A Geology for Engineers

to
Margaret Blyth and Mary de Freitas
for their support and patience during
the preparation of this 7th Edition.

A Geology for Engineers

Seventh Edition

F.G.H. Blyth Ph.D., D.I.C., F.G.S.

Emeritus Reader in Engineering Geology,
Imperial College of Science and Technology, London

M.H. de Freitas M.I.Geol., Ph.D., D.I.C., F.G.S.

Senior Lecturer in Engineering Geology,
Imperial College of Science and Technology, London

OXFORD AUCKLAND BOSTON JOHANNESBURG MELBOURNE NEW DELHI

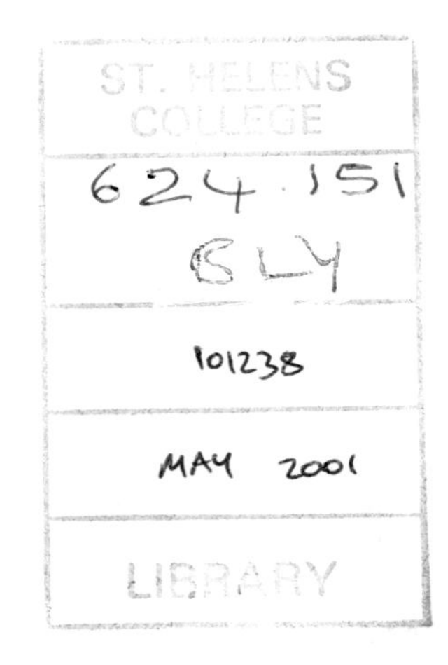

Butterworth-Heinemann
Linacre House, Jordan Hill, Oxford OX2 8DP
225 Wildwood Avenue, Woburn, MA 01801-2041
A division of Reed Educational and Professional Publishing Ltd

A member of the Reed Elsevier plc group

First published by Arnold 1943
Seventh edition published by Arnold 1984
Reprinted with amendments by Arnold 1986
Reprinted by Butterworth-Heinemann 2001

British Library Cataloguing in Publication Data
Blyth, F G H
A geology for engineers.—7th ed
1. Engineering geology
I. Title II. De Freitas, MH
551'.024624 TA705

ISBN 0 7131 2882 8

Printed and bound in Great Britain by The Bath Press, Bath

Preface

The Seventh Edition of *A Geology for Engineers* has been written to provide students of engineering with a recent text in geology for use during their first degree in Civil Engineering or Mining. As with previous editions, we have attempted to explain both the subject of geology and its relevance to engineering work in rock and soil. This edition also provides a text which will complement other courses that student engineers attend, such as those in rock and soil mechanics, ground-water flow, and urban development. To achieve these ends the text has been completely revised and much extended. Three new chapters have been written and the structure and content of former chapters have been substantially changed. Much attention has been devoted to the quality of illustrations and tabulated data, and most of the artwork has been redrawn. SI units have been used throughout the text.

For the teacher we have provided the Seventh Edition with three features which we hope will be of assistance to a course of geology for engineers that is restricted to a limited number of lectures and practicals, and the occasional visit to an engineering site.

The first feature is the structure of the book, which has been divided into two parts of approximately equal length. Chapters 1 to 8 inclusive are concerned with fundamental aspects of Earth geology, its processes and products, as would normally be presented to student engineers in the first part of their tuition in geology. Chapter 9 reviews the mechanical properties of geological materials and is designed to supplement the more extensive courses of soil and rock mechanics that the students will be attending at this stage in their degree studies. We hope this Chapter will be of assistance in illustrating the range of behaviour that may be exhibited by rocks and soils; it does not show how these properties are incorporated into engineering design, such considerations being more appropriately dealt with by conventional courses of rock and soil mechanics. Chapters 10 to 18 inclusive represent the second part of the book and consider subjects where the influence of geology upon engineering work may be clearly demonstrated. These chapters are intended to support the lectures on ground-investigation, slope stability, excavations and hydrology that students will be attending as part of their course in engineering. Numerous references have been provided to assist the teacher locate further details.

The second feature of the book concerns its illustrations. Most are line drawings of a type which can be reproduced easily as transparencies for projection during lectures and practicals. Many of the drawings illustrate in a simple manner the fundamental aspects of complex geological processes and materials. This material has been designed for teachers who wish to use the text either to introduce particular subjects of a lecture or to precede the projection of their own transparencies of real situations and materials. Many of the line drawings contain more information than is revealed in either their caption or the text and will enable a variety of topics to be illustrated to a class.

The third feature we hope will be of help to the teacher is the support the text provides for practical work in the laboratory and in the field. The chapters devoted to minerals, rocks and geological maps have been carefully structured and illustrated to assist students with their independent work, so that they may proceed with the description and identification of minerals and rocks, with map reading and interpretation, and with the construction of cross-sections, after they have received initial guidance from their tutor. Visits to site may be introduced with the aid of the chapters describing ground investigation and laboratory testing, and much of the material in the Chapters devoted to ground-water, slopes, dams and reservoirs, excavations and ground treatment, is concerned with illustrating ground conditions that are rarely visible on site but are the cause of much engineering work.

For the student we have incorporated into the Seventh Edition three features that are in addition to those mentioned above.

The first is the general form of the text. All editions of *A Geology for Engineers* have been written for students who are studying geology to become good engineers. We have tried to select those aspects of geology which are likely to be most relevant for both an appreciation of the subject and the safe practice of civil and mining engineering. Scientific terminology has been moderated to provide a comprehensive vocabulary of geological terms which will satisfy the requirements of most engineers. Each geological term is explained and indexed and many terms describing geological processes, structures and materials are illustrated. By these means we hope that the Seventh Edition will enable the student engineer to communicate with his tutors and with geologists and geotechnical engineers, and to understand the terminology that is commonly used in geological and geotechnical literature.

The second feature we have provided to aid the student is a comprehensive system of headings and sub-headings. Many readers will know nothing of geology and will require clear guidance on the scope and content of its various parts. Each chapter therefore contains a system of headings that will reveal the content and extent of the subject and the relationship between its components. Personal study may therefore commence with a rapid assessment of a topic, gained by turning the pages and reading the headings.

The third feature is the provision of material that will assist the student to become acquainted with other sources of geological and geotechnical information. Each chapter concludes with a Selected Bibliography of texts which a student engineer should find of interest and be able to comprehend. Because some of these texts will prove more difficult to understand than others many of the illustrations in the present edition have been drawn to explain the subject of a chapter and to assist an appreciation of more advanced work recorded in the Proceedings of Professional and Learned Societies to which the student engineer may subscribe as either a Junior or Associate Member. Reference to selected case histories has also been given as far as space has permitted.

In completing the Seventh Edition we wish to acknowledge the help we have received from our many colleagues around the world. In particular we want to thank the staff of Imperial College for their assistance with so many matters. We must also record our appreciation of the work undertaken by the staff of Edward Arnold, who have been our Publisher for so many years. The double column format of this Edition has contained within reasonable bounds a text much enlarged on previous Editions.

London, 1984

F.G.H. Blyth
M.H. de Freitas

Contents

1

The Earth: Surface, Structure and Age

Introduction

The science of Geology is concerned with the Earth and the rocks of which it is composed, the processes by which they were formed during geological time, and the modelling of the Earth's surface in the past and at the present day. The Earth is not a static body but is constantly subject to changes both at its surface and at deeper levels.

Surface changes can be observed by engineers and geologists alike; among them erosion is a dominant process which in time destroys coastal cliffs, reduces the height of continents, and transports the material so removed either to the sea or to inland basins of deposition. Changes that originate below the surface are not so easily observed and their nature can only be postulated. Some are the cause of the slow movements of continents across the surface of the globe; others cause the more rapid changes associated with volcanic eruptions and earthquakes.

The changes result from energy transactions, of which the most important are listed in Table 1.1 (Smith, 1973):

Table 1.1

	Joules year^{-1}
(1) Solar energy received and re-radiated; responsible for many geological effects generated within a depth of about 30 m of ground level, especially weathering and erosion.	10^{25}
(2) Geothermal heat loss from the Earth's interior; responsible for many deep-seated movements that affect the elevation and relative position of continents and oceans.	10^{21}
(3) Energy lost by slowing down of Earth's rotation.	10^{19}
(4) Energy released by earthquakes.	10^{18}

The last three items together account for many of the changes that originate below the Earth's surface, and indicate the importance of internal processes in controlling the behaviour of the planet. These processes are thought to have operated for millions of years and geologists believe that processes working at present are fundamentally similar to those that operated in the past. The effects produced by geological processes may appear to be too slow to be significant in engineering, but many of them operate at rates similar to those found in engineering practice. For example, continents drift laterally at a rate of between 1 and 3 cm per year, or at about 10^{-7} cm per second, which is the approximate value for the hydraulic conductivity of good concrete used in dams.

Geological processes such as those which operate at the present day have, during the very large span of geological time, left their record in the rocks – sometimes clearly, sometimes partly obliterated by later events. The rocks therefore record events in the long history of the Earth, as illustrated by the remains or marks of living organisms, animals or plants, when preserved; all rocks make their contribution to the record. In one sense geology is Earth-history.

The term *rock* is used for those materials of many kinds which form the greater part of the relatively thin outer shell, or crust, of the Earth; some are comparatively soft and easily deformed and others are hard and rigid. They are accessible for observation at the surface and in mines and borings. Three broad rock groups are distinguished, on the basis of their origins rather than their composition or strength:

(*i*) *Igneous rocks*, derived from hot material that originated below the Earth's surface and solidified at or near the surface (e.g. basalt, granite, and their derivatives).

(*ii*) *Sedimentary rocks*, mainly formed from the breakdown products of older rocks, the fragments having been sorted by water or wind and built up into deposits of sediment (e.g. sandstone, shale); some rocks in this group have been formed by chemical deposition (e.g. some limestones). The remains of organisms such as marine shells or parts of plants that once lived in the waters and on the land where sediment accumulated, can be found as *fossils*.

(*iii*) *Metamorphic rocks*, derived from earlier igneous or sedimentary rocks, but transformed from their original state by heat or pressure, so as to acquire conspicuous new characteristics (e.g. slate, schist, gneiss).

Rocks are made up of small crystalline units known as *minerals* and a rock can thus be defined as an assemblage of particular minerals, and named accordingly. For engineering purposes, however, the two terms 'rock' and 'soil' have also been adopted to define the mechanical characters of geological materials. 'Rock' is a hard material and 'soil' either a sediment which has not yet become rock-like, or a granular residue from rock that has completely weathered (called a *residual* soil). Neither of these terms is strictly adequate and descriptive qualifications are required to distinguish weak rocks from hard soils. Rocks and soils contain pores and fissures that may be filled either with liquid or with gas: e.g. water or air. Such voids may be very small but can make up a considerable proportion of a rock or soil mass.

In the present chapter we consider the Earth as a whole, its general structure, its larger surface features – the oceans and continents, and its age and origin.

The surface of the Earth

Dimensions and surface relief

The radius of the Earth at the equator is 6370 km and the polar radius is shorter by about 22 km; thus the Earth is not quite a perfect sphere. The planet has a surface area of 510×10^6 km^2, of which some 29 per cent is land. If to this is added the shallow sea areas of the shelf which surrounds the continents, the total land area is nearly 35 per cent of the whole surface. In other words, nearly two-thirds of the surface is covered by deep ocean.

Surface relief is very varied; mountains rise to several kilometres above sea level, with a maximum of 8.9 km at Everest. The average height of land above sea level is 0.86 km and the mean depth of the ocean floor is about 3.8 km. In places the ocean floor descends to much greater depths in elongated areas or trenches (p. 12); the Marianas Trench in the N.W. Pacific reaches the greatest known depth, 11.04 km. The extremes of height and depth are small in comparison with the Earth's radius, and are found only in limited areas. The oceans, seas, lakes and rivers are collectively referred to as the *hydrosphere*; and the whole is surrounded by a gaseous envelope, the *atmosphere*.

Ocean floors

The topography of the deep oceans was known, from soundings, only in broad outline until 50 or 60 years ago. Advances in measurement techniques have made possible much more detailed surveys, particularly with the use of seismic refraction methods, which enable a profile of the ocean floor to be drawn. Methods of coring the floor at great depths have also been developed and, from the core samples obtained, the distribution and composition of the hard rocks that form the floor and its cover of softer sediments have been recorded in many areas. The topographical features of a continental margin, such as that of the North Atlantic, are shown in Fig. 1.1. The *continental shelf* is a submerged continuation of the land, with a gentle slope of 1 in 1000 or less, and is of varying width. It continues to a depth of about 100 fathoms (183 m), where there is a marked change in slope known as the *shelf break*, the gradient becoming 1 in 40 or more. The shelf break marks the beginning of the *continental slope*, which continues until the gradient begins to flatten out and merges into the *continental rise*, which is often several hundred kilometres wide as in the North Atlantic, with a diminishing gradient. Continental slopes in many places show erosional features known as *submarine canyons*, which are steep-sided gorge-like valleys incised into the sea floor (Fig. 1.2). Some lie opposite the mouths of large rivers, as at the Hudson Canyon opposite Long Island. Many of the canyons have been excavated by turbidity currents, i.e. submarine movements down the slope, similar to landslides. They carry much suspended sediment and are thus denser than normal sea water. In some instances they continue down to the continental rise.

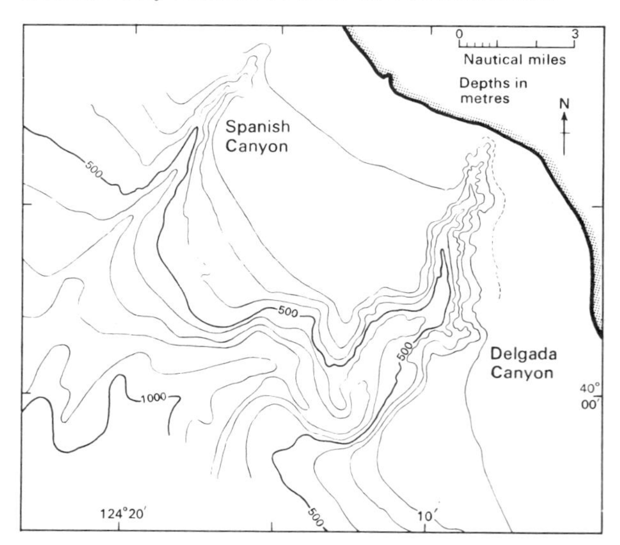

Fig. 1.2 Map of submarine canyons off the West coast of California.

At depths greater than about 2700 fathoms (or 5 km) the deep *abyssal plain* is reached. This is the ocean floor and from it rise submarine volcanic islands, some of which may be fringed with coral reefs. Volcanoes that no longer break the ocean surface are called drowned peaks or *sea mounts*. The volcanoes are related to *oceanic ridges*

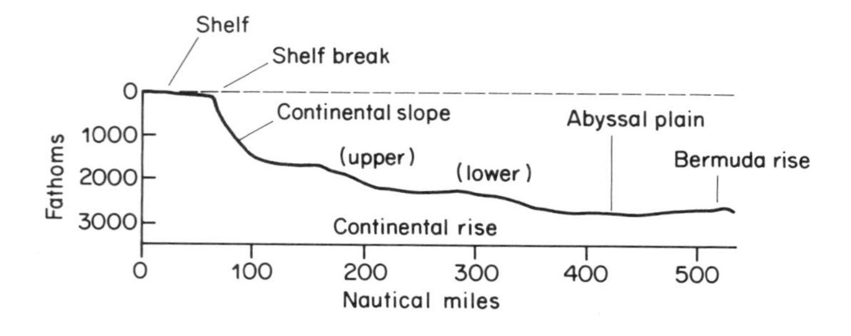

Fig. 1.1 Profile of a continental margin from the continental shelf to deep ocean floor, based on data from the North Atlantic (after Heezen, Tharp & Ewing, 1959). Horizontal scale in nautical miles (1 nautical mile = 1185 km); vertical scale exaggerated (1 fathom = 1.82 m).

which form submarine chains of mountains. The mid-Atlantic ridge rises to a height of 2 to 4 km above the ocean floor and is several thousand kilometres wide. Oceanic ridges are discussed further on p. 8.

The interior of the Earth

Our knowledge of the Earth's interior is at present based on those direct investigations that can be made to depths of a few kilometres from the surface, together with extrapolations to lower levels. Studies of heat-flow, geostatic pressure, earthquakes, and estimations of isostatic balance (p. 6) reveal much about the interior of the Earth.

Temperature gradient and density

It is well known from deep mining operations that temperature increases downwards at an average rate of 30°C per km. This rate is higher near a source of heat such as an active volcanic centre, and is also affected by the thermal conductivity of the rocks at a particular locality. Assuming for the moment that the temperature gradient continues at the average rate, calculation shows that at a depth of some 30 km the temperature would be such that most known rocks would begin to melt. The high pressure prevailing at that depth and the ability of crustal rocks to conduct heat away to the surface of the Earth result in the rock-material there remaining in a relatively solid condition; but there will be a depth at which it becomes essentially a viscous fluid and this defines the base of the *lithosphere* (Greek: lithos = stone), Fig. 1.3.

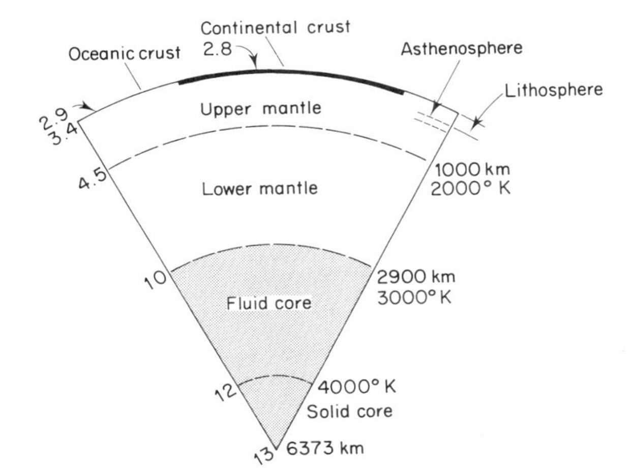

Fig. 1.3 Composition of the Earth (after Bott, 1982); depths from surface in km; temperature scale in degrees K; figures on left are mass density in 10^3 kg m^3.

The mean mass density of the Earth, which is found from its size and motion around the Sun, is 5.527 g cm^{-3}. This is greater than the density of most rocks found at the surface, which rarely exceeds 3; sedimentary rocks average 2.3, and the abundant igneous rock granite about 2.7. In order to bring the mean density to 5.5 there must therefore be denser material at lower levels within the Earth. This has been confirmed from the study of the elastic waves generated by earthquakes, in particular from research into the way in which earthquake waves are bent (by diffraction at certain boundaries) as they pass through the Earth: our knowledge of the Earth's interior comes mainly from such studies. These have shown that our planet has a *core* of heavy material with a density of about 8. Two metals, iron and nickel, have densities a little below and above 8 respectively, and the core is believed to be a mixture of these composed mainly of iron. Surrounding this heavy core is the region known as the *mantle* (Fig. 1.3); and overlying that is the *crust*, which is itself composite. In continental areas the average thickness of the crust is about 30 km: in the oceans it is 10 km. The mantle has a range of density intermediate between that of the crust and the core, as indicated in the figure. In order to discuss further the evidence from seismic work for this earth structure we turn to the subject of earthquakes.

Earthquakes

The numerous shocks which continually take place are due to sharp movements along fractures (called faults) which relieve stress in the crustal rocks. Stress accumulates locally from various causes until it exceeds the strength of the rocks, when failure and slip along fractures occur, followed usually by a smaller rebound. A small movement on a fault, perhaps a few centimetres or less, can produce a considerable shock because of the amount of energy involved and the fault may 'grow' by successive movements of this kind. Earthquakes range from slight tremors which do little damage, to severe shocks which can open fissures in the ground, initiate fault scarps and landslides, break and overthrow buildings, and sever supply mains and lines of transport. The worst effects are produced in weak ground, especially young deposits of sand, silt and clay. These sediments may shake violently if their moduli of elasticity and rigidity are insufficient to attenuate adequately the acceleration imparted to their particles by an earthquake. The bedrock beneath them may be little affected by reason of its strength. Lives and property may be saved if earthquake resisting structures are built (Rosenblueth, 1980). These have frames of steel or wood that are founded directly onto rock whenever possible, and will remain intact when shaken. Dams, embankments, slopes and underground excavations can be designed so as to function whilst shaking (Newmark and Rosenblueth, 1971).

Prior to a major earthquake, strain in the crust

accumulates to the extent that small changes may be noticed in the shape of the land surface, in water levels, in the flow, temperature and chemistry of springs, in the magnetic properties of the strained crust and the velocity with which it transmits vibrations, and in the frequency and location of very small (micro-) earthquakes. These precursors are studied in an attempt to predict location and time of major earthquakes.

When a major earthquake at sea rapidly changes the elevation of the ocean floor, a volume is created that has to be filled by sea-water. Sea-level drops, sometimes causing beaches in the region to be exposed, and large waves, called *tsunamis*, may be generated as sea-level re-establishes itself: these can devastate coastal areas when they strike a shore-line.

Most of the active earthquake centres at the present day are located along two belts at the Earth's surface: one belt extends around the coastal regions of the Pacific, from the East Indies through the Philippines, Japan, the Aleutian Isles, and thence down the western coasts of North and South America; the other runs from Europe (the Alpine ranges) through the eastern Mediterranean to the Himalayas and East Indies, where it joins the first belt (Fig. 1.4). These belts are mainly parallel to the younger mountain chains (p. 15), where much faulting is associated with crumpled rocks; numerous volcanoes are also situated along the earthquake belts. It is estimated that 75 per cent of all earthquake activity occurs in the circum-Pacific belt, and about 22 per cent in the Alpine area. Many smaller shocks also occur in zones of submarine fault activity associated with the oceanic ridges, such as the mid-Atlantic Ridge (p. 9); and others in fault-zones on the continents, e.g. the Rift Valley system of Africa. In areas remote from these earthquake zones only small tremors and shocks of moderate intensity are normally recorded; for example, earthquakes in Britain include those at Colchester (1884), Inverness (1901, 1934), Nottingham (1957), Dent (1970), and Lleyn (1984). All earthquakes are generated in the outer 700 km of the Earth (Fig. 1.3) and all destructive earthquakes, wherever they occur, originate at depths less than 70 km. The deeper earthquakes are discussed on p. 11.

The intensity of an earthquake can be estimated from the effects felt or seen by an observer, and such observations are collected and used to determine the centre of the disturbance. They are graded according to a *Scale of Intensity* such as the Mercalli Scale, which has twelve grades:

- I Detected only by instruments.
- II Felt by some persons at rest; suspended objects may swing.
- III Felt noticeably indoors; vibration like the passing of a truck.
- IV Felt indoors by many, outdoors by some; windows and doors rattle.
- V Felt by nearly everyone; some windows broken; pendulum clocks stop.
- VI Felt by all, many frightened; some heavy furniture moved, some fallen plaster; general damage slight.
- VII Everyone runs outdoors; damage to poorly constructed buildings; weak chimneys fall.
- VIII Much damage to buildings, except those specially designed. Tall chimneys, columns fall; sand and mud flow from cracks in ground.
- IX Damage considerable in substantial buildings; ground cracked, buried pipes broken.
- X Disastrous; framed buildings destroyed, rails bent, small landslides.
- XI Few structures left standing; wide fissures opened in ground, with slumps and landslides.
- XII Damage total; ground warped, waves seen moving through ground, objects thrown upwards.

The observed intensity at points in the area affected

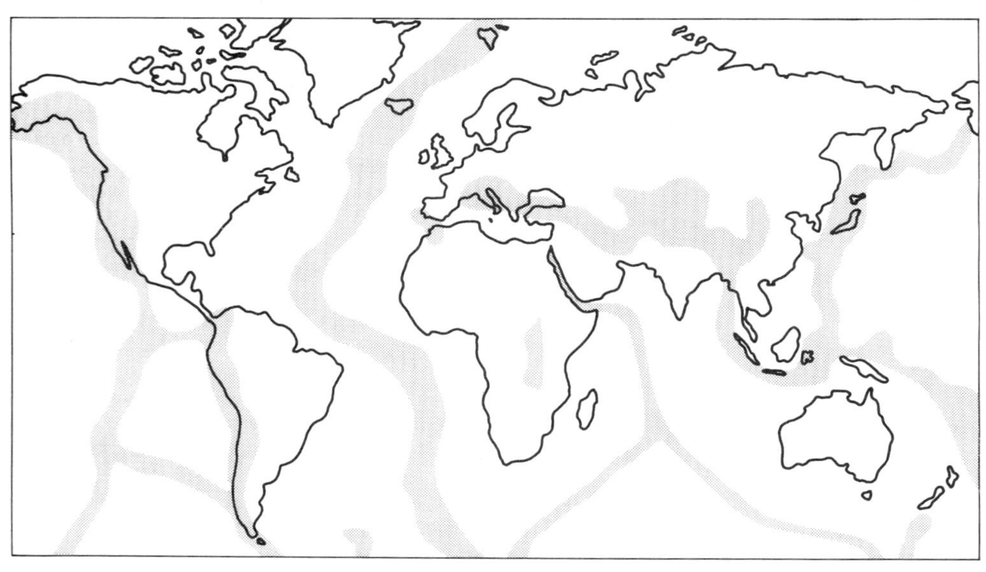

Fig. 1.4 Distribution of earthquakes; the shaded areas are zones of active epicentres.

can be marked on a map, and lines of equal intensity (isoseismal lines) then drawn to enclose those points where damage of a certain degree is done giving an *isoseismal map*.

A more accurate measure of earthquake activity is provided by the amount of seismic energy released in an earthquake; this defines its *magnitude*, for which the symbol M is used. The *Scale of Magnitudes* due to C. F. Richter (1952) and now in general use is based on the maximum amplitudes shown on records made with a standard seismometer. The scale is logarithmic and is related to the elastic wave energy (E), measured in joules (1 erg $= 10^{-7}$ joules), an approximate relationship being $\log E \approx 4.8 + 1.5\,M$, M ranges from magnitude 0 to magnitude 9. The smallest felt shocks have $M = 2$ to $2\frac{1}{2}$. Damaging shocks have $M = 5$ or more; and any earthquake greater than $M = 7$ is a major disaster. The Richter Scale of Magnitudes and the Mercalli Scale of Intensities are not strictly comparable; but $M = 5$ corresponds roughly with Grade VI (damage to chimneys, plaster, etc.) on the Mercalli Scale. The historic record of earthquakes reveals that shocks of large magnitude occur less frequently than those of lesser magnitude. A relationship exists between the magnitude of an earthquake that is likely to occur at a location and its *return period*, and this relationship is used to select the accelerations that must be resisted by the earthquake resisting structures for the locality.

When an earthquake occurs elastic vibrations (or waves) are propagated in all directions from its centre of origin, or *focus*; the point on the Earth's surface immediately above the earthquake focus is called the *epicentre:* here the effects are usually most intense. Two kinds of wave are recorded: (i) *body waves*, comprising of compressional vibrations, called primary or P waves, which are the fastest and the first to arrive at a recording station, and transverse or shear vibrations, called S waves, a little slower than the P waves; and (ii) *surface waves*, (or L-waves) similar to the ripples seen expanding from the point where a stone is dropped into water, and created by Love-wave (LQ) and Rayleigh-wave (LR) ground motions. Surface waves are of long period that follow the periphery of the Earth; they are the slowest but have a large amplitude and do the greatest damage at the surface: M is calculated from their amplitude. The vibrations are detected and recorded by a seismograph, an instrument consisting essentially of a lightly suspended beam which is pivoted to a frame fixed to the ground, and which carries a heavy mass (Fig. 1.5*a*). Owing to the inertia of the heavy mass a movement is imparted to the beam when vibrations reach the instrument, and the movement is recorded on a chart on a rotating drum (Fig. 1.5*b*). On this record, or *seismogram*, time intervals are marked, from which the times of arrival of the vibrations can be read off.

Using known velocities of transmission for the vibrations, the distance of an epicentre from the recording apparatus can be calculated. Two instruments are needed to record north-south and east-west components of the vibrations, and a third instrument to detect vertical movements. Note that large explosions, which are also detected by seismographs, can be distinguished from earthquakes.

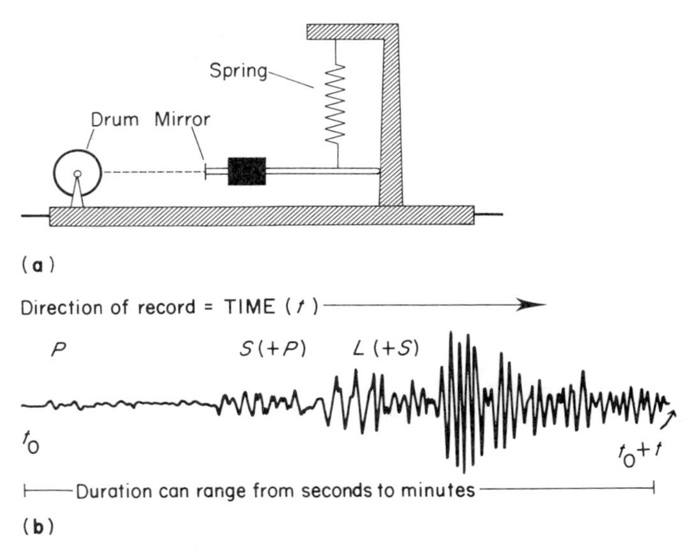

Fig. 1.5 (**a**) Diagram of a seismograph for recording vertical gound movement. (**b**) Record (or seismogram) of a distant earthquake showing onsets of P, S, and L waves in the order of their arrival.

For a distant earthquake, seismographs situated at distances up to 105° of arc from the epicentre record the onsets of P, S, and L waves (Fig. 1.6). Between 105° and 142° of arc, the region known as the 'shadow zone', no P or S waves arrive, but from 142° onwards the P waves are again received. They have, however, taken longer to travel and hence must have been slowed down over some part of their path through the Earth. This was interpreted by R.D. Oldham in 1906 as being due to the presence of

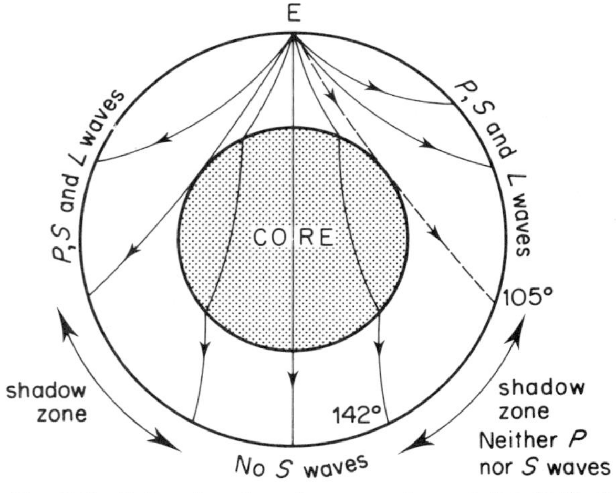

Fig. 1.6 Paths of earthquake waves through the Earth. A few paths only are shown out of the many that radiate from the epicentre (E). Note the refraction that occurs when waves cross the boundary between mantle and core.

a central Earth *core*, of such composition that *P* waves penetrating to a greater depth than the 105° path enter the core and move there with a lower velocity. The transverse *S* vibrations are not transmitted through the core, indicating that it has the properties of a fluid (which would not transmit shear vibrations). Modern work suggests that, while the outer part of the core is fluid, the innermost part is probably solid (Fig. 1.3) and is composed mainly of iron in a densely-packed state. The core extends to within 2900 km of the Earth's surface, i.e. its radius is rather more than half the Earth's radius (Fig. 1.3). There is a sharp discontinuity between the core and the overlying mantle; the latter transmits both *P* and *S* vibrations.

Records obtained from near earthquakes (within about 1000 km of an epicentre), as distinct from distant earthquakes, monitor seismic waves that have travelled for their greater distance through crustal rocks and such records have yielded information about the crust of the Earth. The Serbian seismologist, A. Mohorovičič, in 1909, noticed that *two* sets of *P* and *S* waves were sometimes recorded, the two sets having slightly different travel times. This, he suggested, indicated that one set of vibrations travelled by a direct path from the focus and the other set by a different route. In Fig. 1.7, the set P_g and S_g follow the direct path in an upper (granitic) layer, while the set *P* and *S* are refracted at the boundary of a lower layer and travel there with a higher velocity because the material of the lower layer is denser. This boundary may be considered to mark the base of the crust and is called the *Mohorovičič discontinuity*, or 'the Moho'. Later a third set of vibrations was detected on some seismograms; they are called P* and S* and have velocities lying between those of the other two sets. They follow a path in the layer below the granitic layer (Fig. 1.7). The velocities of the three sets of waves, as determined by H. Jeffreys from European earthquake data, are as follows:

P_g 5.57 km s^{-1} S_g 3.36 km s^{-1}

P* 6.65 km s^{-1} S* 3.74 km s^{-1}

P 7.76 km s^{-1} *S* 4.36 km s^{-1}

These values correspond to those derived from elasticity tests in the laboratory on the igneous rocks *granite*, *basalt*, and *peridotite* respectively. Peridotite is a rock whose mineralogy is formed at pressures and temperatures similar to those expected in the upper mantle. Thus the fastest waves, *P* and *S*, travel for the greater part of their course in material of peridotite composition, in the upper part of the mantle just below the Moho. Above the Moho is the basaltic crust, in which the P* and S* waves travel. The granitic layer, which forms the upper part of the continental crust, transmits the P_g and S_g vibrations. The granitic layer itself is mainly covered by sedimentary rocks, in which velocities of transmission are lower, from about 2 to 4 km s^{-1}. The thicknesses of the crustal layers varies considerably in different situations. The average thickness of the crust in a continental area is about 30 km, but beneath a mountain mass it may thicken to 40 km or more as discussed below. In an oceanic area the crust is

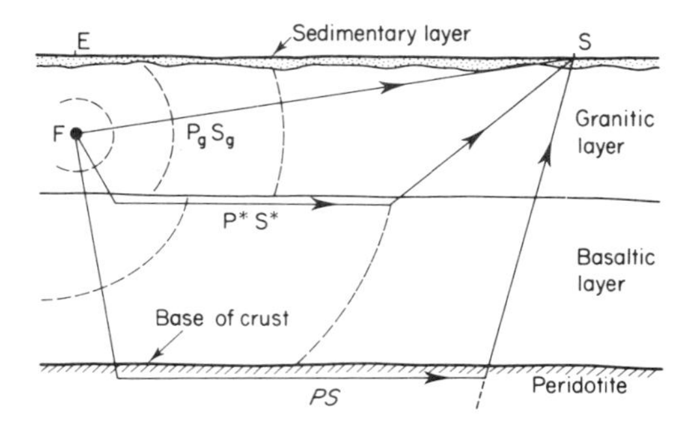

Fig. 1.7 Seismic waves radiating from the location where crustal fracture has occurred, the focus (F), and travelling through the continental crust and uppermost mantle at velocities P_gS_g: P*S* and *PS*. E=epicentre, situated above the focus, S=seismograph.

thinner, 5 to 10 km, and is composed of basalt with a thin sedimentary cover and no granitic layer. This distinction between continental crust and oceanic crust is referred to again on p. 10. The study of earthquake waves has demonstrated that the Earth consists of concentric shells of different density, the lightest being the outer lithosphere. This contains the oceanic and continental crust which rests upon the heavier rock at the top of the upper mantle, whose character is in part revealed by the vertical and horizontal movements of the lithosphere. These movements require the presence of a weaker layer at depth; the asthenosphere. To explain the vertical movements of the lithosphere the theory of isostacy was proposed: horizontal displacements required the theory of continental drift for their explanation. The new theory of Plate Tectonics unifies both these concepts.

Isostasy

This term (Greek, meaning 'in equipoise') is used to denote an ideal state of balance between different parts of the crust. The continental masses can be visualized as extensive blocks or 'rafts' essentially of granitic composition supported by underlying sub-crustal material. The difference in the density between these two implies that the continents are largely submerged in denser sub-crustal material rather like blocks of ice floating in water. A state of balance tends to be maintained above a certain level called the *level of compensation*. Thus in Fig. 1.8 the weight of a column of matter in a mountain region, as at A, equals that of a column B, where the lighter crust is thinner and displaces less of the underlying denser layer. The columns are balanced at a depth (namely the level of compensation) where their weights are the same.

The concept of isostatic balance has been tested by gravity surveys, which reveal excess or deficiency of density in the make-up of the crust below the area surveyed. From all the evidence it is probable that very large topographical features at the Earth's surface such as a range of mountains are isostatically compensated on a regional

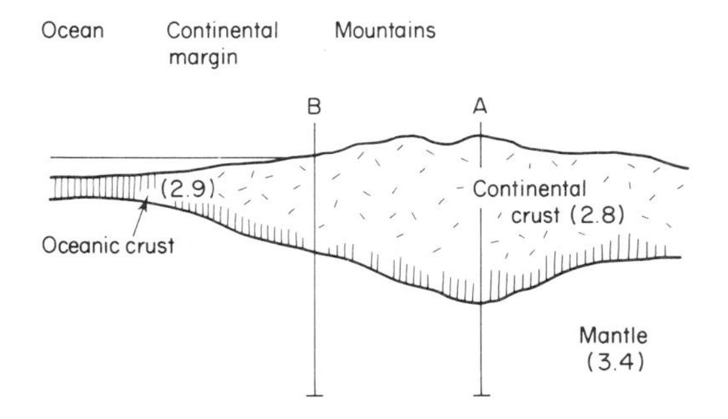

Fig. 1.8 Diagrammatic section through part of a continent. Density in $10^3\,kg\,m^{-3}$.

scale, and probably bounded by faults. The Alps, for example, are balanced in this way, their topographical mass above sea level being continued downwards as a deep 'root' of granitic continental material (Fig. 1.9). For smaller masses local isostatic compensation is unlikely to be complete because their weight is partly supported by the strength of the surrounding crust, i.e. smaller mountains and valleys exist because of the crust's rigidity. Geophysical surveys have also shown that continental margins at the present day are largely compensated and in near isostatic balance.

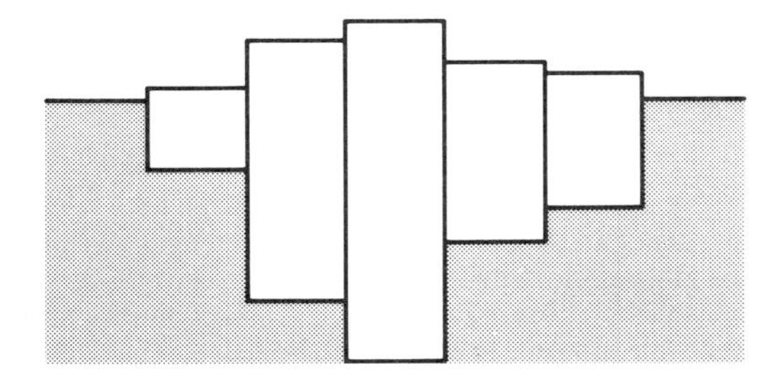

Fig. 1.9 Isostatic balance according to Airy's hypothesis; ideal columns of crust of different lengths are largely submerged in heavier sub-crustal matter which is displaced to a greater depth by the higher columns, corresponding to the 'roots' of mountains.

Isostasy requires that below the relatively strong outer shell of the Earth, the lithosphere, there is a weak layer (or Earth-shell) which has the capacity to yield to stresses which persist for a long time. This weak zone is called the *asthenosphere* (Greek: *a*, not, and *sthene*, strength). It lies in the uppermost part of the mantle (Fig. 1.3) and its distinctive feature is its comparative weakness. Isostasy implies that for a land area undergoing denudation, there is a slow rise of the surface as it is lightened, with an inflow of denser material below the area. Because of the different densities (2.8 and 3.4) the removal of, say, 300 m of granitic crust will be balaced by the inflow of about 247 m of the denser material; the final ground level when isostatic adjustment is complete will thus be only 53 m lower than at first. It is thought that the height of the Himalayas, for example, has been maintained by this mechanism during the erosion of their many deep gorges, which involved the removal of great quantities of rock.

During the Glacial epoch, when thick ice-sheets covered much of the lands of the northern hemisphere (p. 29), the load of ice on an area resulted in the depression of the area. With the removal of the load as the ice melted, isostasy slowly restored the balance by re-elevating the area. In this way many beaches, such as those around the coasts of Scandinavia and Scotland, were raised as the land was elevated in stages, bringing the raised beaches to their present positions above existing sea level, after the melting of the ice (Fig. 2.21).

Continental drift

The possible movement of the continents relative to one another in the geological past was first outlined at length by Alfred Wegener in 1912, and it became a matter of controversy for many years. During the 1960s, however, new evidence came to light which conclusively demonstrated that drifting had taken place; the evidence came largely from the study of magnetism in the rocks of the Earth's crust and from detailed surveys of the ocean floors. These demonstrated that the continents have not remained in their same relative position and that the ocean floors are much *younger* than the continents they separate.

Wegener and others pointed out the similarity of the coastlines of Africa and South America which, although separated at the present day by the Atlantic, would be explained if the two continents were originally adjacent and parts of a single land mass. He postulated a supercontinent to which he gave the name 'Pangaea'. There are also geological features in the two continents that correspond, such as belts of strongly-folded rocks in South Africa and North Africa which run out to the coast and have their counterparts in South America. Other similarities are shown by fossil faunas, one example being the remains of the early horse (*Hipparion*) found on either side of the Atlantic (Fig. 1.10). These features were set out in detail by A.L. du Toit (1937) as evidence that the two continents, originally adjacent to one another, had drifted apart. Modern work shows that there is an accurate fit of Africa and South America at the margins of their continental shelves (Fig. 1.11). The figure also shows that during the separation there was a rotation of one continent relative to the other. In a similar way, North America and Europe possess features that were once adjacent before the opening of the North Atlantic. When their positions are restored, mountain ranges such as the Palaeozoic folds of eastern North America become continuous with the Caledonian folds of Norway and Scotland, both of which have similar geological structures (Fig. 1.10).

Lands in the southern hemisphere including South America, Africa, Antarctica, Australia, and peninsular India formed a large continent, called *Gondwanaland* (Fig. 1.12), some 400 my ago in Carboniferous times;

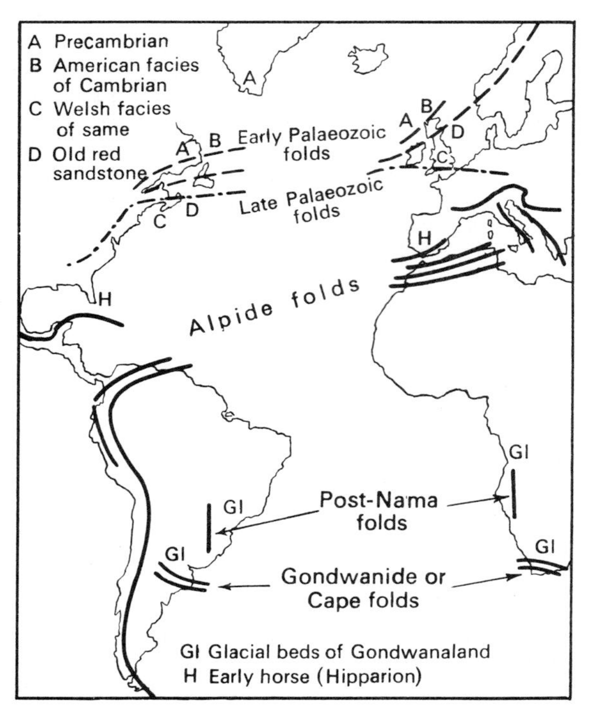

Fig. 1.10 Geological resemblances across the Atlantic (after A. L. du Toit, 1937).

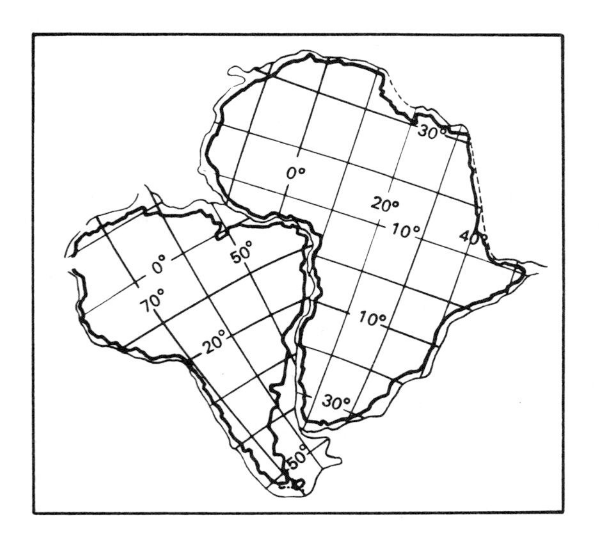

Fig. 1.11 Fit of South America and Africa at the 1000 fathom line (Bullard, 1965; after S.W. Carey, 1958).

they have since moved apart to their present positions. When Antarctica and Australia (with New Zealand) lie together as shown in the figure, certain geological features (g) of the two continents become aligned; also the west side of India and Sri Lanka when alongside east Africa show a correspondence of particular rocks (a) (see caption).

An extensive glaciation in Carboniferous times affected what is now southern Africa, India, south Australia, and parts of Brazil and Argentina, as evidenced by glacial deposits found in all those areas (Chapter 2, p. 21). This glaciation is readily explained if the glaciated lands were originally parts of Gondwanaland, the Earth's south pole at the time being situated at about the centre of the area shown in Fig. 1.12. When the continent broke up and its

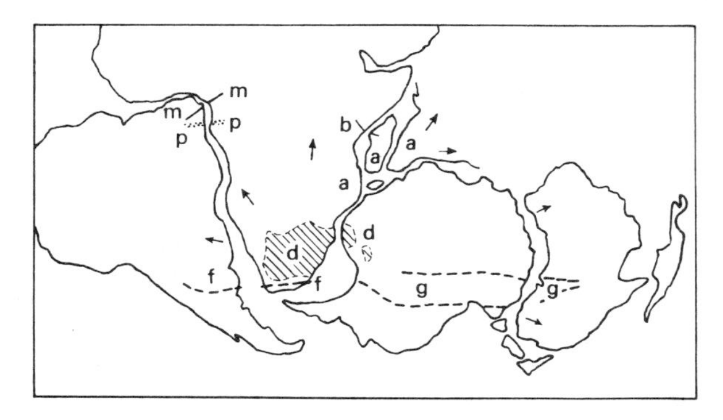

Fig. 1.12 Reconstruction of Gondwanaland (after G. Smith and Hallam, 1970). Matching features include: (a) Precambrian anorthosites; (b) limit of Jurassic marine rocks; (d) Mesozoic dolerites; (f) fold-belt; (g) geosynclinal (early Cambrian); (m) mylonites; (p) Precambrian geosyncline. Arrows, with radial arrangement, show directions of ice movement.

several parts began to separate, some 200 my ago, Africa and India moved northwards and eventually impinged upon the southern margin of the Eurasian continent, where great fold-mountain systems – the Atlas, Alps, and Himalayas – were ridged up in early Tertiary times. It is estimated that the Indian block moved northwards at a rate of some 20 cm per year to reach its present position. By comparing Figs. 1.10, 1.11 and 1.12 with Figs 1.13 and 1.14 it will be noticed that between the drifting continents lie the oceanic ridges. These, and the ocean floor on either side of them, provide evidence that explains the mechanism for continental drift.

Oceanic ridges

These structures, mentioned briefly on p. 2, resemble submerged mountain ranges and are found in all the oceans. The existence of a large rise below the North Atlantic had been known for a long time; surveys have now shown that a ridge extends from Iceland southwards through the North Atlantic, and thence continues into the South Atlantic about midway between Africa and South America (Fig. 1.13). After passing Tristan da Cunha the oceanic ridge turns east and continues into the Indian Ocean. Other ridges lie below the East Pacific, as shown in the figure, and between Australia and Antarctica; and in the

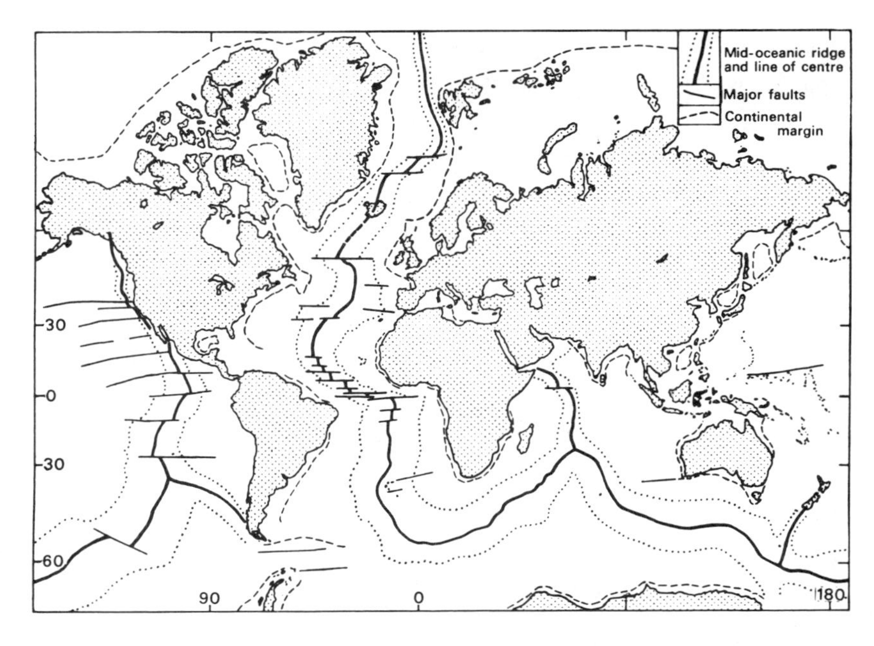

Fig. 1.13 Map of the oceanic ridges (after Heezen, 1963, *The Sea*). Heavy lines show the position of the centre of a ridge; thin lines show displacements by transcurrent faults. Mercator projection.

Indian Ocean a ridge runs northwards to the Red Sea. The mid-Atlantic Ridge (Fig. 1.14) rises some 2.3 km above the deep ocean floor, and to within 2.2 km (1200 fathoms) of the ocean surface. Along the line of its summit a deep cleft called the *median rift* extends to a depth of over 450 m (900 fathoms). Rock samples taken in the vicinity of this rift are mainly volcanic rocks such as

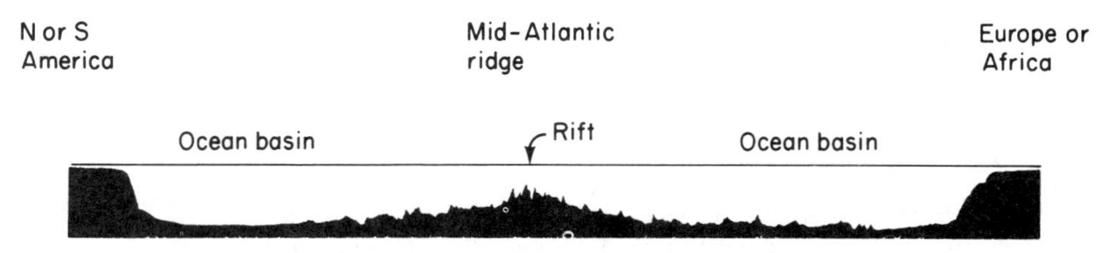

Fig. 1.14 Profile across the Mid-Atlantic Ridge (after Heezen, 1959).

basalts, and are interpreted as material that emerged from fissures along the line of the rift and accumulated on the ocean floor. From radiometric dating it is known that the basalts become older with increasing distance on either side of the rift. The volcanic material is envisaged as rising along the line of median rift and being pushed aside laterally, in either direction away from the rift, by subsequent eruptions, thus forming new ocean floor. The process is termed *ocean floor spreading*. The upper part of the mantle above the asthenosphere, which is in a semi-molten state, is involved in these processes.

Rock magnetism

Studies of the magnetism found in basaltic rocks have yielded independent evidence for ocean floor spreading and continental drift. Minerals which have magnetic properties, such as magnetite (p. 85), are found in basaltic rocks; when the crystals were formed they acted as small magnets and became lined up in the Earth's magnetic field of that time. This *palaeomagnetism* (or 'fossil magnetism') is retained in the rocks and in many instances its direction does not agree with that of the Earth's present magnetic field. This is evidence of a change in the rock's position since it acquired its magnetism, which in turn may be attributed to continental drift or some other cause.

The palaeomagnetism in the basaltic rocks of an oceanic ridge, e.g. the North Atlantic Ridge, shows a pattern of stripes parallel to the median rift, alternate stripes having a reversed magnetism corresponding to the periodic reversal of the Earth's magnetic poles (Fig. 1.15). It is found that matching patterns of stripes have the same sequence in opposite directions away from the median rift, and this was taken as independent evidence for ocean floor spreading. New basalt rising to the ocean floor in successive stages and cooling there acquired a magnetism of the same polarity as that of the Earth's magnetic field at the time. Reversals of the Earth's field have occurred

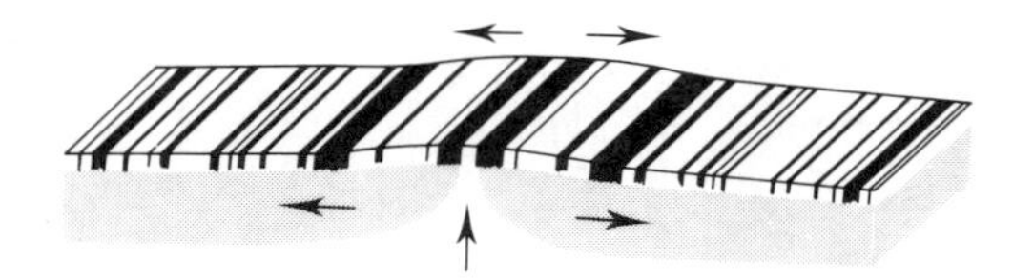

Fig. 1.15 Symmetrical pattern of magnetic stripes in oceanic crust at a spreading ridge (diagrammatic). Black stripes, normal polarity; white stripes, reversed polarity. The rocks at the ends of the diagram are about 8 my older than basalt rising in the centre.

at irregular intervals, and the patterns can be dated by radiometric determinations of age. The last reversal was about 700 000 years ago, and others took place about 1.75 and 2.5 million years ago.

Samples of basaltic rocks from the North Atlantic floor increase in age with distance from the median rift; those near it are only 13 000 years old, while rocks 64 km to the west of it are about 8 million years old. The palaeomagnetic zones closest to the American side of the ridge match those closest to the European side, and those near the American continent match those near the European continent. Radiometric dating thus shows that ocean floor spreading in the North Atlantic has gone on for millions of years in the recent geological past, with the formation of new ocean floor. The present rate of spreading is between 1 cm and 3 cm per year, though it may have varied in the past. The separation of the American continents from Eurasia and Africa probably began in late Jurassic or early Cretaceous times.

Mechanism of drift

Continental drift is associated with the opening and extension of the ocean floor at the oceanic ridges. The temperatures of rocks near the centre of a ridge are higher than on either side of it, because material from the mantle rises towards the surface in the hotter central part of a ridge. The cause of this upward flow is believed to be the operation of slow-moving convention currents in the Earth's mantle (Fig. 1.16). The currents rise towards the base of the lithosphere and spread out horizontally, passing the continental margins and descending again. The hotter rock-material in the rising current is less dense and

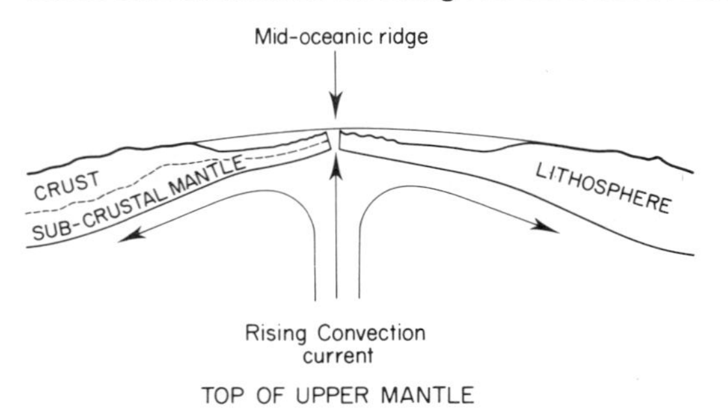

Fig. 1.16 Concept of convection currents and its relationship to continental drift.

possesses buoyancy, which is the driving force of the mechanism. Differences in the rate of movement of adjacent masses away from the oceanic ridges are accommodated by displacement on fractures called transcurrent faults (see Fig. 1.13).

The recognition of extensive fracture systems, with horizontal displacements of hundreds of kilometres, has shown that large fault movements form part of the architecture of the Earth's crust. Thus the Great Glen Fault in Scotland (Fig. 2.21) continues past Caithness and the Orkneys, as shown by geophysical surveys, towards the Shetlands; the San Andreas Fault along the coast of California has a length of over 1200 km; and the great Alpine Fault of New Zealand, along the north-west side of the Southern Alps, has a similar extent. All these are transcurrent faults involving horizontal movements parallel to the line of the fault; similar extensive fractures are located in the ocean floors, e.g. in the east Pacific (Fig. 1.17).

Plate tectonics

When the validity of continental drift became accepted, in the mid-1960s, the idea was advanced that the outer shell of the Earth, the lithosphere, could be considered as a mosaic of twelve or more large rigid *plates* (Fig. 1.17). These plates were free to move with respect to the underlying asthenosphere, and could also move relatively to one another in three ways: (*i*) by one plate sliding past another along its margin; (*ii*) by two plates moving away from one another; (*iii*) by two plates moving together and one sliding underneath the edge of the other. The first of these is expressed at the Earth's surface by movement along major transcurrent faults, such as the San Andreas fault. The second type of movement is shown by the formation of oceanic ridges, (see Figs. 1.16 and 1.18). The third kind of movement is expressed by the deep ocean trenches (Fig. 1.19), where the edge of one plate has moved downwards under the other and is dispersed in the mantle, a process known as *subduction*. The main trenches include the Aleutian trench, the Kuril–Japan–Marianas trench, and the Philippines and Indonesian trenches (Fig. 1.20).

A distinction must be made between continental plate and oceanic plate. The former is capped by continental crust, i.e. the continents 'ride' on the underlying plate. Six of these major plates are distinguished, namely the North and South American, Eurasian, African, Indo-Australian, and Pacific Plates (Fig. 1.17); there are many other smaller plates whose movements are more difficult to determine. Oceanic plate is covered by a thin oceanic crust, mainly basaltic in composition and having a thin covering of sediments (Fig. 1.18).

The term *plate tectonics* came to be used to denote the processes involved in the movements and interactions of the plates ('tectonic' is derived from Greek *tekton*, a builder). Where two continental plates have converged, with the formation of a belt of intercontinental fold-

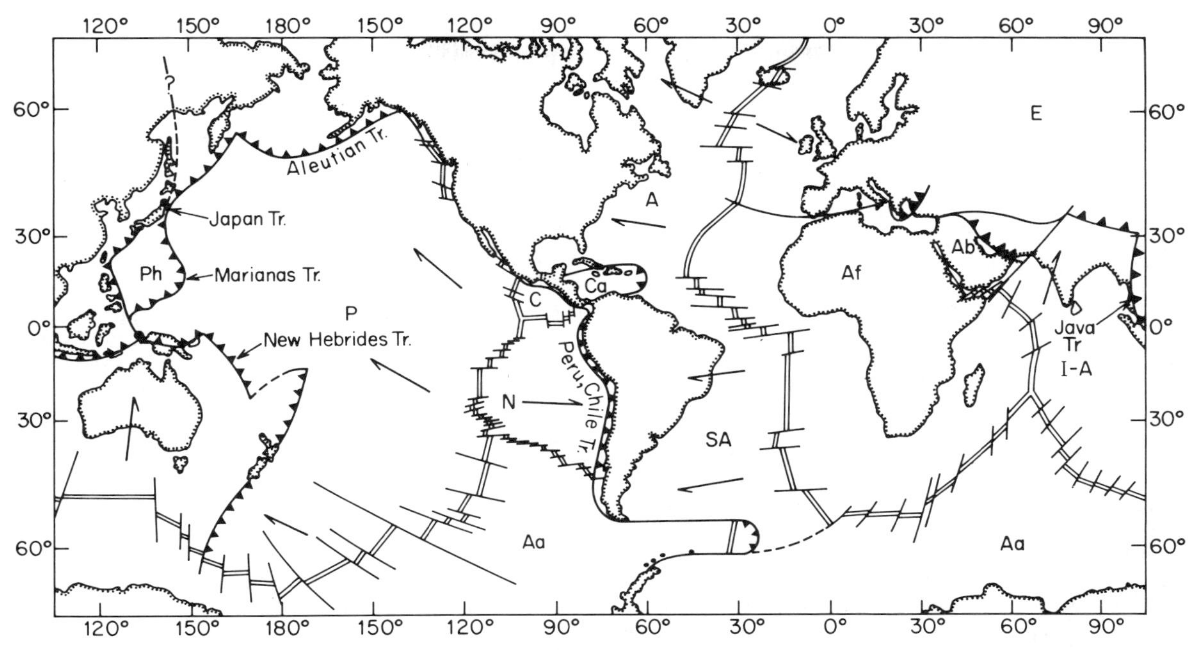

Fig. 1.17 Plate boundaries in the Earth's crust. P, Pacific Plate. A, North American Plate. SA, South American Plate. Af, African Plate. E, Eurasian Plate. I-A, Indo-Australian Plate. Aa, Antarctica. Ph, Philippine. Ca, Caribbean. N, Nazca. C, Cocos. Ab, Arabian. (After Oxburgh, 1974, with modifications.) The plate boundaries largely coincide with zones of seismic and volcanic activity (Fig. 1.4). Oceanic ridges shown by double lines, transcurrent faults by single lines. ▲▲▲ = zones of subduction.

mountains such as the Alpine–Himalayan orogenic belt (p. 17), the term *collision zone* can be used.

The validity of plate tectonics theory received strong support from precise seismic data collected through a period of years by the world-wide seismic network that was set up in the late 1950s. The data showed that the zones in which most of the world's earthquakes occur are very narrow and sharply defined, suggesting that most recorded earthquakes (apart from minor tremors) result from the movements of plates where they impinge on one another. Thus seismic data can be used to map plate boundaries.

The formation of new sea-floor at oceanic ridges, discussed earlier, involves the separation of continents and thus an increase of the area of ocean floor. This increase is balanced by the destruction of plate by subduction, where oceanic crust is carried into the mantle and consumed (compare Figs 1.19 and 1.20). It has been shown that at a subduction zone, earthquakes are generated at deep foci (more than 300 km below the surface) and are related to inclined planes dipping at angles around 30° to 40° beneath the continental margin (Benioff, 1954). Such planes intersect the ocean floor at the deep trenches bordered by island arcs (*Benioff zones*, Fig. 1.20). Volcanic

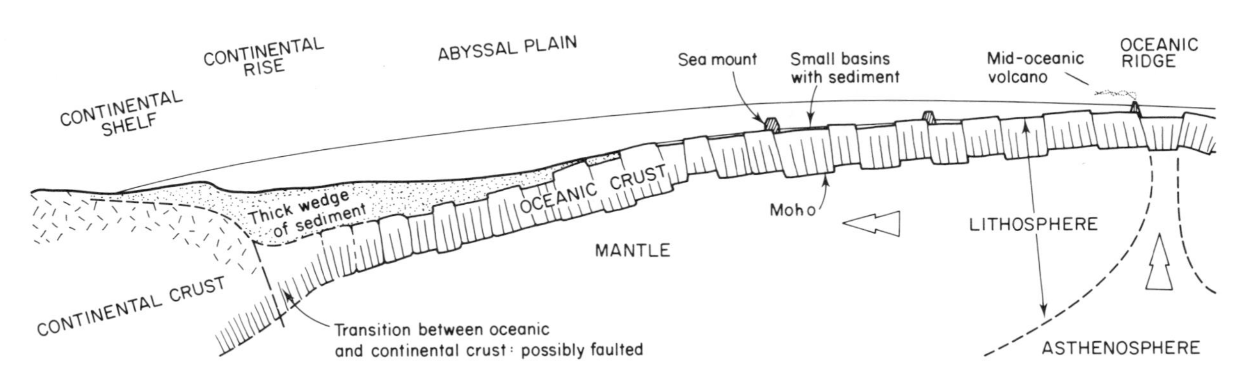

Fig. 1.18 Generalized cross-section across the western Atlantic: based on Dewey and Bird (1970).

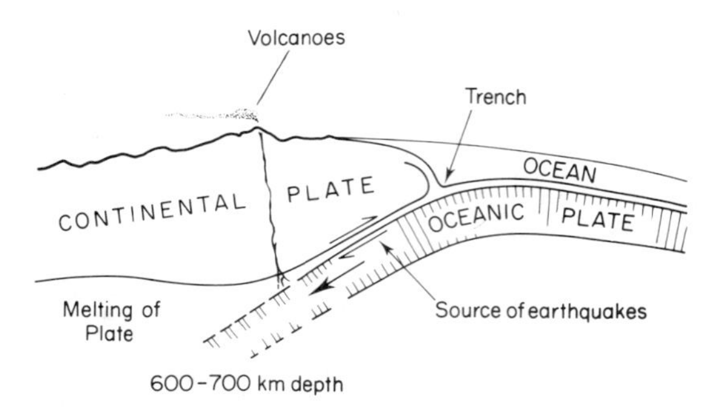

Fig. 1.19 Diagrammatic section through an ocean trench and its relation to subduction of an oceanic plate.

activity is also associated with the island arcs in these zones, as in the Kurile Islands; and the volcanoes of Sumatra and Java which border the deep Indonesian trench on its north side.

An additional hypothesis to that given above is held by a relatively small group of geologists, who accept that the Earth is also slowly expanding and its surface area increasing; the separation of continents by the growth of new oceans need not, therefore, be entirely compensated by the destruction of crust by subduction, if expansion is allowed for.

Is the Earth expanding? Evidence of several kinds has been put forward since the 1930s to suggest that, during geological time, the Earth has expanded from an originally small size. One line of evidence derives from studies of maps showing the distribution of land and sea in earlier geological periods (palaeogeographical maps); the maps show that the extent of water-covered continental areas in the past was greater than their present areas. From these studies it was estimated that the Earth's radius has increased by about 141 km in the last 600 million years, since the early Cambrian; that is, by 2.7 per cent, or a rate of 0.24 mm per year. If this rate of expansion has gone on during the much longer time since the formation and initial cooling of the planet, some 4500 my, the total expansion would amount to about 20 per cent. Another approach takes account of astronomical evidence for the slowly increasing length of the day (as recorded in the growth rings of plants and corals now preserved as fossils), calcuated at 2 seconds in 100 000 years. After allowing for the effect of tidal friction at the Earth's surface, the lengthening of the day could be explained by an increase in the radius of the Earth of the same order of magnitude as above, with a consequent slower rate of rotation. Further evidence comes from palaeomagnetism. The magnetic field of minerals should be aligned to that of the Earth at their time of formation, and when corrected for movements resulting from continental drift, should all point to the existing poles. But they do not and there appears to be a consistent difference that can be explained if the polar radius increases with time, i.e. if the Earth has expanded.

Earth age and origin

The Earth and other members of the Solar System are believed to have been formed about 4600 million years ago by condensation from a flattened rotating cloud of gas and dust. This contracted slowly, giving rise to the primitive Sun at its centre – a new star – surrounded by a mass of cosmic gases in which local condensations generated the planets. They, and other bodies such as the asteroids and meteorites, all revolve in the same direction in orbits around the Sun. The cold primitive Earth became gradually heated as its interior was compressed by the increasing weight of accumulated matter and by the decay of natural radioactive materials. Heat was produced more quickly than it could escape from the compressed mass, resulting in the melting of some constituents and heavier matter being drawn by gravity towards the Earth's centre. The planet thus gradually acquired a core, surrounded by a mantle of less dense material, and an outer crust.

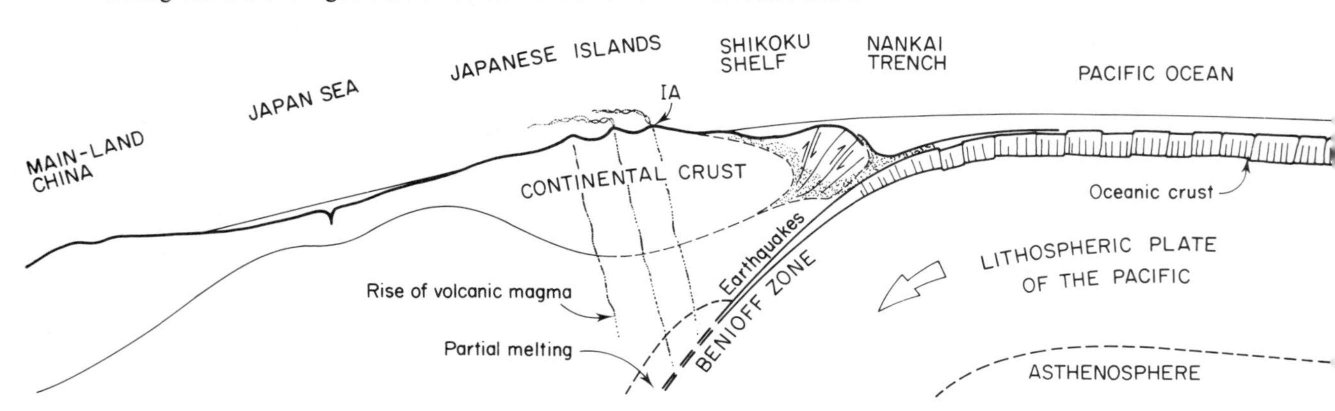

Fig. 1.20 Generalized cross-section across Japan: based on Miyashiro (1970). IA = Island arc of volcanoes. Note: sediments of the Shikoku Shelf are accreted as a wedge against the continental crust of the Eurasian Plate by the movement of the Pacific Plate. (See Fig. 1.17 for the location of similar trenches.)

The oldest rocks so far discovered are dated at about 3900 million years, and as rock samples from the moon range in age from 4400 to 3200 million years, it is probable that a primitive crust formed on Earth about 4400 to 4500 million years ago. Stony meteorites (chondrites) which have fallen on the present surface of the Earth also give ages of the same order. These results together suggest an age of 4600 million years for the Earth and its moon.

The primitive crust was probably basaltic, and was cracked and re-melted, with the separation of lighter (granitic) fluids, which accumulated and eventually contributed to the material of the continents. As the Earth's surface continued to cool, water began to collect on the surface to form the embryo oceans. The atmosphere that we know was formed much later, perhaps within the last 1000 million years, when plant life had become established and contributed oxygen to the volcanic emanations of an earlier stage.

Modern estimates of the ages of rocks are based on determinations made on radioactive minerals contained in the rocks. Before these *radiometric methods* were developed, earlier estimates of age had been made from data such as the amount of salt in the oceans and its estimated rate of accumulation. These gave results that were too low because of innacurate assumptions. The discovery of radioactivity was made by Pierre and Marie Curie, who in 1898 first isolated compounds of radium. This element is found, together with uranium, in the mineral pitchblende, a nearly black pitch-like substance occurring in certain igneous rocks and veins. Uranium during its lifetime undergoes a transformation into an isotope of lead, and radium is formed at one stage in the process. The rate at which this radioactive change takes place is constant. Similarly the element thorium undergoes a transformation into another isotope of lead. The known rates of these changes together with determinations of the amounts of uranium and thorium in a pitchblende, and of the lead content of the mineral, give data for calculating its age, i.e. the length of time that has elapsed during the formation of the lead.

Radioactive transformations that are used for the calculation of age also include potassium into argon, particularly useful for dating igneous rocks (because potassium is a constitutent of many feldspars found in the rocks); and rubidium into strontium, for metamorphic rocks. The greatest terrestrial age so far determined is about 3900 million years, for a mineral in a rock from the ancient Precambrian group.

For much smaller ages, the radioactive isotope of carbon (C^{14}), which becomes converted into nitrogen (N^{14}), is used for dating materials such as wood and plant remains that are enclosed in deposits younger than about 70,000 years.

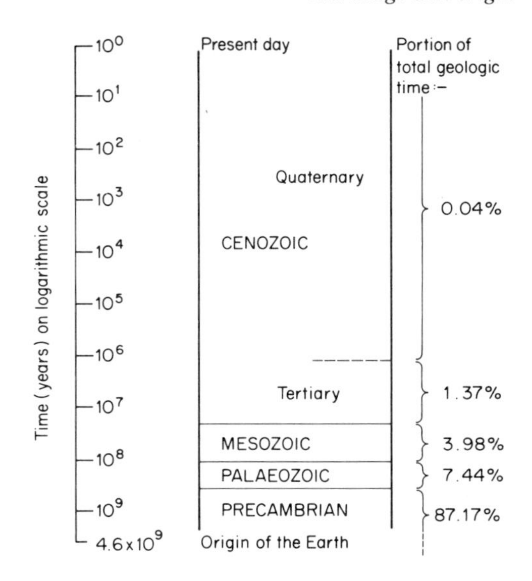

Fig. 1.21 Geological time (after Harland *et al.*, 1982: see text).

The large geological groups into which rocks are divided according to their age are as shown in Fig. 1.21. This has been drawn to emphasize the importance of the last ten-thousandth part of geological time to engineering in rock and soil. The mechanical character of most geological materials in which engineering is conducted will have been affected by this portion of geological history which should always be studied for every site. The great majority of rocks that are present at or near ground level formed during the last one-eighth of geological time. Approximately seven-eighths of geological history is described as Precambrian and is poorly known.

Selected bibliography

Gass, I.G., Smith, P.J. and Wilson, R.C.L. (Eds.) (1971). *Understanding the Earth*. Artemis Press, Sussex, and M.I.T. Press, Cambridge, Massachusetts.

Smith, P.J. (1973). *Topics in Geophysics*. Open University Press, Milton Keynes, England.

Wyllie, P.J. (1971). *The Dynamic Earth: Textbook on Geosciences*. J. Wiley & Sons, New York.

Bott, M.H.P. (1982). *The Interior of Earth*, 2nd edition. Edward Arnold, London and Elsevier, New York.

Costa, J.E. and Baker, V.R. (1981). *Surficial Geology*. J. Wiley & Sons, New York.

2

Geological History

Of the three broad rock groups described in Chapter 1, igneous, sedimentary and metamorphic, it is with the sedimentary rocks that the principles used to study the history of the Earth can be demonstrated most clearly. Sedimentary rocks were deposited in layers, the youngest being at the top. They contain the remains of organisms, i.e. fossils, which represent the life of past geological times and permit the age of the sediment to be defined. Each layer of a sedimentary rock represents a particular event in geological time and the sequence of layers in a pile of sediments thus records a series of events in geological history.

James Hutton in his *Theory of the Earth* (1795), stated an important principle which he called *uniformitarianism.* The principle states that events recorded in the rocks can be understood by reference to the present day activities of geological agents. Thus sedimentary deposits were formed in the past by the action of running water, wind and waves in the same ways as they are at present. Put briefly, 'the present is the key to the past'. Hutton can properly be regarded as the founder of modern geology.

At about this time Fuchsel had shown that in the coal-bearing sedimentary rocks of Thuringia (central Germany) particular fossils were characteristic of certain layers: he introduced the term *stratum* for a layer of sedimentary rock. Cuvier and Brogniart, early in the nineteenth century, examined the sediments of the Paris Basin and worked out the order in which successive deposits there were laid down. In England, William Smith (1769–1839) noted that layers of rock could be traced across country, and he described them as resembling 'superimposed layers of bread and butter'. He also noted that 'the same strata are always found in the same order of superposition, and contain the same peculiar fossils'.

Order of superposition implies that in an undisturbed series of beds, the stratum at the bottom of the series is the oldest (i.e. the earliest formed) and successively younger beds lie upon it. The idea of a *succession* of strata was thus developed.

Following the idea of superposition, which was easily observed in undisturbed horizontal strata, it was necessary to be able to determine the 'way-up' of a sequence of beds when they were steeply inclined as a result of folding or overturning. Various tests that can be applied include observations of internal structures formed during the deposition of the sediments (Fig. 2.1).

(*i*) *Current-bedding* where the tops of beds are truncated by *younger beds.*
(*ii*) *Graded-bedding* where grains of different sizes have settled at different velocities, coarse grains being at the bottom grading to finer grains at the top.
(*iii*) *Included fragments* where inclusions of rock (such as pebbles) have been derived from an *older* formation.
(*iv*) *Fossils* which indicate the relative age of strata in which they are found.

SEDIMENTARY STRUCTURE

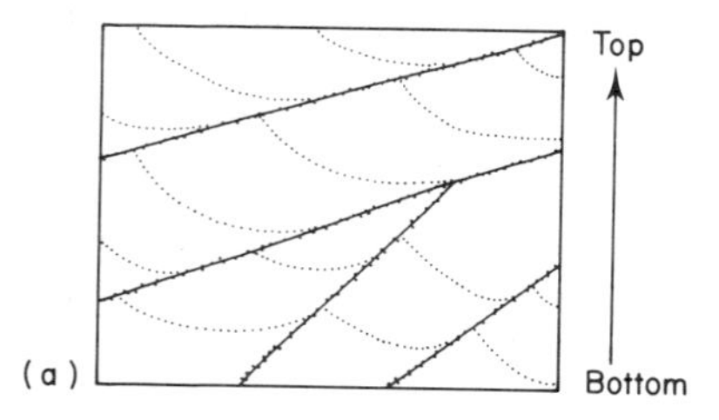

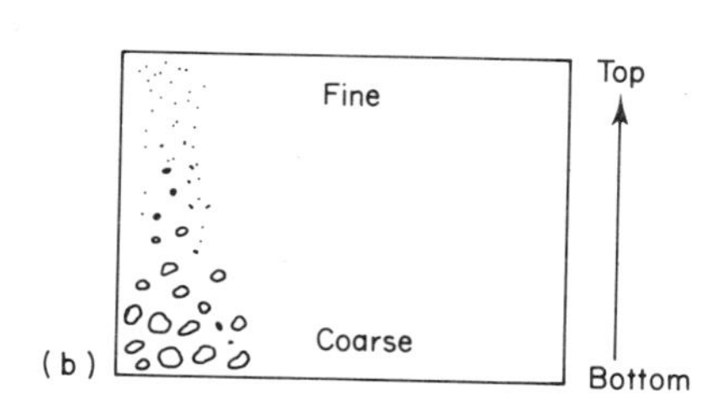

Fig. 2.1 Sedimentary structure. Indications of 'way-up' from small scale structures preserved in sediments: (**a**) current-bedding in a sandstone; heavy lines are erosion surfaces which truncate the current-bedded layers (dotted); the latter are asymptotic to the surface on which they rest (see also Fig.3.19). (**b**) Graded bedding (see text). Note, the grains that constitute a sediment must be derived from rock that is older than the sediment the grains are forming.

Table 2.1 The stratigraphical column showing the divisions of geological time and the age of certain events.

EON	ERA	PERIOD	EPOCH (here given only for the CENOZOIC ERA)	DURATION (10⁶ years)	AGE (10⁶ years)	MOUNTAIN BUILDING (Orogenies) (EUROPE)	MOUNTAIN BUILDING (Orogenies) (N. AMERICA)	NOTES
PHANEROZOIC (= 'evident life')	CENOZOIC (= 'recent life')	Quaternary	Holocene (last 10,000 yrs)	0.01	PRESENT		Coast Ranges	At least 3 major glaciations in N. Hemisphere and changes in sea level from +10 m to −100 m
			Pleistocene	1.99	2	(Himalayan)		
		Tertiary (Neogene)	Pliocene Miocene	23	25	Britain emerges		← First record of hominids ← Red Sea opens
		Tertiary (Paleogene)	Oligocene Eocene Paleocene	40	65	Alpine	Laramide	Australia separates ← from Antarctica ← Primates appear
	MESOZOIC (= 'middle life')	Cretaceous Jurassic Triassic	Each of these Periods is divided into numerous Epochs which can be recognized throughout the world.	79 69 35	144 213 248		Nevadan	S. Atlantic opens. ← Indian & southern oceans open ← C. Atlantic opens. Gondwanaland separates
	PALAEOZOIC (= 'ancient life')	'NEWER': Permian Carboniferous Devonian		38 74 48	286 360 408	Hercynian	Appalachian	
		'OLDER': Silurian Ordovician Cambrian		30 67 85	438 505 590	Caledonian Assyntic	Acadian Taconian	also development in the southern hemisphere of the Samfrau fold belt. ← First appearance of exoskeletal tissue
PROTERO-ZOIC	SINIAN RIPHEAN	All rocks older than the Palaeozoic can be collectively described as PRECAMBRIAN	Many Precambrian rocks are severely deformed and metamorphosed, but large areas of undisturbed Precambrian strata are known. Epochs that can be correlated throughout the world have not been defined	1910	~2500	← 2000 First stable crustal plates		Diamonds, Sn, Cu, NaCl (first appearances) ← Oxygenic atmosphere established, Fe ores ← Great Dyke, Zimbabwe
ARCHEAN				2100	4600	Numerous orogenies probably not involving plate tectonics prior to the development of stable crustal plates		Concentration of dispersed elements to form metalliferous accumulations Cr, Au, U, Pt Origin of the Earth

The stratigraphical column

The sequence of rocks which has formed during geological time is represented by the stratigraphical column, which lists the rocks in their order of age: the oldest rocks are at the base of the column and the youngest at the top (see Table 2.1). The rocks are grouped into Periods many of which are named after the areas where they were first studied in Britain; thus the Cambrian (after *Cambria*, Wales), the Ordovician and Silurian (after the territory of the *Ordovices*, and the *Silures*, both ancient tribes of Wales). Others are named from some characteristic part of their content; thus the Carboniferous refers to the coal-bearing rocks, and Cretaceous to those which include the chalk (Latin *creta*). The Permian was named after Perm in Russia and the Jurassic after the Jura Mountains in Switzerland.

From radiometric dating the absolute ages of many rocks has been found (see column 5 of Table 2.1).

Breaks in the sequence

In many places one series of strata is seen to lie upon an older series with a surface of separation between them. Junctions of this kind are called *unconformities*, some are of local extent, others extend over large areas. The older strata were originally deposited in horizontal layers but often they are now seen to be tilted and covered by beds that lie across them (Fig. 2.2). The upper beds are said to be unconformable on the lower, and there is often a discordance in dip between the younger and older strata. The unconformity represents an interval of time when deposition ceased and denudation took place during an uplift of the area. This sequence of events therefore records a *regression* of the sea prior to uplift and erosion, and the later *transgression* of the sea over the eroded land surface.

Fig. 2.2 Unconformity between horizontal Carboniferous Limestone and steeply dipping Silurian flagstones, Horton-in-Ribblesdale, Yorkshire.

Fold-mountain belts

Mountain building has taken place at intervals throughout geological time and the major periods of mountain building are shown in Table 2.1, column 6. The term *orogeny* is used for this mountain building activity (Greek *oros*, a mountain). In the fold belts the rocks are now seen, after denudation, to have been thrown into complex folds. They are zones of instability in the crust, or *mobile-belts*. The parts of continents adjacent to them are relatively stable areas but subject to vertical or *epeirogenic* movements (Greek *epeiros*, a continent). Epeirogenic and orogenic movements are related to changes in the relative positions of plates of the lithosphere: Fig. 2.3 illustrates an example of this relationship.

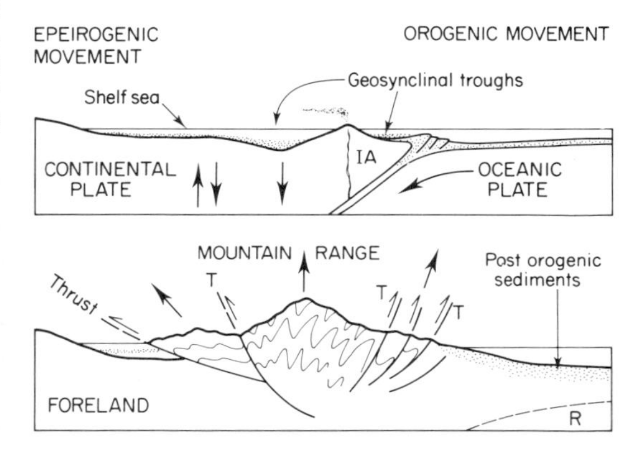

Fig. 2.3 An example of the relationship between epeiric seas, geosynclines and plate tectonics. IA = Island Arc; R = remnant of oceanic plate; T = thrust. Large arrows represent the directions of movement.

Geosyncline

A geosyncline is a large, elongate trough of subsidence. On the subsiding floor of the trough marine sediments are deposited over a long period, in relatively shallow water, as the down warping of the trough proceeds. Later, much coarser sediment derived by rapid weathering of nearby land areas is poured into the trough. Volcanic activity adds igneous material to the accumulation in the trough and the basement is dragged down into hotter depths. In these ways thousands of metres of sediment is concentrated into a comparatively narrow stretch of the crust.

Finally, the contents of the geosyncline are crumpled and broken by thrusts as the sides begin to move together,

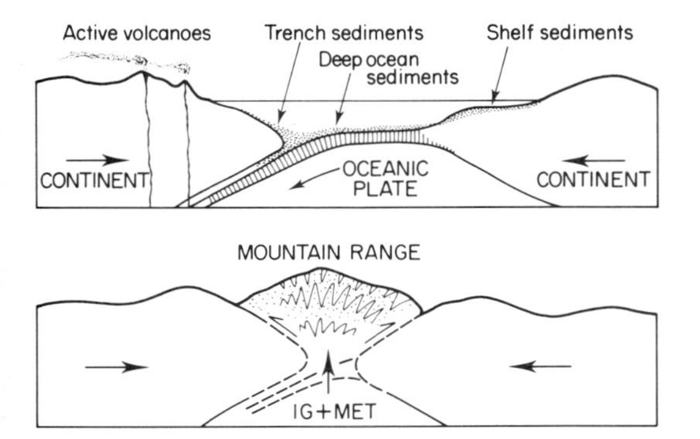

Fig. 2.4 Diagrammatic illustration of mountain building resulting from the collision of continental plates. Rocks at depth are metamorphosed (MET) and may be partially converted to rising igneous material (IG).

and heaved up to form a range of mountains on the site of the earlier trough. This process is illustrated in Figs 2.3 and 2.4.

Precambrian

The ancient rock assemblage of the Precambrian represents some 3600 million years of the Earth's history and comprises all rocks that are older than the Cambrian. To illustrate the immense duration of Precambrian time - if the Earth's age is called 1 hour then the Precambrian would occupy 52 minutes and all other geological periods the remaining 8 minutes.

The Precambrian rocks are largely igneous and metamorphic but also include virtually undisturbed sedimentary deposits which lie in places upon older much altered rocks. Within these sediments can be seen sedimentary structures similar to those formed in present day deposits (Fig. 2.1): they offer convincing evidence for belief in the concept of uniformitarianism. Some of the sediments found in Canada, Norway, South Africa and Australia, are glacial deposits and demonstrate that glacial conditions developed more than once in Precambrian times. In Finland, for example, varved clays (p. 57), now metamorphosed, have been preserved; they were deposited 2800 my ago in a glacial lake of that time.

Many of the metamorphic gneisses were formed at depth below the surface under conditions of high temperature and pressure (discussed in Chapter 7), and are now seen in areas where uplift and erosion have exposed rocks that were once deeply buried. They are penetrated by intrusions of igneous material such as granites, contain mixtures of igneous and sedimentary material (migmatites, *q.v.*), and are frequently traversed by zones in which the rocks are severely deformed.

The large continental areas where Precambrian rocks are close to the present-day surface are called *shields* and are old, stable parts of the present continents that have not been subjected to orogenic folding since the end of Precambrian times. In places undisturbed Cambrian strata lie on their margins and give evidence of their relative age. The location of Precambrian shields is shown in Fig. 2.5.

Valuable economic deposits are found in Precambrian rocks, including the great magnetite ore bodies of Kuruna and Gellivare in north Sweden, and the important iron ores of the Great Lakes area of Canada and the nickel ores of Sudbury. The bulk of the world's metalliferous ores - iron, copper, nickel, gold and silver - come from the Precambrian. Diamonds are mined from the Precambrian at Kimberley, South Africa, and in Brazil.

Precambrian of N.W. Scotland (see British Isles map p. 27)

Many of the features which are characteristic of Precam-

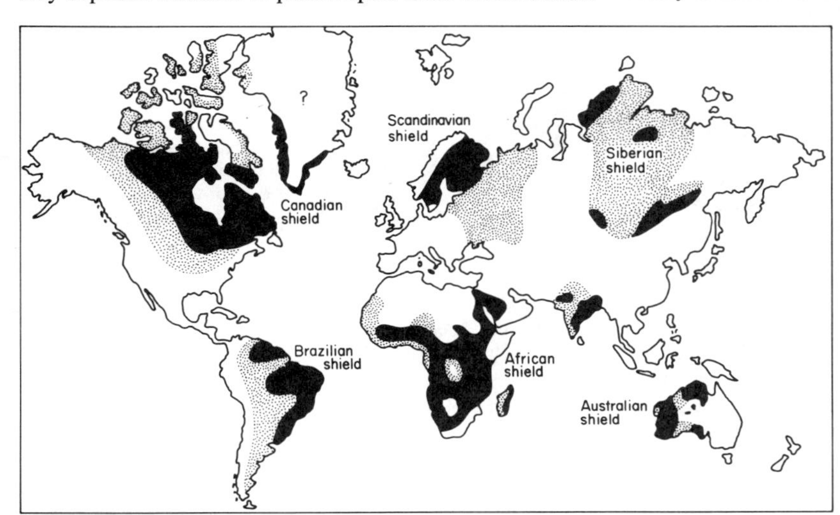

Fig. 2.5 Precambrian shield areas of the world; rocks exposed at the surface are shown in black; platform areas where the Precambrian is covered to a shallow depth by unfolded sediments, stippled.

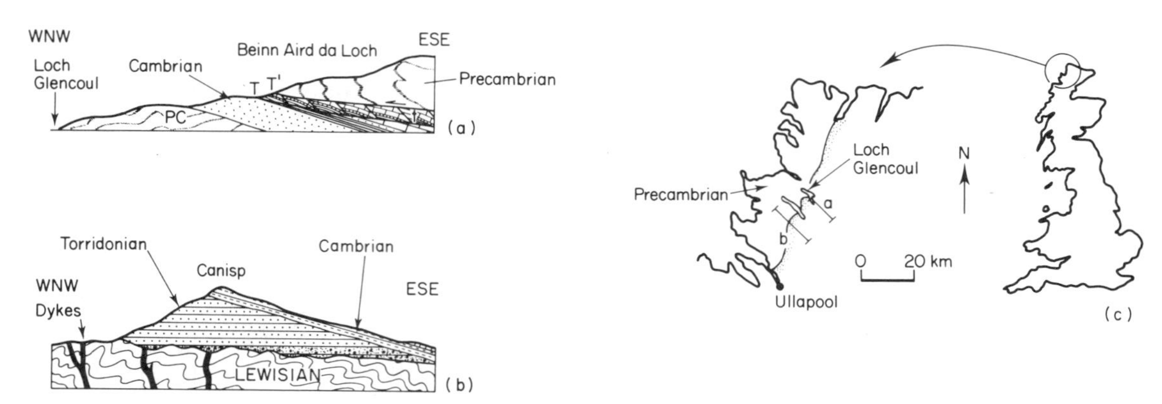

Fig. 2.6 The Precambrian of N.W. Scotland (Glencoul) (**a**) Precambrian thrust (T and T′) over Cambrian that is unconformable on Precambrian. (**b**) Cambrian unconformable on Precambrian Torridonian, itself unconformable on Precambrian Lewisian. Two types of unconformity are shown: the old land surface at the base of the Torridonian, and the tilted plane of marine denudation on which the Cambrian rests. (**c**) Map of location.

brian rocks can be seen in rocks of that age in Scotland: these are shown in Fig. 2.6. The oldest, called Lewisian after the Hebridean island of Lewis which is composed of these rocks, are 3000 to 1700 my old, have been deformed by at least two periods of mountain building and consist of metamorphic rocks cut by igneous intrusions (dykes). They have been eroded and overlain unconformably by the Torridonian series, named from the type locality for these rocks, i.e. Loch Torridon, which consist of brown sandstones with subordinate shales. Boulder beds at the base of the formation fill hollows in the Precambrian topography of the Lewisian land surface (Fig. 2.7). The Torridonian sediments are 800 my old and are themselves overlain unconformably by sediments containing fossils of Cambrian age. Another group the *Moinian*, are upper Precambrian granulites and schists (metamorphosed sediments) which occupy a large area east of the Moine Thrust-zone (Fig. 2.18).

Large thrusts, of which the Moine thrust is an important member (see map on p. 29), have driven slices of Precambrian rock over younger Cambrian strata; Fig. 2.6*a*. These movements were associated with orogenic deformation which occurred 150 my later and built the great range of ancient mountains called the Caledonides.

Careful study of the Precambrian rocks in Scotland has shown that they are closely related to rocks of similar age in the Appalachians, Newfoundland, and Greenland. It is believed these areas were joined as one continent at some time during the Precambrian, although the present location of the shields which contain these remnants (Fig. 2.5) does not represent their original location.

Fig. 2.7 Eroded surface in Lewisian gneisses, Sutherland; in distance a Torridon Sandstone hill, Suilven, rising above the Lewisian floor. Ice from the Pleistocene glaciation has eroded along belts of weaker rock more deeply than in stronger rock giving the Precambrian land surface that was exposed to Pleistocene glaciers a topography of ridges and troughs. T=Torridonian. (Air photograph by Aerofilms Ltd.)

Phanerozoic

The last 13 percent of geological history is represented by the Phanerozoic (Table 2.1) and is distinguished by the development of life. The remains of organisms which

Table 2.2 Development of Phanerozoic life. Ages in 10^6 years

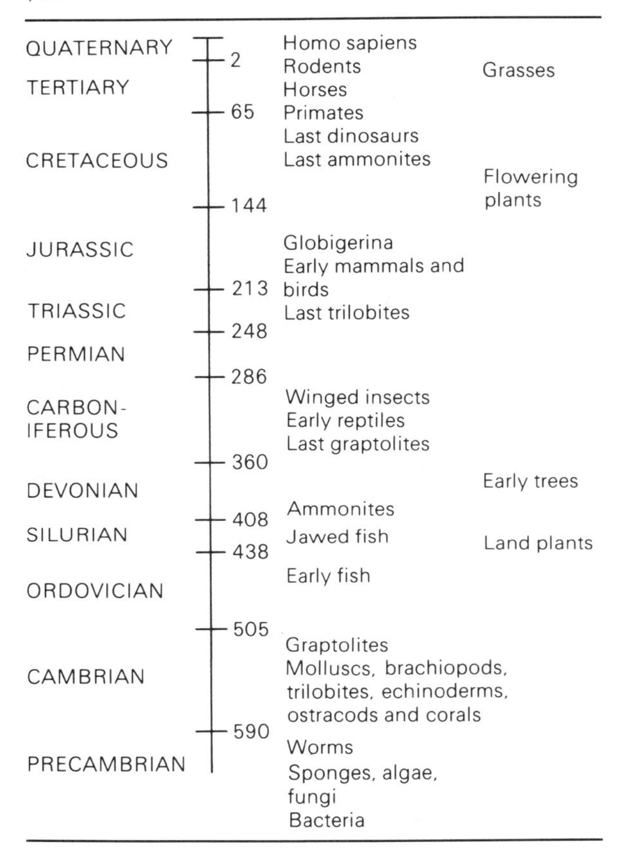

Period	Age	Life	Plants
QUATERNARY	2	Homo sapiens Rodents	Grasses
TERTIARY	65	Horses Primates	
CRETACEOUS	144	Last dinosaurs Last ammonites	Flowering plants
JURASSIC	213	Globigerina Early mammals and birds	
TRIASSIC	248	Last trilobites	
PERMIAN	286		
CARBON-IFEROUS	360	Winged insects Early reptiles Last graptolites	
DEVONIAN	408	Ammonites	Early trees
SILURIAN	438	Jawed fish	Land plants
ORDOVICIAN	505	Early fish	
CAMBRIAN	590	Graptolites Molluscs, brachiopods, trilobites, echinoderms, ostracods and corals	
PRECAMBRIAN		Worms Sponges, algae, fungi Bacteria	

lived when the sediments containing them were formed, are called *fossils* (p. 19) and the history of Earth life has been deduced from studies of the fossil record (Table 2.2).

Older Palaeozoic

Three geological periods, the Cambrian, Ordovician and Silurian make up the Older Palaeozoic rocks, and together cover a span of some 182 my (Table 2.1). They record a long period of marine sedimentation in the oceans between the continents of Precambrian rock and in the shelf seas along their margins.

The number of continents then on the surface of the Earth is not known but four, or more, are believed to have existed, each separated by oceans. One comprised the shield of N. America and Greenland; another the shields of Scandinavia and the Baltic; and a third the shields of Russia and Asia, and a fourth, the shields of S. America, Africa, Antarctica, India and Australasia.

Movement of the oceanic plates against these continents and collisions between the continents, as illustrated in Figs 2.3 and 2.4, produced three extensive mountain ranges. Along the edge of the continent formed by the S. American, African, Antarctic, and Australasian plates was raised the Samfrau fold belt, remnants of which can be found extending from N.E. Australia to Tasmania, in the Ellsworth Mountains of Antarctica, in the Cape Fold Belt (Figs 1.10 and 1.12) and the Sierra de la Vantana of Buenos Aires. The Baltic and Russian shields collided along the line now occupied by the Urals which is believed to have extended into the Franklin range of N. America.

Between N. America and Scandinavia were formed the Caledonides, an ancient range like the others, whose remnants are now found in Scandinavia, the northern part of the British Isles, Newfoundland and the Appalachians. The rocks of the Caledonides were formed in seas at the margins of these converging continents and their stratigraphy records the events of this collision, the period of associated deformation being the Caledonian orogeny. This is described later, because it is an example of mountain building.

Older Palaeozoic fossils

Many forms of life existed and the remains of those that had hard skeletons can be found in profusion: a selection is illustrated in Fig. 2.8.

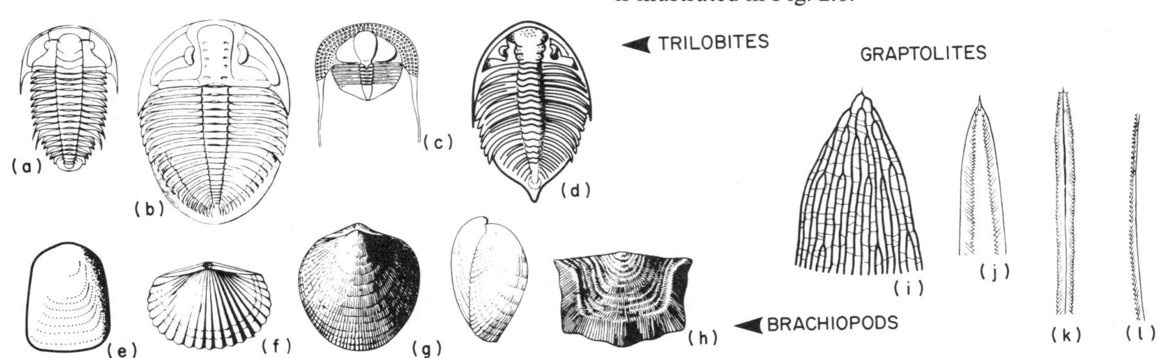

Fig. 2.8 Older Palaeozoic fossils (approximately half life size). Trilobites: **(a)** *Olenus* (Cam); **(b)** *Ogygia* (Ord); **(c)** *Trinucleus* (Sil); **(d)** *Dalmanites* (Sil). Brachiopods: **(e)** *Lingula* (Cam); **(f)** *Orthis* (Cam); **(g)** *Atrypa* (Sil); **(h)** *Leptaena* (Sil). Graptolites: **(i)** *Dictyonema* (Cam); **(j)** *Didymograptus* (Ord); **(k)** *Diplograptus* (Ord); **(l)** *Monograptus* (Sil).

Trilobites lived in the mud of the sea floor and had a segmented outer skeleton consisting of a head, thorax and tail: it was made of a horny substance and divided parallel to its length into three lobes (hence the name).

Brachiopods had a bivalve shell, the two parts being hinged together to form a chamber in which the animal lived. The shells of some early brachiopods (e.g. *Lingula*, Fig. 2.8*e*) were of horny material but as the concentration of calcium and CO_2 in the seas increased, shells of $CaCO_3$ were formed.

Graptolites were small floating organisms comprising colonies of simple hydrozoa which occupied minute cups attached to a stem, the whole resembling a quill pen (hence named from *graphein*, to write). Because graptolites could float, their distribution over the seas was much greater than that of trilobites and brachiopods whose dwelling on the sea floor restricted them to the shelf seas around the continents.

Caledonides

The American and Scandinavian continental areas can be visualized as the left and right continents respectively in Fig. 2.4. Fossils from the shelf sea deposits that fringed them tell us that the continents were situated in the tropics (for example, they contain coral, Fig. 2.9), were drifting northwards, the southern (Scandinavian) more quickly than the northern (American), and converging. The Older Palaeozoic sediments of N.W. Europe and the N.E. America accumulated in the intervening ocean.

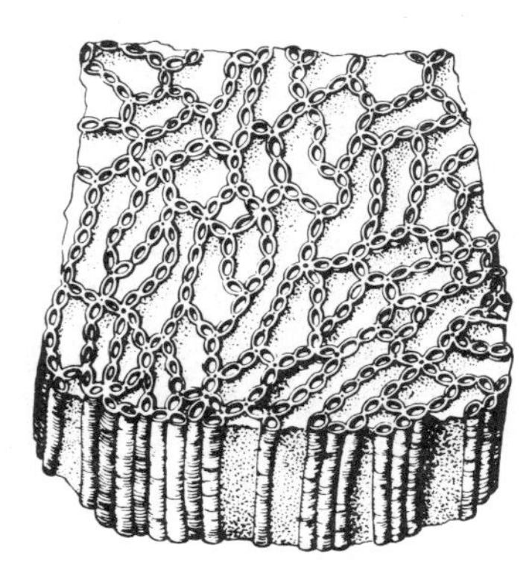

Fig. 2.9 Older Palaeozoic coral (Halysites) from Silurian. Many other coral types flourished at this time.

The fossils from the shelf sea sediments deposited in Scotland and N. America are similar and demonstrate that these two areas were located on the northern margin of the ocean. They differ from the fossils in England and Wales which were located on the southern margin of the ocean. The deep oceanic waters between the two continents acted as a barrier to life forms that inhabited the shallower shelf seas. Only the graptolites could cross the ocean, making them excellent fossils for providing stratigraphic correlation. Towards the end of the Older Palaeozoic the fossil faunas of England and Wales became more closely related to those in America. From this it is concluded that subduction had narrowed the ocean sufficiently for its deeps to be filled with sediment and to no longer provide a barrier to the migration of animals on the sea floor.

The sediments reflect this closure. Shallow water Cambrian sediments formed during the inundation of the continental margins, are overlain by considerable thicknesses of deep sea Ordovician sediments that had been scraped off the oceanic plate as it descends beneath the leading edge of the continent (*cf.* Figs 2.3 and 2.4); much volcanic material is included within them. The Silurian sediments are characteristically shallow water deposits formed when the constricted ocean was almost full of sediment.

Intense mountain building movements began towards the end of the Silurian period and the sediments which had accumulated in the geosyncline between the continents were ridged up into a mountain range, the denuded remnants of which are now seen in the Appalachians, Ireland, Scotland, Wales and Norway. Sediments were thrust over the continents on either side (*cf.* Fig. 2.3) and the Moine and Glencoul thrusts (Figs 2.6, 2.18) are two of many such surfaces: similar thrusts are found in Newfoundland. Others facing in the opposite direction exist in S.W. Sweden. Compression of the sediments created in many a slaty cleavage (p. 133) and the fine grained deep sea deposits so affected were converted into the familiar slate used for roofing. The orogeny continued into the Devonian.

Rising from the root of the mountain range were granitic and other igneous intrusions which became emplaced within the fold belt: they include the large granite and granodiorite masses of the Central Highlands of Scotland. The metamorphism which occurred in the root of the Scottish mountains is shown in Fig. 7.6. Similar intrusions and metamorphism occurred along the length of the mountain range.

To the north of the Caledonides lay the continent of Laurasia, i.e. the shields of N. America (or Laurentia), Greenland, Scandinavia, Baltic, Russia and part of Asia. South of the Caledonides an extensive plane sloped down to the ocean which separated Laurasia from the southern continent of Gondwana. This was centred over the southern pole and contained the shields of S. America, Africa, Antarctica, India (from whence the name Gondwana came) and Australasia.

Newer Palaeozoic

Rocks of three periods, the Devonian, Carboniferous and Permian make up the Newer Palaeozoic and represent some 160 my of geological time. They record the gradual northerly drift of Gondwana and its collision with Laurasia to produce the Hercynian fold belt of N. America and Europe.

In Laurasia the great mountain chains were being denuded and their debris spread across the continent during the Devonian to form continental sediments of *land facies*. The southern part of the continent lay in the tropics and large areas of the continent were desert. Much debris was spread by sudden flash floods from the mountains and collected into deposits of coarse red sands and breccias along the foot of the mountain slopes. Finer material was laid down in lakes and deltas. These deposits are often referred to as the Old Red Sandstone. At the margins of the continent *marine* deposits were accumulated, often as shales and sandstones. Sedimentation continued for about 50 my but then the southern edge of Laurasia began to sink and a marine transgression covered the land. This marked the beginning of the Carboniferous, the period during which the coal basins of N. America and Europe were formed.

This transgression is believed to represent downwarping of a continental margin, caused by subduction of an oceanic plate beneath it. With reference to Fig. 2.4, the left continent can be visualized as Laurasia and the right as Gondwana, which was moving north. Sediments accumulated in a series of trenches that extended from the Baltic to the Appalachians. To their north existed a shallow shelf sea (as in Fig. 2.3) in which thick deposits of limestone were formed (e.g. the Carboniferous Limestone of the British Isles). Further north large deltas were flooding across the shelf bringing coarse sand and grit from the denudation of the Caledonian mountains inland (the Millstone Grit is such a deposit). On these deltas developed and flourished the swamps which supported dense growths of vegetation that later were compressed under the weight of overlying sediment to become coal, so forming the Coal Measures. These basic divisions are shown in Table 2.3.

Throughout this time Gondwana had drifted northwards and by the end of the Carboniferous the ocean separating them had almost disappeared: mountain building had commenced and the period of deformation that followed is called the Hercynian orogeny.

With this Laurasia and Gondwana were joined to form one huge continent called Pangaea (all lands), which marked the end of the Newer Palaeozoic and the Palaeozoic Era (Table 2.1). In the south Gondwana was under the ice of the Gondwana glaciation (Fig. 1.12) as recorded by the Dwyka Conglomerate (a tillite *q.v*). Farther north, in the tropics, were accumulating the red desert sands of the Permian, derived from denudation of the Hercynian Mountains. They are similar to those deposits formed during the Devonian and are called the New Red Sandstone (Fig. 3.36). Extensive and thick deposits of saline sediment and salt formed in the tropical gulfs and embayments of the continent. These saline conditions did not permit aquatic life to flourish and many species became extinct. Around the margins of the continent between latitudes in which temperate climate existed, there accumulated sequences of more normal marine sediments. Those now in the Russian province of Perm provide the standard marine sequence for the period, which is named after the province.

Table 2.3 Basic divisions and global correlation of Carboniferous strata. Ages in millions of years

SUB-PERIOD	*DIVISIONS*				
	N. AMERICA	W. EUROPE			RUSSIA
286					
PENNSYLVANIAN	UPPER PENNSYLVANIAN	Stephanian 296		SILESIAN	UPPER CARBONIFEROUS
	LOWER PENNSYLVANIAN	Westphalian 315	COAL MEASURES	SILESIAN	MIDDLE CARBONIFEROUS
320					
MISSISSIPPIAN	UPPER MISSISSIPPIAN	Namurian 333	MILLSTONE GRIT	SILESIAN	LOWER CARBONIFEROUS
	LOWER MISSISSIPPIAN	Visean 352	CARBON-IFEROUS LIMESTONE	DINANTIAN	
		Tournaisian 360		DINANTIAN	
360					

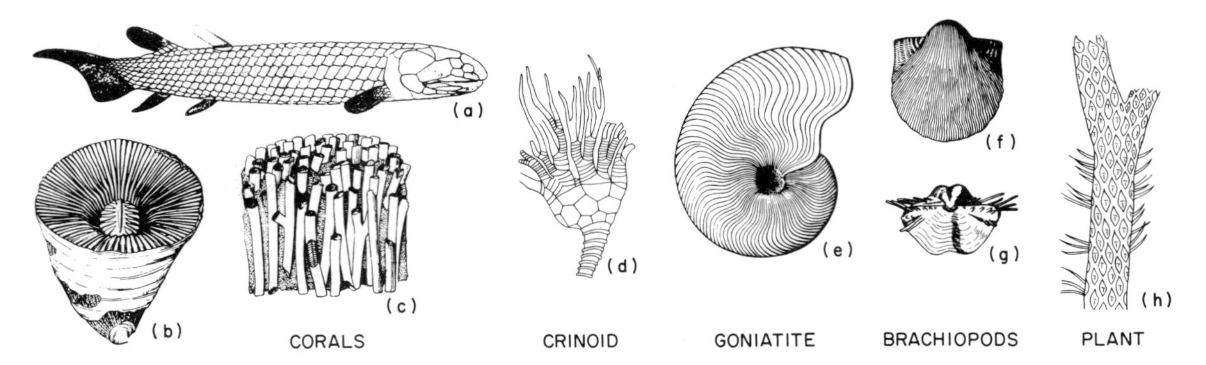

Fig. 2.10 Newer Palaeozoic fossils (not shown to scale). Fish: (**a**) *Thursius* (Devonian); Corals: (**b**) *Dibunophyllum*, (**c**) *Lithostrotion* (both Carboniferous); Crinoid (**d**), Goniatite (**e**) (both Carboniferous); Brachiopods: (**f** and **g**): *Productus* (Carboniferous and Permian); Plant: (**h**) *Lepidodendron* (Carboniferous: Coal Measures).

Newer Palaeozoic fossils

A selection of Newer Palaeozoic forms is illustrated in Fig. 2.10. *Fish*, which appeared in the Older Palaeozoic (Table 2.2) are abundant in the Old Red Sandstone and have continued their line to the present day. *Brachiopods* continued to prefer the shelf seas and occur in abundance in the Carboniferous Limestone. *Corals* also developed in these seas and could grow as solitary forms (*b*), but many grew in colonies (*c*) and built reefs. Their presence permits us to assume that the Carboniferous shelf seas were clear, warm and shallow, resembling those in the south Pacific at the present day. Remains of ancient coral reefs occur as mounds in some limestones, known as *reef-knolls* or *bioherms*. As the reef is porous it can become a reservoir for oil and gas which may subsequently enter it. *Crinoids* (*d*) had a calcareous cup that enclosed the body of the animal, with arms rising from it: the cup was supported on a long stem made of disc-like ossicles. *Cephalopods* had a shell coiled in a flat spiral and divided into chambers by partitions at intervals around the spiral (*cf.* the modern *Nautilus*). The *goniatites* (*e*) are an important group and used as zone fossils in the Coal Measures; particular species permitting correlations to be made between coal seams in different coalfields. The remains of plants such as *Lepidodendron* (*h*) are preserved in some coals together with pollens and spore cases. All manner of insects and reptiles also developed.

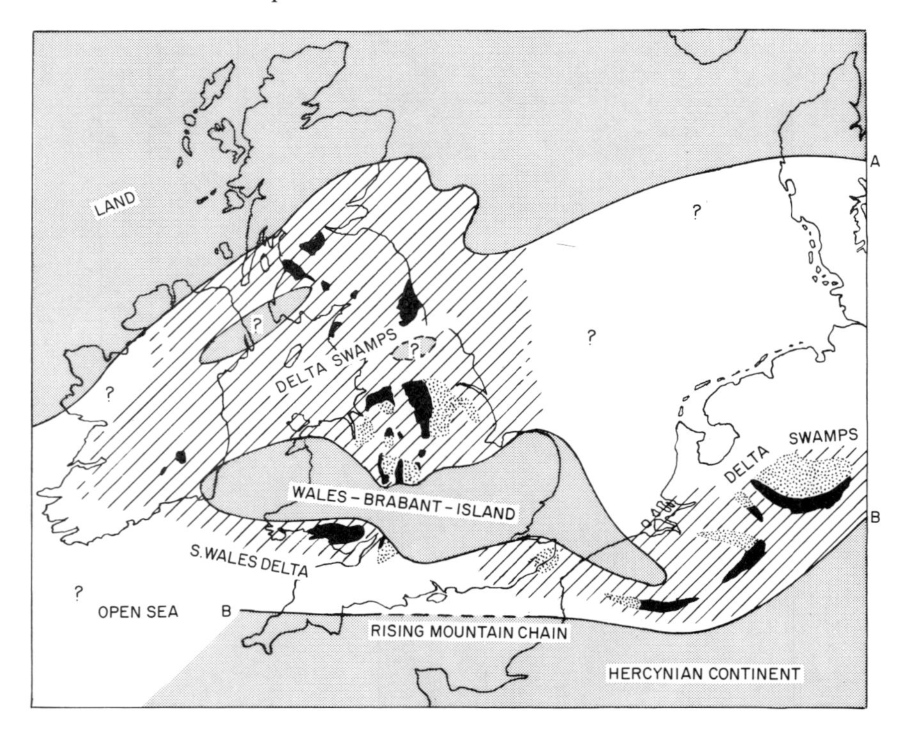

Fig. 2.11 Probable geography of north-west Europe in Coal Measure times (after L.J. Wills and D.V. Ager, 1975). Present day coalfields shown in black, underground extensions dotted; they lie to the north of the Hercynian mountain front.

The Hercynian orogeny

The map shown in Fig. 2.11 illustrates a reconstruction of part of the Hercynian geosyncline as it may have appeared in Coal Measure times. From the account of its development given earlier, it will be realized that the rising mountain chain was located between Gondwana to the south and Laurasia to the north. Coal forest swamps covered the deltas.

From time to time submergency of the swamps took place and the growth of vegetation was buried by incoming sand and mud; the swamp forest then grew again at a higher level. This was repeated many times during oscillations of level to produce a series of sandstones and shales with coal seams at intervals. Beneath each coal seam a layer of fire-clay or ganister (*q.v.*) represents the ancient *seat-earth* in which the vegetation grew (see also Fig. 6.18). In some shales above the coal seams marine fossils are present; such layers are called *marine bands* and show that subsidence had been rapid enough to drown the forest growth temporarily beneath the sea.

The oscillations which produced the repeated sequences of coal and marine bands have been attributed to sudden movements of the continental edge, each period of subsidence recording a small downwarping of the crust as the leading edge of the continent buckled under the lateral forces of the orogeny.

To the south lay the main trough of the geosyncline where thick deposits of marine sediment accumulated, to be raised up as the Hercynian mountain chain (named after the Harz Mountains of Germany). The coal basins to the north were folded less severely. This mountain chain lay to the south of the Caledonides and extended from Romania into Poland, Germany and France. It continued into the southern part of the British Isles and the Appalachians, and in these areas it overprints its structure upon the earlier structures of the Caledonian orogeny.

Granites were intruded into the root of the mountain chain, those of S.W. England, Brittany and Saxony being examples. Further north basic igneous rocks were intruded at higher levels in the crust forming sills and dykes of dolerite: the Whin Sill, which underlies much of northern England, is of this age. Igenous activity reached the surface in Scotland where many volcanoes were pouring forth basalt lavas and ash in the Midland Valley of Scotland, including the basalts of Arthur's Seat in Edinburgh.

The orogeny also created extensive deposits of valuable minerals and veins of lead, zinc, copper and massive sulphide deposits are located along the fold belt.

Mesozoic

The Triassic, Jurassic and Cretaceous periods comprise Mesozoic era and account for approximately 183 my of geological history. The era begins with a single continent and ends with it divided from north to south, its two parts separated by the oceanic ridge of the embryo Atlantic, and drifting apart to east and to west.

Initially, conditions were similar to those in the Permian. In many places the continental deposits of the Triassic are indistinguishable from those of the Permian and are collectively called Permo-Trias. As the continental uplands were eroded the sandy and pebbly deposits derived from them flooded across the adjacent lowlands. From time to time salt lakes were formed. These conditions existed across Laurasia, from Arizona to New York, Spain to Bulgaria and on into China. Much of Gondwana was also being buried under continental deposits. In S. Africa, coal forests were flourishing and demonstrate that the northerly drift of Gondwana had carried S. Africa away from the glaciers of the S. Pole and towards the tropics.

Other movements, of considerable extent, must have also been occurring, for along the entire western edge of the continent there developed the fold belt of the Cordillera, stretching 10 000 km from Alaska to New Zealand. The eastern side extended in two wings of land as if shaped similar to the letter C. To the north were the shields of eastern Russia and Asia, and to the south the shields of Africa, India, and Australia (see Fig. 2.12). Between them lay the ocean called Tethys in which was gathered the sediments that were to be folded by the Alpine orogeny to form the Alps and Himalayas. These eastern wings began to close like a nut-cracker as the shields of Gondwana continued their northerly drift.

In Europe a shallow sea advanced slowly across southern Laurasia depositing a thin sequence of clays (the Rhaetic); this marks the end of the Triassic. A shelf sea developed in which were deposited the extensive European sequences of Jurassic limestones and clays. Laurasia had started to split apart along an opening that was to become the Atlantic. N. America began to separate from Laurasia and move westward, and resulted in great thicknesses of marine sediment and volcanic rock becoming condensed in an elongate trough which extends from Alaska to Mexico. Conditions on this margin were similar to those illustrated in Fig. 1.19. From this accumulation the Sierra Nevada was later formed and into its folds large bodies of granite were intruded: the continuation of their line forms the Baja California.

These conditions continued into the Cretaceous and by the Middle Cretaceous the Atlantic had fully opened, with the Americas to the west and Europe and Africa to the east. N. America continued its westward drift, and this movement assisted the formation of a trough in which the sediments of the Rocky Mountains were gathered, to be raised as a mountain chain by a later stage in the history of Coridilleran mountain building, called the Laramide orogeny (Table 2.1). The western margin of S. America was the site of similar activity. Active volcanism extended along the length of the margin (Fig. 2.13) and immense intrusions of granite began to invade the roots of the Andes. Many of the conditions in western America which developed during the Mesozoic, have continued to the present day, the entire west coast being an area of considerable instability (see Figs 1.4 and 1.17).

The end of the Mesozoic in Europe was heralded by a

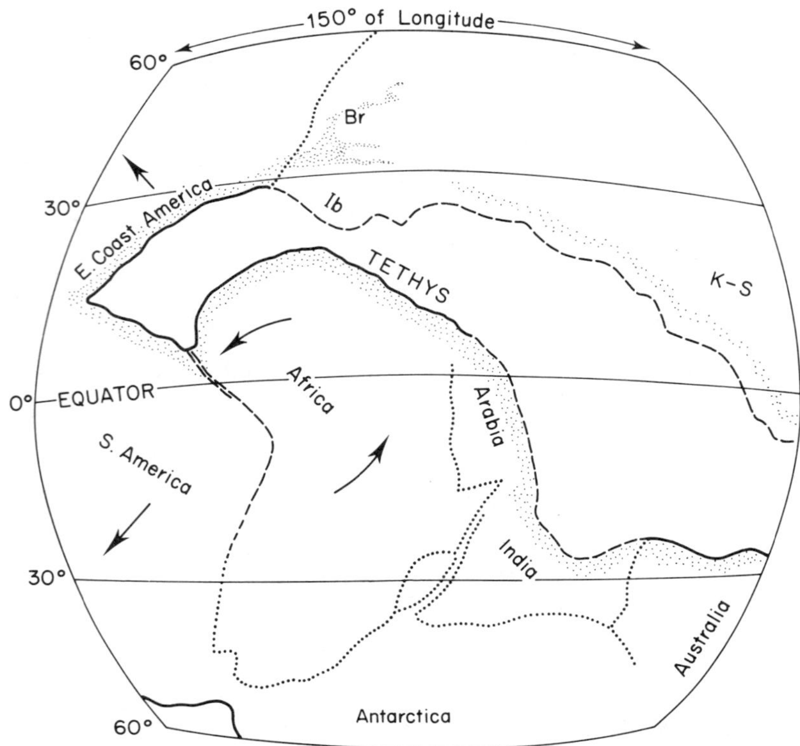

Fig. 2.12 Late Jurassic framework for the Alpine orogeny (based on Smith, Hurley and Briden, 1981; Ager, 1975; Jenkyns, 1980). Ib, Iberia; Br, Britain; K–S, Kazakhstan–Sinkiang.

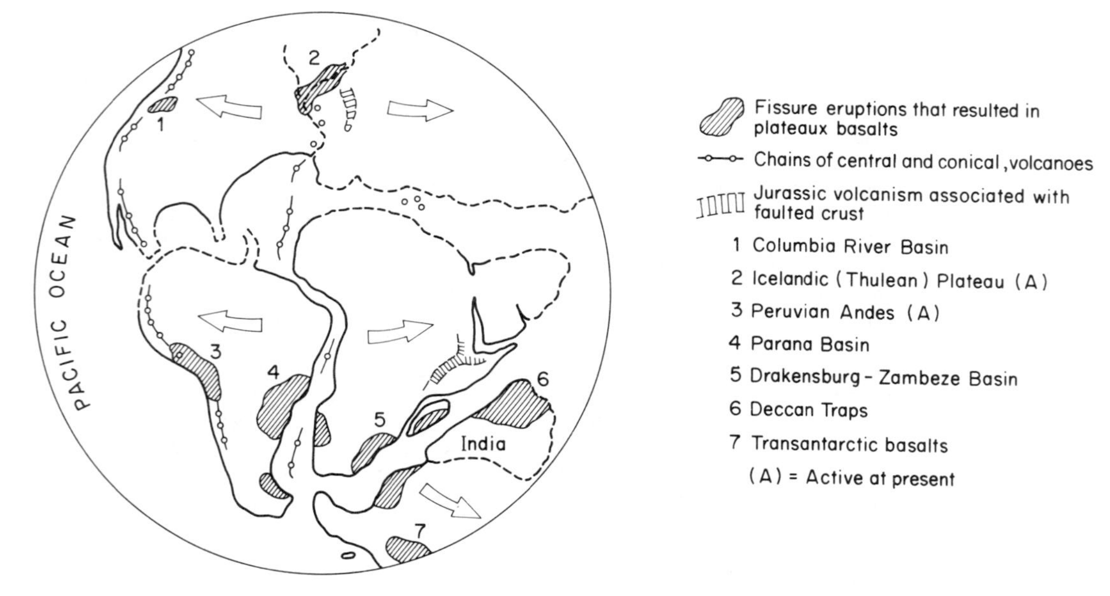

Fig. 2.13 Major volcanic centres of Late Mesozoic and Cenozoic age. (Position of continents is that of Late Jurassic: 150 my ago.)

great transgression of the sea over the land north of Tethys, called the *Cenomanian transgression*. In this sea was deposited a white limestone, the Chalk, over a wide area of Europe. Even greater changes were to follow, as the Alpine orogeny had commenced.

Mesozoic fossils

Some Jurassic and Cretaceous aquatic forms are shown in Fig. 2.14. Among them, the *Ammonites* (*a*) had a coiled planar shell; they are considered to have been one group of the cephalopods. The *Echinoids* (*f, g*) or 'sea-urchins', lived on the sea floor together with many forms of brachiopod (two are shown, *c, d*) and lamellibranch (three are shown, *h, i, j*). *Gastropods* (*e*), resembling the modern snail, also developed many species.

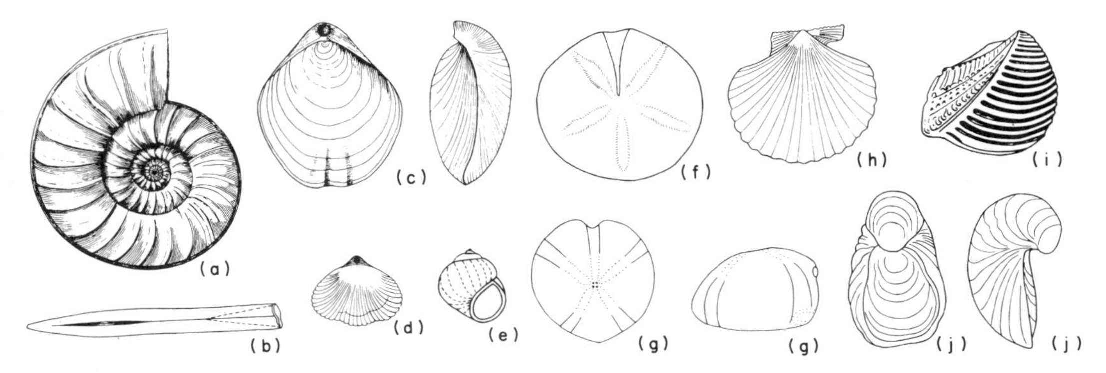

Fig. 2.14 Mesozoic fossils: ×$\frac{1}{8}$ except (**a**). Ammonite: (**a**) *Asteroceras* (Jur), (**b**) A belemnite. Brachiopods: (**c**) *Terebratula* (Cret), (**d**) *Rhynchonella* (Jur). Gastropod: (**e**) *Paludina* (Cret). Echinoids: (**f**) *Clypeus* (Jur), (**g**) *Micraster* (Cret). Lamellibranchs: (**h**) *Pecten* (Jur), (**i**) *Trigonia* (Jur), (**j**) *Gryphaea* (Jur).

The aquatic forms are commonly found as fossils in the Mesozoic sediments, but in addition to them there is ample fossil evidence to demonstrate that throughout the Mesozoic there developed large reptiles including dinasaurs, primitive birds such as *Archaeopteryx*. The earliest mammals also appeared.

By the end of the Mesozoic the continents were supporting modern types of deciduous trees, such as fig, poplar and magnolia, many types of flowering plants and grasses (Table 2.2).

Cenozoic

The Cenozoic Era takes in the last 65 my of geological time and is divided into Tertiary and Quaternary periods, with sub-divisions as shown in Table 2.1. Deposits of this age are a common occurrence close to ground level, often forming soils that cover, unconformably, harder and older rock at depth.

During this era the continents drifted to their present position. The Americas had already broken away from Pangaea, their line of separation having been the site of tension in the crust and fracture throughout the late Jurassic. From this zone had erupted great quantities of basalt lava from depths in the crust, to form piles of basalt lava flows which covered a considerable area: these are the fissure eruptions shown in Fig. 2.13. In the N. Atlantic a big lava field was accumulated the relics of which are seen today in Greenland, Iceland, the Faroes, Antrim and the western islands of Scotland. In the S. Atlantic fissure eruptions had created the basalts of the Parana Basin, in Brazil, their continuation being found in southern Africa (Fig. 2.13). Similar outpourings of basalt lava also occurred at about this time in the Zambeze Basin of Africa and the Deccan of India (Fig. 2.13), and indicate that the shields of India, Antarctica and Australia were detaching themselves from Africa along lines of tensile failure similar to that which developed in the mid-Atlantic. These continents were now free to wander in their separate ways and enabled India to commence its remarkable journey north (Fig 2.15). Africa and Arabia were also moving in that direction and the three

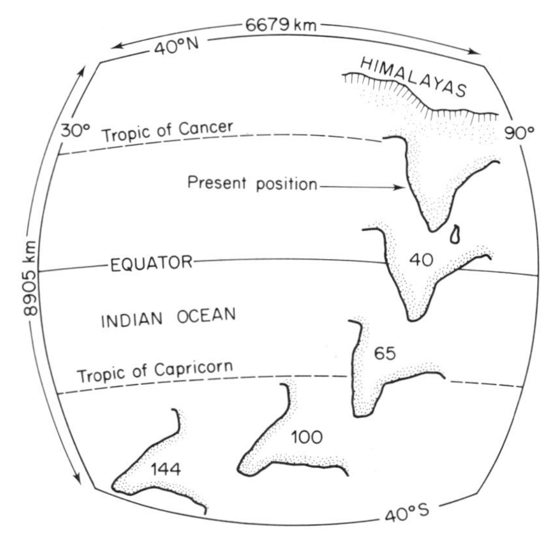

Fig. 2.15 Northward drift of India (based on Smith, Hurley and Briden, 1981). Ages in millions of years.

continents combined to deform the sediments which had accumulated in Tethys (Fig. 2.12). This compression of the sediments was the culmination of the Alpine orogeny and produced the Alps and the Himalayas.

Alpine orogeny

This is the most recent orogeny to affect Europe and seismicity in the Mediterranean (Fig. 1.4) demonstrates that it remains in progress. The lateral forces which compressed the sediments in Tethys produced severe deformation of the crust along the arcuate fold belt of the Alps, as shown in Fig. 2.16. This began in the mid-Cretaceous and reached its climax in the Miocene. The greater part of Europe was affected, with the formation of gentle folds and basins in the north, and complex folding in the south. Two vertical sections are shown in Fig. 2.17 to illustrate this difference in the severity of deformation; their location is shown on Fig. 2.16.

The Alpine fold belt continues east into Turkey, Iran,

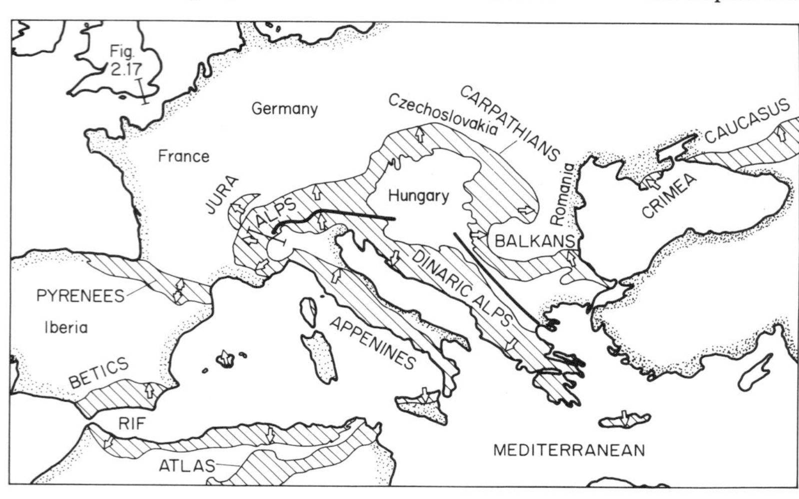

Fig. 2.16 The Alpine fold-belts of Europe; in the eastern chains the main movements (arrows) began in the middle Cretaceous while in the west they were largely Tertiary (Eocene to Miocene). Heavy black line is the suture where much oceanic crust was probably consumed. (Based on D.V. Ager, 1975, by permission.)

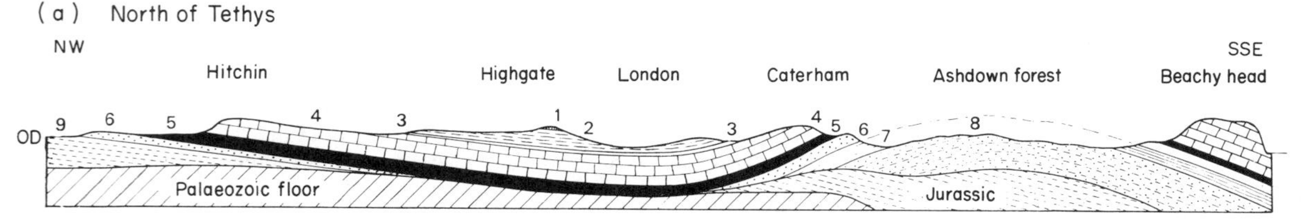

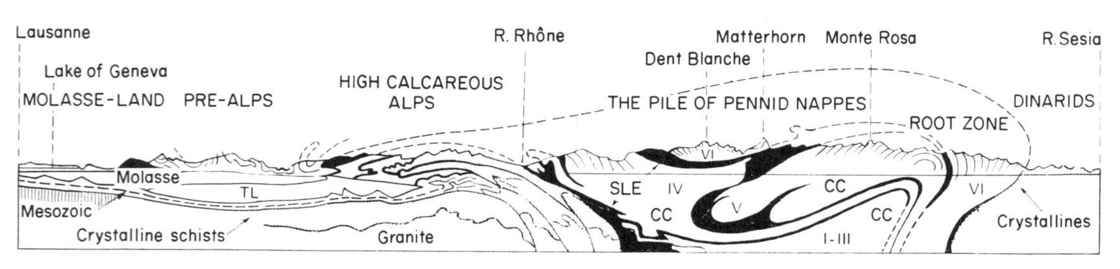

Fig. 2.17 Vertical sections illustrating folding of Alpine age. (**a**) Section across the London Basin and the Weald showing the Palaeozoic floor beneath London. Length of section about 160 km; vertical scale exaggerated about 20 times. (1–3: Tertiary): (1) Bagshot Sands; (2) London Clay; (3) Woolwich and Reading Beds and Thanet Sand. (4–8: Cretaceous): (4) Chalk; (5) Gault and Upper Greensand (the latter is not continuous under London); (6) Lower Greensand; (7) Weald Clay; (8) Hastings Sands; (9) Jurassic rocks.
(**b**) Section across the Alps from N.W.–S.E., distance about 140 km. The molasse (Oligocene sandstones and conglomerates) is formed of debris eroded from the rising Alps and is overthrust by Mesozoic sediments (centre of section). The Pennid Nappes contain big recumbent folds and are named after localities: Simplon (I–III), Great St Bernard (IV), Monte Rosa (V), Dent Blanche (VI). Much structural detail was elucidated from deep railway tunnels in the region. TL=Triassic 'lubricant' (rock-salt); CC, crystalline core; SLE, schistes lustrés envelope, surrounding the core.

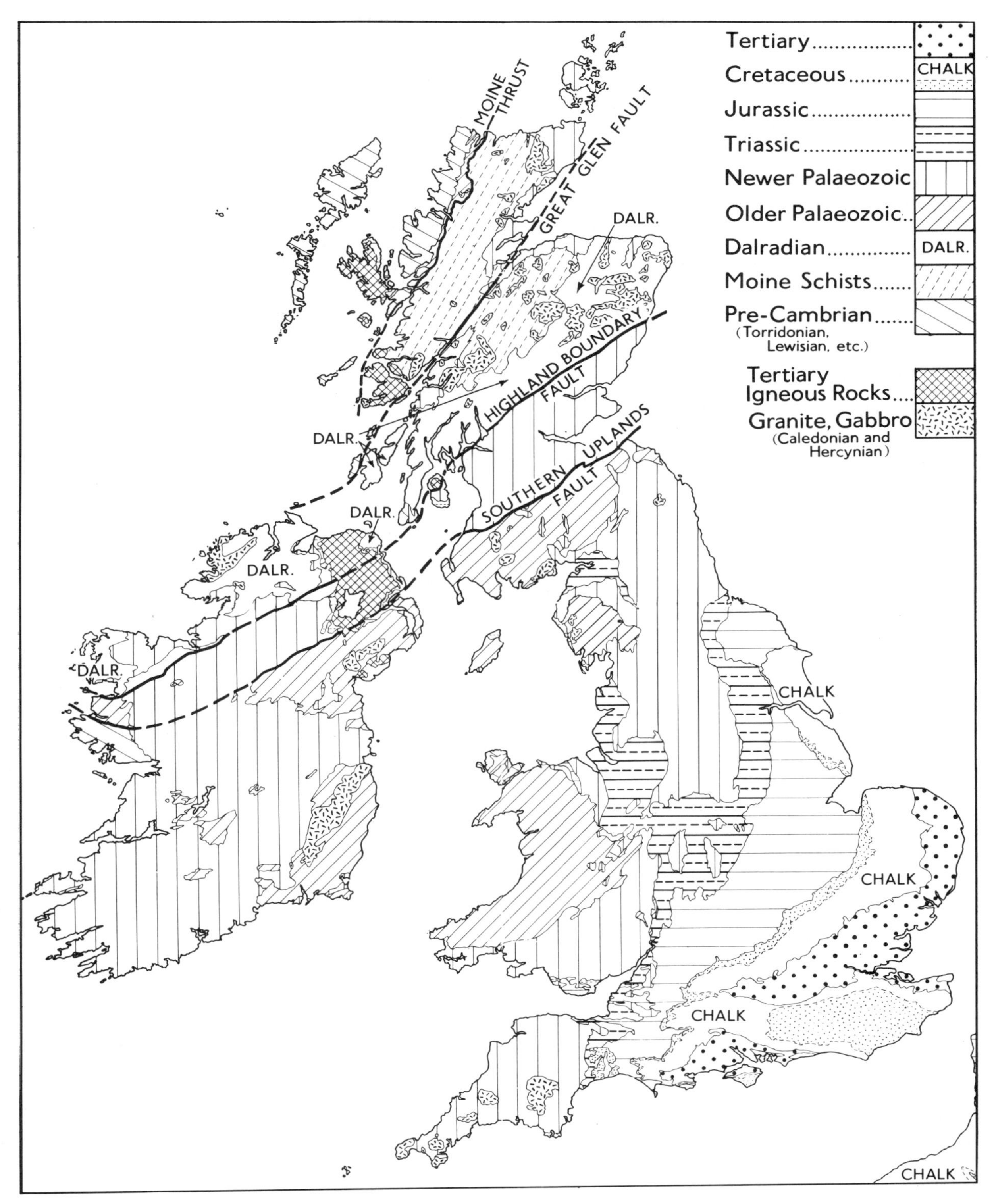

Fig. 2.18 Outline geological map of the British Isles. Scale approximately 1:5 000 000.

Afghanistan, Tibet and Nepal, and on to merge with folds of similar age in Burma, Thailand, Sumatra and Java. Many features of the fold belt are similar to those already described for the Caledonian and Hercynian mountains, for example, the overthrusting of folded sediments, the metamorphism at depth and the intrusion of granite into the roots of the mountain chain.

Orogenies in America

The orogenic movements that were located along the entire length of the western margin of America and which had started in the Mesozoic, were sustained during the Cenozoic by the continued western drift of the American plates. In N. America the deformation of sediments to form the Rocky Mountains, i.e. the Laramide phase of the orogeny, was virtually complete by the Oligocene (Table 2.1). To their west were developing the Cascades and Coast Ranges, and the San Andreas fault system. In S. America the western Andes continued to develop along the junction between the S. American plate and the oceanic Nazca plate which had been sinking beneath it by subduction. Much volcanism has accompanied the building of this mountain range, which also contains extensive intrusions of granite.

Quaternary

Quaternary deposits are more widespread than those for any other period. Some are subaerial, e.g. the screes from rock slopes and the laterites from tropical weathering of ancient land surfaces; others are aquatic such as the sands and gravels in river valleys, and beach and lagoon deposits on shore lines. Others may be volcanic, formed of ash and lava and many will have been made by glacial action. The deposits created during the Quaternary are frequently so different from the harder, more solid rocks they cover, that they are described as 'Drift'. The map of Britain in Fig. 2.18, p. 27 shows the distribution of rocks as they would appear when all Quaternary deposits had been removed.

Table 2.4 Major glacial stages (in capitals) for Europe and N. America. Sea levels, with reference to metres above (+) or below (−) present sea level, are shown in circles

AGE from PRESENT	STAGES N. AMERICA	EUROPE N. EUROPE	EUROPE ALPS		Sea level
		Flandrian (Temperate)		20 000	−5
50 000	WISCONSIN	DEVENSIAN	WURM~WEICHSELIAN	80 000	−100
100 000			RISS~WURM	100 000	
	Sangomanian	Ipswichian (Temperate)			
150 000			RISS~SAALE		
200 000				200 000	
250 000		WALSTONIAN	MINDEL~RISS		
300 000	ILLINOIAN			300 000	
350 000		Hoxnian (Cool)			
400 000		ANGLIAN	MINDEL~EISTER		
450 000		470 000			−150
500 000	Yarmouthian	Supra~Waltonian	GUNZ~MINDEL	500 000 700 000	
1 000 000	KANSAN			900 000	
1 500 000	Aftonian NEBRASKAN	Waltonian	DONAU	1 400 000 1 600 000	
2 000 000					

Pleistocene glaciations

In the Pleistocene an increasingly cold climate became established and large ice sheets developed over the polar regions and adjacent lands. N. America, Greenland, N. Europe and Russia were covered by ice, and many mountain ranges such as the Alps, supported local ice-caps (Fig. 2.19). The Antarctic was also covered by ice, of which the present ice-cap is a dwindling relic.

The glaciation was not a continuous process but interrupted by several *interglacial* episodes during which the climate temporarily improved and the ice fronts retreated. Table 2.4 lists the major glacial stages.

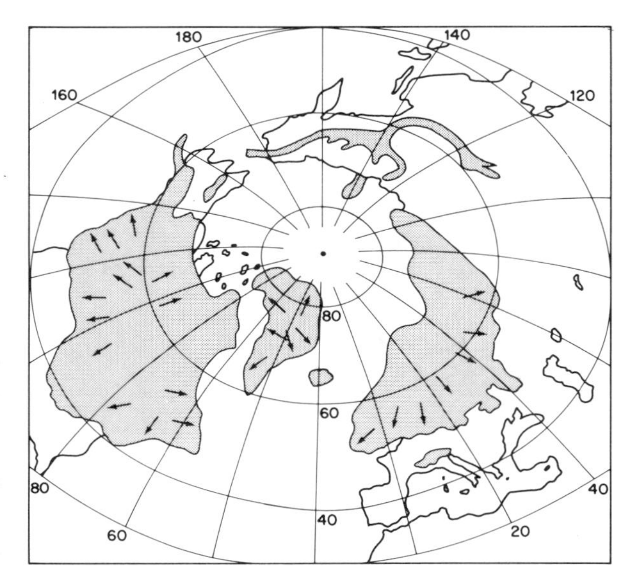

Fig. 2.19 The Pleistocene glaciation in the northern hemisphere (after E. Antevs).

The weight of thick ice on an area resulted in the land being depressed – an *isostatic* effect (p. 6). But with the melting and retreat of the ice as conditions ameliorated the land areas were re-elevated. In Scandinavia there is evidence for a maximum rise of 215 m since the end of the Ice Age some 10 000 years ago (Fig. 2.20).

Another effect of the glaciation, during which vast quantities of water from the oceans were locked in the ice-sheets, was to produce a world-wide fall in sea level, estimated at almost 100 m. With the melting of the ice at the end of the Ice Age much water was gradually released, leading to a word-wide rise in sea level. This *eustatic* change in water level continues at the present day as the remaining ice-caps slowly melt.

Isostatic compensation was not uniform but locally re-elevated the land in stages. Where the isostatic rise has been greater than the eustatic change, coastal land has emerged above sea-level for a time with the formation of *raised platforms* or 'beaches'. These fringe many continents and mark stages in the process when the two effects were nearly equal for a long enough time for a beach to be formed by marine erosion (Fig. 2.21).

Cenozoic fossils

Fossils within the Tertiary sediments record a great diversification of life as new species developed on the isolated continents that were drifting across the latitudes. Mammalian life flourished and plants and animals came to resemble present day forms by the late Neogene (Tables 2.1 and 2.2). Some commonly found Tertiary fossils are shown in Fig. 2.22. Microfossils of Protozoa are of special importance because they indicate the temperature and salinity of the water they inhabited. Remains of microscopic animals can be found, in various states of preservation, in all the marine sediments of the Phanerozoic

Fig. 2.21 Raised marine platform, Isle of Islay, Scotland, formed in late-Glacial times after isostatic uplift, and covered by gravels of the '100-foot' beach (from a Geological Survey photograph).

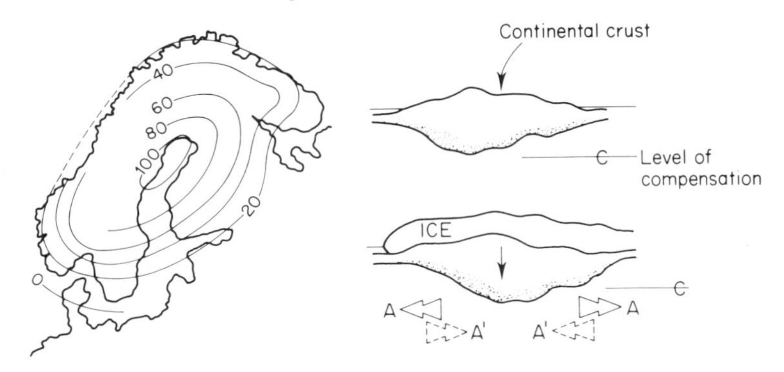

Fig. 2.20 Post-glacial uplift of Scandinavia (in metres). A, signifies the migration of material in the asthenosphere following ice loading. When the ice load is removed isostasy restores the level of compensation at a rate permitted by the return of material in the asthenosphere (A'). (From Woldstedt, 1954.) The present rate of uplift is 2 mm/year in southern Sweden and 9 mm/year in central Sweden.

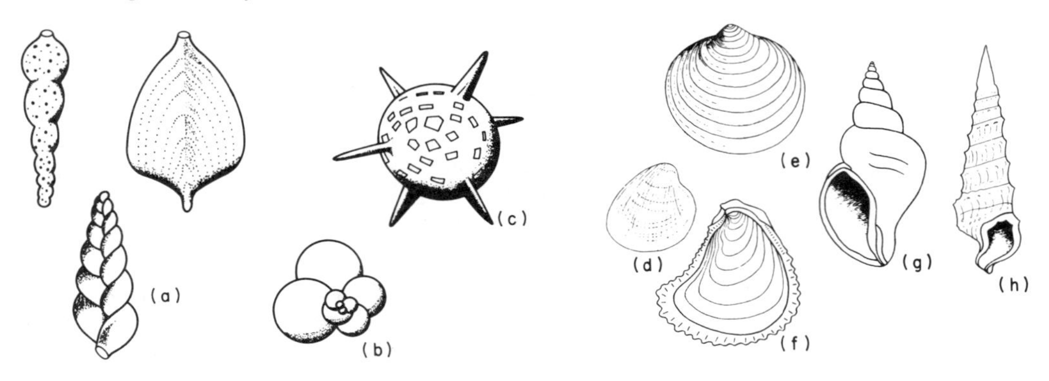

Fig. 2.22 Tertiary fossils. Microfossils (×40): (**a**) foraminifera, (**b**) globigerina, (**c**) radiolarian. Macrofossils (× ×$\frac{1}{3}$) Lamellibranchs: (**d**) *Cardita*, (**e**) *Pectunculus*, (**f**) *Ostrea* (an oyster); Gastropods: (**g**) *Neptunea*, (**h**) *Cerithium*.

and are excellent for dating and correlating strata because of their extensive distribution and rapid evolution. They are particularly valuable in oil field drilling operations where they can be extracted from either the core or the drilling muds and used to identify layers in the sediment that have been drilled through.

Selected bibliography

Ager, D.V. (1975). *Introducing Geology*. Faber and Faber, London.

Institute of Geological Sciences. (1977). *The story of the Earth*. H.M. Stationery Office, London (for the Geological Museum).

Pomerol, C. (1982). *The Cenozoic Era: Tertiary and Quaternary*. Ellis Horwood Pub. Ltd. & J. Wiley & Sons.

Thackray, J. (1980). *The Age of the Earth*. H.M. Stationery Office, London.

Faul, H. (1977). A History of Geologic Time. *American Science*, **66:** 159–65.

3

Surface Processes

The wide range of conditions found in different parts of the globe at present reflect the variety of natural processes that operate to shape the surface of the Earth.

Land areas are continually being reduced and their shape modified by weathering and erosion, and the general term for this is *denudation*. Rocks exposed to the atmosphere undergo *weathering* from atmospheric agents such as rain and frost. *Chemical weathering*, or *decomposition*, is the break-down of minerals into new compounds by the action of chemical agents; acids in the air, in rain and in river water, although they act slowly, produce noticeable effects especially in soluble rocks. *Mechanical weathering*, or *disintegration*, breaks down rocks into small particles by the action of temperature, by impact from raindrops and by abrasion from mineral particles carried in the wind. In very hot and very cold climates changes of temperature produce flaking of exposed rock surfaces. In areas of intense rainfall soil particles may be dislodged and the surface of the soil weakened by raindrops. In arid areas landforms are shaped by sand blasted against them during storms. *Biological weathering* describes those mechanical and chemical changes of the ground that are directly associated with the activities of animals and plants. When present, microbial activity can change the chemistry of the ground close to ground level. Burrowing animals and the roots of plants penetrate the ground and roots produce gasses which increase the acidity of percolating rain water.

By these processes a covering layer of weathered rock is formed on a land surface. Normally the upper layers of this cover are continually removed, exposing the fresher material beneath it to the influence of the weathering agents; in this way the work of denudation continues. In some circumstances the weathered material may remain in position as a *residual deposit* or soil (p. 1) that retains many characters of its parent rock and differs significantly in its mechanical properties from soils formed by the deposition of sediment.

Agents of erosion - rivers, wind, moving ice, water waves - make a large contribution to the denudation of the land. They also transport the weathered material, the *detritus*, away from the areas where it is derived; the removal of material is called *erosion*. The deposition of detritus transported by erosion produces features such as deltas at the mouths of rivers, beaches on shore lines, screes at the base of slopes and sand dunes in deserts. These are examples of the *constructive* or depositional aspect of surface processes.

Weathering

Chemical weathering

The processes most commonly involved in chemical weathering are listed in Table 3.1. Their rate of operation depends upon the presence of water and is greater in wet climates than in dry climates.

Except where bare rock is exposed, the surface on which rain falls consists of the soil which forms the upper part of the weathered layer. This 'top-soil' ranges in depth

Table 3.1 Some commonly occurring processes in chemical weathering

SOLUTION	Dissociation of minerals into ions, greatly aided by the presence of CO_2 in the soil profile, which forms carbonic acid (H_2CO_3) with percolating rainwater.
OXIDATION	The combination of oxygen with a mineral to form oxides and hydroxides or any other reaction in which the oxidation number of the oxidized elements is increased.
REDUCTION	The release of oxygen from a mineral to its surrounding environment: ions leave the mineral structure as the oxidation number of the reduced elements is decreased.
HYDRATION	Absorption of water molecules into the mineral structure. Note: this normally results in expansion, some clays expand as much as 60%, and by admitting water hasten the processes of solution, oxidation, reduction and hydrolysis.
HYDROLYSIS	Hydrogen ions in percolating water replace mineral cations: no oxidation–reduction occurs.
LEACHING	The migration of ions produced by the above processes. Note: the mobility of ions depends upon their ionic potential: Ca, Mg, Na, K are easily leached by moving water, Fe is more resistant, Si is difficult to leach and Al is almost immobile.
CATION EXCHANGE	Absorption onto the surface of negatively charged clay of positively charged cations in solution, especially Ca, H, K, Mg.

from only a few centimetres to a metre or more, according to the climate and type of rock from which it has been derived. In temperate climates, in general, it is a mixture of inorganic particles and vegetable humus and has a high porosity, i.e. a large proportion of interstices in a given volume: it has a high compressibility and a low strength and for these reasons is removed from a site prior to construction. The soil grades down into 'sub-soil', which is a mixture of soil with rock fragments, with decreasing organic content, and then into weathered rock and finally unweathered rock. The terms A-horizon (for soil), B-horizon (for sub-soil) and C-horizon (for the rock at depth), are used in pedology; in most places the materials have been derived from the underlying horizons during weathering. A vertical column showing this sequence is called a *soil-profile* (Fig. 3.1).

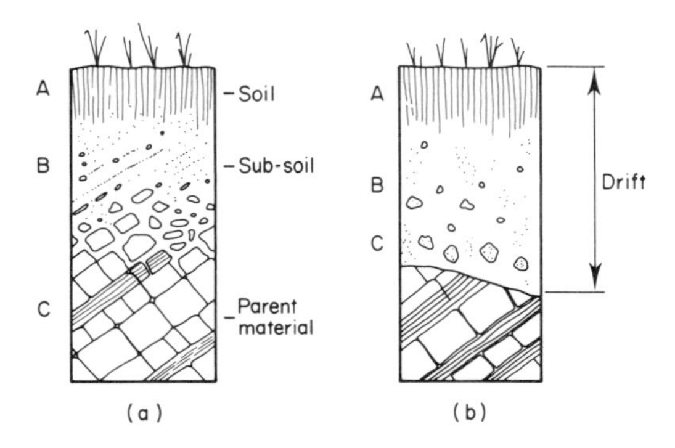

Fig. 3.1 Soil profiles that are: (**a**) related to bed-rock, (**b**) unrelated to rock at depth.

The speed and severity of weathering in wet climates depends essentially upon the activity of the root zone, i.e. the rate of growth of vegetation and production of CO_2 in the root zone, and the frequency with which percolating rainwater can flush weathered constituents from the weathering profile.

Chemical weathering is seen most readily in its solvent action on some rocks, notably limestones and those rocks containing the minerals Halite (NaCl), Anhydrite ($CaSO_4$) and Gypsum ($CaSO_4\ 2H_2O$). In limestones the process depends on the presence of feeble acids, derived from gases such as CO_2 and SO_2 which enter into solution in percolating rainwater. The calcium carbonate of the limestone is slowly dissolved by rainwater containing carbon dioxide, and is held in solution as calcium bicarbonate, thus:

$$CaCO_3 + H_2O + CO_2 \rightleftharpoons Ca(HCO_3)_2.$$

The ground surface of a limestone area commonly shows solution hollows, depressions that may continue downwards as tapering or irregular channels. These may be filled with sediment such as sand or clay, derived from overlying deposits which have collapsed into a solution cavity at some time after its formation. The upper surface of the chalk shows many features of this kind, known as 'pipes' and seen where intersected by quarry faces. Solution cavities are difficult to locate during a programme of ground investigation conducted prior to construction of foundations and their presence should be suspected even if none is identified by boring. Vertical joints in the rock are widened by solution as rain passes down over their walls, and are then known as *grikes*. Continued solution may lead to the formation of *swallow holes*, rough shafts which communicate with solution passages at lower levels along which water flows underground. A swallow hole is often situated at a point where vertical joints in the rock intersect and provide easier passage for surface water. Underground caverns are opened out by solution aided by the fall of loosened blocks of limestone from the cavern roof (Fig. 3.2). Large systems of caves and solution channels are found in the French Pyrenees, the limestone plateau of Kentucky, U.S.A., the Transvaal in S. Africa and elsewhere, most notably in S.E. Asia: China, Vietnam and New Guinea. The Karst area of Istria in Yugoslavia and the Dalmatian coast of the Adriatic has given the name *karst topography* to landforms which are characteristic of chemically weathered limestone; karst topography is developed most spectacularly and most extensively in southern China, where one-seventh of the country, i.e. 500 000 km^2 is karstic.

Water circulating underground helps to extend channels and caverns by solution, particularly in limestone formations; streams which once flowed on the surface disappear down swallow holes and open joints, Fig. 3.3, and continue their journey by flowing along bedding planes and joints below ground. In the Cheddar caves of the Mendips of S.W. England the former surface stream now has its course 15 to 18 m below the floor of the caverns, which are now dry. The Cheddar Gorge, 128 m deep at one point, is probably a large cave system which has become exposed at the surface by the collapse of its roof. As water charged with calcium bicarbonate trickles over the walls and drips from the roofs of caves, part of

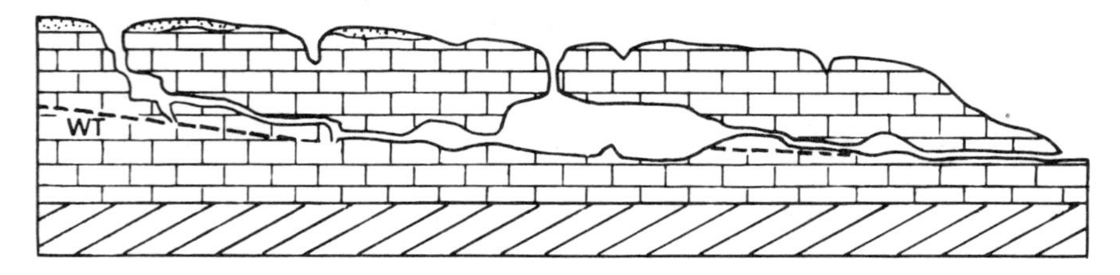

Fig. 3.2 Section through a limestone plateau to show solution features. WT = water table.

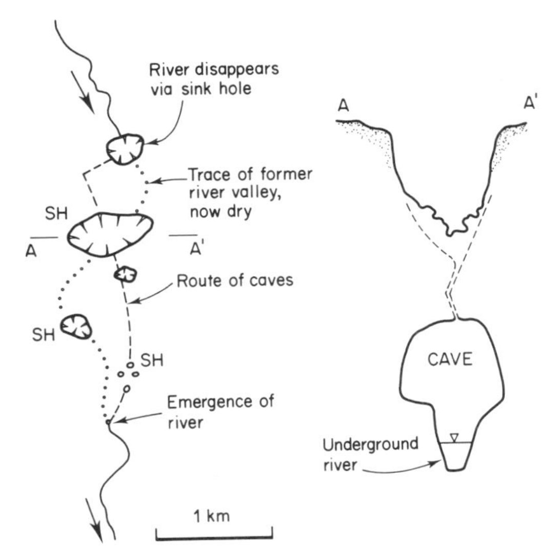

Fig. 3.3 Map showing a river that disappears from the surface to continue its course below ground (SH = sink hole), and a vertical section illustrating the nature of an underground river in Karstic regions.

it evaporates and calcium carbonate is slowly re-deposited as loss of carbon dioxide occurs (i.e. the equation given on p. 32 is reversible). In this way masses of *stalactite*, hanging from the roof or coating the walls of a cave, are formed, sometimes making slender columns where they have become united with *stalagmites* which have been slowly built up from the floor of the cave, onto which water has dripped over a long period of time. Sheet stalactite coats the walls of many caverns and may be coloured by traces of iron and lead compounds.

Chemical weathering is not restricted to easily soluble rocks but attacks all rock types. The most easily weathered are limestones; of greater resistance are sandstones and shales: igneous rocks (excluding certain volcanic rocks that weather rapidly) and quartzites are the most resistant. The effect of chemical weathering on these more resistant rocks can be clearly seen where *deep weathering* has occurred and a thick cover of rotted material lies above the irregular surface which bounds the solid rocks beneath the weathered zone. In areas of jointed igneous rocks, such as granite, groundwater made acid during its passage through the soil, penetrates along joints and can reduce the granite to a crumbly, rotted condition often many metres in depth. In parts of Australia depths of 60 to 90 metres of weathered granite have been recorded in engineering works. On Dartmoor, England, weathered rock up to 10 m deep has been encountered in excavations, especially in the proximity of faults, sometimes enclosing hard masses of unweathered rock (corestones). In Hong Kong considerable depths of weathering exist and may grade from completely weathered granite at ground level to fresh granite sometimes at depths of up to 60 m (Fig. 3.4).

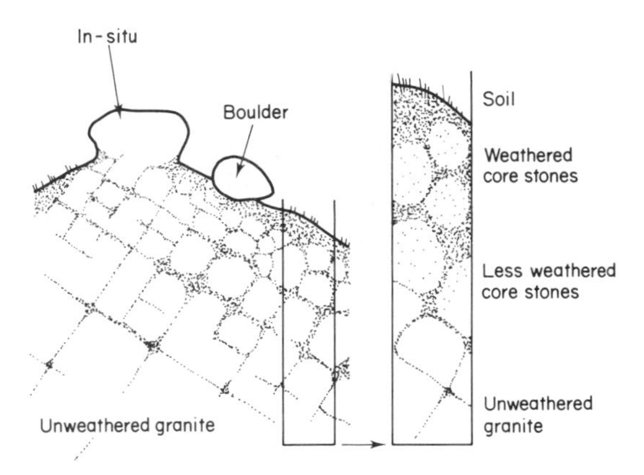

Fig. 3.4 Corestones of less weathered rock within a weathering profile.

The formation of granite *tors* is related to the frequency of jointing in the rocks, illustrated in Fig. 3.5; the tors are upstanding masses of solid granite, preserved where the spacing of joints is wider, in contrast to the adjacent rock; the latter has been more easily and therefore more extensively denuded owing to the presence in it of closely-spaced joints, which have resulted in more rapid weathering above the water-table (the broken line in the figure).

In sub-tropical climates where the gound is covered

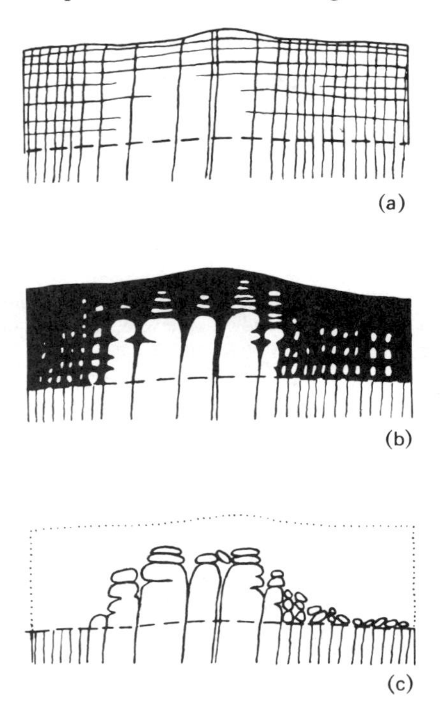

Fig. 3.5 Stages in the formation of tors under humid conditions of weathering (after Linton, 1955). (**a**) Original jointing (in vertical section). (**b**) Weathered rock (black). (**c**) Final form when erosion removes weathered rock.

with dense vegetation and subject to heavy rainfall – as in the monsoon areas of East Asia – very deep zones of weathered rock have been formed, e.g. in the Cameron Highlands, Malaya, depths of weathered rock up to several hundreds of metres may be encountered.

In dry climates chemical weathering is superficial and much retarded by the lack of water, producing thin zones of weathered rock. In very dry climates mechanical processes are the dominant weathering agents.

Mechanical weathering

The processes most commonly involved in mechanical weathering are listed in Table 3.2.

Table 3.2 The processes of mechanical weathering

MECHANICAL UNLOADING	Vertical expansion due to the reduction of vertical load by erosion. This will open existing fractures and may permit the creation of new fractures.
MECHANICAL LOADING	Impact on rock, and abrasion, by sand and silt size windborne particles in deserts. Impact on soil and weak rocks by rain drops during intense rainfall storms.
THERMAL LOADING	Expansion by the freezing of water in pores and fractures in cold regions, or by the heating of rocks in hot regions. Contraction by the cooling of rocks and soils in cold regions.
WETTING AND DRYING	Expansion and contraction associated with the repeated absorption and loss of water molecules from mineral surfaces and structures: (see HYDRATION: Table 3.1).
CRYSTALLIZATION	Expansion of pores and fissures by crystallization within them of minerals that were originally in solution. Note: expansion is only severe when crystallization occurs within a confined space.
PNEUMATIC LOADING	The repeated loading by waves of air trapped at the head of fractures exposed in the wave zone of a sea cliff.

Unloading

One result of denudation is to reduce the load on an area as the removal of rock cover proceeds, leading to relief of stress in the rock below. The unloading allows a small vertical expansion which gives rise to the formation of 'sheets' of rock by the opening of joints parallel to the ground surface. This is frequently seen in igneous rocks such as a granite intrusion, where the sheet-jointing is developed in the upper part of the mass; the 'sheets' or slabs of rock are commonly up to a metre or so in thickness. In valleys the surfaces of parting often lie parallel to the valley sides and approach the horizontal at the bottom of a valley. The frequency of the sheet-joints diminishes with depth below the surface; but during deep quarrying in an igneous body, parting surfaces may open with a loud cracking noise as stress in the rock is relieved. The production of smaller platy fragments in a similar manner is known as *spalling*. The formation of other fractures by release of strain energy is further described in Chapter 9.

Frost action

In cold climates repeated freezing breaks off flakes and angular fragments from exposed rock surfaces, a process referred to as the operation of the 'ice-wedge'; it leads to the formation of screes on mountain slopes and produces the serrated appearance of a high mountain sky-line. Water enters rocks by pores, cracks, and fissures; the ice formed on freezing occupies nearly 10 per cent greater volume, and exerts a pressure of about 13.8×10^6 N m^{-2} if the freezing occurs in a confined space. The freezing is thus like a miniature blasting action and brings about the disintegration of the outer layers of rock. The loosened fragments fall and accumulate as heaps of *scree* or *talus* at lower levels, material which may later be consolidated into deposits known as *breccia*. By the removal of the fragments the surface of the rock is left open to further frost action, and the process continues. Some well-known screes in the English Lake District are those along the eastern side of Wastwater, where the mountain slope falls steeply to the water's edge. Joints and cleavage planes in rocks assist the action of frost and to some extent control the shape of the fragments produced. Very little material of smaller size than 0.6 mm (the upper limit of the silt grade) is produced by the freezing.

The term *permafrost* (due to S. W. Muller, 1945) is used to denote perennially frozen ground. Permafrost areas have remained below 0°C for many years and in most cases for thousands of years. Pore spaces and other openings in such rocks and soils are filled with ice. Conditions of this kind at the present day are found within the Arctic Circle, but extend well south of it in continental areas in northern Canada and in Siberia, in some instances as far south as 60°N. It is estimated that one-fifth of the Earth's land surface is underlain by permafrost. Conditions that exist in sediments with temperatures below 0°C and contain no ice are described as *dry permafrost*.

An impervious permafrost layer is formed in the ground a little below the surface, and may range in thickness from less than 1 m to hundreds of metres; thicknesses of over 400 metres have been reported from boreholes in Alaska and Spitzbergen (Fig. 3.6). Permafrost requires a cold climate with a mean annual air temperature of − 1°C or lower for its formation and maintenance.

The upper limit of continuous permafrost in the soil is generally situated between about 10 and 60 cm below the surface; well-drained soils may thaw annually to a depth of one or two metres during the milder season. This shallow zone which freezes in winter and thaws in summer is known as the *active layer*. Thawing of the near-surface soils means the release of large volumes of water from

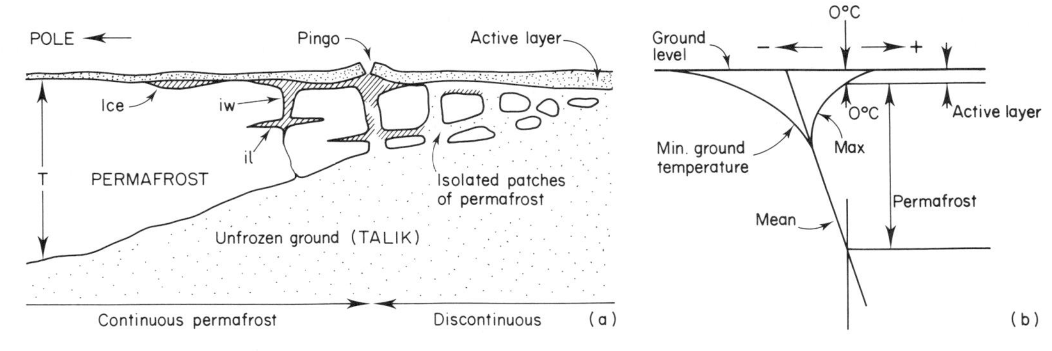

Fig. 3.6 Permafrost. (**a**) Vertical profile through ground containing permafrost. Thickness of permafrost (T) may be 1000 m+. (**b**) Temperature distribution with depth. As ground temperature fluctuates between summer maximum and winter minimum, so a depth of ground (0.5 to 3.0 m+) is seasonally frozen and thawed (the active layer). iw = ice wedge; il = ice lense. Based on data from Brown (1970) and Stearns (1966).

within their mass. The water cannot drain away through the still frozen ground below and it accumulates in the soil, which then flows readily down slopes, even those as low as 3 degrees. The amount of moving soil involved is large and flow is comparatively rapid. A landscape under these conditions is reduced to long smooth slopes and gently rounded forms.

Near the margins of permafrost areas, where slopes are covered with a mixture of fine and coarse rock fragments, the slow movement of this surface material over the frozen ground beneath it takes place by repeated freezing and thawing (Fig. 3.7). The process is known as *solifluction* (=soil-flow).

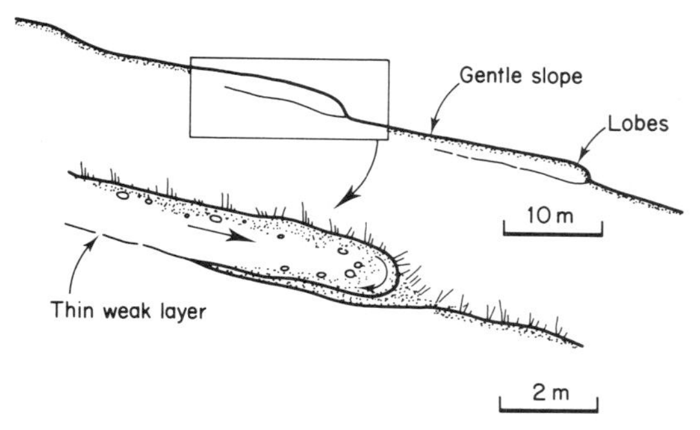

Fig. 3.7 Vertical section through slope affected by solifluction, illustrating lobate character of soil flow.

Frost-heaving

This occurs when the freezing of the soil results in the formation of layers of segregated ice at shallow depths. Each lens of ice is separated from the next by a layer of soil, whose water content freezes solid. The ice lenses vary in thickness from a few millimetres to about 30 mm. The total heave of the surface is approximately equal to the aggregate thickness of all the ice layers.

The frost-heaving of foundations of buildings is also caused by forces originating in the active layer and is a common problem in the Arctic.

Many engineering problems arise from this cause in areas of permafrost, as in connection with bridge foundations, and rail and road construction and maintenance in Alaska, Canada, N. Europe and northern Russia.

Buildings which are heated can be placed a little above ground level with a large air space beneath them. Cold air in winter then circulates under the building and counteracts the heating effect from it. Piped services to the buildings are also placed above ground level to prevent their rupture by ground movement.

Insolation

In hot climates, when a rock surface is exposed to a considerable daily range of temperature, as in arid and semi-arid regions, the expansion that occurs during the day and the contraction at night, constantly repeated, weaken the structure of the rock. The outer heated layers tend to pull away from the cooler rock underneath and flakes and slabs split off, a process known as *exfoliation*. This weathering is called *insolation*.

A large range of temperature occurs daily in deserts, commonly 30°C and sometimes as high as 50°C; the daily range for rock *surfaces* is often higher than for air. Strain is set up in a rock by the unequal expansion and contraction of its different mineral constituents and its texture is thereby loosened. A more homogeneous rock, made up largely of minerals having similar thermal expansion, would not be affected so much as a rock containing several kinds of minerals having different rates of expansion.

Under natural conditions, insolation of rock faces may result in the opening of many small cracks - some of hair-like fineness - into which water and dissolved salts enter; and thus both the decomposition of the rock and its disintegration are promoted. The crystallization of salts from solution in confined spaces, such as cracks and interstices (pore spaces) may hasten the process of weathering in deserts and coastal areas in arid regions, e.g. as along the Arabian and Persian coasts.

Biological weathering

Weathering effects which are small in themselves but noticeable in the aggregate can be attributed to plants and animals (*biotic weathering*). Plants retain moisture and any rock surface on which they grow is kept damp, thus promoting the solvent action of the water. The chemical decay of rock is also aided by the formation of vegetable humus, i.e. organic products derived from plants, and this is helped by the action of bacteria and fungi. Organic acids are thereby added to percolating rain-water and increase its solvent power. Bacteria species may live in the aerobic and anaerobic pore space of the weathering zone, and mobilize C, N, Fe, S and O, so assisting the process of weathering and sometimes attacking concrete and steel. Their mineral by-products can accumulate and cause expansion of the ground if not washed away by percollating water.

The mechanical break-up of rocks is hastened when the roots of plants penetrate into cracks and wedge apart the walls of the crack.

Global trends

Present distribution of the weathering described is illustrated in Fig. 3.8. Different distributions existed during the Quaternary (p. 28) because much of the continental area between the tropics was drier than at present and to the north and south lay cooler and colder conditions peripheral to the extended glaciers: the former position of permafrost is shown in the figure. Weathering profiles produced under these earlier conditions are frequently preserved beneath more recent drift and the present weathering profiles. Excavations and foundations of moderate depth may therefore expose rock and soil weathered under a previous, and perhaps more severe, weathering regime.

In previously hot, semi-arid regions, where evaporation from the ground had been rapid and nearly equal to the rainfall, chemical decomposition of the rocks will have proceeded to great depths and a hard, superficial crust formed by the deposition of mineral matter just below the soil. The water from the occasional rains carries dissolved salts only a short distance below the surface, where they are retained by capillarity, with the result that as evaporation proceeds a mineral deposit is built up. If solutions are saturated with calcium carbonate the deposit will be a calcareous one (*calcrete* or *kankar*), like that which covers large areas in India. With ferruginous solutions, such as would result from the decomposition of basic igneous rocks, a red concretionary deposit may

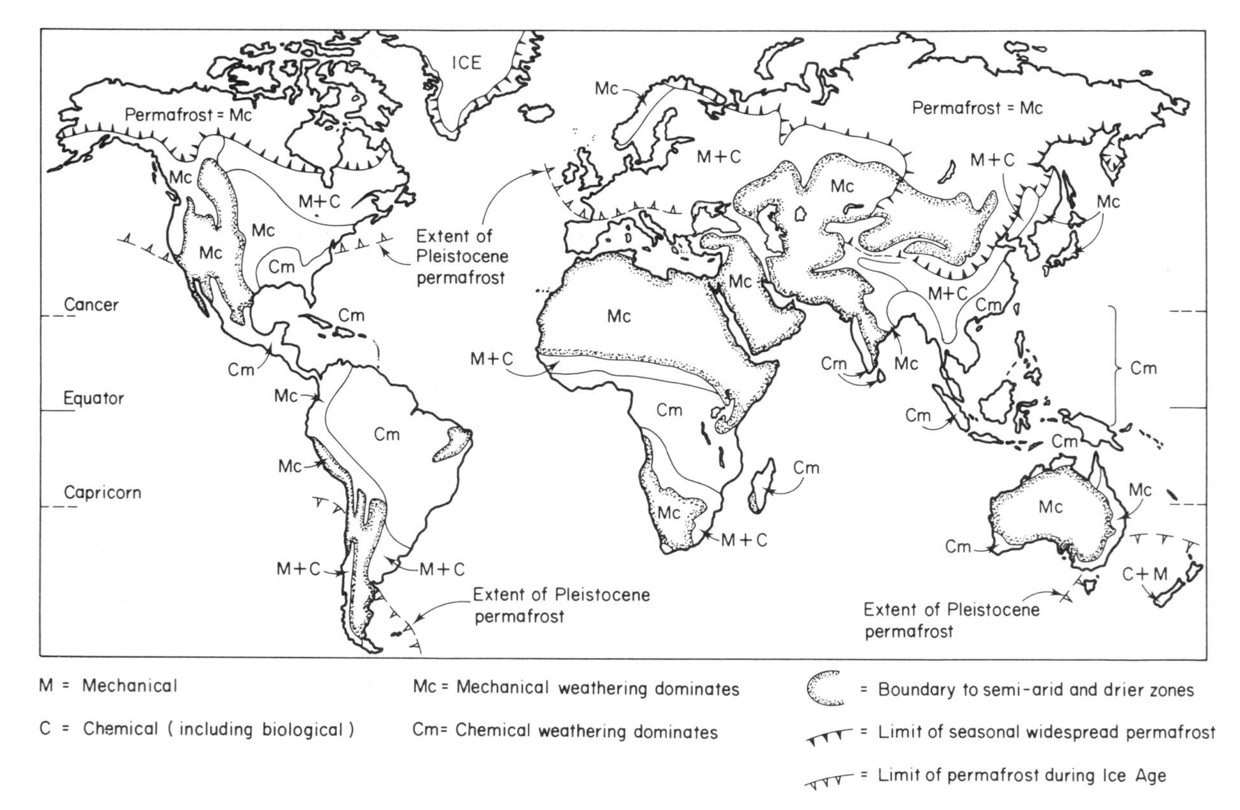

Fig. 3.8 Present distribution of weathering types. Based on data from Washburn (1979); Meigs (1953); Flint (1957); and Strakhov (1967).

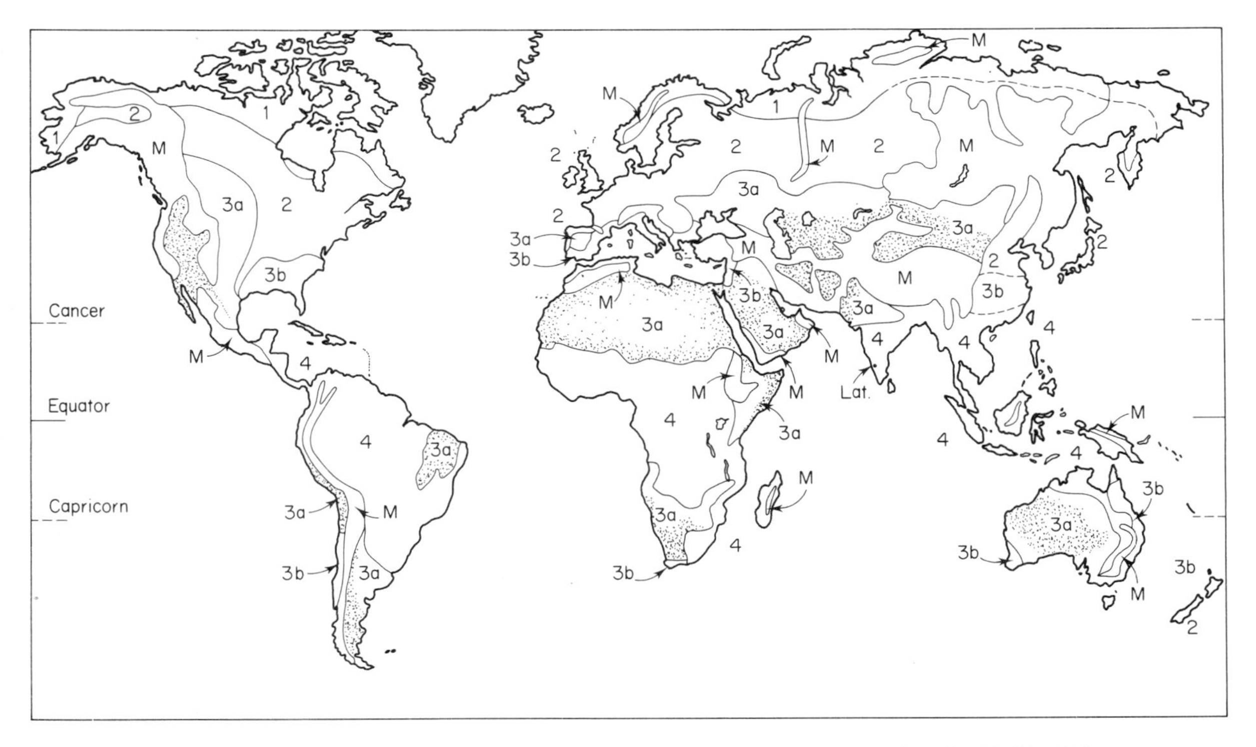

1 Tundra soils (black-brown soils)

2 Taiga and Northern Forest soils (Podzols and grey-brown soils)

3a Prairie, or Steppe long-grass soils (Chernozems) to Steppe short-grass (Chestnut-brown) soils and Desert soils (red-grey)

3b Mediterranean woodland soil (red-yellow) and scrub

4 Savannah grass and scrub soils (ferallitic-red and yellow-and dark grey-black soils) and Tropical rain forest soils (red-yellow)

M = Mountain conditions control soil generation: mainly thin soil profiles

Lat = Type locality for Laterite (for distribution see Chapter 20)

(stippled symbol) = Red-grey desert soils of (3a)

Fig. 3.9 General distribution of soils. Based on data from Bridges (1970); Kellog (1950); Thornbury (1969); and Walton (1969).

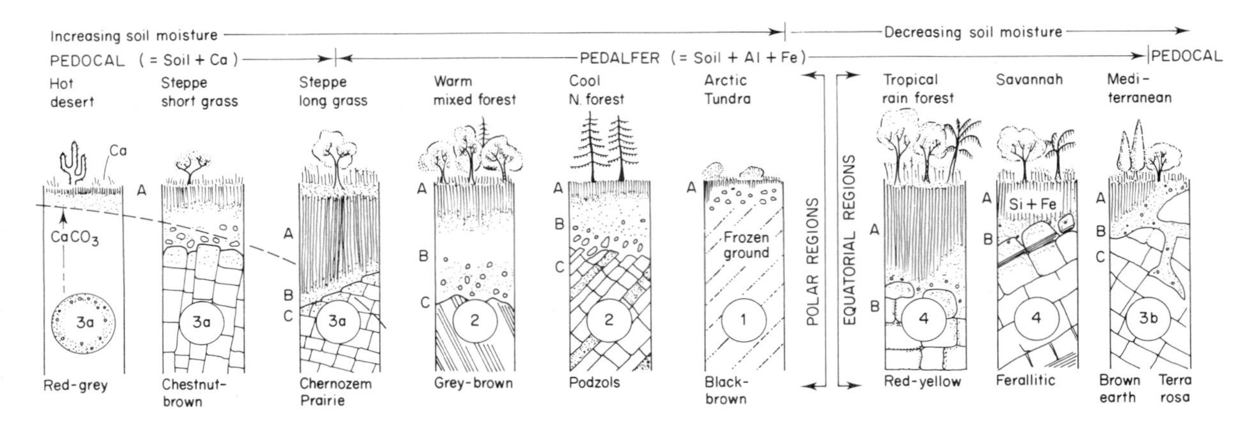

Fig. 3.10 Soils of the major climatic vegetation regions. Numbers refer to Fig. 3.9. $CaCO_3$ found in soil profiles of warm dry regions. Ca, Si, Fe = Calcite, Silica, Iron precipitate, often as a hard durable crust (*duricrust*): see text.

be formed, as in many parts of Africa. Major accumulations of silica occur as *silcrete* duricrusts in Australia and S. Africa. They form very resistant horizons up to 5 m thick and are often found capping prominent plateaux and mesas (Fig. 3.11*e*). It is believed that the silcrete formed during a period of warm, humid climate that existed at the end of the Tertiary.

The distribution of pedological soils is shown in Fig. 3.9. A basic relationship exists between soil type and weathering type (Fig. 3.8) because both reflect climate. Soils can be classified according to their climatic zone and soil profiles usually reflect their latitude (Fig. 3.10).

A special deposit, called *laterite*, exists in certain areas of S. America, Central and Southern Africa, India, Sri Lanka, S.E. Asia, East Indies and Australia. It is rich in oxides of aluminium and iron and may have formed under previous climatic conditions. The term 'laterite' was originally used to describe material which could irreversibly harden when cut from the weathering profile and air dried, so that it could be used as bricks (from the Latin, *later* – a brick). The word is now used to describe many other forms of red soil which contain hard bands and nodules (concretions) but are not self-hardening.

Erosion and deposition

Rivers, wind, moving ice and water waves are capable of loosening, dislodging and carrying particles of soil, sediment and larger pieces of rock. They are therefore described as the agents of erosion.

The work of rivers

The work of erosion performed by rivers results in the widening and deepening of their valleys. The rate of erosion is greatly enhanced in times of flood. Rivers are also agents of transport, and carry much material in suspension, to re-deposit part of it along their course further downstream, or in lakes, or in times of flood as levees (p. 42) and over the flood plain; ultimately most of the eroded material reaches the sea. Some matter is carried in solution and contributes to the salinity of the oceans. The energy which is imparted to sediment moved by a stream, the finer particles in suspension and the coarser (including boulders) rolled along the bed during floods, performs work by abrading the channel of the river. Hollows known as *pot-holes* are often worn in the rock of a river-bed by the grinding action of pebbles which are swirled round by eddying water. Such a water-worn rock surface is easily recognized, and if observed near but above an existing stream it marks a former course at a higher level.

Valleys

A drainage system is initiated when, for example, a new land surface is formed by uplift of the sea floor. Streams begin to flow over it and excavate valleys, their courses mainly directed by the general slope of the surface but also controlled by any irregularities which it may possess. In many instances present day lowland valleys have been shaped by the streams that occupy them; these were most active during the interglacial periods when their discharge was considerably greater than it is at present. Valleys in mountains and glaciated areas have been modified by other agencies, such as the action of avalanches, landslides and moving ice. In course of time a valley becomes deepened and widened, and the river is extended by tributaries. The area drained by a river and its tributaries is called a catchment, or river basin. Stages of *youth*, *maturity*, and *old age* may be distinguished in the history of a river, and topographical forms characteristic of these stages can be recognized in modern landscapes. Thus there is the steep-sided valley of the youthful stream; the broader valley and more deeply dissected landscape of the mature river system; and the subdued topography of the catchment of a river in old age.

Youthful rivers cut gorges in hard, jointed rocks and V-shaped valleys in softer rocks, Fig. 3.11*a*, and are characteristic of many upland areas. A youthful valley frequently follows a zig-zag course, leaving overlapping spurs which project from either side of the valley. Debris, loosened by frost, rain or insolation, falls from the valley sides, to be carried away by the stream and assist in its work of abrasion. Rain-wash and soil creep (p. 59) also contribute material from the slopes, especially in more mature stages of valley development, when a mantle of soil has been formed on a land surface. Large movements in the form of landslides also contribute to the erosion of a valley by transporting material to the river (Fig. 3.11*b*). Gradually as the headwaters of a river cut back, increasing the length of the river's course, its valley is deepened and widened into a broader V; slope movement causes the valley sides to recede and deposits debris on the valley floor (Fig. 3.11*c*). Small scree slopes form at the base of the valley sides. The deepening and widening, if uninterrupted, continue as shown in the figure until the stage of maturity is reached, when there is maximum topographical relief (i.e. the difference in height between valley floor and adjacent ridge tops). Eventually the tributaries of one catchment will be separated by only a thin ridge from those of its neighbour, and when denudation reduces the height of the ridges the river can be considered to have entered its old age: the valley comes to have a wide, flat floor over which the river follows a winding course, the upper slopes may be convex and the depth of relief is less. Such a sequence would be followed by a river in a temperate climate, if uninterrupted. Under other conditions, as in a drier climate, the valley slopes stand at a steeper angle because they are less affected by atmospheric weathering and slope instability, particularly landslides: the ridges between valleys are then sharper. Flat-topped hills may be left in partly denuded horizontal strata where a layer of hard rock such as sandstone, or a dolerite sill, forms the capping of the hill; the term *mesa* (= table) is used for this topographical form. Continued weathering and erosion reduce the mesa to isolated, steep-sided, pillar-like hills (Fig. 3.11*e,f*).

Youth, maturity and old age constitute a 'cycle' of

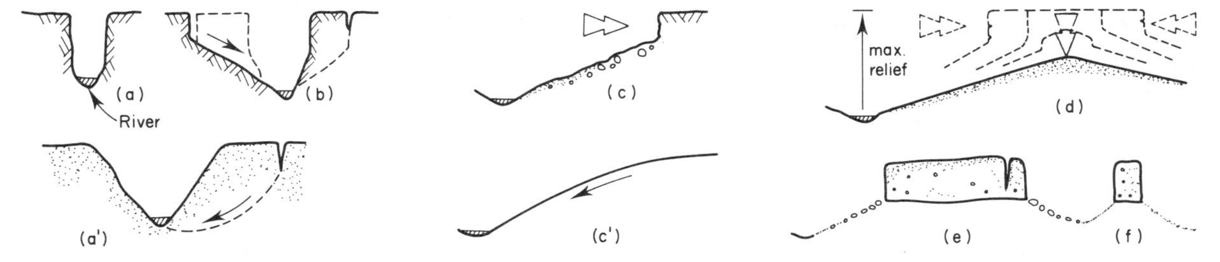

Fig. 3.11 Valley profiles: stages (**a–c**) mainly eroding back, stage (**d**) mainly eroding down. (**a'**) valley in weak rock. (**c'**) convexo–concave slope (see text). (**e**) mesa. (**f**) rock pillar. No scale implied.

erosion which, to be completed, requires the river to remove all the material supplied to it by the degrading slopes. If the rate of supply is equal to its rate of removal, linear slopes of constant angle can be expected (Fig. 3.11*c*). When the rate of removal exceeds the rate of supply convexo-concave slopes result.

The profile of a valley also depends on the kind of rocks which have been eroded; when alternate hard and soft layers are present, as in Fig. 3.12, erosion in the softer rock is more rapid than in the harder rock and terraced slopes are developed. If the rock layers or strata are inclined in one direction and the river flows parallel to the hard layers instead of across them, long hollows or vales will be carved in the softer beds, separated by ridges of harder rock which form escarpments.

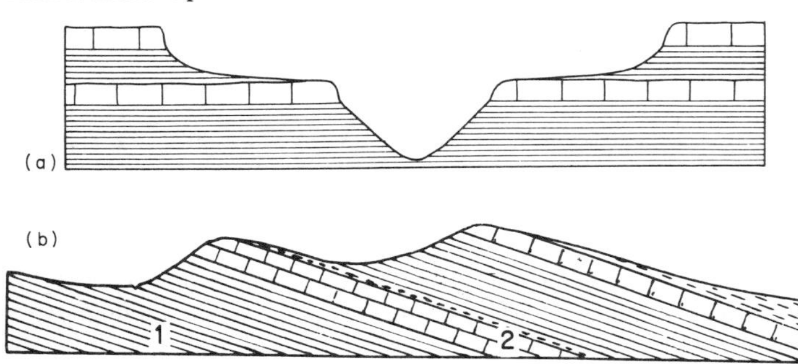

Fig. 3.12 (**a**) Terraced profile. (**b**) Valleys eroded into strata of unequal strength. (1) softer strata, (2) harder strata. (Based on a vertical cross-section through Wenlock Edge where (1) = Wenlock Shales and (2) = Wenlock Limestone; both are of Silurian age.)

Grade and rejuvenation

The profile taken *along* the course of a river also changes during the river's evolution. For a young stream, actively eroding, the long-profile is an irregular curve that is steeper where the river crosses more resistant rocks, perhaps forming rapids or waterfalls, and flatter where it flows over more easily eroded rocks. If left undisturbed by movements such as uplift, or other factors causing change, the river continues to reduce irregularities of gradient until in maturity it is said to be *graded* or at grade. Its longitudinal profile is then independent of the kind of rock over which it passes, and tends towards a smooth curve (Fig. 3.13). Many rivers have developed long-profiles of this kind. Larger rivers such as the Colorado and the Mississippi are essentially graded in their lower courses.

The *base-level* of a river is the level of the sea or lake into which it discharges; in Fig. 3.13 the river profile is asymptotic to base-level. For a tributary, base-level is the level of the main stream at its point of entry, and as this changes in course of time the tributary is continually adjusting its grade to a new level. The cutting power of a river which has reached a graded course in old age may, however, be revived by uplift or tilting of the land, by recession of the coast-line due to marine erosion (p. 45), or from other causes; as a result the stream is given a new fall to the sea. It is said to be *rejuvenated*, and begins to cut back again and lower its bed by the newly acquired energy. Owing to such interruptions in their cycle of activity not all streams become completely graded.

The rejuvenation of a river leaves its mark on the long-profile, where the break in slope at the junction of the old course with the new (deeper) cut is called the *nick-point* (Fig. 3.13).

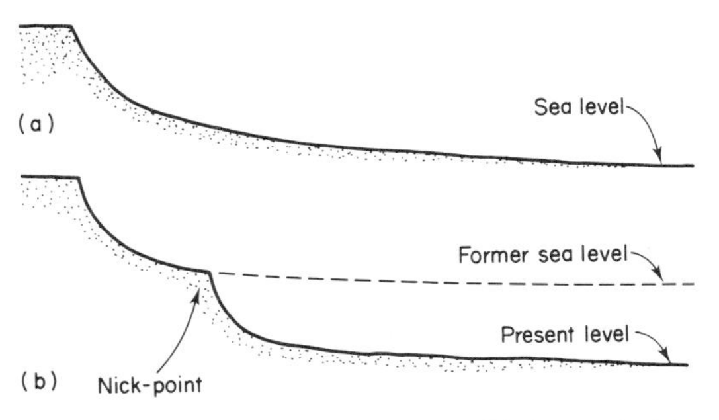

Fig. 3.13 Profile along the course of (**a**) a graded river; (**b**) a rejuvenated river.

Waterfalls and gorges

Rapids and waterfalls are formed where a stream flows over rocks of differing hardness. A hard layer is worn away less quickly than softer rock, with the result that the river's gradient is locally steepened where the outcrop of the harder band is crossed. As the river flows over it, less resistant material below the harder becomes undercut by the eddying of the water and in time an overhanging ledge is formed, over which the stream falls (Fig. 3.14). The weight of the ledge eventually becomes greater than can be supported by the strength of the rock, the ledge breaks away, and by successive stages the waterfall gradually recedes upstream.

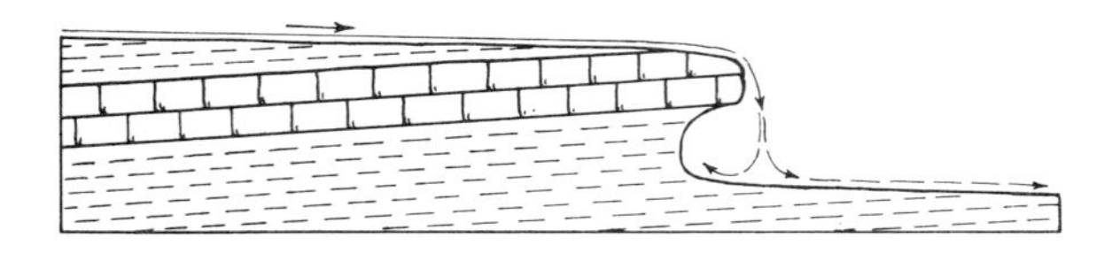

Fig. 3.14 Waterfall formed at a hard layer below which softer beds are eroded.

At Niagara on the Canada–United States border, the water from Lake Erie flowing north to Lake Ontario crosses a hard limestone formation, the Niagara Limestone; this lies above softer shales and the river here makes a drop of 55 m. The limestone forms a broad ledge which is undercut by the river, and the falls have retreated upstream with successive collapses of the ledge. Below the falls the river flows through a gorge 11 km long which has been cut during the retreat of the falls; the rate of recession was formerly about 1.2 m per year. Taking this as an average figure, the time taken to erode the Niagara gorge would be about 10 000 years; thus the gorge has been formed since the end of the Pleistocene glaciation (p. 29). The rate of recession of the Niagara falls was reduced to about a quarter of its former amount when hydro-electric power generation began, greatly reducing the flow of water over the ledge.

Many deep gorges are found in mountains, and in general, gorges are cut by youthful streams that erode rapidly downwards. In many instances a slow uplift of the area has contributed to the formation of the gorge by maintaining the downcutting power of the river. Where prominent joints or other lines of weakness in the rocks are present they largely control the shape of the sides; if the rocks are sufficiently strong they will stand with steep or vertical walls. In less strong rocks the gorge becomes widened out by weathering. When erosion is aided by solution, as in a limestone formation such as the Carboniferous Limestone of the Pennines, the solution effects also play a part in the shaping of the gorge.

The Grand Canyon of the Colorado, a vast erosional feature over 400 km in length and with a maximum depth of about 2 km, has been cut in limestones and shales (mainly of Carboniferous age) by the Colorado River flowing from the Rocky Mountains. The river was rejuvenated by a geologically recent (late Tertiary) uplift of more than 2 km, its rapid downward erosion being assisted by a rising land surface. The upper part of the canyon has been widened out and stands with nearly vertical limestone walls, below which are flatter shale slopes, giving a terraced profile.

River patterns

A river and its tributaries constitute a *network* whose pattern can be influenced by the position and shape of boundaries separating the various rocks within a catchment: Fig. 3.15. With continued down cutting a river may expose older rock of different structure at depth (see Fig. 3.15*e*) and *superimpose* its drainage pattern upon it.

A youthful energetic river which is cutting back vigorously in an easily eroded formation, and thus extending its catchment, may approach the course of a neighbouring stream and divert the headwaters of the latter into its own channel. The process described is known as *river capture*, and the stream that has lost its headwaters is said to be beheaded and dwindles in size. It is then too small for the valley in which it flows and is called a 'misfit'. Detailed field work has demonstrated that many misfit streams

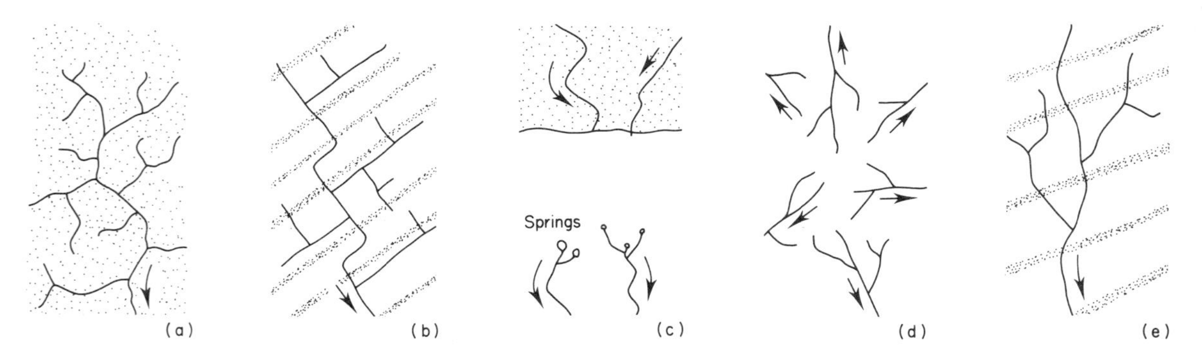

Fig. 3.15 Examples of river patterns. (**a**) *Dendritic*, uniform conditions of drainage. (**b**) *Rectangular*, or *trellis*, linear geological structures (indicated by ornament). (**c**) *Disappearing*, soluble rocks (*cf.* Fig. 3.3). (**d**) *Radial*, uplift of the area. (**e**) *Superimposed*.

have resulted from the diversion of their former river and not from its capture.

A river that swings from side to side of its valley and eventually flows in big loops is said to *meander*. Early stages in their formation are shown in Fig. 3.16*a*. The flow of the stream crosses from one side of its channel to the other, and the bank on the concave side is eroded and undercut, contributing sediment which is moved by the river. As the river flows round the bend (Fig. 3.16*b*) the main current (f) impinges on the concave bank and the channel there is deepened. This flow near the stream bed is crossed by a return current (g) moving towards the convex bank, and thus a helical motion is imparted to the water in the stream. The scouring effect of the bed current results in the deepening of the channel on the concave side of the bend, as shown by the asymmetric profile in Fig. 3.16*b*. Alluvium is deposited on the convex side, in slack water at the foot of the spur, as a bank of sand or shingle. As this process goes on, each bend of the river tends to migrate downstream and the meanders as whole gradually move downstream. The undercutting and collapse of banks is not, however, the cause of meandering.

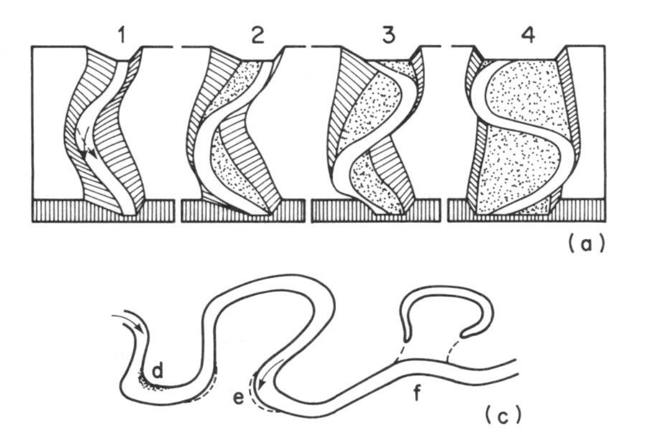

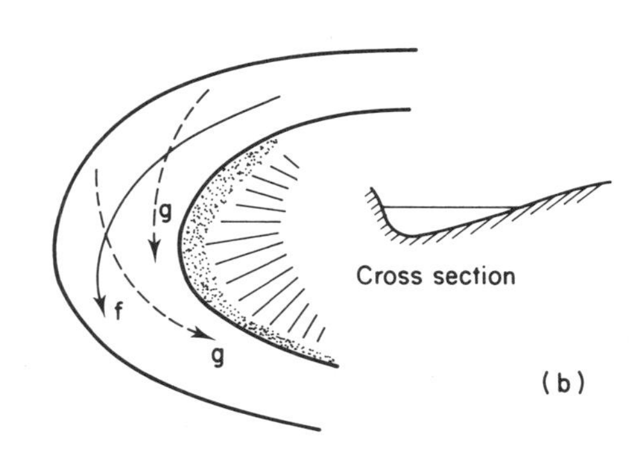

Fig. 3.16 (**a**) Stages in the widening of valley floor and development of meanders; dotted areas indicate flood-plain alluvium. (**b**) Flow in a river at a bend (plan), and profile across the river bed. (**c**) Fully developed meanders, with deposition (d) and erosion (e) at bends, and ox-bow lake (f).

The 'wavelength' of fully formed meanders is generally between 8 and 12 times the width of the stream, with 11 as a mean value. As the meanders shift bodily downstream the river erodes its flood-plain deposits and rebuilds them behind it, and thus it works over a large area of land. Comparison of old and new maps of a river will often show the shift and change in shape of meanders that has occurred. Instances of this are found on early maps of the Mississippi below Vicksburg; some meanders of this river move distances up to 20 m per year. Most rivers have developed meanders and those around which cities have been built, such as the Thames and the Seine, now have parts of their lower courses trained between walls and embankments, thus preventing further changes in position, and flooding.

When a meandering river in times of flood breaks through the neck of land at the end of one of the loops, as in Fig. 3.16*c*, it leaves a crescent-shaped lake or 'ox-bow' in the abandoned meander; examples may be seen from many maps and air photographs.

Meanders can become *incised* when fresh energy imparted to a stream results in downcutting; thus the River Wye on the Welsh border has preserved its meandering course in the limestone gorge below Goodrich (Kerne Bridge) during a slow rejuvenation resulting from uplift. On a much larger scale the San Juan River, Utah, is an example of a river that has deeply entrenched its course during a geologically recent uplift.

River deposits

River deposits take four basic forms, namely: (*i*) those in the river channel during periods of normal flow, (*ii*) those spread over the plain on either side of the river during periods of flood, (*iii*) those deposited across the floor of the estuary and interbedded with sediments carried into the estuary by the sea, and (*iv*) those of deltas.

Alluvium

This is the general term given to deposits laid down by rivers; it may include fine material such as silt and mud and coarse sand and gravel. The transporting power of a stream increases at the rate of the fifth or sixth power of its velocity. Thus, if the normal speed of flow is trebled, as may be the case after heavy rains, the carrying capacity is increased several hundred times. The result is seen in boulder-strewn tracts in hill country. Large boulders which would not be moved under normal conditions of flow are trundled along by a stream in spate, and become partly rounded by the buffeting they receive. Large quantities of smaller boulders, gravel, and sand are also transported at such times, the larger fragments being moved on the stream bed and the smaller carried in suspension.

Transported sediment is dropped by a stream whenever its velocity is reduced. A river emerging from a mountain valley onto flatter ground, such as the edge of a plain, builds up a heap of detritus known as an *alluvial cone* where the change of gradient occurs.

In the lower course of a mature river the finer alluvium is spread out to form an *alluvial flat* (Fig. 3.17); it is subject to periodic flooding and a fresh layer of alluvium is deposited at each flood. The coarser particles are dropped nearest the stream, and gradually build up a bank or *levee* on each side of it, which is only overtopped by water in times of flood. In some cases the growth of an alluvial flat may involve the burial of the lower valley slopes, and the alluvium then abuts directly onto the upper valley slopes. Alluvial deposits usually have only a small thickness, but may go up to 10 m or more. Many valleys were deepened during the Pleistocene glaciations when sea levels were lower and rivers cut down into their valley floor to re-establish their grade. At the end of the last major glaciation (Table 2.4) sea level was restored as the glaciers melted and rivers had to adjust to a higher base-level causing old valleys to be filled with sediment (Fig. 3.17). Buried valleys should be suspected beneath the alluvium of most lowland rivers.

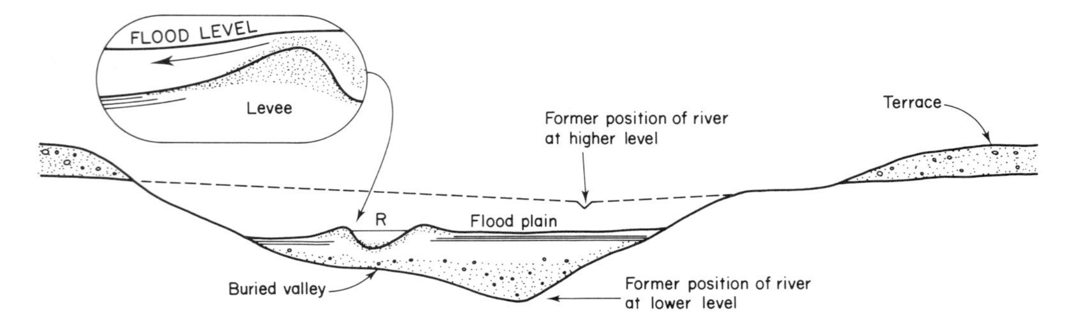

Fig. 3.17 Section across a river valley with inset to illustrate the formation of levees at times of flood. R = present river. Note: levees permit the river level to be raised above the level of the flood plain.

Alluvium is usually very porous and will be compressible if rich in clay, and permeable if composed mainly of silt, sand or gravel. In excavations the zone of saturation may be met at a small depth below the surface of the alluvium. Running sands and other forms of instability can occur under these conditions unless groundwater is controlled (Chapters 16 and 17).

The rapid variation in the nature of alluvial deposits is well illustrated by exposures in dock and harbour works: layers of mud, silt, and sometimes very compressible peat alternate and thin out in irregular fashion, and may fill hollows ('wash-outs') that have been scoured out by floods in earlier deposits.

When a river, after having deposited alluvium on a flood plain (Fig. 3.17), is rejuvenated and cuts down its channel to lower levels, the remnants of earlier deposits may be left on the old valley as *terraces* at different levels (Fig. 3.17). In the valley of the River Thames near London three such terraces are found, known as the Boyn Hill, Taplow, and Flood Plain terraces respectively (Fig. 3.18). Valuable deposits of gravel and sand occur in these terrace gravels and are excavated on a large scale for supplies of aggregate and sand for concrete and mortar, as in the area west of London.

Alluvial mud may be used as one of the raw materials

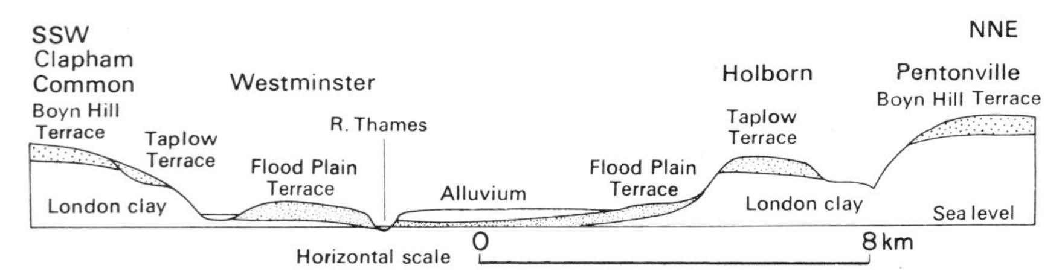

Fig. 3.18 Section across the Thames valley at central London to show river terraces with gravels (stippled).

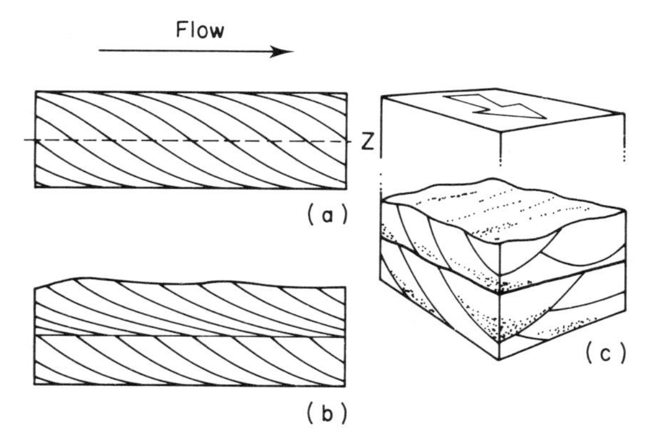

Fig. 3.19 (**a**) Current bedded deposit; (**b**) after erosion to letter *z*, with new cover on truncated layers; (**c**) note cuspate form in direction orthogonal to flow.

in the manufacture of cement, as on the River Medway and at Lewes, Sussex; in both cases the mud is mixed with Chalk (a limestone) in the required proportions.

River deposited and other shallow-water sediments frequently show structures known as *current-bedding* or *cross-bedding*, that are formed during their deposition (Fig. 2.1). Thin, inclined layers of sediment (sand or silt) are laid down on gentle slopes or ledges in the bed of a river or in other shallow water. The laminae are lenticular and gently curved. Their upper parts may be eroded by a subsequent current flowing over them, forming a surface which truncates the inclined laminae (Fig. 3.19). Current bedding formed by wind action and found in eolian sands, is known as *dune-bedding* (p. 53).

Deltas

A river entering a body of water, such as a lake or the sea, drops much of its load of sediment as its velocity is reduced and forms a delta which is gradually built forward into the water. Lake deltas are frequently seen, e.g. in many mountain lakes, where lateral streams that enter build up flat cones of debris. During floods these deposits may accumulate to a level above that normal for the lake and in time become grassed over to stand as small promontories.

A situation analogous to the deposition of alluvium in lakes is found in the silting up process that goes on in reservoirs. Streams flowing into a reservoir deposit in it the sediment that they carry, and the storage capacity of the reservoir is gradually reduced. The Colorado River in the western United States carries a large load of sediment especially in times of flood; when the 222 m high Boulder Dam (re-named the Hoover Dam, in 1947) was completed in 1936 it was estimated that the reservoir would be filled with sediment in about 200 years. Its life has been extended by the construction upstream of the Glen Dam (1964), which intercepts the sediment that would otherwise be destined for the reservoir of the Hoover Dam.

At a coast larger deltas may be formed. Near a shoreline much of the muddy sediment brought in suspension by a river is deposited, though some is carried farther out by weak currents. Settlement of the very fine particles of

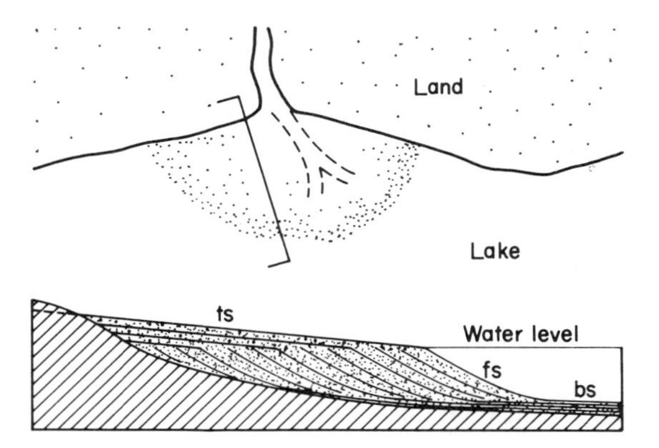

Fig. 3.20 Section through a small lake delta ts, topset beds; fs, foreset beds; bs, bottomset beds.

mud is promoted by *flocculation* (the aggregation of fine particles into clusters or flocks), e.g. where fresh and salt water mix.

When a delta is being formed the river characteristically divides so that the deltaic deposits in time come to cover a large area which may have a roughly triangular shape (like the Greek letter Δ, from which the name *delta* is derived). A section through the deposits of a typical small delta is given in Fig. 3.20; as each flood brings down its load of sediment the coarser material is dropped in front of the growing pile and comes to rest at its angle of repose in water, building up the *foreset beds*. The slope of these layers varies from about 12° to 32°, larger (sandy) particles standing at higher and smaller (clayey) particles at lower angles. Ahead of the foreset beds and continuous with them the finer material is deposited as the *bottomset beds*. As the delta is built forwards, foreset layers come to rest on earlier bottomset deposits, as shown in the figure. Foreset beds may periodically slump and slide from the upper levels of the delta slope to be re-deposited at its toe. The upper surface of the delta is composed of gently sloping *topset beds* of coarser material, which are a continuation of the alluvial plain of the river and cover successive foreset deposits.

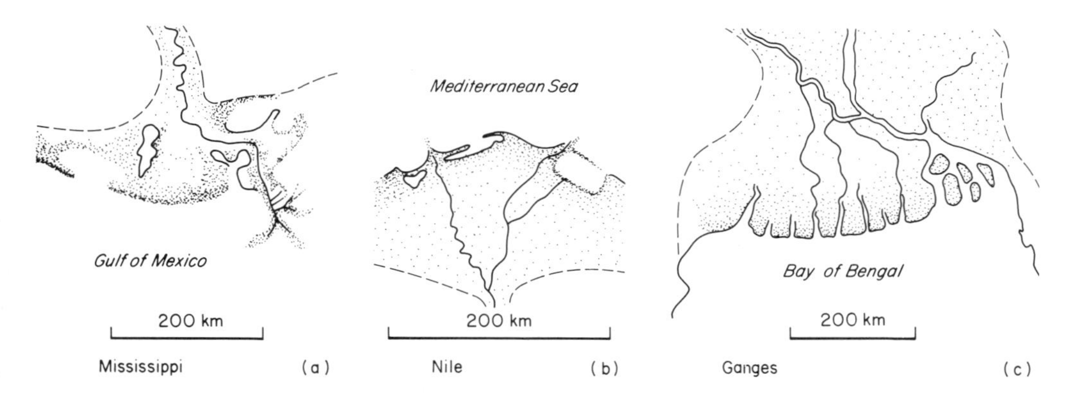

Fig. 3.21 Maps of three large deltas. (**a**) Birdsfoot. (**b**) Classical. (**c**) Lobate.

The sediments of ancient deltas can be recognized in many of the sandstones beneath the coal-bearing rocks of Carboniferous age. Those of the Millstone Grit of Yorkshire in England, are extensive deltaic and flood-plain deposits laid down by a large river or group of rivers from the north-east. Pebbles in the sediments show that the detritus was probably derived from rocks in the Scottish Highlands.

Within a large delta such as that of the Nile, Niger, Ganges, or Indus, all of which are over 160 km across and have a great underwater extent beyond the present coast-line, is a thick pile of sediment. The Nile delta-front (Fig. 3.21*b*) extends northwards under the sea for some 100 km, and its slope increases from about 1 in 1000 near the shore to 1 in 40 at its outer edge (this steeper slope representing the foreset beds). These submarine deposits consist of Nile silts and clays overlying sands with layers of gravel. Borings in them have found thicknesses up to 700 m of shallow-water deposits, which are all geologically recent, indicating that the great deltaic pile must have accumulated on a subsiding floor.

The Mississippi delta (Fig. 3.21*a*), about 30 000 km^2 in extent is an area that is continuing to extend seaward. It differs from those discussed above in possessing a 'bird's foot' projection called the Balize sub-delta, which has been formed in recent times, and across which run levees built of coarse sediment. Between the levees finer material has been deposited on which swamps have formed. Beyond this sub-delta earlier deltaic deposits extend below sea-level, and other shallow-water deposits are accumulating.

The low-lying Indo-Gangetic plains which separate the Himalayas to the north from peninsular India on the south are another area where hundreds of metres of alluvial deposits, brought by rivers from the mountains, have accumulated in the vast delta of the Ganges (Fig. 3.21*c*). Deep borings here have penetrated many successive beds of sand and clay with intercalated layers of peat and kankar (p. 36), and the deposits are of similar type throughout. Their thickness was estimated at nearly 2 km.

The work of the sea

The waves which break on a shore erode the land margin by the force of their impact and, especially in storms, by the impact of the debris they carry forward. The debris itself comprises rock fragments derived either from local cliffs or brought to an area from adjoining beaches. The fragments become rounded and reduced in size by the battering they receive from the waves, and make up the familiar pebbly and sandy beach deposits.

The continental shelf around the margins of a large land mass has been described in Chapter 1. Relatively shallow seas such as the North Sea or the Baltic, which lie on the continental shelf, are called *epicontinental seas*: they are to be distinguished from the deep *oceans* such as the Atlantic and Pacific.

The terms used by different writers in describing the parts of an area where land and sea meet vary; those employed here may be defined as follows. The margin of the land is called the *coastline*, and the land zone adjacent to it is the *coast*, which is frequently bounded by a line of cliffs or dunes. The *shore* is the zone extending from the base of the cliffs down to low-water mark (Fig. 3.22); it may be subdivided into the *foreshore*, which is that part lying between ordinary high- and low-tide marks, and the *backshore* or area between high tide level and the foot of the cliffs. When no backshore is present, high tide then extends up to the cliff base.

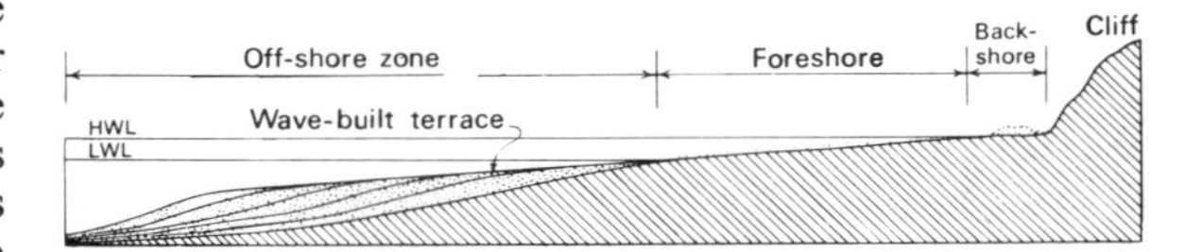

Fig. 3.22 Nomenclature of shore zones.

The shore is a wave-cut platform and on it lie beach deposits, which are moved by waves and covered and uncovered daily by the tides. In the off-shore zone beyond low tide level (Fig. 3.22), thicker deposits of land-derived sediment form a terrace whose seaward slope continues that of the wave-cut platform. The deposits of this terrace grade from coarser to finer material with distance outwards from the shore.

Tides and currents

The periodic rise and fall of the sea, or *tide*, is due to the pull exerted by the sun and moon on the globe; the bulge of water thus produced moves round the Earth as the latter rotates. The highest ordinary tides are called Spring Tides, and occur at the times of full and new moon, when the sun and moon and Earth are in line. The smallest or Neap Tides occur at the moon's first and third quarters, when its pull is at right angles to that of the sun. When high tides are accompanied by heavy seas in severe storms the coarser deposits of the backshore, which are normally above the level of Spring Tides, may be piled up to form a ridge or storm-beach.

In shallow seas and narrow channels water is heaped up with the rise of the tide, so that a tidal current is generated; in the English Channel for instance the tidal current flows eastwards at about 3 km per hour on a rising tide, and westwards to the Atlantic on a falling tide. Tidal currents may be strong enough to move the less coarse sediment in a shore zone but rarely move the coarser shingle.

Tidal surges

A surge is defined, simply, as a water movement which is quickly generated and soon over. One example is the surge which occurred in the North Sea on the night of 31 January 1953. A severe storm swept south past Scotland and into the North Sea; it caused extensive flooding and damage at many places on the east coast of England, from the Humber to Kent, and also affected the coasts of Holland and other countries of north Europe.

Much new construction of defence works was subsequently put in hand, and also measures to restore the productivity of agricultural land that had been flooded with sea water and sand.

Waves

Wave motion is produced when a water surface is swept by wind. It is an oscillatory motion, and any particle near the surface moves in a vertical circular orbit, as may be observed by watching the movement of a floating object. Particles below the surface move in phase with those at the surface but in smaller orbits. Waves in an area where they are being driven by wind are known as *forced waves*; when they move out into water where they are unaccompanied by wind they are called *free waves* or *swell*.

The height of waves from trough to crest commonly varies from a few metres to much more than this in heavy seas. The wavelength (distance from crest to crest) of forced waves is from 60 to 200 m in the open ocean, but the wavelength of a swell may be greater. Wave motion diminishes with depth from the surface, and at a depth equal to the wavelength the movement almost ceases. As a wave runs into shallow water (less deep than a wavelength), the sea floor interferes with water circulating in the lower orbits of the wave while the top of the wave runs on. The wave-front thus increases in steepness and ultimately the crest falls over in front of the wave, and the wave is said to break. This forward movement of the water constitutes the means by which a wave becomes an agent of erosion. The breaking of the wave is followed by the *backwash*, which combs down the components of the beach, especially the finer material, towards deeper water where they are temporarily deposited.

With an onshore wind blowing at right angles to the shore-line, water is heaped up against a coast; this is compensated by a return current away from the land, called the *undertow*, which may be concentrated into narrow channels. The undertow can transport finer sediments out to sea, and in storms a strong undertow can remove large quantities of beach.

Waves running up to a line of cliffs fronting moderately deep water are partly reflected, and an up-and-down movement is imparted to the water at the cliff face. By contrast, where waves run in over a long distance in gradially shallowing water, much of their energy is dissipated by interference with the sea floor and friction, and they become reduced in size (and hence in erosive power) before they break. A flat sandy foreshore is therefore to a large extent its own protection against erosion. When the slope of the foreshore is steeper, low frequency waves may have a constructive effect, resulting in aggradation of the beach deposits.

The pressures exerted by waves were measured by Stevenson in 1849, at Scottish localities, who found a variation from about 28 $kN\,m^{-2}$ up to 96 $kN\,m^{-2}$ in winter; and in one gale 288 $kN\,m^{-2}$ was recorded. The effectiveness of storm waves in breaking massive stone and concrete structures has been demonstrated many times around harbours and it should be noted that storm waves do far more damage in a short time than normal seas acting over a longer period. The action of waves in moving and modifying the form of beach deposits has been studied extensively by the U.S. Beach Erosion Board.

Coastal erosion

At the base of many coastal cliffs a *wave-cut platform* is formed, which slopes gently seawards (Fig. 3.22). The rocks of the foreshore are bevelled off and exposed on the platform, and may be partly covered by beach deposits. The sea exerts a 'sawing' action at the base of cliffs, cutting a horizontal notch which gradually weakens the cliff base. While this goes on, atmospheric denudation slowly wears away the upper part of the cliff and gravity induced shear stresses weaken it internally; slides and debris which fall to the foot of the cliff is broken up by the waves and largely removed into the offshore zone. Thus the cliff face recedes as it is undercut by waves and denuded by rain and frost; and as the wave-cut platform is widened the erosion at the cliff foot is slowed down. Streams which drain to the coast may be rejuvenated by the steepening of their lower courses, as cliffs are cut back; if this takes place faster than the streams can achieve a new grade, waterfalls are formed where the truncated valleys meet the cliff. Instances of truncated valleys are

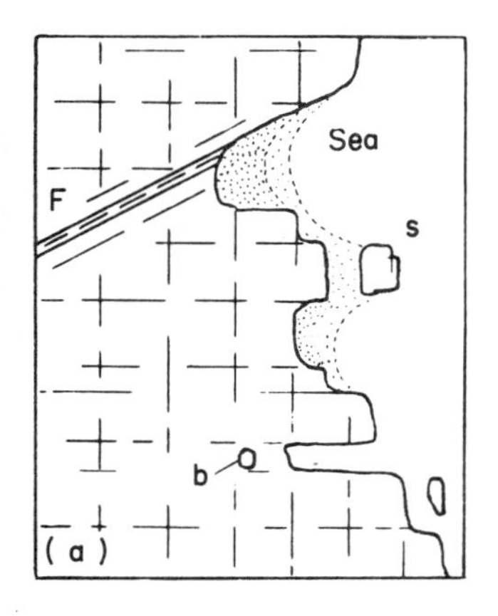

Fig. 3.23 (**a**) A map illustrating joint-control of erosion, with formation of stacks; joint system shown by broken lines. b, blowhole; s, stack; F, fault, controlling local direction of cliff. (**b**) Arch in a coastal cliff; when the crown of the arch falls a stack is formed.

seen in the famous Chalk cliffs on the Sussex coast, e.g. the Seven Sisters.

The influence of jointing on erosion is frequently seen, particularly in hard rocks that stand with a vertical or near vertical face; the sea uses any joints and other surfaces of weakness to remove loosened joint-blocks (Fig. 3.23), and in this way caves and tunnels in the cliffs may be carved out. The process is greatly assisted by the repeated compression of air in the joints by incoming waves; the pressure thus generated acts at the back and sides of joints blocks, which become more readily loosened. When the roof of a cave collapses a narrow inlet or *geo* can be formed; and such a cave is sometimes connected with the surface by a vertical shaft or *blow-hole*, in which the water rises and falls as waves run in and recede. Erosion in jointed rocks will in time leave isolated pillars of rock, or *sea-stacks*, sometimes of fanciful shape and *arches*. Many stacks of hard rock are found along the northern coasts of Scotland and the Orkney and Shetland islands.

Where bands of hard rock alternate with softer, and intersect a coast, the more resistant rocks stand out as headlands and the softer are hollowed out into bays; this is seen for example in South Wales near St. Davids. As waves approach such a shore, with bays and headlands, the wave-front becomes curved so that it is concave towards a headland and convex towards a bay. The wave energy is then concentrated on the headlands which are attacked from both sides, while the bays are largely protected and beach deposits are accumulated in them (Fig. 3.24).

Dipping strata of different resistance may lie parallel to a coast, as at Lulworth Cove in Dorset; here the breaching of a harder outer formation (the Portland Beds) has given the sea access to softer rocks (Purbeck and Wealden Beds), which have been hollowed out into the nearly circular cove behind the Portland Limestone. Poorly consolidated sediments, such as sands and clays offer little resistance to erosion and form low cliffs often scarred by landslides (Fig. 3.24*c*).

Types of coastline

Some coasts at the present day are the result of *submergence*, i.e. a rise of sea-level relative to the land. When this occurs the sea rests against hill and valley slopes of the land surface; ridges stand out as headlands, and cliffs are formed around them by erosion (Fig. 3.25). A coast of *emergence* may result from a rise of the land relative to the sea, eventually giving a nearly smooth coastline bordering an area of newly exposed sea-floor. Many areas, however, have been affected by both rising and falling sea-levels in the geologically recent past.

Coastal marine deposits

Part of the work of the sea is seen in the deposition of material derived from the land by denudation. After being sorted by the action of waves and currents, the sediments contribute to the shallow-water deposits bordering a coast. The fragments are broadly graded according to size, from the shingle and coarse sand of the beach to finer sands and muds; the finest particles that are carried by feeble currents out to sea accumulate there in deeper water. The sediments are laid down on the continental shelf.

The gradation from coarse to fine with increasing distance from the shore and deepening water, is a sequence that can often be recognized in consolidated sedimentary rocks.

Littoral drift

Waves which approach a shore obliquely carry beach material forward up the shore as they advance and break, but the backwash as each wave recedes drags back pebbles and sand along a path nearly at right angles to the shore-line (Fig. 3.26). In this way an alongshore movement is imparted to the sediment by the incoming waves, known as *littoral drift* or 'long-shore' drifting. It may be supplemented to some extent by coastal currents, but these can carry only the finer sediment, and it is the action of waves that is the main cause of the drifting. Transport of coarse material alongshore by wave action in one direc-

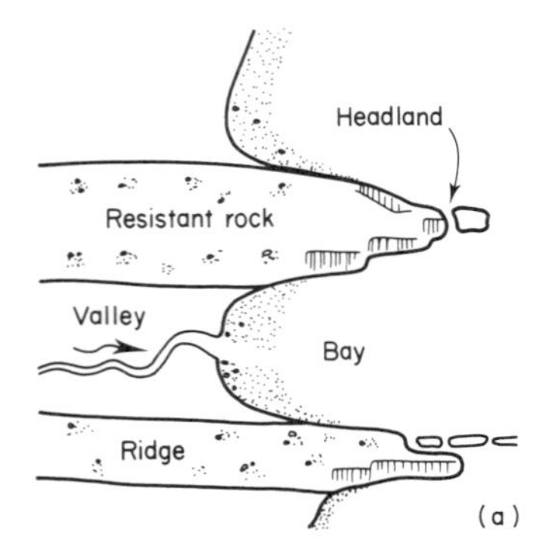

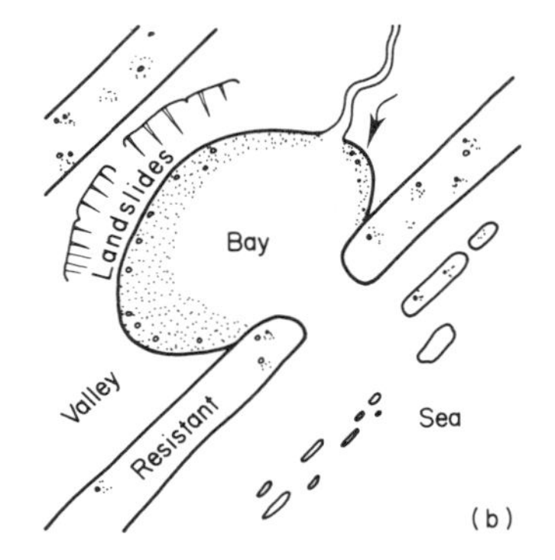

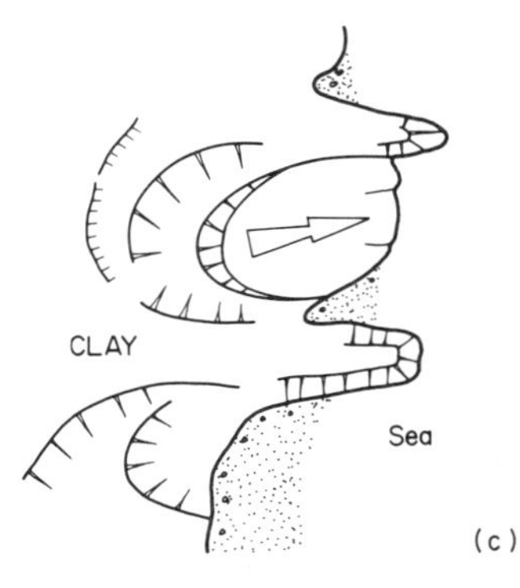

Fig. 3.24 Maps of coastal shapes. (**a–b**) Preferential erosion of weak strata. (**c**) Landslides on a coastline made of clay. (See also Chapter 14.)

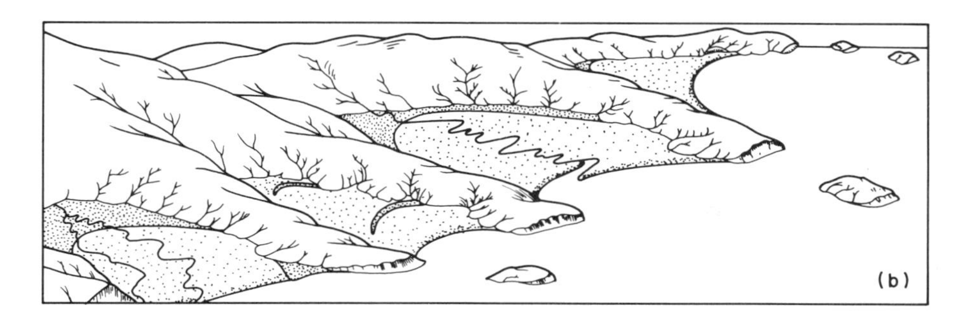

Fig. 3.25 Perspective view of a submerged coastline (**a**), with outline of former coastline indicated. (**b**) Erosion of headlands and deposition of sediment in bays, so smoothing the coastline (see Fig. 2.21 for emerged coastline).

tion and of fine material by currents in the opposite direction has been observed.

Movement of sediment should be expected along coast-lines. On the east coast of England the prevalent drift of shingle is southward, and along the south coast it is eastward; but there are local exceptions to this general pattern, as on the Norfolk coast west of Sheringham where the drift is westward towards the Wash. Since a good shingle beach affords the land a considerable measure of protection from erosion, it is generally an advantage to preserve beach deposits and to encourage their accumulation by suitably placed groynes, across the foreshore. On the other hand, interference with the normal process of littoral drift by engineering works may lead to undesired results. The construction of harbour works, for example, may arrest the natural travel of beach and cause shingle to accumulate upstream of the construction. This can deprive the shores downstream of their protective beach and so expose the coastline there to attack from the sea.

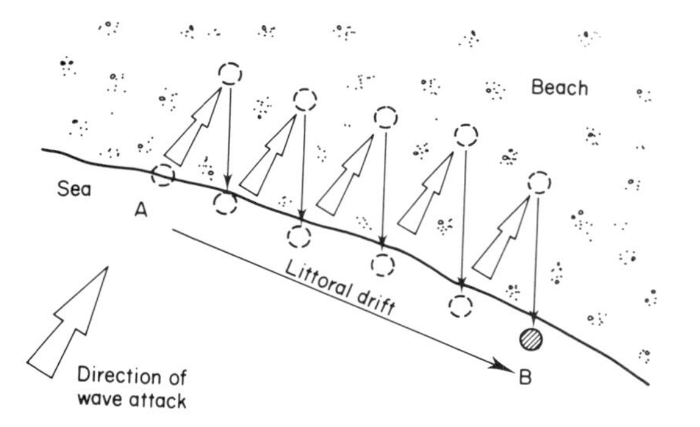

Fig. 3.26 Littoral drift resulting from the movement of sediment up and down a beach. Drift can be measured by monitoring the movement of marked pebbles.

Spits and bars

These accumulations of shingle and sand are brought about by longshore drifting in situations where there is a change in direction of the coastline. A *spit* is a sandy ridge formed by the longshore drift, extending out into open water from a bend in the coast as at the mouth of a river or bay or from the leeward side of a headland (Fig. 3.27*a*). With the latter, erosion of the headland may contribute rock material to the building of the spit. On the east coast of England, Spurn Head, Blakeney Point, and Orfordness are spits built by the southward movement of beach material.

Orfordness, in Norfolk, deflects the mouth of the River Alda, which once entered the sea at Aldeburgh; the river now turns southwards there and flows parallel to the coast for some 16 km, separated from the sea by a shingle barrier. It is estimated that the spit grew 9 km in length, from near Orford to its termination, during 700 years from the founding of Orford Castle in the twelfth century.

Cuspate forelands

These are, in effect, compound spits and may develop a prominence on a coast that gradually builds out to sea, as at Dungeness in Sussex. This large shingle structure now extends some 16 km beyond the original coastline and its outer part consists of shingle storm-ridges that face the dominant waves. Seas approaching the foreland from the two main directions have built up successive storm beaches, now seen as parallel shingle ridges, in the course of centuries. Longshore drifting may further shape such headlands, as occurs at Dungeness where sediment is swept round the point (Fig. 3.27*b*).

is probably the longest known stretch of this type of coastline.

Mud flats

A bay-bar when formed protects the coast locally from wave attack and the area behind it becomes silted up with river-transported sediment, giving rise to mud-flats lying between low and high water levels. Mud and silt trapped in this way can in some situations become new land by

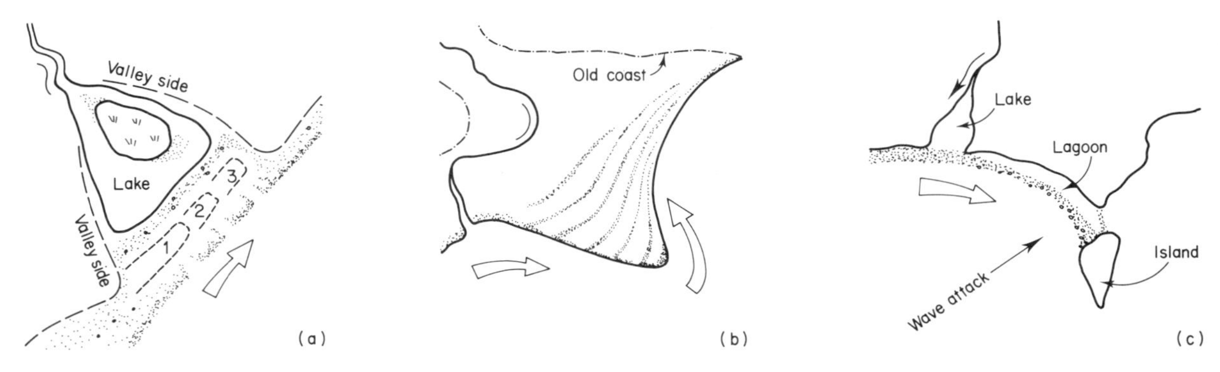

Fig. 3.27 Constructive coastlines: broad arrow indicates direction of longshore drift. (**a**) Bar produced by spit growth (1→ 2→3→), damming a river valley and impounding a fresh water lake, with marsh and mud flats. (**b**) Cuspate foreland (modelled on Dungeness). (**c**) Bar and barrier creating lake and lagoon (modelled on Chesil Beach).

Barriers

When a spit extends across a bay it forms a *bay-bar* or *barrier*, which may lie across the mouth of a small estuary, with a lagoon behind it. The barrier at Looe Pool in Cornwall is one example on the south coast of England. The river flow can percolate through the permeable shingle of the barrier (Fig. 3.27).

The long shingle barrier known as Chesil Beach, in Dorset, extends for about 22 km from near Abbotsbury to the Isle of Portland, and encloses a lagoon (the Fleet) at its eastern end. The size of the pebbles composing it grades from small to large from west to east, and the barrier lies broadly at right angles to the direction from which the longest waves approach this part of the coast. The eastern end, where the shore is steeper, is a storm beach that reaches a height of some 13 m, and here the largest pebbles and also rocks locally derived from the Portland Limestone have been thrown up by wave action. At the western end, longshore drifting from the west has contributed the smaller beach material, mainly sand and small pebbles. The even grading of the pebbles from fine to coarse along the extent of the Beach is considered to have been brought about by wave action, which has helped to drift the larger material westwards.

Barrier islands are composed essentially of sand and, separated by inlets, form series of off-shore islands that are elongated parallel to a coast. They are characteristic of many flat coasts which have a low gradient and a good supply of sediment. The eastern coast of the U.S.A., from about latitude 40°N to Florida and the Gulf of Mexico, is fringed by spits, bars and barrier islands, and

the growth of *salt-marsh* vegetation. The upward growth of a salt-marsh accumulates further sediment, and the rate at which this occurs has been found by spreading a layer of coloured material over the surface, and subsequently measuring the thickness of silt deposited above it. Rates of accumulation may range from about 1/3 to 1 cm per year. Extensive flats exist in the English Fenlands of Norfolk and Lincolnshire, in the Dismal Swamps of Norfolk, Virginia, in N. America, and in the Polders of the Netherlands.

The English fens of East Anglia, are a flat expanse of silt and clay, with layers of peat at certain levels. The area, some 600 km^2, was once a bay or estuary of which the Wash is now a remnant, formed by the submergence of a broad valley which was hollowed out in soft Jurassic clays. The fen deposits are sandy silts (locally called *warp*) and smooth soft clays (*buttery clay*); they have accumulated as sea level has risen slowly relative to the land, with several oscillations in level, since the end of the Pleistocene glaciation. Slight emergence of the area resulted in an extension of the growth of peat seawards; submergence caused silts and clays to be deposited *over* the peat as the sea re-advanced. The deposits reach a thickness of about 18 m and rest on glacial clay. Much of the lowlying fenland is now protected from the sea by artificial embankments.

Very different mud flat deposits accumulate along coastlines exposed to hot dry weather and prolonged evaporation. Behind the coastal barrier islands which fringe the Arabian shores of the Persian Gulf are mud flats that contain extensive deposits, brown in colour, of wind-blown sand mixed with crystals of salt and sulphates of calcium. These are the *sabkhas* (the Arabian term for salt flats). Similar deposits exist at the head of the Baja California and in the Rann of Kutch in N.W. India. The ground water within these mud flats is usually hypersaline, its salt content having been concentrated by the repeated infiltration of sea water when the flats are inundated at high tide, and its evaporation during periods of lower tide. The Bahamas, and Shark Bay in West Australia, are also sites of barrier reefs, extensive lagoons and mud flats. In the Bahamas the muds are mainly of silica and calcium carbonate but contain limestones and minerals precipitated by evaporation (see *evaporites*, p. 127).

Coral reefs

Coral reefs are built by colonies of calcium carbonate secreting animals and plants that enjoy minimum temperatures of 23 to 24°C, sunlight and hence depths of water usually no greater than 50 m, and sea water of normal salinity. They are found extensively developed between the Tropics in the warm waters of the Indian and Pacific oceans, the Caribbean and the Red Sea. No more than 10% of a reef need be the product of reef building colonies, and within the voids that account for the remaining 90% shelter a miriad of animals which make the reef their home. Coral reefs are therefore largely composed of the remains of organisms other than coral.

The large reefs take two principal forms:

(*i*) barriers which grow parallel to the coast but are separated from it, e.g. the Great Barrier Reef of E. Australia, almost 2100 km long; and fringe reefs which grow along the coast as in the Bahamas and Florida.

(*ii*) atolls, which result from the colonization of volcanic sea mounts (p. 2). As mounts move away from the oceanic ridges (Fig. 1.18), they sink into deeper water and may become submerged. If the reef can build at a rate equal to the rate of submergence an atoll is formed with a central lagoon (Fig. 3.28).

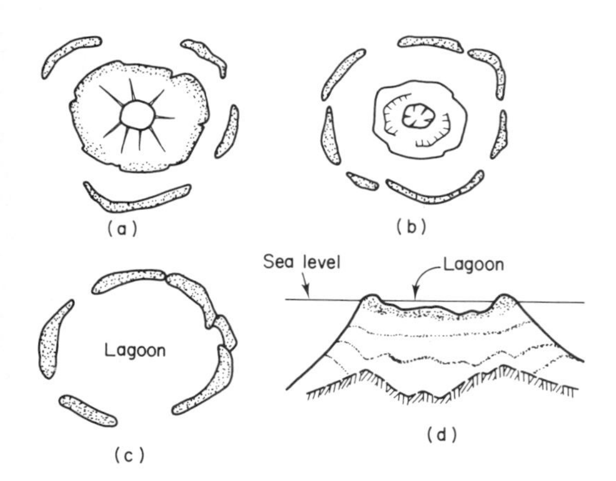

Fig. 3.28 Formation of an atoll. (**a**) Volcanic island (plan) with fringing reefs. (**b**) Sinking. (**c**) Lagoon and atoll. (**d**) Vertical section through (c) showing submerged volcano and successive development of reefs.

Many reefs were exposed to sub-aerial weathering during the Pleistocene glaciations when sea levels were lower and the reefs stood above the sea as islands of porous limestone. As a result many reefs have irregular karstic features with caves, solution cavities and areas which have collapsed into such voids.

Open sea deposits

Beyond the continental shelf lies the open sea in which fine grained deposits of oceanic 'clay' (*ooze*), wind blown sand and volcanic dust slowly accumulate on the ocean floor at rates of between 1 and 50 mm per 1000 years. These are collectively described as *pelagic* sediments (i.e. of the open sea). Much of the ooze is derived from the skeletons of microscopic organisms (see Fig. 2.22), and is either *calcareous* (of calcium carbonate) or *siliceous* (of silica). Calcium carbonate cannot accumulate at great depths because its rate of supply from microscopic skeletons is less than its rate of dissolution by sea water. The depth at which a balance exists between supply and dissolution rates is called the calcite compensation depth; it is 4.5 km below sea level at the equator and shallows to 3.0 km to the south and north. Only siliceous skeletons survive at greater depths.

The volcanic and wind derived material accumulates to produce distinctive deposits of red clay. Towards the continents land derived (*terrigenous*) sediments are found. Much of the sediment in the polar oceans has been deposited by glaciers (Fig. 3.29). Also found on the ocean floor are nodules rich in manganese oxide, particularly in areas of deep sea clay.

Topographic surveys of the deep ocean floor and photographs reveal that the sediments deposited there are moved into wave-like forms by deep ocean currents. The general terms applied to these features are given in Table 3.3.

Table 3.3 Ocean bed forms

Order	Name	Wavelength	Amplitude
First	Mud waves	2000 to 6000 m	20 to 100 m
[Second	Sand waves	30 to 500 m	1.5 to 25 m]*
Second	Ribbons	>10 to >100 m	0.1 to 1.0 m
Third	Ripples	<1.0 to 30 m	0.04 to 1.5 m
Fourth	Lineations	<0.001 to 0.01 m	<0.005 m

* Much more common on the continental shelf than in the deep ocean.

The work of wind

Much of the Earth's solar energy is gained between the tropics and lost at the poles, resulting in a system of global winds which are indirectly responsible for much erosion. They create waves on the oceans which attack the continental coastlines and carry water vapour from

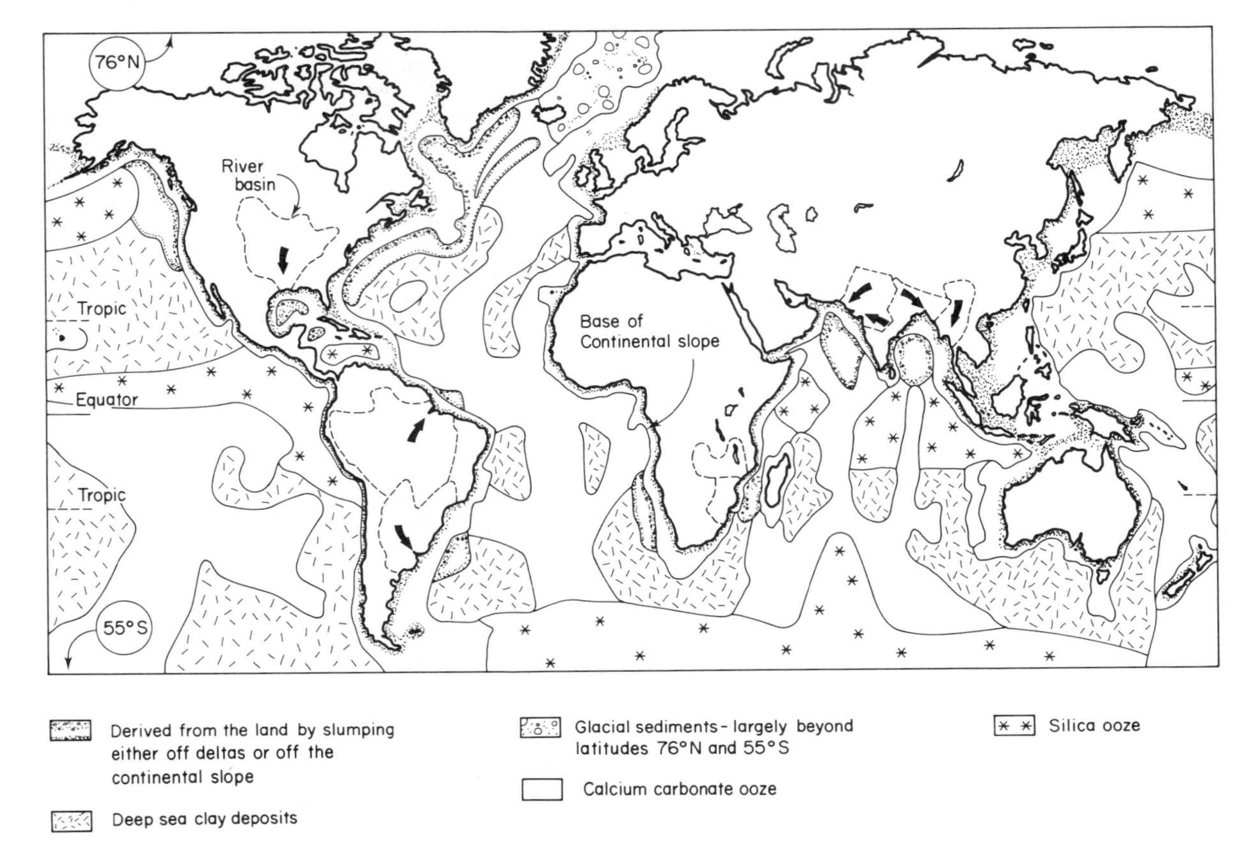

Fig. 3.29 Distribution of sediments on the ocean floor. (Based on data from Davis *et. al.*, 1976, Times Newspapers Ltd., 1980.)

the oceans to fall as rain and snow on land to support the rivers and glaciers of the world. The direct effect of wind on land is greatly reduced by vegetation and thus it is bare ground that is most influenced by this agent of erosion.

Wind erosion

The work of denudation by wind is seen most prominently in regions that have a hot, dry climate (Fig. 3.8). Blowing over weathered surfaces it removes small loose particles of dry and decayed rock, both in deserts and in more temperate regions. In dry desert areas sand of even grade is carried near the ground, at heights up to about 10 cm. Dust particles, however, are blown to greater heights in large quantities, giving rise to repeated duststorms. Most deserts have a rocky floor and any rock surfaces standing above the level of the floor are abraded by the blown sand. The surfaces are smoothed, sharp corners are rounded, and softer bands of rock become more deeply etched than harder layers. These features are typically seen in rocks that have been subjected to wind erosion over a period of time. Larger masses of rock standing above the floor of the desert are undercut by eddying sand near the ground (to heights of little more than a metre), giving characteristic 'mushroom-shaped' rocks with a narrowed base (Fig. 3.30). Blown sand ac-

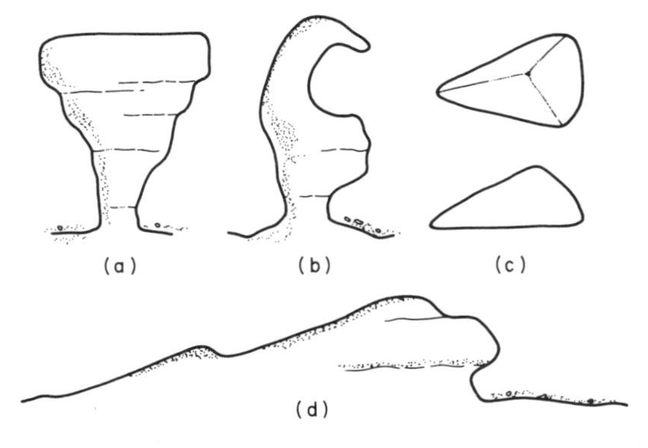

Fig. 3.30 Small scale desert landforms. (**a**) Zeugen (3–5 m high), (**b**) Hollow (3–5 m high), (**c**) Driekanter (3–4 cm long), (**d**) Yardang (10+ m high).

cumulates when wind can no longer keep particles in suspension, and *sand dunes* are formed.

Wind-blown sand (or *eolian sand*) grains, dominantly composed of quartz, become worn down to well-rounded, nearly spherical forms with frosted surfaces. This rounding is more perfect than for water-worn sand, and is diagnostic; the grains are also well graded, i.e. of nearly uniform size, since wind of a given velocity cannot move particles larger than a certain diameter. Grains of quartz 1 mm in diameter need a wind velocity of $8\,m\,s^{-1}$ to move them, a wind of $4\,m\,s^{-1}$ would lift grains of about 0.35 mm diameter. Most wind-blown grains of the deserts are between 0.15 and 0.3 mm in size. By the winnowing action of the wind the finer particles of dust are separated from the sand and carried over large distances, to be dropped far away from their source and form deposits of *loess*. Finer particles are lifted high into the atmosphere, and have been found at heights of 1 and 2 km above ground level. There they are carried over the oceans and continents to be deposited thousands of kilometres from their source. Sand from the Sahara frequently falls with the rain over N.W. Europe.

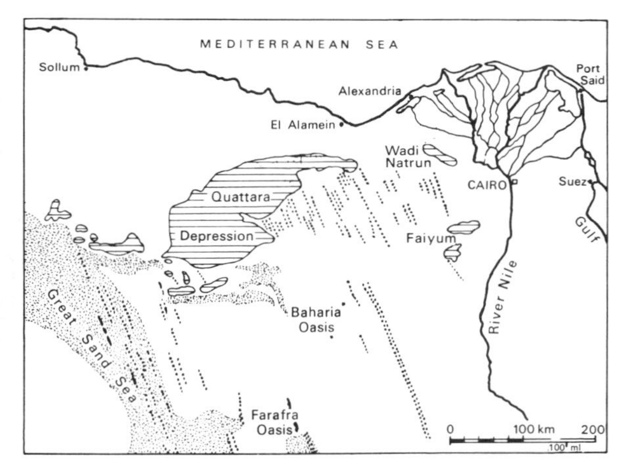

Fig. 3.31 Map of part of the Egyptian desert showing sand accumulations (dots), pattern of dunes, and depressions (based on Holmes, 1978). The Quattara depression reaches a local level of 128 m below sea level. Salt marshes have formed.

Pebbles and boulders lying on the hard floor of a desert (and too large to be moved by the wind) become smoothed and faceted by exposure to the blown sand, and frequently present three-cornered shapes whence they are known as *dreikanters* (Fig. 3.30*c*). They lie with their lengths parallel to the wind direction. When found in old sandstones such as parts of the Precambrian Torridon Sandstone of Scotland, dreikanters are evidence of past desert conditions.

The surface of a desert may be worn down locally until a level is reached where ground-water is exposed. Many hollows in which *oases* are situated have been eroded down to the level of the water table, and so support vegetation. The wet sand is the lower limit of wind action, as in the deep depressions of the Egyptian desert (Fig. 3.31). Lines of communication may be buried by wind-blown sand in arid and semi-arid countries and must be designed to reduce this hazard.

Serious damage has been caused by the denuding action of wind removing vast quantities of dry soil in many regions of the world. The 'Dust Bowl' areas of Kansas and Nebraska are an example of widespread *soil erosion* in a district of low rainfall. A century ago this region was a short grass country supporting large herds of buffalo, whose bones remain as evidence of their former numbers. After intensive cultivation and the felling of trees and draining of swamps the soil became dry and exhausted. A series of droughts followed, and the loose dusty soil was blown by winds, reducing large areas to desert.

Other instances of soil erosion, for example in Kenya and Nigeria, have arisen largely through the overstocking of sparse grasslands with cattle, sheep and goats. Depletion of the scanty cover of grass soon follows and soil erosion begins. A notable feature associated with arid and wind-swept landscapes are residual peaks of hard rock left upstanding and wind polished above the general level; these are termed *inselbergs* (or 'island mounts'). Resistant rocks such as granite and gneiss compose the larger inselbergs; they have undergone long periods of exfoliation, during which successive shells of heated rock

Fig. 3.32 Large inselberg at Spitzkop, Namibia (courtesy of Dr A.O. Fuller).

have fallen away. Outstanding examples occur in parts of Africa, including Mozambique and Namibia (Fig. 3.32), and in northern Australia. The level ground or *pediment* from which they stand up is mainly the result of denudation in alternate dry *and wet* periods. Although seen best in arid landscapes they can occur in more humid regions.

Wind-formed deposits

These include the *coastal sand-hills* of temperate regions; and in deserts *barchan dunes*, crescentic in shape (in plan); *seif dunes*, which form long ridges; *whalebacks*, built of coarser sand, perhaps the relics of linear dunes; and *sand-sheets*, the larger sand accumulations of desert areas, e.g. the Sand Sea of the Egyptian Desert (Fig. 3.31). Wind-blown or *eolian* sands that are deposited on the floor of a desert undergo movements known as *saltation* (Latin *saltare*, to leap) (Fig. 3.33). Sand grains are lifted from the sand-surface by the impacts of other moving grains, and the trajectories which they follow meet the ground at a small angle, often about 15°. Small ridges and hollows formed in the sand by the movement may develop into *ripples*; the length of the saltation path then corresponds with the ripple wavelength.

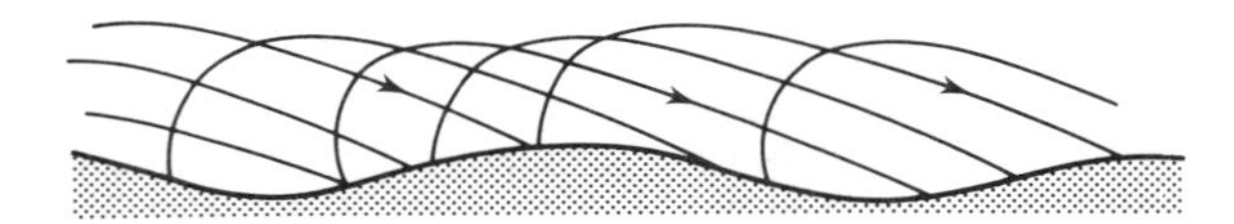

Fig. 3.33 Saltation paths and ripples on a sand surface (Bagnold, 1941).

Coastal sand hills

Mounds of blown sand are piled up in some coastal regions by the prevailing wind, in temperate climates. Thus the western coasts of Europe are liable to dune formation from south-west winds, for example on the west coast of France south of the Gironde estuary. Small accumulations of blown sand are found on the East Anglian coasts and in other parts of Britain. These coastal dunes are not stationary but move slowly and may overwhelm land areas, the dune migrating in the direction of the prevailing wind. Evidence for this movement is sometimes provided by old records and maps, as on the coast of Norfolk at Eccles, where the church was almost completely buried by sand in 1839; 23 years later the tower was uncovered as the sand-hills moved inland, and later its foundation was exposed on the shore. The rate of movement of the sand here can be estimated at about 1 m per year; it varies in different localities and at times may be rapid enough to warrant steps being taken to arrest the process if valuable areas are threatened. The stabilization of coastal dunes is frequently effected by the planting of marram grass (*Ammophila arenaria*), or rice grass, whose long roots bind the surface layers of sand and so hinder its removal by wind. A larger scale method of dealing with the same problem is by afforestation.

Desert dunes

Dunes of many sizes develop in deserts, Table 3.4, and it is common to find lower order forms covering those of higher order. Thus 'ripples' can be found on 'dunes' and 'dunes' can develop on 'draas'. The largest forms generally contain the coarsest sand and it is apparent that as wind speed increases so the wavelength of the dune increases.

Table 3.4 A classification of desert dunes

(Based on Wilson 1972.) Compare with Table 3.3.

Order	Name	Wavelength	Amplitude
First	Draa	300 to 5500 m	20 to 450 m
Second	Dune	3 to 600 m	<1 to 100 m
Third	Ripple	2.5 to 20 m	0.05 to 1.0 m
Fourth	Ripple	<0.01 to 0.15 m	<.001 to 0.05 m

The largest features are called *draas*: they form long ridges similar in shape and continuity to a long, low rounded hill. They are the dune form least susceptible to major change.

The elongated *seif-dunes* of the deserts are long ridges of coarse sand, several hundreds of metres in height and surmounted by crests of finer sand at intervals. The ridges extend in straight lines for great distances in some cases up to 100 km or more; they are built by gentle winds that bring in supplies of sand, and stronger winds from another direction which form the sand into ridges (Fig. 3.34). On the northern margin of the Sahara west of Cairo, in the region of the great Quattara Depression, the sand-bearing winds are from the south-west and the predominant strong winds from the north-north-west; these together have given rise to many seifs. The flat floor of the desert between the ridges is swept clear of the sand. Dunes in the Sahara are presently moving south at one km per year.

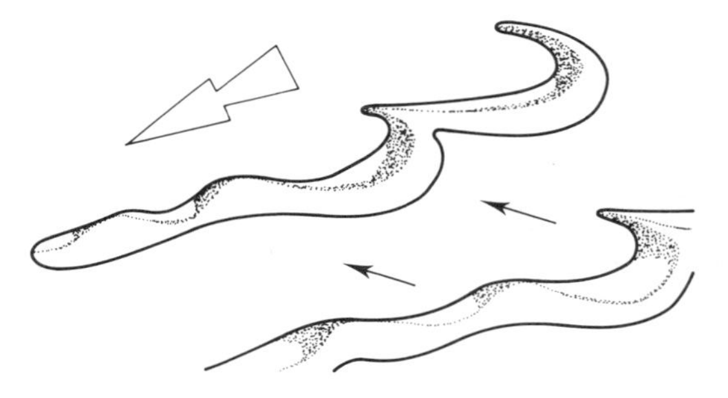

Fig. 3.34 Map of sief dunes: a strong prevailing wind (large arrow) alternates with a gentle wind carrying sand (smaller arrows).

Barchan dunes appear to develop from small heaps of sand not less than 30 cm in height, which increase in extent with continued impacts of moving grains. These desert dunes form most easily on flat surfaces of erosion without large obstructions. The barchans have a crescent shape (Fig. 3.35) and are found in areas where the wind blows from a nearly constant direction, with a limited supply of sand. The sand blown onto the windward slope builds up thin layers (accretion layers) and the rate of deposition is greatest at the summit of the dune, where

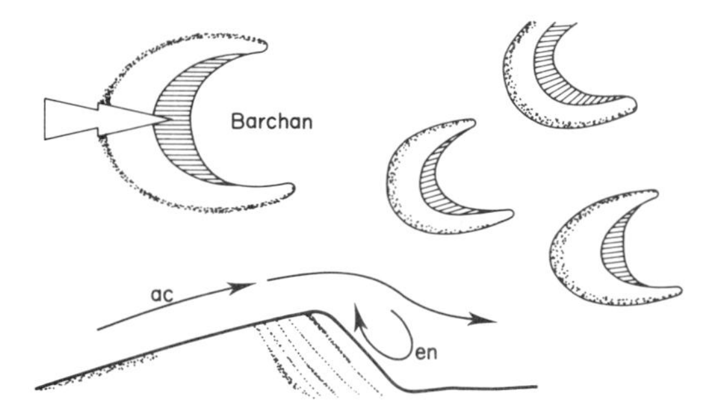

Fig. 3.35 A map of barchan dunes advancing in the wind: slip face shaded. Lower: structure of the dune: ac, accretion layer; en, encroachment layer.

some sand falls over the crest onto the leeward slope. The latter becomes steeper until its slope is greater than the angle of repose for the dry grains, about 32°; when this value is exceeded the sand on the leeward slope shears away along a *slip-surface*. As more sand from windward builds up, another slip-surface is formed ahead of the first, and the process continues. The barchan thus comes to have a structure of thin inclined layers of sand between successive slip-surfaces, which make an angle of 30 to 32° to the horizontal (Fig. 3.35). The dune moves forward in the direction of the wind as successive layers are added to it on the lee side. Barchans range in height from about 10 to 30 m; the larger ones have a width of some 400 m or more.

Ancient dune-bedding is found in certain sandstones formed in past geological ages, e.g. the Nubian Sandstone of North Africa, and the Permian sandstones of Ayrshire, Scotland (Fig. 3.36). The orientation of dunes reflects the wind direction of the desert.

Loess

The wind-blown deposit known as *loess*, a fine-grained calcareous clay or loam, extends from Central Europe through Russia into Asia, and covers large areas in China where it reaches its greatest thickness. It consists of the finer particles that are blown out from the deserts to distant areas, forming a porous deposit which may be traversed by networks of narrow tubes that once enclosed the roots of grasses; these during their growth bound the particles of dust and silt in their grip. Loess is moderately resistant to weathering and can support steep slopes, and where dissected by the action of streams stands as low vertical cliffs. This is aided by the presence of closely-spaced vertical jointing, which is a diagnostic feature of undisturbed loess. The deserts of North Africa have contributed much of this material to Europe and the steppes of Russia. Generally of buff colour, loess is darkened by admixture with vegetable matter and in this condition forms the 'black earth' of the Russian steppes (Figs 3.9 and 3.10). An essentially similar deposit which is developed in the semi-arid regions of the western U.S.A. and in the pampas country of South America is called

Fig. 3.36 Dune-bedding in Permian sandstone, Ballochmyle, Ayrshire.

adobe, a yellowish calcareous clay. A particular feature to note is that loess has a tendency to lose strength and collapse if saturated. It may also flow like a liquid if violently disturbed as during a prolonged earthquake.

The work of ice

A land surface whose topographical features have been fashioned by the action of rivers and atmospheric agents is considerably modified when it becomes covered by an ice-sheet or by glaciers. Valleys are deepened and straightened, rock surfaces smoothed by erosion, and when the ice melts away it leaves behind a variety of deposits which mark its former extent. The main features of these processes are discussed here, and glacial deposits formed during the Pleistocene glaciation when large ares of the British Isles, north-west Europe, North America and other northern lands were under a load of ice, are summarized.

Ice is formed by compaction of snow in cold regions and at high altitudes, where the supply of snow exceeds the wastage by melting. In an intermediate stage between snow and ice the partly compacted granular mass is called *néve*. Ice of sufficient thickness on land will begin to move down a slope and such a moving mass is called a *glacier*. It may occupy a valley, as a *valley glacier*, of which many examples are found in the Alps, the Rockies, the Himalayas and other mountain regions; they are the relics of larger ice-caps (Fig. 3.37). Where several valley glaciers meet on low ground in front of a mountain range a *piedmont glacier* is formed, e.g. the Malaspina Glacier of Alaska. The accumulations of thick ice much larger than those of valley glaciers, constitute the *ice-sheets*, and cover great areas. The Greenland ice-sheet extends over about 1.73×10^6 km^2; drill cores have been obtained from the ice at depths up to 1400 metres. The Antarctic ice-sheet is more than six times greater in extent. Rock peaks protruding through an ice cover are called *nunataks*. When land ice meets the sea it begins to float and break up into *icebergs*; any land-derived debris held in the ice is carried by the bergs and dropped as they become reduced by melting.

Movement of ice

The movement of ice is seen best in valley glaciers. At the head of a valley glacier the ice breaks away from the parent snow-filled by a big crevasse known as the *bergschrund*, and in moving over the ground it behaves like a viscous body, flowing over the irregularities of its course. It has a granular texture, similar to the crystalline structure of metals, and movement within the mass takes place by melting and refreezing at the surfaces of crystals and along glide planes within them. There is also movement along larger surfaces in the ice, which breaks by shearing as it rises and falls over irregularities in its bed. Thus the upper part of the mass undergoes movement by displacement on inclined surfaces and the lower part of the ice moves by slow flowage. Failure in tension also occurs and opens crevasses which can extend across or along the glacier (Fig. 3.38).

An average rate of movement for an Alpine glacier is 0.6 m per day, but this figure varies greatly, depending on the steepness of the slope over which the ice is moving, the thickness of the glacier, and the air temperature. Rates up to 18 m per day have been measured in Greenland. At lower elevations, where melting balances the supply of ice coming down a valley, the thickness of the glacier is gradually reduced until it ends in a 'snout', from which issues a stream of melt-water. Such glacial streams may supply man-made reservoirs constructed to store water

Fig. 3.37 Glaciers in south-west Renland, Scoresby Sound, East Greenland. The ice-dammed lake is 1.5 km wide and lies 820 m above sea level; the cliffs on the left rise a further 550 m above the lake level. (Courtesy of the Director, Geological Survey of Greenland. Photography by Dr. Brian Chadwick.)

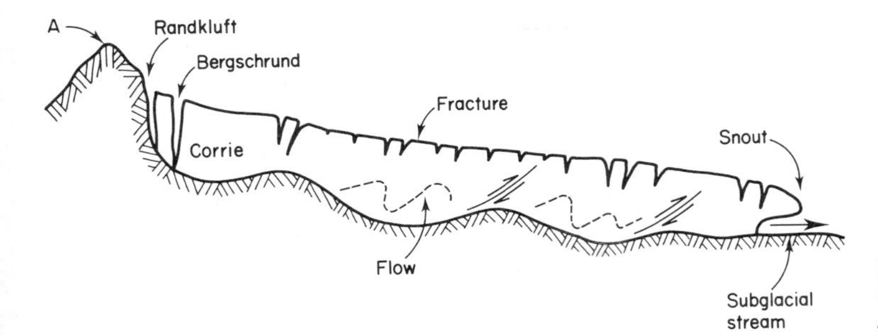

Fig. 3.38 Section along a valley glacier: note ridges and hollows along valley floor. A, arête.

for hydro-electric power generation; the glacier is in effect a natural frozen reservoir.

Transport by ice

A glacier carries along boulders and stones of all sizes, which fall onto its surface from the valley walls on either side; superficial (or *supraglacial*) debris of this kind is called *moraine*. It usually lies in two marginal bands, or *lateral moraines*, parallel to the sides of the glacier; in some cases the ice is completely covered by stones and dirt. The confluence of two glaciers, where a tributary enters the main valley, results in the formation of a *medial moraine* from the two laterals that become adjacent when the two ice-streams meet. The united glaciers below their point of junction thus possess one more line of moraine than the number of ice-streams that have converged (see Fig. 3.37).

Rock debris which falls into crevasses is carried within the ice, and blocks under these conditions can penetrate by their weight to the sole of the glacier. Other fragments are plucked from the underlying rocks as the ice moves, and are held in the lower part of the mass, together with loose surface debris that is picked up by the base of the ice. All this assorted material, much of which is protected from wear as it is enclosed in ice during transport, is known as *englacial material*.

Glacial erosion

Ice moving over a land surface removes soil and loose rock, exposing the bed-rock below. Englacial material held in the lower part of the moving ice acts as an abrasive and the rock floor over which it is rubbed becomes smoothed and a *glaciated surface* is thus formed. Sharp corners of blocks held in the sole of a glacier cut grooves or *striae* in the surfaces over which they are carried, and the striae indicate the direction of movement. Grinding occurs at such points of rock contact and produces fine particles, called *rock-flour* whose movement *polishes* the glaciated surface. Protrusions of rock from the valley floor and sides are sheared off by the force of rock blocks pressed against them by the moving glacier. Large rock-obstructions in the path of a glacier are smoothed on their iceward slopes and plucked on the lee side; when exposed to view after the disappearance of the ice they show rounded forms known as *roches moutonnées*, which are typical of glaciated upland regions.

'Plucking' is thought to occur when rocks, which become lodged in or frozen to the base of the glacier, are separated from their parent mass by glacier movement. A similar plucking action may occur behind the *randkluft* (Fig. 3.38) on the rock wall at the head of a valley glacier, and in time a steep backed hollow, or *corrie* is excavated (also called a *cirque* or *cwm*). Later this is often occupied by a lake (see Fig. 3.39) as at Glaslyn on Snowdon, in Wales.

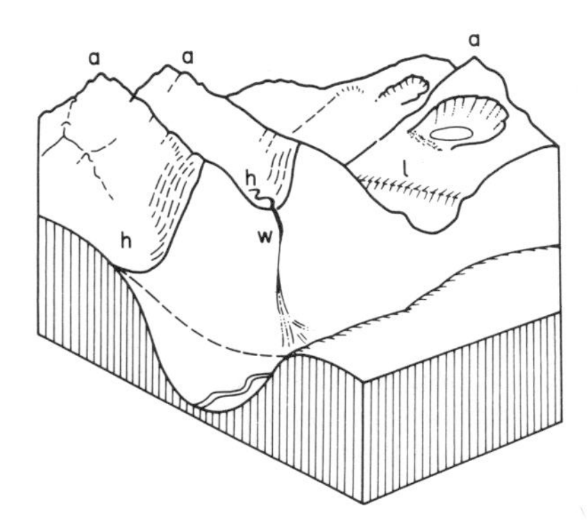

Fig. 3.39 U-shaped profile of a glaciated valley, with hanging valleys (h), waterfall (w) and truncated spurs between tributaries to main valley. Corrie with lake in distance, and shoreline sediments of a previous glacial lake (l). (a) = arête. Original river valley profile shown by broken line.

The base of a glacier need not follow a graded profile like that of a river (Fig. 3.13) but can be most uneven and contain deep hollows separated by ridges (Fig. 3.38). Hollows ground out by the ice to form *rock basins* are often occupied by lakes at the present day. In places the floors of such lakes have been gouged out by the ice to great depths below sea level, and are said to be *over-deepened*. Loch Morar in western Scotland is more than 300 m in depth although its surface level is only 15 m above present sea level; and many of the fjords of Norway and Canada are similarly over-deepened.

The River Devon, a tributary of the Forth of Scotland, flows over the flat flood-plain of the Forth below Stirling. The flood-plain falls gently along its length (42 km) from 15 m above datum near Stirling to 8 m at Grangemouth, and beneath it the pre-Glacial valley of the Devon is filled with glacial deposits. Borings put down to reach Coal Measures had encountered the old valley floor at a depth of 107 m.

The profile of a valley down which a glacier has flowed is also changed, from a V to a U-shaped form giving a broader profile (Fig. 3.39). Not only is the valley widened and deepened, but youthful spurs that formerly projected into it are truncated, and sharp bends removed (Fig. 3.40). Tributary valleys that were once graded to the

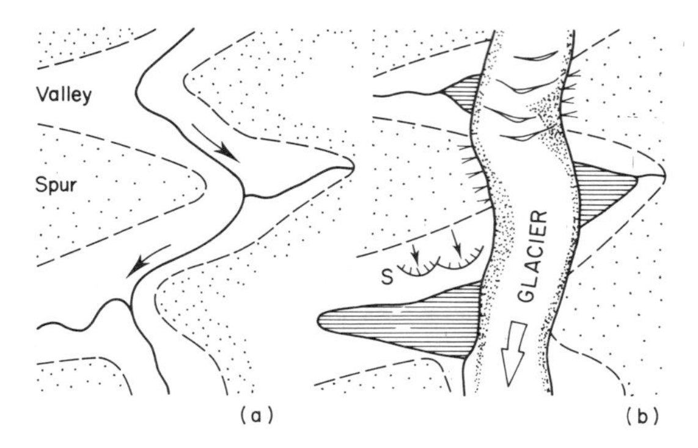

Fig. 3.40 Map of a river valley (a) before and (b) during glaciation. Open ornament = high ground. Lakes are shaded. S = solifluction lobes (see Fig. 3.7).

main valley have been cut off, and end at some height above the new floor. They are called *hanging valleys* and streams from them form waterfalls as they drop to the new level (Fig. 3.39). The U-shaped profile at the base of many glaciated valleys is obscured by glacial deposits the upper levels of which are reworked, by meandering rivers, into flat alluvial plains (Fig. 3.41).

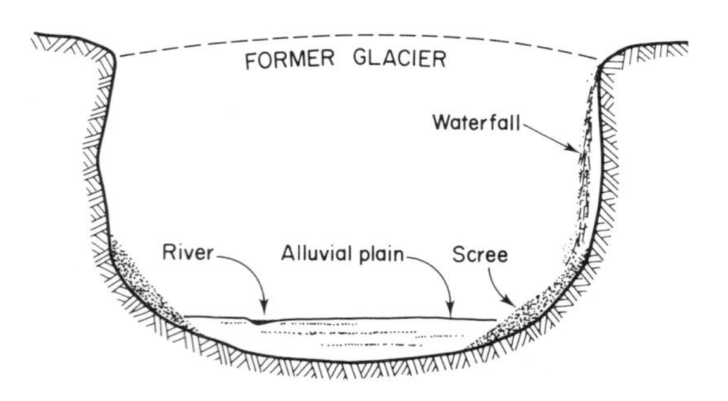

Fig. 3.41 Vertical section through a U-shaped glaciated valley the bottom of which is concealed by alluvium and colluvium (= scree).

Glacial deposits

When a glacier or ice-sheet has retreated from an area, it leaves behind it characteristic deposits of sediment to which the term *glacial drift* is given. This is irregularly distributed, generally thickest in valleys, and shows a lack of sorting and arrangement which distinguishes it from water-borne sediment. More than one-third of the Earth was glaciated during the Pleistocene and a greater area was covered by glacial drift. That part of the drift which comes directly from a glacier is called *till* (Fig. 3.42). Drift deposited by melt water from the glacier produces *outwash deposits*, which may be spread over a wide area or concentrated into rivers and lakes. Supraglacial and englacial debris which melts from the glacier at its snout, slides off the glacier to produce a *flow till*. This often covers the deposits of subglacial debris, called *lodgement till*. *Boulder clay* is an example of a lodgement till and is formed underneath an ice-sheet; the 'clay' content is largely composed of ground-up debris (fine rock-flour) contributed by the melting out of englacial material and by abrasion of the rocks over which the ice has passed. Embedded in this clay are boulders and rock fragments of all sizes. It is deposited as an irregular layer over the surface of the ground, and is unstratified except where later modified by water. The colour of the clay derives from that of the rocks from which the ice obtained raw material during its passage. Thus a red boulder clay would be formed when the ice had passed over red rocks, as with the Irish Sea ice which crossed the red Triassic rocks of Cheshire on its way into the Midlands, carrying also far-travelled boulders from the Lake District and southern Scotland. A feature of particular note is the hardness of lodgement till, sometimes it is very hard: lodgement till encountered during the construction of the St. Lawrence Seaway in Canada, had to be excavated on occasions by drilling and blasting as if it were rock.

Boulder clays and other lodgement tills may give rise to difficulties in construction because of their extreme variability. They may consist of a high proportion of stones and boulders, concentrations of boulders or pockets of sand and gravel in a clay matrix, or may be nearly entirely of clay. When very large boulders are present – and some reach tens of metres in size – excavation is impeded; the large boulders hinder the use of excavation machinery and may have to be broken by blasting and removed piecemeal. Boulder clay also masks the true nature of the underlying rock surface, concealing hollows and sometimes deep valleys; buried channels have been proved by borings through drift in the valleys of most north European, Russian and N. American rivers.

Any large blocks transported by the glacier are left stranded when the ice has melted wherever they happen to be. They are called indicator boulders or *erratics*, since they are different from the local rock on which they come to rest. Studies of the distribution of erratics have made it possible to draw conclusions as to the extent and direction of travel of ice, such as that which during the Pleistocene glaciation of the British Isles moved out from the higher mountain regions of the north. Blocks of a distinctive microgranite from Ailsa Craig, for example, are found as far south as North Wales; and boulders of laurvikite from Norway are found along the Yorkshire coast, indicating transport by Scandinavian ice across what is now the North Sea (Fig. 3.43).

Many of the outwash deposits consist of boulders and

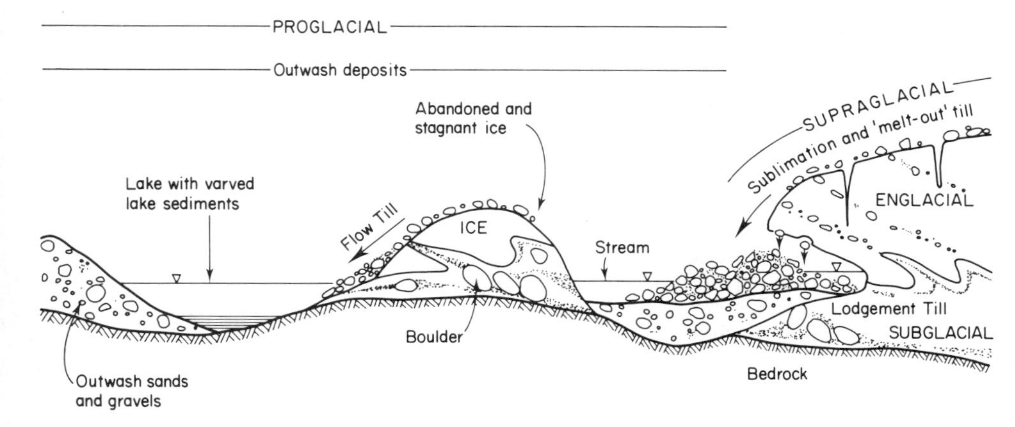

Fig. 3.42 Glacial drift: a diagrammatic vertical section through the sediments deposited beneath a glacier and at its snout. Note outwash deposits may erode previously deposited lodgement till to lie directly on bedrock near a glacier. Upper surface of lodgement till is often deformed. (Terminology based on Boulton, 1980.)

smaller particles that are capable of being transported by flowing water. Accumulations of silt, sand and gravel are abundant. Finer particles can be carried furthest from the glacier. The fine-grained laminated clays known as *varves* are deposited in horizontal layers in still water impounded in front of a retreating glacier or in glacial lakes. They show alternate coarser (silty) and finer (muddy) layers; the mud, which settles slowly in water, lies above the more quickly sedimented silt, and each pair of layers (coarse and fine) represents a season's melting of the ice. Varved clays thus provide a time scale, in which a pair of layers counts as one year. In an engineering excavation at Lake Ragunda, Sweden, a thickness of 24 m of undisturbed varves was exposed, above a deposit of moraine. From the number of laminae in the varves it was estimated that 7000 years had elapsed since the ice margin retreated past the region; other counts near by, when added in, showed that 9000 years had passed since Stockholm became free from ice, giving a date for the end of the Pleistocene glaciation. The method has also been applied extensively to varves in N. America, northern Europe and Russia.

Landforms

Some of the processes which deposit glacial drift differ sufficiently for the shape of their accumulations to be diagnostic of their mode of formation. Distinctive landforms can be identified of which the most common are now described. Debris dropped at the front of a glacier as the ice melts forms a hummocky ridge or *terminal moraine*; in the British Isles these end-moraines are found for example in north Yorkshire, South Wales, and southern Ireland. Extensive terminal moraines exist in Canada, Scandinavia and Russia. Other mounds of sand and gravel which were derived from the ice along its margins during pauses in its retreat, and lie parallel to the ice-

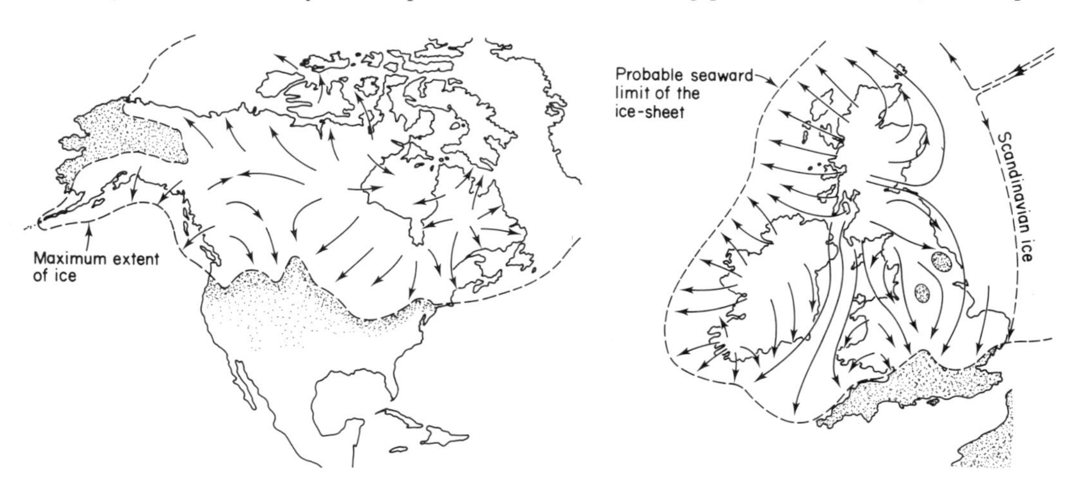

Fig. 3.43 Patterns of ice sheet movement: stipples = unglaciated areas subject to permafrost. Based on data from Legget (1976), Wright (1937), Boulton *et. al.* (1977).

front, are known as *kames*. Good examples of kames in Britain can be seen near Carstairs, Lanarkshire, and in the Clyde valley, and at places in the north of England.

Streams of melt-water flow in tunnels in the base of the ice and move much sediment, fragments of which become partly rounded in the process. When these streams emerge along the ice-front their speed is checked and the transported debris is spread out on land as an *outwash fan* of gravel and sand. If the debris is discharged into still water it builds up a delta. Ridges and mounds of gravel deposited by such sub-glacial streams, with their length roughly at right angles to the ice-front, are termed *eskers*; one broad belt of eskers extends east and west across the Central Plain of Ireland from Galway to Co. Dublin. In Finland some roads and railways are built on esker gravel ridges, which rise above the level of adjacent lakes.

Smooth oval-shaped mounds known as *drumlins*, and commonly 30 to 35 m in height, are composed of boulder clay or englacial debris, sometimes moulded over a roche moutonnée. The long axis of the oval is parallel to the direction taken by the ice. Groups of drumlins are well developed in country bordering the Lake District, as in the Vale of Eden, and in parts of Galloway. Clew Bay in north-west Ireland contains a remarkable drumlin archipelago, formed by partly submerged drumlin country.

In low-lying country a large rock-mass in the path of advancing ice may serve to divide the ice, which passes on either side of the crag, and a 'tail' of drift is preserved on the lee side of the crag. A well-known example of this *crag and tail* topography is seen at Edinburgh, where the crag is Castle Rock from which the High Street and Canongate, running eastwards to Holyrood, lie on the gentle slope of the tail (Fig. 3.44).

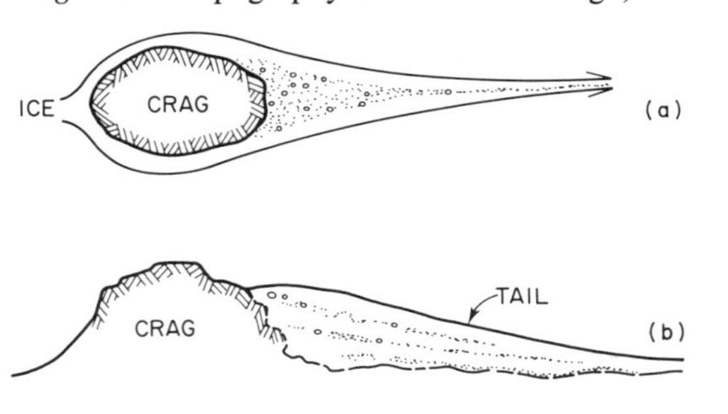

Fig. 3.44 Crag and tail in plan (**a**) and vertical section (**b**).

Landform associations

The grouping of landforms associated with upland valley glaciers tends to differ from that found in flatter lowland areas. Commonly found associations are illustrated in Figs 3.45 and 3.46. In addition to these features there are

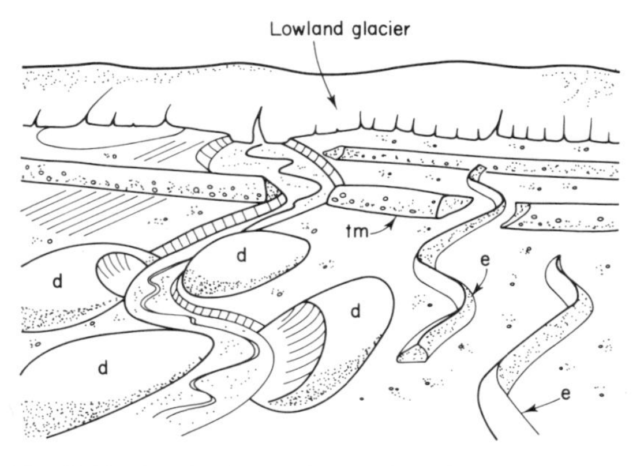

Fig. 3.46 Lowland association of glacial sediments. Lodgement till, sometimes grooved by forward movement of the glacier, covers the ground and is cut by glacial streams depositing outwash sand and gravel. Terminal moraine (tm), esker, (e), drumlin (d).

glacial lakes. Upland lakes often developed in hanging valleys dammed by the glacier of the main valley (Fig. 3.40). In Switzerland the Marjelen See is impounded in a tributary valley by the Aletsch Glacier which fills the main valley. During the Pleistocene Ice Age temporary lakes of this kind existed in many glaciated regions. One well-known example is the lake that occupied Glen Roy, north-east of Ben Nevis; three shore-lines lie at different levels along the slopes of the Glen – the Parallel Roads of Glen Roy – and mark three levels at which the water stood for a time, draining by successive overflow channels as the ice diminished in stages (see Fig. 3.39).

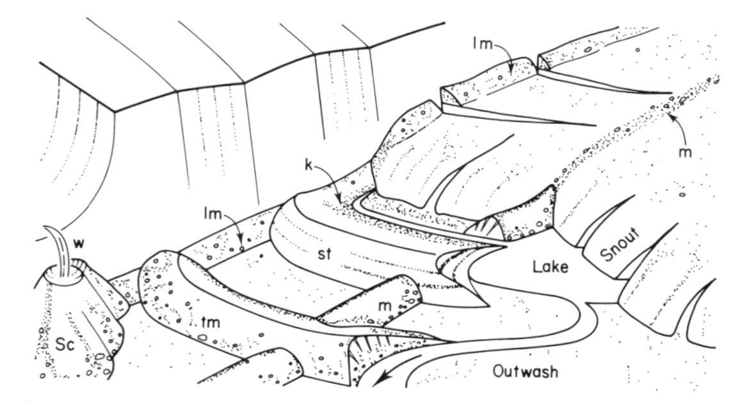

Fig. 3.45 Upland association of glacial sediments. Two valley glaciers meet and create a medial moraine (m). River deposits accumulate on stagnant ice (st) abandoned by glacier recession and form kames (k). Terminal moraine = tm. Lateral moraine = lm. A waterfall (w) discharges from a hanging valley and a scree (sc) accumulates at its base.

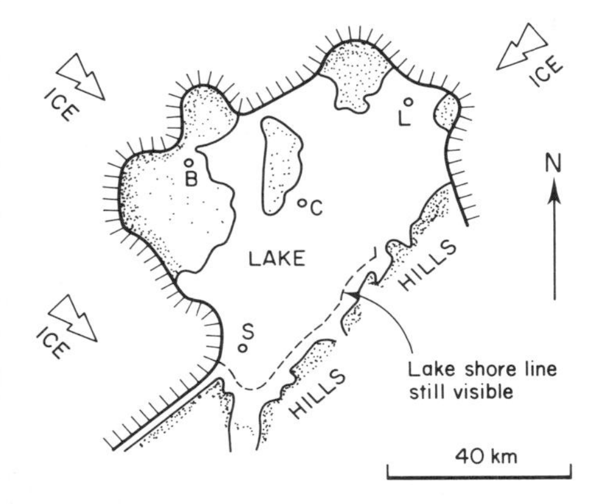

Fig. 3.47 Glacial Lake Harrison. B = Birmingham, L = Leicester, C = Coventry, S = Stratford. Unglaciated high ground is stippled. (Based on Shotton, 1953; Shore-line from Dury, 1951.) See also Fig. 3.37.

Overflow channels from former glacial lakes have sometimes been cut deeply, having acted as a drainage outlet for a long time. They then make noticeable topographic features at the present day.

The Grand Coulee in Washington, U.S.A., a canyon over 64 km in length and 3 to 5 km wide, is an instance of a glacial channel being used for large engineering works. The canyon was eroded by the overflow from a glacial lake, impounded by ice blocking the gorge of the Columbia River, and water overflowed past the margin of the ice for long enough to cut the canyon, in places over 183 m deep, in basaltic lavas. After the recession of the ice the river recovered its former channel. The Grand Coulee is now the site of a reservoir, water for irrigation purposes being pumped up to it from the dammed Columbia River and impounded between two low dams built within the Coulee. In lowland areas glacial lakes often developed against the snout of an ice-sheet. Many large lakes covered England (Fig. 3.47) and overflowed cutting gorges, e.g. the gorge at Ironbridge, Shropshire, which was eroded by the overflow from a large lake impounded by ice that covered the Cheshire plain. This deep overflow served to divert the River Severn from its original north-westerly course into its present channel through the Ironbridge Gorge, on the retreat of the ice.

Post-depositional changes

Till may be overriden by a subsequent advance of a glacier, or be reworked by its outwash streams, or be subjected to repeated freezing and thawing. These and other post-depositional changes alter the character of glacial drift. When a glacier melts the vertical (overburden) load on its subglacial till is reduced and sub-horizontal fissures open within the till, as it expands. Subglacial till covered by moving ice will normally be sheared by the stresses transmitted to it from the ice. The shear surfaces will be planar, often grooved in the direction of movement (slickensided), polished and of low shear strength. The water content of such surfaces may also be much greater than that of the undeformed till, and this will make the shear surfaces very weak (and often soft) when the till thaws. Repeated freezing and thawing of till and its eventual desiccation can produce sets of vertical fractures that cut the till into six-sided (hexagonal) columns; sometimes this is so well developed the till has the appearance of a columnar basalt (p. 92). Weathering will leach till and deposits now buried by later drift may contain weathered profiles beneath horizons that were once ground level.

Other mass transport

Rivers, the sea, wind and ice collectively move more material than other natural processes and are the principal agents of mass transport. Here we mention three others which are of considerable importance to engineering, namely rainfall, ground-water and slope instability.

Rainfall

The *mechanical* action of rain falling continually on a ground surface is to dislodge loose particles of soil and rock dust, which are gradually washed down slopes to lower levels, a process known as *rain-wash*. The muddy appearance of a river after rainfall is evidence that the land surface is slowly but continually being lowered. The denuding effects of heavy showers and storms can be severe, especially in regions where a covering of vegetation is lacking. The rain may then be channelled into courses which collect into deeper channels or *gullies*, especially when heavy storms occur. Cloudbursts (a high rate of precipitation over a relatively small area) can produce deep gullies, entrenched in the surface, and cause local damage by the destruction of roads and livestock and the undermining of farm buildings. The running water thus derived from rainfall then becomes an agent of erosion (p. 31). Gulleying is common in sub-tropical countries, but is occasionally found in more arid climates.

A cover of vegetation protects the ground from the immediate effects of the impact of rain and over-grazing

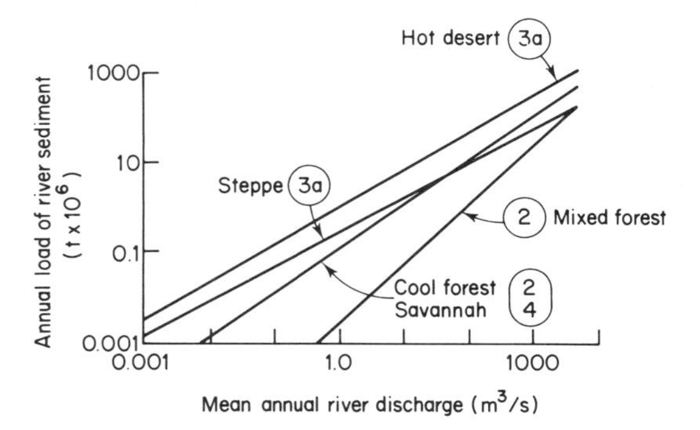

Fig. 3.48 Form of relationship between mass transport of suspended sediment in rivers and vegetation: numbers refer to Fig 3.10. (Data from Fleming, 1969.)

and the clearing of wooded areas has been followed in many places by considerable denudation of the bare ground surface. If this proceeds for some time the entire covering of soil may be removed. Fig. 3.48 illustrates the typical trend of soil erosion by rainfall.

In areas of thick soil containing embedded boulders, *earth pillars* may be formed by rain denudation. They are tapering columns of soil each capped by a large stone or boulder which has preserved the material below it from being washed away.

Ground-water

Chemical weathering associated with the infiltration of rainfall enriches ground-water in dissolved constituents (see Leaching, Table 3.1). By this process the slow but continuous flux of mineralized ground-water can remove from a land mass the equivalent of tonnes of solid matter each year. The chemistry of ground-water is described in Chapter 13. When evaporated, ground-water will precipitate its dissolved constituents, as can occur in hot deserts. Many hot deserts have developed where latitude and distance from the ocean combine to shield a topographic basin from rain-bearing prevailing winds. Drainage is towards the centre of the basin and in wet periods there accumulate lakes, which by evaporation of their water, become rich in salt. These are *playa* lakes and their deposits create extensive, smooth and horizontal *salt-flats* (Fig. 3.49). The Lake Eyre Basin of Australia, and the

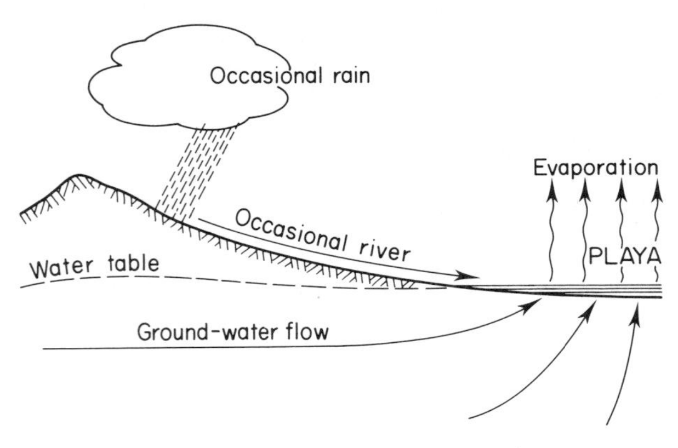

Fig. 3.49 Vertical section through an inland basin showing the movement of salt-bearing waters to a Playa.

Great Salt Lake Desert of N. America, are well known examples, but many others exist as near the borders of southern Iran with Afghanistan and Pakistan, in the northern Nafud of Saudi Arabia and across the Sahara Desert from Egypt to Mauritania.

Slope instability

Mass-movement is a general term used to describe the transport of material down a slope: many mechanisms can operate. Rain soaking into an upland area of deep soil over a long time builds up a higher water content in the soil, which may lead to the development of *mud-flows*. An unstable water-logged mass of this kind becomes mobile when sufficient pressure is developed to burst any lateral restraint at a weak point on its margin, and the mass then begins to move down slopes as a flow of mud. It may engulf or displace buildings and lines of transport before coming to rest at a lower level, and valleys have been temporarily blocked by such flows, leading to local flooding.

Avalanches of snow and rock have a similar mechanism. The very slow movements of surface layers on slopes, such as *soil-creep* and *rock-creep* (see Fig. 14.1), which continue over a long period and may ultimately displace trees, fences, and lines of communication come under this heading. Although these effects are not called landslides they arise from like causes, and are included in the general category of *mass-movements*. The cumulative effect of creep in bringing soil and rock fragments within the sphere of action of transporting agents such as rivers is an important contributory factor in the process of denudation. Repeated, but small scale, mass-movement on hills of hard rock creates escarpments at the base of which may accumulate deposits of loose boulders, called *colluvium* (for example, see Fig. 3.41). Larger mass-movements which transport monolithic masses of rock and soil are called *landslides* and are described in Chapter 14.

Selected bibliography

Ollier, C.D. (1969). *Weathering*. Longman, London.
Holmes, A. (1978). *Principles of Physical Geology*, 3rd Edition. Nelson, London.
Gregory, K.J. and Walling, D.E. (1973). *Drainage Basin Form and Process*. Edward Arnold, London.
King, Cuchlaine, A.M. (1972). *Beaches and Coasts*. Edward Arnold, London.
Cook, R.U. and Warren, A. (1973). *Geomorphology in Deserts*. B.T. Batsford Ltd., London.
Washburn, A.L. (1979). *Geocryology*. A survey of periglacial processes and environments. Edward Arnold, London.
Flint, R.F. (1971). *Glacial and Quaternary Geology*. J. Wiley & Sons, New York.
Sugden, D.E. and John, B.S. (1976). *Glaciers and Landscape*. A geomorphological approach. Edward Arnold, London.
Williams, P.J. (1982). *The Surface of the Earth: an introduction to geotechnical science*. Longmans, London and New York.
Trudgill, S.T. (1983). *Weathering and Erosion*. Butterworths, London.
Press, F. and Siever, R. (1982). *Earth*. W.H. Freeman and Co. Ltd., Oxford.

4

Minerals

Minerals are the solid constituents of all rocks, igneous, sedimentary and metamorphic (p. 1), and occur as crystals. A *mineral* can be defined as a natural inorganic substance having a particular chemical composition or range of composition, and a regular atomic structure to which its crystalline form is related. Before beginning the study of rocks it is necessary to know something of the chief rock-forming minerals.

The average composition of crustal rocks is given in Table 4.1 and has been calculated from many chemical analyses. The last item includes the oxides of other metals; and also gases such as carbon dioxide, sulphur dioxide, chlorine, fluorine, and others; and trace elements, which occur in very small quantities.

Table 4.1 The average composition of crustal rocks

	%
SiO_2	59.26
Al_2O_3	15.35
Fe_2O_3	3.14
FeO	3.74
MgO	3.46
CaO	5.08
Na_2O	3.81
K_2O	3.12
H_2O	1.26
P_2O_5	0.28
TiO_2	0.73
rest	0.77
Total:	100.0

Table 4.2 lists eight *elements* in their order of abundance in crustal rocks. Thus silicon and oxygen together make up nearly 75 per cent of crustal rocks, and the other elements over 98 per cent.

Since silicon and oxygen preponderate in the rocks, the chief rock-forming minerals are silicates. Although over three thousand different minerals are known to the mineralogist, the commonly occurring silicates are relatively few.

Table 4.2 The most abundant elements

Oxygen	(46.60%)
Silicon	(27.72%)
Aluminium	(8.13%)
Iron	(5.00%)
Calcium	(3.63%)
Sodium	(2.83%)
Potassium	(2.59%)
Magnesium	(2.09%)

Physical characters

Included under this head are properties such as *colour*, *lustre*, *form*, *hardness*, *cleavage*, *fracture*, *tenacity*, and *specific gravity*. Not all of these properties would necessarily be needed to identify any one mineral; two or three of them taken together may be sufficient, apart from optical properties (p. 65). Other characters such as *fusibility*, *fluorescence*, *magnetism*, and *electrical conductivity* are also useful in some cases as means of identification, and will be referred to as they arise in the descriptions of mineral species. In a few instances *taste* (e.g. rock-salt) and *touch* (e.g. talc, feels soapy) are useful indicators.

Colour

Some minerals have a distinctive colour, for example the green colour of chlorite, but most naturally occurring minerals contain traces of substances which modify their colour. Thus quartz, which is colourless when pure, may be white, grey, pink or yellow, when certain chemical impurities or included particles are present.

Much more constant is the colour of a mineral in the powdered condition, known as the *streak*. This may be produced by rubbing the mineral on a piece of unglazed porcelain, called a streak-plate, or other rough surface. Streak is useful, for example, in distinguishing the various oxides of iron; haematite (Fe_2O_3) gives a red streak, limonite (hydrated Fe_2O_3) a brown, and magnetite (Fe_3O_4) a grey streak.

Lustre

Lustre is the appearance of a mineral surface in reflected light. It may be described as *metallic*, as in pyrite or galena; glassy or *vitreous*, as in quartz; *resinous* or greasy, as in opal; *pearly*, as in talc; or *silky*, as in fibrous minerals such as asbestos and satin-spar (fibrous gypsum). Minerals with no lustre are described as *dull.*

Form

Under this heading come a number of terms which are commonly used to describe various shapes assumed by minerals *in groups or clusters* (Fig. 4.1); the crystalline form of *individual* minerals is discussed on page 63.

Acicular – in fine needle-like crystals (also described as *filiform*). e.g. schorl, natrolite.

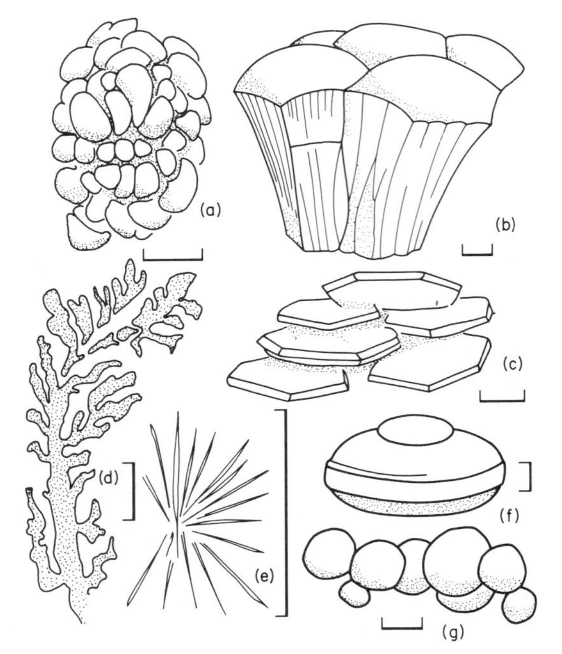

Fig. 4.1 Common shapes of mineral clusters: scale bar=1 cm. (**a**) Botryoidal. (**b**) Reniform. (**c**) Tabular. (**d**) Dendritic. (**e**) Acicular. (**f** and **g**) Concretionary.

Botryoidal – consisting of spheroidal aggregations, somewhat resembling a bunch of grapes; e.g. chalcedony. The curved surfaces are boundaries of the ends of many crystal fibres arranged in radiating clusters.

Concretionary or nodular – terms applied to minerals found in detached masses of spherical, ellipsoidal, or irregular shape; e.g. the flint nodules of the chalk.

Dendritic – moss-like or tree-like forms, generally produced by the deposition of a mineral in thin veneers on joint planes or in crevices; e.g. dendritic deposits of manganese oxide.

Reniform – kidney-shaped, the rounded surfaces of the mineral resembling those of kidneys; e.g. kidney iron-ore, a variety of haematite.

Tabular – showing broad flat surfaces; e.g. the 6-sided crystals of mica.

Note that the above terms do not apply to rocks.

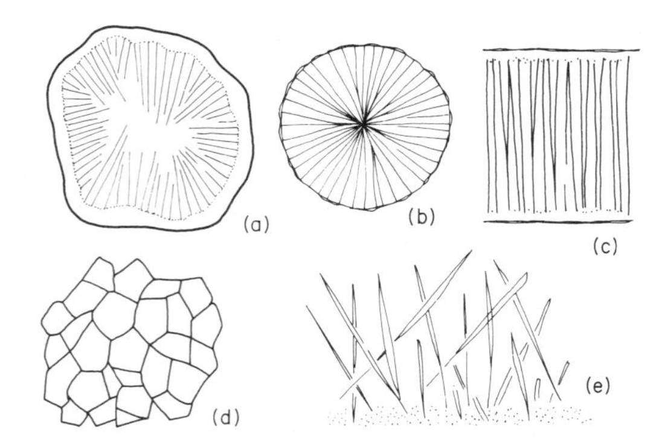

Fig. 4.2 Common relationship between minerals in clusters. (**a**) Drusy. (**b**) Radiated. (**c**) Fibrous. (**d**) Granular. (**e**) Reticulated.

Figure 4.2 illustrates five commonly occurring mineral relationships, as follows.

Drusy – closely packed small crystals growing into a cavity, such as a gas bubble preserved in solidified lava.

Radiated – needle-like crystals radiating from a centre: the illustration is of a pyrite (FeS_2) concretion.

Fibrous – consisting of fine thread-like strands; e.g. asbestos and the satin-spar variety of gypsum.

Granular – in grains, either coarse or fine; the rock marble is an even granular aggregate of calcite crystals.

Reticulated – a mesh of crossing crystals.

Table 4.3 Mohs' Scale of Hardness

1. Talc	6. Orthoclase Feldspar
2. Gypsum	7. Quartz
3. Calcite	8. Topaz
4. Fluorspar	9. Corundum
5. Apatite	10. Diamond

Hardness

Hardness, or resistance to abrasion, is measured relative to a standard scale of ten minerals, known as Mohs' Scale of Hardness: Table 4.3. These minerals are chosen so that their hardness increases in the order 1 to 10. Hardness is tested by attempting to scratch the minerals of the scale with the specimen under examination. A mineral which scratches calcite, for example, but not fluorspar, is said to have a hardness between 3 and 4, or H=3–4. Talc and gypsum can be scratched with a finger-nail, and a steel knife will cut apatite (5) and perhaps feldspar (6), but not quartz (7). Soft glass can be scratched by quartz. The hardness test, in various forms, is simple, easily made, and useful; it is a ready means for distinguishing, for example, between quartz and calcite.

Cleavage

Many minerals possess a tendency to split easily in certain regular directions, and yield smooth plane surfaces called *cleavage planes* when thus broken. These directions depend on the arrangement of the atoms in a mineral (p. 70), and are parallel to definite crystal faces. *Perfect, good, distinct*, and *imperfect* are terms used to describe the quality of mineral cleavage. Mica, for example, has a perfect cleavage by means of which it can be split into very thin flakes; feldspars have two sets of good cleavage planes. Calcite has three directions of cleavage.

Fracture

The nature of a broken surface of a mineral is known as *fracture*, the break being irregular and independent of cleavage. It is sometimes characteristic of a mineral and, also, a fresh fracture shows the true colour of a mineral. Fracture is described as *conchoidal*, when the mineral breaks with a curved surface, e.g. in quartz and flint; as *even*, when it is nearly flat; as *uneven*, when it is rough; and as *hackly* when the surface carries small sharp irregularities (Fig. 4.3). Most minerals show uneven fracture.

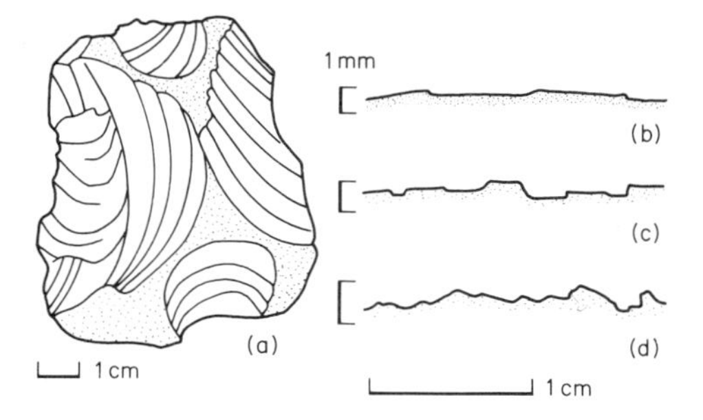

Fig. 4.3 Mineral fracture. (**a**) Conchoidal, which can occur on a smaller scale than shown. (**b**) Even. (**c**) Uneven. (**d**) Hackly.

Tenacity

The response of a mineral to a hammer blow, to cutting with a knife and to bending is described by its tenacity. Minerals that can be beaten into new shapes are *malleable*; e.g. the native metals of gold, silver and copper. Most minerals are *brittle* and fracture when struck with a hammer. A few brittle minerals can be cut with a knife and are described as *sectile*. Flakes of mica can be bent and yet return to their flat tabular shape when free to do so: they are both *flexible* and *elastic*: cleavage flakes of gypsum are flexible but *inelastic*.

Specific gravity

Minerals range from 1 to over 20 in specific gravity (e.g. native platinum, 21.46), but most lie betwen 2 and 7 (see Table 4.4). For determining this property a steelyard

Table 4.4 Specific gravity of common minerals

halite	2.16	muscovite	2.8–3.0	rutile	4.2
glauconite	2.3	apatite	3.2	zircon	4.7
gypsum	2.32	hornblende	3.2 (av.)	haematite	4.72
feldspar	2.56–2.7	tourmaline	3.0–3.2	ilmenite	4.8
clays	2.5–2.8	sphene	3.5	pyrite	5.01
quartz	2.65	topaz	3.6	monazite	5.2
calcite	2.71	kyanite	3.6	magnetite	5.2
dolomite	2.85	staurolite	3.7	cassiterite	6.9
chlorite	2.6–3.3	garnet	3.7–4.3		

apparatus such as the Walker Balance can be used, for crystals or fragments which are not too small. The mineral (or rock) is weighed in air and in water, and the specific gravity, G, is calculated from the formula: $G = w_1/(w_1 - w_2)$, where w_1 is the weight in air and w_2 the weight in water. Other apparatus is described in appropriate textbooks.

The specific gravity of small mineral *grains* is estimated by the use of heavy liquids, of which the chief are bromoform ($CHBr_3$), G = 2.80, and methylene iodide (CH_3I_2), G = 3.33, both of which may be diluted with benzene, and Clerici's solution (G = 4.25), a mixture of thallium salts which may be diluted with water.

The separation of fragments of a particular mineral from a sediment containing many minerals may be achieved by placing them in a liquid solution of known specific gravity equal to that of the mineral to be separated. Grains of greater specific gravity than the liquid will sink whilst those of lower specific gravity will float. Magnetic minerals such as magnetite and ilmenite may be separated from non-magnetic minerals, such as cassiterite, by an electro-magnet.

Crystalline form

Minerals occur as *crystals*, i.e. bodies of geometric shape which are bounded by *faces* arranged in a regular manner and related to the internal atomic structure (p. 70). When a mineral substance grows freely from a fused, liquid state (or out of solution, or by sublimation), it tends to assume its own characteristic crystal shape; *the angles between adjacent crystal faces are always constant for similar crystals of any particular mineral*. Faces are conveniently defined by reference to *crystallographic axes*, three or four in number, which intersect in a common origin within the crystal and form, as it were, a scaffolding on which the crystal faces are erected. The arrangements of faces in crystals possess varying degrees of symmetry, and according to their type of symmetry, crystals can be arranged in seven Systems, which are summarized below and illustrated in Fig. 4.4. A *plane of symmetry* divides a crystal into exactly similar halves, each of which is the mirror image of the other; it contains one or more of the crystallographic axes. The number of planes of symmetry stated in Fig. 4.4 is that for the highest class of symmetry in each of the Systems. Two other forms of symmetry exist, namely *axes* of symmetry and *centres* of symmetry (Fig. 4.5). Planes, axes and centres of symmetry define 32 classes of symmetry which are grouped into the seven Systems. These correspond to the seven types of *unit cell* determined by X-ray analyses of the three-dimensional patterns of the atoms in crystals. The unit cell is the smallest complete unit of pattern in the atomic structure of a crystal.

Crystal faces

In the Cubic System many crystals are bounded by faces which are all similar; such a shape is called a *form*. Two forms are illustrated in fig. 4.4*a*, the cube (six faces) and the octahedron (eight faces); other forms are the dodecahedron (twelve diamond-shaped faces, fig. 4.35*a*), the trapezohedron (24 faces, Fig. 4.35*b*), and with lower symmetry the pyritohedron (12 pentagonal faces, fig. 4.40); among others the tetrahedron has four triangular faces, the smallest number possible for a regular solid. Crystals

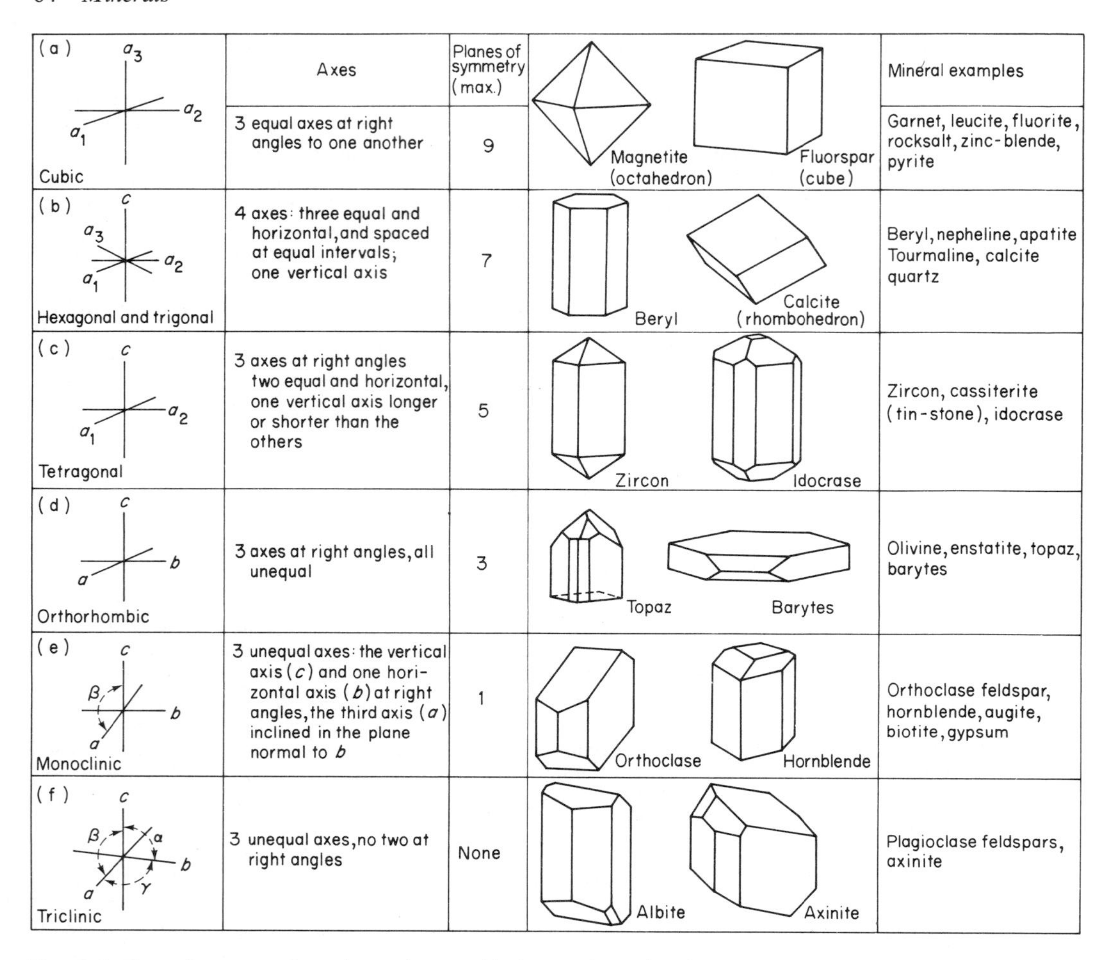

System	Axes	Planes of symmetry (max.)	Crystals	Mineral examples
(a) Cubic (a_1, a_2, a_3)	3 equal axes at right angles to one another	9	Magnetite (octahedron); Fluorspar (cube)	Garnet, leucite, fluorite, rocksalt, zinc-blende, pyrite
(b) Hexagonal and trigonal (a_1, a_2, a_3, c)	4 axes: three equal and horizontal, and spaced at equal intervals; one vertical axis	7	Beryl; Calcite (rhombohedron)	Beryl, nepheline, apatite Tourmaline, calcite quartz
(c) Tetragonal (a_1, a_2, c)	3 axes at right angles two equal and horizontal, one vertical axis longer or shorter than the others	5	Zircon; Idocrase	Zircon, cassiterite (tin-stone), idocrase
(d) Orthorhombic (a, b, c)	3 axes at right angles, all unequal	3	Topaz; Barytes	Olivine, enstatite, topaz, barytes
(e) Monoclinic (a, b, c, β)	3 unequal axes: the vertical axis (c) and one horizontal axis (b) at right angles, the third axis (a) inclined in the plane normal to b	1	Orthoclase; Hornblende	Orthoclase feldspar, hornblende, augite, biotite, gypsum
(f) Triclinic (a, b, c, α, β, γ)	3 unequal axes, no two at right angles	None	Albite; Axinite	Plagioclase feldspars, axinite

Fig. 4.4 Crystal systems. Crystals are drawn with the *c*-axis vertical (the reading position).

may grow as one form only, or as a combination of two or more forms; for example, Garnet (Fig. 4.35*a, b*) occurs as the dodecahedron, as the trapezohedron, or as a combination of the two.

In the Orthorhombic, Monoclinic, and Triclinic Systems, where the axes are all unequal, faces are named as follows: a face which (when produced) would cut all three axes is called a *pyramid*; there are eight such faces in a complete form, one in each octant formed by the axes. Faces which cut two lateral axes and are parallel to the vertical axis are known as *prisims*, and make groups of four, symmetrically placed about the axes. A *pinacoid* is a face which cuts any one axis and is parallel to the other two. A *dome* cuts one lateral and the vertical axis, and is parallel to the other lateral axis. These faces are illustrated in Fig. 4.6.

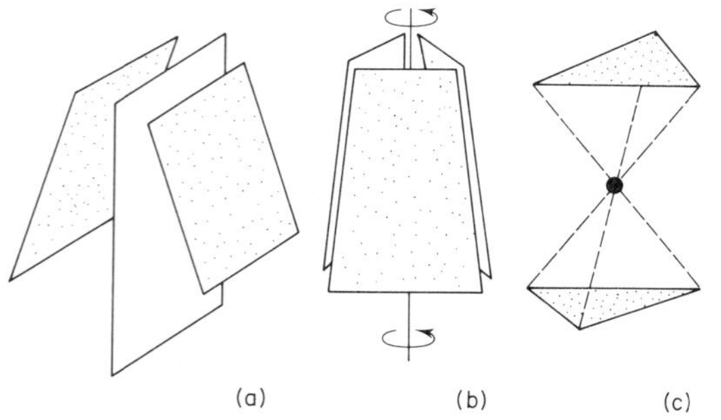

Fig. 4.5 Crystal symmetry. (**a**) Plane. (**b**) Axis (in this case a three-fold axis). (**c**) Centre.

Lastly, in the Hexagonal and Trigonal Systems the names pyramid and prism are used as before for faces which cut more than one lateral axis; but since there are now three lateral axes instead of two, six prism faces (parallel to the *c*-axis) are found in the Hexagonal System (Fig. 4.4*b*, beryl), and 12 pyramid faces. In Trigonal crystals, some faces are arranged in groups of three, equally spaced around the *c*-axis. This is seen, for example, in a rhombohedron of calcite (Fig. 4.4*b*), a form having six equal diamond-shaped faces. Calcite also occurs as 'nail-head' crystals, bounded by the faces of the

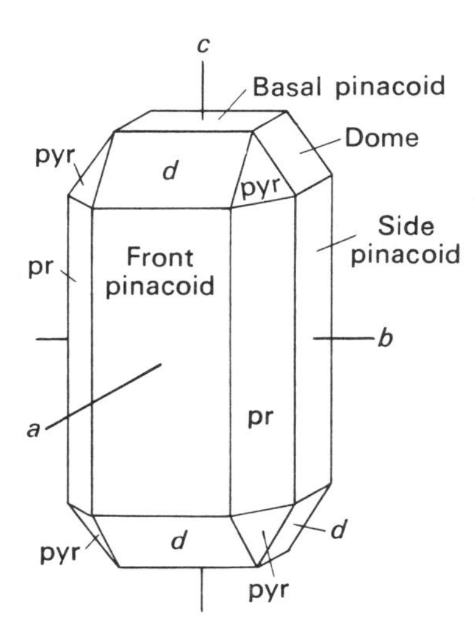

Fig. 4.6 Olivine crystal. To show crystal axes and faces in the orthorhombic system. pr = prism. *d* = dome. pyr = pyramid.

rhombohedron combined with the hexagonal prism (Fig. 4.7*a*); and in 'dog-tooth' crystals (Fig. 4.7*b*) where the pointed terminations are bounded by the six faces of the scalenohedron, each face of which is a scalene triangle. In quartz crystals the six-sided terminations, which sometimes appear symmetrical, are a combination of two rhombohedra whose faces alternate, as shown by shading in Fig. 4.7*c*. This fact is shown by etching a quartz crystal with hydrofluoric acid; two different sets of etch-marks appear on the alternate triangular faces indicating the arrangement of the atomic structure within the crystal to be in trigonal and not hexagonal symmetry. Trigonal (or three-fold) symmetry of quartz is also shown by extra faces which are sometimes present.

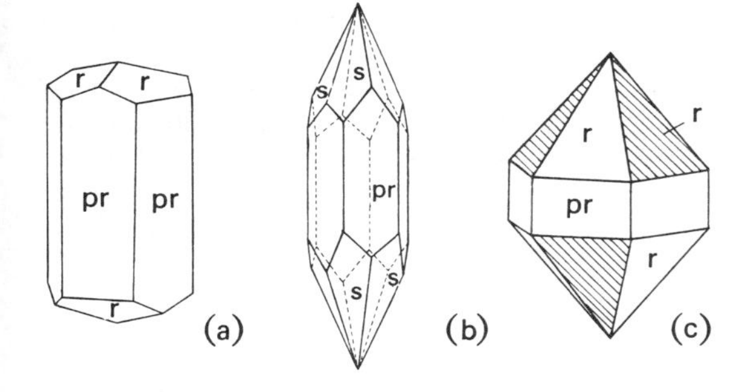

Fig. 4.7 Calcite and quartz. (**a**) Nail-head spar. (**b**) Dog-tooth spar. (**c**) Quartz, with negative rhombohedron shaded. r = faces of rhombohedron. pr = prism. s = scalenohedron.

Symbols are given to crystal faces and are based on the lengths of the intercepts which the faces make on the crystallographic axes, or the reciprocals of the intercepts (Miller symbols). The Miller notation is set out fully in textbooks of Mineralogy.

Twin crystals

When two closely adjacent crystals have grown together with a crystallographic plane or direction common to both, but one reversed relative to the other, a *twin crystal* results. In many instances the twin crystal appears as if a single crystal had been divided on a plane, and one half of the crystal rotated relative to the other half on this plane. If the rotation is 180°, points at opposite ends of a crystal are thus brought to the same end as a result of the twinning and re-entrant angles between crystal faces are then frequently produced; they are characteristic of many twins. Examples are shown in Fig. 4.8.

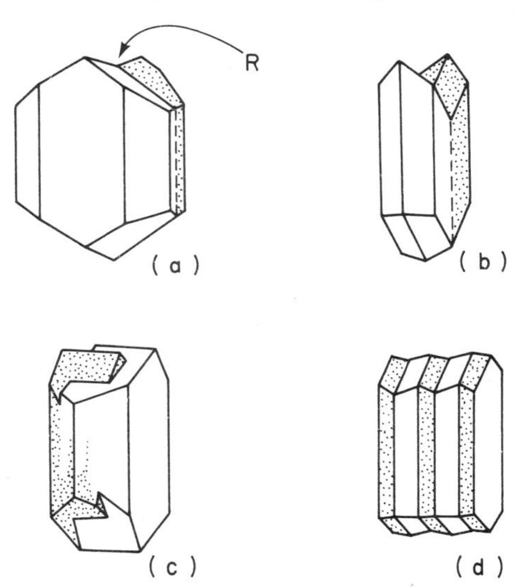

Fig. 4.8 Twin crystals. (**a**) Augite, showing re-entrant angle R; twin plane parallel to front pinacoid. (**b**) Arrow-head twin of Gypsum. (**c**) Carlsbad twin of Orthoclase. (**d**) Multiple twin of Plagioclase (Albite twinning).

Optical properties of minerals

Rock slices

In making a rock slice, a chip of rock (or slice cut by a rotating steel disc armed with diamond dust) is smoothed on one side and mounted on a strip of glass 75 × 25 mm (Fig. 4.9). The specimen is cemented to the glass strip by means of Canada balsam, a gum which sets hard after being heated, or a synthetic resin. The mounted chip of rock is then ground down with carborundum and emery abrasives to the required thinness, generally 30 μm (1 micrometre = $\frac{1}{1000}$ millimetre). It is now a transparent slice, and is completed by being covered with a thin glass strip fixed with balsam. Surplus balsam is washed off with methylated spirit. The surfaces of the specimen have been smoothed in making the slice, and they are free from all but very small irregularities. The effects observable when

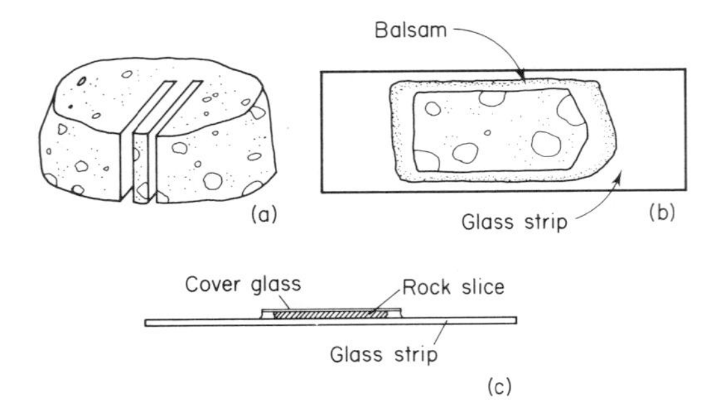

Fig. 4.9 Preparation of a thin section for the microscope. (**a**) Section sliced from sample of rock. (**b**) Slice mounted on glass strip. (**c**) Final section ready for use. Slice thickness = 30 μm.

light is transmitted through such a slice of crystalline material are now described.

Refraction and refractive index

A ray of light travelling through one medium is bent or *refracted* when it enters another medium of different density. Figure 4.10 shows the path of such a ray (RR), which makes angles i and r with the normal (NN) to the surface separating the two media. The angle between the ray and the normal to the surface is smaller in the denser medium, i.e. a ray is bent towards the normal on entering a denser medium, and conversely.

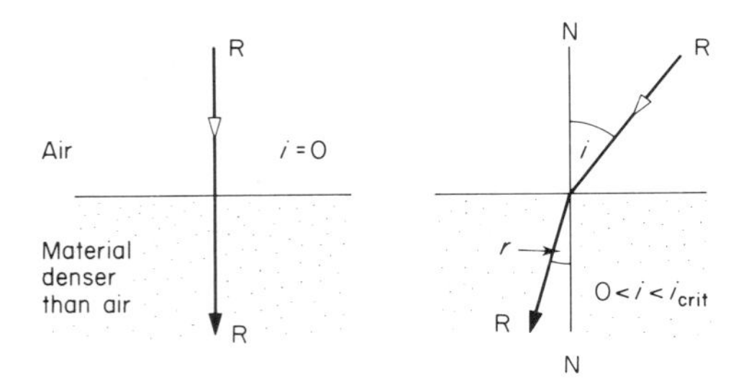

Fig. 4.10 Refraction of light at the interface between two media (see text).

If the angle of incidence (i) is measured for air, as is usual, then the ratio sin i/sin r is called the *refractive index*, n, for the other medium, and is constant whatever the angle of incidence. It can be shown that sin i and sin r are proportional to the velocities of light (v_1, v_2) in the two media, i.e. $n = \sin i/\sin r = v_1/v_2$ (Snell's Law). The refractive index of a substance is therefore inversely proportional to the velocity of light through the substance.

For Canada balsam, and similar cements in which rock slices are mounted, $n = 1.54$. Minerals with a much higher or lower refractive index than this appear in stronger outline, in a thin section under the microscope, than those which are nearer in value to 1.54; garnet ($n = 1.83$) is an example of a mineral which shows a very strong outline (Fig. 4.34). Minerals whose refractive index is near to 1.54 appear in weak relief, as in the case of quartz ($n = 1.553$ to 1.544). Some minerals have an index less than 1.54, e.g. fluorspar ($n = 1.43$).

Polarized light

According to the Wave Theory of Light, a ray is represented as a wave motion, propagated by vibrations in directions at right angles to the path of the ray. In ordinary light these vibrations take place in all planes containing the direction of propagation; in plane polarized light the vibrations are confined to one plane (Fig. 4.11).

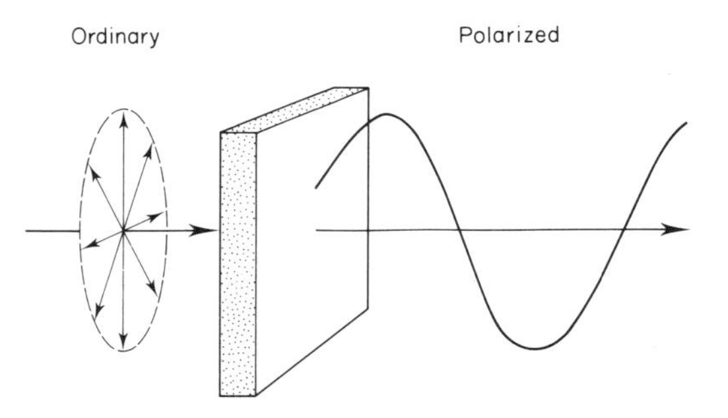

Fig. 4.11 Ordinary light (diagrammatic), composed of many vibrations in planes containing the direction of the ray (double-headed arrows represent double amplitude of vibrations) enters a crystal and is polarized, vibrations in one plane only being transmitted.

Light which passes through a crystal is, in general, polarized. While much can be learned from a microscopic examination of minerals in ordinary transmitted light, polarized light enables minerals to be identified with certainty.

Double refraction

Crystals other than those in the Cubic System have the property of splitting a ray of light which enters them into two rays, one of which is refracted more than the other. A cleavage rhomb of clear calcite does this effectively, as is shown if it is placed over a dot on a piece of paper; two dots are then seen on looking down through the crystal, i.e. two images are produced (1 and 2 in Fig. 4.12). On

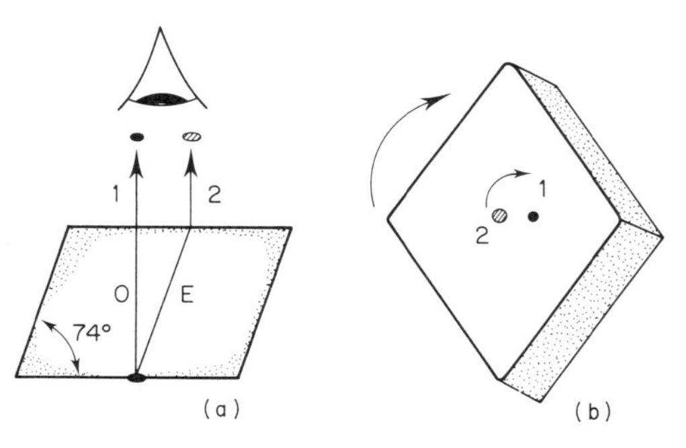

Fig. 4.12 Double refraction of calcite. O = ordinary ray. E = extraordinary ray. 1, ordinary image. 2, extraordinary image. (**a**) In vertical section. (**b**) When viewed from above.

turning the crystal, one dot appears to move round the other. The light passing through the calcite is split into two rays, called the ordinary ray (O) and the extraordinary ray (E). The orientation of the plane of vibration for the O-ray is 90° from that for the E-ray. The two rays travel at different velocities in the crystal, the E-ray being the faster, and both rays are plane polarized.

A mineral which has this property of dividing a ray of light into two is said to be *doubly refractive* or *birefringent*. Since the two rays within the mineral travel at different velocities, there are two values of refractive index, one for each ray. The difference between these two values is known as the *birefringence* of the mineral.

Minerals which have the same refractive index for light which enters in any direction are called *isotropic*; they do not divide a ray entering them and are therefore *singly refracting*. All Cubic crystals are isotropic, and also all basal sections of Hexagonal and Tetragonal crystals.

Optic axis

There is *one* direction in a calcite crystal, for example, along which light entering it is not split into two rays but passes through the crystal undivided. This direction is called the *optic axis* of the crystal, and in calcite it coincides with the crystallographic c-axis. Such a mineral is called *uniaxial*; all Hexagonal and Tetragonal minerals are uniaxial. Orthorhombic, Monoclinic, and Triclinic crystals all have *two* optic axes, i.e. two directions along which light can pass without being doubly refracted. They are therefore called *biaxial*.

Petrological microscope

For obtaining polarized light a synthetic material, *polaroid*, is used in which are embedded many minute needle-shaped crystals, oriented parallel to one another, i.e. in one direction. The material polarizes light which passes through it. Two discs of polaroid are mounted in the microscope (Fig. 4.13), one (the *polarizer*) below the stage, the other (the *analyser*) above the stage. The two polaroids are so set that the directions for their polarized light are at right angles. This setting is known as 'crossed polars'. The stage of the microscope can be rotated and is graduated in degrees.

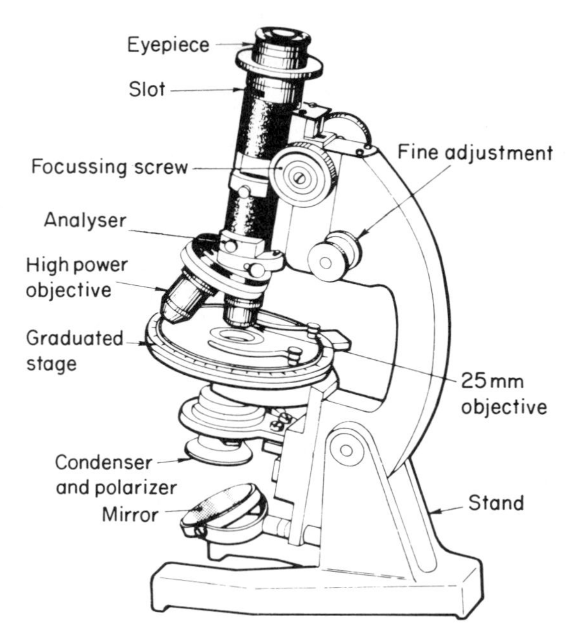

Fig. 4.13 A petrological microscope.

Passage of light through the microscope

Light is reflected from the mirror (Fig. 4.13) up through the polarizer, where it is plane polarized. If no mineral or rock section is placed on the stage, the light from the polarizer passes up through the objective and so enters the analyser, vibrating as it leaves the polarizer. Since the analyser only transmits vibrations at right angles to those from the polarizer, the polars being 'crossed', no light emerges from the eyepiece of the microscope. This fact should be tested as follows: with polarizer in, analyser out, look through the eyepiece and adjust the mirror so that the light is reflected up the tube. Slide in the analyser; a completely dark field of view should result if the polaroids are in adjustment and of good quality.

But when a slice of a doubly refracting mineral is placed on the stage of the microscope between crossed polars, a coloured image of the mineral is usually seen. A thicker

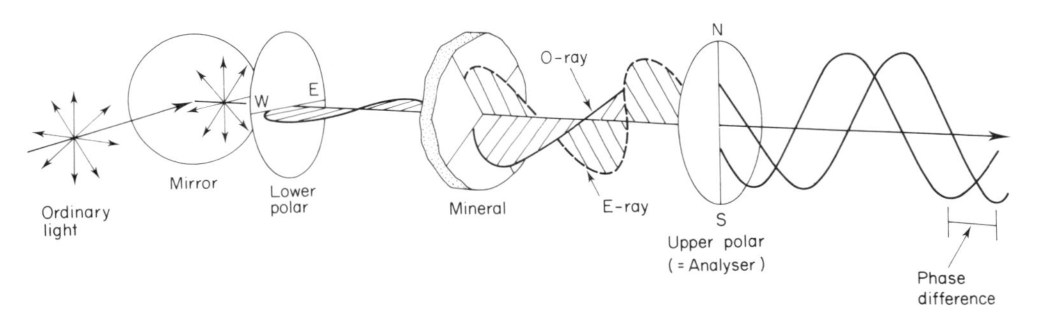

Fig. 4.14 Passage of light through microscope (diagrammatic): plane of polarization is 'E–W' for lower polar, and 'N–S' for upper polar, but may also be 'N–S' and 'E–W' respectively.

or thinner slice of the same mineral produces a different colour. These effects are explained as follows:

The ray of light which leaves the polarizer vibrating in one plane enters the thin section of the mineral on the stage of the microscope, and is divided into two rays (since the mineral is birefringent), an O-ray and an E-ray (Fig. 4.14). One of these travels faster than the other through the mineral slice; they can therefore be called the 'fast' ray and the 'slow' ray. They emerge from the mineral slice vibrating in two planes at right angles, and there is a phase difference between them, since they have travelled at different speeds and by different paths through the mineral slice.

The two rays enter the analyser, but only the component of their movement in the plane of polarization is transmitted (Fig. 4.15). The rays emerge from it *vibrating in the same plane* but out of phase with one another. This results in interference between the emergent vibrations, and the eye sees an interference colour, also called the *polarization colour*, which is of diagnostic value.

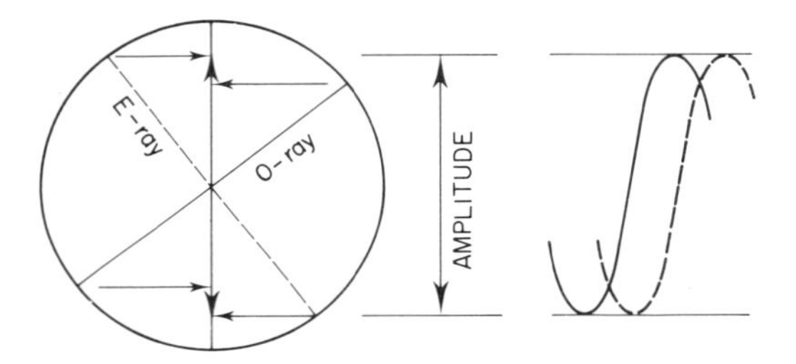

Fig. 4.15 Resolution of waves in the analyser (=upper polar). On left, rays arrive at analyser which transmits their components in the plane of polarization. On right, form of resolved waves in the plane of polarization.

Polarization colours

White light is made up of waves of coloured light, from red at one end of the spectrum to violet at the other. Each colour has a different wavelength; that of red light, for example, is 0.00076 mm, and of violet light 0.00040 mm. These lengths are also expressed as 760 and 400 nm respectively.

The polarization colour obtained with a particular mineral slice depends on (1) the birefringence of the slice, which in turn depends on the refractive indexes of the mineral and the direction in which it has been cut; and (2) the thickness of the slice. These facts are now illustrated by reference to the mineral quartz.

Polarization of quartz

A crystal of quartz has a maximum refractive index of 1.553, for light vibrating parallel to the *c*-axis of the crystal, and a minimum value of 1.544 for light vibrating in a direction perpendicular to the *c*-axis (Fig. 4.16). The difference in these two values is 0.009, which is the maximum birefringence for the mineral. This birefringence holds only for a longitudinal slice of the quartz crystal; the polarization colour of such a slice is a pale yellow (if the slice is of standard thickness). As explained above, the polarization colour is produced by the interference of two rays which have acquired a phase difference in traversing the mineral; the birefringence of 0.009 represents a phase difference of 0.009 μm, or 9 nm, per μm thickness of slice. Since the usual thickness is 30 μm, the full phase difference between the two rays after passing through the slice of quartz is $30 \times 9 = 270$ nm, which corresponds to a pale yellow.

Consider now a basal section of quartz, i.e. one cut perpendicular to the *c*-axis (the direction *aa* in Fig. 4.16). Such a slice has only one value of refractive index for light vibrations traversing it. It has therefore no birefringence, i.e. it is *isotropic* and appears completely black between crossed polars.

Between the above extremes, the pale yellow colour (maximum) and black (or nil), a slice of quartz cut obliquely to the *c*-axis will show a white or grey polarization colour, the tint passing from dark grey to pale grey and white as the orientation of the slice approaches parallelism with *c*-axis.

Thus, in general, a crystal slice gives a characteristic colour between crossed polars according to the direction in which it has been cut from the mineral. The maximum

Table 4.5 Polarization colours for some common minerals in sections of standard thickness

Mineral	Max. Biref.	nm Colour
Muscovite	0.043	1290 = delicate green
Olivine	0.035	1050 = bright red
Augite	0.025	750 = bright blue
Quartz	0.009	270 = pale yellow
Orthoclase	0.007	210 = pale grey

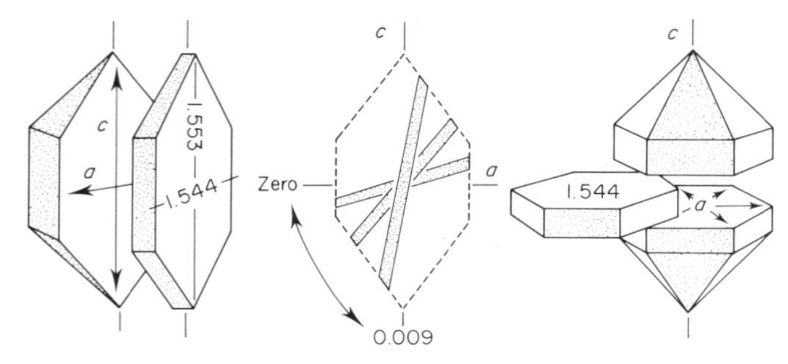

Fig. 4.16 Vibration directions and refractive indices in vertical and horizontal sections of quartz, and variation of birefringence with direction of slice through the crystal.

birefringence and polarization colour of some common minerals, for a 30 μm thickness of slice, are given in Table 4.5.

Extinction

When a birefringent mineral slice is rotated on the microscope stage between crossed polars, one or other vibration direction in the mineral can be brought parallel to the vibration plane of the polarizer. This occurs four times in each complete rotation. In such positions the light vibrations from the polarizer pass directly through the mineral slice to the analyser, where they are cut out (because it is set at right angles to the polarizer), so that no light emerges; the mineral thus appears completely dark at intervals of 90° during rotation. This effect is known as *extinction*. Half-way between successive extinction positions the mineral appears brightest.

The polars of the microscope are set so that their vibration planes are parallel to the cross-wires of the diaphragm, one 'east-west', the other 'north-south'. If now a mineral is found to be in extinction when some crystallographic direction such as its length or a prominent cleavage is brought parallel to a cross-wire, the mineral is said to have *straight extinction* with regard to that length or cleavage. If extinction occurs when the length of the mineral makes an angle with the cross-wire, it is said to have *oblique extinction*; the extinction angle can be measured by means of the graduations around the edge of the stage (Fig. 4.17).

Opaque minerals

The composition of some minerals prevents them transmitting light and they appear dark brown or black when viewed through the microscope. These are called *opaque* minerals: common examples are listed in Table 4.6.

Table 4.6 Some commonly occurring opaque minerals

Mineral	Colour of reflected light
Graphite (C)	Black
Pyrite (FeS_2)	Brass yellow
Chalcopyrite ($CuFeS_2$)	Strong brass yellow
Haematite (Fe_2O_3)	Blue, red and black
Limonite ($(2Fe_2O_3)(3H_2O)$)	Yellow-brown
Ilmenite ($FeTiO_3$)	Strong purple-black
Magnetite (Fe_3O_4)	Blue-black
Chromite ($FeCr_2O_4$)	Brown-black
Leucoxene (TiO_2)	White

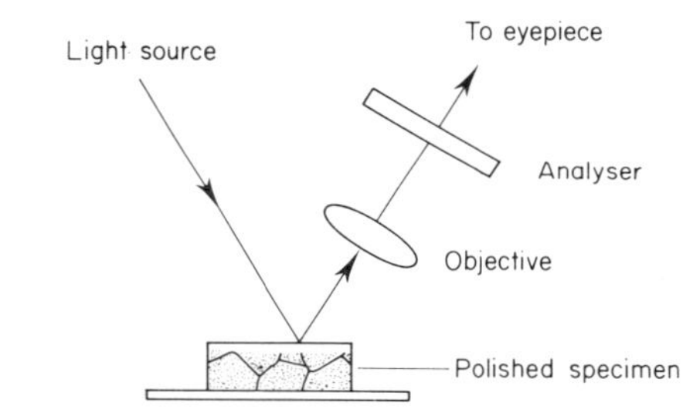

Fig. 4.18 Study of opaque minerals by reflected light.

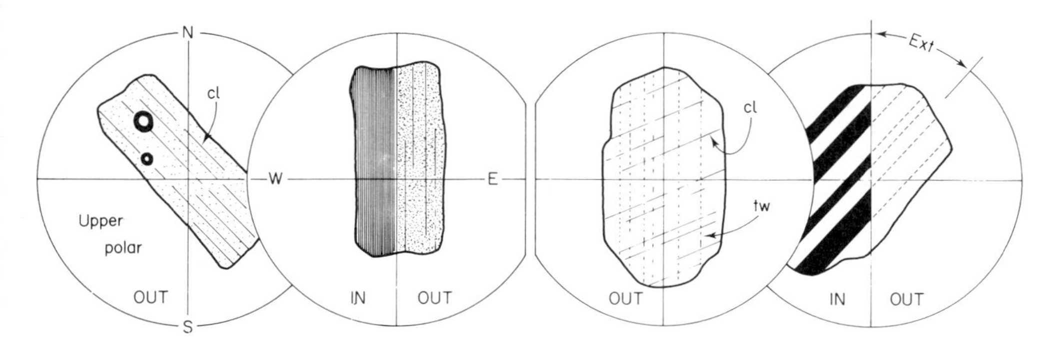

Fig. 4.17 Extinction observed through the microscope. Straight for Biotite mica (left). Oblique for Plagioclase feldspar (right). In all cases the lower polar is in position, its plane of polarization being N–S. Polarization of upper polar is E–W. cl = cleavage. tw = twin plane (Fig. 4.8). Ext = extinction angle.

Pleochroism

In some minerals a change of colour is seen when only the lower polarizer is used and the mineral is rotated above it on the stage of the microscope. This *pleochroism* is due to the fact that the mineral absorbs the components of white light differently in different directions. It is shown strongly by the mica biotite, which when oriented with its cleavage direction parallel to the vibration plane of the polarizer appears a much darker brown than when at 90° to that position (Fig. 4.17). Hornblende and tourmaline are two other strongly pleochroic minerals.

Opaque minerals must be illuminated from above to be studied with the microscope (Fig. 4.18). Light is directed so that the mineral surface reflects it into the objective and thence up the microscope to the eyepiece where it can be seen. When ores consisting mainly of opaque minerals are to be studied a piece of the rock (about 20 mm^3) is mounted on a glass plate and its upper surface ground flat, parallel to the glass plate. The ground surface is then polished to enhance its reflection of light and enable the fine details of mineral structure to be observed. Physical and chemical tests, such as scratching the surface of minerals and etching them with acid, may be conducted on the polished surface and their result observed through the microscope.

Table 4.7 summarizes the observations to be made with a thin slice of rock.

Table 4.7 Summary of observations with a thin slice of rock

(1) With hand lens:	observe form of minerals, their relative size, any preferred orientation of minerals and presence of opaque minerals.
(2) With microscope:	(ordinary light), look at mineral form, colour, inclusions and alteration products. Study opaque minerals under reflected light.
(3) Insert lower polar:	test for pleochrosim.
(4) Insert upper polar:	(analyser), observe isotropic minerals (cubic). In other (anisotropic) minerals observe the polarization colour, extinction (straight or oblique), twinning (present or absent) and alteration if any.

Note; The magnification of the microscope for the combination of lenses used may be assessed by observing a scale with small graduations, e.g. mm.

Atomic structures

The atomic structure of crystals can be investigated by methods of X-ray analysis. The arrangement and spacing of the atoms of which a crystal is composed control its regular form and properties. For example, the atoms of sodium and chlorine in a crystal of common salt (NaCl) are arranged alternately at the corners of a cubic pattern (Fig. 4.19), which is repeated indefinitely in all directions. Salt crystals grown from solution are cubes, a shape which expresses the internal structure.

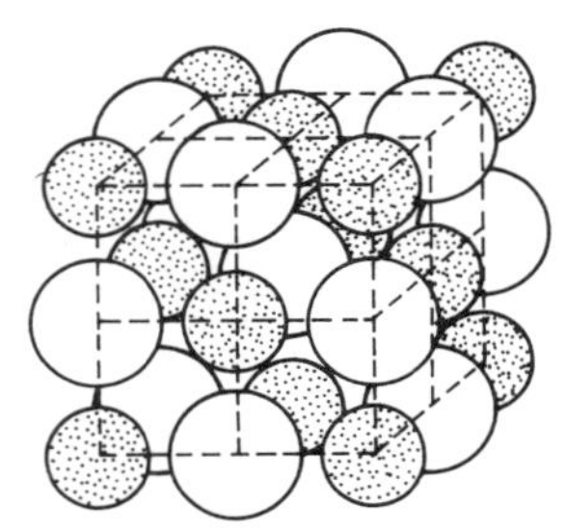

Fig. 4.19 Arrangement of atoms in unit cell of NaCl (small circles, with ornament, Na; large circles, Cl).

When the *silicate minerals* were investigated it was found that they could be placed in a very few groups, according to the arrangement of the silicon and oxygen atoms. The silicon atom is tetravalent, and is always surrounded by four oxygen atoms which are spaced at the corners of a regular tetrahedron (Fig. 4.20). This SiO_4-tetrahedron is the unit of silicate structure, and is built into the different structures as follows:

1. *Separate* SiO_4*-groups* are found in some minerals, are closely packed in regularly spaced rows and columns throughout the crystal structure, and linked together by metal atoms (Mg, Fe, Ca, etc.) situated between the tetrahedra. Since each oxygen has two negative valencies and silicon four positive valencies, the SiO_4-group has an excess of four negative valencies; these are balanced when it is linked to metal atoms contributing four positive valencies, as in olivine, Mg_2SiO_4. Some Mg atoms may be replaced by Fe. (*Note:* in this account the word atom is used instead of *ion*, which is more strictly correct.)

2. *Single Chain Structures* (Si_2O_6) are formed by SiO_4-groups linked together in linear chains, each group sharing two oxygens with its neighbours (Fig. 4.20*e*). This structure is characteristic of the Pyroxenes, e.g. diopside, $CaMg(Si_2O_6)$. The chains lie parallel to the *c*-axis of the mineral, are bonded together by Mg, Fe, Ca, or other atoms, which lie between them. An end view of a chain is given at (f) in the figure. The vertical cleavages in the mineral run between the chains, as shown in (f′), and intersect at an angle of 93°. Aluminium atoms, since they have nearly the same size as silicon, may repace silicon in the structure to a limited extent, and may also occur among the atoms which lie between the chains; in this way aluminous pyroxenes such as augite are formed.

Ring structures are built up by groups of three, four, or six tetrahedra, each of which shares two oxygens with its neighbours. Figure 4.20*d* shows a ring of six linked tetrahedra, which is found in the mineral beryl, $Be_3Al_2Si_6O_{18}$.

3. *Double Chain Structures* (Si_4O_{11}), in which two single chains are joined together side by side (Fig. 4.20*g*), are found in the Amphiboles, e.g. tremolite, $Ca_2Mg_5(Si_4O_{11})_2(OH)_2$. The double chains run parallel to the *c*-axis of the minerals, and are linked laterally by metal atoms lying between the chains. The cleavage directions are as shown at (*h*′), and intersect at an angle of 56°. In hornblende, aluminium replaces part of the silicon (cf. augite above). Hydroxyl groups (OH), are always present in amphiboles to the extent of about one to every eleven oxygens.

4. *Sheet Structures* (Si_4O_{10}) are formed when the SiO_4-tetrahedra are linked by three oxygens each, and lie with their bases in a common plane (Fig. 4.20*i*). Sheets of this kind are found in the Micas and other flaky minerals (e.g. chlorite, talc, the clay minerals), whose perfect cleavage is parallel to the silicon-oxygen sheets. In the mica muscovite, $KAl_2(AlSi_3)O_{10}(OH)_2$, aluminium replaces about one quarter of the silicon, and hydroxyl is always present. The Si_4O_{10}-sheets are arranged in pairs, with aluminium atoms between them, and each pair is separated from the next pair by a layer of potassium atoms (Fig. 4.26). Other structures involving silicon-oxygen sheets are discussed under Clay Minerals, p. 80.

5. *Three-dimensional Frameworks* (SiO_2) are formed when each tetrahedron is linked by all four oxygens, sharing the oxygens with adjacent groups. The mineral quartz, SiO_2, has a framework in which the SiO_4-groups

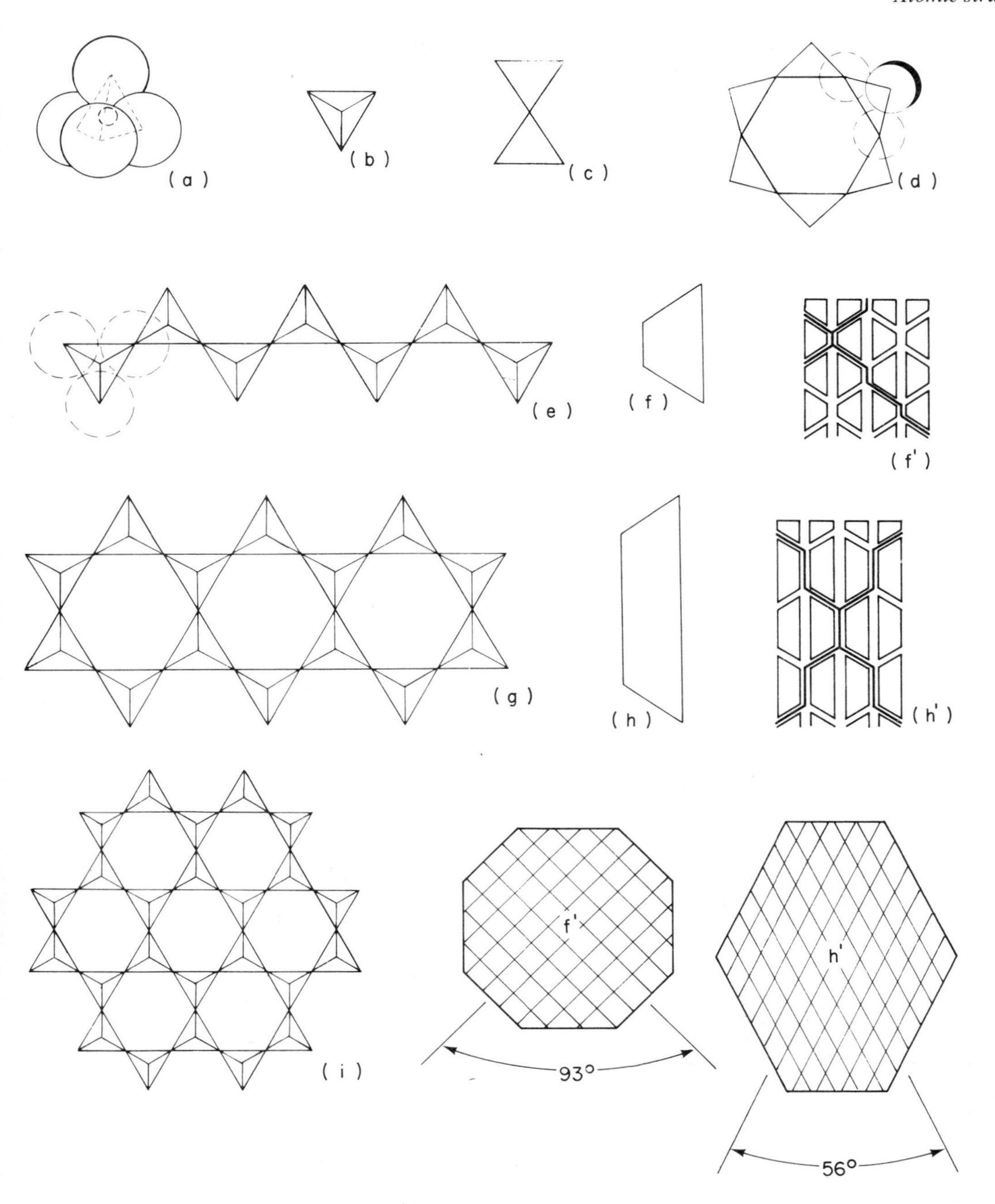

Fig. 4.20 Silicate structures. (**a**) SiO_4-tetrahedron: large circles represent oxygen ions, small circle silicon. (**b**) The tetrahedron in plan, apex upwards (circles omitted). (**c**) Two tetrahedra sharing one oxygen, Si_2O_7. (**d**) Ring of six tetrahedra (the centres of the two outer oxygens lie one above another), Si_6O_{18}. (**e**) Single chain, Si_2O_6; tetrahedra share two oxygens each; (**f**) conventional end-view of single chain. (**g**) Double chain, Si_4O_{11}, and (**h**) end-view of double chain. (**i**) Sheet of tetrahedra, Si_4O_{10}; each shares three oxygens with adjoining tetrahedra. (**f′**, **h′**) Stacking of single and double chains respectively (viewed along *c*-axis); resulting cleavage directions shown by heavy lines. The typical appearance of single and double chain minerals cut 90° to vertical (c) axis shown in bottom right: (f′) a pyroxene, (h′) an amphibole.

form a series of linked spirals (Fig. 4.32). In the Feldspars another type of framework is found, and Al replaces part of the Si. Thus in orthoclase feldspar, one silicon in four is replaced by Al; the substitution of trivalent aluminium for a tetravalent silicon releases one negative (oxygen) valency, which is satisfied by the attachment of a univalent sodium or potassium atom, thus:

Orthoclase $= KAlSi_3O_8$; Albite $= NaAlSi_3O_8$; and Anorthite $= CaAl_2Si_2O_8$.

The K, Na, or Ca atoms are accommodated in spaces within the frameworks. The structure of the silicate minerals is summarized in Table 4.8.

Table 4.8 Structure of silicate minerals. Simple structures, such as that of olivine, form at higher temperatures than those more complex structures such as that of quartz.

Type of structure	Si:O ratio	Repeated pattern	Mineral group
Separate SiO_4-groups	1:4	SiO_4	Olivine
Single chain	1:3	Si_2O_6	Pyroxenes
Double chain	1:2.75	Si_4O_{11}	Amphiboles
Sheet	1:2.5	Si_4O_{10}	Micas
Framework	1:2	$(Al,Si)_nO_{2n}$	Feldspars
		SiO_2	Quartz

The rock-forming minerals

It is convenient to distinguish between minerals which are *essential* constituents of the rocks in which they occur, their presence being implied by the rock name, and others which are *accessory*. The latter are commonly found in small amount in a rock but their presence or absence does not affect the naming of it. *Secondary* minerals are those which result from the decomposition of earlier minerals, often promoted by the action of water in some form, with the addition or subtraction of other material, and with the formation of mineral by-products.

Identification of minerals in hand specimen

It should be possible to identify the common rock-forming minerals in the hand specimen with a pocket lens where one dimension of the mineral grain is not less than about 1 mm. With practice much smaller grains can be determined. The most useful characteristics for this purpose are:

(1) General shape of grains, depending on the crystallization of the mineral; the faces of well-formed crystals can often be observed, but where grains have been modified (e.g. by rounding) other characters may be needed.
(2) Colour and transparency.
(3) Presence or absence of cleavage.
(4) Presence or absence of twinning, and type of twinning.
(5) Hardness.

In the following descriptions of minerals, notes are included to aid identification in the hand specimen on the above lines with abbreviations for specific gravity (G) and hardness (H). They are followed by notes on the simpler optical properties, headed ***in thin sections***, and include refractive index (*R.I.*) and birefringence (*Biref.*).

Silicate minerals

The olivine group

Olivine

Common olivine has the composition $(MgFe)_2SiO_4$, in which Fe^{2+} replaces part of the Mg^{2+}.

Crystals: Orthorhombic (Fig. 4.6); pale olive-green or yellow; vitreous lustre; conchoidal fracture. H $= 6\frac{1}{2}$. G $= 3.2$ to 3.6.

Olivine occurs chiefly in basic and ultrabasic rocks. Since it crystallizes at a high temperature, over 1000°C, it is one of the first minerals to form from many basic magmas. (*Magma* is the molten rock-material from which igneous rocks have solidifed, p. 91.) Alteration to green serpentine is common (p. 80).

In thin section: Porphyritic crystals (which are large compared with the grain size of the matrix in which they are set; they are generally well-formed crystals) commonly show 6- or 8-sided sections, the outline generally somewhat rounded. Cleavage rarely seen; irregular cracks common (Fig. 4.21).

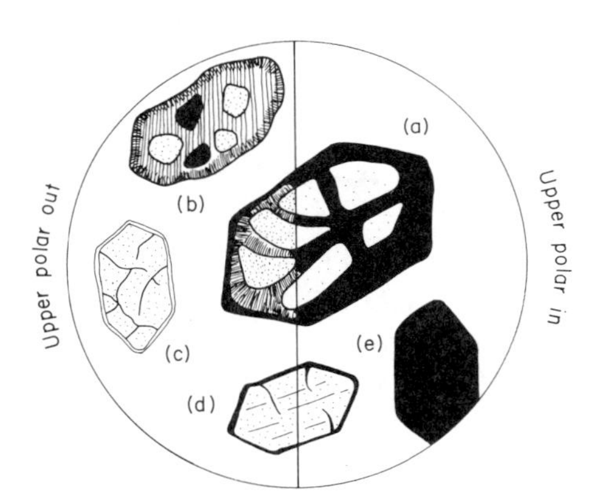

Fig. 4.21 Olivine in thin section (×12): (**a**) with serpentine filled cracks, (**b**) almost completely altered, (**c**) fresh, (**d**) exhibiting weak cleavage, (**e**) in straight extinction.

Colour: None when fresh. Alteration to greenish serpentine is very characteristic, this mineral being often developed along cracks and around the margins of olivine crystals. Some olivines have been entirely converted to serpentine; or relics of original olivine may be preserved as isolated colourless areas in the serpentine. Magnetite (Fe_3O_4) may be formed during the alteration, from iron in the original olivine, and appears as small black specks in the serpentine.

Mean R.I. = 1.66 to 1.68, giving a bold outline.

Biref: Strong (max. = 0.04), giving bright polarization colours.

Extinction: Straight, parallel to crystal outlines and traces of cleavage.

The pyroxene group

Minerals of this group belong to two systems of crystallization:

1. Orthorhombic, e.g. *enstatite*, *hypersthene*.
2. Monoclinic, e.g. *augite*, *diopside*.

They possess two good cleavage directions parallel to the prism faces of the crystals (*prismatic cleavage*); the intersection angle of the cleavages is nearly 90°, a characteristic feature of the group (see Fig. 4.20f′). They form 8-sided crystals, and being silicates of Fe and Mg they are dark in colour (except diopside, CaMg).

Orthorhombic pyroxenes

Enstatite, $MgSiO_3$. ***Hypersthene*** $(MgFe)SiO_3$

The names 'enstatite' and 'hypersthene' have Greek derivations which refer to colour changes in pleochroism: *enstates*, weak; *sthene*, strong.

Crystals: Usually dark brown or green (hypersthene nearly black), 8-sided and prismatic. In addition to the prismatic cleavages mentioned above there are poorer partings parallel to the front and side pinacoids; lustre, vitreous to metallic. H = 5 to 6. G = 3.2 (enstatite), increasing with the iron content to 3.5 (hypersthene).

The minerals occur in some basic rocks such as norite (*q.v.*), as black lustrous grains interlocked with the other constituent minerals; also in some andesites, usually as small black porphyritic crystals; and in certain ultrabasic rocks.

Monoclinic pyroxenes

Augite $(CaMgFeAl)_2(SiAl)_2O_6$

An aluminous silicate whose formula can be written as above in conformity with the Si_2O_6 pattern of the atomic chain structure. The relative proportions of the metal ions (Ca, Mg, Fe, Al) are variable within limits, giving a range of composition and different varieties of the mineral. Some Al^{3+} is substituted for Si^{4+}.

Crystals: Commonly 8-sided and prismatic, terminated by two pyramid faces at each end (Fig. 4.22); brown to black in colour, vitreous to resinous lustre. Twin crystals (Fig. 4.8*a*) show a re-entrant angle. H = 5 to 6. G = 3.3 to 3.5. The two vertical cleavages may be observed on the end faces of suitable crystals.

Augite occurs chiefly in basic and ultrabasic rocks; e.g. in gabbro, where it appears as dark areas intermingled with the paler feldspar. In fine-grained basic rocks it is not distinguishable in the hand specimen

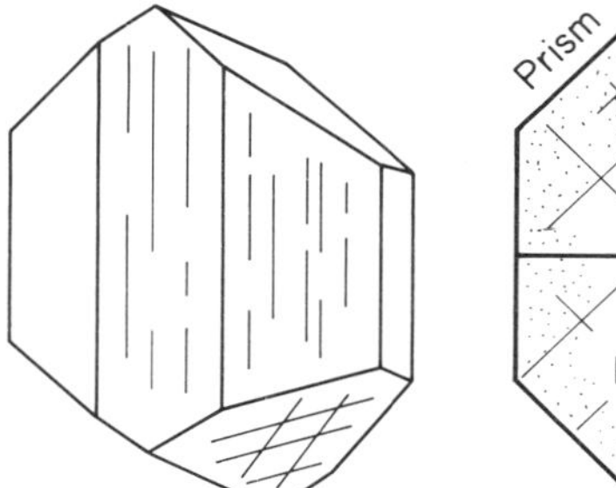

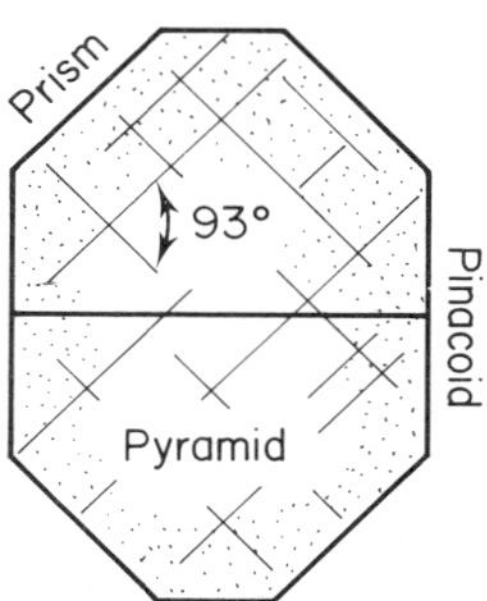

Fig. 4.22 Crystal of augite: oblique view and vertical view showing cleavage and principal crystal faces.

unless it is porphyritic. Augite is also a constituent of some andesites and diorites, and occasionally of granites.

In thin section: idiomorphic (i.e. well formed) crystals show characteristic 8-sided transverse sections, bounded by prism and pinacoid faces, with the two prismatic cleavages intersecting at nearly 90°. Longitudinal sections show only one cleavage direction (Fig. 4.23).

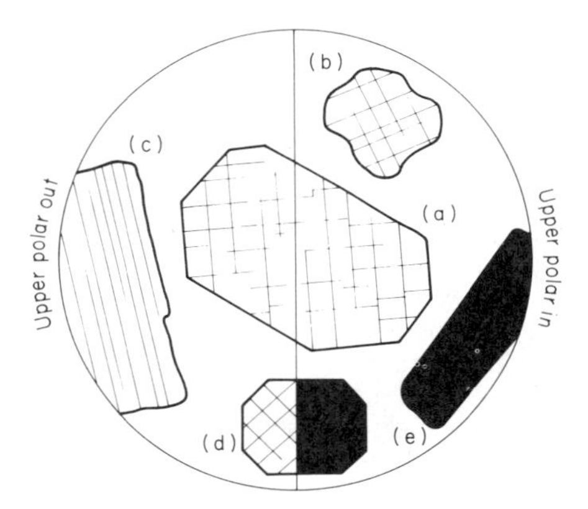

Fig. 4.23 Augite in thin section (×12): (**a**) regular cross-section with 93° cleavage, (**b**) irregular fragment displaying diagnostic cleavage, (**c**) vertical section, (**d**) symmetrical extinction, (**e**) oblique extinction for long section.

Colour: Pale brown to nil. Pleochroism generally absent or weak.

Mean R.I. = about 1.70, giving strong relief in balsam.

Biref: Hypersthene: weak (0.106) giving grey, white or yellow polarization colours. Augite: strong (0.025) giving bright blues, reds and browns.

Extinction: Hypersthene: parallel to cleavage in long section. Augite: oblique to cleavage in long section.

The amphibole group

Minerals of this group are mainly monoclinic. The crystals are elongated in the *c*-direction and usually bounded by six vertical faces, of which the prism faces intersect at an angle of 124°, Fig. 4.24. This is also the

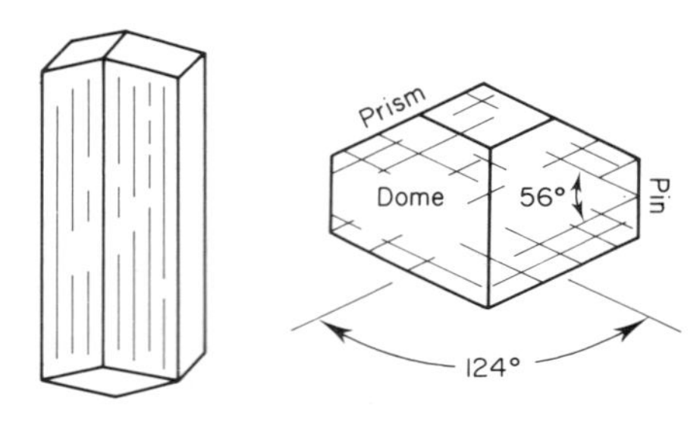

Fig. 4.24 Crystal of hornblende: oblique view and vertical view showing cleavage and principal crystal faces. Pin = pinacoid.

angle between the two cleavage directions, parallel to the prisms (Fig. 4.20g, h, h′). Only the most common monoclinic amphibole, *hornblende*, is given here in detail.

Hornblende $(CaMgFeNaAl)_3$-$(AlSi)_4O_{11}(OH)$

An aluminous silicate whose relative proportions of the metal ions vary within the limits shown, giving a range of composition; the (OH)-radicle is found in all amphiboles. Al^{3+} is substituted for Si^{4+} in about one in four positions.

Crystals: Monoclinic, dark brown or greenish black; usually 6-sided, of longer habit than augite, with three dome faces at each end (Fig. 4.24). Vitreous lustre. H = 5 to 6. G = 3 to 3.4.

Common hornblende is found in diorites and some andesites as the dark constituent; in andesite it is porphyritic and may be recognized by its elongated shape, the length of a crystal being often several times its breadth. It is also found in some syenites and granodiorites, and in metamorphic rocks such as hornblende-schist.

In thin section: 6-sided transverse sections, bounded by four prism and two pinacoid faces, are very characteristic (Fig. 4.25), and show the prismatic cleavages intersecting at 124°. Longitudinal sections are elongated and show one cleavage direction parallel with the length.

Colour: Green to brown; pleochroism strong in shades of green, yellow, and brown.

Mean R.I. varies from 1.63 to 1.72.

Biref: Strong (max. = 0.024).

Extinction: Oblique in most longitudinal sections, at angles up to 25° with the cleavage; sections parallel to the front pinacoid show straight extinction. Symmetrical extinction in transverse section. Twinning is common.

Alteration is to chlorite, often with the formation of by-products.

Hornblende is to be distinguished from augite by its colour and lower maximum angle of extinction; and from biotite, which it may resemble in sections showing one cleavage, by the fact that biotite always gives straight extinction.

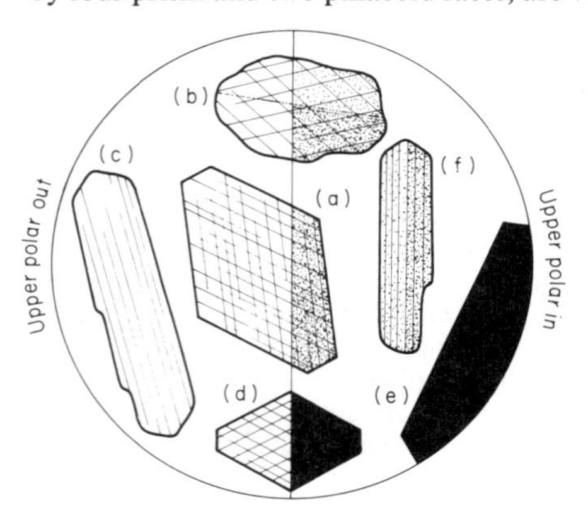

Fig. 4.25 Hornblende in thin section (×12). (**a**) regular cross section with 124° cleavage, (**b**) irregular fragment displaying diagnostic cleavage and twinning, (**c**) vertical section, (**d**) symmetrical extinction, (**e**) oblique extinction for long section, (**f**) twinning in vertical section.

Asbestos

The fibrous form of the amphibole *tremolite*, in which crystals grow very long and are flexible. In commerce the term 'asbestos' also includes other fibrous minerals such as *chrysotile* (fibrous serpentine, *q.v.*) and *crocidolite* (a soda-amphibole). These minerals are useful because of their resistance to heat and because of their fibrous nature, which enables them to be woven into fireproof fabrics, cord, and brake-linings, and made in to boards, tiles, and felt.

The mica group

The micas are a group of monoclinic minerals whose property of splitting into very thin flakes is characteristic and easily recognized. It is due to the

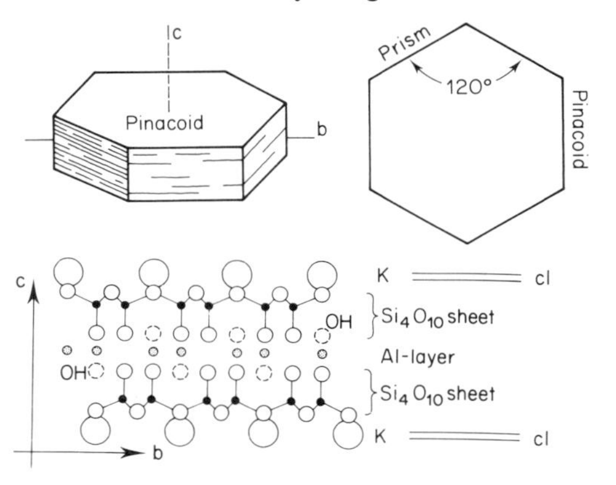

Fig. 4.26 Crystal of mica: oblique view showing cleavage parallel to basal pinacoid, and a vertical view of pseudo-hexagonal symmetry and principal crystal faces. Below is the structure of muscovite (diagrammatic). The silicon-oxygen tetrahedra are linked to form 'sheets', two of which are shown, with aluminium ions lying between them. Layers of potassium ions are situated between successive pairs of the Si_4O_{10} sheets. cl = cleavage.

perfect cleavage parallel to the basal plane in mica crystals, which in turn results from the layered atomic structure of the minerals, as shown in Fig. 4.26. It is along the K^+ layers, where the bonding is weak, that the cleavage directions lie.

The commonly occurring micas, *muscovite* (colourless or slightly tinted) and *biotite* (dark brown to nearly black), are described below.

Mica crystals are six-sided, with pseudo-hexagonal symmetry. Their cleavage flakes are flexible, elastic, and transparent. A six-rayed percussion figure is produced when a flake is struck with a pointed instrument, and one of the cracks thus formed is parallel to the plane of symmetry of the monoclinic mineral.

Muscovite, $KAl_2(Si_3Al)O_{10}(OH)_2$

Form and cleavage as stated above. White in colour, unless impurities are present to tint the mineral; pearly lustre. H = 2 to 2½ (easily cut with a knife). G = about 2.9 (variable).

Muscovite occurs in granites and other acid rocks as silvery crystals, from which flakes can be readily detached by the point of a penknife; also in some gneisses and mica-schists. It is a very stable mineral, and persists as minute flakes in sedimentary rocks such as micaceous sandstones. The name *sericite* is given to secondary muscovite, which may be produced by the alteration of orthoclase. The mica of commerce comes from large crystals found in pegmatite veins (p. 106).

In thin section: vertical sections (i.e. across the cleavage) are often parallel-sided and show the perfect cleavage (Fig. 4.17); basal sections appear as 6-sided or irregular colourless plates. Alteration uncommon.

Mean R.I. = 1.59.

Biref: Strong (max. = 0.04), giving bright pinks and greens in vertical sections.

Extinction: Straight, with reference to the cleavage.

Biotite, $K(MgFe)_3(Si_3Al)O_{10}(OH)_2$

Crystals are brown to nearly black in hand specimen; single flakes are pale brown and have a sub-metallic or pearly lustre. Form and cleavage as stated above. H = 2½ to 3. G = 2.8 to 3.1.

Biotite occurs in many igneous rocks, e.g. granites, syenites, diorites, and their lavas and dyke rocks, as dark lustrous crystals, distinguished from muscovite by their colour. Also a common constituent of certain gneisses and schists.

In thin section: Sections showing the cleavage often have two parallel sides and ragged ends (Fig. 4.17). In some biotites, small crystals of zircon enclosed in the mica have developed spheres of alteration around themselves by radioactivity. These spheres in section appear as small dark areas or 'haloes' around the zircon and are pleochroic.

Colour: Shades of brown and yellow in sections across the cleavage, which are strongly pleochroic; the mineral is darkest (i.e. light absorption is a maximum) when the cleavage is parallel to the vibration direction of the polarizer. Basal sections have a deeper tint and are only feebly pleochroic.

Mean R.I. = about 1.64.

Biref: Strong, about 0.05 (max.). Basal sections are almost isotropic.

Extinction: Parallel to the cleavage.

Alteration to green chlorite is common, when the mineral loses its strong birefringence and polarizes in light greys (see under Chlorite, p. 80).

The feldspar group

The feldspars form a large group of monoclinic and triclinic minerals, and are the most abundant constituents of the igneous rocks. The chief members of the Feldspar family can be classified by composition as *orthoclase*, $KAlSi_3O_8$; *albite*, $NaAlSi_3O_8$; and *anorthite*, $CaAl_2Si_2O_8$. They are represented in Fig. 4.27 at the corners of a triangle.

Orthoclase, $KAlSi_3O_8$. Potassium feldspar

Crystals: Monoclinic; white or pink in colour, vitreous lustre; bounded by prism faces, basal and side pinacoids, and domes (Fig. 4.28). Simple twins are frequent: Carlsbad twins unite on the side pinacoid (Fig. 4.8b), Manebach twins on the basal plane, and Baveno twins on a diagonal plane (dome) parallel to the *a*-axis (Fig. 4.29). Two good cleavages, parallel to the side and basal pinacoids, intersect at 90°; the name of the mineral is due to this property (from the Greek *orthos*, straight or rectangular, and *klasis*, breaking). H = 6. G = 2.56. The colourless glassy variety, *sanidine*, is a high temperature form found in quickly-cooled lavas.

Orthoclase occurs in granites and syenites as hard, cleaved white or pink crystals, generally constituting the greater part of the rock. In the dyke rocks related to granites and syenites, orthoclase may occur as porphyritic crystals (see p. 106). Found also in some gneisses and felspathic sandstones.

In thin section: shape often nearly rectangular if crystals are idiomorphic, irregular if interlocked with other minerals. Cleavage not always seen.

Colour: None when fresh, but shows frequent alteration to kaolin (p. 80), when the mineral appears 'cloudy' and looks white by top light (reflected from the surface of the slice).

Mean R.I. = 1.52.

Biref: Weak (max. = 0.008), giving grey and white polarization colours, in sections of normal thickness.

Extinction: Oblique, up to 21° on the side pinacoid; sections perpendicular to the side pinacoid show straight extinction. Simple twins common (Fig. 4.29); distinguished from plagioclase by the absence of multiple twinning.

Alteration to kaolin is common (p. 80); sometimes alters to an aggregate of very small flakes of *sericite* (secondary white mica), revealed by their bright polarization colours.

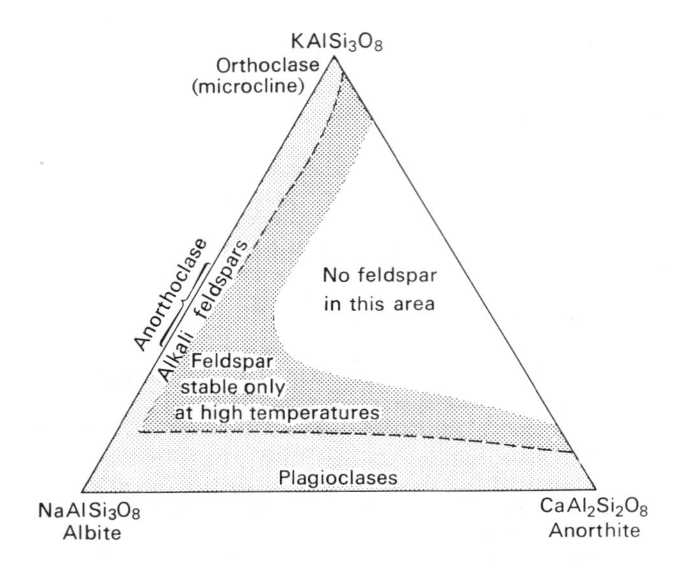

Fig. 4.27 Feldspar composition diagram. For a feldspar represented by a point within the triangle, the proportions of the three quantities Orthoclase, Albite, Anorthite are given by the lengths of the perpendiculars from the point onto the sides of the triangle. Each corner represents 100% of the component named there. With increasing Ca the plagioclases grade from Albite (0–10% $CaAl_2$) to Oligoclase, Andesine, Labradorite, Bytownite and Anorthite (90–100% $CaAl_2$).

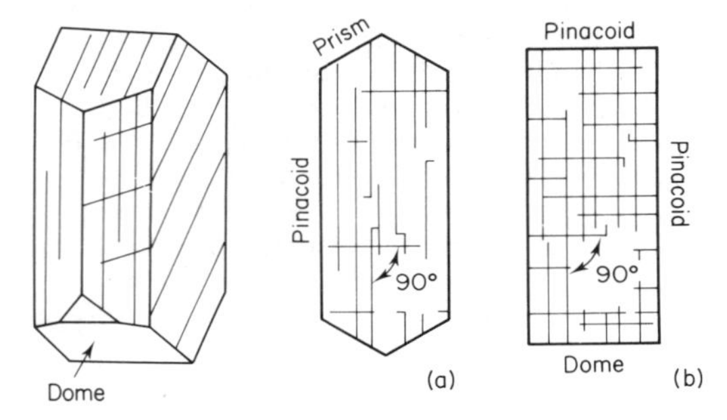

Fig. 4.28 Crystal of orthoclase (left) showing cleavage; (**a**) cross section and (**b**) vertical section showing cleavage and principal crystal faces.

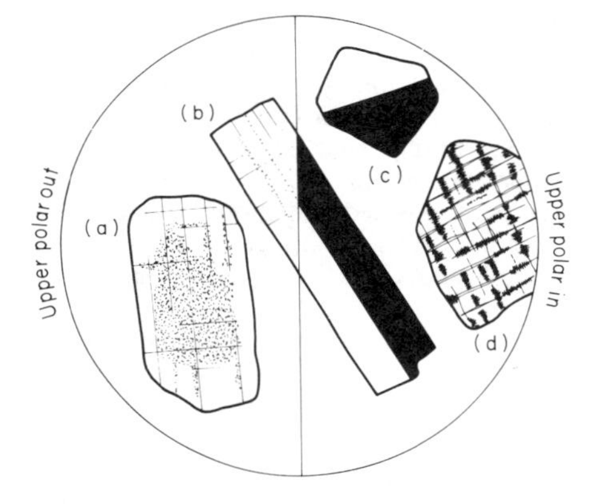

Fig. 4.29 Orthoclase in thin section (×3). (**a**) alteration to kaolin, (**b**) vertical section exhibiting Carlsbad twin, (**c**) Baveno twin, (**d**) Microcline.

Microcline, $KAlSi_3O_8$. Triclinic; the low temperature form of potassium feldspar.

Crystal form is similar to that of orthoclase, but the two cleavages (parallel to the basal and side pinacoids) intersect at about 89° instead of 90°.

Colour: White, pink, or green.

R.I. and *Biref:* Similar to orthoclase. Multiple twinning, with two intersecting sets of gently tapering twin-lamellae; in thin sections this is seen as a characteristic 'cross-hatching' effect between crossed polars, and distinguishes microcline from other feldspars (Fig. 4.29).

Anorthoclase Triclinic. (Fig. 4.27.) The beautiful grey schillerized feldspar which occurs in *laurvikite* (a Norwegian syenite, p. 107); it is also found as rhomb-shaped crystals in *rhomb-porphyry* in the Oslo district.

Schiller is the play of colours seen by reflected light in some minerals, in which minute platy inclusions are arranged in parallel planes; the effect is produced by the interference of light from these platy inclusions.

The Plagioclases

Feldspars of this series are formed of mixtures (solid solutions) of albite, $NaAlSi_3O_8$, and anorthite, $CaAl_2Si_2O_8$, in all proportions. The range of composition is divided into six parts, which are named. A plagioclase, for example, containing 40 per cent albite and 60 per cent anorthite would be called labradorite (written $Ab_{40}An_{60}$).

Crystals: Triclinic; white or colourless (albite) to grey (anorthite), bounded by prisms, basal and side pinacoids, and domes (Fig. 4.30). Vitreous lustre. Cleavages parallel to the basal and side pinacoids meet at an angle of about 86° (hence the name; Greek *plagios*, oblique; *klasis*, breaking). Multiple twins parallel to the side pinacoid (Albite twinning, Figs 4.8c and 4.31a) are

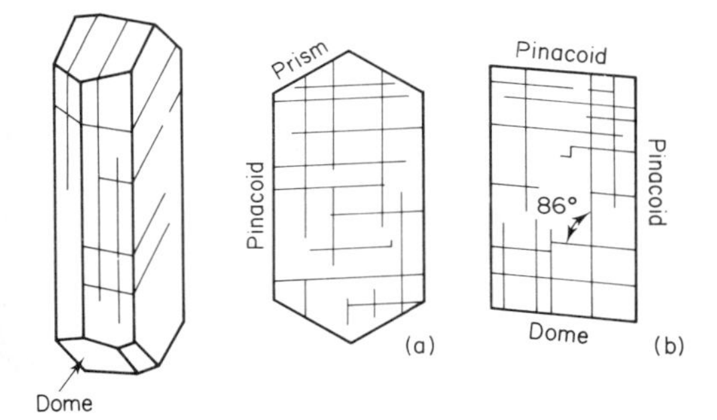

Fig. 4.30 Crystal of albite (a plagioclase): oblique view showing cleavage, (**a**) cross section and (**b**) vertical section, showing cleavage and principal crystal faces.

characteristic of plagioclases; the closely spaced twin-lamellae can often be seen with a lens as stripes on the basal cleavage and other surfaces. Another set of twins is sometimes developed on planes, parallel to the *b*-axis, which make a small angle with the basal pinacoid, Fig. 4.31 (Pericline twinning). H = 6 to 6½. G = 2.60 (albite), rising to 2.76 (anorthite).

Plagioclase feldspars occur in most igneous rocks, and in some sedimentary and metamorphic rocks. They appear as white or grey cleaved crystals in the coarse-grained igneous rocks, where their multiple twinning may be seen with a hand lens in suitable crystals.

In thin section: idiomorphic crystals (e.g. in lavas) commonly show rectangular sections; parallel-sided 'laths', with their length several times as great as their breadth, are seen when the crystals sectioned are flat and thin parallel to the side pinacoid (see Fig. 5.19). Cleavage not often visible. The minerals are normally colourless but may be clouded with alteration products: these are mainly kaolin in Na-rich, and epidote in Ca-rich varieties.

The characteristic multiple twinning appears as light and dark grey parallel stripes between crossed polars

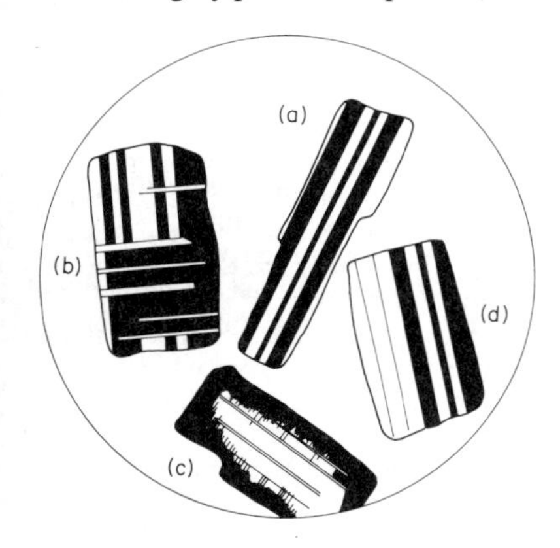

Fig. 4.31 Plagioclase in thin section (× 2) (upper polar in) – see also Fig. 4.17. (**a**) Common albite twin, (**b**) pericline twin, (**c**) zoned, (**d**) combined carlsbad-albite twin.

(Fig. 4.31), and sets of alternate stripes extinguish obliquely in different positions. Average values for the optical properties of the plagioclases are: *R.I.* = 1.55, birefringence weak (0.009) which gives grey and yellow polarization colours, and oblique extinction of long sections that varies from approximately 15° for albite to 40° for anorthite.

Zoned crystals Many crystals possess internal variations of composition which are expressed (*i*) by colour, e.g. tourmaline, Fig. 4.36; (ii) by inclusions, small particles locked up in a crystal at some stage in its growth, e.g. quartz, Fig. 4.33, or (*iii*) by zones of different composition. This last is particularly evident in some plagioclases, where it results from differences in successive layers of material acquired during a crystal's growth. For example, a plagioclase may have begun to grow as anorthite; but because of changes in the relative concentrations of constituents in the melt that was breeding the crystal, further growth may have used material containing less anorthite and more albite. A slice of such a crystal seen between crossed polars does not show sharp extinction in one position, but the zones of different composition extinguish successively at slightly different angles as the slice is rotated (Fig. 4.31c). Augite and hornblende may also show compositional zoning of this kind.

The feldspathoid group

Minerals of this group resemble the feldspars chemically, and have 3-dimensional framework structures; they differ from the feldspars in their lower content of silicon. Stated in another way, the Al:Si ratio is higher in the feldspathoids than in the corresponding feldspars. The two chief minerals of the group, which are not discussed here because their occurrence is somewhat limited in nature, are:

Leucite, $K(AlSi_2)O_6$ (cf. Orthoclase)

Nepheline, $Na(AlSi)O_4$ (cf. Albite), usually with a little K.

Feldspathoids occur in certain undersaturated lavas, which have a low silica- and high alkali-content, such as the leucite-basalts from Vesuvius.

Forms of silica

Silica is found uncombined with other elements in several crystalline forms of which *quartz*, one of the most common minerals in nature, is of special importance. As the quartz content of a rock increases so may its strength and also its abrasiveness to machinery used for drilling and excavating. When drilling quartz-rich rock fine dust may be created which should not be inhaled as it may damage lung tissue. Other forms of silica include the high temperature *tridymite* (see below); *chalcedony*, aggregates of quartz fibres; and the cryptocrystalline forms *flint* and *opal*, and *chert* (p. 126).

Quartz, SiO_2

In the structure of quartz the silicon-oxygen

tetrahedra build up a three-dimensional framework in which each oxygen is shared between two silicons. There are no substitutions of other ions in the silicon positions.

Crystals: Trigonal with 6-sided prisms and rhombohedral terminations (Figs. 4.32 and 4.7); faces sometimes unequally developed; occasionally other faces belonging to trigonal forms are present. Vitreous lustre; conchoidal fracture. Colourless when pure (e.g. '*rock crystal*'), but many coloured varieties occur, the colour being due to traces of impurities, e.g. *rose quartz* (pink), *smoky quartz* (grey), *milky quartz* (white), *amethyst* (violet). Some quartz contains minute inclusions or liquid-filled cavities, which may be arranged in regular directions in a crystal. No cleavage; twins rare. H = 7, cannot be scratched with a knife. G = 2.66.

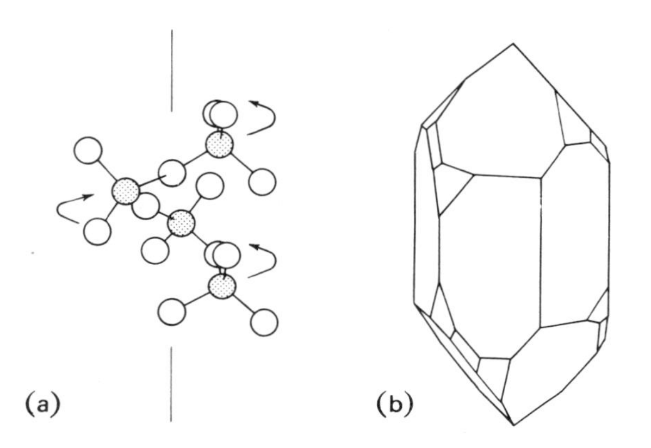

Fig. 4.32 (**a**) Spiral structure in atomic framework of quartz. (**b**) Quartz crystal with extra (trigonal) faces.

Quartz is an essential constituent of granites, and can be recognized in the rock as hard, glassy grains of irregular shape and without cleavage. It occurs in smaller amount in granodiorite and quartz-diorite, and is present as well-shaped porphyritic crystals in acid dyke-rocks and lavas. *Vein quartz* is an aggregate of interlocking crystals of glassy or milky appearance, filling fractures in rocks; the boundaries of the crystals may be coated with brown iron oxide. Well-shaped quartz crystals are found in cavities (*druses*) in both veins and granitic rocks. Most sands and sandstones have quartz as their main constituent; the grains have a high resistance to abrasion and thus persist over long periods during erosion and transport. The mineral is also found abundantly in gneisses, quartzites, and in some schists and other metamorphic rocks.

In thin section: basal sections are regular hexagons (Fig. 4.16) when the crystals are well formed; also see Fig. 4.16 for longitudinal section. When the mineral has crystallized among others, as in granite, its shape is irregular (Figs. 4.33 and 5.24).

Colourless. Never shows alteration, but crystals in lavas sometimes have corroded and embayed margins.

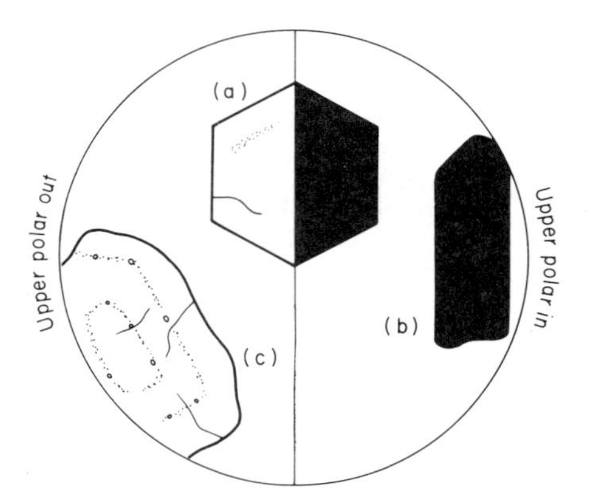

Fig. 4.33 Quartz in thin section (×10). (**a**) Cross-section normal to *c*-axis; (**b**) vertical section parallel to *c*-axis; and (**c**) irregular (from an igneous rock such as granite), lines of minute inclusions (dots) may be present.

R.I. = 1.553 (max.), 1.544 (min.); weak outline in Canada balsam.

Biref: = 0.009 (max.) (see p. 68).

Extinction: Straight in longitudinal sections.

Quartz is distinguished from orthoclase by the absence of twinning and by its entire lack of alteration; it always appears fresh, although inclusions may be present. The mineral may show strain polarization in rocks which have been deformed and then has no sharp extinction position.

Chalcedony, SiO_2

Radiating aggregates of quartz fibres, their ends often forming a curved surface; white or brownish colour and of waxy appearance in the mass. Chiefly found in layers lining the vesicles of igneous rocks (Fig. 4.2a). In thin section, such layers show a radiating structure, of which the crystal fibres have straight extinction and give an extinction 'brush' which remains in position as the stage is rotated. *R.I.* = 1.54; polarization colours are light greys.

Flint Cryptocrystalline (i.e. made of a large number of minute crystals which are too small to be distinguished separately except under very high magnification) silica, possibly with an admixture of opal, representing a dried-up gel; occurs in nodules in the chalk (see p. 127). Often black in colour on a freshly broken surface, with conchoidal fracture. Split flints were much used in the past as a decorative facing to buildings.

Opal Hydrated silica, SiO_2nH_2O; amorphous. White, grey, or yellow in colour, with a pearly appearance (opalescence), and often displaying coloured internal reflections. Conchoidal fracture. H = about 6. G = 2.2. Occurs as a filling to cracks and cavities in igneous rocks. When it replaces woody tissues it preserves the original textures and is known as *wood opal.* The microscopic organisms known as *diatoms* (p. 30) which live in oceans and lakes, are also composed of opaline silica. Opal is an undesirable constituent in rocks used for concrete aggregates, owing to the possibility of reaction occurring between it and alkalis in the cement (*q.v.*).

In thin section, colourless and isotropic, with low *R.I.* (1.44).

Accessory minerals

Minerals that occur as small crystals and in limited quantities are described as accessory. Occasionally they may develop to form large crystals and in certain circumstances can be concentrated to become a major rock constituent.

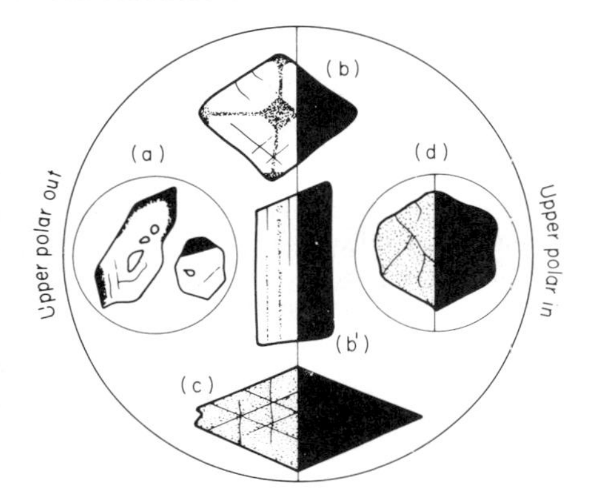

Fig. 4.34 Some accessory minerals in thin section (×15). (**a**) Zircon (at greater magnification than other minerals shown). (**b**) Chiastolite cross-section, (b') long section. (**c**) Sphene. (**d**) Garnet, showing isotropic character under crossed polars.

Zircon, $ZrSiO_4$

Tetragonal. H = 7.5, G = 4.7. Zircon occurs in granites and syenites as an original constituent. Crystals are usually very small and in thin section are recognized chiefly by their shape: Fig. 4.34*a*.

Andalusite, Al_2OSiO_4

Orthorhombic. Colour, pink or grey. H = 7.5, G = 3.2. Andalusite occurs in contact metamorphosed shales and slates. Crystals are prismatic and have a nearly square transverse section. The variety *chiastolite* contains inclusions of carbon (Fig. 4.34*b*). Two other forms of Al_2OSiO_4 are *sillimanite* (formed at high temperature as in an aureole) and *kyanite* (a pale blue mineral formed under high stress and moderate temperature).

Sphene, $CaTiSO_4$ (O, OH, F)

Monoclinic. H = 5, G = 3.5. Sphene occurs as small wedge-shaped crystals (Gr. *sphene*, a wedge) in granite, diorites and syenites. Fig. 4.34*c*.

Garnet $(Ca, Mg, Fe^{2+}, Mn)_3\ (Al, Fe^{3+}, Cr)_2\ (SiO_4)_3$

Cubic (Fig. 4.35). Colour, pale pink. H = 6.5 to 7.5,

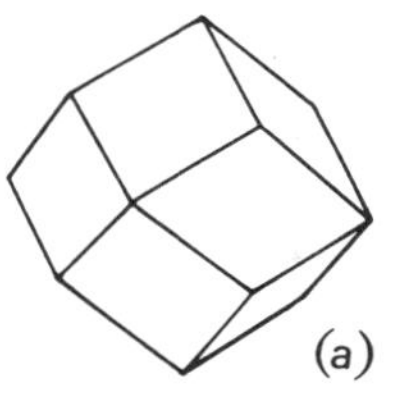

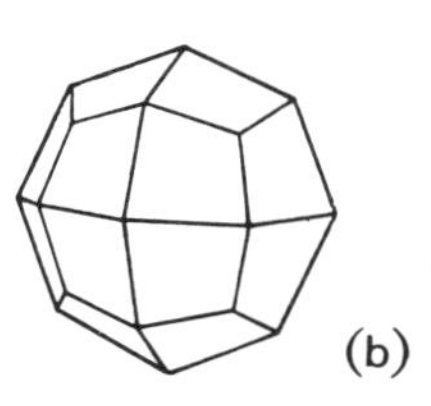

Fig. 4.35 Crystals of garnet oblique views. (**a**) rhombdodecahedron; (**b**) trapezohedron.

G = 3.5 to 4.0 according to species of mineral. Garnet occurs in metamorphic rocks such as mica-schist. In thin section crystals are isotropic. Fig. 4.34d.

Tourmaline. Complex silicate of Na, Mg, Fe, Al with Si_6O_{18} rings.

Trigonal. Colour, black, red, green, blue. H = 7, G = 3.0. Tourmaline occurs in granites and vein rocks such as pegmatites. The variety *schorl* commonly grows in radiating clusters (Fig. 4.1e) e.g. in pneumatolysed granites. Yellow in thin section with extinction parallel to length. Fig. 4.36*a–c*.

Cordierite, $Mg_2Al_3(AlSi_5)O_{18}$

Orthorhombic but commonly appears nearly hexagonal in shape when crystals grow together as twins (Fig. 4.36*d*). Occurs in metamorphic rocks such as hornfels (p. 136).

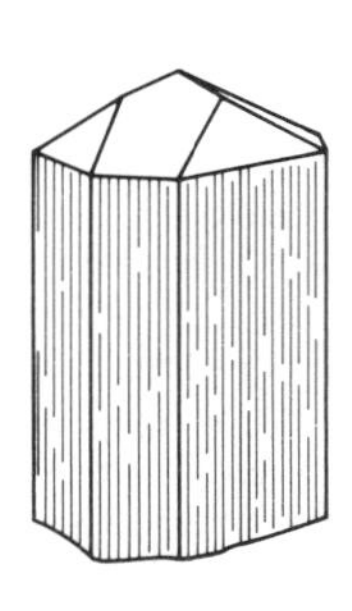

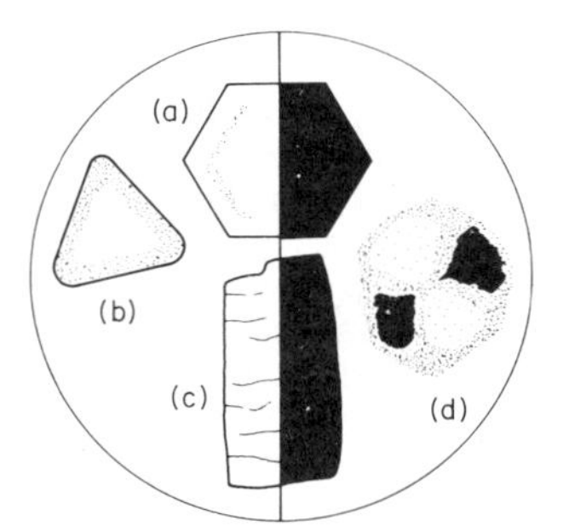

Fig. 4.36 Crystal of tourmaline. Tourmaline seen in thin section (×5), (**a**) and (**b**) cross-sections; (**c**) vertical section. Also shown is cordierite (**d**).

Secondary minerals

Described under this head are the minerals *chlorite*, *serpentine*, *talc*, *kaolin*, *epidote* and *zeolite*, all of which result from the alteration of pre-existent minerals. These minerals have little mechanical strength and small angles of friction. Their presence on fractures can significantly reduce the strength of a rock mass.

Chlorite $(MgFe)_5Al(Si_3Al)O_{10}(OH)_8$, variable

The chlorites (Greek *chloros*, green) form a family of green flaky minerals which are hydrous silicates of magnesium and aluminium. Some Fe replaces Mg and gives colour to the chlorite. Like the micas, they have a perfect cleavage, due to the atomic sheet structure (Fig. 4.26). Different kinds of chlorite are given distinctive names (e.g. penninite, clinochlore); these are not distinguished in the following general description.

Crystals: Monoclinic, frequently 6-sided in shape, with a perfect cleavage parallel to the basal plane; the mineral splits into hexagonal flakes which are flexible but not elastic (cf. mica). H = 2 to $2\frac{1}{2}$ (often soft enough to be scratched by the finger-nail). G = 2.65 to 3.0.

Chlorite is found in igneous rocks, as described below, and in metamorphic rocks such as chlorite-schist, and in some clays.

In thin section: Chlorite occurs as an alteration product of biotite, augite, or hornblende; it may replace these minerals completely, forming a *pseudomorph* (= 'false form') in which the aggregate of chlorite flakes and fibres retains the shape of the original mineral. Together with other minerals such as calcite, chlorite also forms an infilling to cavities in basalts (*q.v.*).

Colour: Shades of bluish-green and yellowish-green, sometimes very pale; noticeably pleochroic; cleavage often seen.

Mean R.I. = about 1.58.

Biref: Weak grey.

Serpentine, $Mg_6Si_4O_{10}(OH)_8$, some Fe replaces Mg, in part

Serpentine is an alteration product of olivine, of orthorhombic pyroxene, or of hornblende. This reaction takes place in an igneous rock while it is still moderately hot (*hydrothermal action*), the source of the hot water being magmatic; it is thought that the change from olivine to serpentine may also be brought about by the action of water and silica.

Serpentine grows as a mass of green fibres or plates, which replace the original mineral as a pseudomorph. A fibrous variety is called *chrysotile*, and is worked in veins for commercial asbestos. In the mass, serpentine is rather soapy to the touch, and may be coloured red if iron oxide is present. H = 3 to 4. G = 2.6. Serpentine is found in basic and ultrabasic rocks (p. 101), and in serpentine-marble.

In this section: as a pseudomorph after olivine, serpentine appears as a matte of pale green fibres, weakly birefringent, and having a low *R.I.* (1.57). Specks of black magnetite, the oxidized by-product from iron in the original olivine, are often present. The change to serpentine involves an increase in volume, and this expansion may fracture the surrounding minerals in the rock, fine threads of serpentine being developed in the cracks so formed.

Talc, $Mg_3Si_4O_{10}(OH)_2$

A soft, flaky mineral, white or greenish in colour, which occurs as a secondary product in basic and ultrabasic rocks, and in talc-schist (p. 140). It is often associated with serpentine. Flakes are flexible but not elastic, and are easily scratched by the finger-nail. H = 1.

Kaolin (china clay)

This substance is largely made up of the mineral *kaolinite*, $Al_4Si_4O_{10}(OH)_8$, one of the group of Clay Minerals which, like the micas, are built up of silicon-oxygen sheets (Fig. 4.26).

Kaolin is derived from the breakdown of feldspar by the action of water and carbon dioxide; the chemical equation for the change is given and the kaolinization of granite masses is described on p. 138. It is white or grey, soft, and floury to the touch, with a clayey smell when damp. G = 2.6. In thin section it is seen as a decomposition product of feldspar (Fig. 4.29), which when altered appears clouded and looks white by top light (i.e. by light reflected from the surface of the slice and not transmitted through it).

Epidote, $Ca_2(AlFe)_3(SiO_4)_3(OH)$

The monoclinic crystals of this mineral are typically of a yellowish-green colour. Often in radiating clusters; vitreous lustre. H = 6 to 7. G = 3.4.

Epidote occurs as an alteration product of calcic plagioclases or of augite; also as infillings to vesicles in basalts, and as pale green veins traversing igneous and metamorphoc rocks.

Zeolites

These form a group of hydrous aluminous silicates of calcium, sodium, or potassium; they contain molecular water which is readily driven off on heating, a property to which the name refers (Greek *zein*, to boil). They occur as white or glassy crystals clusters, filling or lining the cavities left by escaping gases (amygdales, p. 100) in basic lavas, or filling open joints, and are derived from feldspars or feldspathoids by hydration.

Two commonly occurring natural zeolites are:

Analcite, $NaAlSi_2O_6H_2O$

Cubic; crystallized as trapezohedra (as in Fig. 4.35b), white in colour. G = 2.25. Occurs in the amygdales of basalts.

Natrolite, $Na_2Al_2Si_3O_{10}2H_2O$

Forms white, acicular orthorhombic crystals, generally in radiating clusters. G = 2.2.

Clay minerals

Clays can form as either primary or secondary minerals. Here they are grouped under one heading because of their economic importance, their presence in most profiles of weathering and their influence upon the mechanical character of rocks and less well consolidated sediments. Being minute they can be seen using only an electron microscope (Fig. 4.37), but their presence may be revealed by placing a few particles of material that is

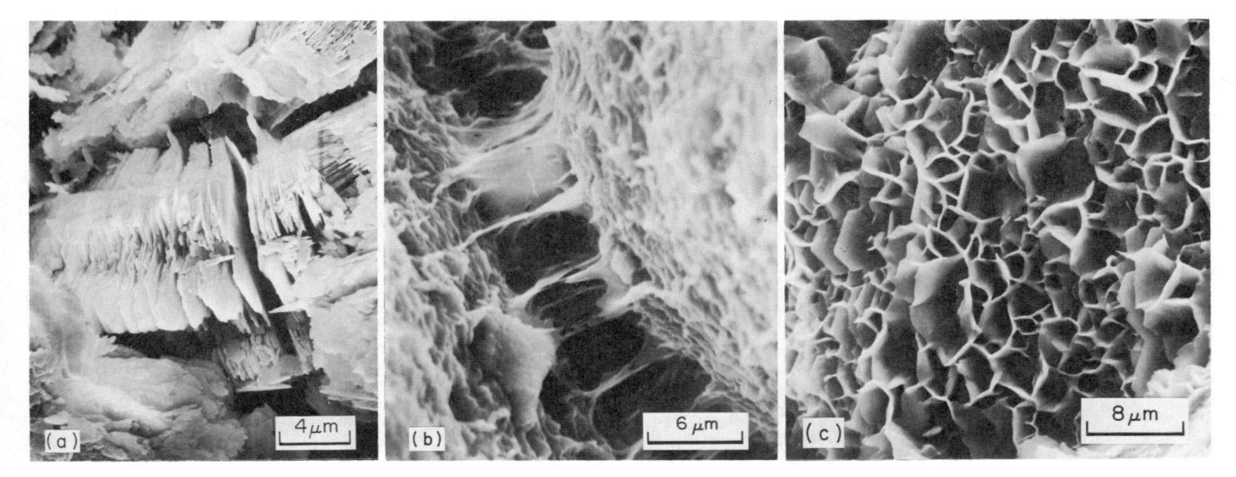

Fig. 4.37 Clay minerals. (**a**) Kaolinite, (**b**) montmorillonite and (**c**) illite, as seen at high magnification with an electron microscope (provided by L. Dobereiner).

suspected to contain clay, on the teeth. The clay component, if present, will feel smooth and the non-clay component will be gritty. The dominant clay mineral present can be identified inexpensively by measuring its *activity*, as described on p. 195.

In common with other flaky minerals such as the micas, chlorites, and talc, clay minerals are built up of two-dimensional atomic layers or sheets (p. 74) which are stacked one upon another in the *c*-direction. The layers are of two kinds:

(*i*) silicon-oxygen sheets, each formed by the linking together of tetrahedral SiO_4-groups as described on p. 70, and generally referred to as a *tetrahedral layer*. The composition of this layer is a multiple of Si_2O_5, or with attached hydrogen, $Si_2O_3(OH)_2$.

(*ii*) *octahedral layers*, in which a metal ion (Al or Mg) lies within a group of six hydroxyls which are arranged at
the corners of an octahedron (Fig. 4.38). Adjoining octahedra are linked by sharing hydroxyls. Such an octahedral layer has the composition $Al_2(OH)_6$ or $Mg_3(OH)_6$ (the minerals *gibbsite* and *brucite* respectively); in addition, some substitution may take pace for the metal ion, for example Fe^{3+} for Al^{3+}, or Fe^{2+} for Mg^{2+}

Different arrangements of the above layers build up the units of which the clay minerals are composed, the flat surfaces of the minute crystals being parallel to the layers. Some clay minerals have two-layer units, as in Fig. 4.38 (kaolinite); others (e.g. montmorillonite) have three-layer units in which an octahedral layer lies between two tetrahedral layers, one of which is inverted relative to the other so that apexes of the tetrahedra point inwards in the unit (*cf.* Fig. 4.26). Layers of molecular water may lie between these units, as in montmorillonite.

The commoner clay minerals include:

(*a*) ***Kaolinite***, $Al_4Si_4O_{10}(OH)_8$, made up of alternate tetrahedral and octahedral-layers; each pair, $Si_2O_3(OH)_2 + Al_2(OH)_6$, with loss of water becomes $Al_2Si_2O_5(OH)_4$.Kaolinite occurs in hexagonal flakes of minute size, and forms the greater part of kaolin (china clay) deposits; it is also found in soils and sedimentary clays, of which it forms a variable and often small proportion. It is the main constituent of fire-clays (*q.v.*).

The mineral *dickite* has the same composition as kaolinite, but the layers in the structure have a different arrangement relative to one another. *Halloysite*, $Al_2Si_2O_5(OH)_4{\cdot}2H_2O$, may be included in a group with dickite and kaolinite; it occurs as minute tubes, the rolled-up 'sheets' of silicon-oxygen and Al-hydroxyl composition. Certain clays having a high content of halloysite possess special properties with regard to porosity and water content, which are discussed in Chapter 18 (p. 289).

(*b*) ***Montmorillonite***, which has important base-exchange properties, is built up of 3-layer units comprising two tetrahedral layers separated by an octahedral layer, and has the ideal formula $Al_4Si_8O_{20}(OH)_4$. Some aluminium is usually replaced by magnesium or iron, and small amounts of sodium or calcium are then attached, as ions lying between the 3-layer units or around the edges of the minute crystals. These alkali ions are exchangeable, and give rise to the high base-exchange capacity of the mineral. In addition, layers of molecular water may occur between the 3-layer units. A typical Ca-montmorillonite would be represented by the formula: $Ca_{0\cdot5}(MgAl_3)Si_8O_{20}(OH)_4{\cdot}xH_2O$; the calcium is replaced by sodium in Na-montmorillonite. The proportion of water is variable, and the absorption of water between the 3-layer units gives rise to the considerable swelling properties possessed by clays containing much montmorillonite.

The mineral occurs sparsely in soils along with kaolinite; it is the chief component of clays such as

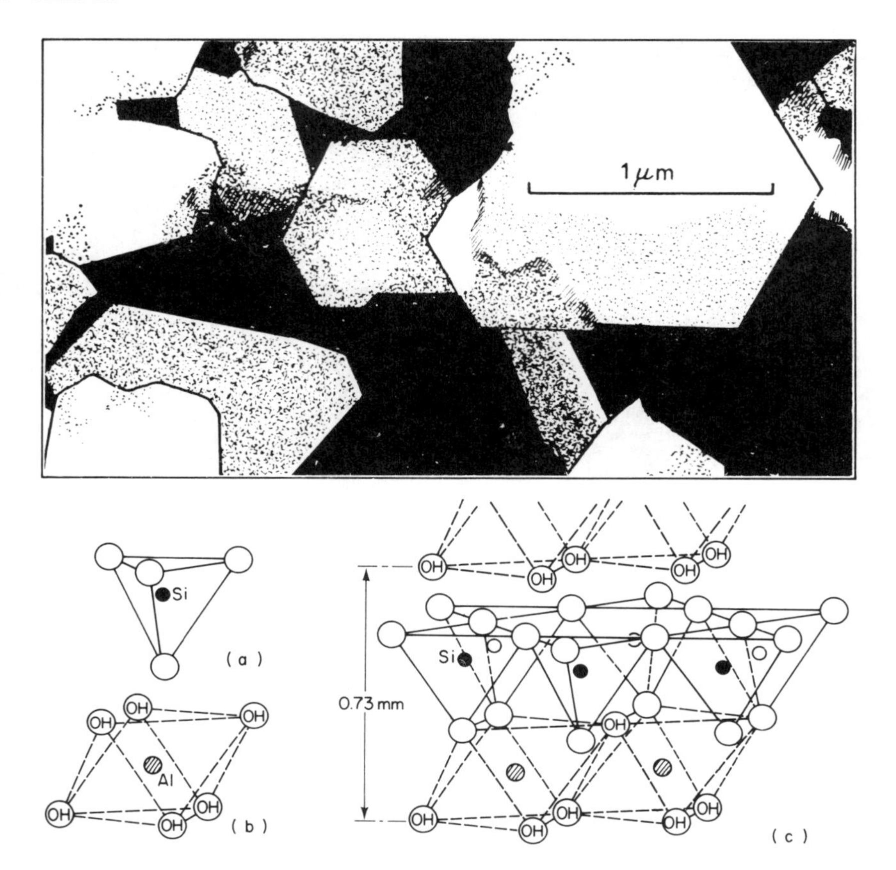

Fig. 4.38 Kaolinite. *Upper.* Electron micrograph (×35 000). *Lower.* Atomic structure of kaolinite (open circles represent oxygen atoms). (**a**) Tetrahedral group; (**b**) octahedral group; (**c**) unit of structure, consisting of a silicon-oxygen layer of linked tetrahedral groups, combined with an aluminium-hydroxyl, layer. The *c*-axis of the mineral is vertical.

fuller's earth and bentonite, which are described briefly on p. 122, with a note on their uses.

(*c*) ***Illite*** (named after Illinois by R.E. Grim, 1937) is similar in many respects to white mica, but has less potassium and more water in its composition. It has a much lower base-exchange capacity than montmorillonite. Illite is built up of units comprising two tetrahedral layers separated by an octahedral layer, and forms minute flaky crystals in a similar way to montmorillonite. Some of the silicon is replaced by aluminium, and atoms of potassium are attached, giving a general formula of the type: $K_xAl_4(Si_{8-x}Al_x)O_{20}(OH)_4$, the value of x varying between 1.0 and 1.5. The OH-content may exceed 4, out of a total of 24 for O+OH.

Sedimentary clays are mostly mixtures of illite and kaolinite, with some montmorillonite, and shales have illite as the dominant clay mineral. Illite is probably the most widely distributed clay mineral in marine argillaceous sediments.

An early use (1940) in engineering of the property of base-exchange in clays was to render impervious a leaky clay lining to the artificial freshwater lake constructed at Treasure Island, San Francisco. The material used for the lining was a sandy clay, having a small content of calcium which was probably attached as ions to aggregates of colloidal particles, by virtue of which the clay was 'crumbly' and to some extent permeable. By filling the lake with salt water the clay was enabled to take up sodium in exchange for the calcium; this resulted in a considerable decrease in its permeability and a 90% reduction in the seepage losses. The colloidal aggregates were dispersed by the exchange of bases, thus changing the physical properties of the clay and filling the voids with a sticky gel which rendered it largely impervious to water. This treatment is the reverse of the common agricultural process of adding calcium (in the form of lime) to a heavy, sticky soil in order to improve its working qualities.

Non-silicate minerals

The more common, or economically important non-silicate minerals are listed in Table 4.9, each with

Table 4.9 Examples of important non-silicate minerals

Class	Example	H	G	Colour	Streak	Lustre	Notes
NATIVE ELEMENTS	Copper Cu	2.5	8.9	red	red	metallic	malleable
	Silver Ag	2.5	10.5	grey-white	white	″	″
	Gold Au	2.5	19.3	yellow	yellow	″	″
	Antimony Sb	3.5	6.7	grey-white	grey	″	sectile
	Diamond C*	10	3.5	variable	—	greasy	sparkles
	Graphite C*	1	2.2	black	black	metallic	marks paper
	Sulphur S	2	2.1	yellow	white-yellow	resinous	odour on burning
SULPHIDES	Chalcocite Cu_2S	2.5	5.7	grey-black	black	metallic	blue-green tarnish
	Bornite Cu_5FeS_4	3	5.1	brown	grey	″	iridescent
	Galena PbS*	2.5	7.5	lead grey	″	″ (dull)	cubic cleavage
	Sphalerite ZnS*	3.5	4	dark brown	brown	resinous	translucent
	Chalcopyrite $CuFeS_2$*	3.5	4.2	brass yellow	black	metallic	iridescent
	Covellite CuS	1.5	4.7	dark blue	grey-black	″	purple tarnish
	Cinnabar HgS*	2.5	8.1	red	red	vitreous	sectile
	Pyrite FeS_2*	6.5	5.1	brass yellow	black	metallic	strikes fire with steel
	Molybdenite MoS_2*	1	4.6	grey	grey	″	marks paper
HALIDES	Halite NaCl*	2.5	2.2	variable	white	vitreous	saline taste
	Fluorite CaF_2*	4	3.2	″	″	″	translucent
OXIDES & HYDROXIDES	Cuprite Cu_2O	3.5	6.1	red	red	metallic	soluble in HCl
	Corundum Al_2O_3	9	4	brown-blue	—	vitreous	v. hard
	Haematite Fe_2O_3*	6	5.3	brown-black	red	metallic	magnetic when heated
	Ilmenite $FeTiO_3$*	5.5	4.7	black	black	″	non-magnetic
	Magnetite Fe_3O_4*	6	5.2	″	″	″	magnetic
	Chromite $FeCr_2O_4$*	5.5	4.6	″	brown	″	no cleavage
	Cassiterite SnO_2*	6.5	7	brown-black	white-brown	vitreous	translucent
	Goethite FeO(OH)*	5	4.4	brown	brown	dull	crystalline
	Bauxite $Al_2O_3(2H_2O)$*	2	2.1	red-brown	″	″	earthy, concretionary
	Limonite $2Fe_2O_3(3H_2O)$*	5	3.8	brown	yellow-brown	″	earthy
CARBONATES	Calcite $CaCO_3$*	3	2.7	colourless	white	vitreous	effervesces in cold HCl
	Siderite Fe_2CO_3*	3.5	3.9	brown	″	″	—
	Dolomite $CaMg(CO_3)_2$	3.5	2.9	white-brown	″	″	effervesces in warm HCl
	Aragonite $CaCO_3$*	3.5	2.9	colourless	″	″	stained with $MnSO_4$
	Witherite $BaCO_3$	3.5	4.3	white-grey	″	resinous	rough striations common
	Cerussite $PbCO_3$	3	6.6	″	″	vireous	effervesces in warm HNO_3
	Azurite $Cu_3(OH)_2(CO_3)_2$	3.5	3.8	dark blue	blue	″	—
	Malachite $Cu_2(OH)_2CO_3$*	3.5	4	bright green	green	silky-dull	—
T & P	Wolframite $(FeMn)WO_4$*	4.5	7.3	black	red-brown	submetallic	dull on fractures
	Apatite $Ca_5(PO_4)_3(F,Cl,OH)$*	5	3.2	green	white	vitreous	sub-resinous also
SULPHATES	Barytes $BaSO_4$*	3	4.5	colourless-white	white	vitreous	tabular crystals common
	Celestine $SrSO_4$*	3.5	4	white-blue	″	″	v. brittle
	Anglesite $PbSO_4$	3	6.3	white	″	″	often with galena
	Anhydrite $CaSO_4$*	3	3	white-brown	″	pearly	splintery fracture
	Gypsum $CaSO_4 2H_2O$*	2	2.3	colourless	″	vitreous	flexible, inelastic

T & P = Tungstates and phosphates.
H = Hardness in Mohs' scale.
G = Specific gravity.
* See text.

diagnostic characters that can be defined easily. A selection is briefly described here.

Native elements

Minerals in this class have one type of atom in their structure and can be metals (Cu, Ag, Au, Pt, Fe), semi-metals (As, Sb, Bi) and non-metals (C, S).

Diamond and graphite

The element carbon (C) is found in two crystalline forms: *diamond* (cubic), and *graphite* (hexagonal). Amorphous carbon also occurs, as charcoal and soot.

Diamond and graphite have entirely different atomic structures (Fig. 4.39) which account for their different physical properties. The diamond structure permits carbon atoms to be linked in tetrahedral groups and the strength of this structure is reflected in the hardness of the mineral.

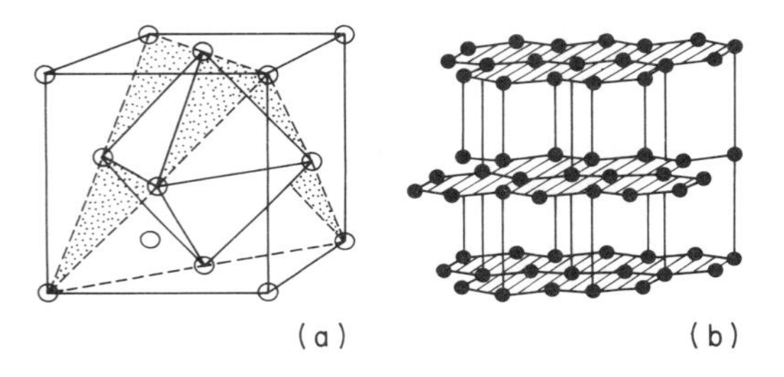

Fig. 4.39 The atomic structure of diamond (**a**) and graphite (**b**). Carbon atoms in diamond are shown clear and those at the centre of the tetrahedra are omitted for clarity. Carbon atoms in graphite are shown black and the layers formed by them are shaded.

Diamond occurs as octahedral crystals in ultrabasic rocks (e.g. serpentinite), as in volcanic pipes at Kimberley, South Africa, and also in related rocks in Brazil and Russia. It is also found in alluvial deposits along with other hard minerals of high specific gravity. Diamonds that are colourless or nearly so are valuable as gems. In the form of granular aggregates (*bort*) they are much used as abrasives, and for the cutting edges of diamond drills, and in emery wheels. (Natural abrasives, in the order of their hardness, are diamond, corundum (Al_2O_3), emery (a grey-black variety of corundum containing magnetite and haematite, and garnet.)

In graphite the carbon atoms are in layers and there is only a weak bonding between the layers; in consequence graphite is a very soft mineral and has important uses as a lubricant. The mineral occurs in veins (as at Borrowdale, Lake District), or in lenticular patches in certain metamorphic rocks such as schists and gneisses (Sri Lanka); and in some crystalline limestones (e.g. in eastern Canada).

Sulphides

Compounds of the large atom sulphur, are important ore minerals and often concentrated to form economic accumulations of Cu, Zn and Pb. As shown in Table 4.9, they are relatively soft, heavy and dark. In thin section they are opaque.

Galena

Generally with some contant of silver sulphide; when enough silver is present to be worth extracting the term 'argentiferous galena' is used. Galena occurs in lodes in association with sphalerite (*q.v.*), calcite, quartz, and other gangue minerals (p. 88), often filling fracture zones; or replacing limestone along joints as in the small occurrence in Derbyshire, Cumbria, and Cornwall, England. The well-known lode at Broken Hill, New South Wales, yields argentiferous galena. At Leadville, Colorado, the mineral replaces limestone. Galena is the chief ore of lead and an important source of silver. Crystals are cubic with faces of the cube and octahedra: perfect cleavage parallel to the cube faces.

Sphalerite (also called Zincblende, or Blende)

Blende is the principal ore of zinc, and often occurs in lodes with galena and gangue minerals. Apart from the small British examples (see galena), rich deposits are found at Broken Hill, New South Wales; and at the large Sullivan Mine, British Columbia, where the lead-zinc ore-bodies are replacements in Precambrian quartzites. The United States is the chief zinc-producing country. Crystals are cubic, commonly as tetrahedra, with perfect cleavage.

Chalcopyrite (also called Copper pyrites)

Chalcopyrite may contain 34% Cu and is the principal commercial source of copper. It is mainly formed in association with igneous rocks especially during the late stages of intrusion when Cu rich hydrothermal fluids permeate the cooling magma and surrounding country-rock (see p. 109). Important deposits occur world-wide but especially in fold belts, e.g. Norway (Caledonian), Cornwall, Rio Tinto in Spain (Hercynian), western America (from Alaska to California) and in Japan. Crystals are tetragonal but cleavage is poor and the mineral normally occurs in a massive state.

Pyrite (also called 'Fool's gold')

Pyrite occurs massive, and in ore-veins, and as nodules with a radiating structure; sometimes replacing calcite in fossil shells. Deposits of massive pyrite may contain a small percentage of copper and gold. The important deposits at Rio Tinto, Spain, are worked for sulphur and copper. Others in Norway, Sweden, Italy, Russia, and Japan are also worked on a large scale. By decomposition and oxidation, pyrite gives rise to sulphuric acid (H_2SO_4). The iron released during natural decomposition helps to form limonitic coatings on rock surfaces (p. 86). Crystals are cubic in the form

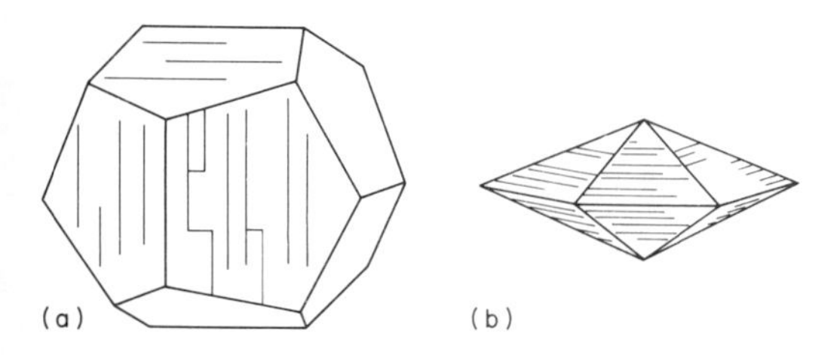

Fig. 4.40 (**a**) Pyrite (pyritohedron), note striations on faces of the crystal. (**b**) Marcasite (orthorhombic form of FeS_2), composition as for pyrite.

of cubes or pyritohedron (Fig. 4.40), normally with striated surfaces.

Pyrrhotite, describes the variety Fe_nS_{n+1}, where *n* is between 6 and 11. Colour brownish or coppery; often occurs massive, tarnishes on exposure. Streak, dark greyish-black; lustre metallic. H = 3.5 to 4.5, G = 4.6 (for Fe_7S_8). Magnetic (a distinction from pyrite), and soluble in HCl.

In the important pyrrhotite ore-deposits at Sudbury, Canada, the mineral is accompanied by the nickel sulphide *pentlandite* (Fe,Ni)S. These deposits yield the greater part of the world's nickel supply; the ore-bodies occur at the margins of a basic igneous mass (gabbro or norite).

Cinnabar

The chief ore of mercury, usually the product of volcanism as in the orogenic belts of California, Peru and at Almada in Spain. Commonly associated with chalcopyrite, pyrite, quartz, opal and calcite. Crystals are hexagonal but the mineral normally occurs in massive or granular form.

Molybdenite

The chief ore of molybdenum in which two sheets of sulphur atoms sandwich a single sheet of molybdenum atoms to produce a layered atomic structure similar to that of graphite (Fig. 4.39). This gives the mineral important lubricant properties. Found near igneous intrusions of granite and often associated with tourmaline. Important occurrences have been located in Colorado, New Mexico, Norway, Queensland and New South Wales. Crystals are hexagonal often in the form of shiny scales.

Halides

This class has many members, the most important often occurring as chemically deposited sediments formed by evaporation (p. 127) and as vein minerals in igneous rocks.

Halite (also called Rock-salt)

Rock-salt occurs, together with gypsum and other salts, as a deposit from the evaporation of enclosed bodies of salt water (see p. 127). Deposits are worked in the Triassic beds of Cheshire; at Stassfurt, Germany; in Ontario and Michigan, and in other countries. Occurrences in salt-domes are discussed on p. 151. Crystals are cubic with perfect cleavage parallel to the faces of the cube (Fig. 4.19). Commonly colourless but may be white, yellow or brown.

Fluorite (also called Fluorspar)

The mineral occurs in hydrothermal veins, often associated with blende and galena, or with tinstone (cassiterite, p. 86). Fluorspar is used in the manufacture of hydrofluoric acid. A massive purple or blue variety from Derbyshire is called Blue John. Crystals are cubic, commonly as cubes, sometimes zoned from green at their centre through white to purple at their outermost zone. Cleavage perfect and parallel to octahedron faces (see Fig. 4.4*a*).

Oxides and Hydroxides

These occur mainly as alteration products of sulphide ores (see above) and as accessory minerals in igneous rocks (p. 100). The most remarkable oxide, SiO_2 (silica dioxide), is not included here as it is the fundamental component of silicate minerals and described on p. 77.

Haematite (also spelled Hematite)

An important ore of iron. The great haematite deposits of the Lake Superior area (the Mesabi and Marquette iron-ranges) contain 50–60% Fe, and are an extensive source of iron ore. At Hamersley, western Australia, banded haematite and chert form an important economic deposit. Residual haematite deposits are found in Cuba; and in Brazil large accumulations are metamophosed sedimentary ores. It is also found in large or small 'pockets' in limestone, as a replacement, as at Ulverston in Cumbria, England.

Haematite is the cementing material in many sandstones (p. 120), and is a very common cause of the red staining seen in many rocks.

Crystals are Trigonal rhombohedral in form, often in thin tabular forms, with brilliant metallic lustre; this variety of haematite is called *specular iron*. *Kidney ore* is a massive, reniform variety (Fig. 4.1*b*) with an internal structure of radial fibres, and is the common form of red haematite.

Ilmenite

The chief ore of titanium. Occurs as an accessory mineral in basic igneous rocks; large, massive segregations of the mineral are found in association with such rocks, as in Norway and Canada (Quebec and Ontario). Ilmenite grains are found in many beach-sands; important deposits of this type are worked in India (Travancore), Australia, Tasmania, Florida, and elsewhere. Alters to white *leucoxene*, the presence of which distinguishes ilmenite from magnetite.

Magnetite

An important iron ore. Occurs in small amount in many igneous rocks; when segregated into large masses it forms a valuable ore of iron, as at Kiruna and Gellivaare in north Sweden, and in the Urals. Lenses of magnetite are found in schists in the Adirondacks of eastern United States, and elsewhere. Magnetite grains are commonly found in the heavy residues obtained from sands. Crystals are cubic (Fig. 4.4*a*) but magnetite often occurs in massive form.

Chromite

The only commercial source of chrome. Occurs as a primary mineral in basic igneous rocks accumulating during the early stages of magma crystallization. Being extremely resistant to weathering it also accumulates as a secondary mineral in sediments derived from the erosion of chromite-rich rocks. Principal sources are in the Ural Mountains, Zimbabwe, the Bushveld (S. Africa), Kiruna (Sweden), Turkey, Cuba and India. Crystals are cubic but the mineral is normally massive.

Cassiterite

The most important source of tin. Found in acid igneous rocks; in quartz veins and pegmatites (*q.v.*), often in association with tourmaline, topaz, and fluorspar in granite areas, as in Cornwall, Saxony and Tasmania. Occurs also in alluvial deposits, as water-worn grains which can be recovered from many stream-sands ('placer' deposits) in granite areas. Deposits of this kind were worked extensively in Malaya for commercial supplies of tin, but are now becoming exhausted. Similar deposits have been located at Yunnan (China), Nigeria, Congo and Bolivia. Crystals are tetragonal and frequently twinned into 'knee-shaped' or 'L-shaped' forms.

Bauxite

An important source of aluminium being a mixture of three minerals, *gibbsite* $Al(OH_3)$, *diaspore* AlO(OH) and *boehmite* AlO(OH) which is the major constituent of bauxite. Generally produced by intense chemical weathering of Al-rich silicate minerals in tropical climates and may occur as residual deposits capping less weathered parent rock at depth, or as sedimentary deposits derived from the erosion of such weathered profiles. Important reserves occur in Guiana, Jamaica, the southern states of N. America, southern Europe especially France, Russia, India and Australia.

Limonite

The name limonite is used as a field term to denote hydrated oxides of iron which are poorly crystallized. They are brown in colour, yellowish brown to yellow when earthy, and form coatings on rock joints and weathered surfaces, and concretions. The principal constituent is *goethite* which results from the alteration of Fe-bearing minerals, and in residual deposits may be mixed with clay and other materials (*cf.* laterite, p. 38). Goethite produces the 'iron-hat', or gossan that mantles a weathered exposure of sulphide ores (see p. 89). It may form the iron-pan in bogs and tropical soils.

Carbonates

In this group, atoms of carbon are surrounded by three atoms of oxygen to form a planar triangle, the resulting structure being arranged in sheets. Calcite consists of these carbonate sheets interlayered with sheets of calcium which facilitate perfect cleavage. In dolomite the calcium sheets also contain magnesium. Many hydrous carbonates are the weathered products of other minerals.

Calcite

The principal source of carbonate of lime, invaluable to manufacturing and chemical industries. Limestones are essentially composed of calcium carbonate, of which crystalline calcite may form a large part. Calcite also occurs as a secondary mineral in many igneous rocks, e.g. in the amygdales of basalts. Open fractures in rocks are often filled by calcite veins, which are recognized by their colour and distinguished from quartz veins by the much lower hardness. Calcite is commonly associated with sulphide ores such as blende and galena in mineral veins (p. 109); it forms the material of stalactites and stalagmites in caverns (*q.v.*).

Crystals are Trigonal (Figs. 4.4b and 4.7), often as well-formed crystals. Generally colourless or white, but may have various tints. Cleavage is perfect, parallel to the rhombohedral faces, and twinning is common on rhombohedral planes. H = 3; the ease with which the mineral can be scratched with a knife affords a useful index to identification. Calcite dissolves in dilute acids, including acid groundwater.

In thin section: Crystals in rocks often interlocking, irregular in shape, colourless, the rhombohedral cleavage seen in some sections.

R.I. = 1.658 (max.), 1.486 (min.); these values are respectively above and below the R.I. of balsam. In consequence, the mineral shows changes in relief and strength of outline as it is rotated on the microscope stage (a *twinkling* effect).

Biref. = 0.172 (max.); polarization colours are whites or delicate pinks and greens. Lamellar twinning appears as bands of colour along the diagonals of rhomb-shaped sections.

Aragonite, orthorhombic, is another crystalline form of calcium carbonate; it is less common and less stable than calcite, and is often associated with gypsum, as in evaporites (p. 127).

Siderite (also called Chalybite)

Siderite occurs in clay-ironstone beds and nodules in the Coal Measures; formerly worked as a source of iron in British coalfields. The mineral is also found in rocks where Fe-carbonate has replaced the original calcium carbonate of a limestone (p. 126) and occurs in many of the marine sedimentary oolitic iron ores of Mesozoic age in Britain and continental Europe. In bog-iron ores siderite was precipitated direct from the water in lakes (p. 129). Found as brown rhombhedral crystals (trigonal), and also massive.

Malachite

A valuable ore of copper and an ornamental stone. Produced by weathering of copper-rich minerals, usually sulphides, by percolating ground-water. Famous deposits exist in the Urals, southern Zaire and northern Zambia. Crystals are monoclinic but malachite normally occurs in botryoidal form (Fig. 4.1*a*).

Tungstates and Phosphates

Molybdates, Arsenates, and Vanadates may also be considered with these classes. All generally occur in hydrothermal fluids that develop late in the crystallization of a granite magma and permeate the cooling igneous intrusion.

Wolframite

The principal ore of tungsten. Occurs in veins associated with the intrusion of granite and found world-wide, e.g. in Bolivia, Portugal and Britain (in St Austell, Cornwall, and Carrock in Cumbria). Frequently associated with cassiterite, and like that mineral, resists weathering: often found concentrated in alluvial deposits, as in Burma. Crystals are monoclinic and prismatic. Often occurs as massive or bladed forms.

Apatite

An important source of phosphorus. Apatite occurs as a common accessory mineral in many igneous rocks, the small crystals being usually visible only with a microscope. Large crystals are found in coarse-grained veins (pegmatites, *q.v.*), which yield commercial supplies from which phosphate is obtained for use as fertilizers, as in Canada, Norway, and the Kola peninsula.

Sulphates

Most sulphates form at low temperatures either during weathering, especially of mineral veins, or by evaporation in arid and semi-arid climates. Barytes is an exception to this general occurrence and forms in hydrothermal veins, often in association with galena, fluorite and quartz.

Barytes (also called 'Heavy-spar')

Barytes is of common occurrence in ore-veins of lead and zinc in association with galena, blende, fluorspar and quartz, and such veins are worked in the north of England, in Germany, the United States, and in many other countries. The mineral also occurs alone (or with calcite) in veins traversing limestones; and as a cement in certain sandstones (p. 119).

Crystals are orthorhombic and elongated parallel to the *b*-axis (Fig. 4.4d). Perfect cleavage is developed parallel to the prism faces and basal plane, so that crystals break into flat diamond-shaped cleavage fragments. Also occurs massive and in granular form.

Celestine

Principal source of strontium. Often occurs as an evaporite, as at Yate in Bristol, and associated with halite and gypsum; found in salt domes (p. 151). Also occurs with sulphur deposits as in Sicily. Crystals are orthorhombic, tabular (like those of barytes, Fig. 4.4d): often occurs in massive form.

Gypsum

Gypsum is formed chiefly by the evaporation of salt water in shallow inland seas, the calcium sulphate in solution being precipitated, as at the southern end of the Dead Sea; extensive deposits of Permian age, hundreds of metres thick, are worked at Stassfurt in Germany. Gypsum is also formed by the decomposition of pyrite (FeS_2) in the presence of calcium carbonate, e.g. crystals of selenite found in the London Clay are due to this reaction. Gypsum is much used in the building industry in the manufacture of plasters and plasterboard, and as a retarder of cement. *Selenite* is the transparent variety of gypsum. *Alabaster* is white or pink massive gypsum; and the form known as *satin-spar* is composed of silky fibres, occurring in veins.

Crystals are monoclinic flattened parallel to the side pinacoid and often twinned (Fig. 4.8b). Perfect cleavage parallel to side pinacoid.

Anhydrite

Frequently associated with gypsum in evaporites (p. 127), and also occurs in the cap-rock of salt domes (p. 151). Crystals are orthorhombic, tabular, white or greyish in colour; Anhydrite expands on hydration.

Mineral accumulations

The outer part of the lithosphere, the crust, is largely composed from only 13 of the 100 or so elements known, namely O, Si, Al, Fe, Ca, Na, K, Mg, Ti, P, H, C, and Mn. Many of the elements of value to industry exist in a diffuse state within the crust and must be concentrated by the natural processes of mineralization before they form a resource that can be worked economically. The product of such concentrating processes reflects the chemical affinities of the minerals involved and certain minerals typically occur together to form 'mineral associations' (e.g. gold and silver, or copper, lead, zinc, gold and silver, or nickel and copper). Such associations can be of value when exploring for minerals, as a mineral difficult to find may be located by tracing its more easily identified associates. The natural processes that concentrate minerals of economic value are here described after certain commonly used terms are defined.

Ore mineral
A desired metal generally bound with other elements, e.g. chalcopyrite, Cu Fe S_2 (80% metal).

Gangue mineral
A non-metallic mineral e.g. quartz, calcite mixed with ore minerals. Many gangue minerals are valuable resources and are mined, e.g. fluorite, apatite, arytes and gypsum.

Ore
All ore mineral and gangue that has to be mined.

Hypogene mineral
One formed with the ore and not from the later alteration of existing ore: may be considered as *primary*.

Supergene mineral
Formed by the alteration of existing minerals: may be considered as *secondary*.

Tenor
The metal content of an ore (also called the *grade*) expressed either as a percentage or a weight of ore (g per tonne). The higher the tenor, the richer the ore.

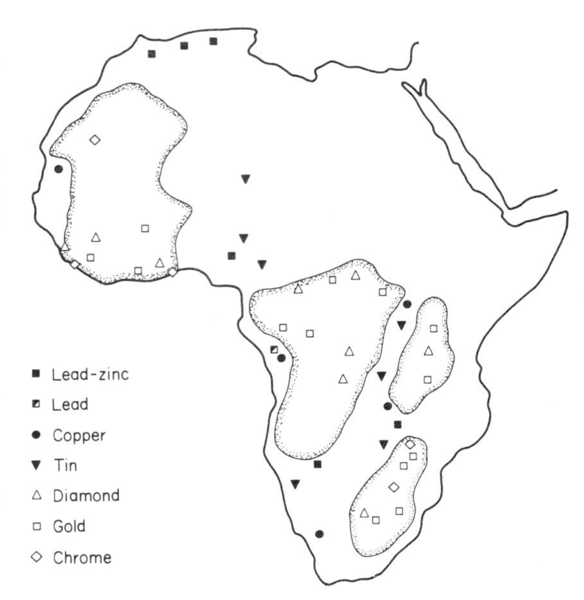

Fig. 4.41 Distribution of some major mineral deposits in Africa. Stippled areas have been stable for the past 1.5×10^9 years. Other areas have been deformed during the past 1.2×10^9 years. (Data from Clifford 1971.)

The concentration of minerals

Deposits of ore and other valuable non-metals are formed by the processes associated with igneous activity, sedimentation and metamorphism, and examples of them are described in Chapters 5, 6 and 7. Igneous rocks such as basalt, whose origin is closely related to mantle magmas at the base of the lithosphere, may bring to the near surface metals rare in the crust. Other igneous rocks, especially granites, form at depth in the lithosphere. They are produced by heat and pressure which create chemical gradients that may affect hundreds of cubic kilometres of country rock. Such gradients encourage the mobilization of elements and their reassembly in concentrated form as new minerals, some of which are valuable. Such migration from surrounding ground to particular centres creates *metallogenic provinces*, e.g. the Canadian Shield gold belt extending 3200 km from Quebec to the Great Slave Lake; the Chilean copper belt; the lead-zinc-tin province of Cornwall; the pyrite belt of Portugal, Spain and Central Europe (p. 23); the silver-lead-zinc belt of Broken Hill in Australia. Numerous provinces can be identified in Africa and are illustrated in Fig. 4.41. The location of many such provinces appears to be related to the position of plate boundaries, and the crustal deformation that has occurred or is occurring along them (see Chapters 1 and 2).

Many ores of igneous origin extend to considerable depth (p. 109 and p. 110). Ores so formed are enriched by chemical weathering and some of the richest deposits have developed in this way (Fig. 4.42).

Sedimentary processes concentrate valuable metals and non-metals by the mechanical sorting of minerals weathered from existing deposits, according to their specific gravity (such accumulations are called *placer deposits*); by the precipitation of minerals in solution during the evaporation of water (to produce *evaporites*) and through precipitation promoted by the activity of aerobic and anaerobic bacteria. The former live in the presence of oxygen and are thought to be responsible for many sedimentary deposits of iron; the latter live in environments devoid of oxygen and precipitate sulphides and sulphur. Sedimentary processes are also responsible for the burial of organic matter from which are derived coal, oil and petroleum gas. Many sedimentary deposits extend laterally for considerable distances unless geometrically condensed by later folding and faulting.

In all these processes the migration and concentration of valuable substances is intimately associated with the movement of water in the crust, as a hot hydrothermal fluid, or as a low temperature brine, or as a vapour. The position and shape of many valuable deposits has been governed by the former pathways available for fluid flow, especially faults, joints and other fractures, and the presence of impermeable rock that acted as a barrier to such movement (see Figs 5.27 and 5.28).

The search for minerals

Most near surface deposits reveal their presence by the colour of soil and nature of vegetation developed upon them (e.g. the iron-hat of sulphide veins, Fig. 4.42), and have been located by prospectors. Nowadays the deeper reserves must be found that are hidden from sight. To locate these large areas of country, or sea, are systemati-

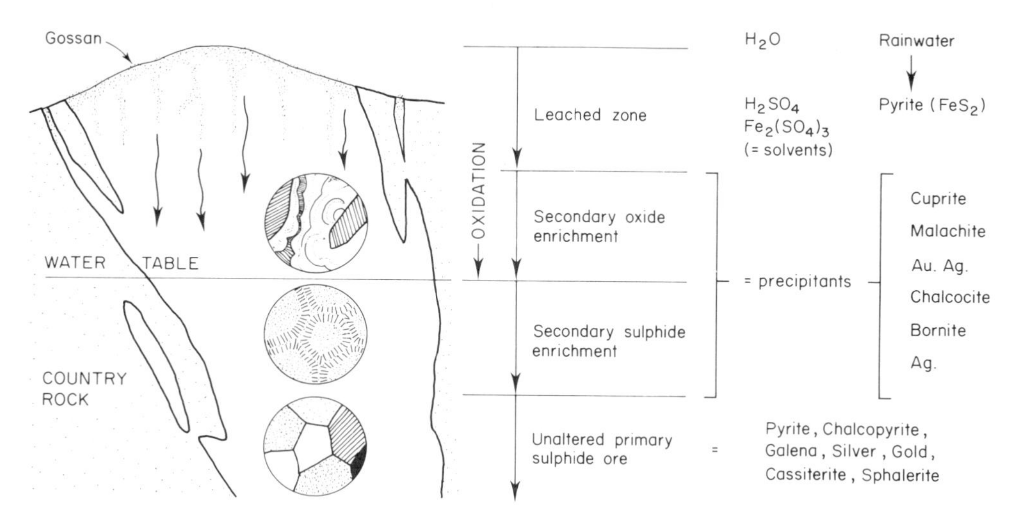

Fig. 4.42 Secondary enrichment of primary sulphide vein by weathering. Rainwater reacts with oxidized pyrite and creates ferric sulphate and sulphuric acid which leach the vein leaving insoluble iron hydroxides (limonite) as a gossan, or iron-hat. Botryoidal precipitates accumulate in the zone of oxide enrichment filling fractures and other voids. This zone is terminated by the water table. Below the water table primary sulphides are progressively replaced in order of solubility. Examples of precipitants and primary ore minerals are shown: many other minerals also exist in such veins. The insets illustrate diagrammatically the microscopic appearance of samples from the zones.

cally searched to identify areas which exhibit a noticeable change in some geophysical or geochemical character. For example, the magnetic field of the Earth follows a smooth path along the lines of longitude from pole to pole unless deflected from this path by the presence of other magnetic sources such as metallic ores. Hence anomalies in the magnetic field of a land area may indicate the presence of ore at depth. Much initial surveying for land resources can be conducted from the air, including magnetic surveying, and utilizes air photographs and remotely sensed images.

Areas where anomalies have been identified or where, for other reasons, ores are suspected to exist, are searched in detail using ground surveys. These have the following objectives.

(*i*) To map the geology of the area.
(*ii*) To improve the map with results from detailed geophysical surveys (see Table 10.3).
(*iii*) To sample the area, collecting specimens for mineralogical analyses. Panning stream deposits in the traditional manner of the old prospector is one method of sampling and remains an essential part of many surveys. Alluvium, colluvium and soil profiles on hillsides will reveal the presence of thin veins of ore that are hidden beneath a weathered profile and whose location may indicate richer reserves at greater depths (Fig. 4.43).
(*iv*) To collect samples for chemical analyses. This is geochemical prospecting and is an exploration technique of equal importance as geophysical prospecting. Many elements concentrated in ores at depth can migrate to the surface as gas, to be adsorbed onto mineral surfaces in the soil profile, absorbed into plants or lost to the atmosphere. The analyses of soil samples, of leaves and of gases in soil can reveal the presence of ore at depth.

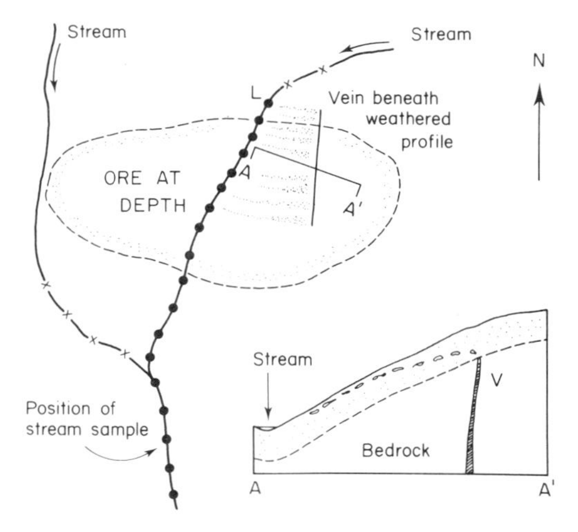

Fig. 4.43 Airborne regional geophysical surveys identify an anomaly in the area of two streams. Stream sampling reveals appropriate minerals (● = minerals present × = absent) and locates the northern limit (L) of a possible vein (V) coming from the ore body. Sampling on the A–A′ traverse locates the vein. Deep drilling follows.

Off-shore a different approach is required and surveys rely heavily upon seismic geophysical surveys and sampling with dredgers.

Eventually promising areas are drilled to confirm predictions. Once a deposit has been located it must be evaluated before being accepted as an economically workable resource.

Selected bibliography

Battey, M.H. (1972). *Mineralogy for Students*. Oliver and Boyd, Edinburgh.

Mackenzie, W.S. and Guilford, C. (1980). *Atlas of Rock Forming Minerals in Thin Section*. Longman, London.

Dixon, C.J. (1979). *Atlas of Economic Mineral Deposits*. Chapman Hall, London.

Bates, R.L. (1969). *Geology of the Industrial Rocks and Minerals*. Dover, New York.

Evans, A.M. (1980). *An Introduction to Ore Geology*. Blackwell Scientific Publishers, Oxford.

Peters, W.C. (1978). *Exploration and Mining Geology*. John Wiley and Sons, New York.

Reedman, J.H. (1979). *Techniques in Mineral Exploration*. Applied Science Publishers, London.

Parasnis, D.S. (1973). *Mining Geophysics*. Elsevier Publishing Company, Amsterdam.

Read, H.H. (1970). *Rutley's Elements of Mineralogy*, 26th edition. George Allen and Unwin, London.

5

Igneous Rocks

Geological processes due to the natural agents which operate at the Earth's surface have been discussed in Chapter 3. Other processes, however, originate below the surface and these include the action of volcanoes, or volcanicity. Molten rock material which is generated within or below the Earth's crust reaches the surface from time to time, and flows out from volcanic orifices as *lava*. Similar material may be injected into the rocks of the crust, giving rise to a variety of igneous *intrusions* which cool slowly and solidify; many which were formed during past geological ages are now exposed to view after the removal of their covering rocks by denudation. The solidified lavas and intrusions constitute the *igneous rocks*.

The molten material from which igneous rocks have solidified is called *magma*. Natural magmas are hot, viscous siliceous melts in which the chief elements present are silicon and oxygen, and the metals potassium, sodium, calcium, magnesium, aluminium, and iron (in the order of their chemical activity). Together with these main constituents are small amounts of many other elements, and gases such as CO_2, SO_2, and H_2O. Magmas are thus complex bodies and the rocks derived from them have a wide variety of composition. Cooled quickly, a magma solidifies as a rock-glass, without crystals; cooled slowly, rock-forming minerals crystallize out from it.

The content of silica (as SiO_2) in igneous rocks varies from over 80% to about 40% and results in some, e.g. granites, containing visible quartz (SiO_2) and others, e.g. gabbros, having no quartz. SiO_2 is a non-metal oxide and the basic component of silicate minerals. These were considered to be 'salts' of silicic acids and rocks containing much silica were originally called *acid*, and those with less silica and correspondingly more of the metallic oxides were called *basic*. The concept has long been abandoned but its nomenclature remains because the broad distinction it creates is a useful one. Basic magmas are less viscous than acid magmas; the temperatures at which they exist in the crust are incompletely known, but measurements at volcanoes indicate values in the neighbourhood of 1000°C for basic lavas, a figure which may be considerably lowered if fluxes are present. (A flux lowers the melting point of substances with which it is mixed; the gases in magma, for example, act as fluxes.)

Volcanoes and extrusive rocks

A volcano is essentially a conduit between the Earth's surface and a body of magma within the crust beneath it (see Figs 1.18 to 1.20). During an eruption lava is extruded from the volcanic vent and gases contained in the lava are separated from it; they may be discharged quietly if the lava is very fluid and the gas content small, but commonly they are discharged with explosive violence. In a submarine eruption the lavas flow out over the sea floor; a volcanic pile may be built up which can eventually rise above sea-level to form an island. The new island of Surtsey, off the south coast of Iceland, was formed in this way in 1963 (Figs 5.1 and 1.18).

Different styles of volcanic action are distinguished as (*i*) *fissure eruptions* and (*ii*) *central eruptions*.

Fig. 5.1 Surtsey Volcano, Iceland, in eruption 1963. The dark cloud near the cone is a dense ash emission. (Photograph by G. P. L. Walker.)

Fissure eruptions

Lava issues quietly from long cracks in the Earth's surface, with little gas emission. It is basic and mobile, with a low viscosity, and spreads rapidly over large areas. In past geological times vast floods of basalt have been poured out in different regions, and are attributed to eruptions from fissures. These *plateau-basalts* at the present day include the Deccan Traps (*trap* is an old field term for a fine-grained igneous rock, possibly derived from an even older Swedish term for a staircase, so describing the stepped escarpments eroded at the edges of the plateaux formed by the fine-grained basalts), which cover 10^6 km² in peninsular India; built up of flow upon flow of lava they reach a thickness in places exceeding 1800 m. The plateau-basalts of the Snake River area in North America cover 500 000 km² and resemble the Deccan Traps in composition. The basalts of Antrim and the Western Isles of Scotland, including the hexagonally-jointed rocks of the Giant's Causeway and of Staffa (Fig. 5.2), are the remnants of a much larger lava-field which included Iceland and Greenland. The basalt flows of the world (see Fig. 2.13) are estimated to cover in all 2.5 million km²; and the source of these widespread lava flows is the basaltic layer situated at the base of the lithosphere. Active fissures in which the rock is still hot can be observed in Iceland; the island, which lies across the mid-Atlantic rift (Fig. 1.13), is built of lava and has continued to grow as the rift has opened.

Fig. 5.2 Hexagonal columns of basalt at Giant's Causeway, Antrim, N. Ireland.

Central eruptions

A central eruption builds a volcano that has a cone with a summit crater connected to the volcanic 'pipe', through which are ejected lava, gases, and fragments of exploded lava (ash) and broken rock. Vesuvius, Etna, and Stromboli in the Mediterranean region, Popocatapetl, and Cotopaxi in the Andes, the Mounts St Helens, Rainier, Crater Lake, Shaster and Lassen Peak in the Cascades of N. America, and other active volcanoes of the present day belong to this type. When the dissected volcanic cones of central eruptions that were active in past geological ages are now exposed to view after long denudation, their structure can be studied in detail; Arthur's Seat at Edinburgh and Largo Law in Fife are examples of such 'fossil volcanoes'.

In the upper part of the magma chamber beneath the volcano (Fig. 5.3) gases accumulate and build up pressure in the pipe of the volcano. When eruption occurs the expanding gases burst the lava into countless small fragments of dust, ash, or pumice, which ultimately fall around the vent or are blown to a distance by wind. Larger fragments (lapilli) and still larger lumps of magma (bombs) may also be ejected, together with fragments and blocks of rock torn by the force of the eruption from the walls of the volcanic vent.

Among the central volcanoes are those which have steep-sided cones and erupt stiff lavas derived from *acid* magmas whose large gas content gives rise to violent explosions when suddenly released (described as *paroxysmal* eruptions). Incandescent lava-fragments, suspended in a cloud of hot gas, can flow rapidly downhill as *nuées ardentes*, or glowing clouds. An eruption of this type occurred at Mont Pelée on Martinique, in the West Indies, in 1902, and brought about the destruction of the town of St Pierre.

The eruption of the Mount St Helens volcano in the state of Washington, in May 1980, was an explosive discharge of ash and dust most of which was ejected horizontally to devastate a segment of surrounding country,

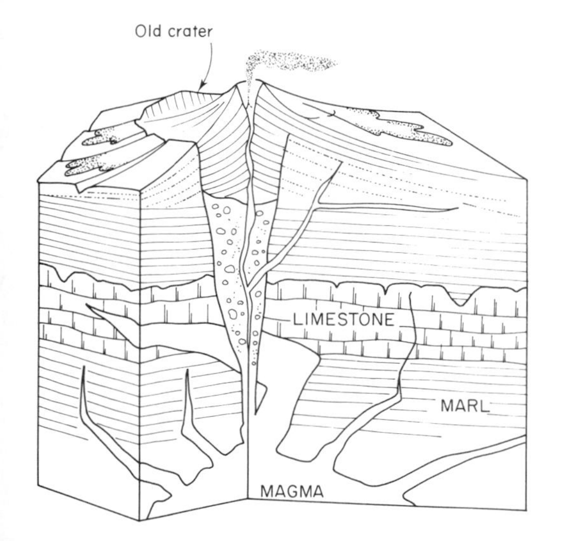

Fig. 5.3 An idealized cross section through a central volcano (based upon the geology at Vesuvius). A volcanic pipe connects with magma at a depth of about 5 km.

20 km by 30 km wide. Much volcanic dust, instead of rising into the upper atmosphere, was spread over a large area in the north-western United States, and affected agricultural conditions there.

The volcanic island of Krakatoa in the East Indies, part of the rim of a former crater, underwent four paroxysmal explosions in August 1883. After two centuries of quiescence the eruption began with the emission of steam, followed by the explosions which blew away most of the island, estimated at 18 km^3 of rock. Shocks were felt to great distances and the tidal waves generated did much damage to low coasts. The dust from Krakatoa floated around the Earth in the upper atmosphere for a year afterwards.

Not all central eruptions produce the high conical volcanoes associated with acid magmas: some are characterized by large flat cones called *shield volcanoes*, formed by successive flows of mobile basaltic lava derived from basic magmas. At the top of the pile and connected with the volcanic conduit is a large pit or *caldera*, in which the lava column, fed from below, is maintained under pressure and from time to time overflows into the caldera. Flows of lava are also emitted from fissures on the slopes of the cone.

Mauna Loa, a great shield volcano, is the largest of four in the Hawaiian Islands of the South Pacific (Fig. 5.4). The summit caldera is an elliptical pit about 5.6 km in length and nearly 3 km across. Jets of molten rock are expelled from within the pit at times, their spray being blown by the wind to fall as glassy threads around it.

A more accessible caldera, Kilauea, is situated on the eastern slope of the mountain. Within the caldera is a lava lake in which the level of the molten rock rises and falls. At intervals the lake becomes crusted over, and at other times fountains of lava are ejected by the discharge of gases. The temperature of the basaltic lava at the surface of the lake is about 1050°C but is 100°C lower 6 m below, because at the surface burning gases keep the lava at a higher temperature.

Other shield volcanoes are found in Iceland, where they were active in geologically recent (Pleistocene) times, and in Java.

Descriptions of past volcanic outbursts include Pliny's account of the violent eruption of Vesuvius in A.D. 79, which overwhelmed Pompeii and Herculaneum. Modern studies have been made from observatories sited on volcanoes, as at Vesuvius or Kilauea; aerial reconnaissance using photographic, radar, and geophysical equipment has made possible comprehensive studies that monitor volcanic activity on local and regional scales.

Waning phases

In areas of waning volcanic activity the emission of steam and hot gases (HCl, CO_2, H_2S and HF) at high temperture from *fumaroles* (gas-vents) may continue for a long time. Sulphur is deposited around gas-vents, called *solfataras*, as in the Vesuvian area, and commercial supplies of sulphur were formerly obtained from these deposits. *Geysers* (Icelandic = roarer) are eruptive springs of boiling water and steam. The Yellowstone Park region of Wyoming, U.S.A., is famous for its geysers and hot springs. One geyser, named Old Faithful, which erupts regularly at intervals of about an hour and shoots a

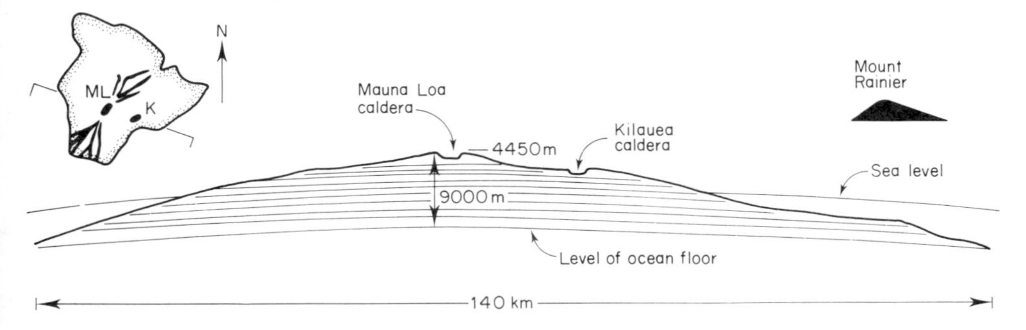

Fig. 5.4 A vertical section through the shield volcano of Mauna Loa, Hawaii, with silhouette of Mount Rainier drawn to the same scales. Sketch map of Hawaii shows recent lava flows and line of vertical section. (Data from Stearns, 1966.)

column of water to a height of 46 m, was investigated. Heat from the surrounding rock raises the temperature of the underground water. The upper layers of water begin to boil, and the boiling spreads downwards in the fissure network until enough steam pressure has formed to eject the water in the upper part of the geyser. The time interval between successive eruptions of Old Faithful decreased from 64 to about 60 minutes just before the Hogben Lake earthquake in 1959, and increased again to 67 minutes after the earthquake. It fell again before the onset of the more distant Alaska earthquake in March 1964. A possible link between the behaviour of the geyser, stress in the crust and earthquake activity was thus indicated.

Geysers and hot springs are common in Iceland, where the hot water is used for domestic heating and cooking, and for laundries. The temperature of *hot springs* is generally lower than that of geysers, and the former flow constantly whereas geysers are intermittent. Around the orifice of a spring or geyser a cone or small mound of *sinter* is deposited; sinter is a siliceous substance which may be a sublimate (i.e. deposited from gaseous emanations in a cooler environment), or thrown out of solution from the hot spring. It is usually white in colour but may have other tints. In New Zealand large sinter deposits are found, sometimes forming terraces in the Hot Springs district of North Island.

Pyroclastic rocks

The deposits formed by the consolidation of fragments ejected during an eruption are called *pyroclastic* (literally, 'fire broken'). The largest ejected masses of lava (bombs) fall in and around the vent and become embedded in dust and ash; a deposit of this kind is called an *agglomerate* (or volcanic agglomerate). The smaller particles of ash and dust may be blown by wind and spread over large areas in layers; they become hardened into rocks that are called *tuffs*.

The tuffs formed on land from a nuée ardente typically have their component fragments welded together by the heat involved in their formation; such welded tuffs are called *ignimbrites* (Latin *ignis*, fire; *nimbus*, cloud).

Poorly consolidated tuffs from the Naples area were used by the Romans for making 'hydraulic cement', and were called *pozzolana*; mixed with lime they harden under water. Similar material from the Eifel district of Germany has also been used; tuffs known as *trass*, when mixed with an equal amount of limestone, form a cement.

When an eruption takes place on the sea floor through the opening of a submarine vent, or on land adjacent to the sea, the ejected dust, ash, and other fragments form pyroclastic layers interbedded with normal aqueous sediments. Many examples of this are found in the Lower Palaeozoic rocks of Wales, the Welsh Borderland, and the Scottish Lowlands; the considerable thicknesses of bedded tuffs in those areas point to the extensive volcanic activity of that time (see Chapter 2, p. 20).

Extrusive rock associations

The products of volcanism can produce complex associations when they accumulate on land: some are shown in Fig. 5.5. The oldest rock shown is a lava flow (1) whose top was once the land surface on which developed a soil profile (stippled). This and the vegetation it supported (crossed) were buried by ash (2) from a later eruption that also generated mudflows of liquefied ash (3) before culminating in lava flows (4a and b). The upper surface of lava (4b) rapidly solidified to form a crust beneath which liquid lava continued to flow, forming *lava tunnels* (5) which are now found partially filled. Deep river valleys (6) eroded the land surface and became the routes followed by minor lava flows at a later date. The initial flows buried the river deposits in the base of the valleys, the later flows being restricted to lava channels, like rivers (6a). Ash falls (7 and 8) blanket the area.

The land surface is typically free from running water and often dry as rain soaks readily into the loose deposits except during periods of intense rainfall when gulleys (9) are eroded. The pyroclastic rocks are normally porous and the more massive lava flows are fractured by shrinkage cracks on cooling. Most of the rocks freely transmit ground-water which is often some distance below ground level. In regions of active volcanism springs of hot water may issue from fractures, accompanied by sulphurous gases. Strong lavas may be separated by layers of weak ash; horizons which appear continuous can be abruptly terminated by other rocks; zones of weathering can be found buried beneath unweathered rock; sediments can

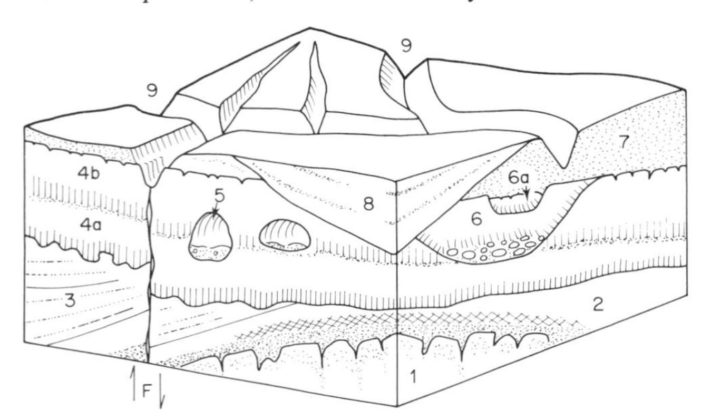

Fig. 5.5 Example of extrusive rock associations: for explanation, see text. F = fault.

be interlayered with lava flows, and changes in the thickness of members that constitute such extrusive associations must be expected.

Intrusive rocks and rock forms

A body of magma in the Earth's crust may rise to higher levels and penetrate the rocks above it without reaching the surface; it is then said to be *intrusive*. During the process of intrusion it may incorporate within itself some of the country-rocks with which it comes into contact, a process known as *assimilation* (p. 98), and it is thereby modified. Some magmas may give off hot fluids that penetrate and change the rocks in their immediate neighbourhood. A large mass of magma, many cubic km in volume, is a *major intrusion* and cools slowly because of its size. Large crystals are able to form and the rocks so formed are coarsely crystalline. When the magma rises and fills fractures or other openings in the country-rocks it forms *minor intrusions*, i.e. smaller igneous bodies. These include *dykes*, which are wall-like masses, steep or vertical, with approximately parallel sides (Fig. 5.6); and *sills*, sheets of rock whose extent is more or less horizontal, and which lie parallel to the bedding of sedimentary rocks into which they are intruded. Dyke and sill rocks commonly have a fine to medium-grained texture (see p. 98 for textures). *Veins* are smaller injections of igneous material and are often thin and irregular, filling cracks which have been opened up in the country-rocks around an intrusion.

Fig. 5.6 Basic dyke cutting folded Damara rocks, Brandberg West area, Namibia. (Courtesy of Dr A.O. Fuller.)

Minor intrusions

Dykes

Dykes, (Fig. 5.7) defined above, vary in width from a few centimetres to many metres, but most are not more than 3 m wide. They commonly outcrop in nearly straight lines over short distances (Fig. 12.17), but exceptionally may run for many km across country. A remarkable example, several km in width (formerly known as the Great Rhodesian Dyke), extends 450 km through Zimbabwe into South Africa. In the north of England the Armathwaite–Cleveland dyke of dolerite can be traced for 209 km. If the dyke-rock is harder than the country-rocks into which it is intruded, it will stand above the general ground level and appear as a linear feature (Fig. 5.7). The term 'country-rock' is used for rocks (of any kind) which are invaded by an igneous mass. A fine-grained *chilled margin* is often formed by the rapid cooling of the igneous body at its contact with the country-rock.

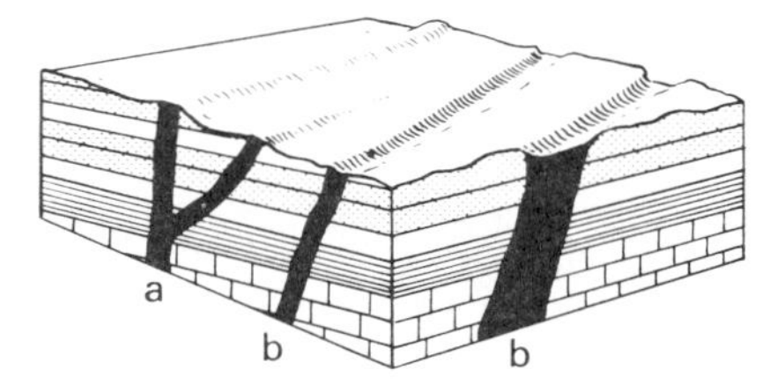

Fig. 5.7 Dykes (**a**) more resistant to weathering, (**b**) less resistant than the country-rock.

A *dyke swarm* is a group of parallel or radiating dykes (Fig. 5.8). In the island of Mull the well-known *parallel swarm* intersects the south coast of the island over a distance of 20 km. Some individual dykes of the swarm extend across to the Scottish mainland, where they can be traced for many km farther. The stretching of the crust that was needed to open the dyke fissures points to the operation of tensile stresses across the area. A *radial swarm* occurs on the island of Rhum, off the west coast of Scotland, where 700 basic dykes are grouped about a centre in the south of the island.

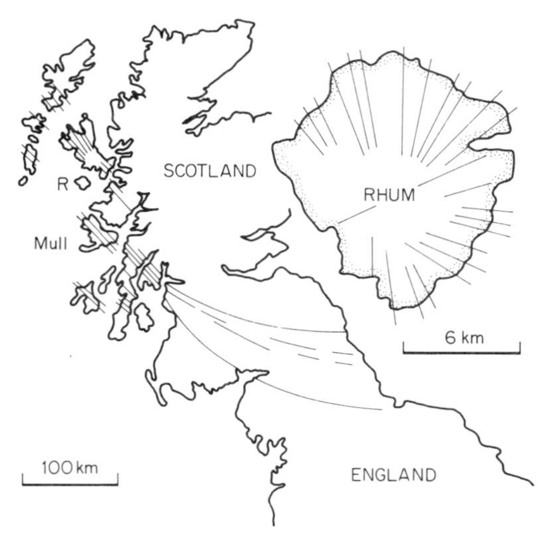

Fig. 5.8 Tertiary dyke swarms of the British Isles, with inset showing the radial dykes of Rhum (R). (Data from *Geol. Survey. G.B.* 1961.)

Sills

Sills, (Fig. 5.9), in contrast to dykes, have been intruded under a flat cover or 'roof' against a vertical pressure due to the weight of the cover. A columnar structure is often developed in such an igneous sheet by the formation of sets of joints which lie at right angles to its roof and floor; the joints in general result from the contraction of the igneous body. This is seen, for example, in the well-exposed dolerite sill at Salisbury Craigs, Edinburgh, and at other localities. Many lava-flows also show a joint pattern (Fig. 5.10). The sediments above and below a sill are baked by the heat of the intrusion, as also occurs either side of a dyke, and jointing develops by the cooling and lateral contraction of the sill-rock; similar jointing develops in dykes.

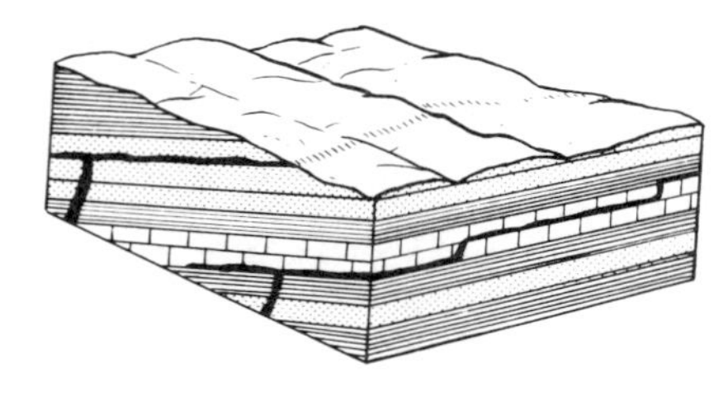

Fig. 5.9 Sills fed by dykes.

A sill is sometimes stepped up from one level to another, the two parallel parts being connected by a short length of dyke (Fig. 5.9). If the country-rocks into which the sill is intruded are later tilted or folded the igneous body will be tilted or folded with them. Sills may be of considerable area seen in plan. The Whin Sill of the north of England is a sheet of basic rock (quartz-dolerite) with an extent of some 3900 km^2 and an average thickness of about 30 m. It is exposed along the sides of many valleys and in escarpments, where denudation has cut down through it, and in coastal cliffs. It yields a high quality road-stone, locally called 'whinstone'.

Sills and dykes may be *composite*, with contrasted margins and centre due to successive injections of different material (e.g. basic and acid); if there are successive injections of *similar* material the term *multiple* sill or dyke is used.

The Palisades Sill, New Jersey, is an olivine-dolerite (or diabase) with a thickness of 303 m and fine-grained chilled margins at the upper and lower contacts. The mineral composition varies from top to bottom of the sill as a result of the setting of early-formed heavy crystals, olivine and pyroxene, during the slow cooling of the mass. Most of the olivine is concentrated near the base of the sill at a level known as the 'olivine ledge'. Crystal settling is referred to again on p. 109.

Ring structures

Ring dykes are intrusive masses filling curved fractures, sometimes appearing as a complete circle or loop in plan. They are formed when a detached plug of country-rock, which occupies a cylindrical fracture in rocks above a body of magma, sinks, and the magma then rises and fills the annular space around the plug (Fig. 5.11). This process is called *cauldron subsidence*; it was first described from Glen Coe, Scotland, where andesite lavas within a ring fault are surrounded by the Cruachan granite. Such igneous bodies are also referred to as *permitted intrusions*, which are 'allowed' to fill spaces formed, e.g. by the subsidence of a plug (in the case of ring fractures), or filling open joints or other spaces in the rocks. *Cone-sheets* are fractures having a conical shape, with the apex of the cone pointing downwards, and filled by magma (Fig. 5.12). Both cone-sheets and ring dykes are seen in the Ardnamurchan peninsula in west Scotland, where they are composed of quartz-dolerite.

Laccolith and phacolith

A laccolith is a small intrusion having a flat floor and domed roof (Fig. 5.13a), the roof having been arched by the pressure of incoming magma. Laccoliths in the Henry Mountains, Utah, were intruded into mainly horizontal strata and are now exposed after denudation. Others are found in Iceland, in Skye (gabbro laccoliths), and elsewhere.

A phacolith is a somewhat similar body but has both a

Fig. 5.10 Dolerite sill with columnar jointing, intrusive into jointed sandstone. Erosion has removed the rock that overlay the sill.

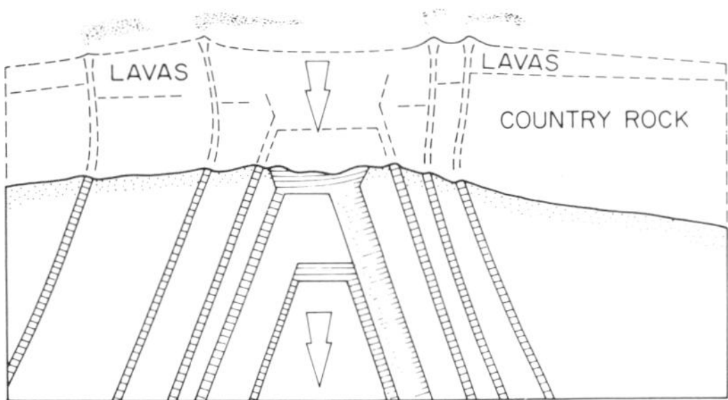

Fig. 5.11 The lower levels of a volcanic centre, now exposed by erosion, containing annular ring dykes produced by cauldron subsidence. Repeated downward movements of a central plug were associated with injections of magma which moved up the annular fractures to occupy the space created by each period of subsidence.

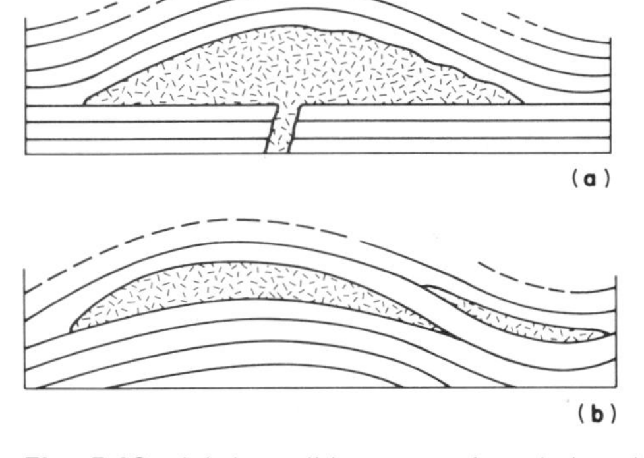

Fig. 5.13 (**a**) Laccolith, magma intruded under pressure produces arching of roof. (**b**) Phacolith, emplaced in an existing antiformal structure.

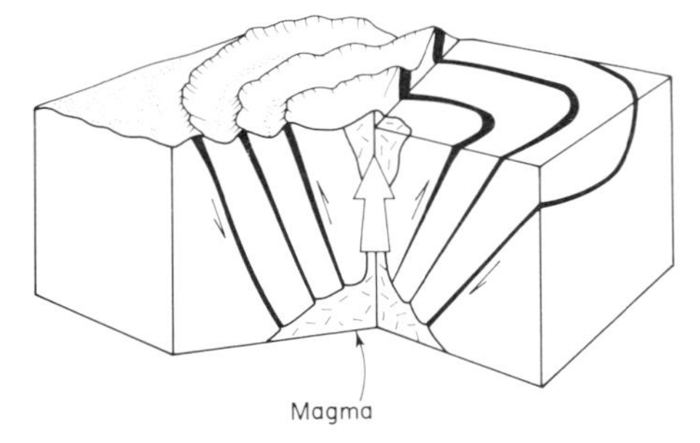

Fig. 5.12 An igneous source exposed by erosion and revealing cone-sheets, where rising magma has filled conical fractures opened by vertical pressure acting over a small area.

curved floor and roof (Fig. 5.13b), magma having been intruded into rocks that were already folded. The Corndon phacolith, S. Shropshire, is 2 km in length and occupies the crestal region of a dome in folded mudstones.

Major intrusions

Plutons, stocks, and batholiths

The term *pluton* is used to denote a moderately large body of magma which is intruded essentially at one time and is contained within a single boundary. Plutons have various shapes but are commonly nearly circular in cross-section; an average area for many granitic plutons is about 150 km^2 but many are larger. Those emplaced early during an episode of intrusive activity may be larger than others that come later; some reach about 1000 km^2 in outcrop area and are composite.

The granites of Devon and Cornwall (Fig. 2.18) may be considered as a group of six steep-sided plutons, possibly (though not necessarily) connected at a low level. They are now seen at the surface where their sedimentary roof rocks have been removed by denudation.

The older term *stock* was introduced by R.A. Daly (1912) to denote a vertical nearly cylindrical body of igneous rock, cutting across the rocks into which it is intruded, with a cross-sectional area up to 100 km^2.

A *batholith* was formerly defined as a large bottomless igneous mass (the word means 'depth-rock') rising as an irregular projection into sedimentary and other rocks of the crust. Recent research, however, has shown that most batholiths consist of a cluster of plutons, located by some structural control during their intrusion. Examples in Britain are the Donegal batholith, Ireland, which consists of 8 separate plutons; and the Leinster granite batholith with 5 separate plutons which form elongated domes aligned along faults.

The Leinster granite batholith, situated south-west of Dublin, occupies 1000 km^2 at the surface and is the largest granite body in the British Isles. It is elongated in a NNE–SSW direction (Fig. 2.18). Larger batholiths elsewhere are also found to consist of multiple units (plutons) which have risen into the country-rocks during intrusion without necessarily reaching the surface.

A buried granite underlies the northern part of the Pennines between Alston and Stanhope, Co. Durham, where mineral veins are found injected into the sedimentary rocks of the area. After its presence was confirmed by a geophysical survey, the buried (Weardale) granite was reached by a boring at Rookhope in 1961, at a depth of 369 m.

In the Grampian Highlands of Scotland large granite and granodiorite masses were emplaced during the Caledonian orogeny (p. 20). Many of them are bounded by steep walls, and during the process of intrusion appear to have pushed aside the rocks which they invade. They are *forceful intrusions*, as distinct from the permitted intrusions referred to earlier (p. 96), and represent magma formed at depth and intruded at a higher level. They include the Moor of Rannoch, Etive, Cairngorm, Lochnagar, and Hill of Fare granites, and the Aberdeenshire granites; the shapes of these large igneous bodies are not completely known.

Stoping

A contributory process by which magma rises into country-rocks during a process of intrusion is known as stoping, or magmatic stoping. The term 'stope' has long been used in mining for the opening up of a vertical or inclined shaft by the piecemeal removal of the rock. Stoping as applied to magma was discussed by Billings (1925). 'Fingers' of magma penetrate under pressure into cracks and joints in the country-rocks, and blocks of rock are wedged off by them. If the density of the blocks is greater than that of the magma they will sink in the viscous fluid, which is then displaced and rises. Shales and many metamorphic rocks, for example, are heavier than granitic magma and would sink in this way. When the stoped rocks (especially the smaller ones) remain for some time in such an environment they are softened and may be streaked out by viscous flow in the magna; some are partly or wholly assimilated, i.e. dispersed in the magma, thereby changing its composition locally. Others remain as enclosures or *xenoliths* (Fig. 5.14), often with their mineral constituents re-crystallized. Such enclosures are often seen in the marginal parts of granites and other igneous bodies.

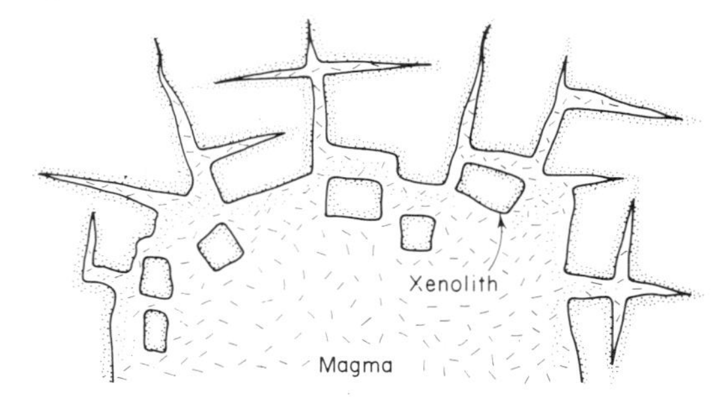

Fig. 5.14 Schematic vertical cross-section of a plutonic intrusion at depth, stoping country-rock. Xenolith = slab of country-rock enclosed in the mass near its walls and roof. Small inclusions of country-rock have been softened and dispersed in the igneous rock as it cooled slowly, locally modifying its composition near its contacts. Thus a granite may grade into quartz-diorite and diorite at its margin.

Basic sheets

Certain large intrusions have the form of *sheets* which are much thicker in proportion to their extent than sills,

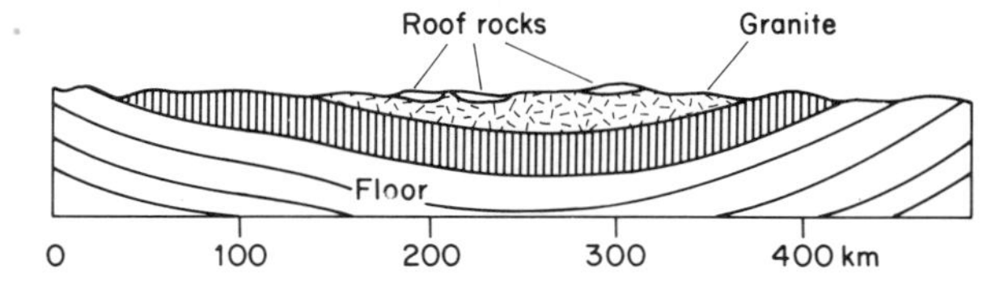

Fig. 5.15 Lopolith: a generalized section across the great Bushveld Lopolith of S. Africa; note that length of section is about 400 km.

Table 5.1 A grouping for igneous rocks

Mode of formation		Rock types	Rock textures
EXTRUSIVE	(*Volcanic*)	Lavas	Glassy or fine-grained
INTRUSIVE	Minor intrusions	Dykes and sills, laccoliths	Fine to moderately coarse texture
	Major intrusions (*Plutonic*)	Batholiths, stocks, bosses, and sheets	Coarsely crystalline rocks

and are often basic in composition (gabbro or dolerite). They occupy many square kilometres of ground, over much of which the igneous rock may not be exposed at the surface because covered by sedimentary rocks. The Duluth gabbro of Minnesota is one such basic sheet, with an extent of some 6000 km² and a thickness of about 6 km. Another example is the complex layered intrusion of the Bushveld, S. Africa, which has a flat basin-shaped form described as a *lopolith* (Fig. 5.15).

We can now construct the following grouping for igneous rocks (Table 5.1), based on their mode of occurrence as described above, and leading to the classification on p. 100.

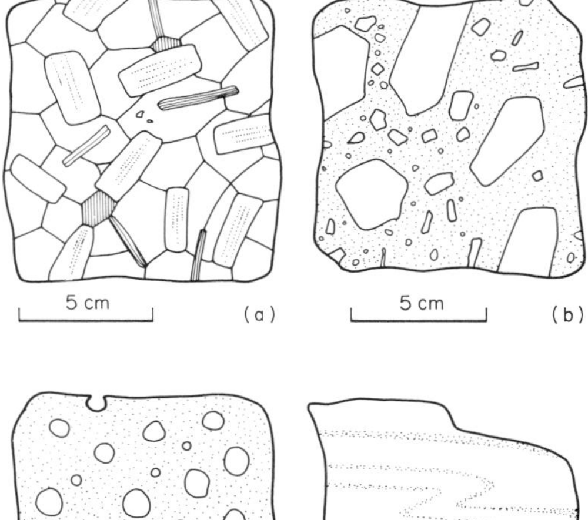

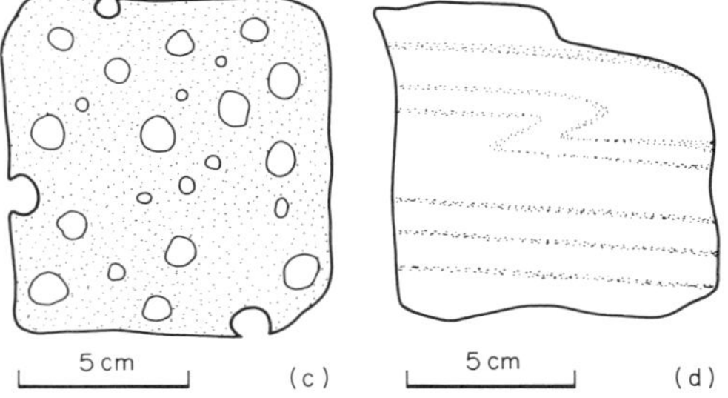

Fig. 5.16 (**a**) Equigranular texture. (**b**) Porphyritic texture. (**c**) Vesicular texture. (**d**) Flow-structure.

Texture and composition

Texture

The texture, or relative size and arrangement of the component minerals, of an igneous rock corresponds broadly to the rock's mode of occurrence. Many descriptive terms are used by geologists and a selection of the more common is presented in Table 5.2.

Plutonic rocks, which have cooled slowly under a cover perhaps several kilometres thick, are coarsely crystalline or *phaneritic*; their component crystals are large (2 to

Table 5.2 Commonly used terms for describing igneous rock texture: all may be used as field terms and require no greater visual aid than a ×10 hand lens.

1. *Degree of crystallinity* (commonly reflects speed of cooling)
   Holocrystalline = entirely composed of crystals (cooled slowly)
   Hypocrystalline = composed of crystals and glass
   Hyalocrystalline = no crystals, i.e. glassy (cooled quickly)
2. *Visible crystallinity* (commonly reflects speed of cooling)
   Phaneritic = individual crystals can be distinguished (cooled slowly)
   Aphanitic = granularity from the presence of crystals can be seen, but individual crystals cannot be distinguished
   Glassy = entirely glass-like, or some crystals set in glass (cooled quickly)
3. *Crystal size* (commonly reflects speed of cooling)
   Coarse = >2 mm
   Medium = 2–0.06 mm   Phaneritic
   Fine = <0.06 mm
   (Microcrystalline = must be observed using a microscope)
   (Cryptocrystalline = cannot be seen with a microscope but can be sensed from birefringence of groundmass)
4. *Relative crystal size* (commonly reflects abundance of ions and uniformity of cooling history)
   Equigranular = all crystals are approximately of equal size
   Inequigranular = some crystals are clearly larger than others
   Porphyritic = large crystals surrounded by much smaller crystals
5. *Crystal shape* (commonly reflects sequence of crystallization)
   Euhedral = well defined regular shape (crystallized early from melt) (= idiomorphic)
   Anhedral = poorly defined and irregular outlines (crystallized late from melt)

5 mm *or more*) and can easily be distinguished by the naked eye (Fig. 5.16*a* and *b*). Rocks of medium grain often have crystals *between* about 1 and 2 mm, and in fine-grained rocks crystals may be considerably less than 1 mm across. When the texture is so fine that individual crystals cannot be distinguished without the aid of a microscope it is called *aphanitic*, or *microcrystalline*. For extremely fine-grained rocks, when their crystalline character is only revealed by viewing a rock slice through crossed polars, which enables the birefringent colours of each embryo crystal to be displayed, the term *cryptocrystalline* is used. These textures are all even-grained, or

equigranular, i.e. having crystals of much the same size (Fig. 5.16*a*); but some rocks show the *porphyritic* texture (Fig. 5.16*b*), in which a number of larger crystals are set in a uniformly finer base (or groundmass). The large, conspicuous crystals are called porphyritic crystals or *megacrysts*. Porphyritic feldspars in some granites, for instance, may be 5 to 10 cm long.

Extrusive rocks (lavas) which have cooled rapidly at the Earth's surface are often entirely *glassy* or *vitreous* (without crystals), or partly glassy and partly crystalline. Within a single lava flow the outer part, i.e. its upper and lower boundaries, may be glassy, because rapidly chilled, and the inner part crystalline. Expanding gases in a magma during its extrusion give rise to cavities or *vesicles*, resulting in a *vesicular texture* (Fig. 5.16*c*), the vesicles may subsequently be filled with secondary minerals, when they are called *amygdales* (a name derived from 'almond-shaped' cavities with secondary infilling). *Flow-structure* in lavas (Fig. 5.16*d*) is a banding produced by differential movement between layers of the viscous material as it flowed (see rhyolite = 'flow-stone'). Another kind of banding is formed by crystal-settling (p. 109).

Composition

The mineral composition and colour of rocks are related to their chemical composition. When the chemical analyses of an acid rock, e.g. granite, and a basic rock, basalt, are compared, marked differences are seen: such as the greater proportion of silica (SiO_2) and of alkalies (Na_2O, K_2O) in the acid rock, and the higher content of lime, magnesia, and iron oxide in the basic. Table 5.3 records the averages of a large number of analyses and illustrates this broad distinction. The higher alkali content in granite corresponds to a greater proportion of feldspar in the rock; conversely, basalt has more dark minerals containing Fe, Mg, and Ca (mafic minerals, below).

Table 5.3 The average composition of granite and basalt

	Average granite	Average basalt
SiO_2	70.2%	49.1%
Al_2O_3	14.4	15.7
Fe_2O_3	1.6	5.4
FeO	1.8	6.4
MgO	0.9	6.2
CaO	2.0	9.0
Na_2O	3.5	3.1
K_2O	4.1	1.5
H_2O	0.8	1.6
rest	0.7	2.0
Total =	100	100

During the cooling of a magma its different constituents unite to form crystals of silicate minerals, such as pyroxenes, amphiboles and feldspars, and oxides such as magnetite. In a basic magma, minerals such as olivine

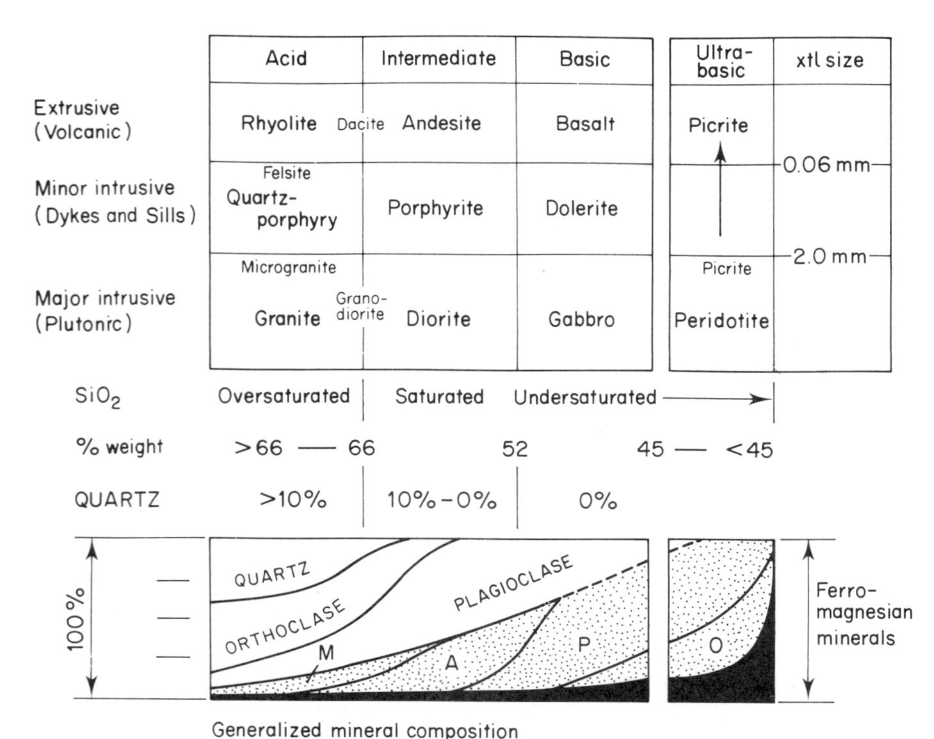

Fig. 5.17 A classification of igneous rocks (excluding alkaline rocks). xtl size = usual size of crystals. M = micas. A = amphiboles (e.g. Hornblende). P = pyroxenes (e.g. Augite). O = olivine. Solid black = opaque minerals.

and magnetite are often the first to crystallize, using up some of the silica, magnesium, and iron; the remaining Mg and Fe, together with CaO and Al_2O_3, is used up later in augite (pyroxene), hornblende (amphibole), and dark mica. Such minerals, on account of their composition, are called *ferromagnesian* or *mafic* (a word coined from *ma* for magnesium and *fe* for iron). In contrast to these dark and relatively heavy minerals the alkalies and calcium, together with Al_2O_3 and SiO_2, form light-coloured or *felsic* minerals, which include feldspars, feldspathoids, and quartz. Most of the calcium in a basic magna would go to form a plagioclase, a little contributing to augite. In acid rocks felsic minerals predominate and give the rock a paler colour, in contrast to the darker basic rocks. Between *acid* and *basic* types there are rocks of *intermediate* composition.

Classification

A convenient scheme of classification can be constructed for these more common varieties of igneous rocks; it does not include all igneous types, but some of the less common rocks are briefly mentioned in the descriptions that follow. In Fig. 5.17 the common rocks are arranged in three columns headed Acid, Intermediate, and Basic, containing the Granite, Diorite, and Gabbro groups respectively. A transitional type, Granodiorite, between Granite and Diorite, is also indicated. Rocks of the Syenite family (alkaline) fall outside this grouping and are treated separately on p. 107 for the reasons stated there. The range from left to right in the figure corresponds to a decreasing silica content. In each column there are three divisions, the lowest giving the name of the coarse-grained plutonic member, the middle division the dyke or sill equivalent of the plutonic type, and the top division the extrusive or volcanic rocks. The main minerals which make up the plutonic rocks are shown in the diagram below the figure. The columns therefore give a grouping based on mineral composition (which also expresses the chemical composition), and the horizontal divisions of the figure are based on texture, which usually reflects mode of occurrence, as this largely governs the rock texture. In the lower diagram, intercepts on any vertical line give approximately the average mineral composition for the corresponding plutonic rock in the figure. Granite, andesite, gabbro and basalt account for more than 90% by volume, of all the igneous rocks in the upper levels of the crust. Of these, granite is the most common intrusive rock and basalt the dominant extrusive.

Ultrabasic rocks

Picrite and peridotite

These ultrabasic rocks consist mainly of mafic minerals and contain little or no feldspar. They are coarse-grained, mostly dark in colour, and have a high specific gravity (3.0 to 3.3). Their content of SiO_2 is around 40 to 42 per cent, lower than average gabbro (p. 100).

Ultrabasic rocks have relatively small outcrops at the Earth's surface and often form the lower parts of basic intrusions: the heavy crystals of which they are composed have sunk through a body of magma before it fully crystallized, and have accumulated to form an olivine-rich layer.

Picrite

Contains a little feldspar, up to about 10–12%; the bulk of the rock is made of olivine and augite or hornblende, and olivine crystals may be enclosed in the other mafic mineral. By increase in the feldspar content and corresponding decrease in other constituents, picrite grades into olivine-gabbro and gabbro. The Lugar Sill in Ayrshire contains a band of picrite 7.6 m thick, which merges downwards into peridotite and upwards into a feldspathoidal gabbro.

Peridotite and Serpentinite

Olivine is the chief constituent of peridotite (from French *peridot*, olivine); other minerals include pyroxene, hornblende, biotite, and iron oxides. Felsic minerals are absent. A variety composed almost entirely of olivine is called *dunite*, from the Dun Mountain, New Zealand; pale green in colour, it has been used as a decorative stone on a small scale.

Serpentinite (or serpentine-rock) results from the alteration of peridotite by the action of steam and other magmatic fluids while the rock is still hot. Red and green coloured serpentinite bodies occur, e.g. in the Lizard district, Cornwall, and in the Cabrach area near Aberdeen, in Scotland.

The fibrous variety of serpentine, *crysotile*, furnishes one source of commercial asbestos when it is found in veins of suitable size; Canada, Zimbabwe, and U.S.S.R. are main suppliers of the mineral. Diamond is another economic product associated with peridotites of volcanic origin. The famous mines of Kimberley, S. Africa, are located in old volcanic pipes which are filled with a diamond-bearing breccia known as 'blue ground'. The breccia has the composition of serpentine-rock (*q.v.*); it is easily weathered and is crushed for the extraction of the diamonds that are sparsely distributed through it.

Other bodies of ultrabasic rock consist almost entirely of one kind of dark mineral: thus, *pyroxenite* (all pyroxene), and *hornblende-rock* (all hornblende). They are of small volume and are usually associated with basic plutonic rocks: the Bushveld intrusion in S. Africa and the Stillwater complex in Montana, U.S.A., contain such rocks.

Basic rocks

Gabbro, dolerite, and basalt

These basic rocks, some of which are economically important as construction stone, road-stone and aggregate, have a large content of ferromagnesian minerals which give the rocks a dark appearance. Specific gravity ranges from about 2.9 to 3.2. Because basic lavas are relatively fluid, basalt is frequently (though finely) crystalline; it may grade into dolerite with increase of crystal size, and similarly dolerite may grade into gabbro as the texture grows coarser.

Gabbro

Minerals

Essential minerals are a plagioclase (generally labradorite) and a monoclinic pyroxene (augite or diallage). The plagioclase composition reflects the high CaO and low Na_2O content in gabbro (see analysis, p. 100). Other minerals which may be present in different gabbros are hypersthene, olivine, hornblende, biotite, and sometimes nepheline. Ilmenite, magnetite, and apatite are common accessories. Quartz in very small amount may be present in some gabbros (see below). Mafic minerals make up over 50% of the rock, Fig. 5.18a.

Fig. 5.18 Polished surface of gabbro of ring-dyke, Ardnamurchan, Scotland (×1). Plagioclase, light; ferro-magnesian minerals, dark.

Texture

Coarsely crystalline, rarely porphyritic, sometimes with finer modifications. Hand specimens appear mottled dark grey to greenish-black in colour because of the large mafic content. Under the microscope the texture appears as interlocking crystals (Fig. 5.19). Olivine if present is often in well-formed crystals because of its early separation from the magma, and commonly encloses black grains of iron oxide. Serpentine after olivine, and chlorite after pyroxene are common alteration products.

Varieties

If a mafic mineral in addition to augite is present, its name is added to the rock name, thus: hypersthene-gabbro, olivine-gabbro, hornblende-gabbro; nepheline-gabbro, =*essexite*.

Norite is a variety containing essentially hypersthene in-

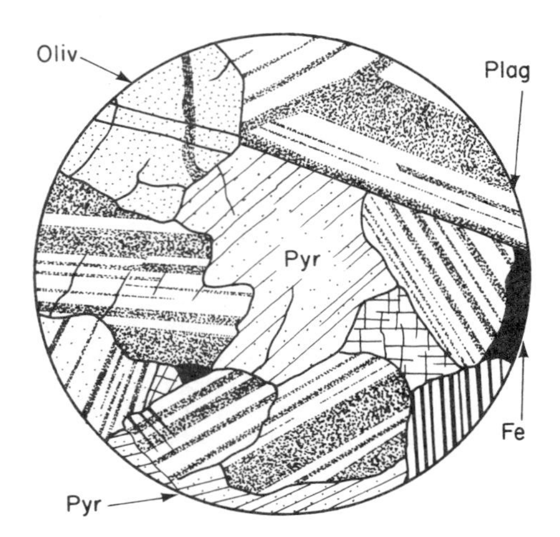

Fig. 5.19 Gabbro seen in thin section; crossed polars (×12). Oliv=olivine; Pyr=pyroxene; Fe=iron-ore; Plag=plagioclase.

stead of augite, i.e. a hypersthene-labradorite rock, and is of common occurrence.

Troctolite has olivine and plagioclase (no augite).

Quartz-gabbro contains a little interstitial quartz, derived from the last liquid to crystallize from a magma with slightly higher silica content than normal; e.g. the Carrock Fell rock, Cumberland.

In Britain gabbros are found in Skye (the Cuillins), Ardnamurchan (ring dykes), Northern Ireland (Slieve Gullion), the Lake District, and the Lizard. The large basic sheets of Aberdeenshire as at Insch, Haddo House, and Huntly contain much norite and hypersthene-gabbro. In other countries much larger gabbro intrusions include those of Skaergard, Sudbury, and the Bushveld, which are now briefly described.

The *Skaergaard intrusion* is shaped like an inverted cone or funnel, with an area of 50 km² at the surface, and is exposed on the east coast of Greenland to a vertical depth of some 2500 m; the funnel-shaped mass may continue below this for a similar distance. The lowest part of the exposed gabbro possesses a nearly horizontal layering, consisting of dark bands rich in olivine (formed by the gravity-settling of olivine crystals in the magma) separated by broader bands of olivine-gabbro and hypersthene gabbro. A small scale layering which is repeated many times is called *rhythmic banding*. The layered rocks pass upwards into gabbros without olivine, succeeded by iron-rich gabbros and finally by quartz-gabbros. The plagioclase in the gabbros ranges from labradorite in the lower rocks to andesine in the upper. At the margins of the mass fine-grained gabbros are present.

The *Bushveld complex* of the Transvaal, S. Africa, is a vast lopolith, one of the largest known igneous bodies, with a width at the surface of about 390 km (Fig. 5.15). The lowest rocks which outcrop along its margins are norites, with bands of ultrabasic rock, some containing important *chromite* deposits. Above them lie hypersthene–gabbros which reach a thickness of some 5 km, and which show layering by mineral composition. They in turn are overlain by a zone of red rocks, alkaline diorites and syenites. Valuable concentrations of *platinum* and *magnetite* are found at certain levels in the gabbros.

The large nickel-bearing intrusion of Sudbury, Ontario, is also of lopolith form (Fig. 5.15) with an overall length and width of some 57 and 24 km respectively. A thick norite sheet occupies the lowest part of the elongated basin, and nickel sulphide ores are concentrated near its base and in mineralized dykes. Above the norite lie granophyric rocks (p. 106) which may be a separate intrusion; they are covered by sedimentary rocks in the centre of the lopolith. The combined thickness of the norite and granophyre is about 2.4 km. It has been suggested that the depression now occupied by the Sudbury norite is a crater formed by the impact of a large meteorite.

Dolerite

The chemistry of this intrusive rock corresponds to gabbro but its texture is finer. Dolerite forms dykes, sills, and other intrusions. The rock is dark grey in colour, except where its content of feldspar is greater than average. Dolerite is important as a road-stone for surfacing because of its toughness, and its capacity for holding a coating of bitumen and giving a good 'bind'. In its unweathered state dolerite is one of the strongest of the building stones and used for vaults and strong-rooms, as in the Bank of England.

Minerals

As for gabbro, but the plagioclase is usually lath-shaped.

Texture

Medium to fine-grained; some dolerites have a coarser texture, when the lath-like shape of the feldspar is less emphatic and the rock tends towards a gabbro. When the plagioclase 'laths' are partly or completely enclosed in augite the texture is called *ophitic* (Fig. 5.20); this interlocking of the chief components gives a very strong, tough rock.

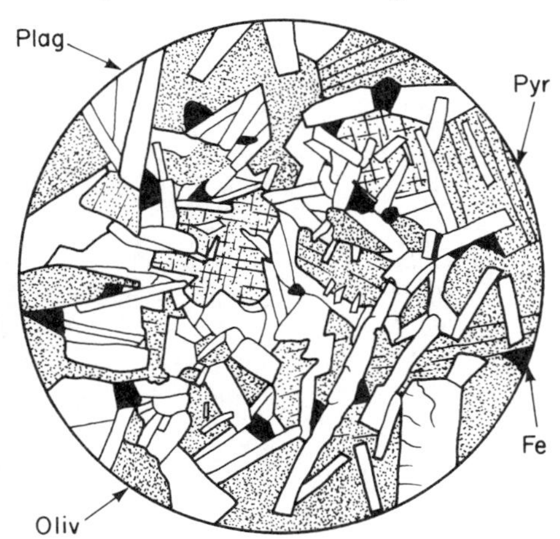

Fig. 5.20 Ophitic dolerite seen in thin section with upper polar out (×12). Plag = plagioclase; Pyr = pyroxene; Oliv = olivine; Fe = iron ore.

Varieties

Normal dolerite = labradorite + augite + iron oxides; if olivine is present the term *olivine–dolerite* is used. Much altered dolerites, in which both the feldspars and mafic minerals show alteration products are called *diabase*, though in America the term is often used synonymously with the British usage of dolerite.

Basalt

Basalt is a dark, dense-looking rock, often with small porphyritic crystals, and weathering to a brown colour on exposed surfaces. It is the commonest of all lavas, the basalt flows of the world being estimated to have five times the volume of all other extrusive rocks together. Basalt also forms small intrusions.

Minerals

Essentially plagioclase (labradorite) and augite; but some basalts have a more calcic plagioclase. Olivine occurs in many basalts and may show alteration to serpentine. Magnetite and ilmenite are common accessories; if vesicles are present they may be filled with calcite, chlorite, chalcedony, and other secondary minerals. Nepheline, leucite, and analcite are found in basalts with a low content of silica.

Texture

Fine-grained or partly glassy; hand specimens appear even-textured on broken surfaces, unless the rock is porphyritic or vesicular (Fig. 5.16); small porphyritic crystals of olivine or augite may need some magnification for identification. Under the microscope the texture is microcrystalline to cryptocrystalline or partly glassy. At the chilled margins of small intrusions a selvedge of black basalt glass, or *tachylite*, is formed by the rapid cooling.

Varieties

Basalt and olivine–basalt are abundant (Fig. 5.21*a*); varieties containing feldspathoids include nepheline–basalt and leucite–basalt (e.g. the lavas from Vesuvius). Soda-rich basalts in which the plagioclase is mainly albite are called *spilites*, and often show 'pillow-structure' in the mass, resembling a pile of sacks; they are erupted on the sea floor, Fig. 5.21b. Their rapid cooling in the sea prevents the minerals crystallized from achieving chemical equilibrium; they are *reactive* and alter readily. Between the pillows are baked marine sediments, often containing chert and jasper (SiO_2). These features of pillow lavas make them a most unsuitable form of basalt for concrete aggregate.

Some of the great flows of basalt in different parts of the world have been referred to earlier; their virtually

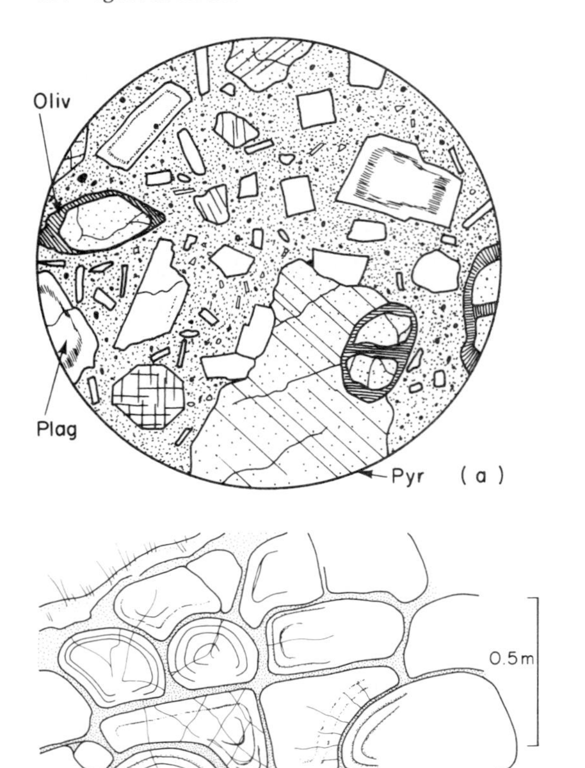

Fig. 5.21 (**a**) Olivine-basalt seen in thin section with upper polar out (×12). Oliv=olivine; Pyr=pyroxene; Plag=plagioclase. (**b**) Pillow lavas exposed in excavation. Note their concentric internal structure and how they sag into the space between underlying pillows.

constant composition suggests a common source, the basaltic layer of the Earth's crust.

Intermediate rocks

Diorite and andesite

The intermediate rocks shown in Fig. 5.17 typically contain little or no quartz. Diorite is related to granite, and by increase of silica content and the incoming of orthoclase grades into the acid rocks, thus: *diorite* → *quartz-diorite* → *granodiorite* → *granite*. The average specific gravity of diorite is 2.87.

Diorite

Minerals

Plagioclase (andesine) and hornblende; a small amount of biotite or pyroxene, and a little quartz may be present, and occasional orthoclase. Accessories include Fe-oxides, apatite and sphene. The dark minerals make up from 15% to 40% of the rock, and hand specimens are less dark than gabbro.

Texture

Coarse to medium-grained, rarely porphyritic. In hand specimens minerals can usually be distinguished with the aid of a lens. Under the microscope minerals show interlocking outlines, the mafic minerals tending to be *idiomorphic* (=exhibit a regular shape).

Varieties

Quartz-diorite (the amount of quartz is much less than in granite) is perhaps more common than diorite as defined above. Fine-grained varieties are called *microdiorite*.

Diorites are found in relatively small masses and frequently form local modifications to granodiorite and granite intrusions; e.g. at the margins of the Newer Granites of Scotland. Small diorite masses occur at Comrie and Garabal Hill, Perthshire.

Andesite

Andesites are fine-grained volcanic rocks, are common as lava flows in orogenic regions and occasionally form small intrusions. They are compact, sometimes vesicular, often brown in colour, and in total extent are second only to basalts. Most andesite flows are found on continental areas, e.g. in the Andes of South America (whence the name is taken), where many volcanoes have emitted ash and lava of andesitic composition; also in parts of the Cordilleras of North America. Certain ores are associated with andesites (p. 110).

Minerals and texture

Essentially plagioclase (mainly adesine) and a mafic mineral (hornblende, biotite, augite, enstatite); the small porphyritic crystals are set in a groundmass that may be glassy, cryptocrystalline, or microlithic (Fig. 5.22). The

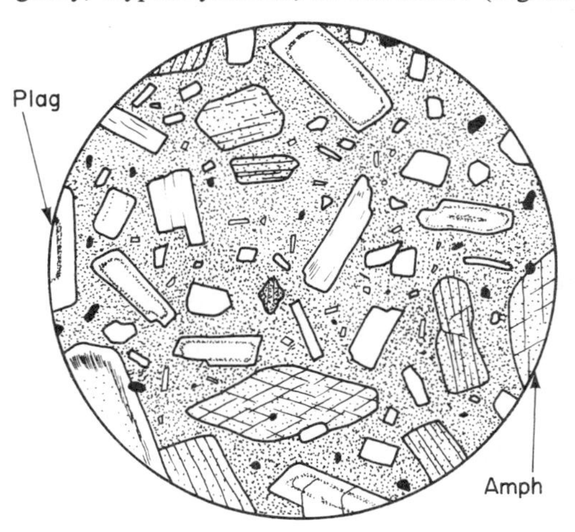

Fig. 5.22 Hornblende-andesite seen in thin section with upper polar out (×12). Plag=plagioclase; Amph=amphibole.

microlithic texture contains many small elongated feldspars, or *microliths*, often having a sub-parallel arrangement, *cf.* the trachytic textures, p. 108). Opaque iron oxides (e.g. magnetite) are common accessories. Hornblende may show dark borders due to reaction with magma.

Varieties Hornblende-andesite, augite-andesite, enstatite-andesite, biotite-andesite, and quartz-andesite (= *dacite*). The pyroxene-bearing varieties are abundant and may grade into basalts. Andesites that have been altered by hot mineralizing waters of volcanic origin (see Fig. 5.28a), with the production of secondary minerals, are called *propylites*.

In Britain, andesite lavas of Old Red Sandstone age are found in the Pentland Hills, Edinburgh; Glencoe and Ben Nevis; the Lorne volcanic plateau, Argyll; the Cheviot Hills; and as small intrusions in the Shelve area, Shropshire.

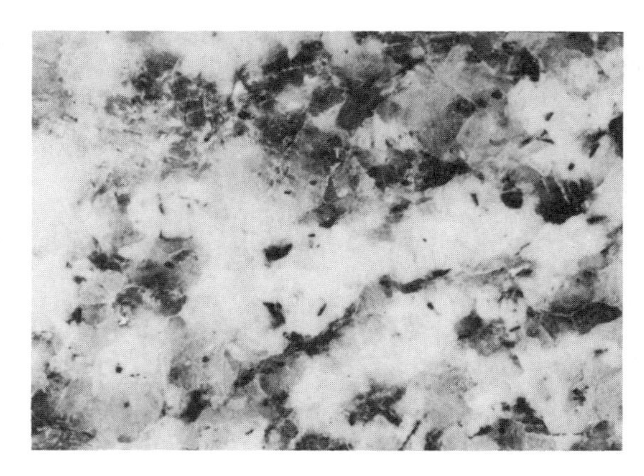

Fig. 5.23 Polished surface of biotite-granite, Merrivale, Dartmoor (×1). Feldspar, white; quartz, grey; biotite, black.

Acid rocks

Granite and granodiorite

The acid plutonic rocks, granite and granodiorite, make the greater part of the large batholiths, found in the cores of mountain fold-belts (p. 17); they also form smaller masses in the upper levels of the Earth's crust, and they are the most abundant of all plutonic rocks. Granite is an important structural stone because of its good appearance, hardness and resistance to weathering (except when crystals of mica are large and weather leaving voids which pit the finished surface), and its strength in compression; the crushing strength of sound granite ranges from about 135×10^6 to 240×10^6 Nm^{-2}. The strength of the rock and its rough fracture are also valuable properties when it is used as concrete aggregate. The average specific gravities are: granite, 2.67, granodiorite, 2.72.

It should be noted that in commerce the term 'granite' is used for some rocks that are not granite in the geological sense.

Granite

Minerals

Quartz, alkali feldspar, a smaller amount of plagioclase, and mica are essential constituents; in some rocks microcline is present. The feldspar may form up to 50% of the rock; the mica is biotite or muscovite, or both. Other minerals found in some granites include hornblende and tourmaline; alkaline types of granite may have Na-rich minerals such as aegirite and riebeckite. Accessory minerals are apatite, magnetite, sphene, zircon, and occasionally garnet (Fig. 5.23).

The average composition of granite is shown in Table 5.3.

Texture

Coarse-grained; individual minerals - cleaved feldspar, glassy quartz, and flaky mica - can be distinguished without a lens. Porphyritic granites contain some much larger feldspar crystals than those in the groundmass; fragments of country-rock that have been enclosed in the granite during intrusion may be present as xenoliths (Fig. 5.14).

Under the microscope the crystals show interlocking outlines (Fig. 5.24). Less coarse varieties include *microgranite*, a fine-grained rock frequently found as chilled margins to a larger mass or as a vein rock.

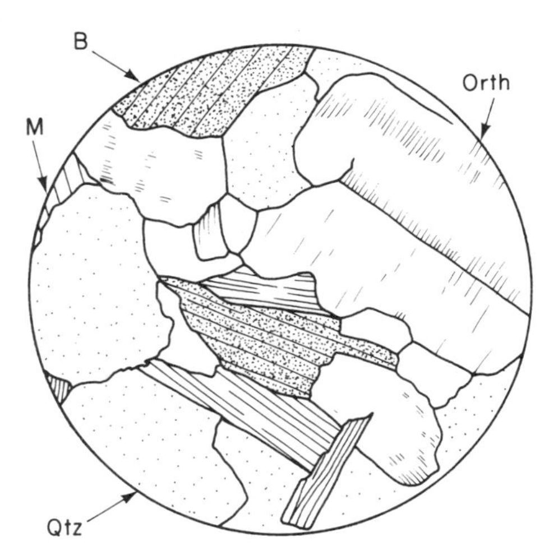

Fig. 5.24 Muscovite-biotite granite seen in thin section, upper polar out (×12). M = muscovite; B = biotite; Orth = orthoclase; Qtz = quartz.

Graphic granite contains intergrowths of quartz and feldspar due to simultaneous crystallization of the two minerals, with angular pockets of quartz in orthoclase or microcline, resembling cuneiform writing. In thin section with crossed polars groups of the quartz areas show simultaneous extinction, indicating that they belong to a single quartz structure interpenetrating the feldspar. When fine-grained the texture is called *micrographic*, and is seen in the rock *granophyre*. The proportion of alkali feldspar to quartz in graphic granite is practically constant, about 70:30, and results from the crystallization of a eutectic mixture of the two minerals. Similar textures on a small scale, however, may sometimes result from the *replacement* of parts of the feldspar by quartz.

Varieties

Varieties of granite are named according to the chief mineral present other than quartz and feldspar, thus: biotite-granite, muscovite-granite, muscovite-biotite-granite, hornblende-granite, or tourmaline-granite.

The worldwide extent of granitic rocks has been mentioned earlier; the larger granite masses of the British Isles are shown in Fig. 2.18.

Granodiorite

Granitic rocks in which the plagioclase content is greater than that of the potash-feldspar are called *granodiorites*; their dark minerals (biotite, hornblende) are usually rather more plentiful than in granite, and the amount of quartz is less. As shown in Fig. 5.17 (p. 100), granodiorite is transitional between granite and diorite; its texture is normally coarse-grained. The name *tonalite* (after the Tonali Pass, Italy) is used for similar rocks in which the plagioclase content is more than two-thirds of the total feldspar; and *adamellite* (after Mt Adamello) for those in which the potash-feldspar and plagioclase are nearly equal in amount.

In Scotland, the 'granites' of Caledonian age are mainly granodiorites with tonalite or adamellite portions; the three rock types often form successive intrusions in a single mass, which in some cases may also have a margin of quartz-diorite. They include the Moor of Rannoch intrusion, a hornblende-granodiorite, south-east of Ben Nevis; the great Cairngorm intrusion (mainly adamellite), and the Lochnagar and Hill of Fare intrusions of Aberdeenshire, as well as many smaller bodies.

The granodiorite from the Mountsorrel quarries, Leicestershire, showed the following mineral composition in thin sections:

quartz, 22.6%, orthoclase 19.7%, plagioclase 46.8%, biotite 5.6%, hornblende and magnetite, 5.1%.

The rock has a medium to coarse texture and is quarried on a large scale for road-stone.

Much larger and more complex igneous masses of broadly granodiorite composition include the Coast Range batholith of British Columbia, with an outcrop length of some 2000 km and an average width about 200 km. Composite bodies of such great extent have considerable variation in composition from one part to another.

Migmatites

There are many instances, as in the Scottish Highlands and elsewhere, where granitic material is seen to have become intimately mingled with the country-rocks, as if it had soaked into them, and the mixed rocks are called *migmatites* (Greek *migma*, a mixture). Zones of migmatite may be formed in areas where the country-rocks are metamorphic and have been invaded by granite; the migmatites pass gradually into the metamorphic rocks and into the (paler) granitic rocks. Structures in the metamorphic rocks may be traceable through the zone of mixing as inherited, or 'ghost', features. Migmatites are developed on a regional scale in fold-belts, as exposed in the Precambrian of southern Sweden and Finland, and in central Sutherland in Scotland.

Quartz-porphyry and acid vein rocks

Quartz-porphyry

The dyke equivalent of granite; contains porphyritic quartz and orthoclase in a microcrystalline matrix of feldspar and quartz (Fig. 5.25); small crystals of mica are also present. Dykes and sills are commonly found in granite areas. A similar rock but without porphyritic crystals is called *felsite*.

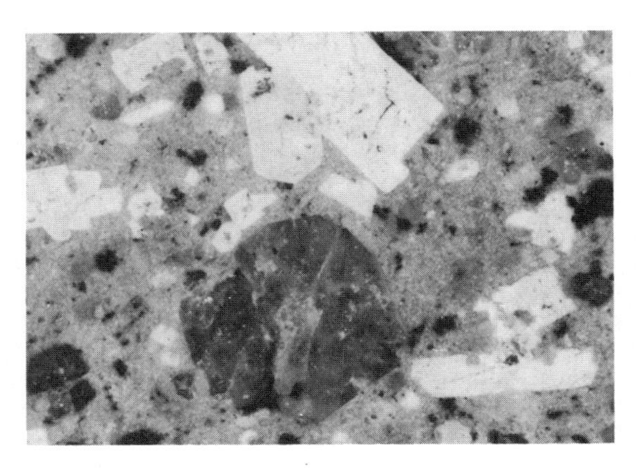

Fig. 5.25 Polished surface of quartz-porphyry, Cornwall (×3). Feldspar, white; quartz, grey; biotite, black.

Pegmatites

Pegmatites are very coarse-grained vein rocks that represent the last part of a granitic magma to solidify. The residual magmatic fluids are rich in volatile constituents, which contain the rarer elements in the magma. Thus in addition to the common minerals quartz, alkali feldspar and micas, large crystals of less common minerals such as beryl, topaz, and tourmaline are found in pegmatities. Also residual fluids carrying other rare elements, e.g. lithium, cerium, tungsten, give minerals in the pegmatites that can be worked for their extraction, such as the lithium pyroxene *spodumene*, the cerium phosphate *monazite* and wolfram. The mica used in industry - mainly muscovite and phlogopite (*q.v.*), is obtained from pegmatites;

individual crystals may be many centimetres across, yielding large mica plates. Canada, India, and the United States produce mica from such sources.

Pegmatites are found in the outer parts of intrusive granites and also penetrating the country-rocks.

Aplites

By contrast to pegmatites, are fine-grained rocks of even texture, found as small dykes and veins in and around granites. They are composed mainly of quartz and feldspar, with few or no dark minerals. Their fine texture points to derivation from more viscous fluids than for pegmatites; but they are commonly associated, and aplites and pegmatites may occur within the same vein (Fig. 5.26). Aplites also contain fewer rare elements than pegmatites.

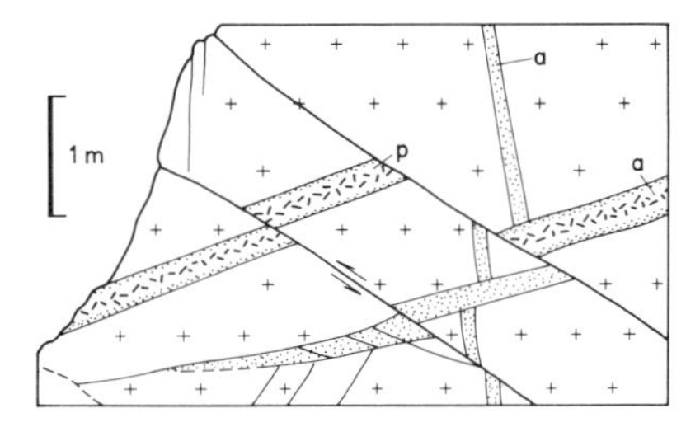

Fig. 5.26 Sketch of a vertical rock face in granite traversed by a 20 cm vein of pegmatite (p) with aplite margins (a). Other aplites (a) are stippled (from Blyth, 1954).

Acid lavas

These include rhyolite, obsidian, and dacite; they have a restricted occurrence and their bulk is very small compared with basic lavas.

Rhyolite (Greek *rheo*, flow) characteristically shows flow-structure, i.e. a banding formed by viscous flow in the lava during extrusion (Fig. 5.16d). The rock may be glassy or cryptocrystalline, and may contain a little porphyritic quartz and orthoclase. Some rhyolites show *spherulitic* structures, which are small spheres of radiating quartz and feldspar fibrous crystals formed by devitrification of the glass, and often situated along flow-lines. In the course of time an originally glassy rock may become entirely cryptocrystalline, and is said to be *devitrified.*

Obsidian is a black glassy rock which breaks with a conchoidal fracture and is almost entirely devoid of crystals. Obsidian Cliff in the Yellowstone Park, U.S.A., is a classic locality.

Pitchstone, another glassy lava, has a pitch-like lustre and general greenish colour; otherwise resembling rhyolite, pitchstone usually contains a few per cent of water in its composition. Small curved contraction cracks, formed around centres during the cooling of the glass, are known as *perlitic structure*.

Pumice is a very vesicular 'lava froth', with a sponge-like texture due to escaping gases, making the rock so light as to float on water. It may have the composition of rhyolite or may be basic in character (black pumice). Pumice is used as a light-weight aggregate for concrete.

Alkaline rocks

Syenite and trachyte

These alkaline rocks, of which syenite (named after Syene, Egypt) is the plutonic type, are placed separately here because they do not form part of the diorite/granodiorite/granite series already described. Syenite is somewhat like granite but contains little of no quartz; it is called an *alkaline* rock because it contains alkali-feldspars, rich in Na and K. Rocks of this group are not abundant by comparison with the world's granites; where they are locally well developed, however, they can be quarried and used for construction; e.g. the syenites and related basic rocks near Montreal, Canada, are worked on a large scale for road metal and aggregate. The *laurvikites* of southern Norway are attractive decorative stones that have been used for facing slabs in buildings. In the Kola (Murmansk) peninsula, U.S.S.R., large bodies of metalliferous ores are related to syenitic rocks; and lenses of apatite-nepheline pegmatite yield supplies of apatite (*q.v.*) for the manufacture of phosphates.

Syenite

Minerals

Orthoclase or other alkali feldspar usually forms over half the rock, with smaller amount of plagioclase (oligoclase); the dark minerals may be biotite, hornblende, or a pyroxene; and apatite, sphene, zircon and opaque iron oxides are accessories. A little quartz may be present, filling interstices between the other minerals. The texture is coarse-grained, sometimes porphyritic. Hand specimens are usually pale coloured.

Varieties

Named after the chief dark mineral, e.g. biotite-syenite, hornblende-syenite, augite-syenite, aegitite-syenite. There are few syenite bodies in Britain - one occurs at Loch Borolan, in Assynt, Sutherland; large masses are found in Canada and norway, and in the Kola peninsula, U.S.S.R., with associated ore bodies.

A soda-rich syenite from Laurvik, Norway, is known as *laurvikite* and consists mainly of the feldspar anorthoclase, together with titanium-rich augite, mica, and iron oxides. The anorthoclase shows blue and green 'schiller' effects which enhance the appearance of the polished stone when used for decorative purposes.

Trachyte

This is typically a pale coloured rough-looking lava (Greek *trachys*, rough) having porphyritic crystals of orthoclase in a groundmass composed mainly of feldspar microliths, with a small amount of biotite or hornblende.

The microliths of the groundmass in trachyte often have a sub-parallel arrangement, the *trachytic texture*, due to viscous flow in the lava, and flow-lines are deflected around the porphyritic crystals.

Varieties

Phonolite, a feldspathoidal trachyte which contains nepheline as well as feldspar, forms the hill of Traprain Law, Haddington. The Wolf Rock, off the coast of Cornwall, is a nosean-phonolite; the name (from *phonē*, a sound) refers to the ringing sound made when the rock is struck with a hammer. The Eifel district of Germany has many leucire-bearing trachytes and tuffs; and the well-known trachyte domes of the Auvergne in Central France are extrusions of stiff, very viscous lavas that solidified around the vents as dome-like masses.

Origin of igneous rocks

This subject has been a matter for discussion for many years, as research has continued to provide new data, and it is only briefly outlined here. The igneous rocks can be held to be derived from two kinds of *magma*, one granitic (acid) and the other basaltic (basic), which originate at different levels below the Earth's surface. The primary basic magma comes from the mantle, at considerable depth whereas bodies of granitic magma are generated in the crust, in the Earth's *orogenic belts*, where the crust becomes hot enough to liquefy. Two different groups of rocks are thus generated: granite and its relatives (diorite, porphyrite, andesite, quartz-porphyry, and some rhyolites) from the granitic magmas; and basalt lavas, dolerite, gabbro, and ultrabasics (such as peridotite and picrite), from the basaltic magma. This grouping corresponds to the way in which igneous rocks are distributed. Granite and its relatives are not found in the oceans but are restricted to the continents in zones of deep folding; basic rocks predominate in the oceans, the oceanic plates being largely composed of them and basic magma is known to issue from deep (oceanic) rifts that communicate with the asthenosphere. Basic rocks also occur in stable continental areas where rifting is thought to have permitted similar lines of communication with the base of the crust (e.g. as with plateau lavas, p. 24). Former belief in the presence of a thickness of liquid basic magma beneath the lithosphere has been disproved by geophysics as shear waves penetrate the upper mantle (Fig. 1.6). It is more correct to visualize these deep layers as being crystalline but containing 1% to 3% liquid as intercrystalline films. This liquid may accumulate and be added to by preferential melting of crystals when pressure and temperature permit. Often this occurs in conjunction with deformation of the plates of the lithosphere.

Modification of an intrusive mass may also come about by the *assimilation* of foreign (xenolithic) material at its contacts; the marginal type thus formed grades locally into the main rock (e.g. see diorite, p. 104).

Much discussion has centred on the origin of *granite*, the most abundant of all plutonic igneous rocks (granodiorite is included here with granite). Various suggestions have been put forward, of which the following are believed to be important.

(*i*) The melting of large amounts of crustal material at depth in high temperature conditions. This process, called *palingenesis*, was proposed by Sederholm (1907) to account for many of the granite and granodiorite masses of Fennoscandia, and was subsequently developed by other investigators.

(*ii*) The *permeation* or soaking of country-rocks by igneous fluids, especially those of alkaline-silicate composition, resulting in the formation of rocks of granitic appearance. Crustal material is 'made over' *in situ* into granite on a large scale, thus obviating the presence of great quantities of intrusive magma. The term 'granitization' is also used. The character of the 'granite' thus produced depends on the composition of the rocks that have undergone permeation; shales and sandstones, for instance, are more readily transformed than some other sediments.

The emplacement of bodies of granite by different mechanisms, at different levels in the Earth's crust, has been discussed by H.H. Read in *The Granite Controversy* (1957), and his concept of a *granite series* is outlined in Table 5.4.

The rise of successive *pulses* of magma from depth form a composite batholith. Recent studies by W.S. Pitcher have shown that the immense Coastal Batholith of Peru,

Table 5.4 The formation and ascent of igneous rock

General features of magma generation and alteration[1]		The Granite Series of H.H. Read.[2]
Intrusion of magma into the upper crust.	Last in time ↑	*High level granite.* Mechanically emplaced as permitted intrusions, e.g. cauldron subsidence (Fig 5.11)
Hot magma ascends because of its buoyancy and is contaminated by the crust through which it passes.	Later in time ↑	*Low level granite.* Forceful intrusions of magma and alteration of surrounding rocks by metamorphism
Generation of magma from parent material.	Early in time	*Deep-seated granite.* Formed *in-situ* by conversion of parent material such as sediments and metamorphic rocks at great depths.

[1] From Best (1982). [2] Read (1955).

with a length of 1600 km, is a multiple intrusion of granite, tonalite, and gabbro. The magmas were intruded as hundreds of separate plutons, with associated ring-dykes, in an area where vertical movements were dominant. Rapid uplift and deep denudation have en-abled a three-dimensional view of the batholith to be obtained.

Ores of igneous origin

Natural processes that concentrate valuable minerals have been described briefly within this Chapter and in Chapter 4. Here we consider three important processes which attend the intrusion of igneous rock.

Magmatic concentration

During the solidification of magma early formed heavy minerals, usually metallic oxides, chlorides, sulphides and certain native metals (e.g. nickel and platinum) sink and become concentrated in layers at or near the base of the mass. Such minerals include magnetite (Fe_3O_4) and chromite ($FeCr_2O_4$). The large magnetite deposits at Kiruna, Sweden, which have yielded millions of tonnes of high grade ore, were formed this way. Similar products occur in the famous examples of layered intrusion at Stillwater, Montana, U.S.A.; Muskox, Mackenzie, Canada; Skeargaard in Greenland and the Bushveld in S. Africa.

Hydrothermal processes

As crystallization progresses within an intrusion a residual magmatic liquid accumulates. Magmas of acid composition (such as granite) are frequently rich in volatile constituents, and may also contain small quantities of many metals. As crystallization proceeds, with the formation of minerals like feldspar and quartz which form the bulk of the resulting rock, the metals - which were originally disseminated throughout the magma and were not incorporated in the feldspar and other crystals - become concentrated in the residual fluids. These also contain the volatile constituents, and are thus able to remain fluid to lower temperatures than would be possible without the fluxing effect of the volatiles. If, then, tensile fractures are formed in the outer (first solidified) part of the granitic mass and its surrounding rocks, as will occur in the stretched roof of an ascending magma or with shrinkage of the intrusion on cooling, they become channels into which the residual fluids migrate, there to permeate and alter rock in contact with the granite. This is *contact metasomatism* (Greek = change of substance). Minerals which crystallize directly from the fluids may be described as (*i*) *pneumatolytic deposits*, due to the action of gaseous emanations at a high temperature (600°C or over), including water in the gaseous state; and (*ii*) *hydrothermal deposits*, due to the operation of hot aqueous fluids, at temperatures from about 600°C downwards. It is difficult to draw a sharp distinction between the higher temperature hydrothermal minerals and those of pneumatolytic origin.

Economic reserves occur when these deposits are concentrated in veins, and other voids. *Veins and lodes* consist of the infillings of fissures and fractures developed in the outer part of an intrusive body or in the surrounding roof and wall-rocks. Veins which contain metalliferous minerals are termed lodes. (See also pegmatites, p. 106.)

Metals which are commonly associated in this way with acid rocks include copper, lead, zinc, arsenic, tin, tungsten, gold, and silver. The volatile constituents, consisting chiefly of water vapour but often also including fluorine, chlorine, boron, and other gases, act as carriers for the metals into the area of mineralization around the intrusion. As the fluids pass outwards into zones where lower temperatures and suitable pressures prevail, they deposit e.g. tin as cassiterite, SnO_2; tungsten as wolfram $(Fe, Mn)WO_4$; and copper as chalcopyrite, $CuFeS_2$ (Fig. 5.27). Thus tin and tungsten lodes may be formed in and around granite masses, as in Devon and Cornwall; they may be regarded as of pneumatolytic origin. Minerals such as quartz, pyrite, topaz, and tourmaline are commonly associated with the metalliferous minerals in such lodes. Iron, lead, and zinc may also be carried outwards and deposited as haematite, galena, and blende in joints and fractures in cooler rocks at somewhat greater distances from the igneous source.

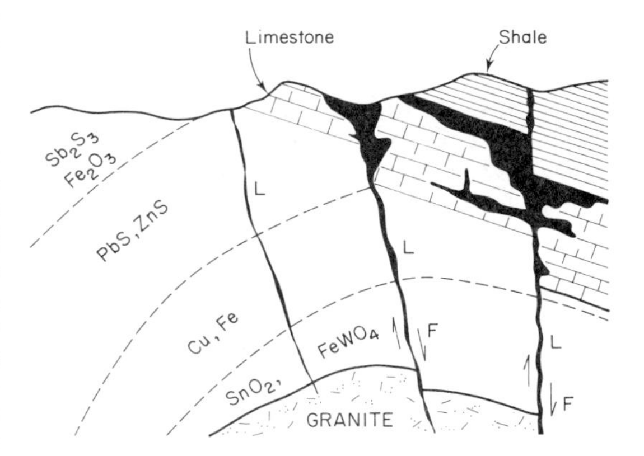

Fig. 5.27 Vertical section diagrammatically illustrating lodes (L) emanating from an intrusion at depth, with zoning of mineral deposits (see text) and solution in limestone F = fault, black = massive mineral deposit.

The latter stages of crystallization involve the movement of abundant quantities of water. Some comes from the magma but a greater volume comes from the surrounding country-rock and is drawn towards the magma by extensive convection cells which surround the chamber, Fig. 5.28. Millions of tonnes of water may be transferred in this way and its dilute content of dissolved solids precipitated to form with time mineral concentrations of great wealth. These merge naturally with the products of contact metasomatism but extend much further into the surrounding rock and have formed some of the largest and most important mineral deposits.

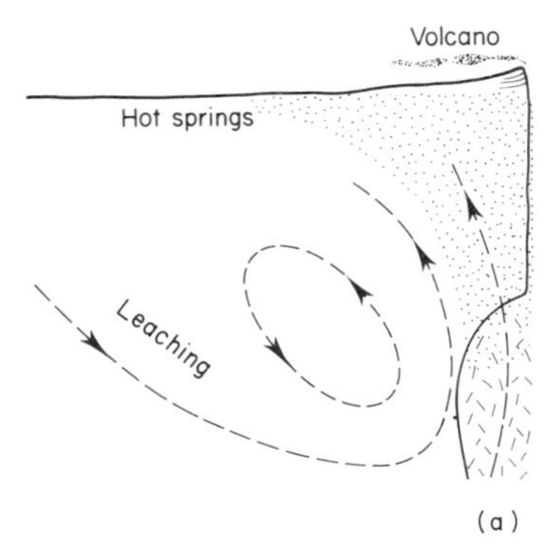

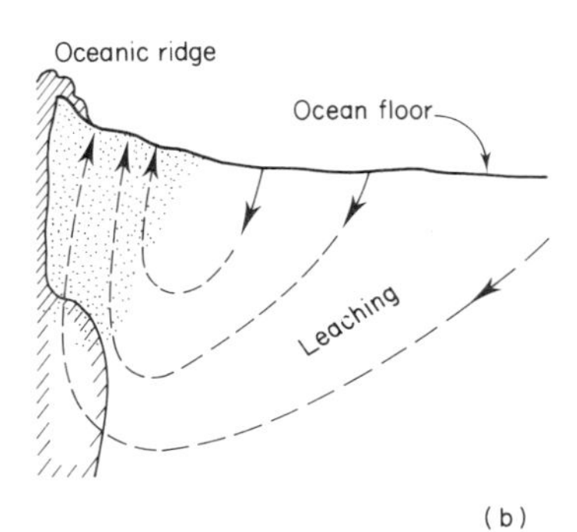

Fig. 5.28 Vertical sections (unscaled) through (**a**) continental intrusion, (**b**) oceanic ridge, to illustrate convection cells. Stippled = area of mineral precipitation.

Of special note is the dissolution of limestone by hydrothermal fluids (see Fig. 5.27). Where fissures and pores are enlarged and filled *disseminated* mineral deposits are formed: when larger fractures are widened and cavities are created and filled, *massive* deposits accumulate. Such concentrated reserves are not restricted to limestone and may fill available voids in any rock, but tend to be remarkably well developed in soluble strata. In England the Carboniferous Limestone is host to many such deposits of galena, blende, and calcite.

Minerals generally formed under high-temperature hydrothermal conditions are mainly the sulphides of iron, copper, lead, and zinc (see p. 84). They occur in association with non-metalliferous minerals such as quartz, fluorite, calcite, or dolomite.

Among the deposits formed at intermediate temperatures are some lead and zinc veins, certain gold-bearing quartz veins, and some copper and pyrite deposits. Lead and zinc commonly occur as the sulphides, galena and blende, but sometimes as compounds with arsenic and antimony, often in association with pyrite, and quartz, calcite, fluorite, or barytes.

Deposits formed under low-temperature hydrothermal conditions occur at shallow depths and are often associated with andesites. They include certain gold occurrences, in which the gold (together with some silver) is combined as a telluride. Gold telluride ores are worked at Kalgoorlie, Australia, and were formerly worked at Cripple Creek, Colorado. Mercury, combined as cinnabar (HgS), and antimony, as stibnite (Sb_2S_3), are two other examples of low-temperature hydrothermal minerals. They are sometimes associated in veins with minerals of the zeolite group, indicating a temperature of formation not greater than 200°C.

Selected bibliography

Francis, P. (1976). *Volcanoes*. Penguin Books Ltd. London and New York.

Green, J. and Short, N.M. (1971). *Volcanic Landforms and Surface Features*. Springer-Verlag, New York.

Volcanic Processes in Ore Genesis. (1977). Proc. Joint Meeting of Geol. Soc. London and Instn. Mining and Metallurgy. Jan. 1966. *Geol. Soc. London, Spec. Pub.*, **7.**

Descriptions of igneous intrusions

Fiske, R.S., Hopson, C.A. and Waters, A.C. (1963) *Geology of Mt. Rainier National Park, Washington*. U.S. Geol. Survey. Professional Paper 444.

Baker, B.H., Mohr, P.A. and Williams, L.A. (1972). Geology of the eastern rift system of Africa. *Geol. Soc. America. Special Paper*, **136.**

Richey, J.E., MacGregor, A.G. and Anderson, R.W. (1961) *British Regional Geology. Scotland: The Tertiary Volcanic Districts.*

Bussell, M.A., Pitcher, W.S. and Wilson, P.A. (1976). Ring complexes of the Peruvian coastal batholith. Canadian *Jl. Earth Sciences*, **13,** 1020–30.

Jacobson, R.R.E., MacLeod, W.N. and Black, R. (1958). Ring-complexes in the younger granite province of northern Nigeria. *Geol. Soc. of London. Memoir* No 1.

Hamilton, W. and Meyers, B. (1967). *The nature of batholiths*. U.S. Geol. Survey. Professional Paper 554-C.

Descriptions of igneous intrusions with mineral deposits

Williams, D., Stanton, R.L. and Rambaud, F. (1975). The Planes – San Antonio pyritic deposit of Rio Tinto, Spain: its nature, environment and genesis. *Trans. Inst. Mining Metallurgy*, **84,** 373–82.

Gustafson, L.B. and Hunt, J.P. (1975). The porphyry copper deposit at El Salvador, Chile. *Economic Geol.*, **70,** 857–912.

Hosking, K.F.G. and Shrimpton, G.J. (1964). (Editors) *Present views of some aspects of the geology of Cornwall and Devon*. 150th Anniversary Edition of the Royal Geol. Soc. of Cornwall, Penzance.

Mayer, C., Shea, E.P., Goddard, C.C. and other staff. (1968). Ore deposits at Butte, Montana, in: *Ore Deposits in the United States*, 1933–67, Editor, J.D. Ridge. Amer. Inst. Mining, Petroleum Eng. Inc. New York.

Hall, A.L. (1932). The Bushveld igneous complex of the central Transvaal. *S. African Geol. Surv. Memoir* 28.

Goudge, M.F., Haw, V.A. and Hewitt, D.F. (1957). (Editors) The geology of Canadian industrial mineral deposits. *6th Commonwealth Mining, Metallurgy Congres*. Montreal (see paper by Allen, C.C., Gill, J.C. and Koski, J.S. p. 27–36).

6

Sedimentary Rocks

Sediments form a relatively thin surface layer of the Earth's crust, covering the igneous or metamorphic rocks that underlie them. This sedimentary cover is discontinuous and averages about 0.8 km in thickness; but it locally reaches a thickness of 12 km or more in the long orogenic belts that are the sites of former geosynclines (p. 16). It has been estimated that the sedimentary rocks constitute little more than 5% of all crustal rocks (to a depth of 16 km); within this percentage the proportions of the three main sedimentary types are: shales and clays, 4%; sandstones, 0.75%; and limestones, 0.25%. Among other varieties of smaller amount are rocks composed of organic remains, such as coals and lignites; and those formed by chemical deposition.

Composition

The raw materials from which the sedimentary rocks have been formed include accumulations of loose sand and muddy detritus, derived from the breakdown of older rocks and brought together and sorted by water or wind, as discussed in Chapter 3. Some sediments are formed mainly from the remains of animals and plants that lived in rivers, estuaries, on deltas, along coast-lines and in the sea. Shelly and coral limestones, coal and many sedimentary iron ores are composed of such remains. Sediments may also be formed by evaporation of water and precipitation of the soluble minerals within it, as in playa lakes (p. 60) and sabkhas (p. 49). When sea water evaporates its components precipitate valuable deposits of chemicals such as sodium chloride (= rock salt = halite). The chemical properties of sea water are due to dissolved matter brought by rivers and contributed by volcanic eruptions, and to the presence of marine organisms and sediments (Table 6.1).

Table 6.1 Concentration of the components of sea water for a salinity of 35%. Many other metal and non-metal ions are present in sea water, including Sr, B, Si, Rb, F, P, I, Cu, and Ba (in decreasing order of abundance). The density of sea water varies between 1.025 and 1.028, depending on the salinity and temperature.

Component	Concentration
Chloride	19.35 g kg^{-1}
Sodium	10.76
Sulphate	2.71
Magnesium	1.29
Calcium	0.41
Potassium	0.39
Bicarbonate	0.14
Bromide	0.067

Development

The components of sediments become hardened into sedimentary rocks such as sandstone, quartzite, limestone and shale by changes which commence soon after the sediment has accumulated. Water percolating through the voids (or pores) between the particles of sediment carries mineral matter which coats the grains and acts as a cement that binds them together (Fig. 6.1*a*). Such processes are known as *cementation*; they may eventually completely fill the pores and are responsible for converting many coarse-grained sediments to rock. The conversion of muddy sediment to rock is mainly achieved by the very small particles of silt and clay of which they are mainly composed being pressed together by the weight of overlying sediment, interstitial water being squeezed out and mineral matter precipitated in the microscopic network of pores (see Figs. 4.37 and 6.1*b*). In course of time mud will become a coherent mass of clay, mudstone, or shale. The process is called *compaction*, it being more than simple consolidation, and it affects the muddy sediments to a much greater degree than the sands. During compaction, while much pore-contained water in the mud is pressed out, some water with its dissolved salts may remain in the sediment; it is known as *connate water* (connate = 'born with').

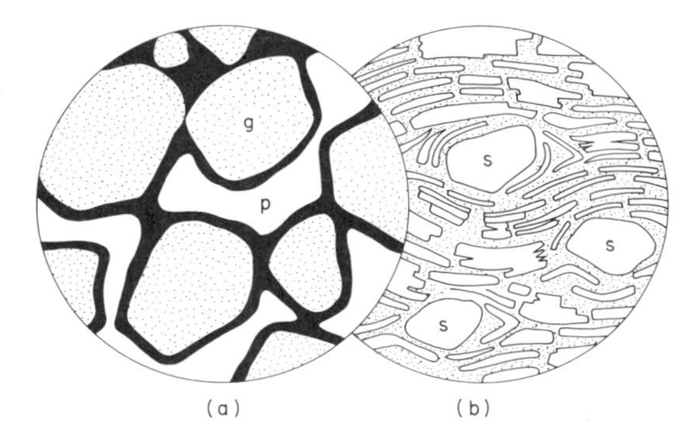

Fig. 6.1 Two sedimentary rocks, in thin section. (**a**) Sandstone with iron cement (black) coating the sand grains (g); p, pore (×30). (**b**) Compacted clay with silt (s). Stipple = smaller crystals of clay mineral and new minerals formed by precipitation from pore water during compaction (× 1000).

The general term *diagenesis* is used to denote the processes outlined above, which convert sediments into sedimentary rocks. Diagenetic processes include not only cementation and compaction but also solution and re-deposition of material, to produce extremely strong, or very weak rocks (Fig. 6.2). Other changes such as replacement, take effect in particular rocks (see e.g. dolomitization). Much salt in marine sediments is removed by leaching. All these changes take place near the Earth's surface at normal temperatures.

When fully-formed rocks come again into the zone of weathering, perhaps after a long history of burial, soluble substances are removed and insoluble particles are released, to begin a new cycle of sedimentation in rivers and the sea.

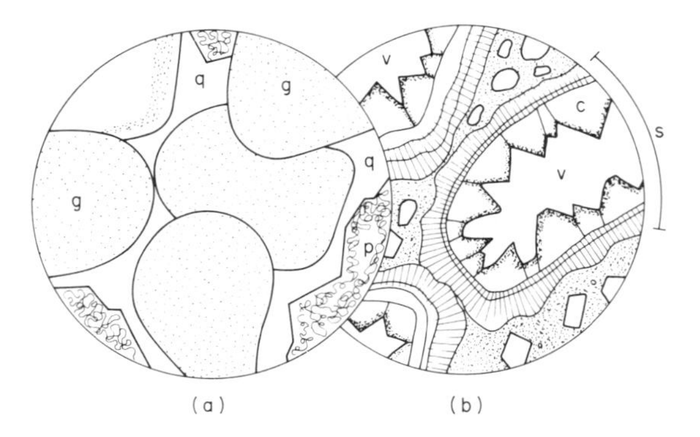

Fig. 6.2 Sedimentary rocks in thin section. (**a**) Sandstone in which compaction has produced dissolution at grain boundaries and an interlocking texture, with precipitation of dissolved material as new cement (q) at grain edges; g, grain, p, pore (× 30). (**b**) Limestone where shell fragments (s) have been partially dissolved leaving voids (v) that have later been incompletely filled with calcite (c). (× 25.)

Texture

The texture of a sedimentary rock reflects the mode of sediment deposition, diagenesis during burial and subsequent weathering when uplifted and exposed at the surface of the Earth. Some common microscopic textures are illustrated in Fig. 6.3. Two important characters of sediments are their porosity and packing.

Porosity

Porosity, n, is defined as the percentage of void space in a deposit (or in a rock), i.e.

$$n = \frac{\text{volume of voids}}{\text{total volume}} \times 100.$$

The ratio of void space to solid matter, by volume, is called the *voids ratio*, e,

$$e = \frac{\text{volume of voids}}{\text{volume of solids}} \times 100.$$

The relationship between these two quantities is given by

$$n = \frac{e}{(1+e)} \text{ and } e = \frac{n}{(1-n)}.$$

Clays have smaller pores than sands but may have much greater porosity.

Packing

In coarse sediments the packing of grains results mainly from the range of grain sizes deposited (Fig. 6.3*f*,*g*). In many sandy deposits (but not all) the grains accumulate in a configuration that can be little altered by later overburden loads. Fine-grained deposits, especially clays,

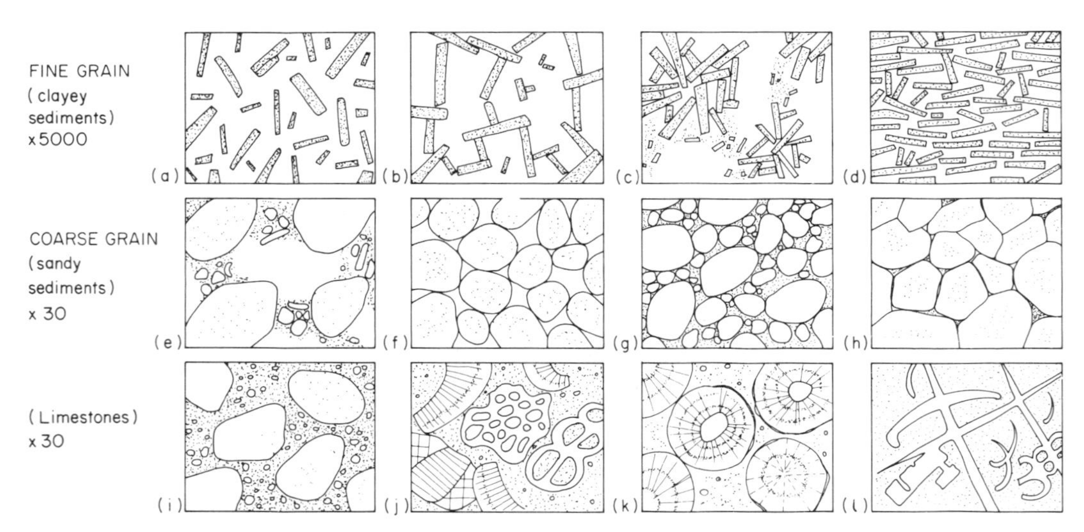

Fig. 6.3 Microscopic texture of some common sediments. Clays, (**a**) dispersed as a colloidal suspension prior to sedimentation; (**b**) aggregated edge to face as often deposited in fresh water; (**c**) flocculated, the form of many newly deposited marine clays; (**d**) compacted with parallel packing, the eventual form of many clays after burial. Sands, (**e**) grains separated by finer matrix; (**f**) grains in contact; (**g**) dense and closely packed grains; (**h**) grain boundaries in contact (*sutured*), as when severely compressed by deep burial. Limestones, (**i**) grains separated by matrix of lime mud; (**j**) shelly limestone; (**k**) oolitic limestone; (**l**) shelly limestone broken by fractures with both shells and fractures enlarged by dissolution to create voids.

generally possess an open structure when they accumulate and are susceptible to considerable change in packing as the overburden load upon them increases (compare Figs 6.3*c* and 6.14). The change in packing that accompanies burial of any sediment results from the process of *consolidation*.

A freshly deposited mud on the sea floor has a high porosity (and high void ratio), represented in Fig. 6.4 by

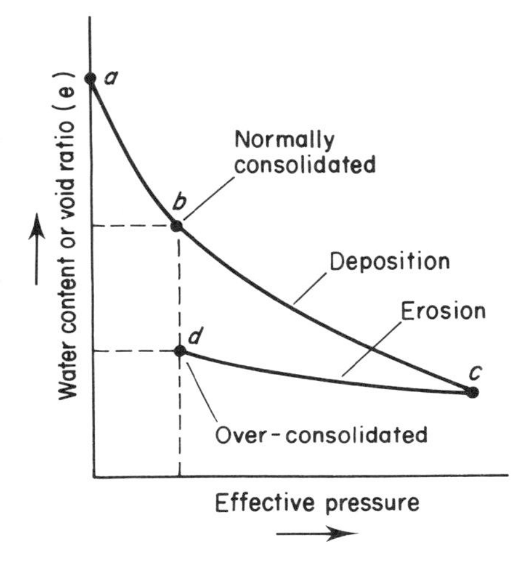

Fig. 6.4 Normally-consolidated and over-consolidated clay (after Skempton, 1970).

the point (*a*). As deposition continues, the overburden pressure increases and the sediment compacts, e.g. to point (*b*) or further to (*c*). The curve *abc* in the figure is called the sedimentation compression curve for that particular clay. The clay is said to be *normally consolidated* if it has never been under a pressure greater than the existing overburden load. If, however, the clay has in the past been subjected to a greater overburden pressure, due to the deposition of more sediment which has subsequently been eroded, it is said to be *over-consolidated*; this is represented by point (*d*) in Fig. 6.4. Thus two clays, at points (*b*) and (*d*), which are under the same effective pressure and are identical except for their consolidation history, have very different void ratios and packing. The construction of curves such as those shown in the figure has been discussed by Skempton (1970). The London Clay is an example of an over-consolidated clay; its consolidation took place under the weight of the overlying Upper Eocene beds (Fig. 2.17), the estimated load in the central London area being about 3400 kNm^{-2}. Uplift, and the erosion which resulted from it, reduced the overburden to its present amount.

Similar effects to consolidation by overburden load may be produced in individual beds of clay if, during deposition in shallow water, they are periodically exposed to the atmosphere and desiccate at an early stage in their geological history. The 'ball-clays' of Devon and clays elsewhere are thought to have been over-consolidated in this manner (Best and Fookes, 1970).

Facies

At any given time different kinds of deposit may be forming in different environments, such as marine, continental, lacustrine deposits. Examination of a particular sedimentary formation (or group of strata) yields information about the materials composing it, their textures, contained fossils, and other characters. All these features together distinguish a rock and are known as its *facies*, and from them deductions can be made about the environment in which the rock was formed. Thus, in marine sediments occur *marine* facies in which there is a transition from pebbly and sandy material (*littoral* facies) near the shore-line to a contemporaneous *muddy* facies in deeper water. Sediments formed on land, such as breccias, wind-blown sands, and wadi deposits, constitute a *continental* facies.

Igneous rocks erupted from volcanic vents may provide a *volcanic* facies contemporaneous with sedimentation elsewhere.

Environments of deposition

The compositions and textures of sedimentary rocks are controlled by the processes that operated during their formation; and these processes are in turn governed by the environment in which sedimentation takes place. Three major environments may be identified, namely continental, shelf sea and open or deep sea.

Continental environments

These develop on land areas where *desert*, *piedmont*, *alluvial*, *lacustrine* and *glacial* deposits are accumulated. Desert or eolian sands have been discussed earlier (p. 000). Piedmont deposits, which are formed during rapid weathering of mountains at the end of an orogenic upheaval and lie at the foot of steep slopes that are undergoing denudation, include scree deposits and arkoses (*q.v.*). Lacustrine clays are slowly deposited in lakes of still water. Where water is impounded in glacial lakes, seasonal melting of the ice leads to the formation of varved clays with alternations of coarser (silty) and finer (muddy) layers. Many of these environments have been described in Chapter 3: *cf.* the work of rivers, wind, ice and other mass transport.

Shelf sea environments

These exist at the margin of a sea on the continental shelf; deposits of pebbles and sand of various grades are formed, together with muds and calcareous material. The rough water caused by wave action along a shore results in the rounding of rock particles into pebbles, and the wearing down of the larger sand grains, which are mainly quartz. Minerals less hard than quartz do not persist for so long, nor do minerals with cleavages (such as feldspars). Still finer particles of silt also consist largely of quartz.

Structures such as current-bedding and ripple-marks (Fig. 2.1) may be preserved in the resulting sandy and silty rocks. Shales are formed from muds deposited in somewhat deeper and less turbulent water at a distance from the shore. Limestone-forming materials derived from calcareous skeletal remains (e.g. shell debris, crinoids, algae) are often associated with such sands and muds. A sequence of fossiliferous layers may thus be formed in a marine area; they cover the older rocks that underlie the area of deposition (e.g. Fig. 2.2). Because of its content of fossil shells and shell fragments such a series may be referred to as a *shelly facies*. Shelf sea deposits are frequently of relatively small thickness (hundreds of metres rather than thousands); but if subsidence has affected the area during the deposition of the sediments, a greater thickness of shallow-water deposits can be accumulated. In these conditions successive pebbly layers on the foreshore come to lie farther and farther inland from the original coast, the foreshore migrating inland as the water slowly deepens over the area of sedimentation.

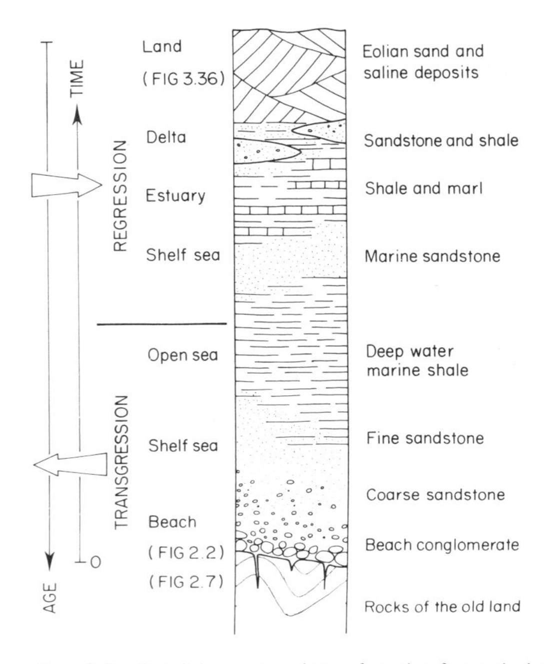

Fig. 6.5 Possible succession of rocks formed during a transgression of the sea and a regression. (After Read and Watson, 1971.) No scale implied.

A succession of sediments which are laid down during the advance of the sea over a land area (or *transgression*) differs from that formed during the retreat (or *regression*) of the sea. A possible sequence of the rock types formed is shown in Fig. 6.5. The figure is read from the bottom upwards, and it may be imagined that at time 'zero' a sea exists to the right of the position represented by the column of strata and land to the left, the column being located at the beach.

The figure records an initial period of trangression, the sea advancing from right to left with deepening water, when pebbles and coarse sand (a littoral facies) come to be covered by finer marine sands, and these eventually by marine muds. By this time the coastline has moved a considerable distance to the left of the column. During the retreat of the sea (from left to right), with shallowing water, the muds are in time succeeded by sands and these in turn give place to estuarine or lagoonal deposits as a land surface begins to emerge. Finally, if conditions are suitable, deltaic deposits may be put down on the earlier sediments and eolian sands accumulate on the new land surface. The shore-line is now a distance to the right of the column.

In mobile belts (p. 16) of orogenic zones the shelf sea forms the margin of a geosynclinal trench (Figs. 2.3 and 2.4). During the development of a geosyncline much sediment of various kinds and of differing grade sizes is contributed to the deepening trough. Coarse detritus is poured into the trough at times, contributing lenses of deposit derived from the land margins by rapid weathering, and often together with material from volcanic eruptions. These sediments are accumulated rapidly, without having undergone long processes of sorting and rounding, and therefore have a high proportion of angular particles. Muddy sediments built up on the slopes of the trough are affected by slumping, and mud-flows run their course to lower levels; when seen now as consolidated rocks (shales) they show slump structures (Fig. 6.6).

Under such conditions *greywackes* are formed, i.e. badly sorted muddy sediments with much coarse clastic material (p. 121). With them are associated breccias, and lenses of poorly graded conglomerates in which partly rounded pebbles are set in a matrix of angular mineral grains; the minerals are often of more than one variety

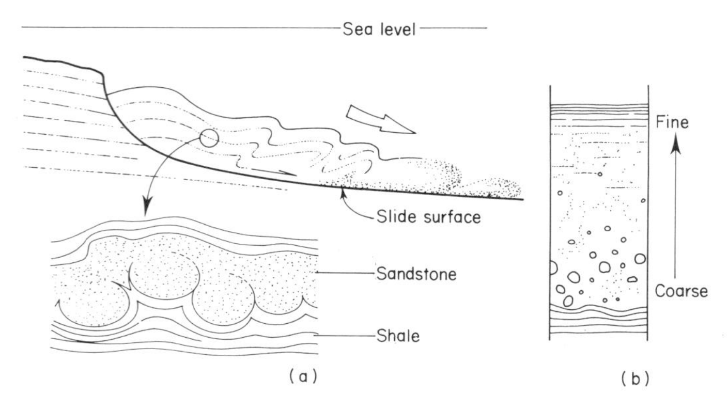

Fig. 6.6 (**a**) A slump: detail shows how sand horizons sink into underlying clays liquefied during slumping. (**b**) Vertical sequence through an horizon deposited from slumped strata.

(in contrast to the predominant quartz of some beach deposits described above).

Limestones if present are generally thin and represent temporary conditions of greater stability that prevailed for a time, especially when the trough was nearly filled. All these sediments suffer severe compression at the end of the geosynclinal phase, as their buckled and fractured condition indicates at the present day (Fig 2.17*b*).

Deep-sea environments

The deep-sea sediments were first studied from the results of the *H.M.S. Challenger* expedition of 1872–76, which obtained many hundreds of samples dredged from the ocean floors of the world. These were classified by reference to predominant constituents, easily visible, with the use of terms such as Globigerina ooze, siliceous ooze, and 'red clay'. The second *Challenger* expedition of 1952, and the more recently developed methods of core sampling in the ocean floors have yielded much new information.

The deep-sea sediments are spread over vast areas of ocean floor (Fig. 3.29), and characteristically contain no large fragments and no features due to surface current or wave action. In places they connect with shallower water (land-derived) sediments by gradual transition. Land-derived muds that lie on the continental slopes are blue, red, or green in colour and are composed of very small clay particles (Table 6.2). The blue muds extend to depths of about 2.75 km (1500 fathoms). At greater depths are found calcareous and siliceous sediments, composed of the skeletons of minute floating organisms which sink and accumulate slowly on the ocean floor. Thus, the Globigerina 'ooze' is a calcareous deposit, composed of the minute tests of coccoliths and of the foram *Globigerina* (Fig. 2.22); it is estimated to cover some 130 million km^2 in the Atlantic and Pacific oceans, at depths of around 3.65 km (2000 fathoms). Siliceous organisms persist to greater depths than those reached by calcareous-tests, which are slowly dissolved as they sink through deep water. Thus *radiolaria* (Fig. 2.22), and small disc-like skeletons of *diatoms*, and a few sponges form a siliceous 'ooze' covering about 7 million km^2 in the South Pacific. From about 4.0 km (2200 fathoms) down to the greatest known depths the deposit described as 'red clay' covers large areas of ocean floor. This soft, plastic material is partly derived from fine volcanic dust from the atmosphere which has slowly settled through a great depth of water, and partly of clay minerals. The hard parts of fishes such as sharks' teeth and the bones of whales are found in samples of the 'red clay' that have been dredged up. Spherules of meteoric iron are also found in the deposit, together with manganese nodules ('sea potatoes') and concretions with manganese coatings.

Radiocarbon dating of the upper part of deep-sea cores containing calcium carbonate tests has shown that the last glacial period gave way to present conditions about 11 000 years ago. This confirms the dating of the end of the Pleistocene glaciation from other evidence (p. 57).

Classification

A classification of the unconsolidated sediments and rocks derived from them is given in Table 6.2, in which the broad groupings are:

DETRITAL

(*i*) *terrigenous*. . . . derived from the land by weathering and erosion, and mechanically sorted: e.g. gravels, sands and muds, conglomerates, sandstones, mudstones and shales.
(*ii*) *pyroclastic*. . . . derived from volcanic eruptions.
(*iii*) *calcareous*. . . . derived mainly from calcareous particles which have been mechanically sorted, as if detritus.

CHEMICAL AND BIOCHEMICAL (organic)
. . . . formed mainly either in place or involving animal or vegetable matter.

This classification is suitable for sandy and muddy sediments which together account for most of the sediments deposited, but is not the most appropriate for limestones, varieties of which appear in both parts. For example, the Chalk of N.W. Europe is mainly composed of a foraminiferal mud i.e. biochemical detritus.

The divisions based on particle size are those commonly used in engineering standards for the description of detrital sediments. They are not precisely those recognized by geologists who place the divisions at 256 mm, 64 mm, 4 mm, 2 mm, 0.063 mm, and 0.004 mm. The three divisions used for chemical sediments are not standards but are conveniently related to those used for describing detrital sediments.

Detrital sedimentary rocks

Detrital (terrigenous) sediments

This group is divided according to the sizes of component particles into *rudaceous* (Latin, gravelly), *arenaceous* (Latin, sandy) and *argillaceous* (French, clayey, although the Latin *lutaceous* or silty may also be used). These groups are further divided into grades which describe more precisely a range of particle sizes.

A particular sediment generally contain particles of several different grades rather than a single grade, e.g. a 'clay' may contain a large proportion of the silt grade and even some fine sand, as in the London Clay. The determination of the grades or sizes of particles present in a sample of sediment, known as the *particle size analysis*, is carried out by sieving and other methods, as discussed in Chapter 11. The results of such an analysis can be represented graphically, grade size being plotted on a logarithmic scale against the percentage weights of the fractions obtained (Fig. 11.3).

Unconsolidated sands, silts, and clays are of considerable importance in civil engineering and are known to the engineer as *soils* (p. 1). Particle size analysis may give

Table 6.2 A classification for sediments and sedimentary rocks. Sediments listed in lighter print, sedimentary rocks in darker print.

PARTICLES (or GRAINS) size (mm)	name	DETRITAL (terrigenous) >50% GRAINS OF ROCK & SILICATE MINERALS: Unconsolidated raw material	Geological Term	**Consolidated as rock**	DETRITAL (calcareous) >50% GRAINS OF CALCIUM CARBONATE: Unconsolidated raw material	**Consolidated as rock**	Collective name	DETRITAL (pyroclastic) >50% GRAINS OF VOLCANIC ROCK: Unconsolidated raw material	**Consolidated as rock**
200	Boulder	Storm beach Colluvium (scree, or talus) Glacial boulder beds		**Comglomerate** **Breccia**	Storm beach Colluvium Coral reef debris	**Conglomerate** **Breccia**	Collective name = CALCIRUDITE	Volcanic bombs & ejected rock blocks	**Agglomerate** **Volcanic breccia**
60	Cobble			**Tillite**					
4	Pebble	GRAVEL: Coarse-alluvium (e.g. wadi)	RUDACEOUS (mainly rock fragments)	**Conglomerate**	GRAVEL: Carbonate gravel: eg. coral reef debris. Shell rich beach gravel	**Carbonate gravel**		Fine strands and droplets of lava ejected into the atmosphere	**Lapilli tuff**
2	Granule			**Grit**		**Pisolithic grit** **Carbonate grit**			
0.06	Sand	SAND: Fine-alluvium Deltaic grits and sands Sand beach Desert sand & dust	ARENA-CEOUS (mainly mineral fragments)	**Sandstone (arkose, grey-wacke,** and other varieties) Loess	SAND: Carbonate sand. eg. shelf and beach	**Carbonate sand** **oolitic limestone**	CALC-ARENITE	Volcanic ash	**Tuff**
0.002	Silt	SILT: Estuarine silt	ARGILLACEOUS (mainly mineral fragments)	**Siltstone**	SILT: Carbonate Silt: eg. lagoon silt	**Carbonate silts**	CALCI-LUTITE	Volcanic dust	
	Clay	CLAY: Glacial silts & clays clay ooze		**Mudstone & Shale**	CLAY: Desert Dust Carbonate mud	**Loess** **Cement-stone** **Chalk**			

PARTICLE (or CRYSTAL)	CHEMICAL & BIOCHEMICAL (=organic)				
	SILICEOUS	CALCAREOUS	CARBONACEOUS	SALINE	FERRUGINOUS
Megacrystalline		Coral reef, shell banks **Coral (reef), crinoidal and shelly limestones**		(Some evaporite sediments become megacrystalline with diagenesis)	
—— 2 mm ——					
Mesocrystalline		**Pisolithic grit** **Oolitic limestone**	Peat & drifted vegetation **Lignite (brown coal)** **Humic (bituminous) coals** **Anthracite**	Mineral crust precipitated in salt lakes and marine mud-flats **Evaporites, anhydrite, halite, potash-salts, Gypsum**	
—— 0.06 mm ——					
Microcrystalline	Silica gel & colloids, radiolaria, diatoms **Flint, chert, jasper, radiolarite, diatomite**	Algal, coccolithic and other forams **Algal and foraminiferal limestones** $CaCo_3$ precipitated from solution, & $CaMg(CO_3)_2$ replacement **Travertine (tuffa) & Dolomite**	Organic muds **Oil shale**		Algal precipitates of iron-rich carbonates and silicates: colloidal ferric hydroxides. **Siderite, chamosite and bog-iron ores**

Note that the terms sand, silt and clay have two usages, viz. as a description of particle size and as a name for sediments made from particles of their size. The term gravel strictly refers to a deposit of pebbles and granules, but is also used to describe a range of particle sizes (*cf.* Fig. 11.14).

valuable information about the physical properties of soils, which are named according to the range of particle size present, e.g. sandy gravel, silty clay: see Chapter 11. The proportion of clay present in a sediment, for instance, affects the engineering properties of the material. A good moulding sand should contain a small proportion of particles of the clay grade; a good sand for glass manufacture would be composed mainly of one grade of clean grains. It is to be noted that in general sand and silt particles are physically and chemically inert, but clays (0.002 mm and less) are often highly reactive. A simple test that distinguishes silt from clay is described on p. 81. *Loam* is a deposit containing roughly equal proportions of sand, silt, and clay.

Rudaceous deposits

Boulders, cobbles, pebbles and granules are generally pieces of *rock*, e.g. flint and granite, and by definition are greater than 2 mm in diameter; most are much larger than this. Sand grains, on the other hand, usually consist of *mineral* particles, of which quartz, because of its hardness and resistance to weathering, forms the greater part of most sands.

Marine rudaceous deposits are formed at the foot of cliffs from the break-up of falls of rock and from material drifting along the coast. Sorting by wave action may result in a graded beach, as at Chesil Beach, Dorset, England (Fig. 3.27). The pebbles consist of the harder parts of the rocks which are undergoing denudation, wastage of any softer material being more rapid; thus, a gravel derived from flint-bearing chalk is composed mainly of rounded flints. Pebbles derived from the denudation of boulder clay deposits, as on the Yorkshire coast, are generally of many kinds (e.g. granite, gneiss, vein quartz), and may have travelled far from their source.

River gravels are laid down chiefly in the upper reaches of streams, after being carried over varying distances, and often contain an admixture of sand; at a later date a river may lower its bed and leave the gravels as terraces on the sides of its valley, marking its former course (Figs. 3.17 and 3.18). In the Thames Valley area such gravels are mainly composed of one constituent, flint; 'Thames Ballast' contains as much as 98% flint. In areas where many different rocks have contributed fragments, gravels laid down by the rivers have a more varied composition; e.g. deposits of the River Tay, Scotland, contain pebbles of granite, gneiss and quartzite.

Some rudaceous sediments are accumulated by the action of sudden storms which wash down much detritus, as during the sudden (flash) floods of wadis in deserts. Lens-shaped beds of sand are often intercalated in deposits of river gravels, the variation in size of material being related to the currents which carried it.

Conglomerate

Deposits consisting of boulders, cobbles, pebbles and granules when cemented are known as *conglomerate*; the spaces between the pebbles are often largely filled with sand. Two groups of conglomerates may be distinguished:

(*i*) the beach deposits, in which rounded pebbles are well sorted, and commonly derived from one kind of rock (such as flint, quartzite). These often occur at the base of a formation which follows above an uncomformity, e.g. the orthoquartzite at the base of the Ordovician in Salop. 'Pudding-stone' is a flint-conglomerate formed by the cementing of pebble beds in the Lower Tertiary sands, and is found locally as loose blocks lying on the surface of the Chalk north-east of London.

(*ii*) those conglomerates in which the pebbles are less well rounded and are derived from several different sources. Pebbly layers of this kind are often found within thick beds of sandstone, as in the Old Red Sandstone of Hereford, the N.W. Highlands, Scotland, and other areas (Fig. 6.7). Such deposits represent rapidly accumulated coarse detritus washed into place by storms. *Banket* is the gold-bearing conglomerate, composed of pebbles of vein quartz in a silicous matrix, from the Witwatersrand System, South Africa; alluvial gold was washed into stream gravels which were later cemented into a hard rock.

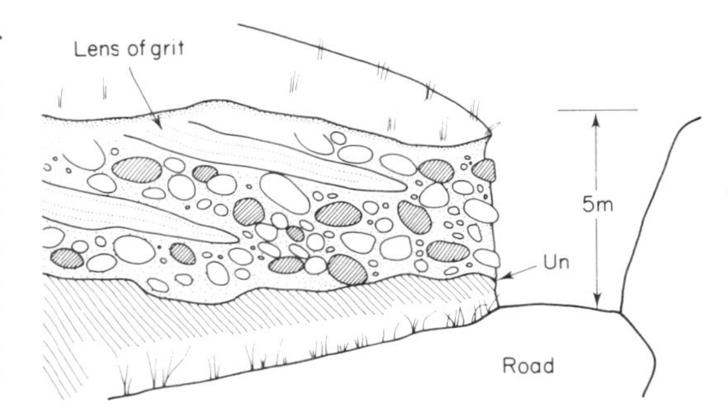

Fig. 6.7 Sketch of a boulder conglomerate of Devonian age resting with unconformity (Un) on much older slate (shaded).

Breccia is the name given to cemented deposits of coarse angular fragments, such as consolidated scree. Such angular fragments, after limited transport, have accumulated at the foot of slopes (Fig 3.41) and have not been rounded, in contrast to the rounded pebbles which are distinctive of conglomerates. Breccias composed of limestone fragments in a red sandy mixture are found in Cumbria, e.g. near Appleby, and are called *brockrams*; they represent the screes which collected in Permian times from the denudation of hills of Carboniferous Limestone. The formation of scree or talus is specially characteristic of semi-arid or desert conditions.

The Dolomite Conglomerate of the Mendips in Somerset is a deposit of Triassic age, containing partly rounded fragments of dolomitic limestone of Carboniferous age and much angular material contributed by rapid weathering; it is a consolidated scree.

Breccias may also be formed by the crushing of rocks, as along a fault zone (p. 151), the fragments there being cemented by mineral matter deposited from percolating solutions after movement along the fault has ceased. These are distinguished as *fault-breccias*. The explosive action of volcanoes also results in the shattering of rocks at a volcanic vent, and the accumulation of their angular fragments as they fall to the ground to form *volcanic breccias*.

Grit

This is a coarse-grained rock, rough to touch and usually composed of quartz grains cemented by silica. Thick and widespread deposits of grit accumulated in Great Britain during the Carboniferous in the deltas upon which were to grow the coal forests. These constitute the Millstone Grit (Table 2.3) so called after their abrasive qualities and uniform character, making them ideal material for use, in the past, as millstones.

Arenaceous deposits

Sands

Most sand-grains, as stated above, are composed of quartz; they may be rounded, sub-angular, or angular, according to the degree of transport and attrition to which they have been subjected. Windblown grains, in addition to being well rounded, often show a frosted surface. Other minerals which occur in sands are feldspar, mica (particularly white mica), apatite, garnet, zircon, tourmaline, and magnetite; they are commonly present in small amounts in many sands, but occasionally may be a prominent constituent. They have been derived from igneous rocks which during denudation have contributed grains to the sediment. Ore mineral grains of high specific gravity are present in stream sands in certain localities, e.g. alluvial tin (cassiterite) or gold.

The many uses of sand in the manufacturing and construction industries has necessitated its detailed sub-division into grades. Two scales are given here (Table 6.3); the Wentworth scale, in common use among geologists, and the Atterberg (or M.I.T.) scale, which is frequently employed in engineering standards and specifications, and used in soil mechanics. In geology a *well graded* sand has particles of mainly one size (e.g. Fig. 6.3*f*): Fig. 6.3*g* illustrates a poorly graded deposit, in geological terms.

Table 6.3 Two commonly used grade scales for sand

Wentworth *Grade*	Wentworth *Size*	Atterberg *Size*	Atterberg *Grade*
	2 mm	2 mm	
v. coarse			
	1 mm		coarse
coarse		0.6 mm	
	$\frac{1}{2}$ mm		
medium			medium
	$\frac{1}{4}$ mm		
		0.2 mm	
fine			
	$\frac{1}{8}$ mm		
			fine
v. fine			
	0.06 mm	0.06 mm	

The porosity of these deposits depends on several factors, including:

(*i*) The grade sizes of the grains; a deposit containing mainly grains of one grade will possess a higher porosity than one consisting of a mixture of grades, since in the latter case the smaller grains will partly fill interstices between the larger and so reduce the pore space (Fig. 6.3*f*, *g*). This applies also to sandstones.

(*ii*) The amount and kind of packing which the grains have acquired. Figure 6.8 shows two kinds of theoretical packing for spherical grains, the pore spaces being interconnected though irregularly shaped. The arrangement in (*a*) is known as hexagonal close packing; in a layer overlying the one shown, each sphere would lie above a group of three below it, and succeeding pairs of layers repeat the arrangement. In (*b*), each grain of the next layer rests directly on the one below it, and the packing is more open, with larger interstices; this is an unstable structure not found in nature.

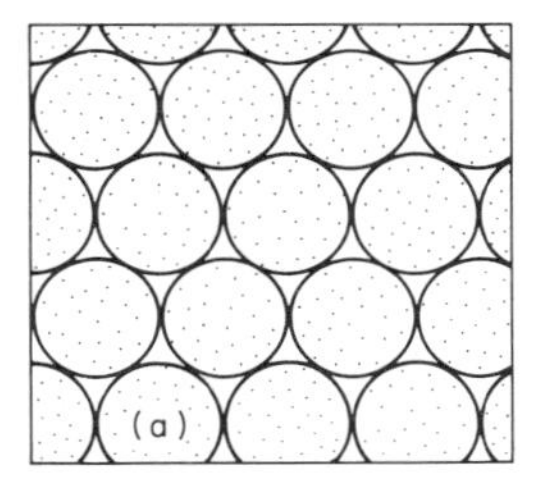

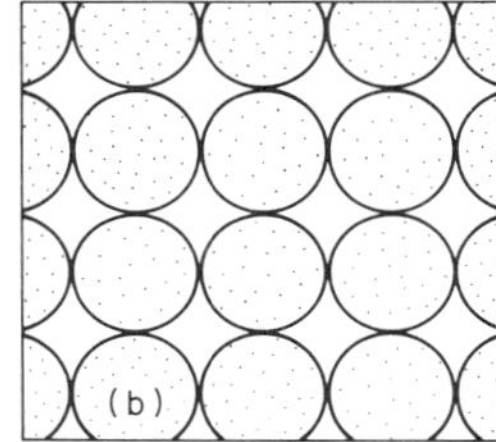

Fig. 6.8 (**a**) Close packing of spheres: porosity=27% (**b**) Open packing: porosity=47%.

(*iii*) The amount of cement present. Interstices may be filled wholly or partly with mineral matter which binds the grains together, and in this way porosity may be considerably reduced (Fig. 6.1).

It will be seen that many different possibilities arise from a combination of the three factors indicated above.

Sandstones

A sandy sediment, after natural compaction and cementation has gone on, is converted into a relatively hard rock called *sandstone*. It has a texture of the kind shown in Fig. 6.3*g*–*h*; there is no interlocking of the component grains, as in an igneous rock, and the mineral composition is often simple. Bedding planes and joints develop in the consolidated rock, and form surfaces of division; these often break the mass into roughly rectangular blocks and are useful in quarrying. Widely spaced bedding planes indicate long intervals of quiet deposition and a constant supply of sediment. Sands laid down in shallow water are subject to rapid changes of eddies and currents, which may form current-bedded layers between the main bedding planes, and this current-bedding is frequently seen in sandstones (see Fig. 3.19).

Cementation occurs in two main ways:

(1) By the enlargement of existing particles; rounded quartz grains become enlarged by the growth of 'jackets' of additional quartz (Fig. 6.2*a*), derived from silica-bearing solutions, the new growth being in optical continuity with the original crystal structure of the grains (a fact which can be tested by observing the extinction between crossed polars). This is seen in the quartz of the Penrith Sandstone (Fig. 6.9*a*), or of the Stiperstones Quartzite of south Salop. (For another use of the term *quartzite* see p. 136.)

Calcareous sandstone has a relatively weak cement of calcite which is easily weathered by acids in rain-water. The 'Calcareous Grit' of Yorkshire, of Corallian age, and the Calciferous Sandstones of Scotland (Lower Carboniferous) are examples. In some rocks the calcite crystals have grown around the quartz grains, which are thus enclosed in them; the crystals are revealed by reflection of light from cleavage surfaces when the rock is handled, and the appearance is described as 'lustre mottling'.

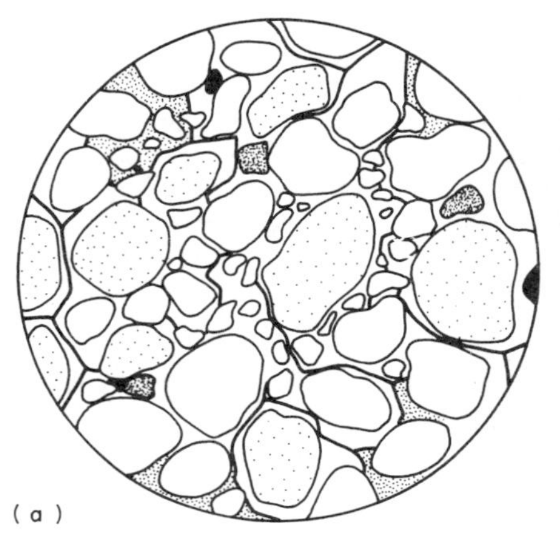

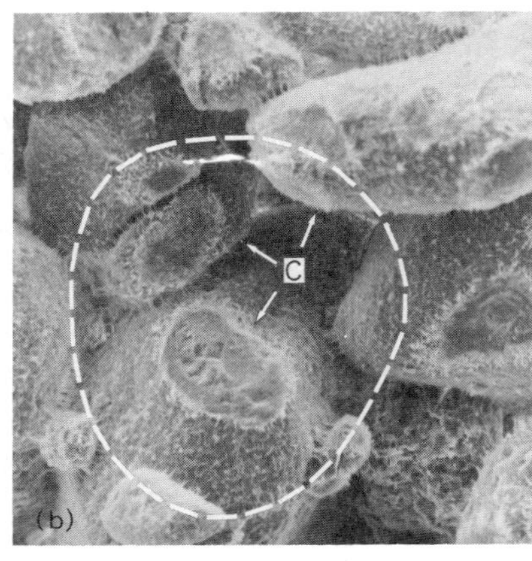

Fig. 6.9 (**a**) Penrith sandstone in thin section, showing enlargement of grains by new growth of quartz. (×12). See also Fig. 6.2. (**b**) Cemented sandstone as seen with an electron microscope. c=cemented contact areas for a grain that has been removed (×75). Photograph by L. Dobereiner.

(2) By the deposition of interstitial cementing matter from percolating waters (Fig. 6.9*b*). The three chief kinds of cement, in the order of their importance, are:

(*i*) silica, in the form of quartz, opal, chalcedony, etc.
(*ii*) iron oxides, e.g. haematite, limonite (Fig 6.1*a*).
(*iii*) carbonates, e.g. calcite, siderite, magnesite, witherite (p. 86).

Sandstones may be named according to their cementing material, or after constituents other than quartz.

Siliceous sandstone, with a cement of quartz or cryptocrystalline silica between the grains of quartz (Fig. 6.9*a*), is a very hard rock because of its high content of silica, which sometimes exceeds 90% of the whole, and is therefore resistant to weathering and abrasive to machines used for excavation. The term *orthoquartzite* is used for nearly pure, *evenly-graded* quartz-sands which have been cemented to form a sandstone. Examples are the Stiperstones quartzite at the base of the Ordovician, and the basal Cambrian quartzite of Hartshill, Warwick; both are the beach deposits of an incoming sea.

Ferruginous sandstone is red or brown in colour, the cement of iron oxide (haematite or limonite) forming a thin coating to the grains of quartz. Many rocks of Old Red Sandstone age (Devonian) in Britain, and Triassic sandstones, have ferruginous cement. Permian sandstones from Ballochmyle, (Fig. 3.36) and Locharbriggs, near Dumfries, are two structural stones of this type.

Argillaceous sandstone has a content of clay which acts as a weak cement, and the rock has a low strength; sandstones of this kind in the Carboniferous of Scotland and elsewhere are crushed to provide moulding sands for use

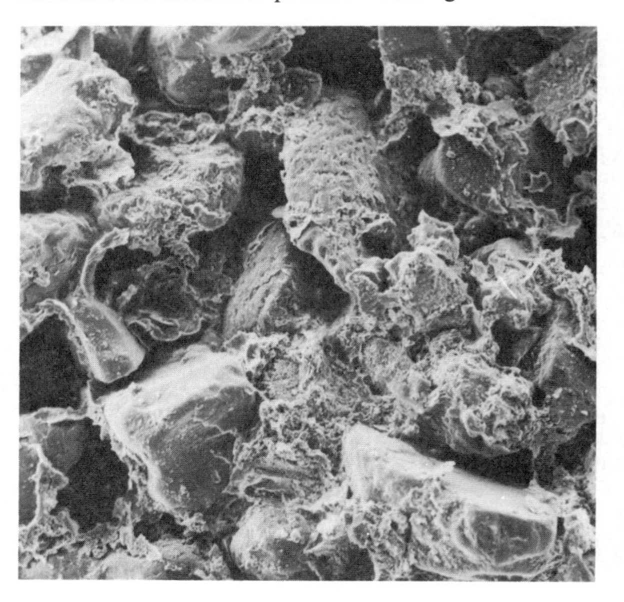

Fig. 6.10 Argillaceous sandstone as seen with an electron microscope: clay occupies the pore space. (×100). Photograph by L. Dobereiner.

in the steel industry. In many argillaceous sandstones there is a growth of clay minerals in the pore spaces between sand grains (Fig. 6.10). This results in a reduced porosity for the rock, and from the point of view of water abstraction it will have a lower permeability.

More rarely, sandstones may be cemented by a sulphate (gypsum or barytes), or a sulphite (pyrite), or occasionally a phosphate. In desert regions like the Sahara cementing by gypsum is observed in wind-blown sands; the gypsum crystals during growth enclose sand grains, thus forming 'sand-gypsum' crystals. Barytes ($BaSO_4$) gives a very hard cementation, as in the Alderley Edge sandstone, Cheshire; it has resulted in the preservation of isolated stacks of the sandstone, such as the Hemlock Stone, which are very resistant to atmospheric weathering.

Sandstones may also be named after prominent constituents other than quartz.

Micaceous sandstone, which has a content of mica flakes, usually muscovite. They may be dispersed throughout the rock, or arranged in parallel layers spaced at intervals of a few centimetres; along these layers the rock splits easily, and is called a *flagstone* (Fig. 6.11). This

Fig. 6.11 Flagstone at Freshwick, south of Wick, Caithness.

structure arises when a mixture of mica and quartz grains is gently sedimented in water; the micas settle through still water more slowly than the quartz because of their platy shape, and are thus separated from the quartz grains to form a layer above them. The process may be repeated many times if the conditions remain stable. The Moher Flags from the coast of Co. Clare, west Ireland, are an example; they (and other occurrences) are used in slabs and possess a useful resistance to weathering. The term *freestone* is used for sandstones *or* limestones which have few joints and can be worked easily in any direction, yielding good building material; e.g. the Clipsham Freestone of Leicester.

Glauconitic sandstone contains some proportion of the green clay-mineral glauconite, a hydrated silicate of iron and potassium formed in marine conditions. The 'Greensands' of the Lower Cretaceous in England are so named from their content of the mineral, which occurs as rounded grains and as infillings to the cavities of small fossils. Parts of the Upper Eocene marine sands in southern England are rich in glauconite, as in the Bracklesham Beds above the London Clay.

Greywacke (or German, 'grauwacke') is dark, often grey-coloured, and containing many coarse angular grains of quartz and feldspar which have been sedimented with little sorting, together with mica and small rock fragments (e.g. of slate) and fine matrix material. Greywackes are formed in the seas adjacent to rapidly uplifted land masses. Such an association commonly accompanies the development and filling of orogenic troughs or geosynclines (Fig. 2.3) when much coarse detritus is washed into the area of deposition. As a result of the eventual compression of the contents of the trough, typical greywackes are now found in areas of sharply folded strata, for example among the Lower Palaeozoic sediments of central Wales and the Southern Uplands of Scotland. Many greywackes show *graded-bedding* (Fig. 2.1), in which the sediment passes from coarser to finer particles from the bottom of a bed upwards; this structure is produced by the settling of a mixture of sand and mud in water, after movement over the sea floor as a turbidity current; a form of slump (Fig. 6.6). (Turbidity currents give rise to a mass of disturbed sediment which on later consolidation forms the rocks known as *turbidites*.) Elongated projections called *flow-casts* or *sole-markings* are frequently found on the undersides of greywacke beds; they show the direction of currents which operated at the time of their formation.

Arkose is the term used to denote typically pale-coloured sandstones, coarse in texture, composed mainly of quartz

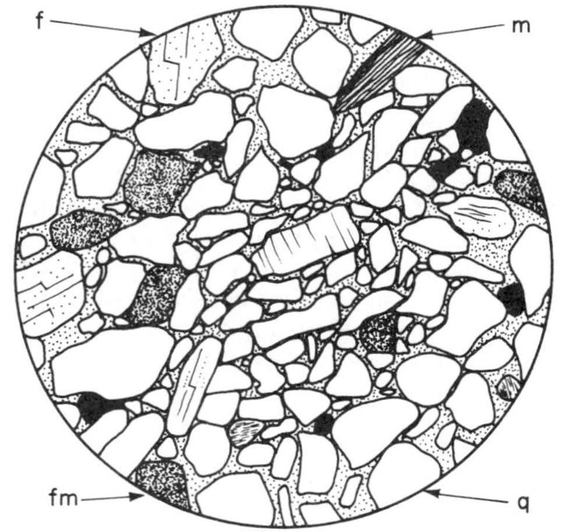

Fig. 6.12 Feldspathic sandstone (arkose) in thin section. m = mica, q = quartz, fm = ferromagnesian, f = feldspar. (× 15.)

and feldspar in angular or partly rounded grains, usually with some mica; the minerals were derived from acid igneous rocks such as granites, or from orthogneisses, which were being rapidly denuded at the end of an orogenic upheaval. The feldspar may amount to a third of the whole rock in some instances; the constituents are bound together by ferruginous or calcareous cement. Much arkose is found in the Torridon Sandstone of Ross-shire (Fig. 6.12).

Ganister is a fine-grained siliceous sandstone or siltstone which is found underlying coal seams in the Coal Measures; it is an important source of raw material for the manufacture of silica bricks and other refractories.

Argillaceous deposits

Silt

Minute fragments of minerals and clusters of microscopic crystals form particles called silt (Table 6.2). They accumulate as laminae interbedded with deposits of clay at periods when the supply of clay to an area of deposition is either diminished, or diluted by an influx of silt particles. Thicker sequences of silt may be formed by the finer particles derived from deltas (p. 43). Clay minerals, when present as in a clayey silt, provide a weak force that binds the silt particles to each other. Without clay minerals a silt composed of silicate fragments (Table 6.2) has little cohesive strength until cemented: in this respect deposits of silt and clay differ significantly. Ground-water flows with greater ease through the pores of silt than it does through those of clay and may precipitate mineral cement converting the silt to *siltstone*.

Clays

These are muddy sediments composed of minute clay minerals (p. 80) of the order of 0.002 mm diameter or less, which were deposited very slowly in still water on the continental shelf or in lakes and estuaries. There are also particles of colloidal size, which in some cases form a high proportion of the clay. The term 'fat clay' is used in some engineering literature for a clay which has a high content of colloidal particles, and is greasy to the touch and very plastic; a 'lean clay', in contrast, has a small colloidal content. Soon after the deposition of a mass of muddy sediment, sliding or slumping may occur on submarine slopes, and the incipient bedding structures are destroyed or disturbed; considerable thicknesses of disturbed shales are seen intercalated between normal beds at many localities, and are attributable to slumping contemporaneous with deposition. Some silt particles are commonly present in these sediments, and by lateral variation, with increase in the proportion of the silt, shales pass gradually into fine-grained sandy deposits.

Compaction and consolidation processes have been described on p. 113, and in argillaceous sediments they result in the progressive change from a soft mud to a stiff clay of greater strength, in which the mineral grains possesses a preferred orientation (see Fig. 6.13).

Sensitive clays

When the strength of an undisturbed sample of clay or silt is compared with the strength of the material after it has been re-moulded, the ratio of the undisturbed strength to the re-moulded strength is called the *sensitivity* of the material. Re-moulding breaks down the structure of the clay (i.e. the clustering and orientation of grains), although its natural water content remains the same. Values of sensitivity up to 35 or more have been recorded; but many clays have sensitivities of about 10, i.e. they lose 90% of their strength on being re-moulded. Such *sensitive clays* when disturbed during engineering construction may undergo great loss of strength, and it is therefore important that their character should be understood when excavations in them are to be made.

Clays which are highly sensitive are also known as 'quick' clays. Certain Norwegian quick clays, of late-glacial age (deposited in lakes) or post-glacial (marine), have been investigated, as well as others from the St Lawrence valley, Ontario. Many of the post-glacial marine clays have been leached by the passage of fresh water through them after the isostatic uplift of the area in which they were deposited (Fig. 2.19), and this has resulted in the removal of some of the dissolved salts in their original pore-water. The connection between reduced salt content of the pore-contained water and increased sensitivity of the clay was demonstrated experimentally for a quick clay from Horten, Norway, by Skempton and Northey (1952): see also Kenney (1964).

Mineral content

Clays and the rocks derived from them have a complex mineral composition, which is the more difficult to investigate on account of the very small size of the particles involved. The *clay minerals* have been described in Chapter 4; they are hydrous aluminium silicates, sometimes with magnesium or iron replacing part of the aluminium and with small amounts of alkalies. They form minute flaky or rod-like crystals, which build up the greater part of most clays and are important in determining their properties. Clays may also contain a variable proportion of other minerals, such as finely divided micaceous and chloritic material, together with colloidal silica, iron oxide, carbon, etc., and a small proportion of harder mineral grains (e.g. finely divided quartz). Organic matter may also be present. In addition, water is an important constituent and on it the plasticity of the clay depends; the water forms thin films around the very small mineral particles and fills the minute pore spaces. These films of water which separate the mineral flakes endow the clay with plastic properties. The water-absorption capacity of clay in turn depends on the nature of the clay minerals. Some particular kinds of clay are now described.

Fuller's earth, a clay largely composed of montmorillonite, has low plasticity and disintegrates in water. It readily absorbs grease and is used for the bleaching and filtration of oils, for cleaning cloth and taking grease out of wool, and for medical purposes. Deposits of Fuller's earth are worked in the Sandgate Beds (part of the Lower Greensand) at Nutfield, Surrey, and in Jurassic rocks at Combe Hay near Bath and at Woburn. Supplies of commercial bentonite (below) are obtained by processing these deposits.

Bentonite is a clay derived from the alteration of volcanic dust and ash deposits, and is mainly composed of montmorillonite. Owing to the capacity of this mineral to absorb water within the crystal lattice (p. 81), as well as acquiring a film of water around each particle, bentonite clays swell enormously on the addition of water, forming a viscous mass. This property renders the material useful for various purposes, such as the thickening of drilling mud in sinking oil wells; it is also used in America as an ingredient of moulding sands for foundries, and as an absorbent in many processes. Commercial deposits are worked in the Black Hills of Wyoming and South Dakota. In civil engineering bentonite is employed as a sealing layer in trenches and cofferdams to prevent the percolation of water, and as a slurry pumped into sands or gravels to fill the voids and render the mass impervious. It is also used, for example, in the construction of the Hyde Park Corner underpass, London, to help support the walls of the excavations in sands and gravel.

Fire-clays are rich in kaolinite, and commonly contain small amounts of quartz and hydromica. They occur beneath coal seams in the Coal Measures, and characteristically have a very low content of alkalies. They can be exposed to high temperatures, 1500 to 1600°C, without melting or disintegrating and are used in the manufacture of refractories.

Ball-clays are plastic, kaolinitic clays formed from transported sediment and generally deposited in lacustrine (=lake) basins. The clays are composed of secondary kaolinite together with some micaceous material (e.g. illite), and may contain a proportion of silt or sand. The kaolinite, which forms 60% or more of the ball-clay, is probably derived from argillaceous rocks such as shales during their weathering and transport, especially under warm temperature conditions; it is less well crystallized than kaolinite derived by hydrothermal decomposition of granite (p. 138), the minute crystals being commonly less than 1 μm in size.

Ball-clays are used in the ceramic industry, and when fired at temperatures of 1150–1200°C, they have a white or near white colour. In England they are worked in the Bovey Basin near Newton Abbot, Devon, where deposits of Oligocene age are associated with lignites; and near Wareham, Dorset. The geotechnical properties of the Bovey Basin deposits have been described by Best and Fookes (1970).

Certain argillaceous deposits are closely associated with the Pleistocene glaciations and include *boulder clays* (a type of till) and *varved clays* (p. 57). Wind-blown deposits such as loess (p. 53) composed of fine clay-sized particles, were widely distributed by dust storms in the cold dry periods during the Pleistocene.

Brickearth is a brown-red silty clay formerly used for making bricks, and is a type of loess.

Apart from uses mentioned above, clays are employed in industry in many ways such as for bricks, tiles, terracotta, tile drains, earthenware and porcelain, firebricks, crucibles and other refractories, puddle for engineering structures (e.g. the cores of earth dams), and in cement manufacture. The suitability of a deposit for any particular purpose depends on the presence of different clay minerals and other constituents, and their relative proportions in the deposit.

Shales

Shales are compacted muds, and possess a finely laminated structure by virtue of which they are fissile and break easily into parallel-sided fragments (Fig. 6.13). This lamination is parallel to bedding-planes and is analogous to the leaves in each book of a pile of books. Starting from a deposit of very fluid mud with a high water content, water has been slowly squeezed out from the sediment as a result of the pressure of overlying deposits, until the mud has a water content of perhaps 10–15%. With further compaction and loss of more water, the deposit has ultimately taken on the typical shaley parting parallel to bedding; there is thus a gradation from a mud to a shale.

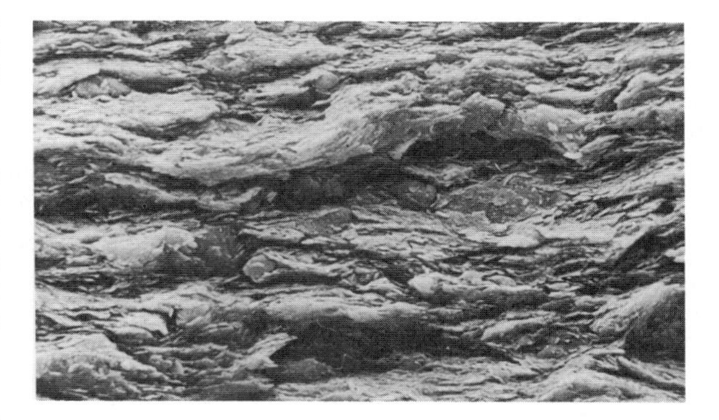

Fig. 6.13 Fissile shale viewed with an electron microscope (×1000).

During the expulsion of water from clay in the process of compaction, mineral reconstitution takes place and the resulting *shale* consists essentially of sericite (*cf.* illite), chlorite, and quartz; small crystals grow out of the raw materials available and form part of the shale, which, however, is not completely crystalline. Chemically, shale is characterized by a high content of alumina and is also generally rich in potash.

Some particular kinds of shale are now described.

Bituminous shale or *oil-shale* is black or dark brown in colour and contains natural hydrocarbons from which crude petroleum can be obtained by distillation. It gives a brown streak and has a 'leathery' appearance, with a tendency to curl when cut; parting planes often look smooth and polished. The oil-shales of the Lothians of Scotland provide raw material for the production of crude oil and ammonia.

Alum-shale is a shale impregnated with alum (alkaline aluminium sulphate): the alum is produced by the oxidation and hydration of pyrite, yielding sulphuric acid which acts on the sericite of the shale.

Mudrock is a general term presently used to describe argillaceous rocks; a review of the mudrocks of the United Kingdom has been edited by de Freitas (1981).

When the shaley parting is not developed an argillaceous rock may be called a *mudstone*.

Marl is a calcareous mudstone. In the Keuper Marl the mineral dolomite is present, in addition to clay minerals (mainly illite), chlorite, and a little quartz. The carbonate content ranges from about 7% to over 20%: for references see Barden and Sides (1970).

Detrital (pyroclastic) sediments

Fine ash and coarser debris (Table 6.2) ejected from an active volcano rain down upon the surrounding land to form a blanket of pyroclastic deposits (p. 94). The eruption of Mt St Helens in May 1980 deposited several metres of pyroclastic sediment within 10 km of the volcano and several centimetres of ash over most of Washington, W. Montana and N. Idaho.

Some pyroclastic deposits exhibit a delicate grading of particles from coarse at the base to fine at the top. Others contain an unsorted mixture of sizes and large rock blocks, and volcanic bombs, may be found embedded in finer ash (Fig. 6.14).

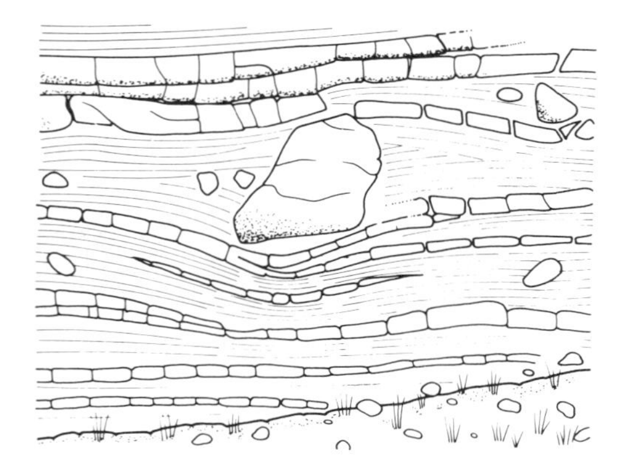

Fig. 6.14 Pyroclastic deposits: a vertical face in bedded tuffs enclosing large block ejected during an eruption (from a photograph).

When pyroclastic deposits settle in water they become interbedded with the other sediments. When deposited on land they may bury previously extruded lavas and be buried by later lavas so that the succession of strata becomes an alternating sequence of extrusive igneous and sedimentary pyroclastic rocks (Fig. 5.5).

Detrital (calcareous) sediments

The Limestones

Limestones consist essentially of calcium carbonate, with which there is generally some magnesium carbonate, and siliceous matter such as quartz grains. The average of over 300 chemical analyses of limestones showed 92% of $CaCO_3$and $MgCO_3$ together, and 5% of SiO_2; the proportion of magnesium carbonate is small except in dolomite and dolomitic limestones. The limestones considered here are those formed mainly by the accumulation of carbonate detritus. They are bedded rocks often containing many fossils; they are readily scratched with a knife, and effervesce on the addition of cold dilute hydrochloric acid (except dolomite). The distance between bedding-planes in limestones is commonly 30 to 60 cm, but varies from a couple of centimetres or less in thin-bedded rocks (such as the Stonesfield 'Slate') to over 6 m in some limestones.

Calcium carbonate is present in the form of crystals of calcite or aragonite, as amorphous calcium carbonate, and also as the hard parts of organisms (fossils) such as shells and calcareous skeletons, or their broken fragments (Fig. 6.2). Thus, a consolidated *shell-sand* (table 6.2) is a limestone by virtue of the calcium carbonate of which the shells are made. On the other hand, chemically deposited calcium carbonate builds limestones (e.g. oolites) under conditions where water of high alkalinity has a restricted circulation, as in a shallow sea or lake. Non-calcareous constituents commonly present in limestones include clay, silica in colloidal form or as quartz grains or as parts of siliceous organisms, and other hard detrital grains. Though usually grey or white in colour, the rock may be tinted, e.g. by iron compounds or finely divided carbon, or by bitumen. Common types of limestone are now described.

Shelly Limestone (Fig. 6.2) is a rock in which fossil shells, such as brachipods and lamellibranchs (Fig. 2.14), form a large part of its bulk. It may be a consolidated shell-sand (*cf.* the 'crags' of East Anglia). *Shell banks*, in which the shells are laid by currents and the spaces between them filled by milky calcite (during diagenesis), are also known as 'conquinites'.

Chalk is a soft white limestone largely made of finely-divided calcium carbonate, much of which has been shown to consist of minute plates, 1 or 2 μm in diameter. These plates are derived from the external skeletons of calcareous algae, and are known as coccoliths. The Chalk also contains many foraminifera, which differ in kind and abundance in different parts of the formation; and other fossils, such as the shells of brachiopods and sea-urchins (echinoids, Fig. 2.14). The *foraminifera* are minute, very primitive jelly-like organisms (protozoa) with a hard globular covering of carbonate of lime (Fig. 2.22); they float at the surface of the sea during life, and then sink and accumulate on the sea floor (p. 113). *Radiolaria*, which have siliceous frameworks, often of a complicated and beautiful pattern, are also found in Chalk but are not so numerous as the foraminifera. Parts of the rock contain about 98% $CaCO_3$ and it is thus almost a pure carbonate rock. It was probably formed at moderate depths (round about 180 m) in clear water on the continental shelf. The Chalk in England is a bedded and jointed rock; the upper parts of the formation contain layers of flint (p. 78) along some bedding-planes and as concretions in joints. Chalk is used for lime-burning and, mixed with clay, is calcined and ground for Portland Cement manufacture.

Limestones which contain a noticeable amount of substances other than carbonate are named after those substances, thus: *siliceous limestone, argillaceous limestone, ferruginous limestone, bituminous limestone*; these terms are self-explanatory. *Cement-stone* (or hydraulic limestone) is an argillaceous limestone in which the proportions of clay and calcium carbonate are such that the rock can be burned for cement without the addition of other material, e.g. the cement-stones of the Lower Lias in England.

Chemical and biochemical limestones, which include oolitic, dolomitic, magnesian, coral, algal and crinoidal limestones, are described in the following section.

Chemical and biochemical sedimentary rocks

Calcareous deposits

The Limestones (cont.)

As explained in the previous section, the calcareous particles of many limestones have organic or biochemical origins but have been sedimented as detritus to give the resulting rock a character that is predominantly detrital. Other limestones are dominantly organic or chemical in nature and are described here: their particles have not been greatly transported although some mechanical sorting may have occurred.

Coral, Algal, and Crinoidal Limestones take their names from prominent fossil constituents (Fig. 6.15). *Algae* are aquatic plants allied to seaweeds; some kinds secrete lime and their remains may build up large parts of some limestones. Some fresh-water algae precipitate calcium carbonate, and other silica. Many corals and crinoids (Fig. 2.10), which have been described earlier (p. 22), are found in parts of the Carboniferous Limestone and limestones elsewhere especially of Silurian and Jurassic age. Reef facies are abundant at the present day, as in the Great

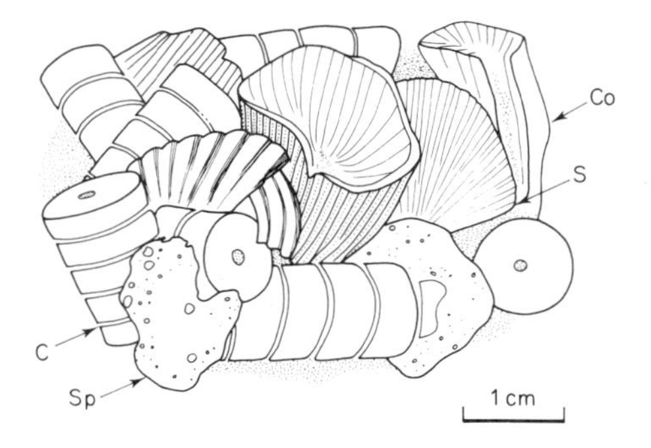

Fig. 6.15 Specimen of limestone containing fragments of crinoid (c), coral (co), shell (s) and sponge-like forms (sp). (And see Fig. 2.10.)

Barrier Reef off the north-east coast of Australia, and the islands of the west Pacific and the Carribean.

Oolitic Limestone, or *oolite*, has a texture resembling the hard roe of a fish (Greek *oon*=egg). It is made up of rounded grains formed by the deposition of successive coats of calcium carbonate around particles such as a grain of sand or piece of shell, which serves as a 'seed'. These particles are rolled to and fro between tide marks in limy water near a limestone coast, and become coated with calcium carbonate; they are called 'ooliths,' and their concentric structure can be seen in thin section (Fig. 6.16*a*). Sometimes the grains are larger, about the size of a pea, when the rock made of them is called *pisolite*. Deposition of $CaCO_3$ may also take place through the agency of algae; each particle, or seed, which is rolled about by currents acquires a coating of calcareous mud, which is surrounded by a layer of blue-green algae. The latter can survive the rolling for a long time. Devonian oolites when carefully dissolved have been shown to contain these concentric organic coatings. Ooliths are forming today in the shallow seas off the coasts of Florida and the Bahamas.

In the course of time the ooliths, often mixed with fossil shells, become cemented, and then form a layer of oolite. When few shells are present the rock has an even texture and is easily worked, and makes a first-class building stone. The Jurassic oolites of the Bath district, and the Portland oolitic limestone, Dorset, are two English examples. Portland stone has been used in many London buildings and other structures, where it weathers evenly over a long period. The cream-coloured stone is quarried from beds of the Portland Series at Portland, Dorset, the best material coming from the Whit Bed and the Base Bed. A coarser shelly layer, the Roach, is used for rough walling.

Dolomitic Limestone and *Dolomite* are rocks which contain the double carbonate $MgCO_3.CaCO_3$, dolomite (the name is used for the mineral as well as for the rock); the mineral occurs as rhomb-shaped crystals (Fig. 6.16b). Dolomite is made entirely of the mineral dolomite, but

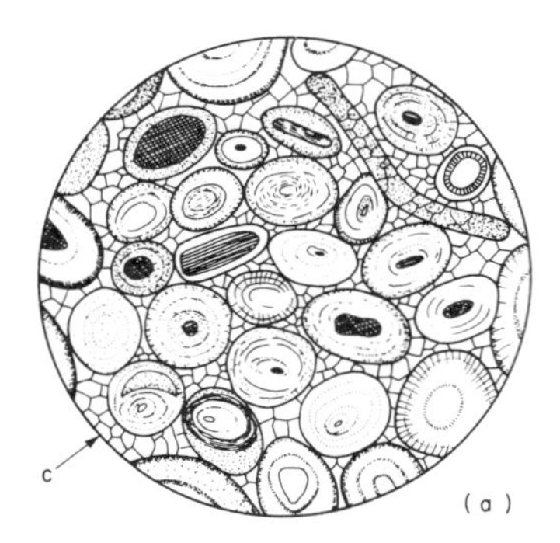

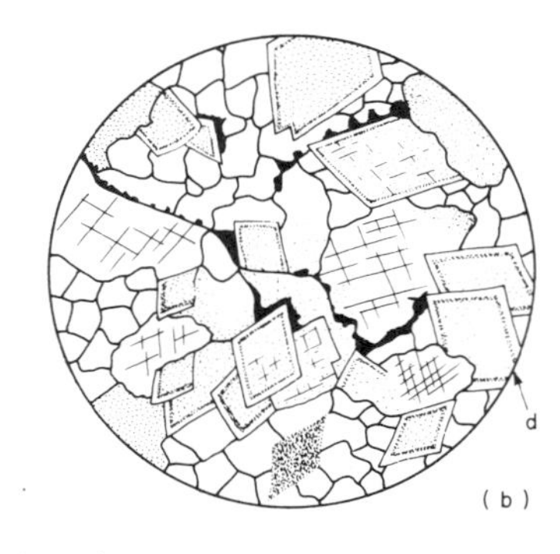

Fig. 6.16 (**a**) Oolitic limestone in thin section: note calcite cement (c) filling original pore space (x10).
(**b**) Dolomitic limestone in thin section: dolomite (d) (x10).

dolomitic limestone has both dolomite and calcite. For a test for dolomite see page 83. *Magnesian limestone* is the term used for a limestone with a small content of magnesium carbonate, which is usually not present as the double carbonate but is held in solid solution in calcite crystals.

The mineral dolomite often replaces original aragonite or calcite in a rock, the magnesium being derived from sea water and introduced into the limestone by solutions passing through it. The replacement process is called *dolomitization*. The change from calcite to dolomite involves a volume contraction of 12.3% and hence often results in a porous texture. Borings in the coral atoll of Funafuti, in the South Pacific, showed progressive dolomitization of the limestone reef with depth, indicating that the change takes place in shallow water. Extensive dolomitization of recently deposited aragonite mud occurs in the tidal zones of sabkhas (p. 49). Sea water carries Mg^{2+} and SO_4^{2-} into the mud flats where Mg^{2+} replaces some Ca^{2+} in the crystal lattice of aragonite. The displaced Ca^{2+} combines with SO_4^{2-} to precipitate as nodules of anhydrite ($CaSO_4$) or gypsum ($CaSO_4.2H_2O$).

Beds of dolomite are found in the Carboniferous Limestone, e.g. in the Mendips, and at Mitcheldean, Forest of Dean; in some cases the process of dolomitization has been selective, the non-crystalline matrix of a limestone being dolomitized before the harder calcite fossils which it contains are affected. The Magnesian Limestone of Permian age is mainly a dolomitic limestone.

The general term for replacement processes, including that outlined above, is *metasomatism* (=change of substance, Greek); original minerals are changed atom by atom into new mineral substances by the agency of percolating solutions, but the outlines of original structures in the rock are frequently preserved. The dolomitization of a limestone thus involves the replacement of part of the calcium by magnesium. Another metasomatic change which takes place in limestones is the replacement of calcium carbonate by silica; in this way silicified limestones and *cherts* are formed, as in parts of the Carboniferous Limestone. Silica from organisms such as sponges and radiolaria is dissolved by water containing potassium carbonate, and re-deposited as chert, which is a form of cryptocrystalline silica. The Hythe Beds of the Lower Greensand in Kent have beds of chert of considerable thickness. Replacement of calcite by siderite ($FeCO_3$) is another important metasomatic change; it is seen in some Jurassic limestones such as the Cornbrash of Yorkshire and the Middle Lias marlstone of the Midlands. The Cleveland Ironstone of Yorkshire, which is oolitic, is believed to have originated in this way.

Travertine, stalactite and *calc-sinter* are made of chemically deposited calcium carbonate from saturated solutions. Calc-sinter (or *tufa*) is a deposit formed around a calcareous spring. Stalagmites and stalactites are columnar deposits built by dripping water in caverns and on the joint planes of rocks (see Chapter 2). Travertine is a variety of calc-sinter, cream or buff-coloured, with a cellular texture due to deposition from springs as a coating to vegetable matter. When the latter decays it leaves irregular cavities. There are deposits of travertine in Italy, in the Auvergne, and elsewhere. The important deposit on the banks of the R. Anio near Tivoli is about 140 m thick, and is soft when quarried but hardens rapidly on exposure. Travertine has been used in many buildings for interior or exterior panelling, with or without a filler; and in slabs for non-slip flooring.

Siliceous deposits

Under this heading are included the organically and chemically formed deposits mentioned in Table 6.2. The deep-sea siliceous oozes are briefly described on p. 49. *Diatomite* is the consolidated equivalent of *diatom-earth*, a deposit formed of the siliceous algae called diatoms, which accumulate principally in fresh-water lakes. It is found at numerous Scottish localities and at Kentmere in the Lake District, where there is a deposit of Pleistocene age. The United States, Denmark, and Canada have large deposits from which considerable exports are made. Diatom-earth is also known as *kieselguhr* or as *Tripoli-powder*, and was originally used as an absorbent in the manufacture of dynamite. It is now employed in various chemical processes as an absorbent inert substance, in filtering processes, as a filler for paints and rubber products, and for high temperature insulation. Deposits of *sponge spicules*, which are fragments of the skeletons of siliceous sponges, are found today on parts of the ocean floor; older deposits of this kind provided the raw material for the cherts which occur in certain limestones.

Flint is the name given to the irregularly shaped siliceous nodules found in the Chalk. It is a brittle substance and breaks with a conchoidal fracture; in thin section it is seen to be cryptocrystalline. The weathered surface of a flint takes on a white or pale brown appearance. In England, flints occur most abundantly in the Upper Chalk where they may lie along bedding-planes or fill joints in the form of tabular vertical layers. The silica of flint may have been derived from organisms such as sponge spicules and radiolaria in the Chalk, which were slowly dissolved and the silica re-deposited probably as silica gel by solutions percolating downwards through the Upper and Middle Chalk. The Lower Chalk generally contains no flints. Other modes of origin for siliceous deposits in limestone have been suggested and it is thought that many such 'deposits' developed during diagenesis by replacement of calcite of aragonite. For *chert* see p. 126.

Jasper is a red variety of cryptocrystalline silica, allied to chert, the colour being due to disseminated Fe-oxide; it is found in Precambrian and Palaeozoic rocks.

Saline deposits

The Evaporites

When a body of salt water has become isolated its salts crystallize out as the water evaporates. The Dead Sea is a well-known example; it has no outlet and its salinity constantly increases. Another instance is that of the shallow gulf of Karabugas in the Caspian Sea, into which salt water flows at high tide but from which there is no outflow, so that salt deposits are formed, in what is virtually a large evaporating basin which is continually replenished. In past geological times great thicknesses of such deposits have been built up in this way in various districts. Deep-sea drilling has revealed 1.5 km of such deposits in the Mediterranean, formed there in Miocene times when it was a subsiding inland basin (p. 24). Greater thicknesses of evaporites (up to 3 km) have been identified in the Red Sead and Danikil Depression of Ethiopia. During evaporation the first salt to be deposited is *gypsum* ($CaSO_4.2H_2O$), beginning when 37% of the water is evaporated: *anhydrite* ($CaSO_4$, orthorhombic) comes next but as explained on p. 126 much of the gypsum and anhydrite formed in natural sediments may have originated from diagenetic changes of previously deposited shallow water calcareous sediments. Continued evaporation eventually precipitates the *halides*. First *rock salt* (NaCl). Pseudomorphs of rock-salt cubes are found in the Keuper Marl (see p. 21 for description of the conditions under which these rocks were deposited). Lastly are formed magnesium and potassium salts such as *polyhalite* ($K_2SO_4.MgSo_4.2CaSO_4.2H_2O$), *kieserite* ($MgSO_4.H_2O$), *carnallite* ($KCl.MgCl_2.6H_2O$), and *sylvite* (KCl). All these are found in the great salt deposits of Stassfurt, Germany, which reach a thickness of over 1200 m, and have been preserved from solution through percolating water by an overlying layer of clay.

Gypsum and rock-salt are found in the Triassic rocks of Britain, e.g. in Cheshire, Worcester and near Middlesbrough, but nothing beyond the rock-salt stage was known in Britain until deep borings were made at Aislaby and Robin Hood Bay, north Yorkshire. These penetrated important potash salt deposits lying above rock-salt and covered by marl, at depths of about 1250 m. The development of such very soluble deposits is rendered difficult owing to the depth at which they occur. Great thicknesses of evaporite exist in the Michigan Basin, in southern New Mexico and in Utah.

Gypsum associated with calcareous sediments is mined at Netherfield, Sussex, where a deposit in the Purbeck Beds is found between 30 and 60 m below the surface. Apart from its use in building industry, gypsum is used as a filler for various materials (e.g. rubber), and for a variety of other purposes (p. 87). Rock-salt, apart from domestic uses, is employed in many chemical manufacturing processes; potash salts are used as fertilizers, as a source of potassium, and in the manufacture of explosives.

Sodium salts. Soda nitre (or 'Chile salt-petre', $NaNO_3$) deposits are found in Atacama Desert of Chile, in sandy beds called *caliche*, of which they form from 14 to 25%. It is exported largely as a source of nitrates, iodine (from sodium iodate) being an important byproduct. Other sodium and magnesium salts were deposited from the waters of alkali lakes, e.g. in arid regions in the western U.S.A., and the Great Salk Lake of Utah. Sodium sulphate ('salt cake') is important in the chemical industry and in paper and glass manufacture. It is produced chiefly in the U.S.S.R., Canada, and the U.S.A.

The evaporites form an economically important group, of which the foregoing is a brief outline.

Carbonaceous deposits

The Coals

The formation of coal through the burial of ancient vegetation is described in Chapter 2 (p. 23). The gradual transformation from rotting woody matter to coal can be represented by peat, lignite (brown coal), humic or bituminous coal (soft coal) and anthracite (hard coal). This sequence is called the *coal series* and the average compositions of several varieties within the series are given in Table 6.4. The quality of a coal is described by its *rank*, whose relationship to the major constituents is shown in Fig. 6.17.

Table 6.4 Average composition of fuels (percentages): sulphur and mineral impurities which create ash are not included

	C	H	N	O
Wood	49.7	6.2	0.9	43.2
Peat	56.6	4.8	1.7	36.9
Lignite	65.2	3.5	1.2	30.1
Bituminous coal	84.5	4.6	1.5	9.4
Anthracite	93.6	2.3	1.1	3.0

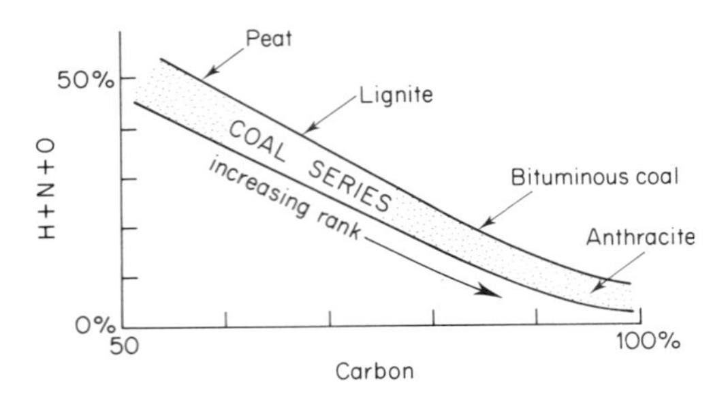

Fig. 6.17 Range of composition for members of the coal series.

Peat is derived from compressed mosses and plants such as sphagnum, heather and cotton-grass. It is a valuable fuel in some areas and is cut and burnt for domestic use; it has a high ash content and burns with a smoky flame. Peat-fired generating stations for electricity are in use in Ireland. Peat is not a coal but the microbial degradation of vegetation that produces peat, and to which the term 'peat stage' refers, is a necessary stage in the formation of coal. In waterlogged environments, as in swamps, the microbes cannot fully decompose vegetable matter and partially digested plant debris accumulates, which can later be converted to coal by compression and diagenesis.

Lignite (or brown coal) is more compact than peat and is the first product of the next phase of coal formation, namely burial. Lignite has lost some moisture and oxygen and its carbon content is consequently higher. Its texture is rather like woody peat, and when dried on exposure to the atmosphere it tends to crumble to powder. Extensive beds of lignite are found in Germany (Cologne), Canada, the southern U.S.A. and Brazil. Average calorific values range from 7×10^6J to 11×10^6J.

Bituminous coal includes the ordinary domestic variety and many coals used in industry: it has a carbon content of 80–90% and a calorific value of 14×10^6J to 16×10^6J. Between it and lignite there are coals called *sub-bituminous* (or 'black lignites') with less moisture than lignite and between 70–80% carbon. *Semi-bituminous* coals are transitional between bituminous coal and anthracite: they have the greatest heating value of any and burn with a smokeless flame.

Typical bituminous coals have a banded structure in which bright and dull bands alternate: they are classified into four types on the basis of their lustre and other physical characters.

(*i*) *Vitrain*, bands or streaks of black coal, several millimetres or more in thickness, representing solidified gel in which the cellular structure from leaves, stems, bark, twigs and logs can be detected.

(*ii*) *Clarain*, less bright and having a satin or silky lustre derived from thin vitrinite sheets alternating with a duller matrix.

(*iii*) *Durain*, dull hard coal, containing spore cases and resins.

(*iv*) *Fusain*, a soft black powdery substance ('mineral charcoal') which readily soils the fingers, occurs in thin bands and is composed of broken spores and cellular tissue which may be the oxidized remains of vegetation.

These coal types reflect the original composition of the coal whereas the rank of coal is governed by its subsequent maturation, especially with burial. 'Bright' coals have a high content of vitrain, which is the best source of coal tar byproducts.

Blocks of coal break easily along the fusain layers, and are bounded by fractures ('cleat' and 'end') which are perpendicular to the layers: a thin film of mineral matter (e.g. pyrite) often coats the surface of these fractures.

Cannel coal is a dense deposit lacking noticeable bedding, dull grey to black in appearance (rather like durain), and is composed mainly of spores. It contains 70–80% carbon, 10–12% hydrogen and ignites with ease to burn with a long smoky flame like a candle (whence its name). Typical cannels have much ash and a calorific value of 12×10^6J to 16×10^6J. They are believed to be formed from transported material (unlike other coals, see later) and the ash is attributable to an admixture of sediment which was originally deposited with the drifted vegetable debris.

Anthracite contains from 90–95% carbon, with low oxygen and hydrogen, and ignites only at a high temperature; its calorific value is about 15×10^6J. It is black with sub-metallic lustre, conchoidal fracture, and banded structure, and does not mark the fingers when handled. Anthracite appears to have been formed when coal-bearing beds have been subjected to pressure or to increased temperatures, during gentle metamorphism; transitions from anthracite to bituminous coals have been traced, e.g. in the South Wales coalfield, where also shearing structures are present in the anthracites. In N. America seams of bituminous coal in Pennsylvania can be traced into the Appalachian fold belt where they are anthracite.

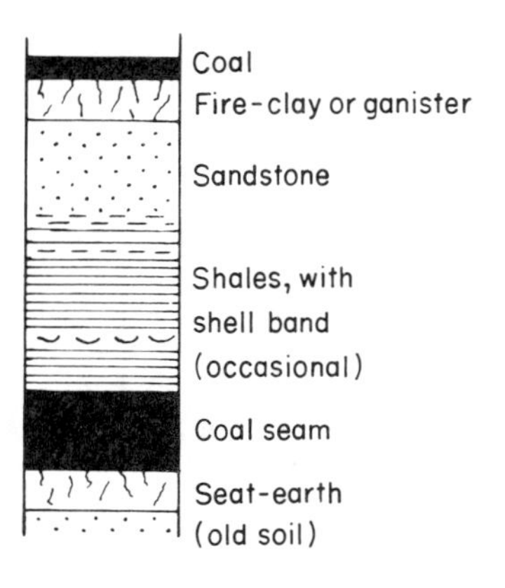

Fig. 6.18 Unit of coal measure rhythm.

Associated with coals are beds of *ganister* (p. 122) and *fireclay* (p. 123) (Fig. 6.18), which lie below the seams and contain the roots of plants in their position of growth; they represent the seat-earth in which the Coal Measure plants grew. Overlying a coal seam is a series of shales, of varying thickness, which grade upwards into sandy shales and then sandstones. These are followed by the next coal seam, with its seat-earth and its overlying shales and sandstones. The repetition of this sequence of rocks over and over again, with minor variations, is referred to as the Coal Measure 'rhythm', each unit of which records the submergence of the vegetation of a coal swamp, and the eventual emergence (as the water shallowed) of sandy shoals on which the swamps grew again.

Oil shale has been described on p. 123. One variety, *boghead coal*, or *torbanite* appears to be closely related to cannel coals. Substantial amounts of oil can be extracted from these shales on heating. The Scottish oil shales yield 90 l^{-t} and those of Colorado, Wyoming and Utah, 125 l^{-t} (Greensmith, 1978).

The existence of most coal fields results from their coal-bearing strata being preserved in a coal basin, Fig. 6.19, that part beneath a cover of younger rock being described as concealed. In the process of folding clay rich horizons within the fire-clays (Fig. 6.18) may be disturbed and lose much of their strength. Slopes in open cast mines may fail on such layers (Walton and Coates, 1980).

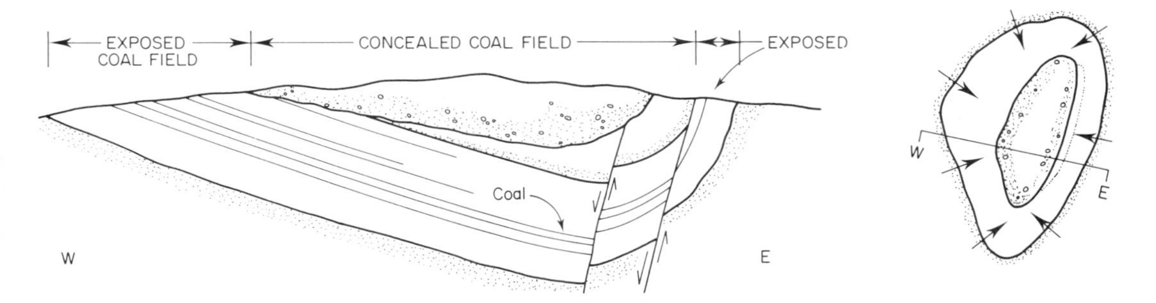

Fig. 6.19 A coal basin in vertical section and plan. Arrows represent the direction of strata inclination towards the basin centre. Such basins may vary in dimension from 10 km to 100s km.

Ferruginous deposits

The Ironstones

Numerous sediments and sedimentary rocks contain iron, some in concentrations that make the deposits valuable ores of iron. The iron may be precipitated as a primary mineral or be locked into crystal lattices during diagenesis.

Blackband ironstone and clay-ironstone are ferruginous deposits associated with the Coal Measures. In Stafford the blackband ironstones contain 10–20% coal, the iron being present as carbonate, and are found replacing the upper parts of coal seams; they are economical to smelt on account of their coal content. Clay-ironstones also consist chiefly of chemically deposited siderite ($FeCO_3$); they occur as nodular masses in the argillaceous rocks of the Coal Measures, and also in the Wealden Beds, where they are of fresh-water origin.

Bog-iron ores are impure limonitic deposits which form in shallow lakes and marshes, as in Finland and Sweden today. The deposition of the iron may be due to the action of bacteria or algae. The ores were much used in the early days of iron-smelting.

Jurassic ironstones Important bedded iron ores of this age in Britain and northern Europe contain the minerals siderite and chamosite (hydrated iron silicate); some of the ironstones are oolitic and others have small crystals of siderite in a matrix of mudstone. The Cleveland ironstone of the Middle Lias is an oolitic rock containing both the above minerals. A rock of the same age, the marlstone of Lincoln and Leicester, has a large amount of calcite as well as chamosite and siderite. The Northampton ironstones are partly oolitic rocks with chamosite as the chief constituent, and partly mudstones with siderite and limonite.

Haematite and limonite ores In some cases iron is present as haematite in a shelly limestone, as at Rhiwbina, South Wales, or as limonite, e.g. the Frodingham ironstone. The haematite deposits of Cumbria are replacements of the Carboniferous Limestone by irregular masses of the iron oxide. The Lake Superior region of North America possesses very extensive haematite deposits; the rocks here are Precambrian sediments in which iron of sedimentary origin has probably been concentrated as the oxide after alteration from its original silicate or carbonate form. The Precambrian iron ores (see Table 2.1) are world wide and extensive deposits exist in the Transvaal of S. Africa and the Hammersly Basin in Western Australia.

Sediment associations

The succession of sediments illustrated in Fig. 6.5 records an orderly sequence of events in which shallow water deposits grade into deeper water sediments, succeeded by shallow water and continental deposits. In this example the shallow water sediments (conglomerates, gravels and coarse sands) can be grouped as an association of similar materials (i.e. a *facies*, p. 113). Other groupings can be made of the sediments accumulated in deeper water, deltas and on land. The facies to which a deposit belongs will indicate the other types of sediment with which it may be associated. For example, a foundation engineer building on alluvium may assume that this facies will contain gravel, sand and clay, and possibly conglomerate, and that care must therefore be taken to ascertain the presence of clay even though sand and gravel may be the only deposits visible at ground level.

Table 6.5 lists some important associations which are commonly encountered.

Continental deposits are often the most variable of all types because they have not endured the long period of mechanical sorting that is associated with the deposition of sediments in the ocean. Boulder conglomerates may pass into finer sandstones and grits within the distance of a few metres (Fig. 6.7). The variations which occur in glacial sediments deposited on land have been described on p. 58. A greater degree of sorting occurs in alluvial sediments.

Shore-line deposits tend to be more uniform than continental sediments, but considerable variability may exist if they accumulated in shallow water. For example a modern mud-flat may contain thin beds of shell debris, nodules of calcium carbonate, horizons of silt and lenses of peat within a deposit mainly composed of clay. The sedimentary rock equivalent of this mud-flat sequence would contain thin bands of shelly and nodular limestone, siltstones, organic mudstones and possibly shale. A delta, which develops in rather deeper water (Fig. 3.20), would be composed mainly of sand with some clay, and compact to produce a clayey sandstone (Fig. 6.10).

Oceanic sediments of the open sea (*pelagic*) are the most uniform of all sediments but when they exist on land it can be assumed they have been placed there by orogenesis; many are found to be much disturbed by folding, faulting and metamorphism. Sequences of dark-coloured shales and siltstones are typical of this association.

Sediments of the shelf sea are less uniform but they will have only gradual changes in thickness and composition

Table 6.5 Commonly encountered associations

DEPOSITIONAL ENVIRONMENT	(Figures)	LIKELY ASSOCIATIONS (see Table 6.2)		RANGE OF PARTICLE SIZE 0.06 m 2.00 mm 60.0 mm
CONTINENTAL				
Sub-aerial	slope (3.7, 3.11, 3.41, 3.49, & see Chapter 14)	Detrital terrigenous:	conglomerate, breccia.	
	volcanic (5.3, 5.5)	Detrital pyroclastic:	agglomerate, volcanic breccia & tuff.	
	desert (3.34 to 3.36)	Detrital terrigenous:	sand	
	cold desert	,, ,, :	loess	in floc
sub-aquatic	river (3.16 to 19)	,, ,, :	conglomerate, grit, sand, clay (+peat)	
	wadi (6.7)	,, ,, :	wadi conglomerate, sand	
	lake (3.20)	,, ,, :	sand, silt (+some peat)	
	desert lake (3.49)	Chemical saline:	silts, muds and evaporites	
	estuary (3.25, 3.27)	Detrital terrigenous:	sand, silt, clay (+peat)	
GLACIAL				
sub-glacial	lodgement till (3.42)	Detrital terrigenous:	sand, silt, clay, with boulders (tillite)	
other glacial	(3.39 to 42, 3.44 to 46)	,, ,, :	as per slopes, cold desert, rivers, lakes.	
SHORE LINE				
	beach (3.22, 3.25, 3.27)	Detrital terrigenous:	conglomerate, breccia, gravel, sand	storm
	coral beach (3.28)	,, calcareous:	,, ,, ,, ,,	
	lagoon (3.27, 3.28)	Detrital terrigenous & calcareous;	sand, silt, clay (+peat)	
		Chemical calcareous	calcareous sand (oolites)	
		& carbonaceous	silts and organic muds:	
		& saline	evaporites	
	reef (3.28)	Biochemical:	coral and associated reef debris	
	delta (3.20, 3.21)	Detrital terrigenous:	sand, silt, clay } + peat & other vegetation	
	coal swamp (6.18)	,, ,,	,, ,, ,, } + peat & other vegetation	
OCEANIC				
	shelf (1.2, 2.4, 6.6)	Detrital terrigenous & calcareous:	grit, sand, silt, clay	
	deep sea (3.29)	,, ,, & chemical siliceous	ooze and deep sea clay	
		calcareous	ooze and deep sea clay	

if deposited on a stable shelf. If deposited on an unstable shelf associated with a geosynclinal trough considerable variation in sediment type or thickness must be expected. Coarse sediments such as grit and sandstone which flood across the shelf during periods of uplift inland, may be interbedded with more uniform and finer sediment such as mudstone and shale, deposited in quieter times.

Sedimentary mineral deposits

The sedimentary processes associated with the mechanical sorting of grains and the evaporation of sea water, the accumulation of organic remains and the activity of bacteria, are responsible for concentrating dispersed constituents into mineral deposits. Many types of deposit exist and their products are collected here under five headings, as follows.

Construction materials

Sand and gravel provide most important resources of natural aggregate, limestone forms the basis of cement, clay is the bulk mineral of ceramic ware, silica sand is the substance of glass and gypsum provides the body of plaster. These sediments sustain the building industry with necessary materials for construction: their main uses have been mentioned on earlier pages and their relevant characters are described further in Chapter 18.

Minerals for chemicals

These are mainly non-metallic and include the following; (from Bateman, 1950).
Salt and brines: from deposits of coastal evaporites.
Borax and borates: from deposits of playa lake evaporites.
Sodium carbonates and sulphates: from deposits of 'alkali' or 'bitter' playa lakes.
Calcium: as a chloride from deposits of evaporite, as a carbonate from limestone strata and as a sulphate from beds of gypsum.
Potash: as a nitrate from deposits of evaporite.
Sulphur: from beds of gypsum and anhydrite that have been altered by reducing bacteria.
Nitrogen: as sodium and potassium nitrate from deposits of evaporite.
Phosphate: from nodular chemically-precipitated beds of phosphate and phosphatic marls and limestones.

The importance of evaporite is obvious and many large deposits are mined. In these excavations the original horizontal bedding is often seen to have been destroyed by faulting and folding. This mixes the layers of evaporite (each originally of different composition) with other strata with which they were interbedded. Evaporites can deform as a ductile material at low temperatures and pressures and beds of salt may flow to form salt-domes which rise to the surface by displacing overlying strata (see Fig. 6.21). The deformation of evaporites can cause severe difficulties to the planning and operation of underground mines.

Placer deposits

Important minerals may be concentrated by the mechanical sorting of sediment in moving water, with lighter minerals being separated from heavier ones which remain as placer deposits. Gold, cassiterite (tin), diamond, zircon, magnetite, ilmenite and chromite have been found as placer deposits, the last four taking the form of 'black sand'.

Sedimentary ores

Two ores are especially associated with sediments; iron ores as described on p. 129, and metallic sulphides which have been precipitated *on* the sea floor from hydrothermal solutions emitted by submarine volcanoes (Fig. 5.28b); these have been called *exhalative* deposits. It is believed that some of the extensive sulphide deposits that are interbedded with marine sediments and volcanic rocks may have formed this way, e.g. the Kuroko ores in Japan.

Organic fuels

Coal, oil and gas are the most important fuels, and coal deposits are described on p. 127. The search for these fuels is only conducted in sedimentary rocks.

Oil and gas

Oil is formed from the putrefaction of organic matter, and organic-rich sediments containing up to 10% organic carbon have been important sources for oilfields: they are called *source rocks*. When a source rock is buried oil and gas are generated from the unoxidized organic matter locked in its pores, sharing the pore space with pore water (Fig. 6.20). The four basic types of organic matter in sediments are *kerogen* (70–80% carbon, 7–11% hydrogen, 10–15% oxygen with some nitrogen and sulphur), *asphalt* or bitumen (80% carbon, 10% hydrogen, 2–8% sulphur with nitrogen and a little oxygen), and the hydrocarbons *crude oil* and *natural gas*. Kerogen and asphalt are solid at normal temperatures and pressures whereas the hydrocarbons are liquid. Kerogen, unlike asphalt, is insoluble in petroleum solvents. A typical crude oil contains 85%

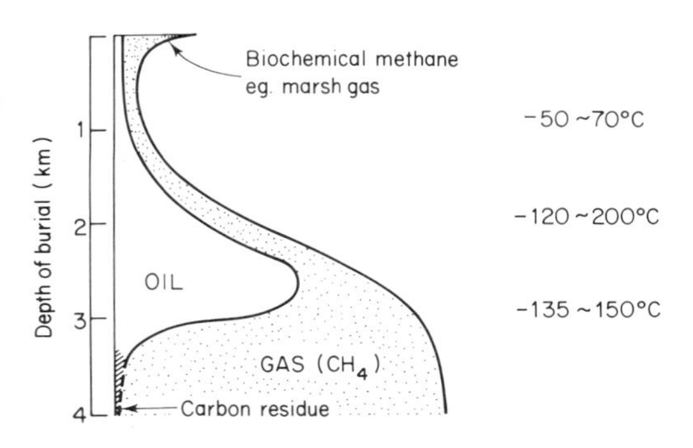

Fig. 6.20 Formation of hydrocarbons and its relationship with temperature and depth of burial (Tissot and Welte, 1978).

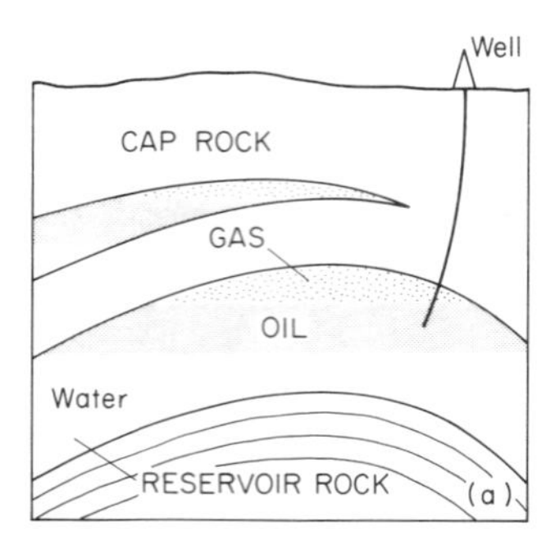

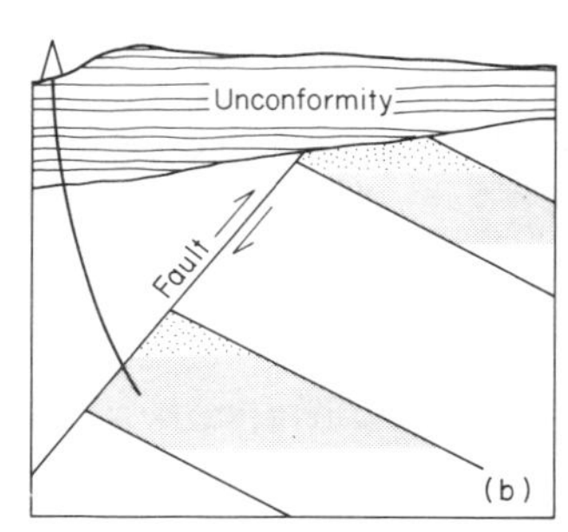

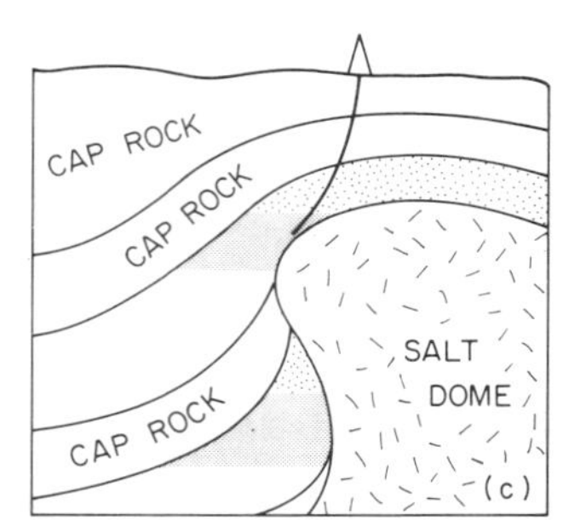

Fig. 6.21 Typical oilfield structures, (**a**) folds and lenses, (**b**) faults and unconformities which terminate reservoir rock against cap rock, (**c**) salt domes.

carbon, 12% hydrogen, 0.7% sulphur and small amounts of nitrogen and oxygen. Natural gas is mainly methane (CH_4) but contains many other hydrocarbons such as ethane, propane and butane. Carbon dioxide, nitrogen and hydrogen sulphide are also commonly present together with water vapour derived from pore water. The pore water associated with oil is usually rich in total dissolved solids (600 000 parts per million is not uncommon) and called a '*brine*'.

Source rocks buried to depths where they are heated to above 100°C generate gas at the expense of oil (Fig. 6.20).

Oil and gas are less dense than the associated brine and will migrate upwards through the overlying deposits with the displaced brine occupying the pore space vacated by the oil and gas. If there are no strata capable of trapping these migrating hydrocarbons they will escape and when this occurs at ground level the low-boiling components will vaporize to leave an asphaltic residue (e.g. a pitch lake). Strata of low permeability capable of halting this upward migration are called *cap rock*: shales, mudstones and evaporites are excellent for this purpose.

The porous rock in which oil and gas accumulate beneath a cap rock is called the *reservoir rock*. Figure 6.21 illustrates typical oilfield structures: note the stratification of gas above oil above brine. Wells drilled into the reservoir rock extract the hydrocarbons by maintaining in each well a fluid pressure slightly smaller than that in the reservoir. The difference between the total head of fluid in the wells and that in the reservoir, together with the expansion of gas which accompanies a reduction in its pressure, drives the oil towards the wells. This is most economically achieved when the reservoir rocks have a permeability that requires a small differential head to sustain an adequate flow. Good reservoir rocks therefore have a porosity sufficiently large to accommodate economic reserves of oil and a high permeability.

Most of the oil discovered occurs in Mesozoic strata, slightly less in Cenozoic strata, and small amounts in the Newer Palaeozoic (see Table 2.1).

Selected bibliography

Greensmith, J.T. (1978). *Petrology of Sedimentary Rocks*, 6th edition. George Allen and Unwin, London, Boston.

Selley, R.C. (1981). *An Introduction to Sedimentology*, 2nd edition. Academic Press, London, New York.

Adams, A.E., MacKenzie, W.S. and Guilford, C. (1984). *Atlas of Sedimentary Rocks under the Microscope*. Longman, Harlow, England.

Fuels

Green, R.P. and Gallagher, J.M. (1980). (Editors) *Future Coal Prospects: Country and regional assessments*. The second and final WOCOL report. World Coal Study. Ballinger Pub. Co. (Harper & Row), Cambridge, Massachusetts.

Murchison, D. and Westoll, T.S. (1968). (Editors) *Coal and Coal-Bearing Strata*. Oliver & Boyd, London.

Hobson, G.E. and Tiratsoo, E.N. (1981). *Introduction to Petroleum Geology*, 2nd edition. Scientific Press, Beaconsfield, England.

Evaporites

Halbouty, M.T. (1967). *Salt Domes: Gulf Region, United States and Mexico*. Gulf Pub. Co., Houston, Texas.

Anon (1980). *Evaporite deposits, illustration and interpretation of some environmental sequences*. Chambre syndicale de la recherche et de la production du petrole et du gaz naturel. Paris.

Clays

Millot, G. (1970). *Geology of Clays: weathering, sedimentology, geochemistry*. Springer-Verlag, New York, London.

Gillot, J.E. (1968). *Clay in Engineering Geology*. Elsevier Pub. Co., London, New York.

Grim, R.E. (1962). *Applied Clay Mineralogy*. McGraw-Hill Book Co. Ltd., New York, London.

7

Metamorphic Rocks

Metamorphism is the term used to denote the transformation of rocks into new types by the recrystallization of their constituents; the term is derived from Greek *meta*, after (signifying a change), and *morphe*, shape.

The original rocks may be igneous, sedimentary or ones that have already been metamorphosed and changes which occur in them result from the addition of heat or the operation of pressure. Heat and pressure are the agents of metamorphism which impart energy to the rocks, sufficient to mobilize the constituents of minerals and reassemble them as new minerals whose composition and crystal lattice are in equilibrium with existing conditions. Such processes transform, or *metamorphose* the rocks and can impose upon them a metamorphic texture that may be entirely different from the texture they originally possessed. The superimposition of textures enables the history of metamorphic rocks to be defined especially when the composition and orientation of new minerals reflect the temperature of metamorphism and the prevailing direction of stress (Fig. 7.1).

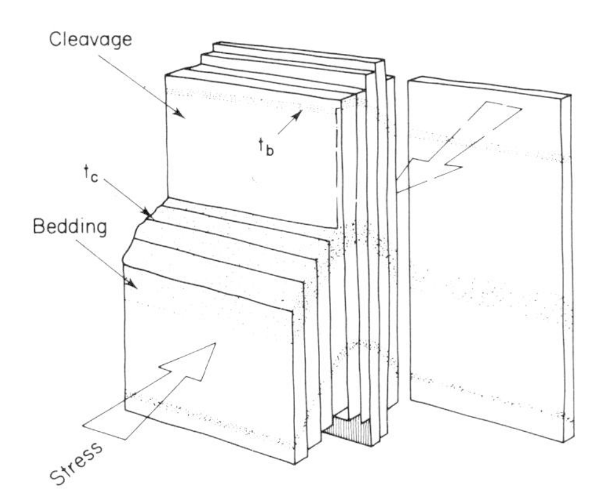

Fig. 7.1 Cleavage in slate, formed at right angles to the maximum compression during metamorphism of the original sediment. t_c = trace of cleavage on bedding, t_b = trace of bedding on cleavage surface.

Metamorphic minerals grow in solid rock, their development being aided by solvents, especially water expelled from remaining pores and the dehydration of clay minerals.

Three broad classes of metamorphism, depending on the controls exercised by temperature and pressure, may be distinguished:

(1) *Thermal or Contact Metamorphism*, where rise of temperature is the dominant factor. Thermal effects are brought about in contact zones adjacent to igneous intrusions; or when sediments are down-folded into hotter regions in the crust.
(2) *Dynamic or Dislocation Metamorphism*, where the dominant control is stress, as in belts of shearing.
(3) *Regional Metamorphism*, where both temperature and pressure have operated over a large (regional) area.

Figures which illustrate the typical occurrence of these classes of metamorphism are listed in Table 7.1.

Table 7.1 Figures illustrating the occurrence of metamorphism

Contact metamorphism adjacent to
small intrusions: 5.6, 5.7, 5.9, 5.10
larger intrusions: 5.3, 5.13, 5.14, 5.27, 7.3
Dislocation metamorphism associated with thrusts:2.3, 2.6, 7.11
Regional metamorphism associated with fold belts:2.3, 2.4, 2.17b, 7.6, 7.11

Contact metamorphism

The intrusion of a hot igneous mass such as granite or gabbro produces an increase of temperature in the surrounding rocks (or 'country-rocks'). This increase promotes the recrystallization of some or all of the components of the rocks affected, the most marked changes occurring near the contact with the igneous body. When recrystallization can develop uninhibited by an external stress acting on the rocks, the new minerals grow haphazardly in all directions and the metamorphosed rock acquires a granular fabric, which is known as the *hornfels* texture (p. 136).

During the metamorphism there may also be a transfer of material at the contact, when hot gases from the igneous mass penetrate the country-rocks, the process known as *pneumatolysis* (p. 109). The country-rocks are not melted, but hot emanations such as carbonic acid, SO_2, water vapour, and volatile compounds of boron and fluorine, percolate into them and result in the formation of new minerals. Temperatures may range from about 500 to 800°C during pneumatolysis (p. 109), and associated hydrothermal emanations may carry ore-metals such as Sn, Zn, and Fe which are deposited as mineral veins in fissures in the country-rocks.

Dislocation metamorphism

Where stress is the principal control and temperature is subordinate, as in zones affected by strong shearing movements, the rocks undergo *dislocation metamorphism*, with the production of cataclastic textures (p. 141); these result from the mechanical breakdown of the rocks under stress, e.g. by shearing or brecciation.

Regional metamorphism

The operation of stresses as well as rise of temperature results in recrystallization, with the formation of new minerals many of which grow with their length or flat cleavage surfaces at right angles to the direction of the maximum compressive stress (Fig. 7.1). High temperatures and stresses are produced in orogenic belts of the crust, and regional metamorphic rocks are found in these great fold-belts where they have become exposed after denudation. Many components of the rocks have acquired a largely parallel orientation, which gives the rocks characteristic textures: the oriented texture produced by platy or columnar minerals is known as *schistosity* (Fig. 7.2), and an alteration of schistose layers with others

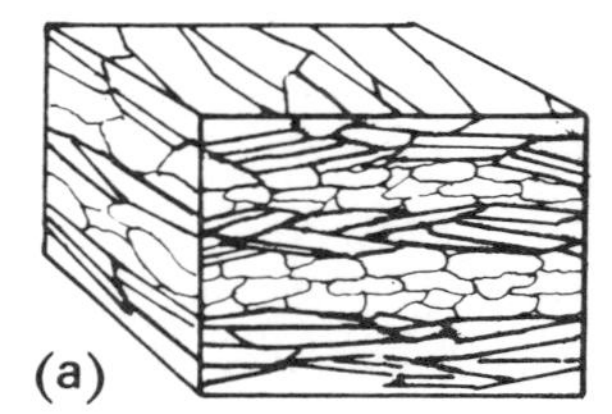

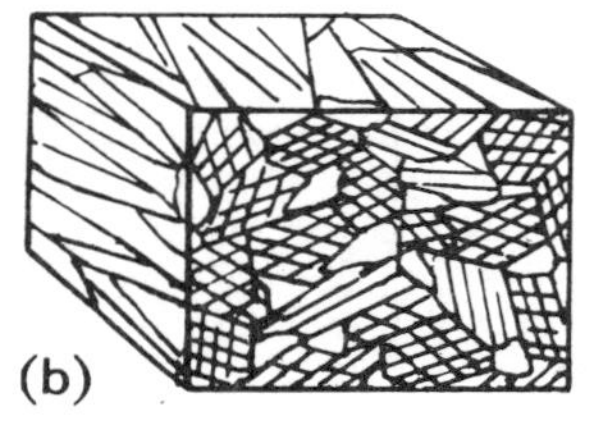

Fig. 7.2 Block diagrams to illustrate schistosity. (**a**) Parallel orientation of mica crystals in mica-schist; quartz grains lie between the micaceous layers. (**b**) Oriented hornblende prisms in hornblende-schist.

less schistose gives the banded texture known as *foliation* (p. 140). Argillaceous rocks under the influence of moderate to low temperature and high stress develop *slaty cleavage* (p. 139).

Crystal shape and fabric

The crystalline shape of a metamorphic mineral partly determines the ease of its growth during metamorphism; thus micas and chlorites, with a single cleavage, grow as thin plates oriented perpendicular to the maximum stress, and amphiboles such as hornblende grow in prismatic forms with length at right angles to the maximum stress. Some minerals of high crystallization strength, e.g. garnet and andalusite, grow to a relatively large size in metamorphic rocks and are then called *porphyroblasts*, i.e. conspicuous crystals in contrast to those which make up the rest of the rock. Feldspars and quartz have low and nearly equal strengths of crystallization, and metamorphic rocks composed of quartz and feldspar show typically a granular texture (granulites).

Fabric

The term fabric is used to denote the arrangement of mineral constituents and textural elements in a rock, in three dimensions. This is particularly useful in metamorphic rocks, as it enables the preferred orientation of minerals, when present, to be described with reference to broader structures. Thus, rocks are either *isotropic*, when there is no orderly arrangement of their components (as in hornfels); or *anisotropic* (as in schists) when there is parallel orientation of minerals, often well developed. Relict textures from an original sediment, such as banding due to variation in composition, may also be preserved in the fabric of the equivalent metamorphic rock (Fig. 7.1). In graded bedding (Figs. 2.1; 6.6) for example, the upper (clayey) part of a graded bed will show conspicuous crystals grown from the clays (e.g. andalusite), in contrast to the equigranular crystals of quartz, formed from the sandy base of the bed.

Terms commonly used to describe the shape and size of metamorphic minerals, and their fabric, are listed in Table 7.2.

Table 7.2 Commonly used terms for describing metamorphic rock texture: all may be used as field terms

1. *Banding.*
   - Foliation = series of parallel surfaces. (Fig. 7.6)
   - Lineation = series of parallel lines as produced by the trace of foliation on a rock surface; e.g. the wall of a tunnel (Figs 7.1, 7.8, 7.10)
2. *Visible crystallinity.*
   - Phaneritic = individual crystals can be distinguished
   - Aphanitic = granularity from the presence of crystals can be seen but individual crystals cannot be distinguished
3. *Crystal size.*
   - Coarse = > 2.0 mm ↑
   - Medium = 2.0–0.06 mm – – Phaneritic – – – – – – – –
   - Fine = < 0.06 mm
4. *Relative crystal size.*
   - Granoblastic = all crystals are approximately the same size
   - Porphyroblastic = larger crystals surrounded by much smaller crystals. (Fig 7.9)

Classification

The visible character most useful for the classification of metamorphic rock is the presence or absence of foliation (see Fig. 7.1). In Table 7.3 the common metamorphic rocks are classified under three headings that describe the anisotropy created by foliation.

The names of most metamorphic rocks are prefixed by the names of the minerals they contain, e.g. quartz-biotite-schist. When the original rock type is known the prefix *meta* may be used, e.g. metasandstone, or metagabbro. Metamorphosed arenaceous sediments (i.e. sands) may be described as *psammitic*, e.g. psammitic gneiss, and metamorphosed argillaceous sediments (i.e. silts and clays) as *pelitic*, e.g. pelitic gneiss.

Table 7.3 A classification for metamorphic rocks.

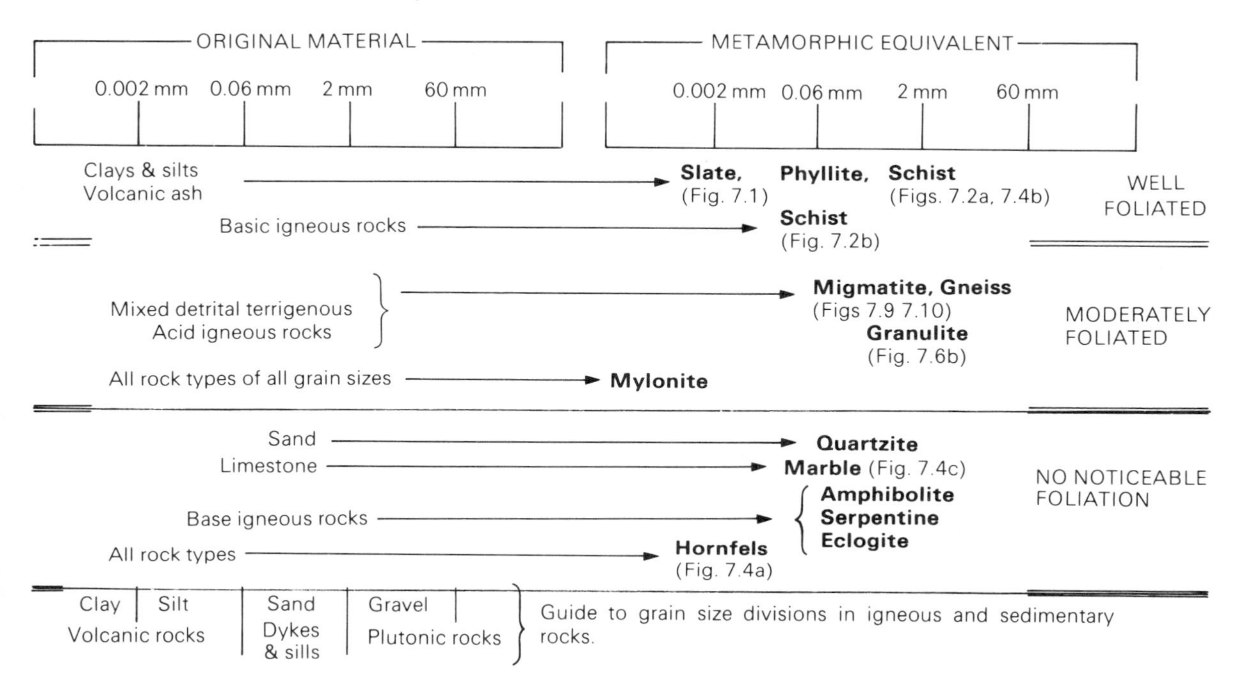

Contact metamorphism

The effects of contact metamorphism on the main sedimentary rock types (shale, sandstone, limestone) are now discussed, assuming rise of temperature to be the dominant and controlling factor. At the contact zone bordering an igneous mass an *aureole* of metamorphism is developed (Fig. 7.3). Within the aureole, zones of increasing metamorphism can be traced as the contact is approached, and they are distinguished by the growth of new minerals. The width of the aureole depends on the amount of heat that was transferred, i.e. on the size and temperature of the igneous body, and is also related to the steepness or otherwise of the contact (Table 7.4).

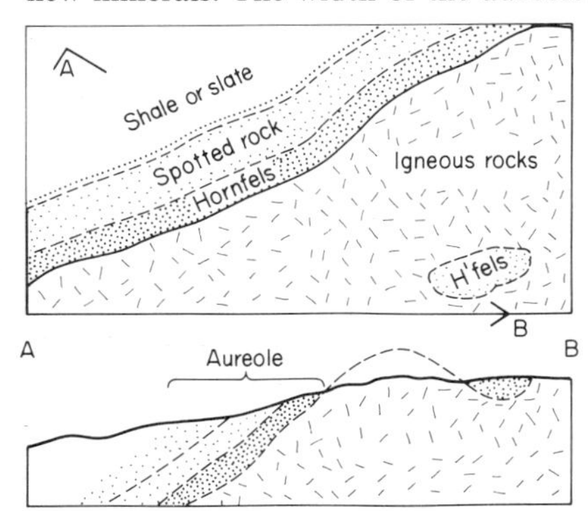

Fig. 7.3 Map of part of a metamorphic aureol around an intrusive igneous mass; small area near B is a relic of denuded roof of the intrusion. Below, section along the line AB; bedding of sediments not shown.

Table 7.4 The reduction in temperature away from an intrusion similar to that in Fig. 7.3

	Basic intrusion (e.g. gabbro)	Acid intrusion (e.g. granite)
magma	1200°C	800°C
contact	900°C	650°C
0.1 km	820°C	600°C
1.0 km	680°C	500°C
10.0 km	350°C	320°C

Contact metamorphism of a shale or clay

An argillaceous rock such as shale is made up of very small particles, most of which are clay minerals and are essentially hydrated aluminium silicates; with them are small sericite (secondary white mica) flakes and chlorite, and smaller amounts of colloidal silica, colloidal iron oxide, carbon, and other substances. The two dominant oxides in a clay or shale, as would be given in a chemical analysis, are thus SiO_2 and Al_2O_3, and when the shale is subjected to heat over a long period the aluminium silicate *andalusite*, or its variety *chiastolite* (Fig. 4.34), is formed. *Cordierite* (Fig. 4.36) is another mineral frequently formed at the same time; it grows as porphyroblasts in the metamorphosed shales.

In the outermost zone of the aureole, farthest from the contact, small but noticeable changes are produced: the shales are somewhat hardened and specks of opaque magnetite are formed. Somewhat nearer the contact incipient crystals of chiastolite and cordierite appear, as small dark spots; and with further progress towards the igneous contact andalusite and cordierite are better developed, as distinct spots, and small flakes of biotite have grown from the chlorite and sericite in the shale. This is the *spotted-rock* zone in Fig. 7.3. The recrystallization is continued in the zone nearest the contact, including the growth of muscovite flakes and quartz and larger magnetites; crystals of rutile (TiO_2, tetragonal) may also be present, derived from titanium in the original sediment. The shale has now been completely recrystallized, and is called a *hornfels*, or to give it its full name, a biotite-andalusite-cordierite-hornfels. It is hard, with a 'horny' fracture, and has a finely granular hornfels texture (Fig. 7.4). In some hornfelses, next to a high temperature contact, needle-shaped crystals of sillimanite may be formed from excess alumina, in addition to the other minerals. The 'zones' described above are not sharply separated but grade into one another; and it is assumed that there has been neither addition nor subtraction of material during the metamorphism.

Localities in Britain where the sequence of rock types - shale (or slate) → spotted rock → hornfels - is found in contact aureoles include those around the granites of Cornwall and Devon (where also mineralized veins have been formed); and in Scottish aureoles as at the Ben Nevis granite or the Insch (Aberdeenshire) gabbro. An interesting case is that of the Skiddaw granite, Cumbria, which is exposed over only three small areas, each less than a kilometre across, in the valleys of streams flowing from the north-east side of the mountain. Over a considerable area of adjacent ground, however, there are exposures of chiastolite-bearing rock, indicating the existence of the granite at no great depth below the present surface, and marking a stage in the process of unroofing of the granite by denudation.

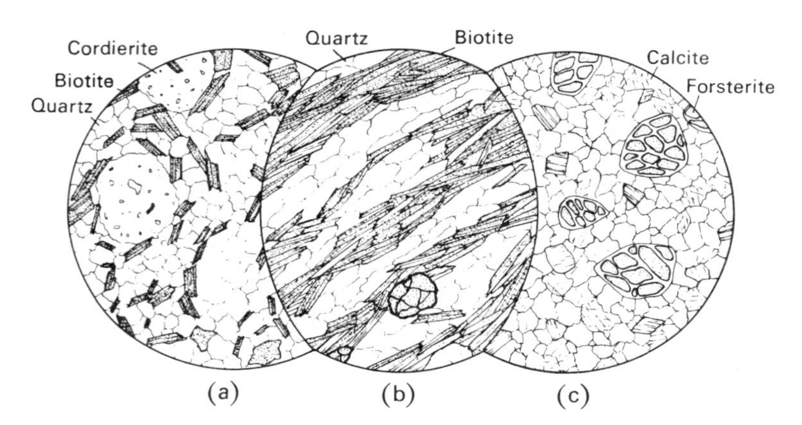

Fig. 7.4 Metamorphic rocks in thin section: (**a**) Biotite-cordierite-quartz-hornfels; the rounded cordierite contains many inclusions. (**b**) Mica-schist with garnets. (**c**) Forsterite-marble (all ×15).

Contact metamorphism of a sandstone

A siliceous rock such as sandstone is converted into a metamorphic *quartzite*. The original quartz grains of the sandstone (and siliceous cement if present) are recrystallized as an interlocking mosaic of quartz crystals. Partial fusion of the mass may occur in special circumstances, but rarely. Constituents other than quartz in the cement between the grains give a rise to new minerals, depending on their composition: e.g. a little biotite (from clay), and magnetite (from iron oxide, as in a ferruginous sandstone). The bulk of the rock consists essentially of quartz.

Contact metamorphism of a limestone

Limestones are made up mainly of calcium carbonate, together with some magnesium carbonate, silica, and minor constituents, and the metamorphic product is a *marble*.

(*i*) In the ideal case of a pure calcium carbonate rock the marble would be composed entirely of grains of calcite. The coarseness of texture of the rock depends on the degree of heating to which it has been subjected, larger crystals growing if the metamorphism is prolonged. Dissociation of the carbonate into CaO and CO_2 is prevented by the operation of pressure. Examples of rocks with a high degree of purity are provided by the statuary marbles.

(*ii*) In the metamorphism of a limestone in which some content of silica is present, as quartz grains or in colloidal form, the following reaction takes place in addition to the crystallization of calcite:

$CaCO_3 + SiO_2 \rightarrow CaSiO_3$ (wollastonite) $+ CO_2$

giving a *wollastonite-marble*, as in parts of the Carboniferous Limestone of Carlingford, Ireland (metamorphosed by gabbro intrusions). The mineral wollastonite forms white or grey tabular, translucent crystals.

(*iii*) During contact metamorphism of a magnesian or dolomitic limestone the dolomite is dissociated, thus:

$CaCO_3.MgCO_3 \rightarrow CaCO_3$ (calcite) $+ MgO$ (periclase) $+ CO_2$

The periclase is readily hydrated to form *brucite*, $Mg(OH)_2$, in colourless tabular hexagonal crystals, and the product is a *brucite-marble*. Blocks of this rock have been ejected from Vesuvius; British occurrences of brucite-marble are in Skye and in the north-west Highlands.

When silica and dolomite are both present, the magnesium silicate *forsterite*, an olivine, (p. 72) is formed together with calcite, CO_2 being lost in the reaction, thus:

$$2CaCO_3.MgCO_3 + SiO_2 \rightarrow 2CaCO_3 \text{ (calcite)} + Mg_2SiO_4 \text{ (forsterite)} + 2CO_2.$$

The metamorphosed rock is a *forsterite-marble* (Fig. 7.4); a British locality for this is at Kilchrist, Skye. If the further change from forsterite to serpentine takes place (by hydration) the rock becomes a *serpentine-marble*. The white calcite streaked with green serpentine in the marble gives an attractive decorative stone, e.g. the Connemara Marble of Ireland.

(*iv*) When clay is present as well as silica in the original rock, minerals such as tremolite and diopside (Ca-Mg-silicates) and CaAl-garnets are formed by the metamorphism, in addition to calcite, and the resulting rock is a *calc-silicate-hornfels*. Its mineral composition varies according to the amounts of different substances in the original limestone; the textures of these rocks tend to be coarse because of the fluxing action of dissociated $MgCO_3$, and the rock is often very hard.

A zone of calc-silicate-hornfels was passed through by the Lochaber water-power tunnel at Ben Nevis; this extremely hard rock together with baked schists at a granite contact, reduced the rate of progress of the tunnel excavation to half the average rate. (Halcrow, W.T., 1930-31.)

Many varieties of marble may arise, some beautifully coloured because of traces of substances in the original sediment. Decorative stones of this kind are exported from Italy and Greece. The rocks are cut into thin slabs, with one surface polished, and are used as panels for interior decoration; advantage may be taken of any pattern due to veining or zones of brecciation by arranging the slabs symmetrically in groups of two or four. The white Carrara Marble (N.W. Italy) is a metamorphosed Jurassic limestone.

Unfortunately the term 'marble' is also used as a trade name for any soft rock that will take a polish easily, and includes many limestones used as decorative stones on account of their colouring or content of fossils. Among these in Britain are the Ashburton 'marble' (a Devonian coral limestone with stromatoporoids), the Hopton Wood stone (a Carboniferous crinoidal limestone), and the Purbeck 'marble' (an Upper Jurassic limestone from Dorset, containing the fossil shell *Paludina*, Fig. 2.14).

Contact metamorphism of igneous rocks

The effects here are not so striking as in the sedimentary rocks, because the minerals of igneous rocks were formed at a relatively high temperature and are less affected by re-heating; but some degree of recrystallization is often evident. A basic rock such as dolerite or diabase may be converted into one containing hornblende and biotite, from the original augite and chlorite, the plagioclase being recrystallized. Secondary minerals that occupy vesicles, as in amygdaloidal basalt, yield new minerals such as calcium-feldspar (after zeolite) and amphibole (after chlorite and epidote).

Igneous rocks which were much weathered before metamorphism may acquire calc-silicate minerals (e.g. Ca-garnet) derived from calcium and Al-silicates in the original rock, and hornblende from chlorite. Andesites and andesitic tuffs during contact metamorphism may develop many small flakes of brown mica and crystals of magnetite, as in the Shap Granite contact zone. *Basic granulites* (rocks of equigranular texture, p. 141) are formed from basic rocks such as gabbro by prolonged high-grade metamorphism, involving high temperature and complete recrystallization; pyroxenes and plagioclase are the main minerals. The term *granulite* is used broadly for rocks in which the main minerals are roughly equidimensional (see further under regional metamorphism).

Pneumatolysis

In the foregoing discussion it has been assumed that there has been no transfer of material from the igneous body across the contact and that metasomatic changes (i.e. changes of substance) have involved only the recombining of original constituents and loss of gas. It frequently happens that the volatile substances accumulated in the upper part of a body of magma as it crystallized, pass into the country-rocks at a moderately high temperature stage in the cooling process of the igneous mass. Their reaction with the rock is called *pneumatolysis* (*pneuma* = gases), and can affect rocks, as indicated on p. 109). The emanating volatiles include compounds of boron, fluorine, carbon dioxide, sulphur dioxide and others; characteristic minerals such as *tourmaline*, *topaz*, *axinite*, *fluorspar* (*q.v.*), and *kaolin* are formed near a contact under these conditions.

Tourmaline

This is formed by the pneumatolytic action of boron and fluorine on mafic minerals (p. 79). It has a high content of alumina (between 30 and 40%) and is found also in rocks of clayey composition adjacent to an igneous contact. When the biotite of a granite is converted into tourmaline the granite itself is often locally reddened by the introduction of iron. The name *luxullianite*, from a Cornish locality, Luxullian, is given to a tourmalinized granite in which the tourmaline occurs as radiating clusters of slender crystals of *schorl* embedded in quartz (Fig. 4.1*e*). Veins of quartz and tourmaline, some carrying cassiterite, are common in certain granite areas.

Axinite

This is a calcium-boron-silicate occurring in contact metamorphic aureoles where boron has been introduced into limestone or altered rocks containing calcite. Axinite crystals are typically flat and acute-edged (Fig. 4.4*f*), brown and transparent with a glassy lustre.

Topaz

This occurs in cavities in acid igneous rocks, often

associated with beryl, tourmaline, and fluorite, and commonly found in *greisen* (see below and Fig. 4.4*d*).

Kaolinization

The term *kaolin* (or China Clay) is used for the decomposition products that result from the alteration of the feldspars of granites, and is partly crystalline *kaolinite* and partly amorphous matter. One equation for the change, in which potassium is removed and silica is liberated, is as follows:

$$\underset{\text{orthoclase}}{4KAlSi_3O_8} + 2CO_2 + 4H_2O = 2K_2CO_3 + \underset{\text{kaolinite}}{Al_4Si_4O_{10}(OH)_8} + 8SiO_2$$

Parts of the Cornish granites, as at St Austell, have been decomposed by this hydrothermal action, giving a soft mass of quartz, kaolin, and mica which crumbles at the touch. In the quarries the kaolinized rock is washed down by jets of water; the milky-looking fluid is then pumped from the sumps into which it has drained and allowed to gravitate through a series of tanks. These remove first the mica and quartz, and the kaolin is allowed to settle out of suspension as a whitish sludge.

Kaolin is an important economic product, and is used as a paper filler, and to a lesser extent in pottery manufacture and for numerous other purposes as an inert absorbent.

Greisen

Composed essentially of quartz, white mica, and accessory amounts of tourmaline, fluorite, and topaz, this is formed from granite under certain pneumatolytic conditions; where, for example, the formation of K_2CO_3 in the above equation is inhibited, white mica (secondary muscovite) is formed from the feldspar of the granite and the name greisen is given to the resulting rock.

China-stone

This represents an arrested stage in the kaolinization of granite; in addition to quartz and decomposed feldspar it frequently contains topaz ($Al_2SiO_4F_2$) and fluorspar, both of which point to incoming fluorine.

Regional metamorphism

Rocks which have undergone regional metamorphism have a much greater extent at the surface than those formed by contact-metamorphism and may be found exposed over many hundreds of square kilometres. A substantial input of energy is required for such metamorphism and occurs at the boundaries of colliding plates. Such a boundary is shown in Fig. 1.20 and the location of others like it, where regional metamorphism is occurring at present, is given in Fig. 1.17. Mountains form from these zones (Figs 2.3 and 2.4) and the deformation that occurs within them is illustrated in Fig. 2.17*b*.

Regional metamorphism develops under the *hydrostatic pressure* (or confining pressure) arising from the weight of overlying rocks and the *shearing stresses* produced by the system of unequal principal stresses associated with plate movement. From Fig. 2.4 it is evident that temperature and pressure increase towards the central root of an orogenic belt and that the *grade* of metamorphism will increase with depth. Likewise, concentric *zones* may be defined around the root in which operate different intensities of metamorphism.

Metamorphic grades

These can be defined in relation to the temperature and stress conditions that prevail (Fig. 7.5); they are named

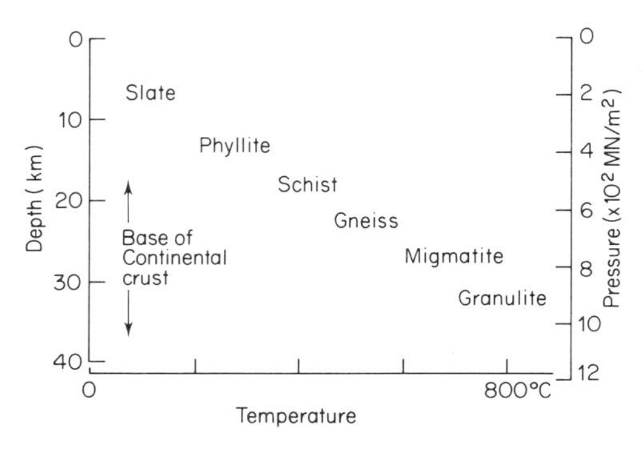

Fig. 7.5 Grades of increasing metamorphism produced by the burial of a muddy sediment.

after the main rock types produced, *viz.* slate, schist, and gneiss. In the *slate grade* of metamorphism, low temperature and high stress differences are the controlling factors (Fig. 7.1). In the *schist grade* moderate temperatures and moderate stress operate during the metamorphism (Fig. 7.2). And in the *gneiss grade* the metamorphism involves high temperatures and low stress differences. With the gneisses can also be included the granulites, which have a texture of nearly equal-sized crystals.

Zones of regional metamorphism

The existence of metamorphic zones is best revealed by the metamorphism of clayey sediments of constant bulk composition. With a progressive increase in temperature, zones of increasing metamorphism are defined by the appearance of *index-minerals* in the rocks. Different index-minerals come into equilibrium at successively higher temperatures, and provide a guide to the temperature reached by the rocks containing them. Most of the metamorphism takes place in the temperature range from about 200° to 700°C.

The sequence of index-minerals in order of rising temperature is: chlorite, biotite, garnet (almandine), staurolite, kyanite, and sillimanite, and the minerals define a series of *metamorphic zones*. In the chlorite-zone (low-grade metamorphism) the chlorite is formed from sericite and other minute mineral particles in the original sediment. At a somewhat higher temperature a red-brown biotite is formed and marks the beginning of the biotite zone. The appearance of abundant pink garnet (almandine, p. 79) defines the next zone; and the rocks are now garnetiferous mica-schists (Fig. 7.4*b*). Following the garnet zone, a staurolite zone before the kyanite zone; but the two minerals commonly occur together. (Staurolite is a Mg-Fe-Al-silicate, and has a crystal structure composed of alternate 'layers' of kyanite and ferrous hydroxide.) Kyanite is stable under stress, and the rocks containing it are kyanite-schists, formed under moderate to high-grade metamorphism. Lastly, sillimanite (p. 79) is formed at higher temperatures; the rocks of the sillimanite zone are schists and gneisses, with which migmatites (p. 141) may be developed. Sillimanite also occurs in the innermost (hottest) parts of some thermal aureoles (p. 136).

The above sequence of zones was demonstrated by G. Barrow (1893) from extensive field mapping of rocks in the south-east Scottish Highlands (the area situated between the Highland Boundary Fault and the Great Glen Fault in Fig. 2.18). Here the roots of the Caledonide mountain chain are exposed (Fig. 7.6): similar zones exist in the Appalachians which are a continuation of this orogenic belt (Fig. 1.10), and in mountains elsewhere. Figure 7.6*b* illustrates some of the folded sediments within this orogenic belt. Their layered structure has been preserved despite changes that the sediments have sustained.

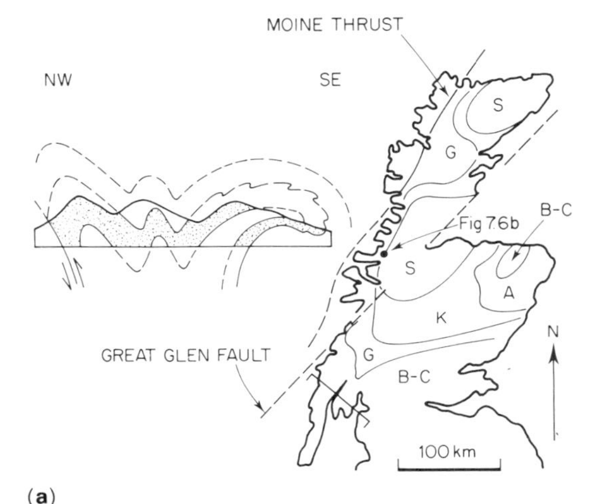

(**a**)

(**b**)

Fig. 7.6(**a**) Metamorphic zones of the Scottish Highlands S=sillmanite, K=kyanite, A=andalusite, G=garnet, B-C=biotite-chlorite. Displacement on the Great Glen Fault occurred after the zones had formed and has been reversed to restore the zones to their original relative positions. Section schematically illustrates structures within the area: compare with Fig. 2.17b. (Based on Dewey & Pankhurst, 1970.) (**b**) Steeply dipping Moine granulites, Loch Moidart, Argyll, Scotland. The rocks retain the bedding planes of the original sediments and are recrystallized quartz-sands.

Slate

Under the influence of high stress combined with low to moderate temperature, argillaceous sediments such as shales are compressed and become slate. Minute crystals of flaky minerals such as chlorite and sericite grow with their cleavage surfaces at right angles to the direction of maximum compression and some original minerals, e.g. quartz grains, recrystallize with their length parallel to this direction (Fig. 7.7). The rock thus possesses a preferential direction of splitting, or *slaty cleavage* (to be distinguished from *mineral* cleavage), parallel to the flat surfaces of the oriented crystals and independent of original bedding (Fig. 7.1). Slaty cleavage is often very regular and perfect, yielding smooth-sided thin plates of rock. Fossils may be deformed by this metamorphism. The cleavage pattern can extend over a wide area and cuts across folds that may be present in the rocks (Fig. 7.8). The essential minerals are chlorite, sericite and quartz.

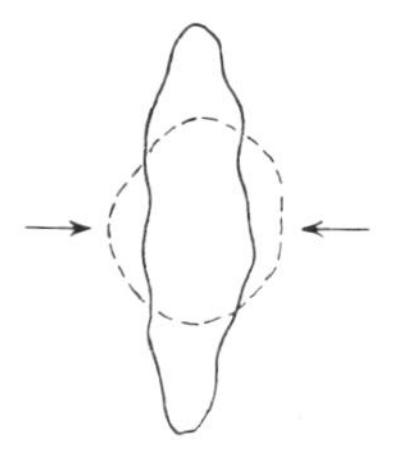

Fig. 7.7 Change of shape of a grain, e.g. quartz, under stress, original grain shown by dotted outline. Stress lowers the melting point, especially where grains touch; solution takes place there with redeposition along directions at right angles to the maximum stress direction (arrowed).

Some slates are derived from fine-grained volcanic tuffs (Table 7.3), e.g. the green Cumbrian slates of andesitic composition, which have become cleaved during a regional metamorphism that affected also other rocks.

The commercial value of slate depends upon the perfection of its cleavage and absence of accessory minerals such as calcite and pyrite which may weather in acid rain to leave holes.

British slate localities include the Bethesda and Llanberis quarries, North Wales (Cambrian age), and Penrhyn and Bangor slates; Ballachulish, Scotland (black slates, some with pyrite); Skiddaw quarries, Cumbria (green slates formed from the fine ash deposits of Ordovician volcanoes); and Delabole, Cornwall (a deep quarry in grey slates of Upper Devonian age).

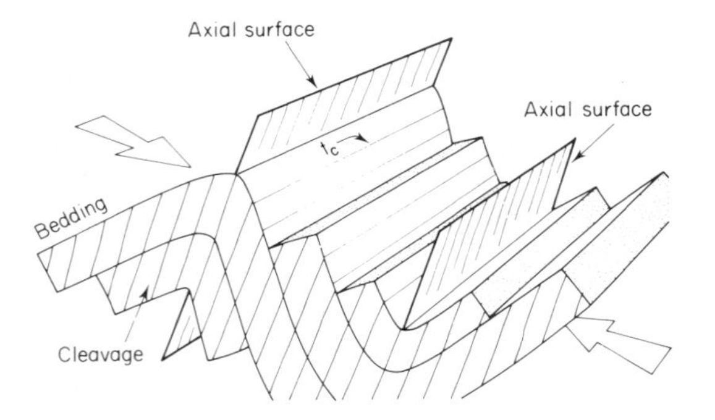

Fig. 7.8 Slaty cleavage in relation to a fold; arrows show deforming forces. Cleavage is developed in the direction of easiest relief, and is often parallel to the axial surfaces of folds. The angle between cleavage and bedding varies, and is greatest at the crest of a fold. tc = trace of cleavage on bedding (see Fig. 7.1).

Phyllite

With stresses continuing and if some rise of temperature ensues, the above mentioned minerals continue to grow to give larger crystals of muscovite and chlorite, and a lustrous, finely crystalline rock called *phyllite* results. With continued metamorphism the minerals increase further in size, leading to the formation of *mica-schists*. There is thus a gradation,

shale→slate→phyllite→mica-schist

the several types being formed from original muddy sediment under conditions of increasing grade of metamorphism (Fig. 7.5).

Schist

This is a crystalline rock of mainly medium-grained texture, whose mineral constituents can be distinguished either by eye or with a hand lens. Different varieties of schist have been formed from sedimentary or igneous rocks during regional metamorphism, under moderately high temperature and pressure. Because of the parallel orientation of their minerals, schists break into more or less flat fragments or *foliae*, which have lustrous surfaces and similar mineral composition. The name 'schist' (from Greek *schistos*, divided) was originally used to denote this property of splitting into foliae (Figs 7.2 and 7.4*b*).

Consider first those schists derived from original sedimentary material:

Mica-schist

Formed from argillaceous or pelitic sediments such as shale or clay, and composed of muscovite and biotite together with quartz in variable amount. Quartz grains under the influence of stress become elongated (Fig. 7.7) and lie with their length in the surface of schistosity. Lenticles of quartz and mica alternate in the rock (Fig. 7.4*b*); garnets may be formed if a higher temperature is reached, and grow as porphyroblasts, tending to push apart the micaceous layers.

Quartz-schist

Derived from sandy sediment, i.e. is psammitic, with a smaller clay content than sediment which forms mica-schist. Certain coarsely foliated rocks of the same derivation, formed at a higher grade of regional metamorphism, are the mica-gneisses (Fig. 7.5).

Schists formed from original igneous rocks include the following:

Chlorite-schist

The metamorphic equivalent of basalt of dolerite, consists predominantly of chlorite crystals in parallel orientation, often with a little quartz and porphyroblasts of magnetite or garnet. It is formed under moderate stress and moderate temperature.

Hornblende-schist

This is also derived from basic rocks such as dolerite, but at a rather higher grade of metamorphism and contains essentially hornblende and quartz (Fig. 7.2*b*). The term *amphibolite* is used as a group name for metamorphic rocks in which hornblende, quartz, and plagioclase are the main minerals. Hornblende-schist is one member of this group. British examples of hornblende-schist are found in the Dalradian rocks of the Scottish Highlands (Fig. 7.6), as in the Loch Tay area, Perthshire; and in the Lewisian near Loch Maree, Ross-shire; and in England at Start Point, Devon, and Landewednack, S. Cornwall.

Derivatives of ultrabasic rocks such as peridotite and dunite (p. 101) include some *serpentine-schists* and *talc-schists*. The latter are composed of talc (a very soft mineral) together with some mica. Soft rocks such as talc-schist and chlorite-schist, and sometimes decomposed mica-shists, may be a source of weakness in engineering excavations, and fail easily along the foliae. Instances are on record where underground works have been hindered when such rocks were encountered in tunnelling and boring: the schist, when unsupported in a working face, may begin to move rapidly (or 'flow') under the pressure exerted by the surrounding rocks (see Chapter 16).

Gneiss

Gneiss has a rough banding or *foliation*, in which pale coloured bands of quartz and feldspar lie parallel with bands or streaks of mafic minerals; the mafic minerals are mainly biotite, hornblende, or in some cases pyroxene. Biotite is often accompanied by muscovite, and garnets are common accessory minerals. A gneiss breaks less readily than a schist and commonly splits across the foliation; it is often coarser in texture than most schists, though some gneisses are relatively fine-grained.

The term *orthogneiss* is used for rocks derived from igneous rocks such as granite by regional metamorphism; and the name *paragneiss* is given to those rocks derived from sediments.

An example of the latter is *biotite-gneiss*, a high grade derivative from argillaceous rocks such as shales (Fig. 7.9). It is composed of bands in which quartz and

Fig. 7.9 Biotite-gneiss, Sutherland, showing foliation and augen-structure.

feldspar are concentrated, and mica-rich bands interspersed with them; the proportions of the felsic (*fe*ldspar, and light coloured) and mafic (*ma*gnesium and *fe*rrous iron, and dark coloured) material may vary according to the composition of the original sediment. Biotite-gneiss thus represents the higher grade of metamorphism in the series:

phyllite→mica-schist→biotite-gneiss (see Fig. 7.5)

Some orthogneisses have the composition of a granite or granodiorite, and show bands of quartz-feldspar composition in parallel with streaks of oriented biotite, or biotite and hornblende. Some of the feldspar in the quartz-feldspar bands may be clustered into lenticular or eye-shaped areas for which the name 'augen' (=eye) is used. Such a rock is an *augen-gneiss*; the foliation is often deflected around the eye-shaped areas (Fig. 7.9). An example of augen-gneiss is the metamorphosed granite of Inchbae, N. Scotland; the porphyritic feldspars of the original granite are now the augen of the gneiss. The granite was intruded into argillaceous rocks, and a contact aureole with andalusite-hornfels was developed around it. When the area underwent a second (regional) metamorphism, the hornfelses were converted into kyanite schists and the granite into augen-gneiss.

Beautifully striped gneisses can result from the injection of thin sheets of quartz-feldspar material (of igneous origin) along parallel surfaces in the parent rock, and present a veined, striped appearance; they are called *injection-gneisses* (Fig. 7.10).

Fig. 7.10 Injection gneiss, near Lochinver, Sutherland (length of hammer head about 12 cm).

Migmatite

The introduction of igneous (e.g. granitic) material into country-rocks of various kinds produces mixed rocks or *migmatites* (p. 106). In some migmatites the mixing is mechanical, by the injection of veins or stripes; in others the mixing is chemical, and arises from the permeation or soaking of the country-rocks by the invading fluids (*cf.* granitization, p. 108). The subject is too extensive for further consideration here, but we may note that migmatites of many kinds are found in the Precambrian rocks of Scandinavia and Finland, the Baltic Shield and similar areas of Precambrian rock (Fig. 2.5); also in areas of high grade metamorphism in many orogenic belts.

Granulite

A rock composed of quartz, feldspar, pyroxene and garnet in nearly equidimensional grains (*granoblastic* texture) in which schistosity is less pronounced because the platy minerals, especially mica, are scarce. Granulites are believed to form in conditions of high temperature and pressure (Figs 7.5 and 7.6*b*).

Dislocation metamorphism

Dislocation metamorphism occurs on faults and thrusts where rock is altered by earth movement (see Figs 1.20 and 2.3). Much energy stored in the surrounding crust is released along these zones and dislocation metamorphism is associated with earthquakes (Chapter 1). Within 10 km of the Earth's surface these movements involve brittle fracture of rock, the mechanical breaking caused by shearing, grinding and crushing being termed *brecciation* or *cataclasis* (=breaking down). Major shear zones continue to great depths and below 10 km pressure and temperature may be sufficient for dislocation to occur by plastic deformation. Fine-grained rocks are produced, called *mylonites* (Greek, *mylon*=mill).

Zones of dislocation metamorphism often contain much greater quantities of the minerals mica and amphibole than occur in adjacent rocks. These minerals require abundant OH^- in their lattice and it is believed they formed when the shear zones acted as conduits for the expulsion of water from the metamorphic belt, water being the major source of OH^-.

Ancient shear zones containing rocks metamorphosed by dislocation exist in the roots of the Caledonian mountains in Scotland (the Moine Thrust is one of many: Figs 2.6 and 2.18), and in ancient mountain belts elsewhere. Modern examples exist in the Alpine–Himalayan chain (Figs 1.17 and 2.17) and in the circum-Pacific orogenic belts (Figs 1.4 and 1.17) in which are exposed the active shear zones of the Franciscan area in the Coast Ranges of N. America, of Honshu Island in Japan, and of the Alpine Fault in New Zealand.

Metamorphic rock associations

The associations that have been described are illustrated in Fig. 7.11. This pattern of metamorphic conditions can vary and an area once at high pressure and low temperature may gradually come under the influence of both high pressure and high temperature. Many metamorphic rocks have a fabric and mineralogy that could only have been produced by more than one phase of metamorphism.

Numerous extensive excavations into metamorphic rock, for mines, tunnels, underground storage chambers and machine halls for power stations, have shown that a mixture of metamorphic rock types must be expected. Regionally metamorphosed rocks may be thermally metamorphosed by granite intrusions and cut by shear zones in which dislocation metamorphism has occurred. Structures such as folds, with cleavage (Fig. 7.8), faults and thrusts produced by one phase of metamorphism, may be refolded by a later metamorphism to create structures of complex geometry and develop a new cleavage. More than one direction of foliation may result, and excavation into rocks of such character is accompanied by the constant risk of slabs of rock becoming detached at a weak foliation surface in the roof or walls, and falling into the excavation. This problem seriously delayed the construction of an underground power station at Kariba Dam in southern Africa, and bankrupted the contractor.

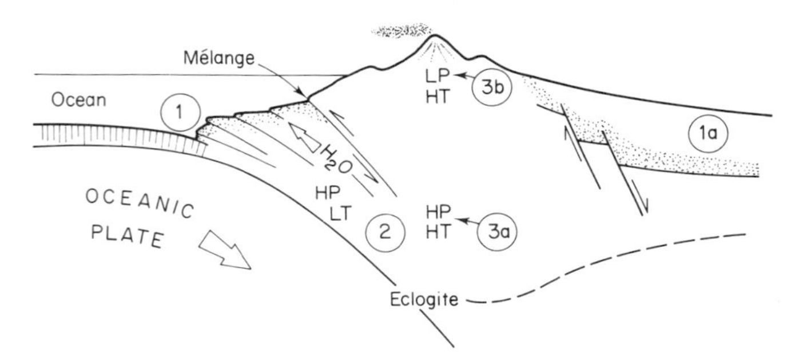

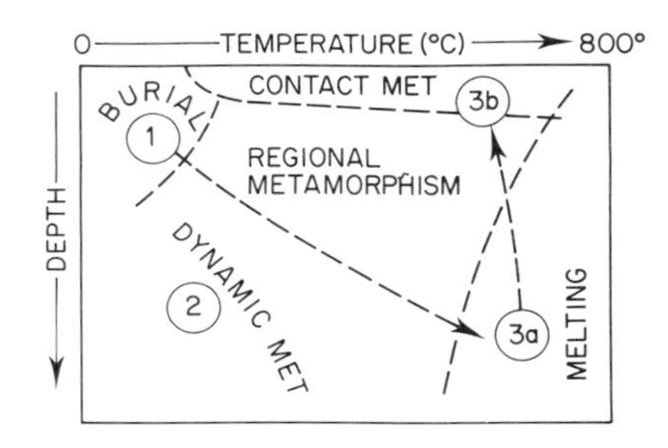

Fig. 7.11 Metamorphic rock associations: HP, LP = high and low pressure: HT, LT = high and low temperature. 1 = burial that results in severe distortion of strata. 1a = burial with the minimum of distortion. Eclogite = rock composed of garnet and pyroxene and quartz. Mélange = a chaotic mixture of rocks associated with major fault zones.

Economic rocks and minerals

Important industries have developed to extract slate for roofing, marble for ornamental stone, migmatite for facing buildings and lining floors, and good quality hornfels for ballast. Of greater importance are special minerals produced by metamorphism.

Asbestos (p. 74), graphite (p. 84), and talc (p. 80) are found associated with metamorphic rocks but are not restricted in their formation to metamorphism.

Sillimanite (p. 79), kyanite (p. 79) and andalusite (p. 79) are metamorphic minerals of great value to the refractory industry. Porcelain containing these minerals is endowed with the ability to withstand very high temperatures and exhibit little expansion. Commercial concentrations of these minerals occur in Kenya, the Appalachians, California, India, S. Africa and Western Australia: all are associated with schists.

Garnet (p. 79) is an important abrasive and mined from gneiss in New York, New Hampshire and N. Carolina. Placer deposits (p. 131) of garnet are worked in many countries where the mineral concentration in rock is too low for economic extraction by mining.

Selected bibliography

Gillen, C. (1982). *Metamorphic Geology*. George Allen & Unwin, London.

Examples of construction in metamorphic rock

Andric, M., Roberts, G. T. and Tarvydas, R.K. (1976). Engineering geology of the Gordon Dam, S.W. Tasmania. *Q. Jl. Engng. Geol.*, **9**, 1–24.

Anderson, J.G.C., Arthur, J. and Powell, D.B. (1977).

The engineering geology of the Dinorwic underground complex and its approach tunnels. *Proc. British Geotechnical Society Confr. on Rock Engineering*. Newcastle-upon-Tyne. pp. 491–510.

Anttikoski, U.V. and Saraste, A.E. (1977). The effect of horizontal stress in the rock on the construction of the Salmisaari oil caverns, in: *Storage in Excavated Rock Caverns*. M. Bergman (Editor) Pergamon Press, Oxford, **3,** 593–8.

Freire, F.C.V. and Souza, R.J.B. (1979). Lining, support and instrumentation of the cavern for the Paulo Alfonso IV power station, Brazil, in: *Tunnelling '79. Instn. Mining and Metallurgy*. London. pp. 182–92.

Kuesel, T.R. and King, E.H. (1979). *MARTA's Peachtree Centre Station*. Proc. 1979. Rapid Excavation and Tunnelling Confr. A.C. Maevis, W.A. Hustrulid (Editors) Amer. Instn. Mining. Engrs. New York, **2,** 1521–44.

Adams, J.W. (1978). Lessons learned at Eisenhower Memorial Tunnel. *Tunnels and Tunnelling*, **10,** 4, 20–3.

8

Geological Structures

Sedimentary rocks, which have been described in Chapter 6, occupy the greater part of the Earth's land surface; they occur essentially as layers or strata and are parts of the stratigraphical sequence of rocks discussed in Chapter 2. A single stratum may be of any thickness from a few millimetres to a metre or more, and the surfaces which separate it from the next stratum above or below are called *bedding planes*. In this chapter we are concerned with the arrangements of sedimentary rocks as structural units in the Earth's outer crust. The ways in which these structures appear on geological maps are considered in Chapter 12.

Dip and strike

Consider a flat uniform stratum which is tilted out of the horizontal (Fig. 8.1). On its sloping surface there is one direction in which a horizontal line can be drawn, called the *strike*. It is a direction that can be measured on beds that are exposed to view and recorded as a compass bearing. At right angles to the strike is the direction of maximum slope, or *dip*. The angle of inclination which a line drawn on the stratum in the dip direction makes with the horizontal is the *angle of dip* (or true dip), and can be measured with a clinometer and recorded to the nearest degree (Fig. 12.8). For example, a dip of 25° in a direction whose bearing is N.140° would be written *25° at 140*. The bearing is taken from north (or, in the southern hemisphere, from south). A line on a sloping rock-surface in any other direction than that of the true dip makes a smaller angle with the horizontal, called an *apparent dip*. Apparent dips are seen in quarry faces where the strike of the face is not parallel to the true dip direction (Fig. 12.12).

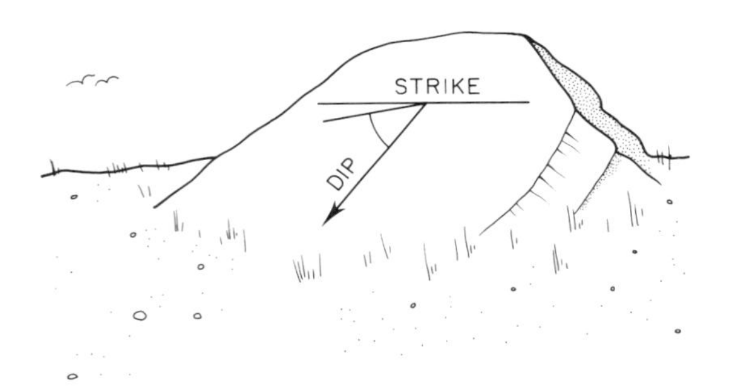

Fig. 8.1 Dip and strike. Dipping strata in an exposure of rock at the surface of the ground. Angle of dip shown.

Strike and dip are two fundamental conceptions in structural geology, and are the geologist's method of defining the attitude of inclined strata. The information is placed on a map as a short arrow (dip arrow) with its tip at the point of observation, together with a number giving the angle of true dip (Fig. 12.1). For horizontal beds the symbol + is used, i.e. where the dip is zero.

Horizontal strata

Small areas where sedimentary rocks are horizontal or nearly so are often preserved as hills capped by a resistant layer; e.g. many are seen along the western margin of the Cotswolds in England. On a larger scale, a flat-topped tableland of *mesa* (Fig. 3.11) has a capping of hard rock from which steep slopes fall away on all sides, as in the Blue Mountains of New South Wales and in Grand Canyon, Arizona. The capping may be a hard sedimentary layer or igneous rock such as a lava flow. Table Mountain, near Cape Town, is a mesa. The distance measured on the ground, between beds of horizontal strata will depend upon their thickness and the slope of the ground (Fig. 3.12).

Dipping strata

A ridge which is formed of hard beds overlying softer, with a small to moderate dip, has one topographical slope steeper than the other and is called an *escarpment* (Fig. 3.12). The length of the ridge follows the strike direction of the dipping beds; the gentler slope, in the dip direction, is the *dip-slope*. Examples in England are Wenlock Edge (Fig. 3.12) and the North and South Downs (Fig. 2.17*a*). When the strata dip more steeply, a ridge is formed with a dip-slope nearly equal in inclination to that of the scarp slope, and the ridge is then called a *hog-back* (p. 146). Steep-sided ridges are also formed by vertical beds, i.e. when the dip is 90°.

An *outlier* is an island of rock left by the erosion of an escarpment and now found entirely surrounded in plan by older rocks, e.g. the hills shown at the top of Fig. 8.2. The converse of this is an *inlier*, i.e. an area of older rocks surrounded by younger. Inliers are often developed in valleys, where streams have cut down and exposed older rocks locally in the bottom of the valley floor.

The distance measured on the ground between beds of dipping strata depends upon their thickness, the shape of the ground *and* their dip (Fig. 12.19*b*). In valleys the strata will 'V' in the direction of dip, Fig. 8.3, an effect recorded

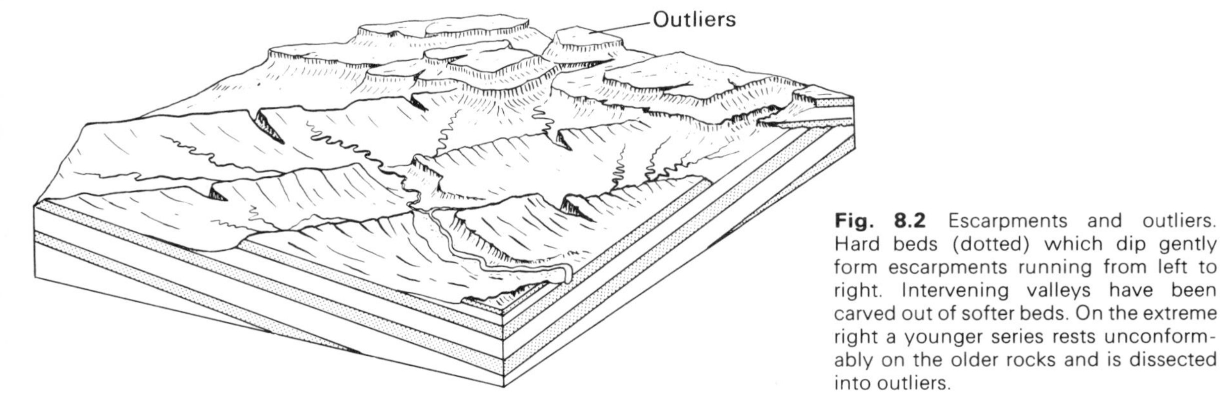

Fig. 8.2 Escarpments and outliers. Hard beds (dotted) which dip gently form escarpments running from left to right. Intervening valleys have been carved out of softer beds. On the extreme right a younger series rests unconformably on the older rocks and is dissected into outliers.

on maps (Fig. 12.17). The only exception to this rule is when the beds dip downstream at a low angle, less than the slope of the valley floor; the 'V' then is elongated and points upstream (Fig. 8.3*c*).

When beds have been upturned to a vertical position their boundaries on a map are straight and independent of topographical rise or fall.

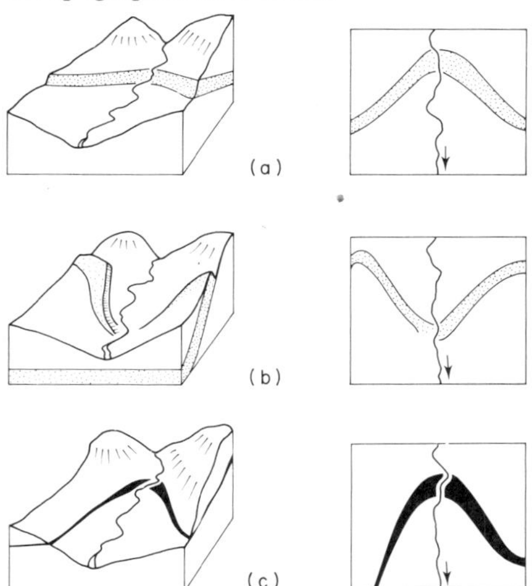

Fig. 8.3 Diagrams to show outcrops of dipping beds which cross a valley, with their form as it appears on a map. (**a**) and (**b**) 'V' in dip-direction. (**c**) Strata dip *downstream* but at an angle that is less than the slope of the valley floor.

Unconformity

Unconformity was described in Chapter 2 (Fig. 2.2). There is frequently a discordance in dip between the older and younger strata, as in Fig. 8.4 where older beds have been folded, and denuded during a rise of the sea floor, before the younger series was deposited on them. The unconformity was later tilted to its present position. As we have seen earlier, an unconformity marks both a break in the process of deposition and an interval of time when no sediments were laid down in the area.

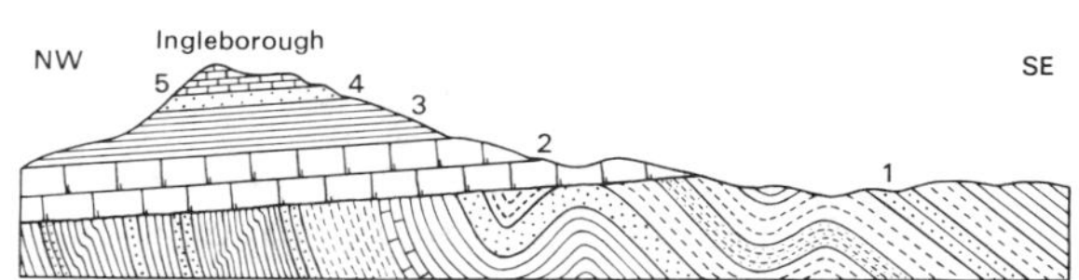

Fig. 8.4 Section through Ingleborough, N. Yorkshire, showing unconformity between Carboniferous and folded older rocks (1). (2) Carboniferous Limestone. (3) Yoredale Series, (4). (5) Millstone Grit. Length of section about 8 km.

Basic geological structures are now described, the processes responsible for their formation are termed *tectonic* (Greek *tekton*, a builder).

Folds

It is frequently seen that strata in many parts of the Earth's crust have been bent or buckled into folds; dipping beds, mentioned above, are often parts of such structures. An arched fold in which the two limbs dip away from one another is called an *antiform*, or an *anticline* when the rocks that form its central part or core are older than the outer strata (Fig. 2.17*a*, the Wealden anticline beneath Ashdown Forest, and Fig. 8.15). A fold in which the limbs dip towards one another is a *synform*, or a *syncline* when the strata forming the core of the fold are younger than those below them (Fig. 2.17*a*, the London Basin, and Fig. 6.19). It is necessary to make this age distinction before naming a fold (which may be turned on its side, as shown in Fig. 8.7, or inverted); if the relative ages of the core and the 'envelope' rocks around it are *not* known, the terms *antiform* and *synform* are used.

The name *monocline* is given to a flexure that has two parallel gently dipping (or horizontal) limbs with a steeper middle part betwen them; it is in effect a local steepening of the dip (Fig. 8.5), and differs from an anticline in that the dips are in one direction only. Large monoclines are sometimes developed in sedimentary rocks overlying a rigid basement when the latter has been subject to fault movement.

The formation of folds has in many cases been due to

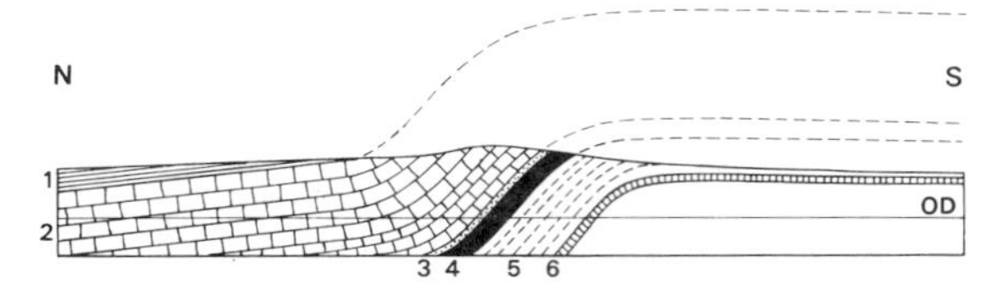

Fig. 8.5 Monocline at the Hog's Back, near Guildford, Surrey. (1) Tertiary beds, (2) Chalk, (3) Upper Greensand, (4) Gault, (5, 6) Lower Greensand. Length of section = 2 km.

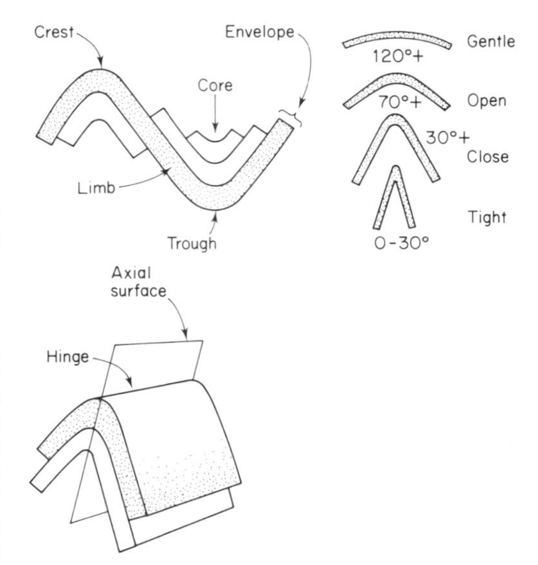

Fig. 8.6 Antiform and synform in upright open folding, the degrees of acuteness in folding, and the hinge of folding.

the operation of forces tangential to the Earth's surface; the rocks have responded to crustal compression by bending or buckling, to form a fold-system whose pattern is related to the controlling forces (Figs 7.1 and 7.8). In other cases forces operating radially to the Earth's surface have caused strata to move and buckle, e.g. by downward movement under gravity (Fig. 6.6) or upward movement above a salt dome (Fig. 6.21). Folding involves brittle and ductile deformation. Strata which have behaved as brittle materials are described as *competent*, whilst strata which flowed as ductile materials are described as *incompetent*.

Flow-folding is an example of incompetent behaviour and arises in beds which offer little resistance to deformation, such as salt deposits (Fig. 8.18), or rocks which become ductile when buried at considerable depth in the crust where high temperatures prevail, as for some gneisses.

Fold geometry

Some simple fold forms are shown in Fig. 8.6, and we now consider their geometry. In the cross-section of an upright fold the highest point on the anticline is the *crest* and the lowest point of the syncline the *trough*; the length of the fold extends parallel to the strike of the beds, i.e. in a direction at right angles to the section.

The line along a particular bed where the curvature is greatest is called the *hinge* or *hinge-line* of the fold (Fig. 8.6); and the part of a folded surface between one hinge and the next is a fold *limb*.

The surface which bisects the angle between the fold limbs is the *axial surface*, and is defined as the locus of the hinges of all beds forming the fold (see Fig. 8.6). This definition allows for the curvature which is frequently found in an axial surface (Fig. 8.7*g*).

The intersection of an axial surface with the surface of the ground can be called the *axial trace* of the fold; in some instances it is marked on a geological map as 'axis of folding' (see Chapter 12).

Two other terms for describing folds, especially useful where dissimilar rocks are involved, are *core* and *envelope*; Fig. 8.6 shows that the core is the inner part of the fold and the envelope the outer part.

Some other fold forms are shown in Fig. 8.7. *Parallel folds* maintain parallel bedding and uniform bed thickness. The beds in *similar folds* thicken towards their hinge and indicate ductile deformation. *Chevron folds* are characterized by their sharp hinge.

The *attitude* of a fold describes the inclination of its axial surface, which can be *upright*, *inclined* or *recumbent*,

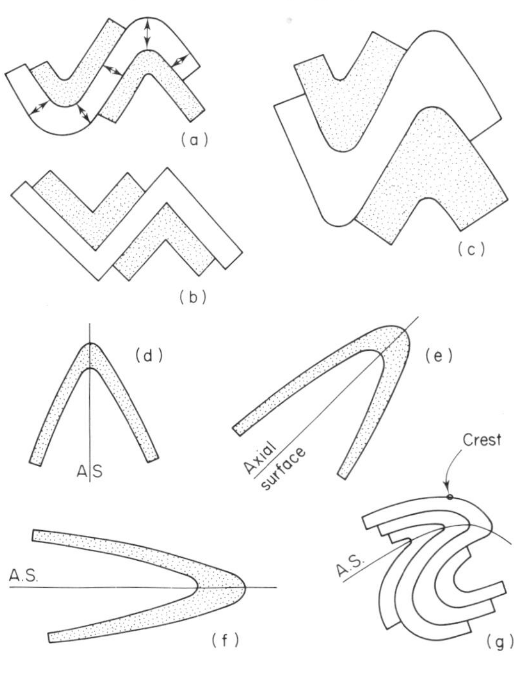

Fig. 8.7 Fold form (**a–c**) and attitude (**d–g**). (**a**) Parallel folds. (**b**) Chevron folds. (**c**) Similar folds. (**d**) Upright, (**e**) inclined, (**f**) recumbent (as in Fig. 8.8), (**g**) curved axial surface.

Fig. 8.8 Recumbent Chevron folds in Carboniferous rocks, Millook, North Cornwall.

Figs 8.7*d–f*; note that the hinge may not coincide with the crest or trough of these folds. Some recumbent folds have developed into flat-lying thrusts by movement of the upper part of the fold relative to the lower. After their formation, folds may be tilted through any angle, or even inverted, by later structural deformations affecting an area (Fig. 2.17b). A recumbent fold does not, from its shape alone, provide a distinction between an anticlinal or synclinal structure (Fig. 8.7). But if the rocks of the core of the fold are *younger* than the envelope rock it is recumbent *syncline*; if they are older it is a recumbent *anticline*. The terms *synform* and *antiform* can be used to describe such folds when the age relation of the rocks involved is not known.

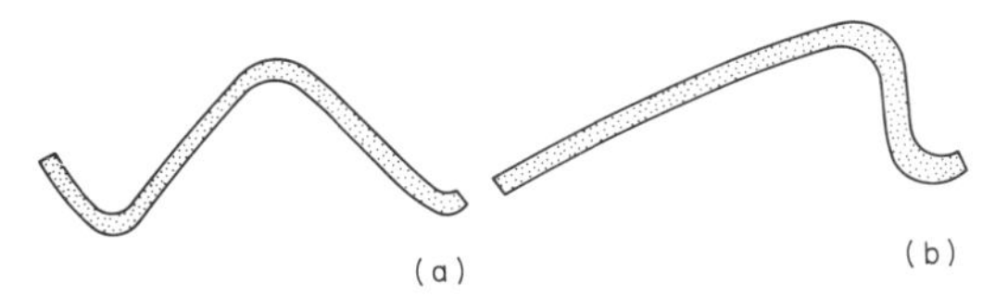

Fig. 8.9 Fold symmetry. (**a**) symmetric, (**b**) asymmetric.

The *symmetry* of a fold describes the relative length of its limbs: when both limbs are of equal length the fold is symmetric; Fig. 8.9.

Plunge

In most instances the fold hinge is inclined to the horizontal, and is then said to plunge. Thus the level of an antiform crest falls in the direction of plunge, and in some cases the antiform diminishes in amplitude when traced along its length in that direction, and may eventually merge into unfolded beds. In a plunging synform (Fig. 8.10) the trough shallows in one direction along the length of the fold and deepens in the opposite direction, which is the direction of plunge. The angle of plunge (or

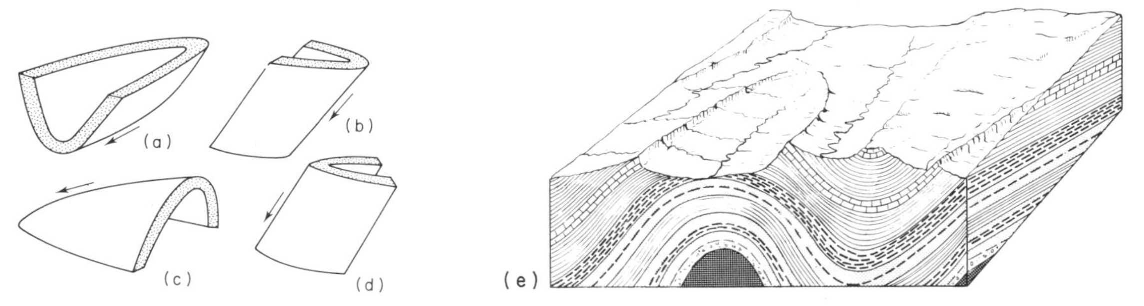

Fig. 8.10 Diagrams to illustrate plunge. (**a**) Upright plunging synform; (**b**) synform with nearly vertical plunge; (**c**) inclined plunging antiform; (**d**) antiform with nearly vertical plunge. (**e**) Block diagram, illustrating plunging folds, with hard beds forming topographical features on a denuded surface. Plunge shown on right hand section.

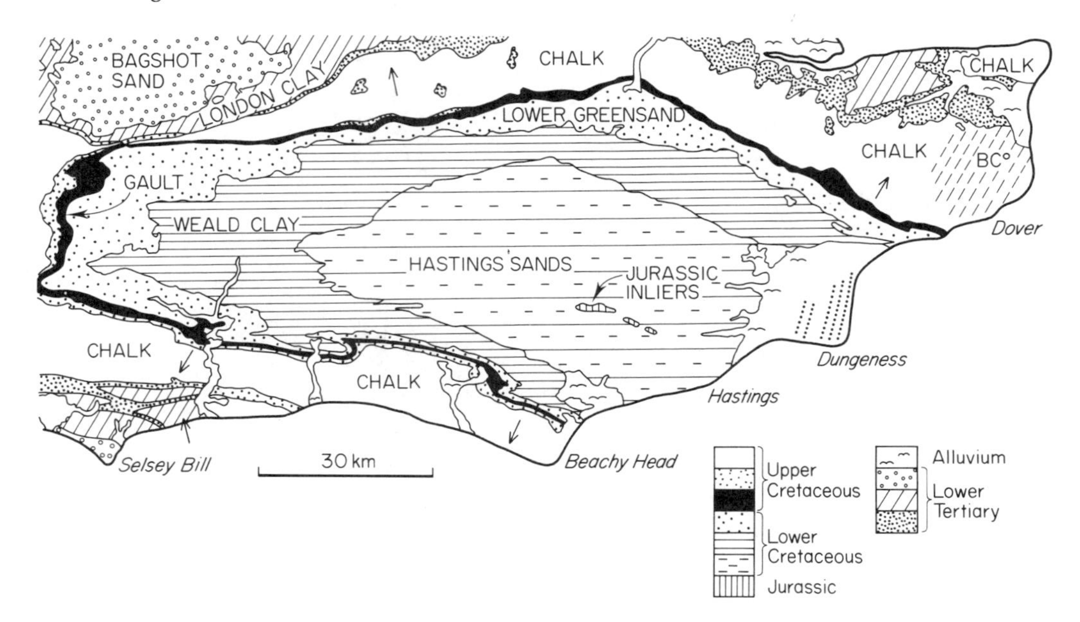

Fig. 8.11 A geological map of the Wealden anticline. For N–S section see Fig. 2.17*a*. Arrows indicate dip of strata away from the fold axis. This structure extends across the English Channel into N. France (see Fig. 2.18). BC = Betteshanger Colliery in the *concealed* (*q.v*) Kent Coalfield (diagonal shading).

inclination of the hinge to the horizontal) may be small, as in the figure, or at a steeper angle (Fig. 8.10), or vertical.

It should be noted that when a plunging antiform is seen intersected by a ground surface, i.e. in outcrop, the strikes of the two limbs converge in the direction of plunge until, at the closure of the fold, the strike is normal to the axial trace (Fig. 8.10*e*). Examples of this are sometimes clearly displayed on a foreshore which has been cut by wave action across folded rocks. The representation of such structures on geological maps often gives characteristic outcrop shapes (Fig. 12.17).

The Wealden anticline of south-east England is a large east-west fold which plunges to the east in the eastern part of its outcrop, and to the west at its western end (Fig. 8.11). The limbs of the fold dip north and south respectively on either side of the main anticlinal axis, and denudation has removed the Chalk and some underlying formations in the crestal area, revealing beds of the Lower Cretaceous Wealden Series in the core of the structure. The Chalk once extended across the present Weald.

Fold groups

The relative strength of strata during folding is reflected by the relationship between folds. Those shown in Figs 8.6 and 8.7 are called *harmonic folds* as adjacent strata have deformed in harmony. *Disharmonic folds* occur when adjacent beds have different wavelengths the smaller folds being termed *parasitic folds* (Fig. 8.12), or 'Z', 'M' and 'S' folds. Box-folds and kink bands are examples of other groups that may occur. Other fold groups, called *nappes*, are illustrated in Fig. 2.17*b* and are discussed on p. 153.

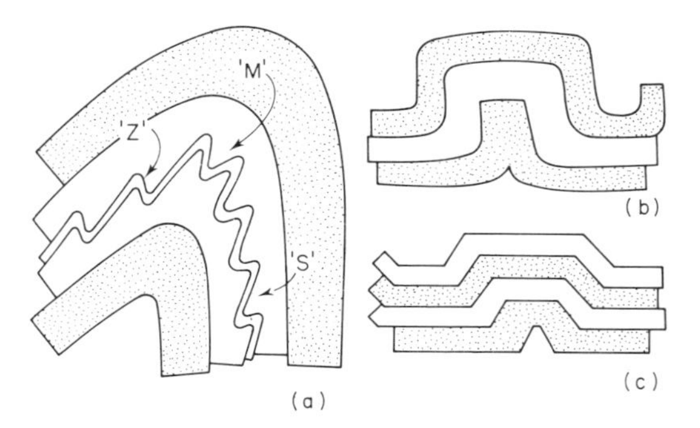

Fig. 8.12 Fold groups. (**a**) Parasitic folds within a larger fold. (**b**) Box-fold, others would occur to the left and right. (**c**) Kink-band; one of a series.

Minor structures

The structures to be considered here are *fracture-cleavage, tension gashes, boudinage* and *slickensides*. Slaty cleavage (or flow cleavage), which results from the growth of new, oriented minerals, has been discussed in Chapter 7 (*q.v.*).

Fracture-cleavage

This is mechanical in origin and consists of parallel fractures in a deformed rock. In Fig. 8.13 a band of shale lies between two harder sandstone beds, and the relative movement of the two sandstones (see arrows) has fractured the shale, which has acquired cleavage surfaces oblique to the bedding. The cleavage is parallel to one of the two conjugate shear directions in the rock. The spacing of such shear fractures varies with the nature of the material, and is closer in incompetent rocks such as shale than in harder (competent) rocks.

Observations of fracture-cleavage in folded rocks are useful in interpreting the overall structure when only part of the fold is open to inspection; e.g. the inclination of the fracture-cleavage on one limb of an antiform will indicate on which side of the limb the fold axis is situated. Thus, in the steeper limb of an overturned fold (Fig. 8.13)

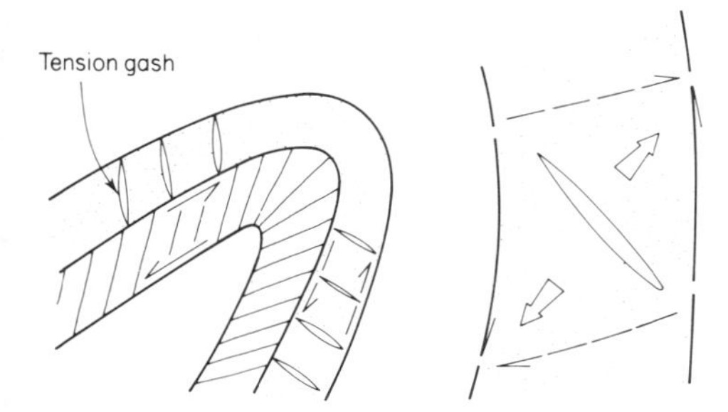

Fig. 8.13 Fracture cleavage in shale folded between stronger beds, with detail of relationship between tension gash and shear stresses.

the cleavage dips at a *lower* angle than the dip of the strata; while in the less steep limb it dips at a greater angle. In an exposure which lies to one side of a fold axis, the position of the axis – either to right or left of the exposure – can be inferred.

Tension gashes

These are formed during the deformation of brittle material and may be related to the shear stress between strata (Fig. 8.13). The gashes are often filled by minerals, usually quartz or calcite.

Boudinage

When a competent layer of rock is subjected to tension in the plane of the layer (Fig. 8.14*a*), deformation by extension may result in fracturing of the layer to give rod-like pieces, or boudins (rather like 'sausages'), with small gaps between them (Fig. 8.14*b*). They are often located on the limbs of folds; softer material above and below the boudin layer is squeezed into the gaps. The length of the boudins generally lies parallel to the fold-axis, and the gaps may be filled with quartz (q). Boudins are found, e.g. in sandstones or limestones interbedded with shales, and also in thick quartz veins in metamorphic rocks such as schists and slates (Fig. 8.14*d*).

Slickensides

These are a lineation associated with the movement of adjacent beds during folding (Fig. 8.14a and c) and occur when weak layers *shear* between more competent beds. Similar features can be produced by the dissolution of minerals under pressure during shear and recrystallization of the mineral matter as streaks oriented in the direction of bed movement. The influence these small structures may have upon the strength of bedded sediments is discussed by Skempton (1966).

Major fold structures

Folds may form large structures 10 to 100 km in size. Examples already considered include fold chains made of many folds (*cf.* Figs 1.10, 2.16, 2.17*b* and 7.6*a*) and large folds (*cf.* Figs 6.19, 6.21*a*, 8.5, 8.10*e* and 8.11). Some terms commonly used to describe these structures are:

Pericline: and elongate antiform or synform, e.g. Figs 8.10*e* and 8.15.

Nappe: a recumbent fold found in mountain belts and cut by thrusts e.g. Fig. 2.17*b* (and see *overthrusts*).

Dome: a circular structure in plan with dips radiating from a central high area: often associated with rising salt domes e.g. Figs 6.21*c*, 8.18.

Basin: the reverse of a dome (Fig. 6.19) often associated with faulting at depth.

Gravity folds

Gravity folds, which may develop in a comparatively short space of time, are due to the sliding of rock masses

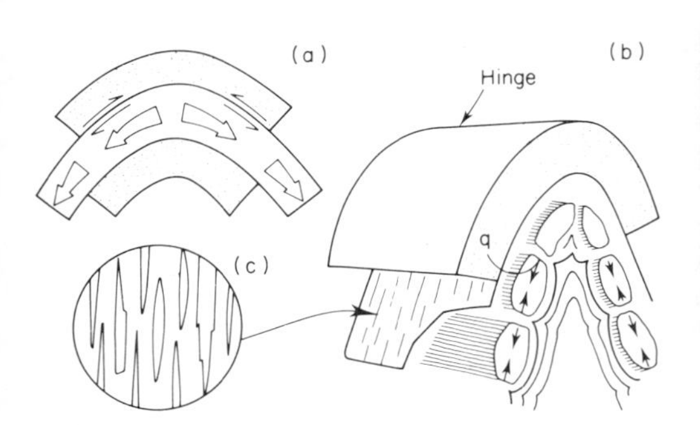

Fig. 8.14 (**a**) Tension within a competent bed; (**b**) boudin structures with quartz (q) between boudins; (**c**) lineations developed on a bedding surface; (**d**) boudins in sandstone and slate, in cliff face near Trebarwith Strand, North Cornwall. Courtesy of the British Geological Survey, London. Crown copyright reserved.

Fig. 8.15 A plunging anticline, or more precisely a plunging anticlinal pericline. Iran. Photo by Aerofilms Ltd. (AB489.)

down a slope under the influence of gravity. Examples of the masses of sediment which move over the sea floor and give rise to slump structures, on a relatively small scale, have been mentioned in Fig. 6.6. Submarine slumping often takes place on slopes greater than 10° or 12° but may also occur on flatter slopes; it is frequently related to submarine fault scarps. Sliding may develop on an inclined stratification surface within a sedimentary series, and the partly compacted sediment is thrown into complex folds.

Larger structures arise where the vertical uplift of an area covered with sedimentary layers has caused the latter to slide off down-dip soon after the uplift developed. An example of such *gravity collapse* structures is the series of folded limestones and marls which lie on the flanks of eroded anticlinal folds in south-west Iran (Fig. 8.16).

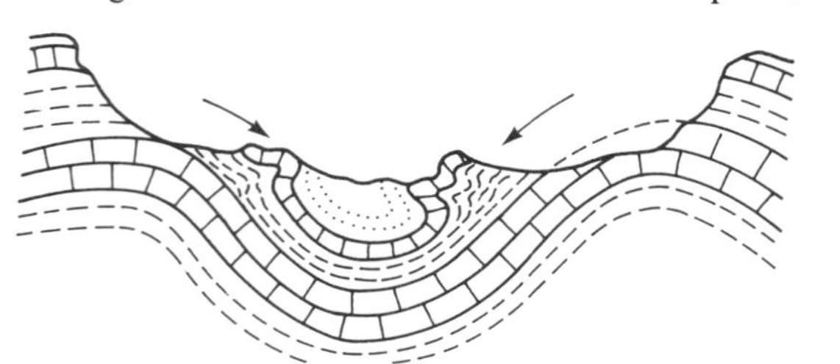

Fig. 8.16 Gravity collapse structures in young sediments (from Harrison and Falcon, 1936).

Valley bulges

Bulges are formed in clays or shales which are interbedded with more competent strata, and are exposed in the

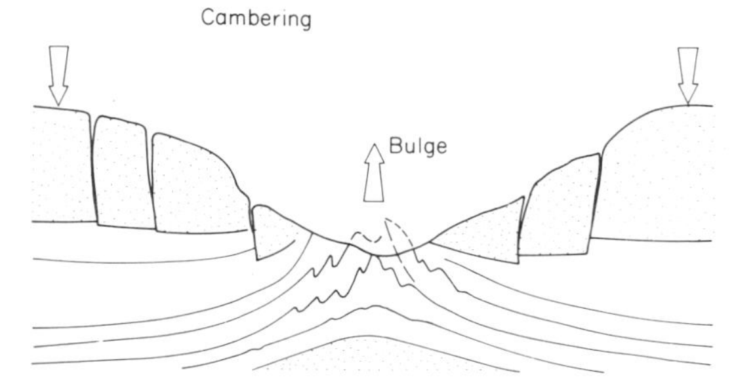

Fig. 8.17 Valley bulging and associated movement (cambering) of the valley sides (see also Figs 14.10 and 16.12).

bottoms of valleys after these have been eroded. The excavation of a river's valley is equivalent to the removal of a large vertical load at the locality. The rocks on either side of the valley exert a downward pressure, which is unbalanced (without lateral restraint) when the valley has been formed; as a result, soft beds in the valley bottom become deformed and squeezed into shallow folds (Fig. 8.17). Residual stress within the sediments often assists in the process (p. 160). Bulges of this kind, and related structures, were described in Jurassic rocks near Northampton by Hollingworth *et al.* in 1944. Other instances are found in the Pennine area, where interbedded shales and sandstones of the Yoredale Series crop out, as at the site of the Ladybower Dam in the Ashop valley, west of Sheffield, England. Many glaciated valleys contain valley bulge (see p. 260).

Salt domes

These are formed where strata are upturned by a plug of salt moving upwards under pressure. A layer of rock-salt is more easily deformed than other rocks with which it is associated, and under pressure can rise as an intrusive plug, penetrating and lifting overlying strata. The doming thus formed (Figs 6.21, 8.18) is often nearly circular in

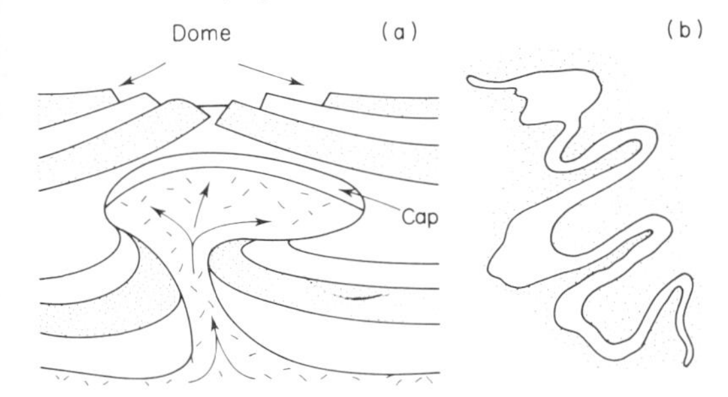

Fig. 8.18 (**a**) A dome created by the ascent of a salt diapir; (**b**) flow folds as may be found in a diapir (see ref. Halbouty, 1967: Chapter 6).

plan. When the salt outcrops at the surface it makes an abrupt change of slope. Many salt domes are found in the Gulf Coast area of Texas, where they are associated with accumulations of petroleum; and in Iran, Germany, and elsewhere. Geophysical exploration of the North Sea basin has revealed the presence of dome-like uplifts which are due to intrusive bodies of salt.

The term *diapir* (= through-piercing) is used for structures produced by materials such as rock-salt or mobile granite which have moved upwards and pierced through the overlying strata. Material ascends as a diapir when it is weak enough to flow and less dense than the overlaying strata.

Faults

The state of stress in the outer part of the Earth's crust is complex, and earlier in this chapter we have seen how some rocks respond to compression by folding and buckling. Commonly also, fractures are formed in relief of stress which has accumulated in rocks, either independently of folds or associated with them. The fractures include faults and joints: *faults* are fractures on which relative displacement of the two sides of the break has taken place; *joints* are those where no displacement has occurred. Groups of faults and sets of joints may both form patterns which can be significant in indicating the orientation of the stresses that resulted in the fracturing, though a clear indication of this is not always forthcoming. In this account our discussion is limited to the geometrical description of fractures and some of the structural patterns to which they give rise.

Brittle fracture

When a brittle material is broken in compression, failure occurs along shear planes inclined to the direction of loading; for example, a prism of concrete will, on breaking, yield a rough cone or pyramid. The angle at the apex of such a cone is theoretically 90°, since the maximum shear stresses should occur on planes inclined at 45° to the direction of maximum principal stress. In practice this angle is generally found to be less than 45°, owing to the operation of frictional forces in the material at the moment of failure. Thus an acute-angled cone is formed; or, in two dimensions, two complementary shear fractures intersect at an acute angle which faces the direction of maximum compression. Many rocks when tested break in this way, as brittle materials, with an acute angle between the shear directions which is often in the neighbourhood of 67° (Fig. 9.17).

Tests have been made on cylinders of rock, with lateral confining pressure as well as an axial compression. In these tests brittle failure occurs at low confining pressures, and ductile failure at high confining pressures; a transitional type of failure which may be called semi-brittle can occur at intermediate pressures, with a stress-strain curve lying between the ductile and brittle curves (Fig. 9.12). Applying these results to natural conditions in the Earth's crust: for a particular rock different types of failure could arise from different conditions, such as the depth at which failure takes place, and the temperature prevailing at that depth. Granite would behave as a brittle material at depths up to several kilometres if surface temperatures prevailed; but there is an average temperature gradient in the outer crust of 30°Ckm^{-1} (see p. 2), and at a depth of 30 km temperature would be about 500°C. At still greater depths and higher temperature the same rock would behave as a ductile material.

Faulting

Near the Earth's surface, hard rocks which have undergone compression may have failed in shear, with the production of single fault, or groups of faults forming a fault pattern. Three main kinds of faults are formed, namely thrust faults, normal faults, and wrench faults.

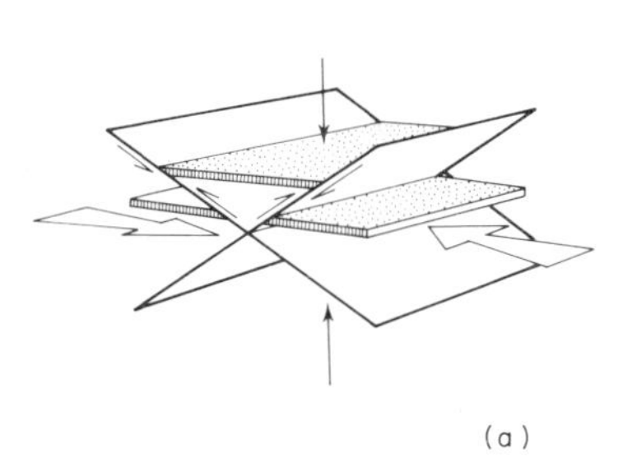

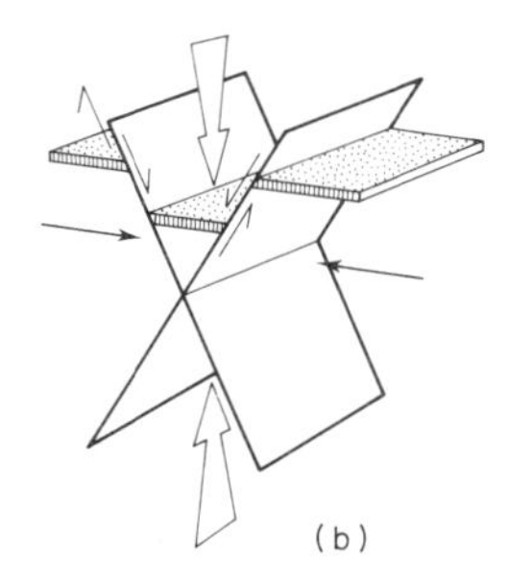

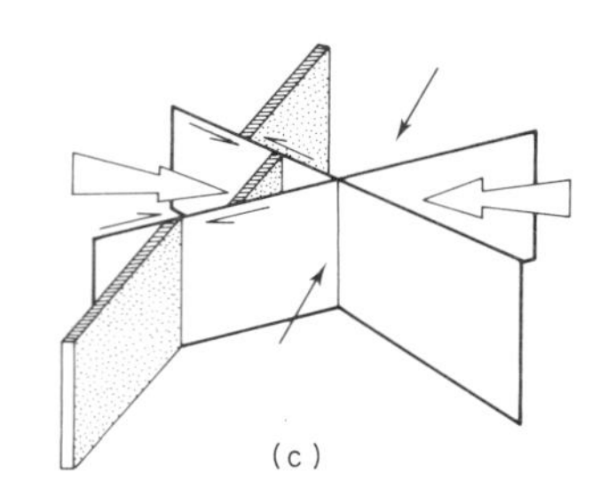

Fig. 8.19 Relationship of faults (shear planes) to axes of principal stress; (**a**) thrusts, (**b**) normal faults, (**c**) wrench faults. Maximum stress, large arrows; minimum stress, small arrows, intermediate stress omitted, but its direction is that of the line along which the shear planes intersect.

Where the dominant compression was horizontal and the vertical load small, the shear fractures formed intersect as shown in Fig. 8.19, the acute angle between them facing the maximum principal stress; faults having this kind of orientation are the *thrusts* or thrust faults. Where the greatest stress was vertical (Fig. 8.19) the shear planes are steeply inclined to the horizontal, and faults formed under such stress conditions are *normal faults*. Thirdly, when both the maximum and minimum stresses were horizontal and the intermediate stress vertical (Fig. 8.19), the resulting fractures are vertical surfaces and correspond to *wrench* faults. The two wrench faults of a pair are often inclined to one another at an angle between 50° and 70°.

Normal and reverse faults

Here the relative movement on the fault surface is mainly in a vertical direction; and the vertical component of the displacement between two originally adjacent points is called the *throw* of the fault.

Two common types of fault are illustrated in Fig. 8.20*a*, which shows a *normal fault* on left, where a bed originally continuous at 'a' has been broken and the side A moved down relative to B, or B moved up relative to A. The side B is called the 'footwall', and A the 'hanging wall.' In a normal fault, therefore, the hanging wall is displaced downwards relative to the footwall. A *reverse fault* is shown in Fig. 8.20*b*. In this case the hanging wall C is displaced upwards relative to the footwall D. Notice that the effect of the reverse faulting is to *reduce* the original horizontal extent of the broken bed; the two displaced ends of a stratum overlap in plan. The effect of a normal fault is to *increase* the original horizontal distance between points on either side of the fracture.

The custom of describing faults as 'normal' or 'reverse' originated in English coal-mining practice. Faults which were inclined towards the downthrow side were met most commonly (e.g. Fig. 6.19). When a seam which was being worked ran into such a fault, it was necessary to continue the heading a short way (as at 'a' in Fig. 8.20*a*), and then sink a shaft to recover the seam; this was the usual or 'normal' practice. When a fault was encountered which had moved the opposite way, that was the reverse of the usual conditions, and the fault was called 'reversed'.

Fault groups

Several normal faults throwing down in the same direction are spoken of as 'step faults' (Fig. 8.20*c*); two normal faults dipping towards one another produce 'trough faulting,' and dipping apart form a pair of 'ridge faults.' Where a stratum approaches a fault it is often bent backwards a little, away from the direction of movement along the fault plane, as shown in the figure. A sunken block, bounded on all sides by faults, is called a *graben*; the Rhine Valley and the Midland Valley of Scotland are examples of this structure on a large scale. A *horst* is a fault-bounded ridge-block, the converse of graben; e.g. the Black Forest and Harz Mountains, Germany, and on a smaller scale the ridge of the Malvern Hills, Hereford. The North Sea contains many horst and graben structures, some of which have provided important traps for oil and gas.

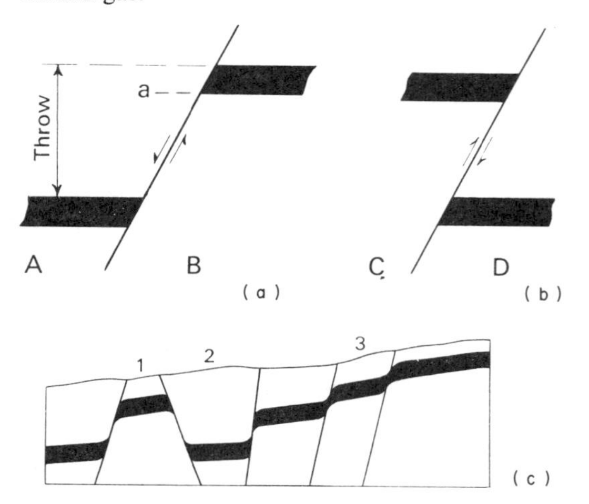

Fig. 8.20 Normal and reverse faults. Vertical cross-sections showing the displacement of a horizontal stratum (in black). (**a**) Normal fault; A=downthrow side (hanging wall), B=upcast side (footwall). (**b**) Reverse fault; C=upcast side (hanging wall), D=downthrow side (footwall). (**c**) (1) Ridge, (2) trough, and (3) step faults. The ends of the black bed are shown slightly dragged round as they approach a fault.

Wrench faults

Faults in which horizontal movement predominates, the other components being small or nil, are called *wrench faults*. The term *transcurrent fault* is also used (and formerly 'tear fault'). Wrench faults are commonly vertical or nearly so (Fig. 8.19*c*) and the relative movement is described as follows: for an observer facing the wrench fault, if the rock on the far side of the fault has been displaced to the right the movement is called *right-handed* or *dextral*; if in the opposite direction, i.e. to the observer's left, it is termed *sinistral* or *left-handed*.

Thrusts

Most of the fractures so far described have a steep inclination to the horizontal; in contrast to this is the group of faults produced by dominantly horizontal compression and known as *thrusts* or thrust planes (Fig. 8.19*a*). They are inclined to the horizontal at angles up to 45°. (If the inclination of the fault surface is greater than 45° it is called a reverse fault.) The development of thrusts from overturned folds is illustrated in Fig 8.21*a*; with continuing compression the middle limb of the fold becomes attenuated (A), then fracture occurs, and further compression results in movement along the thrust (or shear plane) so formed.

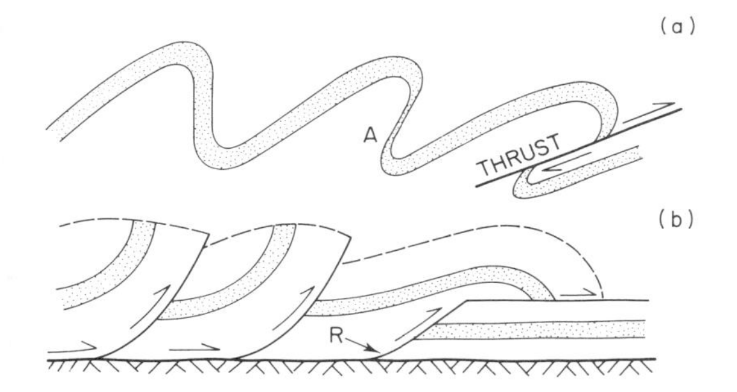

Fig. 8.21 Development of (**a**) thrust and (**b**) over-thrust with imbricate structure. R = ramp (see Fig. 2.6*a*).

Very large thrusts which are virtually horizontal are called *overthrusts*; they do not fit into the classification of faults given above. They are surfaces of large extent, with a small inclination, on which large masses of rock have been moved for considerable distances, Fig. 8.21*b*. In the Moine thrust-zone (p. 139) overthrusting has resulted in a horizontal displacement of some 16 km. The very large displacement on thrusts suggests that the frictional resistance on their sliding surface is greatly reduced by pore water pressure and plastic deformation.

Overthrusts and nappes

In intensely-folded mountain regions such as the Alps, overthrusts associated with recumbent folds occur on a large scale. A recumbent fold driven forward on a thrust surface is called a *nappe*, or thrust-sheet; structures of this kind form the basis of an interpretation of the complex geology of the Alps (Fig. 2.17*b*). The figure illustrates the structural units along a line from Lausanne to Italy. The Swiss plain at Lausanne, betweeen the Jura Mountains and the Pre-Alps, is covered by thick coarse sediments known as *molasse*, mainly of Oligocene age, which were accumulated by rapid denudation of the rising mountain-mass to the east. West of the Pre-Alps the Jura Mountains show a series of upright broken anticlines and synclines that are believed to have slid westwards on a weak layer of Triassic rock-salt (TL in the figure). East of the Pre-Alps rises a pile of gigantic recumbent folds, now deeply eroded; the lower members are covered by higher and later-formed overfolds which have been moved for large distances as nappes. Relics detached from a nappe by later erosion are called nappe-outliers or *klippen* (the plural of *klippe*); the Pre-Alps are probably parts of such folds that have been thrust over the molasse. South-east of the Pre-Alps in the section are the High Calcareous Alps, composed of Mesozoic sediments. They are one of four tectonic units each of which is a group of nappes, and are named in ascending order Helvetid, Pennid, Grisonid, and Dinarid. The Helvetid nappes of the High Calcareous Alps are overridden by the Pennid group of nappes (numbered I to VI in the figure); the Matterhorn pyramid, with a thrust-plane at its base, is an erosion remnant of the Dent Blanche nappe (VI). Beyond the pile of Pennid nappes the rocks turn vertically downwards in a root-zone, beyond which the Dinaric structures show movements towards the south-east. The broad structure of this area is thus asymmetric rather than bilateral as in many orogenic belts (e.g. Fig. 7.6*a*).

A structure involving both major and minor thrusts is called the *imbricate structure* (latin *imbrex*, a tile); the minor thrust masses overlap like the tiles on a roof (Fig. 8.21*b*). It was first described from the district of Assynt, Sutherland, where rock-wedges have been piled up by a series of minor thrusts which are truncated by flatter major thrusts; on these, large heavy masses of rock have been pushed forward over the imbricated zone. Fig. 2.6*a* shows the structure in section at Loch Glencoul in the Western Highlands, first described by B.N. Peach and J. Horne (1907). The rocks immediately below these surfaces of dislocation have been crushed, sheared and deformed by the overriding of the heavy masses, and converted into mylonites (*q.v.*).

Fault components

The movement on a fault may be of any amount and in any direction on the fault surface. The complete displacement along a fault plane between two originally adjacent points can be described by means of three components measured in directions at right angles to one another. The vertical component, or *throw*, has already been mentioned; the two horizontal components are the *heave* (*bc* in Fig. 8.22), measured in a vertical plane at right angles to the fault plane; and the *strike-slip* (*cd*), measured parallel to the strike of the fault plane. The total displacement, *ad*, is called the *slip*, or *oblique-slip*. The other

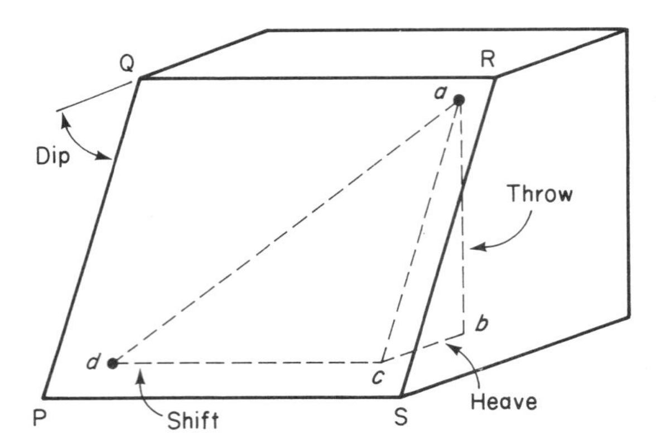

Fig. 8.22 Components of fault displacement. PQRS=fault surface. *ad*=total slip; *ab*=throw; *bc*=heave; *cd*=shift or strike slip being along the strike of the fault surface; *ac*=dip slip being along the dip of the fault. The points *a, c, d* lie on the fault surface.

component on the fault-plane is the *dip-slip* (*ac* in the figure).

The *hade* of a fault plane is its angle of inclination to the vertical, the angle *bac* in Fig. 8.22. It is often more convenient to speak of the dip of a fault, i.e. its inclination to the horizontal, than of its hade: (hade + dip = 90°). When a fault is vertical the hade and heave are zero.

Faults are referred to above as 'planes' only for ease of description; they are more often zones of crushed rock which have been broken by the movement, and range in width from a few millimetres to several metres. Sometimes the crushed rock along the fault surface has been ground very small, and become mixed with water penetrating along the line of the fault, producing a *gouge*. Fault surfaces are often plane only for a small extent; curved and irregular fractures are common.

Strike and dip faults

Faults are also described from the direction of their outcrops on the ground, with reference to the strata which they displace. *Strike faults* outcrop parallel to the strike of the strata; *dip faults* run in the direction of the dip of the beds; and *oblique faults* are those which approximate neither to the dip nor strike direction. These cases are illustrated below.

Effect of normal faulting on outcrop

Many effects can be produced depending upon the inclination of the fault with the strata it intersects, its direction of total slip, and the dip of strata. To illustrate the effects examples of normal faulting are considered first.

(*i*) A dip fault

The general effect of a dip fault is to displace the outcrops of corresponding beds on either side of the fault line. In Fig. 8.23, the side *AB* of the block is parallel to the dip direction of the strata, and the side *BC* shows its strike. The normal fault (FF) throws down to the right. The outcrops of the beds (see on the top of the block and as they would appear on a map of an area of plane ground) are offset on either side of the fault trace; upper (younger) beds on the down-throw side are brought opposite lower (older) beds on the upcast side.

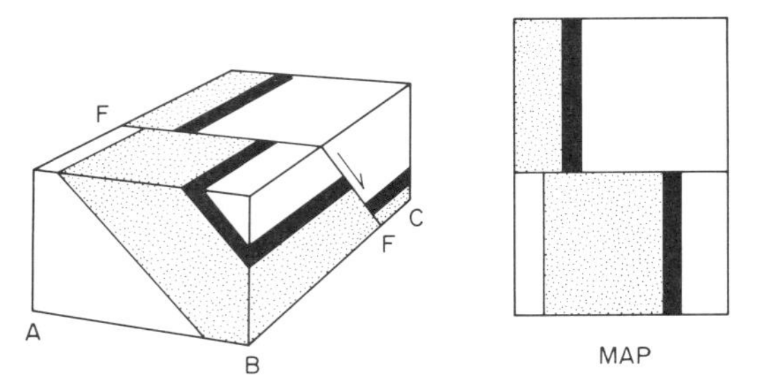

Fig. 8.23 Effect of normal dip fault (FF) on dipping strata, and appearance of this effect as recorded on a map.

(*ii*) A strike fault dipping in the same direction as the strata

Strike faults either *repeat* or *cut out* the outcrop of some parts of the faulted strata. In Fig. 8.24, a series of beds is broken by a normal strike fault (FF) which dips

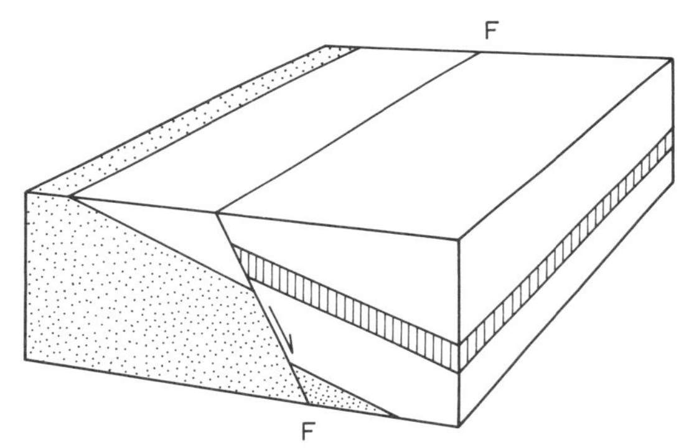

Fig. 8.24 Normal strike fault (FF) with dip in same direction as strata concealing striped bed.

steeply in the same direction as the dip of the beds. The result of this is that one bed (striped) does not appear in outcrop at the surface (the upper surface of the block); it is cut out by the faulting. The case where the dip of such a fault is the same as the dip of the strata is referred to as a *slide*, i.e. the fault movement has taken place along a bedding-plane. Where the dip of the fault plane is *less* than the dip of the beds, and in the same direction, there is *repetition* of some part of the outcrop of the beds.

(*iii*) A strike fault dipping in the opposite direction to strata

Figure 8.25 shows a series of beds broken by a normal strike fault which dips in the opposite direction to the dip of the beds. The effect is to repeat the outcrop of some beds (the black bed, for instance), as seen on the upper surface of the block.

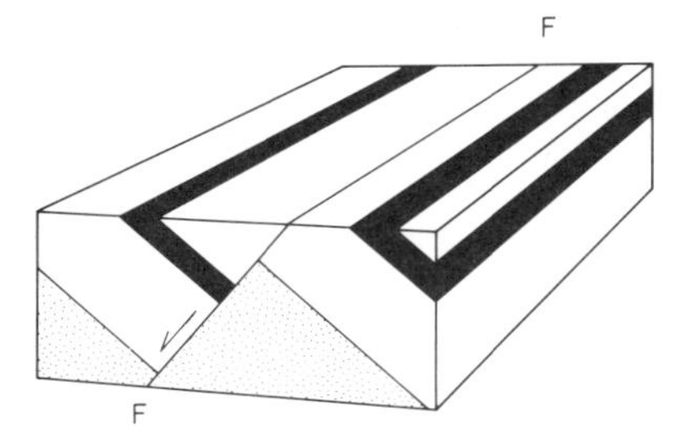

Fig. 8.25 Normal strike fault (FF) with dip in opposite direction to that of strata, and repeating the outcrop of some beds.

Effect of reverse faulting on outcrop

In some instances the effects of reverse faulting on outcrop are not always distinguishable from those of normal faulting. For example, a reverse dip-fault will still bring older beds on the upcast side against newer beds on the downthrow side (*cf.* normal faulting, above). On the other hand, a reverse strike-fault will have the opposite effect to a normal fault in the same direction; in the cases illustrated in Figs 8.24 and 8.25, a little consideration will show that *repetition* of outcrop would be produced by a reversed fault dipping in the same direction as the dip of the strata (Fig. 8.24) and elimination of outcrop in the other case (Fig. 8.25).

Effect of wrench faulting on outcrop

A fault of this type, in which the displacement is essentially horizontal, may shift the *outcrops* of dipping beds in an apparently similar way to a normal fault movement. In Fig. 8.23 for example, the staggering of the outcrops shown could have come about by a horizontal movement of one part of the block past the other, along the fault. It is important therefore not to make rigid assumptions about the nature of a fault solely from the displacement of outcrops, without looking for other evidence.

A wrench fault in the direction of strike will not change the order of the strata in outcrop unless there is some vertical throw as well as the horizontal movement.

Criteria for recognizing wrench faulting are often a matter of field observation and two may be noted here:

(*i*) Parallel grooves have in some cases been cut on a fault surface by projecting irregularities on one block as it moved past the other. These grooves are known as *slickensides* and give valuable information about the direction of the fault movement. If they are horizontal or nearly so, it can be inferred that the last movement on the fault had a mainly wrench component.

(*ii*) When vertical beds are faulted and displaced laterally in outcrop, a wrench fault is indicated. The horizontal shift of a vertical axial surface of a fold is illustrated in Fig. 8.26.

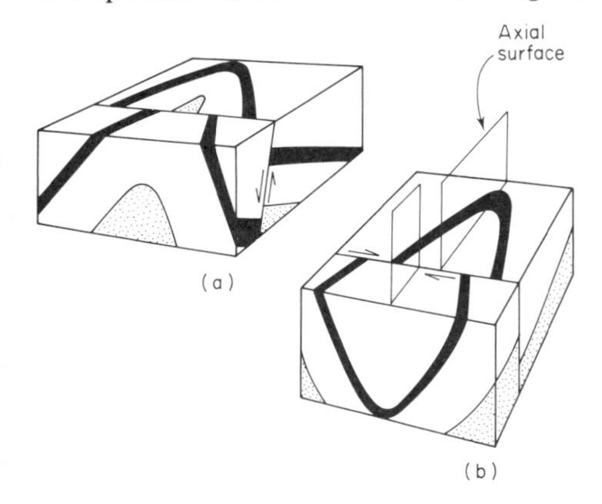

Fig. 8.26 Folded strata broken by faults. (**a**) Faulted antiform, outcrops of limbs are moved apart on upcast side of a normal fault. (**b**) Plunging synform, cut by a wrench fault. Note the horizontal shift of the axial surface.

Effect of faulting on fold outcrop

The principles described in the previous sections can be applied when a fault shifts the outcrop of a fold. Two examples are illustrated, one for a normal fault, the other for a wrench fault (Fig. 8.26).

The diagrams in Fig. 8.26 show outcrops on a plane surface; irregular topography will modify this simple conception, as discussed in Chapter 12. When a fault is vertical it has a straight-line trace, whatever the topography. If a fault is not vertical, the shape of its trace or outcrop on a map gives an indication of its dip, which may show whether it is normal or reverse. But, in general, it is difficult to distinguish between normal and reverse faults solely by their relation to outcrops on a map. Field evidence is more definite.

Joints

Parting-planes known as *joints* are ubiquitous in almost all kinds of rocks, have been referred to earlier in this book (p. 33) and are the most common structure to affect the behaviour of soil and rock in engineering works. The 'fissures' of many overconsolidated sediments (p. 160) are joints. They are fractures on which there has been no movement, or no discernible movement, of one side (or wall) relative to the other. In this way joints differ from faults (p. 151). Groups of parallel joints are called *joint sets*, and for two or more sets which intersect the term *joint system* is used.

Many joints are developed in the relief of tensional or shearing stresses acting on a rock mass. The cause of the stresses has been variously ascribed to shrinkage or contraction, compression, unequal uplift or subsidence, and other phenomena; all are a relief of *in situ* stress.

Joints in young sediments

Shrinkage-joints develop from the drying or freezing, and resultant shrinkage of sedimentary deposits (*cf.* sun-cracks in newly formed estuarine muds and shrinkage cracks in frozen ground). The contraction of a mass of sediment, if evenly distributed, was held by some to be a cause of joint formation; but objections have been put forward to such a process operating in thick sandstones and limestones. Some recently deposited clays found in the beds of lakes, and other wet sediments have, however, been observed to contain joint sets formed relatively early in the history of the sediment.

Joints in folded sediments

Commonly two sets of joints intersecting at right angles or nearly so are found in many sedimentary rocks, and are often perpendicular to the bedding planes (Fig. 8.27).

In dipping or folded sediments the direction of one of these sets frequently corresponds to the strike of the beds, and the other to the dip direction; they are therefore referred to as *strike-joints* and *dip-joints* respectively (Fig 8.27, s and d).

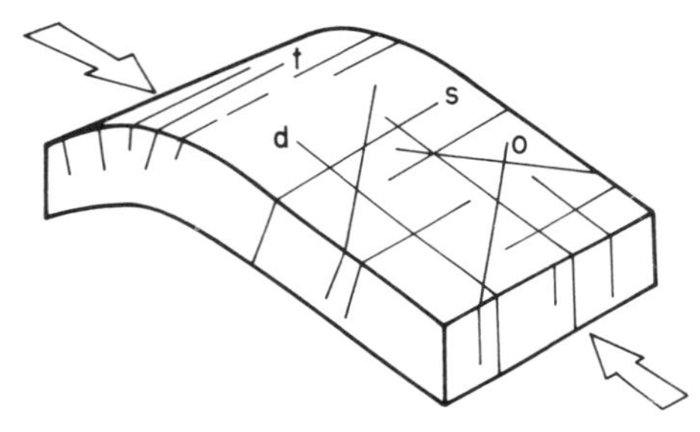

Fig. 8.27 Jointing in a folded stratum. t=tension joints at hinge of fold, s=strike joints, d=dip joints, o=oblique joints (shear joints). Large arrows give direction of maximum compressive stress. (From Price, 1981.)

Others parallel the axis of the fold and are concentrated in the fold hinge (p. 146) where tension is greatest during folding; they are called tension joints (Fig. 8.27, t). Two other joint sets may lie in the planes of maximum shear stress during folding (Fig. 8.27, o).

The joints in sedimentary rocks sometimes end at a bedding-plane, i.e. they extend from one bedding-plane to the next, but others may cross many bedding-planes and are called *master-joints*.

The extent and regular pattern of joints in folded strata demonstrate their relationship to the compression that created the folds and such joints are believed to be a visible expression of the relief of stress that remained in the rocks after deformation.

Joints in igneous rock

In igneous rocks *contraction-joints* are formed as a hot mass cools and contracts. A lava flow often develops a hexagonal joint pattern by contraction around many centres, equally spaced from one another, the contraction being taken up by the opening of tension joints. This gives rise to a columnar structure in the lava sheet, the hexagonal columns running from top to bottom of the mass (Fig. 5.2); the columnar basalts of Antrim and of the Western Isles of Scotland (e.g. Staffa) are familiar examples.

In an intrusive igneous body such as a granite pluton, joint systems develop during the cooling of the mass after its emplacement. Cooling takes effect first near its roof and walls; movement of still liquid or plastic magma at a lower level may then give rise to fracturing in the outer, more solidified, part of the intrusion. Lines of oriented inclusions such as elongated crystals, in the igneous rock, are called *flow-lines*; they have been formed by viscous flow near and often parallel to the roof of the mass (Fig. 8.28).

Cross-joints (or 'Q-joints'), are steep or vertical and lie perpendicular to the flow-lines. These are tension fractures opened by the drag of viscous material past the outer partly rigid shell.

Longitudinal-joints (or 'S-joints') are steep surfaces which lie parallel to the flow-structures and perpendicular to the cross-joints, and are later in formation.

Flat-lying joints (or 'L-joints') of primary origin form a third set of fractures at right angles to the Q- and S-joints. Later formed flat-lying joints, or sheet-jointing (p. 34), are developed during denudation and unloading. Some of these joints act as channels for the passage of residual fluids, which consolidate as veins of aplite or pegmatite; or for the passage of hot gases which affect the walls of open joints (pneumatolysis, p. 109), or fluids which coat them with hydrothermal minerals. All these joints are of early origin. But many granites show other joint systems which are probably tectonic, and often consist of two sets of steep or vertical joints which intersect at 90° or less, and are sometimes accompanied by diagonal joints. *Sheet-jointing*, which crosses such a joint-system and is especially developed near the roof of the igneous body, is probably caused by tensile stresses generated by the unloading of the mass during denudation, as mentioned above, and described on p. 34.

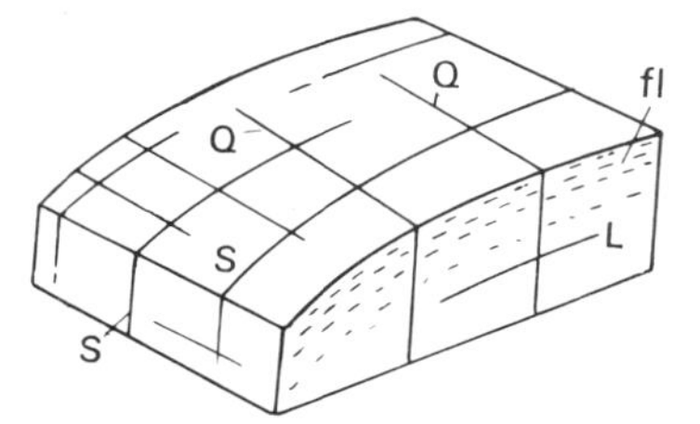

Fig. 8.28 Joint patterns within intrusive plutonic rocks. fl=flow lines. Q, S, L=joint sets named by Hans Cloos (1923). (From Balk, 1937.)

Joints near faults

It is often a matter of observation that, near a visible fault, the rocks are traversed by joints parallel to the fault surface. This suggests that they were formed in response to the same stress system that resulted in the faulting; their frequency diminishes with increasing distance from the fault.

Where shear fractures intersect at a small angle, 20° or less, and one fracture is a fault, the small fractures which lie at the acute angle with it are called pinnate or *feather-joints*. They can sometimes be used to indicate the direction of shearing movement along the main fracture; the acute angle points away from the direction of movement of that side where the pinnate structures occur.

Size and spacing of joints

There is a wide range of size or extent shown by joint planes; master-joints in sediments have been mentioned above and may extend for 100 m or more. Less extensive but well defined joint sets can be called *major joints*, in distinction from smaller breaks or *minor joints*. Minute joints on a microscopic scale are seen in some thin sections of rocks. The spacing or frequency of joints (=number of planes of a particular set in a given direction measured at right angles to the joint surfaces) also varies considerably. These factors are of great importance in quarrying and other excavations; some rocks such as sandstones and limestones, in which the joints may be widely spaced, yield large blocks of stone whereas other rocks may be so closely jointed as to break up into small pieces. The ease of quarrying, excavating, or tunnelling in hard rocks largely depends on the regular or irregular nature of the joints and their direction and spacing. Joints are also important in connection with the movement of water (Chapter 13), and their presence is a main factor in promoting rock weathering (Chapter 3).

Geological structures and economic deposits

Geological structure influences the original location of economic deposits and modifies the shape of previously formed accumulations.

Influence on location of deposits

The influence geological structure may have upon the location of mineral accumulations is most clearly demonstrated by the geometry of deposits formed from mineralized fluids.

Many hydrothermal lode deposits (p. 109) occupy voids created by faults and joints formed when the intruding parent magma deformed surrounding rock (Chapter 5). Joint sets and faults existing early in the hydrothermal phase may contain different minerals from those formed later, if there is a change in the character of the hydrothermal fluids. One joint set may be rich in one mineral association whilst a second may be barren (Fig. 8.29).

Some faults and joints contain mineral deposits that are zoned, the first formed mineral being in contact with

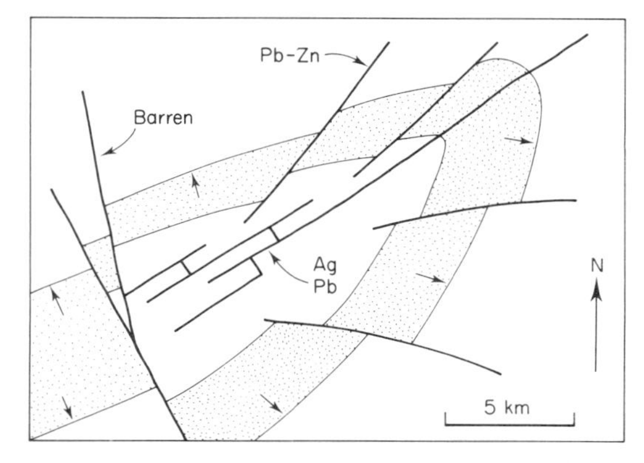

Fig. 8.29 A map of mineral veins and mineralized faults associated with a periclinal antiform. Ag=silver, Pb=lead, Zn=zinc.

the wall of the fracture, with later deposits sandwiched in between. These fractures may have been used many times to conduct fluids away from magma at depth.

Competent and incompetent beds when folded into forms similar to those illustrated in Figs 8.6 and 8.7, will develop voids in the hinge of the folds, as the competent beds ride up over their less competent neighbours. In vertical cross-section these voids have the shape of a saddle and extend in this form along the length of the fold axis: mineral deposits accumulating in them are called 'saddle reefs' and excellent examples exist at Bendigo in Australia (Fig. 8.30). Faults are usually associated with

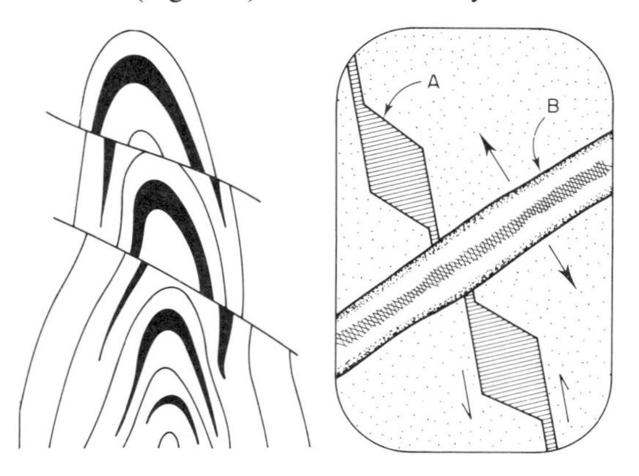

Fig. 8.30 A vertical section through saddle reefs, with detail illustrating successive phases of mineralization in a fault (A) and later fracture (B). Arrows indicate relative movement.

saddle deposits and their relevance to the upward movement of fluids through a fold, and that of joints, can be gauged from an inspection of Figs 8.26 and 8.27.

Tension gashes (Fig. 8.13) are a further example of voids formed by folding. When the space they create is occupied by valuable minerals there results a 'ladder vein', the mineral-rich gashes resembling the treads of a ladder. Similar gashes may occur in dykes (p. 95) stressed by shear displacement on their boundaries and ladder veins have been found within such intrusions.

The importance of faulting and folding to the creation of reservoirs for oil and gas has been described in Chapter 6 (Fig. 6.21).

Modification of deposits

Valuable deposits of economic minerals may have their original proportions modified after formation, by faulting and folding. Faults may truncate and separate valuable seams, or possibly conceal them or duplicate them. The nature of this disruption may be assessed by picturing the seams shown in Figs 8.20 and 8.23–8.26 as being of material for abstraction by surface or underground mining.

Intense folding can distort beds so severely that much of their economic value is dissipated by the complex methods and patterns of extraction required to win them from the ground.

Folding and faulting are both associated with jointing which divides the rock into blocks. Heavy support may be required to prevent an excavation from collapse in ground where jointing is severe. Many joints and faults also provide pathways for the movement of water to excavations.

From these points it may be concluded that the geological structure of an economic reserve is of considerable relevance to an assessment of its value.

Selected bibliography

Park, R.G. (1983). *Foundations of Structural Geology*. Blackie and Son Ltd., London, Methuen Inc., New York

Roberts, J.L. (1982). *Introduction to Geological Maps and Structures*. Pergamon Press, Oxford and New York

Price, N.J. (1981). *Fault and Joint Development in Brittle and Semi-brittle Rock*, second edition. Pergamon Press, Oxford and New York

Anderson, E.M. (1951). *Dynamics of Faulting*, second edition. Oliver and Boyd, Edinburgh

Ragan, D.M. (1968). *Structural Geology: an Introduction to Geometrical Techniques*, second edition. J. Wiley and Sons, New York

9

Strength of Geological Material

The strength of rock, or of less well consolidated sediment, is influenced by the mineralogy of its particles and by the character of the particle contacts. These properties are inherited from the processes that formed the rock, as described in Chapters 5, 6 and 7, and modified by later folding, faulting and jointing as explained in Chapter 8: finally they are affected by the agents of weathering (Chapter 3). Consequently the strength of rocks and sediments will reflect their geological history.

Influence of geological history

Burial and uplift are frequently recurring aspects of geological history (see Chapter 2) and the general effects of these processes are illustrated in Fig. 9.1.

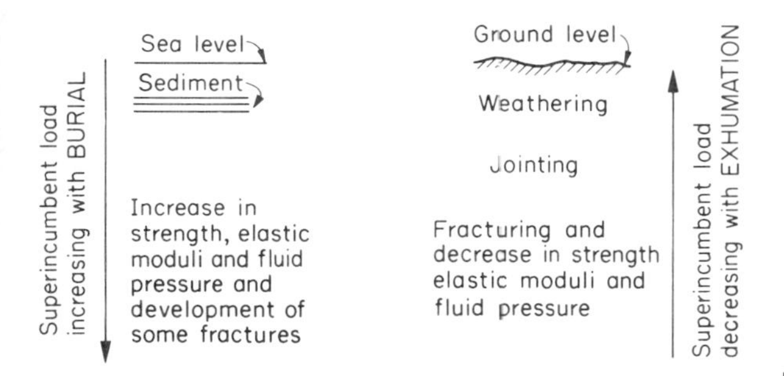

Fig. 9.1 Effect of loading and unloading on the properties of sediments.

Burial

During burial the volume of a sediment is reduced because water is squeezed from its pores. Sometimes the drainage of water is prevented by overlying strata of low permeability, such as a thick layer of mudstone, and water pressure in the pores gradually increases with burial until it equals the strength of the confining layers. Vertical fractures then develop up which the trapped water escapes: this is called *hydrofracturing*. As a sediment dewaters so its grains pack closer together and the strength of the sediment increases. Figure 9.2 illustrates the path of a sediment during burial and the lateral strain that must attend it. These events occur slowly and typical rates are recorded in Fig. 9.3.

The deformation of rock at very slow rates of strain

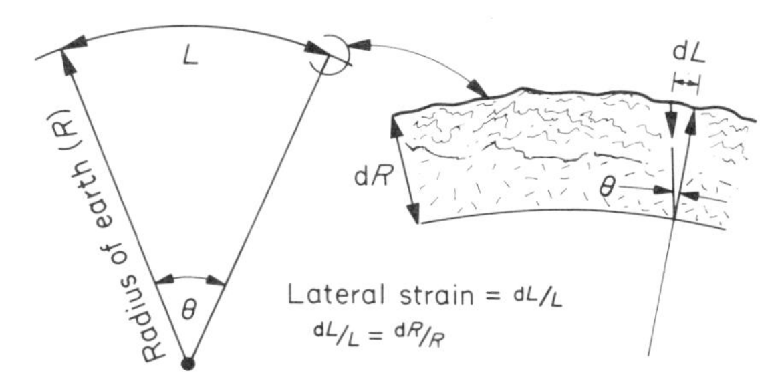

Fig. 9.2 Lateral strains accompanying burial and exhumation of sediments.

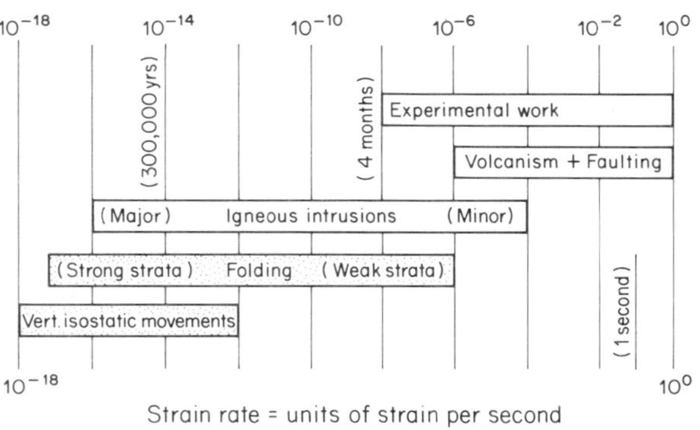

Fig. 9.3 Indicative rates of deformation, with time shown for the development of 10% strain at these rates. (From Price, 1975.)

involves processes collectively described as *creep*: they are illustrated in Fig. 9.4 where the strain that occurs in a sample loaded under a low constant stress is plotted against time. Primary creep is recoverable, and secondary creep is distinguished by the onset of permanent deformation, and tertiary creep by its culmination in failure of the sample. A rock will deform by creep under a lower stress than needed to obtain the same deformation using conventional laboratory testing, in which the normal rate of strain varies from 10^{-2} to 10^{-6} (as shown in Fig. 9.3 by the position of experimental work).

Some aspects of creep are analogous to the deformation of a viscous fluid and the material properties rocks must possess for such an analogy to exist are illustrated in Fig. 9.5. Rock behaves as an elastic material when loaded rapidly and recovers its strain when unloaded,

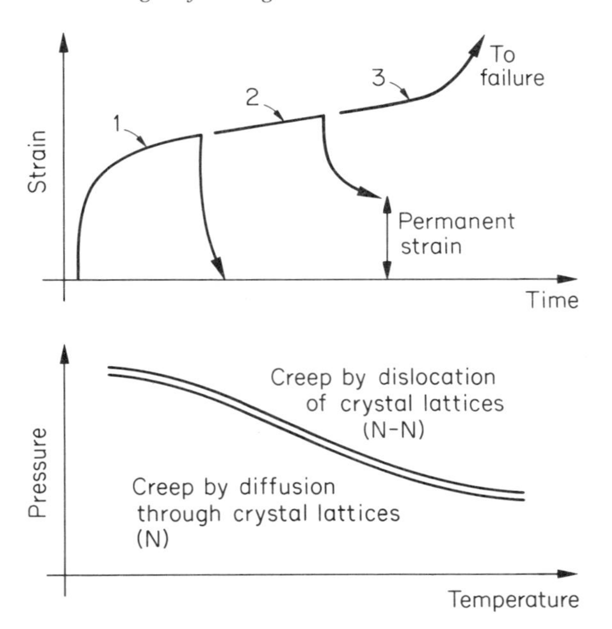

Fig. 9.4 Creep under load that is less than that required for failure in standard laboratory test.
1 = primary creep: recoverable. 2 = secondary creep: permanent strain. 3 = tertiary creep: results in failure. N = Newtonian behaviour: or rate of deformation proportional to applied stress. N–N = Non-Newtonian behaviour.

hence elastic deformation is a part of rock behaviour, Fig. 9.5*a*. Plastic deformation must also exist, Fig. 9.5*b*, because rock loaded slowly recovers only part of its strain when unloaded. This recovery of strain requires considerable time, Fig. 9.5*c*, and many rocks that have been loaded by burial, A–B in Fig. 9.5*c*, and unloaded by uplift, B–C, contain stresses that exceed those calculated

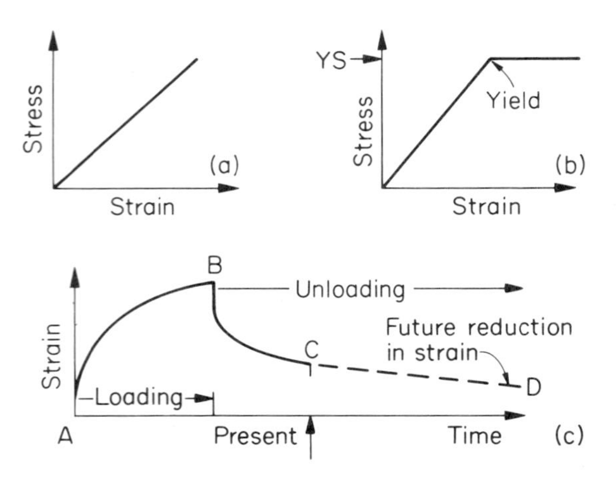

Fig. 9.5 Aspects of the deformation of rock. (**a**) elastic behaviour; (**b**) plastic deformation with viscous strain above the yield stress (YS); (**c**) strain associated with loading and unloading.

from the weight of the overlying column of strata. Some of these stresses influence the behaviour of the rocks when they are sampled for testing in the laboratory and contribute to their failure when the samples are loaded to destruction. *In-situ* stress of this kind may continue to relax, C–D, even after the rock has been further unloaded by excavation for engineering works, and can affect the stability of excavations and buildings constructed in them.

Uplift

The overburden load is progressively reduced above rocks as they are raised towards ground level and this permits them to expand in the vertical direction. Horizontal sets of joints and others of sub-horizontal inclination, will open and bedding surfaces will part (see Fig. 6.11). Figure 9.2 illustrates that uplift is also accompanied by lateral strain which enables vertical and steeply inclined joint sets to develop and open (Fig. 6.11). In addition to the joints that can be seen there are many more fractures, of microscopic size, that open in the 'solid' blocks of rock between the visible joints.

Other microscopic changes occur: crystals and grains begin to move apart as the rock expands, and this movement disrupts the contact between them (see Fig. 6.10). These, and similar processes, gradually convert a rock from the unbroken character it possessed at depth, where its crystals and grains were pressed tightly together, to the broken and porous condition it exhibits at ground level.

Shallow burial and uplift

Many of the younger sediments that are close to the surface of the Earth have not been buried to great depths and are insufficiently consolidated and cemented for them to be described as 'rock'. These are the sediments engineers call 'soil'. Despite their short geological history they exhibit the same trend in their physical character as that described above for rock.

The variation of strength with depth measured in two deposits of clay is illustrated in Fig. 9.6. The normally consolidated clay has never been unloaded and is without fissures or joints. The lateral stress at depth within this clay is similar to that which could be calculated from a knowledge of vertical stress, the theory of elasticity and the use of Poisson's ratio.

The over-consolidated clay has been unloaded by erosion of overlying sediment: it contains fractures, called *fissures*, that decrease in number with distance from ground level. Fissures influence the strength of clay and if present should always be included in the description of a deposit, e.g. 'a stiff fissured clay'. The geological loads under which the clay has consolidated have produced a small amount of visco-elastic deformation and the lateral stresses within the sediment cannot be calculated accurately from a knowledge of vertical stress and the theory of elasticity.

The magnitude of lateral stress *in-situ* is relevant to many aspects of geotechnical engineering, especially to

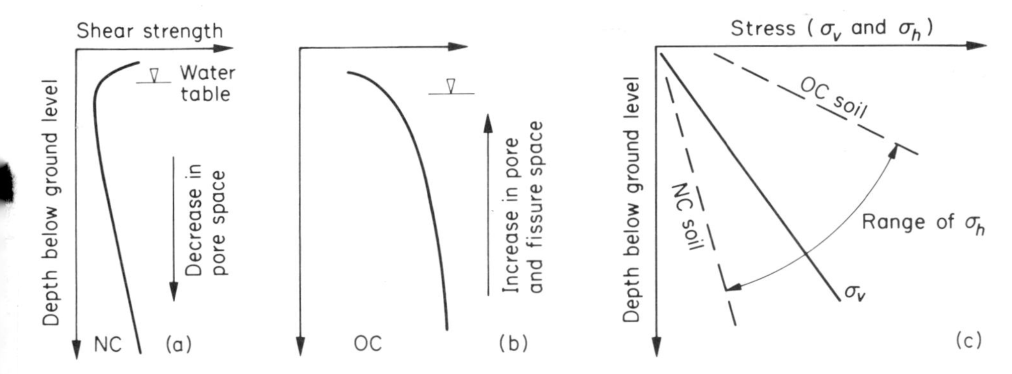

Fig. 9.6 Variation of strength of soils, and stress with depth. Increased strength above water table in (**a**) is associated with capillarity (see Fig. 9.23). NC = normally consolidated, OC = over-consolidated, σ_v and σ_h = *in-situ* stress in the vertical and horizontal directions. The ratio $(\sigma_h/\sigma_v) = K_0$.

the support of excavations in soil. Its value is described as *earth pressure at rest* (K_0); see Fig. 9.6.

Importance of drainage

A porous sediment, if loaded, will deform when its grains move under the influence of an applied load. Thus when a building is constructed upon a sedimentary deposit the sediment consolidates and the building settles. Similarly, when a jointed rock is loaded it will deform as joints and other fractures within it close under the applied load. The closure of voids and fractures, such as pores and joints, is influenced by the ease with which fluids residing in them may be displaced. Ground-water and air are the fluids most commonly encountered, the latter occupying pores and fractures above the level of water saturation.

Effective stress

Rock and soil, whose fractures and pores are completely filled with water are described as being *saturated*. In Fig. 9.7 is shown a saturated sample of soil placed in a cylindrical container, sealed at its base. The top of the sample is covered by a circular plate that can move down the cylinder, like a piston, the small annulus between the two being filled with a water-tight seal. A weight placed on the top plate pushes the plate down and compresses the sample.

The graph in Fig. 9.7 illustrates the response of a soil, such as a clay, if loaded in this apparatus. When load is applied (1) there is a rapid increase in the presure of water within the pores (ΔP). This pressure remains until pore fluid is permitted to drain (see Fig. 9.7), at which time the soil particles move closer together as the sample *consolidates*. Most of the grain movement is irreversible, for when the externally applied load is removed the sample retains its reduced dimensions ($H - \Delta H$). The total deformation (ΔH) is a function of the difference between the total applied load (W) (or *total stress*) and the pressure of

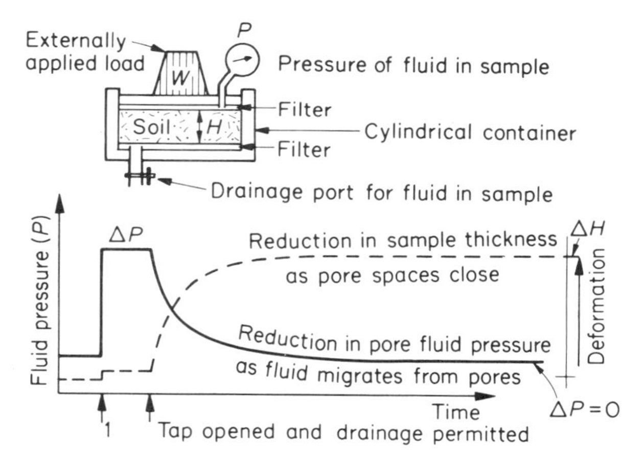

Fig. 9.7 Deformation of soil under load. This test is normally conducted in apparatus called an oedometer (see Fig. 11.6). ΔP = changes in fluid pressure. ΔH = changes in sample thickness. 1 = Load applied whilst tap is shut and drainage prevented. Fluid pressure increased and small, reversible, elastic shortening occurs.

pore water (P). This difference is called *effective stress* and the notation commonly employed to describe it is

$$\underset{\text{effective stress}}{\sigma'} = \underset{\text{total stress}}{\sigma} - \underset{\text{pore fluid pressure}}{u}$$

Laboratory experiments and careful observations of deformation beneath buildings has demonstrated repeatedly that the deformation of soil and rock can only be accurately described in terms of this difference, i.e. in terms of effective stress.

Figures 9.8 and 9.9 illustrate further aspects of this behaviour. In a saturated sample of soil unable to drain, the application of load is accompanied by an increase in pore fluid pressure of equal magnitude. When drainage commences and effective stress increases, (σ-u) in Fig. 9.8,

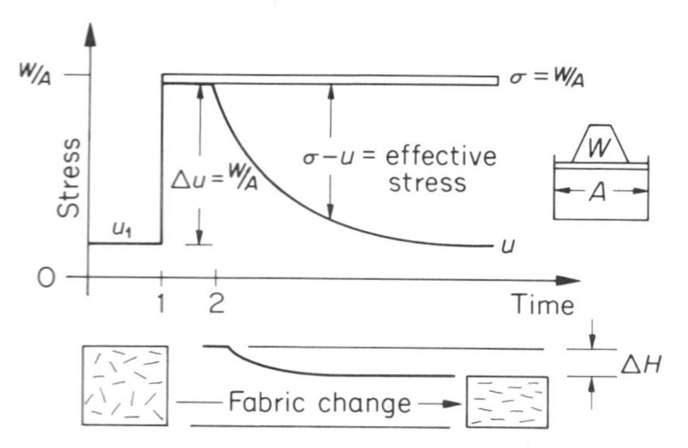

Fig. 9.8 Consolidation of soil (see also Fig. 9.7) 1 = application of load (W). 2 = start of pore fluid drainage.

the fabric of the sample changes and its water content (p. 113) decreases. This increases the strength of the sample. The strength of sediment is increased by consolidation.

A similar response may be observed in rock. Figure 9.9 illustrates the variation in fluid pressure measured in a prominent rock fissure: the manometric heads measured varied between a maximum of h_w in the winter and a minimum of h_s during the summer. The variation of effective stress with time is described in the caption. As effective stress increases the fissures close making them less deformable to further increases in load and more resistant to any shear stress that may develop along them. The strength of rock is thereby increased.

Three parameters have been defined which enable the responses just described to be quantified; namely:

Undrained modulus of elasticity (*Eu*). The reversible deformation that accompanies the application of load to soil or rock that cannot drain and dissipate the pore pressure produced by loading (see Figs 9.7 and 9.8).

$$Eu = \Delta\sigma \Big/ \frac{\Delta H}{H} \qquad \textit{Units}\ \frac{\mathrm{kNm^{-2}}}{1}\ \text{or}\ \mathrm{kNm^{-2}}$$

Coefficient of compressibility (m_v). The strain that accompanies a change in effective stress (see Figs 9.7 and 9.8)

$$m_v = \frac{\Delta H}{H} \Big/ \Delta\sigma' \qquad \text{Units}\ \frac{1}{\mathrm{kNm^{-2}}}\ \text{or}\ \frac{\mathrm{m^2}}{\mathrm{kN}}$$

Coefficient of consolidation (c_v). The relationship between the speed with which consolidation will occur in a deposit of known compressibility in response to an increase in effective stress.

$$c_v = \mathrm{K} \Big/ m_v\gamma_w \qquad \text{Units}\ \frac{\mathrm{m}}{\mathrm{year}} \Big/ \frac{\mathrm{m^2}}{\mathrm{kN}}\,\frac{\mathrm{kN}}{\mathrm{m^3}}\ \text{or}\ \frac{\mathrm{m^2}}{\mathrm{year}}$$

K = coefficient of permeability and is a measure of a material's ability to drain: it is described more fully in Chapter 13.

γ_w = unit weight of the pore fluid, which is usually water.

These parameters are empirical and the values obtained for them will vary according to the condition of the rock or soil tested and the loads applied.

Consolidation is greatly influenced by the presence of strata that permit rapid drainage of ground-water. Two situations are illustrated in Fig. 9.10. In case A a compressible clay lies between less compressible sands of high permeability. In case B a similar clay lies above impermeable rock and is overlain by sand as in case A. When the stockpile is constructed the clay interbedded between the sand (case A) will complete its consolidation before a deposit of identical thickness and compressibility that has only one drainage path (e.g. case B) because of the shorter distances over which fluid pressures have to dissipate ($H/2$ rather than H). Any phenomenon that enhances

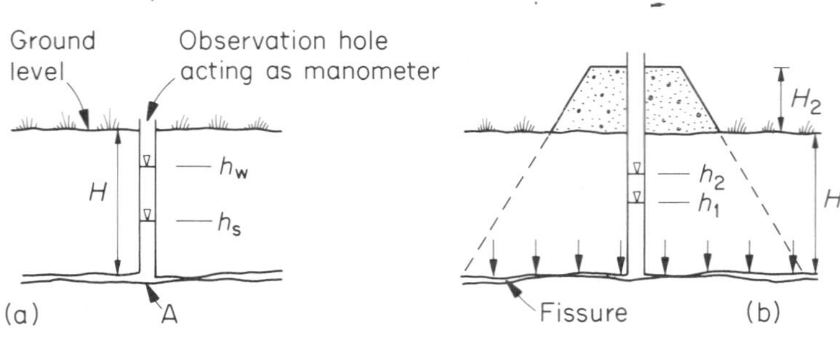

Fig. 9.9 Effect of load (stockpile) placed on rock with water filled joints.
(a) *Prior to construction of stockpile:*
Fluid pressure at A = unit weight of water × h_s in summer
= ,, ,, × h_w in winter
Total stress at A = ,, rock × H in summer and winter
Fluid pressure changes independently of total stress.
(b) *During and after construction of stockpile:*
Fluid pressure at start of construction = unit weight of water × h_1
Total stress at start of construction = ,, rock × H_1
Fluid pressure at end of construction = ,, water × h_2
Total stress at end of construction = ,, (rock × H_1) + (stockpile × H_2)

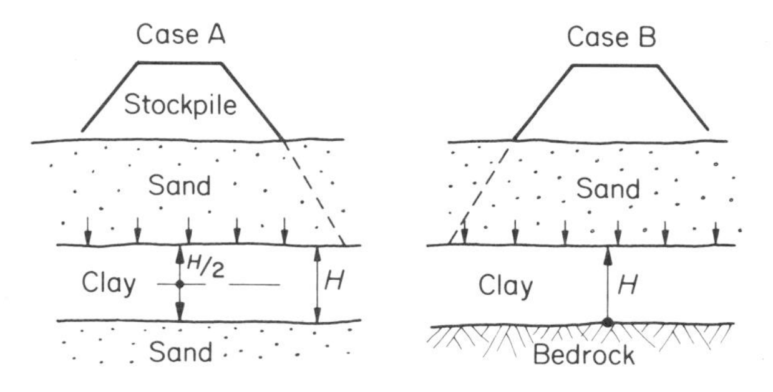

Fig. 9.10 Significance of drainage path length to the rate of consolidation. Sand shown is permeable, the bedrock impermeable.

drainage, e.g. lenses of sand and silt in a clay, root holes etc., or impedes drainage, e.g. the presence of clay layers, frozen ground, etc. is relevant to consolidation (Rowe, 1972).

Behaviour of rock and soil

The behaviour of rock and soil under load may be observed by testing columnar specimens that are representative of the larger body of soil or rock from which they are taken (see Chapter 11). One such specimen is shown in Fig. 9.11. It is sheathed in an impermeable

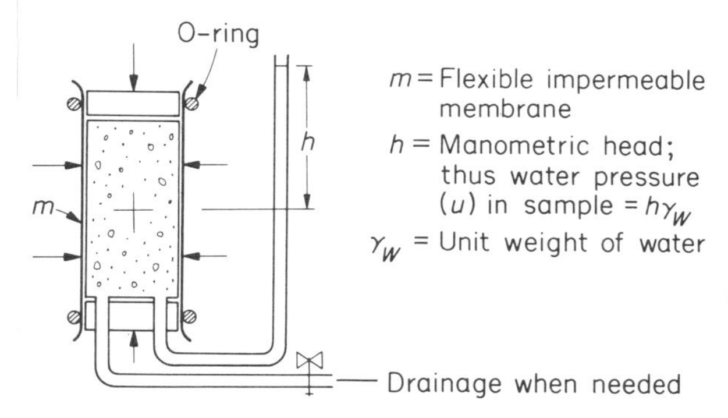

Fig. 9.11 A specimen prepared for testing in a triaxial cell: the specimen is circular in cross section.

flexible membrane that is sealed with O-rings (see Fig 9.11) so that drainage of fluid within the specimen may be controlled and its pressure measured. The specimen is placed in a cell of the type shown in Fig. 11.8 and loaded axially between end plattens, and radially by hydraulic fluid. Three principal stresses are thus imposed, those in the radial direction being equal, and the apparatus is therefore described as a triaxial cell. The apparatus permits consolidation to be studied, but as this behaviour has already been described in the previous section it will not be discussed further.

Stress and strain

Triaxial experiments conducted over a wide range of pressures demonstrate that the behaviour of rock and soil may be brittle or ductile, as illustrated in Fig. 9.12. The behaviour of a specimen is influenced by the radial pressure used to confine it (also called the confining pressure), the pore fluid pressure within the specimen, the speed of loading and other factors, notably the temperature of the specimen.

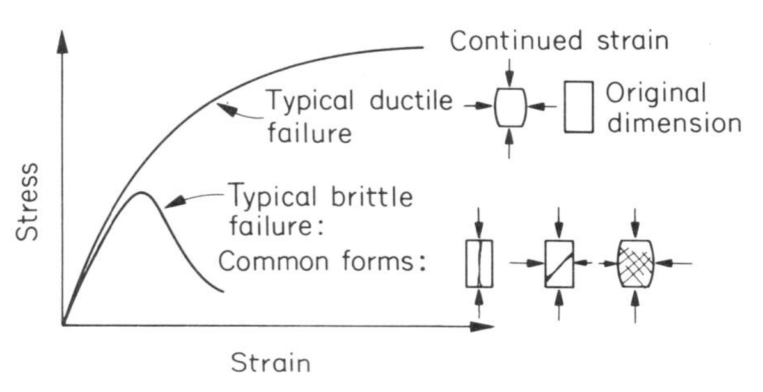

Fig. 9.12 The range of stress–strain relationships exhibited by geological materials.

Rock tested at temperatures and pressures representative of those encountered in most construction sites and mines, strained at moderate rates (see Fig. 9.3) and permitted to drain and thus dissipate pore pressures developed during testing, will exhibit the behaviour shown in Fig. 9.13. Deformation is initially linear and elastic (region 1 in the Fig. 9.13): sound rock typically has a Young's modulus in the range 35 to 70×10^6 kNm^{-2} and

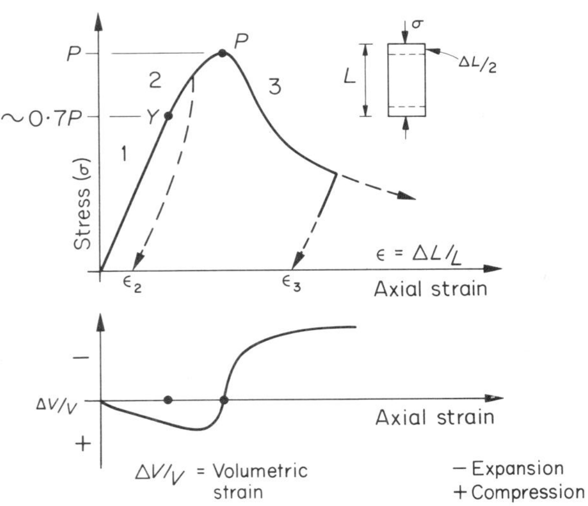

Fig. 9.13 Stress–strain curve for rock.
Note: axial strain for complete failure is usually less than 1%. Y=yield stress, P=peak strength (see text).

a Poisson's ratio of 0.2 to 0.3. When yield occurs (*Y*) non-linear inelastic deformation has commenced (region 2): if the specimen is unloaded it will be observed to have suffered permanent strain (ε_2). The highest, or peak stress (*P*) is that just prior to complete failure after which the stress carried by the specimen cannot be sustained (region 3) and major strains result (ε_3) largely produced by displacement upon failure surfaces, as shown in Fig. 9.12. Axial strain is a component of volumetric strain which is also shown in the figure. The rock initially compresses to a smaller volume then typically expands as failure surfaces develop and movement occurs upon them, one side of each surface riding over the roughness of the other.

If the radial (confining) pressure (σ_2 and σ_3) on the specimen is increased to prevent expansion the axial stress required to cause failure (σ_1) must be increased, i.e. the specimen becomes stronger: Figure 9.14. An increase in either the temperature of the specimen or the time over which loading occurs decreases the stress required to obtain an equivalent strain at lower temperatures using faster rates of loading, i.e. the specimen becomes weaker. Rocks which are brittle at ground level can behave as ductile materials at depth.

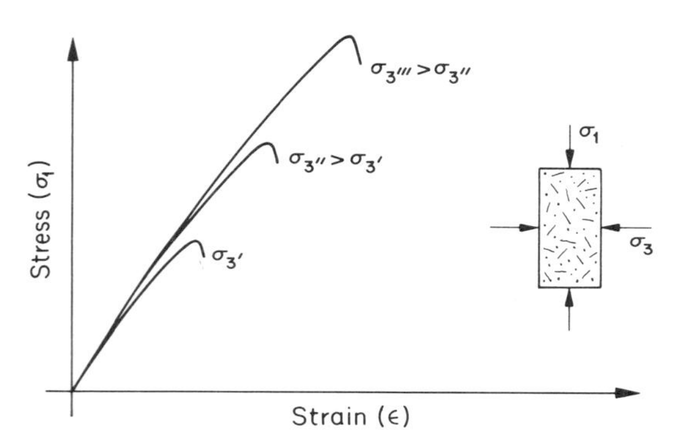

Fig. 9.14 The increase in strength that accompanies an increase in confining stress. (σ_3) = (σ_2.)

The relationship between stress and strain to be expected in soil that is loaded and permitted to drain during testing is shown in Fig. 9.15. Yield usually commences soon after the application of load and failure can be either brittle or ductile. Volumetric strain indicates the reason for this behaviour. Loose soils (e.g. normally consolidated sediments) compress to a smaller volume as their particles rearrange themselves during failure. The particles in dense soils (e.g. over-consolidated sediments) are packed more closely together and must ride over each other when significant strain occurs: such soil expands. As with rock, the strength of soil increases with increased confining pressure. With continued displacement the soil fabric within the zones of failure becomes completely disturbed or *remoulded* (*RM* in Fig 9.15).

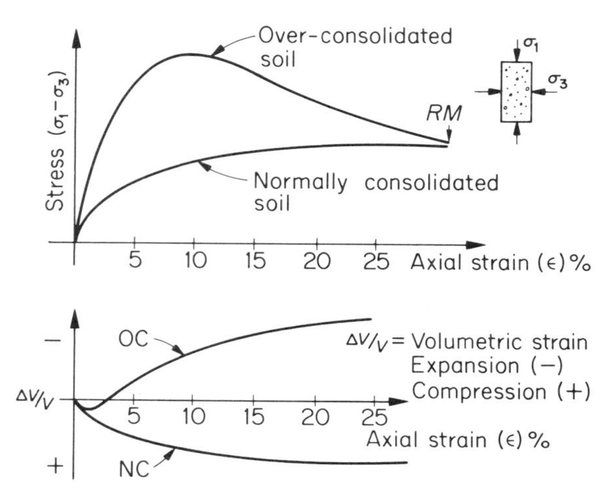

Fig. 9.15 Stress–strain curves for a drained soil, with representative strains: Note large strains required. Normally consolidated soils can exhibit a peak if their fabric is well developed, or strengthened with inter-particle bonds.

Cohesion and friction

Sediment such as clay, has an inherent strength called *cohesion* that must be exceeded for a failure surface to develop: dry uncemented sand has no such strength. The presence of cohesion may be used to divide soils into two classes, namely cohesive, i.e. having cohesion, and non-cohesive. Argillaceous sediments tend to be cohesive (Table 6.2), and arenaceous sediments tend to be non-cohesive unless they are cemented, or contain clay, or have been consolidated and are extremely dense.

Figure 9.16 illustrates the behaviour of a cohesive sediment loaded to failure in three tests, each conducted at a higher confining pressure and requiring a corresponding increase in the vertical stress (σ_1) to cause failure. The resolution of principal stresses (σ_1 and σ_3) into their

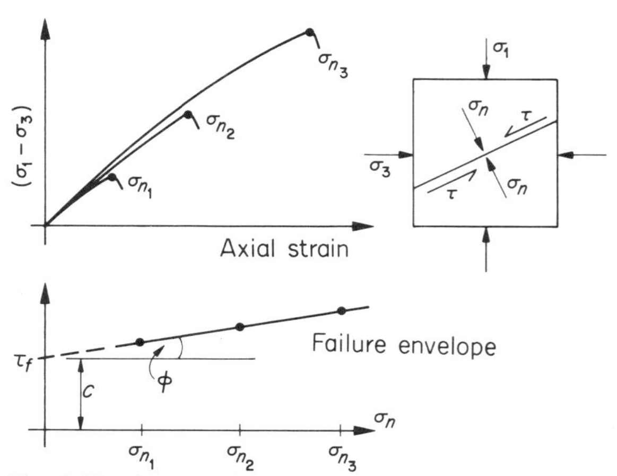

Fig. 9.16 Coulomb failure envelope. σ_n = stress normal to failure surface. τ = shear stress along the failure surface.

components normal and parallel to the shear surface is shown in the figure. A plot of shear stress at failure (τ_f) against the normal load operating at failure (σ_n) defines a relationship described as a *failure envelope* (and used by Coulomb in 1773) whose intersection with the axes is a measure of the cohesion of the soil (c) and whose slope is the angle of shearing resistance (ϕ). Hence if pore pressure (u) is measured the maximum resistance to shear (τ_f) on any plane is given by the expression:

$$\tau_f = c' + (\sigma_n - u) \tan \phi'$$

where prime ′ indicates that the parameter is measured using values of effective stress (p. 161). Cohesion may be derived from the electron attraction between clay particles, from mineral cement, or from an interlocking fabric in which propagation of a shear surface requires the dislodgement or failure of sediment particles. The sediments illustrated in Figs 4.37, 6.9, 6.10 and 6.13 all have cohesion. Sliding friction contributes substantially to the angle of shearing resistance and typical values for both are given in Table 9.1.

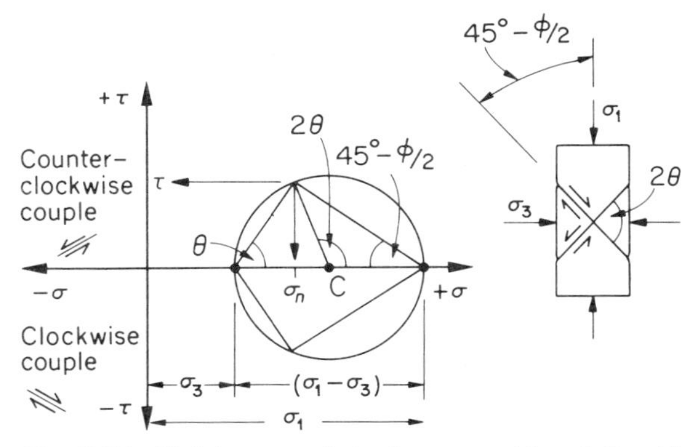

Fig. 9.17 Mohr's stress circle diagram and its relationship to the failure in a specimen of soil or rock. σ_1 and σ_3 = maximum and minimum values of principal stress at failure. ϕ = angle of shearing resistance. C = centre of the circle.

Failure

The failure envelope for many soils and rocks is not entirely linear and numerous failure criteria have been developed to describe their non-linear portions. The criteria proposed in 1900 by Otto Mohr is commonly used to introduce the subject: namely that when shear failure occurs the magnitude of the shear stress is related to that of the normal stress across the failure surface, the relationship being controlled by the strength of the material. His construction for revealing the nature of this relationship is illustrated in Fig. 9.17 and enables the shear stress (τ) and normal stress (σ_n) on a surface of any inclination (θ) to be related to the principal stresses that generated them, σ_1 and σ_3. The surfaces that develop in shear failure are generally inclined at an angle 45°-ϕ/2 to the axis of maximum principal stress.

In soil failure is dominated by sliding of sediment particles past each other and the frictional resistance between their points of contact provides a major contribution to their total strength. Friction is proportional to load and the failure envelope produced by a series of tests conducted using increasing values of principal stress is often linear, and frequently referred to as a Mohr-Coulomb envelope (Figs 9.16 and 9.18).

Rock failure differs from that of soil because of the considerable cohesion that must first be overcome before a continuous failure surface is generated. This cohesion comes from the tight interlocking of crystals and grains, as shown in the numerous illustrations of rock in thin section (see Chapters 5, 6 and 7), and by the bonding along their points of contact. Cohesion provides rock with a noticeable tensile strength.

Rocks crack in tension prior to complete failure and when the principal stresses causing failure are low the cracks, which are microscopic, remain open for much of the time. The failure envelope for this behaviour is often parabolic. In tests where the principal stresses are more compressive the cracks once formed, are later forced to partially close remaining open only at their extremity

Table 9.1 Illustrative values for angles of sliding friction (ϕ_s) of smooth mineral surfaces and shearing resistance after failure of selected rocks and soils in terms of effective stress (ϕ'): note, (ϕ_s) need not equal either (ϕ') or (ϕ).

SILICATES with rocks and soils composed of silicates							NON-SILICATES and rocks composed of non-silicates (see Table 4.9)		
Single and chain lattice minerals (see Fig. 4.20)			Sheet lattice minerals (see Fig. 4.20)						
	(ϕ_s)	(ϕ')		(ϕ_s)		(ϕ')		(ϕ_s)	(ϕ')
Feldspar	25°		Biotite	7°			Diamond	12°	
Quartz	32°		Montmorillonite	8°	M-clay	9°	Graphite	6°	
Quartzite		34°	Talc	9°			Carbon	24°	
Sandstone		37°	Illite	10°	I-clay	12°	Coal		35°
Granite		31°	Kaolinite	10°	K-clay	11°	Calcite	34°	
Gneiss		31°	Chlorite	12°	C-clay	12°	Limestone		35°
Dolerite		34°	Shale			16°	Marble		35°
Gabbro		34°	Muscovite	13°			Rock-salt	35°	
			Serpentine	17°			Fe-oxides	30°	

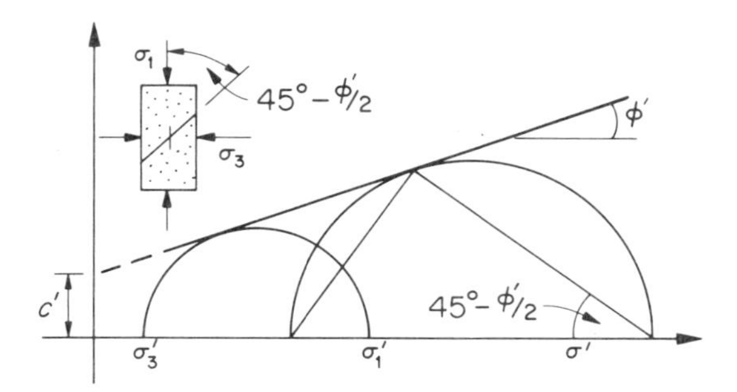

Fig. 9.18 Mohr-Coulomb failure envelope for soil.

where they continue to propagate. Frictional resistance is generated between the crack surfaces in contact and when this process dominates the period of failure the envelope becomes linear. With increasing stress, when σ_1 at failure approaches $3\sigma_3$, ductile failure commences and the envelope again becomes non-linear. Figure 9.19 illustrates

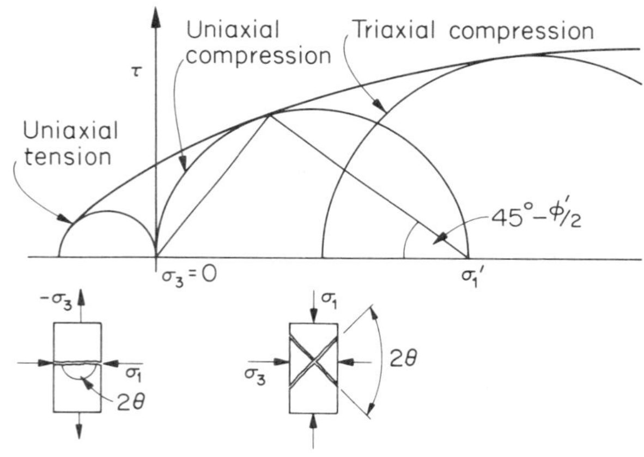

Fig. 9.19 Mohr failure envelope for rock. *Note:* rock is weak in tension. The value (uniaxial compression) ÷ (uniaxial tension) normally ranges from 8.0 to 12.0.

these trends, all of which can be displayed by the Mohr envelope.

Table 9.2 lists representative values of strength expected in uniaxial tension, uniaxial compression and triaxial shear (see Fig. 9.19) for rocks similar to those illustrated in Chapters 5 and 6.

Influence of fabric

The *fabric* of a rock or soil is the pattern formed by the shape, size and distribution of its crystals or sediment particles. Many metamorphic rocks have a banded fabric (*cf.* Figs 7.1, 7.2, 7.4*b*, 7.9, 7.10) and most sediments contain bedding (*cf.* Figs 6.11, 6.13): their fabric is anisotropic and their strength parallel to banding or bedding will differ from that in other directions (see Fig. 11.5). Many igneous rocks have an isotropic fabric (*cf.* Figs 5.18, 5.19, 5.20, 5.23, 5.24) and their strength tends to be similar in all directions.

In soil the influence of fabric is revealed by comparing the strength of a carefully collected sample whose fabric has not been disturbed by sampling, with its strength when remoulded without change in moisture content; remoulding completely destroys the original fabric. This comparison describes the *sensitivity* of the soil. Marine clays are most sensitive because of their open fabric (*cf.* Fig 6.3*b*, *c*), the ratio of their undisturbed and remoulded strength being 10 to 50 plus, sometimes 100 plus. Many late Pleistocene and Recent clays in Canada and Scandinavia are extremely sensitive being marine clays which have been raised above sea level by the post-glacial uplift of the continent on which they were deposited (Fig. 2.19). These clays rapidly lose their strength when disturbed and are aptly described as 'quick' clay (p. 122). Clays having a dispersed but layered structure (Fig. 6.3*d*) and those which have been consolidated (Fig. 6.13) are less sensitive. The sensitivity of normally consolidated clays ranges from 2 to 8 and that of over-consolidated clay, from less than 1 to 4.

Table 9.2 Strength that may be expected from rocks similar to those illustrated in Chapters 5 and 6. T = uniaxial tension; C_u = uniaxial compression with description recommended by Int. Soc. Rock Mechanics. c' = cohesion and ϕ' = angle of shearing resistance, both as measured in triaxial compression and in terms of effective stress. γ = dry bulk unit weight (kNm^{-3}): see also p. 300.

ROCK TYPE	FIGURE	WEIGHT γ	UNIAXIAL STRENGTH ($MN m^{-2}$) T	C_u		TRIAXIAL STRENGTH c' ($MN m^{-2}$)	ϕ' (degrees)
GABBRO	5.18	29	20	150	very strong	30	32°
GRANITE	5.23, 5.25	26	20	150	very strong	30	30°
DOLERITE	5.20	28	35	350	extremely strong	50	35°
BASALT	5.21	27	10	120	very strong	25	34°
SANDSTONE	6.12	25	5	15	weak	5	30°
"	6.9*a*	22	<2	10	weak	<5	30°
"	6.9*b*	21	<1	<10	weak	<5	30°
"	6.10	21	<1	<1	extremely weak	<2	<15°
SHALE	6.13	22	<1	10	weak	5	12°
LIMESTONE	6.16*a*	23	<5	20	weak	<10	35°
"	6.16*b*	24	5	30	moderately strong	10	35°

Influence of water

The strength of a crystal lattice and the energy required to propagate through it a crack is reduced by contact with water, and the presence of water in the pores and fractures of soil and rock lessens the bonds that provide cohesion. Further, the pressure of water within the voids of soil and rock controls the effective stress upon their crystals and particles and the frictional resistance they are able to generate at their points of contact. For these reasons water weakens rock and soil whose strength when saturated is usually less than that when dry.

Drained and undrained strength

The behaviour described so far for rock and soil (Figs 9.18, 9.19) assumed that pore pressures developed during loading may drain and so dissipate: such tests are called *drained tests* and the strengths obtained are *drained strengths*. Silt, sand, gravel, and other sediments and sedimentary rocks of similar permeability to them, normally exhibit drained strengths when loaded *in-situ* by engineering work. The strength of rock and soil whose permeability prevents the rapid drainage of water in their voids, will be reduced by any increase in the pressure of water that develops within them when loaded. Experiments in which drainage is prevented are called *undrained tests* and the strengths obtained are *undrained strengths*. Clay, shale and many rocks of low permeability such as unweathered igneous and metamorphic rocks, and sedimentary rocks whose pores are blocked by minerals and mineral cement, are examples of geological material that will normally exhibit undrained strength when initially loaded *in-situ* by engineering work.

Figure 9.20 illustrates the remarkable influence lack of drainage may have upon the behaviour of rock. When pore pressure is zero and effective stress $(\sigma - u)$ is high, the rock is ductile. When pore pressure is not permitted to dissipate fully and is maintained at the same pressure as the confining stress (σ_3) the rock is brittle, and much weaker.

Figure 9.21 illustrates another effect produced by lack of drainage, this time upon the strength of soil. The results of three tests are shown, conducted upon saturated specimens sealed within a membrane as in Fig. 9.11. When each specimen had been assembled into its triaxial cell the principal stresses upon it were increased by 10, 30 and 60 $\mathrm{kN\,m^{-2}}$; $(\Delta\sigma)$ in the figure. No drainage was permitted from the specimens during this change in all round stress and pore pressure increased by an identical amount (Δu). The vertical stress was then increased steadily until failure occurred $(\sigma_1)_{failure}$ at 50, 70, and 100 $\mathrm{kN\,m^{-2}}$ (see figure). Mohr circles based upon the values of σ_1 and σ_3 reveal the specimens to have no frictional strength, their angle of shearing resistance being zero and the increase in principal stresses producing no increase in strength (compare Fig. 9.21 with Fig. 9.18). The strength present is the product of cohesion and described as the undrained shear strength.

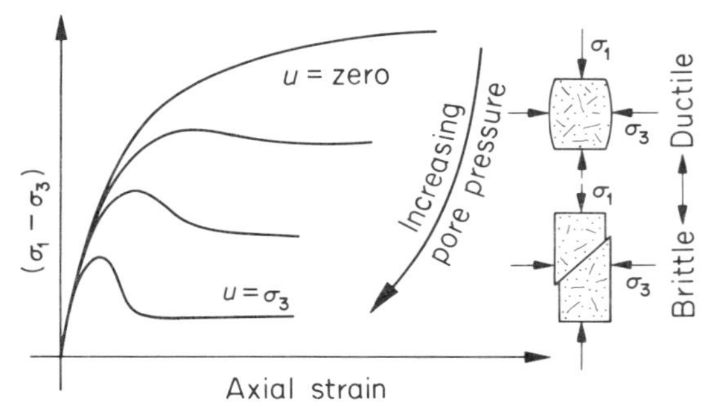

Fig. 9.20 Graphs showing the transition from brittle to ductile failure in rock as a function of pore water pressure (u): σ_3 the same for all tests. (From Robinson, 1959.)

TEST STAGE	Test 1	Test 2	Test 3	
1. $\Delta\sigma_1 = \Delta\sigma_2 = \Delta\sigma_3$	10	30	60	$\mathrm{kNm^{-2}}$
$\Delta u = \Delta\sigma$	10	30	60	—"—
2. $(\sigma_1)_{failure}$	50	70	100	—"—
$\Delta u_{failure}$	30	30	30	—"—
$u_{failure}$	40	60	90	—"—

Fig. 9.21 Triaxial test results of the undrained shear strength of soil (C_u). *Note:* $\theta = 45° - \phi/2$ and when $\phi = 0$, $\theta = 45°$. $(\sigma_1 - \sigma_3)_f$ = difference in principal stresses at failure.

Pore pressure changes

A change in the load applied to a soil or rock produces a change of pore pressure within it whose magnitude may be predicted with the aid of two empirical measurements known as the pore-pressure parameters A and B.

Parameter B describes the change in pore pressure produced in an undrained specimen by a change in all round stress $(\Delta\sigma_1 = \Delta\sigma_2 = \Delta\sigma_3)$. Thus, change in pore pressure = B (change in all round stress). In saturated soil and saturated compressible rock (e.g. Fig. 6.10) B = 1.0. When voids are partially filled with air, as occurs in strata above the zone of saturation, B is less than 1.0. The fabric of strong rock (e.g. Figs 5.20, 5.24) has a low compressibility and offers resistance to changes in all round stress: B is often less than 1.0 in such rock.

Parameter A describes the change in pore pressure produced by a change in deviator stress $(\sigma_1 - \sigma_3)$, and is influenced by the fabric and strength of the specimen (Table 9.3). Hence the change in pore pressure (Δu) that may result from engineering work can be predicted by the expression:

$$\Delta u = B[\Delta\sigma_3 + A(\Delta\sigma_1 - \Delta\sigma_3)]$$

Parameter A may be less than one, or negative, if the deviator stress causing deformation and failure produces a volumetric expansion of pore space sufficient to cause pore pressure to decrease (*cf.* Figs 9.13, 9.15). When this occurs effective stress on the surrounding particles and hence the strength at their points of contact, will increase. This increase in strength remains *until* more water can migrate into the areas of reduced pore pressure and the speed with which this occurs depends upon the permeability of the material. For this reason the strength of consolidated sediments and rocks of low permeability may vary with time.

Table 9.3 Indicative values of pore-pressure parameter A (Skempton, 1954 & 1961). Rocks may have negative values.

Soil	*Parameter A at failure*
Loose sand	2.0 to 3.0
Soft clays	Greater than 1.0
Normally consolidated clays	0.5 to 1.0
Over-consolidated clays	0.25 to 0.5
Heavily over-consolidated clays	0.5 and less

Consistency limits

The volume and strength of most soils varies with their water content and the limit of this variation, i.e. the water content at which the character of the soil is essentially that a solid (as in a brick made from dry mud), or a plastic (as in clay ready for moulding by a potter), or a liquid (as in a slurry) is provided by index tests (p. 195) which define the consistency limits: the limits are illustrated in Fig. 9.22*a*. In these tests the samples are totally remoulded to prevent soil fabric from contributing to their behaviour. The limits therefore reflect the mineralogy of the soil (Fig. 9.22*b*).

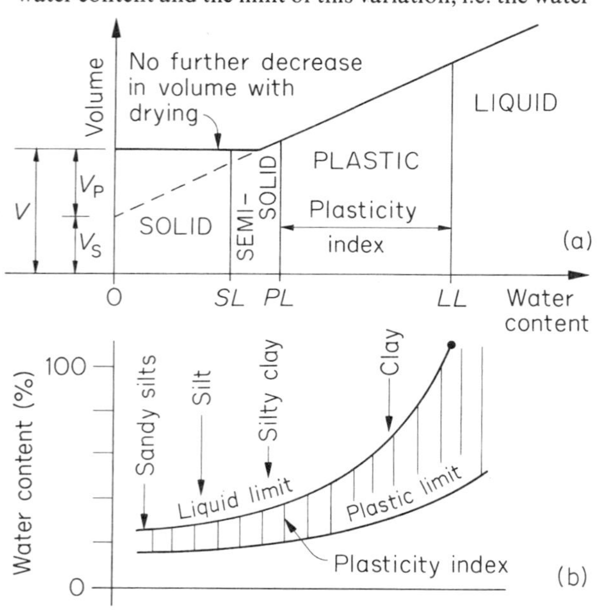

Fig. 9.22 **(a)** Consistency limits: SL = shrinkage limit, PL = plastic limit, LL = liquid limit, V = total soil volume = volume of pores (V_P) + volume of solid grains (V_S) Water content = (Mass of water) ÷ (mass of solid) and can be more than 100%. **(b)** Influence of mineralogy upon consistency limits.

Weak soils, such as loose sand, that are partially saturated may gain strength from the capillary tension of the water meniscus around their areas of grain contact (Fig. 9.23). This increases the effective stress upon the grains and thus the frictional resistance at their points of contact.

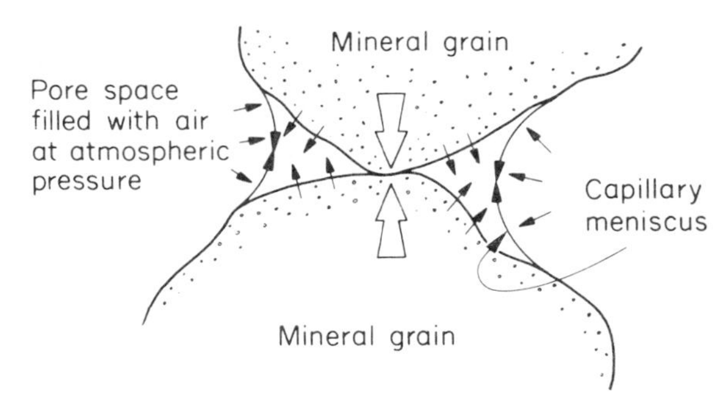

Fig. 9.23 Capillary tension in pore spaces between mineral grains. Large arrows indicate the attraction between grains created by the meniscus.

Elastic moduli

Values of Young's Modulus and Poisson's ratio may be obtained for rock and soil with comparative ease, but the significance of the values obtained must be considered with care as they may vary with time from those applicable to undrained conditions to those for drained conditions. This is especially important for soils, where these moduli should be determined experimentally using conditions that simulate the range of stresses and type of deformation that are expected to operate in the ground during the engineering works.

Behaviour of surfaces

Over-consolidated sediments and all rocks that have experienced unloading and uplift (p. 160) contain microfractures and other failure surfaces such as joints and fissures: many bedding planes will have separated, some containing a visible parting. Such surfaces have a strength that is less than that of the rock or soil in which they occur and they are a major source of weakness: they have formed by failure either in tension or in shear.

Tensile fractures such as joints, tend to be discontinuous (Figs 8.27, 8.28), have rough surfaces, and are open. They have no tensile strength, close when a load is applied normal to them, and have a shear strength that depends upon the resistance provided by the friction and roughness of their adjacent surfaces.

Shear surfaces such as faults (Figs 8.18, 8.23, 8.26) and bedding in folds (Fig. 8.14) are continuous and differ from tensile surfaces in having been sheared. They contain the crushed debris produced by shearing and sometimes this is reduced to clay sized particles. The shear strength of these surfaces under any normal load is usually much lower than that of joints.

All surfaces provide routes for the agents of weathering and many surfaces have the strength of weathered rock or soil. They are difficult to sample and care is required to recover them intact during a ground investigation (Chapter 10). Once recovered, their strength may be measured in a shear box (Fig. 11.9). The specimen containing the surface is placed into the box with the surface oriented horizontally and positioned mid-way between the upper and lower halves of the box. A vertical load normal to the surface is applied and the sample then sheared by applying a horizontal load. Resistance to shear under various normal loads can be measured. The apparatus has no means for controlling drainage within the specimen and rates of testing are chosen to ensure that they are either sufficiently slow to permit drainage or fast enough to prevent it.

Smooth surfaces

The behaviour of a smooth surface such as a bedding plane or one on which shearing has already occurred, is shown in Fig. 9.24. Results from three tests are illustrated; each test was conducted at a speed that permitted drainage to occur and subjected the surfaces to a different normal load (W). Displacement on each surface tested continued until no further loss of strength occurred and this required many centimetres of relative movement. A peak resistance was exhibited early in each test but did not differ significantly from later values.

In soil, any particles whose length is not oriented parallel to the shear surface are aligned to this direction by the relative movement of the surfaces and the strength remaining when all the particles are so oriented, and after all original fabric of different orientation has been destroyed, is called the *residual strength* (ϕ_r). The movement of rock surfaces can produce a different microfabric, more granular in character, and their lowest resistance to shear may be described as ultimate strength (ϕ_u).

The failure envelope for peak, residual and ultimate values of shearing resistance are shown in the figure and from these results it is evident that smooth surfaces may have some apparent cohesion (c') although it is usually small: their greatest strength comes from friction. The linear envelopes enable resistance to shear to be described by the expression:

$$F = c' + W\tan\phi'$$

where ϕ' is the drained angle of shearing resistance.

Rough surfaces

Figure 9.25 illustrates the results from a similar set of tests, but conducted on rough surfaces typically found on joints. As displacement progresses the upper surface must slide over the roughness of the lower and dilatation is associated with failure. This is shown in Fig. 9.26: an initial period when the surfaces come closer together under load is followed by rapid expansion during failure.

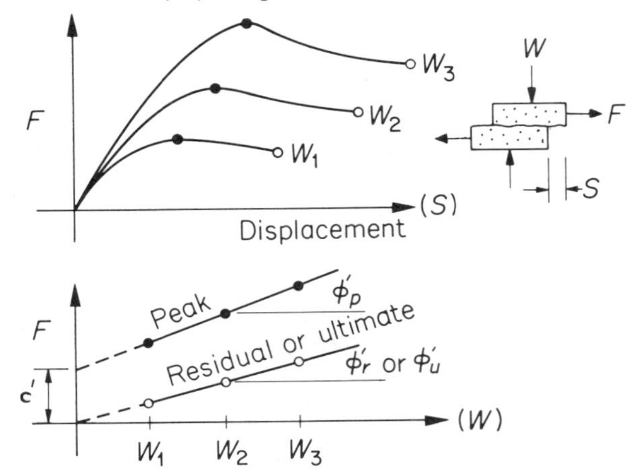

Fig. 9.24 The response of a smooth surface to displacement. F = force required to displace surfaces under normal load (W). Brittleness = $(F_{peak} - F_{residual}) \div (F_{residual})$. c' = cohesion. ϕ = angle of shearing resistance (see text).

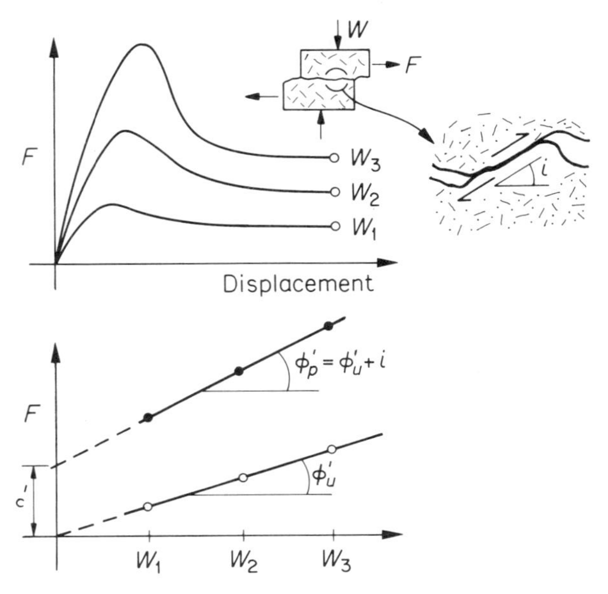

Fig. 9.25 The response of a rough surface to displacement. Note that the inclination (i) of an asperity may increase the resistance to shear (F) and causes surfaces to dilate. c' = cohesion, ϕ' = angle of shearing resistance (see text).

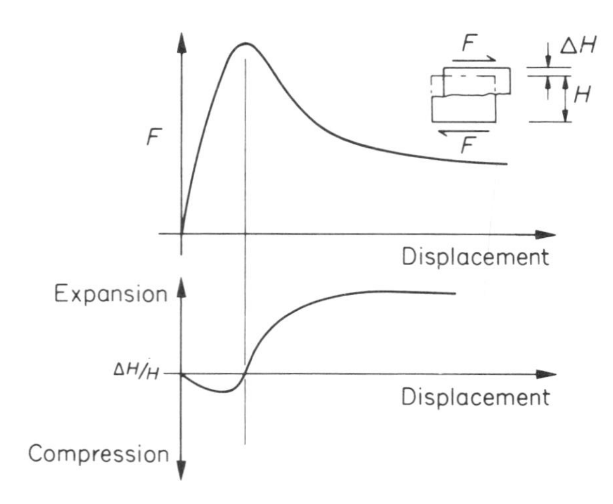

Fig. 9.26 Dilation ($\Delta H/H$) during shear failure.

Peak strength results from the combined resistance provided by roughness and friction. The failure envelope at low normal loads is linear and roughness provides an apparent cohesion.

When normal load across a rough surface is increased (as from W_1 to W_3 in Fig. 9.25) greater shear force is required to overcome friction and move the upper surface over the lower. A joint will 'lock' when the normal load is sufficient to prevent such movement and shear failure occurs through the rock rather than along the joint surface: when this process dominates the failure envelope may become noticeably non-linear.

The strength of a surface reduces to its ultimate value once displacement has removed the influence of roughness.

Variation in the irregularity of a rough surface is usually greater than that which can be sampled for testing in the laboratory and methods for assessing the likely peak shear strength of a rough surface have been proposed by Barton and Choubey (1977).

Lessons from failure

In-situ failure as occurs in landslides on slopes that are too steep, or in foundations that are overloaded, or in excavations that are inadequately supported, provides an opportunity for studying the *in-situ* strength possessed by large volumes of rock and soil. The lessons learnt from such studies are most relevant to successful geotechnical engineering (Mitchell, 1986) and indicate the validity of values for strength and deformation measured from small samples tested in the laboratory (Chapter 11) and from slightly larger volumes tested *in-situ* during ground investigation (Chapter 10).

Indicators of failure

Direct indicators of ground failure are displacement, fracture and water pressure in pores and fractures (i.e. pressure head). Descriptions of instruments which provide such indications are given by the British Geotechnical Society (1973) and the Int. Soc. Rock Mechanics (1977).

Displacement

Ground displacements that cannot be explained entirely by either elastic deformation or consolidation, normally indicate failure. Thus changes in the distance between or the elevation of survey points located on or in the ground can provide a simple indcation of failure. Figure 9.27 illustrates the displacement of a slope prior to its collapse.

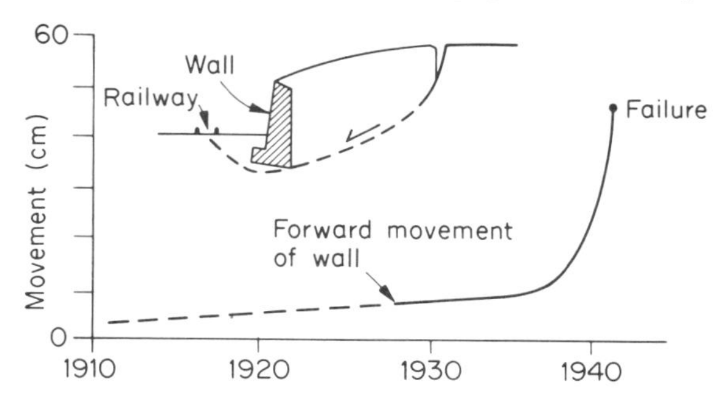

Fig. 9.27 Example of slope movement over a period of 30 years (after Skepton, 1964).

Fractures

Joints may open and new fractures may occur when dis-

Fig. 9.28 Tension Scar. Produced by mining subsidence. (Courtesy of the National Coal Board, London).

placement continues. Figure 9.28 illustrates tensile fractures in ground above mines at depth. Differential movement of the ground on either side of such fractures indicates the presence of shear failure at depth.

Water pressure

Figure 9.9 illustrates that the pressure of ground-water may be gauged by measuring its manometric, or pressure, head and that the vertical effective stress at the level of measurement can be calculated by subtracting the water pressure measured from the vertical stress produced by the overlying strata. The installation used for gauging water pressure is called a *piezometer* (Fig. 10.15). Failure can be anticipated when water pressure reduces the magnitude of effective stress to a value that cannot generate in the ground the strength required.

Indirect indicators

The approach of failure may be indicated by indirect means in areas between those locations where direct measurements of displacement, fracture or water pressure have been obtained. Many indirect methods utilize the change that occurs in the transmissive properties of the ground as it dilates prior to failure (examples of dilatation are shown in Figs 9.13, 9.15, 9.26). The ability of rock and soil to conduct electricity and the velocity with which they transmit seismic waves are sensitive to a change in the volume of voids created by dilatation. The transmission of ground-water is also affected by dilatation as the opening of fractures increases their permeability. Water seepages into underground excavations that were originally dry usually indicate dilation and are often a precursor of failure.

A phenomenon that accompanies failure in rock is the noise (or acoustic emission) made by the cracking prior to ultimate failure. By listening to this it is possible to obtain a clear indication of impending failure (Amer. Soc. Testing and Materials, 1981).

Analyses of failure

Once the shape and position in the ground is known of the surfaces on which failure has occurred, it is possible to calculate the total stress upon them prior to failure and from this their *in-situ* strength, in terms of total stress. If ground-water pressure at the time of failure is known, the *in-situ* strength may be also calculated in terms of effective stress. Assumptions often have to be made concerning the magnitude of the horizontal stress that existed, as this is not easily calculated for rocks and over-consolidated soils, for the reasons explained on p. 160. Similarly, ground-water pressure may not be known accurately and a range of values, varying between the highest and lowest pressure likely to exist, have to be used to complete the analyses. By these means, *in-situ* failure of the ground may be used to obtain values for the strength of large bodies or rock and soil.

Frequency of failure

It is useful to know the frequency with which failure occurs, in particular whether large failures are preceded by smaller failures of limited extent or happen suddenly as catastrophic failures of considerable magnitude. Much can be implied about the strength and behaviour of large bodies of rock and soil from this information: for example, the time required for failure to occur.

A record of failure is often preserved by the stratigraphy of adjacent areas. Figure 9.29 illustrates the fail-

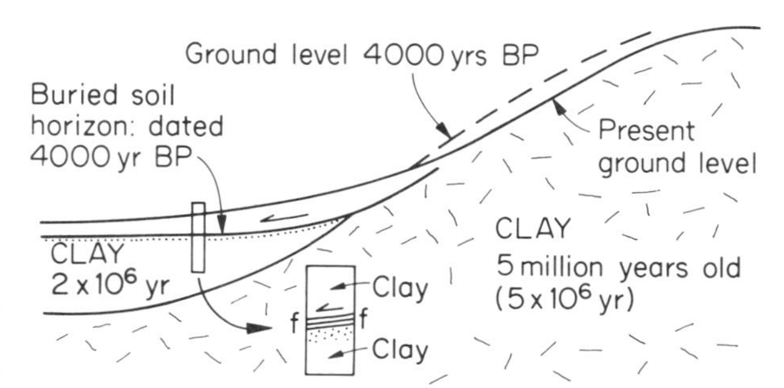

Fig. 9.29 Stratigraphic record of a past failure: *f* = failure, or slip, surface above soil horizon.

ure of a slope recorded in the stratigraphy of an adjacent valley. The stratigraphic principles described at the start of Chapter 2 may be applied, in particular the principle of superposition (p. 14). The slope that failed was formed of clay that is 5×10^6 years old. A valley at the toe of the slope contains alluvium that is 2×10^6 years old on top of which has developed a soil horizon dated as 4000 years old. The landslide debris buries this horizon and thus the slope failure can be dated as occurring within the last 4000 years even though no record of the failure may now be found on the slope itself. Stratigraphic records of this kind can provide a valuable chronology of events covering periods of millions of years (see Fig. 18.15).

Selected bibliography

Farmer, I.W. (1983). *Engineering Behaviour of Rocks*. Chapman and Hall, London and New York.

Price, N.J. (1981). *Fault and Joint Development in Brittle and Semi-brittle Rock*. Pergamon Press, London.

Smith, G.N. (1982). *Elements of Soil Mechanics for Civil and Mining Engineers* (5th Edition). Granada Pub. Co. Ltd, London.

Sowers, G.F. (1979). *Introductory Soil Mechanics and Foundations* (4th Edition). Macmillan Pub. Co. Inc., New York.

Jaeger, J.C. and Cook, N.G.W. (1979). *Fundamentals of Rock Mechanics* (3rd Edition). Chapman & Hall, London.

Mitchell, J.K. (1976). *Fundamentals of Soil Behaviour*. J. Wiley & Sons Inc., New York, London.

10

In-situ Investigations

The satisfactory design and construction of an engineering structure can be accomplished only when the character of the soil or rock, on which or within which it is to be built, is known. For this knowledge to be obtained the ground must be carefully studied by investigations conducted *in-situ*. Engineering structures such as roads, dams, buildings, tunnels and other underground works, are normally constructed according to the requirements of a specific design and from selected construction materials: by observing these requirements the strength of the completed structures is known and their response to load and displacement may be predicted. To obtain comparable information about the soil and rock against which the structure will react it is necessary to understand the geological processes which formed the soils and rocks, this being the only way to reveal their 'design', and the nature of the materials of which they are composed.

Approach

A common difficulty with ground investigation is that most of the ground will remain unseen, but this does not mean that the ground need remain unexplored.

Figure 10.1 illustrates a site considered for a dam: 3% of the foundation rock is naturally exposed and can be seen (Fig. 10.1*a*). Bore-holes and trenches increase the visible evidence by less than 1% (Fig. 10.1*b*) and geophysical investigations, which do not produce visible evidence but merely physical responses that are characteristic of certain ground conditions, assist the extrapolation of visible evidence to other areas. Volumes of ground can be tested *in-situ* (Fig. 10.1*c*) and much smaller volumes, collected as bore-hole core, can be tested in a laboratory. With these data the basic design of the dam is completed and costed. Further investigations may be undertaken if the scheme is feasible but even with these studies construction will commence with only a small fraction of the ground having been seen. As the foundation rocks become exposed during construction (Fig. 10.1*d*) unforeseen ground conditions may be revealed and require further investigation.

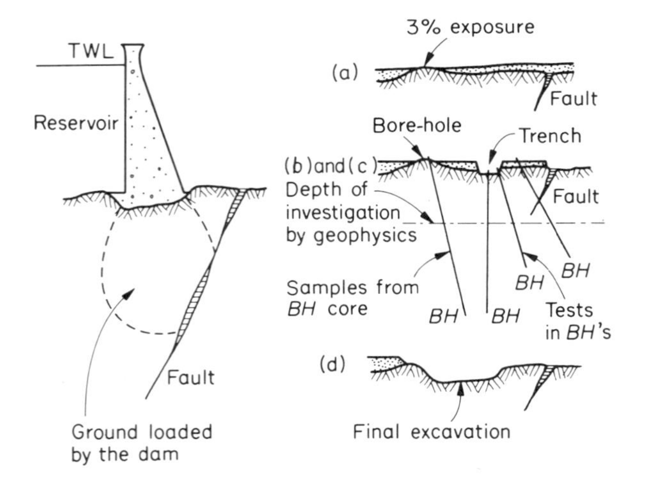

Fig. 10.1 A comparison between the volume of ground affected by a structure and the volume investigated.

There is no agreed approach to ground investigation which may vary from the repetitive programme such as one bore-hole every 1.0 km along the centre line of a road, to the carefully considered exploration that is tailored to site geology and its likely interaction with the engineering structure. Codes of practice exist, which if followed ensure that the methods used in investigation are conducted with an appropriate standard of confidence. They may also advise that certain works be undertaken, such as the excavation of inspection pits, but cannot indicate where best to site them. For this, geological advice is required. The use of probability theory has been considered as an aid to defining an approach to ground investigation, but such theory assumes a knowledge of the geological processes that formed the site, and that is one of the topics the ground investigation is meant to reveal.

Content

Table 10.1 lists five factors that must usually be investigated by every ground investigation. It is therefore useful to ascertain for a site the information already available for each of these factors before commencing further exploration as this will reveal the topics for which insufficient data exists and which must be the subject of further study. To anticipate conditions that might otherwise remain 'unforeseen' it is relevant to consider the geology of the region. For example when it was raised above sea level, how rapidly it has been eroded, the depth of weathering, the likely depth of bedrock in valleys filled with alluvium or colluvium, the possibility of stress relief structures in the valleys and of fossil landslides on the valley sides. Many 'unforeseen' problems can be attributed to a

Table 10.1 Factors involved in behaviour of a mass of rock or soil at an engineering site

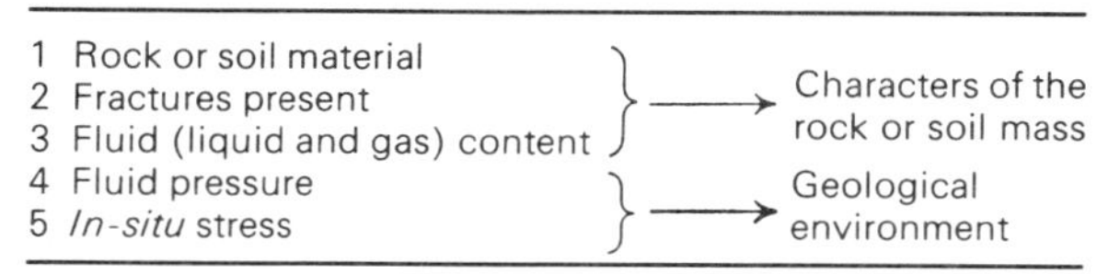

Factor		Group
1 Rock or soil material 2 Fractures present 3 Fluid (liquid and gas) content	→	Characters of the rock or soil mass
4 Fluid pressure 5 *In-situ* stress	→	Geological environment

1 is obtained by observation. 2 to 4 are obtained from measurements of: 2, permeability and storage capacity; 3, saturation; 4, pressure head. 5, is either calculated from the weight of overburden or measured.

ground investigation that has been too restricted in its approach.

Programmes of ground investigation should be flexible in content and phased in execution. Flexibility enables investigation to be amended when unexpected features of importance are encountered. As shown in Fig. 10.1 it is unlikely that every geological feature of note will be revealed prior to construction and a phased investigation permits later studies to be completed. Table 10.2 illustrates an example of a phased investigation.

continue with the investigations conducted in the field and conclude with the maintenance of records of ground exposed during construction. These activities often overlap.

Desk study

This colloquially describes the search through records, maps and other literature relevant to the geology of the area. Information may be disseminated in libraries, Government archives and company files. Dumbleton and West (1976) recommended the following procedure:

(*i*) Locate and (if necessary) acquire any maps, papers, air photographs, imagery and satellite data relating to the site, and interpret as far as possible the geological conditions shown by these sources. In a complex area attempt an analysis of the geology; the preparation of geological sections or block diagrams may help to indicate areas where further information is needed. A short visit to the site is often desirable to confirm observations and predictions already made.

(*ii*) At this stage it is often fruitful to seek additional information from institutions such as the Geological Survey, geological societies, local authorities and libraries, universities, and from engineers who may have been involved in projects in the area.

Table 10.2 Example of phased ground investigations for a dam: rectangles represent periods of work on site (after Kennard, 1979).

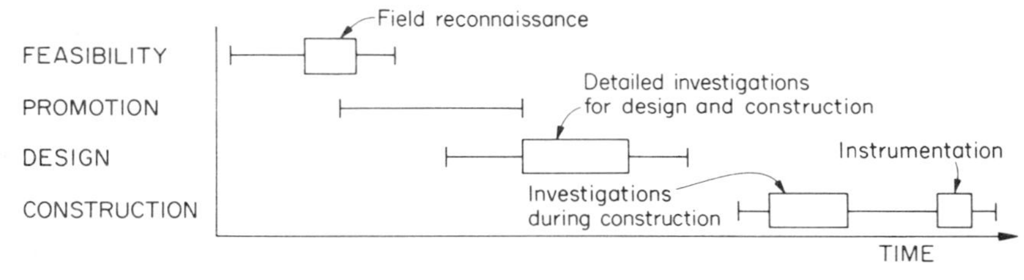

Cost

The cost of most ground investigations is between 0.5% and 1.0% of the cost of the project but this should not be taken as a rule. The principal purpose of ground investigation is to acquire geological and geotechnical data sufficient for the geology of the site to be described and its geological history to be reconstructed, particularly those events occurring during the past 10 000 years, with an accuracy that can be relied upon by design and construction engineers. If the ground is complex, appropriate funds should be made available for its investigation. Ground investigations should never be limited to save money as ignorance of ground conditions can be most dangerous. Money can safely be saved in the design and construction of an engineering structure once engineers can rely upon an accurate appraisal of ground conditions.

Components

A ground investigation contains numerous activities that are here grouped and described as its components. They commence with a 'desk study' to collate existing data,

(*iii*) After these enquiries it may be useful to visit the site again to collate all the data so far obtained, and to identify areas where engineering difficulties may exist and areas where particular investigations are needed.

(*iv*) Compile as good a report as can be made, recording the geological and geotechnical data, the addresses of useful contacts, and references to literature. This preliminary report will assist the ensuing investigation and provide a basis for the final report.

(*v*) To complete this study the construction requirements of the proposed engineeering works at the site should be considered, so that the ensuing investigation will be reasonably suited to both the geology of the area and the requirements of the design.

The above procedure is designed to help in the examination of a new area where there is little information about sub-surface conditions. It can be trimmed to suit the state of knowledge at the time, but vigilance is needed even in areas where the ground is known.

Field reconnaissance

This commences with a preliminary survey to confirm the basic geology of the region and the site: some mapping of geological structure and rock and soil types may be undertaken. For example, a reconnaissance for a tunnel would identify possible sites for its portals that are free from landslides, and topography that may indicate the presence of faults, such as scarps and straight valleys. Exposed rocks and evidence of ground-water, such as springs, would be noted. A certain amount of shallow augering may also be undertaken to indicate the character of the deposits and weathered profiles close to ground level. Sources of aggregate may be considered together with the locations for safe disposal of excavated material (see Fig. 12.9).

Simple tests that can be conducted in the field, require little if any sample preparation and are inexpensive, may be used to provide an index of the geotechnical properties of rocks and soils. They are called *index tests* (p. 195) and are a valuable aid to reconnaissance, permitting sites to be divided into zones within which similar geotechnical properties can be expected.

The geomorphology of the site and the geographical changes that have occurred should also be considered so as to produce an integrated review of ground conditions (Fig. 10.2).

Desk studies and field reconnaissance are the most cost effective components of ground investigation. Much relevant information can be inexpensively gained.

Field investigations

These employ familiar techniques of ground investigation, e.g. bore-hole drilling, and are located in areas where reconnaissance studies indicate that further information is required. The investigations utilize direct methods of study, such as the excavation of trial pits, trenches, shafts and adits, from which the ground can be examined, tested and sampled for further testing in the laboratory. Indirect methods of study may also be used, e.g. geophysical surveys, but these do not expose the ground to view. They are utilized to explore large volumes of rock and soil surrounding and between the smaller volumes of ground studied by direct means. The larger the volume of ground studied within the volume of ground that will be affected by an engineering structure, the better will be the ground investigation, for the character of soil and rock must be expected to vary from place to place.

The particulars of an investigation will vary from one job to another, but in most cases four fairly well defined objectives can be discerned:

(1) To determine, in whatever detail is required, the character of the ground (including both the solid material and the contents of pores and fissures). This accounts for much of the bore-hole drilling that is requested, together with core logging and hole logging, using the variety of optical, electrical, and mechanical devices now incorporated in bore-hole logging instruments.
(2) To determine the variations in the character of the ground within a particular volume and throughout a given time. The former accounts for the position of investigation areas, e.g. the location of bore-holes, their depth and inclination, the core recovery required, the location of geophysical surveys, and so on. The latter controls the length of time instruments such as water level recorders, load cells, and inclinometers, remain in the ground.
(3) To make a quantitative description of the area in terms that are relevant to the programme of testing and the engineering design. This is often difficult because many standard geological descriptions are qualitative and rely on the user correlating some characters of the rock with a range of physical properties.
(4) To determine the response of the ground to certain

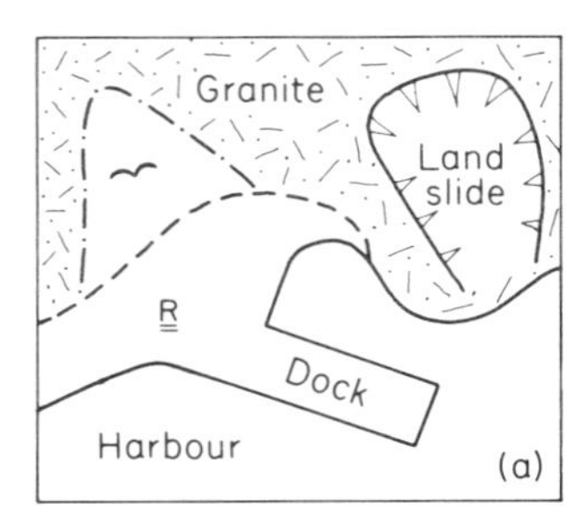

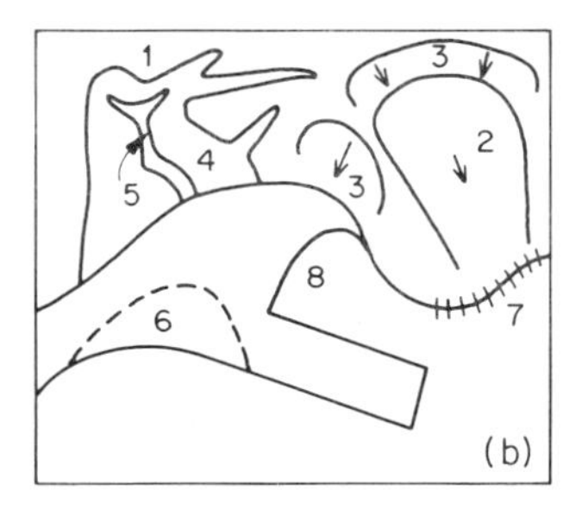

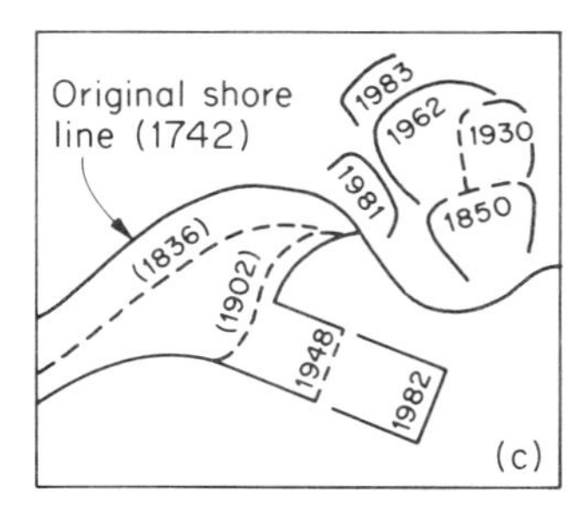

Fig. 10.2 Examples of maps produced by a desk study and a reconnaissance survey.
(**a**) Geological map: ~ = alluvium, R = reclaimed ground

(**b**) Geomorphological map: 1 = Bare rock slopes < 25°; 2 = Former landslide; 3 = Unstable rock; 4 = Wet superficial deposits; 5 = Active river erosion; 6 = Subsidence and flood risk; 7 = Marine erosion active; 8 = Active deposition of sediment.

(**c**) Geographical map.

imposed conditions of stress, strain, drainage, saturation, etc. This is where the majority of the *in-situ* tests are used. They abound in variety and almost anything that needs to be measured can be measured, given both money and time. The many techniques now available for field studies can be grouped into those made (*i*) along a line, (*ii*) over an area, and (*iii*) through a volume.

(i) Linear investigations

These include all kinds of bore-hole work and describe the sampling of a line or column of ground. If the columns are fairly close together a reasonably accurate interpretation of the geology can be made. Most *in-situ* studies are of this type, and it is not surprising that sophisticated methods have been developed for looking at and logging the holes, as well as the cores that come from them. The ingenuity of these methods should not blind the investigator to their limitations. An important point to note here is the orientation of the line within a geological structure. Vertical bore-holes, for example, are unlikely to intersect vertical joints; the same applies for horizontal bore-holes and horizontal joints. The more general case is illustrated in Fig. 10.3. Hence, for each bore-hole there is a blind zone which cannot be sampled.

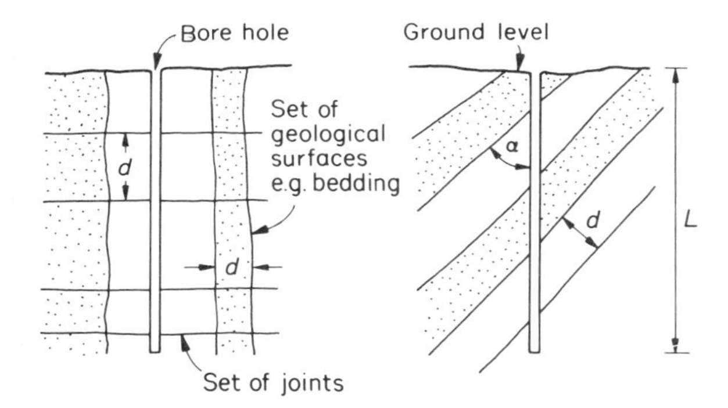

Fig. 10.3 Intersection of geological surfaces by bore-holes. Number of surfaces of a given 'set' intersected $= n = L \sin\alpha/d$ and ranges from L/d for $\alpha = 90°$ to zero for $\alpha = 0°$. (Terzaghi, 1965.)

(ii) Areal investigations

These include most geophysical reconnaissance techniques (except radiometric and single bore-hole logging techniques), all geological mapping and terrain evaluation. They provide a two-dimensional study of the ground and its geological make-up. The areas involved may be exposed on either vertical or horizontal surfaces.

Vertical surfaces can be seen in trenches and pits and are valuable in showing geological variations which occur both laterally and vertically. They are normally a part of any study of near-surface geology, are easily recorded and likely to reveal more information per unit cost than any bore-hole taken to the same depth.

A vertical area of a rather different nature is revealed by certain geophysical surveys, notably those using electrical resistivity and seismic methods. The former measures the potential drop between electrodes, and the latter the time taken by seismic waves (generated by a small explosion) to travel through the ground and return to the surface after reflection or refraction by buried surfaces. Both kinds of survey are conducted along lines laid out at ground level. From the measurements obtained certain deductions can be made concerning the nature of the sub-surface geology and approximate geological sections can be produced. A tabulation of the geophysical methods that can be used is given in Table 10.3.

Geology that is too deep to view in trenches can be seen with the aid of adits, and if these are sufficiently large to accommodate drilling equipment they can be used for making further investigations. Trenches and adits, though valuable, can nevertheless be subject to the effects of stress relief and other disturbances which loosen the ground, and care should be taken when tests are made in them. Attention is also drawn to the support and construction of such excavations (see British Standard Code of Practice, BS 6031:1981 'Earthworks').

Horizontal areas are usually those seen at ground level, either from natural exposures or from sections where the top soil has been removed. They reveal horizontal variations in the ground, and when used in conjunction with trenches and bore-holes can provide a three-dimensional picture of a site. Air photograph interpretation and geological mapping are the techniques most commonly used here, although gravity and magnetic surveys can also be employed (Table 10.3). The interpretation of geology from air photographs, although not strictly an *in-situ* method, is a skilled operation that can be very useful in an investigation.

(iii) Volumetric investigations

These are primarily concerned with determining the 3-dimensional characters of the geology, thereby differing from investigations which study either a local area, or a large area as described above. Many of the large *in-situ* tests fall within this group, e.g. blasting tests, pumping tests, load bearing tests, shear strength tests. The volume of ground involved is an important factor and ideally it should be related to both the geological fabric and the size of the proposed engineering structure.

The ground to be explored can be considered as either a continuous or discontinuous material, depending on the volume that is tested. Many authorities consider that the optimum volume of ground to be tested is that above which an increase in size has no appreciable effect on the results obtained. This can sometimes mean testing exceedingly large samples, particularly in rock masses that have a coarse fabric, e.g. widely spaced joints and bedding surfaces. A more practical procedure, which carries testing to the point where no improvement in test results can be expected by increasing the volume of the samples, can sometimes be adopted. Bieniawski (1969) describes how *in-situ* compression tests on $1\frac{1}{2}$ metre cubes were adequate

Table 10.3 Geophysical methods (after Dunning, 1970 and Griffiths *et. al., 1969*.) Magnetic, Gravity and Radiometric Surveys measure natural forces and are susceptible to such features as buried pipelines, old mine shafts, industrial waste and the like. In built-up areas these can severely interfere with a survey. (See also British Standard BS 5930: 1981.)

Method	Field Operations	Quantities Measured	Computed Results	Applications
Seismic	Reflection and refraction surveys using, *on land*, several trucks with seismic energy sources, detectors, and recording equipment; *at sea*, one or two ships. Data-processing equipment in central office. Two-man refraction team using a sledgehammer energy source.	Time for seismic waves to return to surface after reflection or refraction by sub-surface formations.	Depths to reflecting or refracting formations, speed of seismic waves, seismic contour maps.	Exploration for oil and gas, regional geological studies. Superficial deposit surveys, site investigation for engineering projects, boundaries, material types and elastic moduli.
Electrical and Electromagnetic	Ground self-potential and resistivity surveys, ground and airborne electro-magnetic surveys, induced polarization surveys.	Natural potentials, potential drop between electrodes, induced electromagnetic fields.	Anomaly maps and profiles, position of ore-bodies, depths to rock layers.	Exploration for minerals; site investigations.
Radar	Ground survey 2-man. Portable micro-wave source.	Induced reflections from surfaces in ground.	Depths to reflecting surfaces.	Shallow engineering projects: frozen ground at depth.
Gravity	*Land surveys* using gravity meters; *marine surveys* gravity or submersible meters.	Variations in strength of Earth's gravity field.	Bougner anomaly and residual gravity maps; depths to rocks of contrasting density.	Reconnaissance for oil and gas; detailed geological studies.
Magnetic	Airborne and marine magnetic surveys, using magnetometers. Ground magnetic surveys.	Variations in strength of Earth's magnetic field.	Aero- or marine magnetic maps or profiles; depth to magnetic minerals.	Reconnaissance for oil and gas, search for mineral deposits; geological studies at sites.
Radiometric	Ground and air surveys using scintillation counters and gamma-ray spectrometers; geiger counter ground surveys.	Natural radioactivity levels in rocks and minerals; induced radioactivity.	Iso-rad maps, radiometric anomalies, location of mineral deposits.	Exploration for metals used in atomic energy plant.
Bore-hole Logging	Seismic, gravity, magnetic, electrical and radiometric measurements using special equipment lowered into bore-hole.	Speed of seismic waves, vertical variations in gravity and magnetic fields; apparent resistivities, self-potentials.	Continuous velocity logs; resistivity & thickness of beds; density; gas and oil, and K, Th, U content. Salinity of water.	Discovery of oil, gas, and water supplies; regional geological studies by bore-hole correlation. Applicable to site investigations.

for designing mine pillars that were several times bigger.

The size of the test having been decided, its orientation in relation to the engineering structure should then be considered. Most rock masses are anisotropic and their performance will vary according to the direction in which they are tested. This is most marked when the anisotropy is regular, as in a well-bedded or cleaved rock mass. It is usual to apply loads, promote drainage or initiate displacements in the direction which the proposed structure is likely to load, drain, or displace the ground. Finally it should be remembered that tests are best conducted in ground that has not been disturbed.

The scale of linear, areal and volumetric investigations should be compatible with the scale of the project and characters of the ground. For example, the investigation for a tunnel in rock containing joints spaced at distances greater than the tunnel diameter, will differ from that required for the same tunnel when penetrating intensely broken rock.

Construction records

These record the geology exposed during excavation and usually consist of maps and plans, at a scale of about 1:200, and scaled photographs. From these records models can be simply constructed to elucide the site geology; a cardboard model is illustrated in Fig. 10.4. An example of a construction record is shown in Fig. 10.5, and the basic information that should be recorded is as follows:

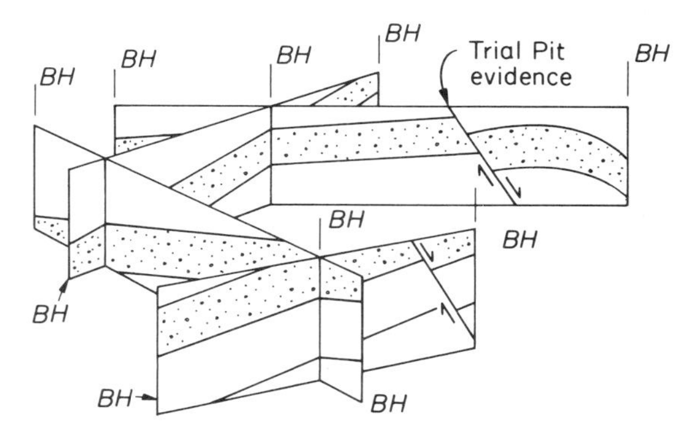

Fig. 10.4 A simple 3-D model made from card. The end of each section and each intersection represents the location of a bore-hole and shows the geology encountered within it. The cards can be cut to further model topography.

(*i*) *rock and soil types*; to ensure consistent descriptions standard terms should be used: see Tables 10.4 and 10.5. Samples may be collected for later reference and their location recorded (see Figs 10.5 and 10.6).

(*ii*) *boundaries*; an indistinct boundary between rock or soil types can be recorded by a broken line; see Figs 10.5 and 10.6.

(*iii*) *structure*; the direction and amount of dip of geological surfaces should be recorded (see Fig. 8.1 and Chapter 12). If possible distinguish bedding from cleavage and fractures such as faults and joints (Fig. 10.6).

Table 10.4 Elements of a description of soil

COLOUR	Use a colour chart (e.g. Munsell Color Co. Inc. *Soil Color Charts*. Baltimore)
BEDDING	Distance between bedding surfaces: >2 m Massive .... or v thick bedding 2 m – 20 mm Bedded: 2000–600 mm thick bedding; 600–200 mm medium ,,; 200–60 mm thin ,,; 60–20 mm v. thin ,, <20 mm Laminated
GRAIN SIZE (see Tables 6.2 & 6.3)	Boulder >200 mm; Cobble 200–60; Pebble 60–4; Granule 4–2; Sand 2–0.06 — Can be seen with unaided eye. Silt 0.06–0.002 Gritty on tongue. Clay <0.002 No grit on tongue.
DENSITY	Loose: can be penetrated with a pencil. Firm: v. slight penetration with a pencil. Dense: no penetration: may break pencil.
STRENGTH	Soft: can be moulded easily in fingers Firm: difficult to mould Stiff: can only be dented by fingers Hard: cannot be dented by fingers
CONDITION (see Fig. 9.22)	Like butter .... Liquid limit Like window putty... Plastic limit Like cheese ....Between Plastic & Shrinkage limits Like chocolate ....Drier than Shrinkage limit
NAME & CLASSIFICATION:	Use one of the recognized systems of classification, e.g. British Standard, or Unified Soil Classification, (for example see Fig 11.14).

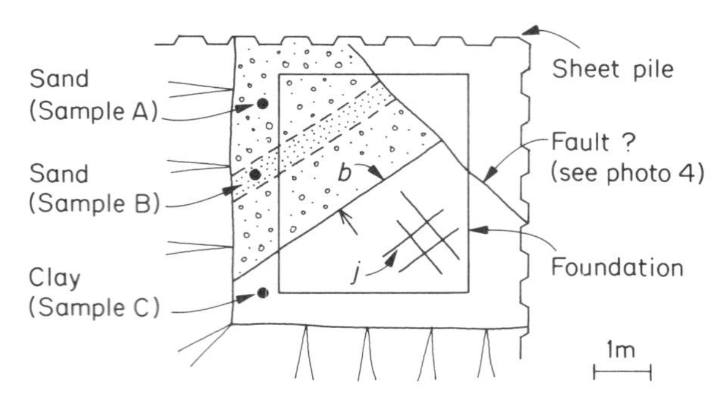

Fig. 10.5 Example of geological details recorded in a foundation excavation. *b*=bedding surface. ←=direction of dip. *j*=sets of vertical fissures.

(*iv*) *water*; springs and similar issues of ground-water should be noted (Fig. 10.6).

Two practical aids for making such records are a polaroid camera, the pictures from which can be annotated on site, and a pro-forma record sheet. The latter is particularly useful when records of exposed geology are required with every shift of work, e.g. as when tunnelling, Fig. 10.7.

Construction records have three important uses:

(*i*) Contractors, who construct the engineering structure, have to price their work prior to construction and must assess the geology of a site from the data provided by the ground investigation. If geological conditions encountered during construction differ significantly from those anticipated the contractor will seek extra payment for any additional costs incurred in coping with the unexpected ground. A record of the geology encountered is therefore a valuable document of reference.

(*ii*) Many engineering structures require the ground beneath or around them to be improved by treatment (Chapter 17) and such an exercise is greatly assisted by accurate records of ground conditions, particularly when the ground has been concealed by the first phase of foundation construction or tunnel lining.

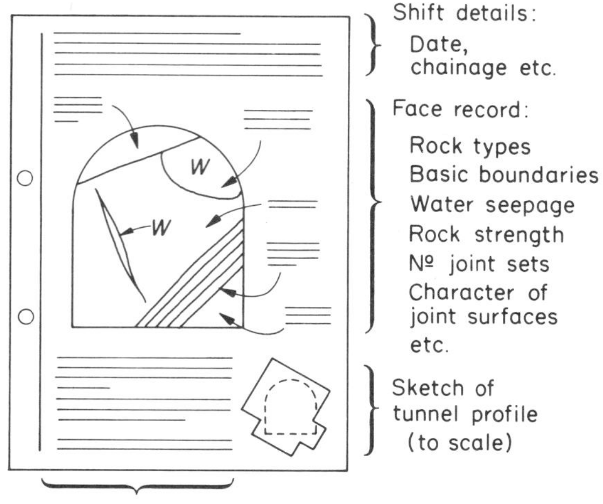

Fig. 10.7 Example of pro-forma for use when tunnelling: details recorded to suit site and contract. See roof and wall logs: Fig. 16.20. In this example *W*= areas where water flowed from the face.

(*iii*) Most large scale constructions, e.g. roads, dams, require their own source of construction material from quarries developed for this purpose. An accurate daily record of the geology exposed, described with the aid of index tests to separate suitable from unsuitable material, will enable control to be exerted on the quality of the rock or soil used.

Methods

When selecting methods of investigation it is necessary to consider those aspects of bore-hole drilling and *in-situ* testing that may affect the ground adversely or damage any samples recovered.

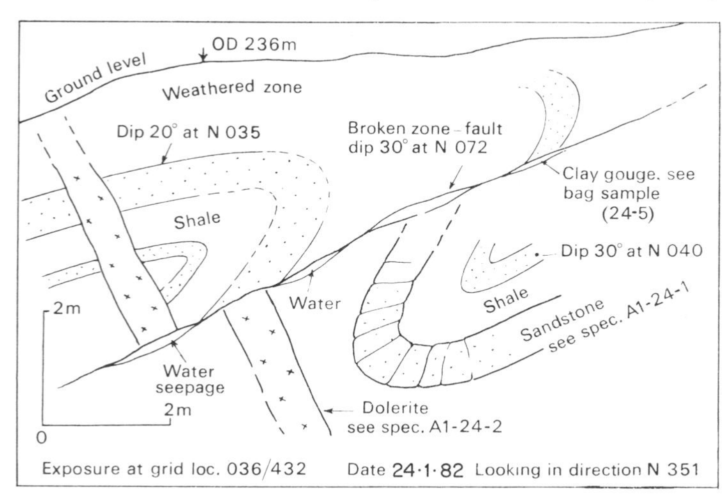

Fig. 10.6 Field sketch of the geology exposed in the vertical face of an excavation.

Table 10.5 Elements of a description of rock

COLOUR	Use a colour chart (e.g. Geological Society of America. *Rock-color Chart*)		
BEDDING	*Sedimentary*	*Volcanic*	*Metamorphic*
>2 m	Massive	Massively banded	Massively foliated
2 m–20 m	Bedded	Banded	Foliated
<20 mm	Laminated	Finely banded	Finely foliated
STRENGTH	Weak: can be scratched with a knife. Strong: breaks with one or two hammer blows. V. strong: difficult to break.		
WEATHERING	Complete: no unaltered rock remains and the rock is weak. Partial: some fresh or discoloured rock is present. On joints only: with slight alteration of adjacent rock. Fresh.		
GRAIN SIZE	*Igneous & Metamorphic*	*Sedimentary*	
>2.0 mm	Coarse } See Tables 5.2 & 7.2	} See Tables 10.4, 6.2 & 6.3.	
2.0–0.06 mm	Medium } See Tables 5.2 & 7.2	} See Tables 10.4, 6.2 & 6.3.	
<0.06 mm	Fine } See Tables 5.2 & 7.2	} See Tables 10.4, 6.2 & 6.3.	
MINERALS	Identify first the dominant minerals, then the subordinate. (cf. *The Rock-forming minerals*: Chapter 4)		
NAME & CLASSIFICATION	For Igneous rocks see	Fig. 5.17 & Table 5.1	
	For Sedimentary rocks see	Tables 6.2 &6.5	
	For Metamorphic rocks see	Table 7.3	

Bore-hole drilling. The common techniques for producing a bore-hole are shown in Table 10.6. A good hole will be drilled and samples recovered with only the minimum of disturbance when an experienced driller is employed to use a drilling machine that is free from vibration. When good core recovery is required the core barrels used should have an internal diameter of between 50 mm and 100 mm, and contain an inner tube (double core barrel) that is linked to the drill-stem by a swivel, to prevent its rotation against the core, Fig. 10.8. Triple-tube core barrels may be required for recovering very weak rock. Core recovery is also enhanced by the correct choice of drilling fluid, which clears the bit of rock debris, lifts the drill cuttings out of the bore-hole, and if correctly chosen will prevent the core from being unduly disturbed. Bore-holes in which *in-situ* tests are later to be performed or water pressures monitored, should be cleaned of any drilling fluid that may influence the tests or the water pressures.

Despite all precautions it is neither possible to drill a hole into the ground where pressure is greater than atmospheric pressure, without disturbing the ground around it, nor feasible to recover a core and bring it to ground level without affecting it in some way.

Disturbance to the walls and immediate vicinity of the hole normally results from the relief of stresses that follow the removal of a column of ground; it may be associated with processes such as desiccation, oxidation, and with the degree of geological stress at depth. The overall effect is to open existing fractures and possibly to generate new ones, so generally loosening the ground.

Table 10.6 Basic methods for drilling holes.

Drilling method	*Application*	*Samples*
AUGER	Soil	Disturbed
PERCUSSION (or cable-tool)	Soil (mainly) & rock	Disturbed to damaged
ROTARY (or core-drill)	Rock (mainly) & soil	Little damage to disturbed

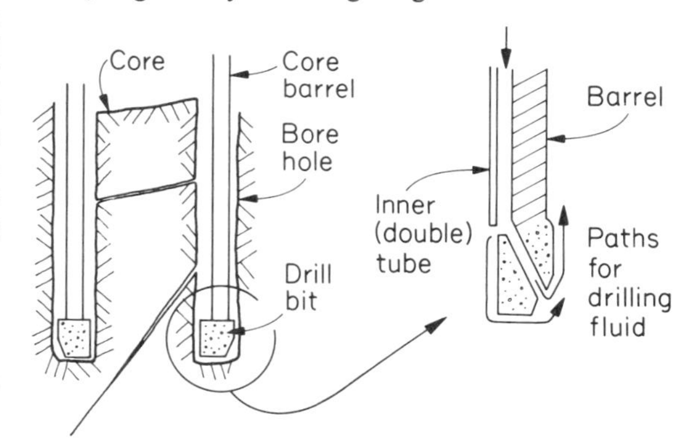

Fig. 10.8 Double-tube core barrel to prevent drilling fluid from unnecessary contact with core whilst in the core barrel.

A certain amount of 'bedding in' or 'settling down' can be expected in the initial stages of *in-situ* tests carried out in holes that have been affected in this way; care should be taken when the results are interpreted, especially when using visual aids such as bore-hole cameras and periscopes, as it is natural to assume that what is actually seen also represents the condition of the unseen ground.

The disturbance of cored samples can vary from skin effects which are restricted to the outer edge of the core to total disarrangement. Helpful guidelines for sampling delicate ground are suggested by Hvorslev (1948). Cores can also be disturbed by stress relief, and shales are notorious in this respect; the intensity of fracturing seen in

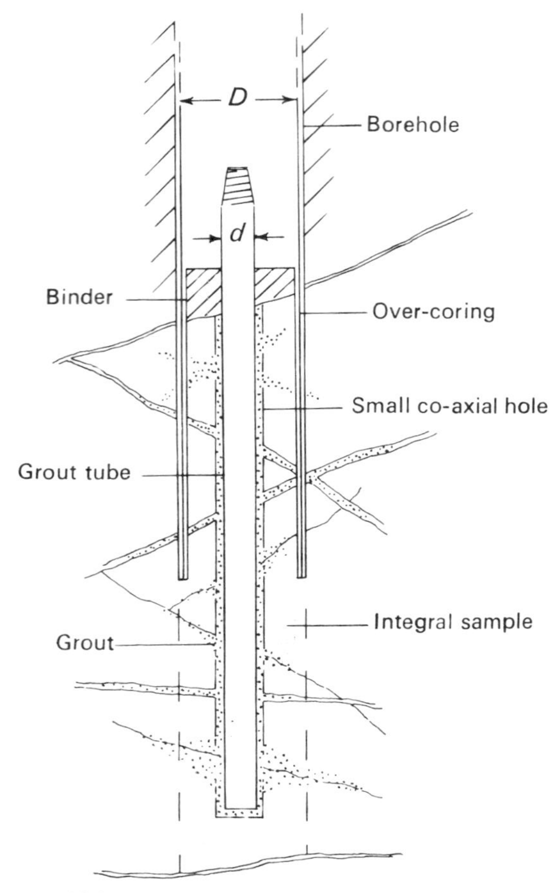

Fig. 10.9 Integral sampling (after Rocha, 1971).

cores of shale is often much greater than that found *in-situ*. A useful, but expensive method for recovering weak and broken rock, called integral sampling, is described by Rocha (1971) and is shown in Fig. 10.9. A hole is drilled to just above the level from which a sample has to be collected. A small hole, coaxial to the first, is then advanced into the sampling area. It is generally recommended that the ratio D/d should be about 3.0. A perforated hollow reinforcing bar of diameter approximately 5 mm less than that of the smaller hole, is next inserted into the smaller hole and bonded to the rock by the injection of grout through its hollow centre. The sample area is then over-cored, using low drilling pressures, and the sample retrieved. Thin-walled bits are recommended in small diameter holes, and double-tube core barrels in weak ground. This method produces beautiful samples because the grout strengthens the weak rock and infills voids, so that the separation between fissure surfaces can be observed. A less expensive alternative involves lowering an inflatable tube (packer), sheathed in a pressure-sensitive skin of plastic, into a bore hole and expanding it so as to force an impression of the fractures intersected by the hole onto its sides: an impression packer. The packer is then deflated, brought out of the hole and its pressure-sensitive skin removed so that the trace of fractures upon it may be studied.

Core logging. The description of core is important and can be of great assistance to both the design engineer and the contractor: it should therefore be completed *on site* by an experienced geologist. Useful logs are not easily produced by inexperienced personnel and there is much to recommend the adoption of methods set out in the report on the logging of rock-cores (Geological Society of London). This report covers the general requirements of a bore-hole log, the factors that determine logging methods, the handling, labelling and preservation of rock cores, and the information to be recorded on the log, including core recovery, descriptive geology and rock grade classification. (See also Stimpson *et al.*, 1970.)

Core logs are most easily made using a scaled proforma of the type shown in Fig. 10.10, supplemented with colour photographs of the cores in their box. The strength of the core at the time of recovery should always be noted, and natural fractures such as joints, should be distinguished from those produced either by drilling or by changes that have occurred later, e.g. through stress relief or desiccation. Three other parameters are normally calculated, namely:

Total core recovery = total core recovered expressed as a percentage of length of hole from which it came.
Example: drilled length = 127 cm
core recovered = 101.6 cm
total core recovery = 80%

Rock Quality Designation (or RQD) = summed length of core sticks greater than 10 cm in length expressed as a percentage of the drilled length.

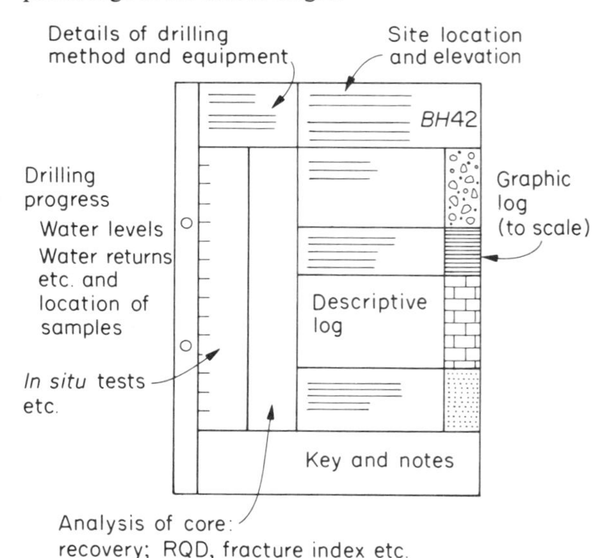

Fig. 10.10 Example of pro-forma for the logging of core: see Geological Socity of London 1970. RQD = rock quality designation (see above).

Example: drilled length = 127 cm
summed length of core sticks greater than 10 cm long = 68.58 cm
RQD = 54%

Fracture index = number of fractures per metre length of core.

Rock quality designation and fracture index are best applied to massive and bedded sedimentary rocks and their igneous, and metamorphic equivalent (see Table 10.5); their application to laminated, finely-banded and foliated rocks such as shale or slate, can produce meaningless values. The formulae for calculating RQD can be written in other ways to suit different geological conditions such as laminated rock, and particular logging requirements (Priest and Hudson, 1976).

After the cores have been described they should be carefully stored in a shed in a manner that will protect them from damage by changes in temperature, moisture content and general rough handling.

A bore-hole provides access to ground at depth and may itself be logged by the techniques listed in Table 10.7.

Table 10.7 Basic-bore hole logs (see also Int. Soc. Rock Mechanics, 1981).

Log. type	Information revealed
1 FRACTURE	Strata boundaries & bedding, joints & other fractures (faults)
2 GEO–DYNAMIC	Elastic modulus & strength
3 HYDRO-DYNAMIC	Strata & fractures containing flowing fluid, fluid temperature & composition, fluid pressure.
4 GEOPHYSICAL	Porosity, lithology, density, seismic velocity.

Geological mapping

This is described in Chapter 12. Such surveys are greatly assisted by trenches to provide additional exposure in areas where it is critical to obtain information: an example is shown in Fig. 10.11. Geophysical surveys may be used to augment this data and Fig. 10.12 shows a bridge crossing where geophysical surveys were used to define the best position for the bridge foundations and to locate the

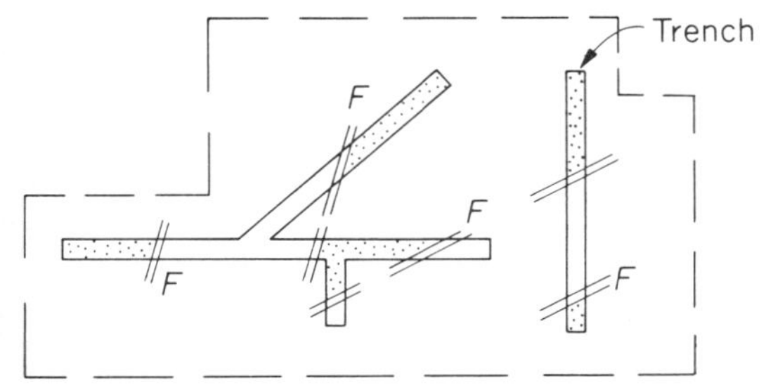

Fig. 10.11 Investigation of faulting (*F*) across a proposed foundation.

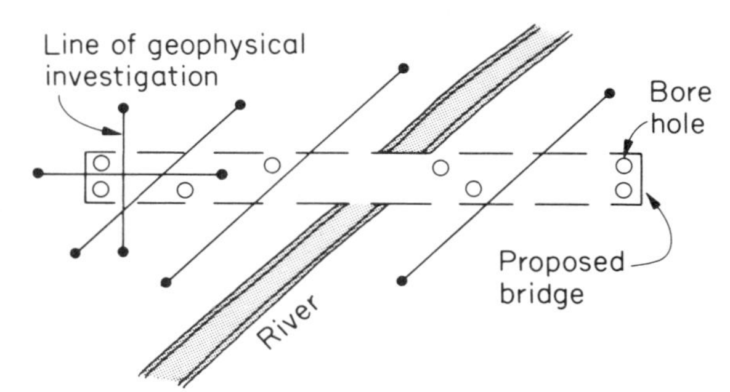

Fig. 10.12 The combined use of geophysical and bore-hole surveys to define the position of a proposed bridge.

bore-holes that were drilled to confirm the geology of the site.

Measurement of stress

Two components of *in-situ* stress often have to be measured, namely the total stress (σ) and the fluid pressure (u) in the ground: these are combined to reveal the value of *in-situ* effective stress ($\sigma - u$): see p 161.

Total stress

This can be measured by inserting into the ground a 'stress meter', located in the base of a bore-hole approximately 30 mm in diameter and measuring the strains that occur within it when over-cored by a larger (e.g. 100 mm) core barrel, Fig. 10.13. The *in-situ* stress required to cause the strains measured may be calculated, but values for the elastic moduli *in-situ* must either be known or assumed

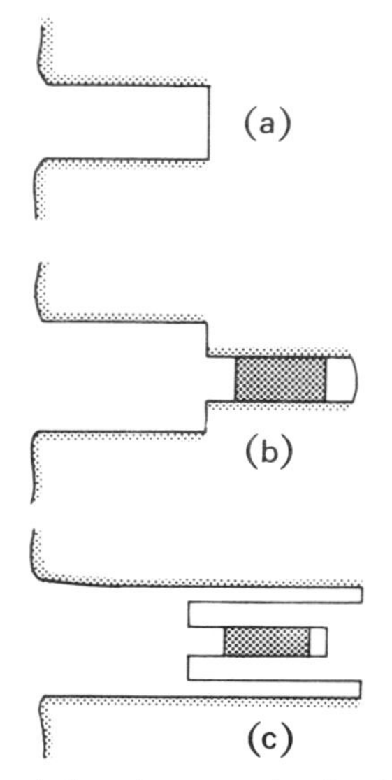

Fig. 10.13 A technique for measuring *in-situ* stress. (**a**) An initial hole is drilled to the required location. (**b**) A co-axial hole is used for housing the deformation meter. (**c**) Over-coring releases the stresses around the meter which measures the resulting deformations.

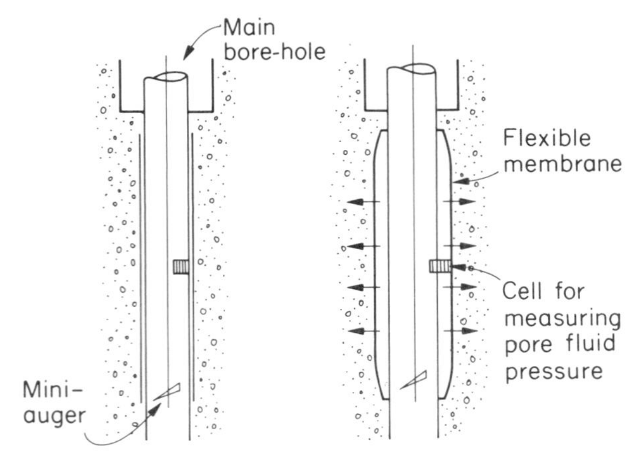

Fig. 10.14 The Camkometer: this instrument bores from the base of a bore-hole into the soil beneath and its flexible membrane may be expanded to deform the surrounding ground (after Wroth *et al.*, 1973).

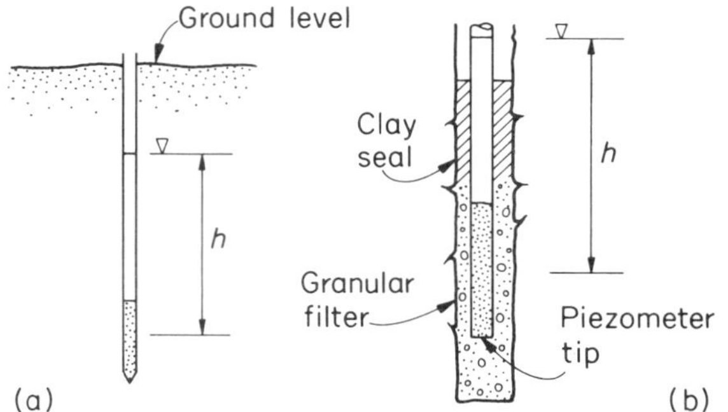

Fig. 10.15 Two examples of a stand-pipe piezometer. (**a**) Can be driven into the ground and need not require a bore-hole; (**b**) is an installation in a bore-hole. The fluid pressure = unit weight of water × *h*.

for the calculation to be completed. This method was designed for use in rock.

An alternative method that does not require a prior knowedge of elastic parameters, cuts into the ground a long slot (1–3 m) which closes slightly under the *in-situ* stress. A thin hydraulic jack, called a *flat-jack*, is inserted into the slot and pressurized to expand and force the slot open. The pressure required to return the slot to its dimensions when cut is taken as a measure of *in-situ* stress: see Hoskins (1966).

A third method injects fluid into a bore-hole, gradually increasing its pressure until the walls of the bore-hole crack, this pressure being a measure of *in-situ* stress. The method can be used in deep bore-holes (Haimson, 1978).

Methods designed for use in soils mainly utilize load cells in which the deformation that occurs under *in-situ* stress is nullified by an applied load whose magnitude is taken as a measure of the *in-situ* stress. Cells fashioned as lances and blades can be pushed into soft soils. The *Camkometer* is a development of such tools, being able to bore itself into the soil, measure *in-situ* stress and expand further to deform the soil and obtain a measure of its strength (Fig. 10.14).

Fluid pressure

Instruments for measuring fluid pressure are called *piezometers*: they are divided into two categories, namely those that require a movement of water and those that do not. The simplest example of the former is a stand-pipe that operates as a manometer, Fig. 10.15. It is suitable for use in permeable ground such as sand, gravel and fractured rock where water can move easily from the ground into the instrument during periods of rising water pressure and the reverse when pressures fall. Such instruments require too great a period of time to respond to changes in pore water pressure that occur in impermeable materials such as clay and their accuracy suffers, the pressure registered by them being that in the ground at some previous date. This phenomenon is called *time-lag* and is overcome by using a piezometer that does not require pore fluid to enter it. These operate as stress gauges: pore water pressure causes them to deform and the hydraulic pressure that has to be provided inside the piezometer to nullify the deformation is a measure of the pore pressure in the ground.

Measurement of deformability

To calculate deformability a static or dynamic load must be applied to the ground and a measurement made of the resulting strain. It is customary to interpret the results on the basis of the theory of elasticity and assign values for Young's Modulus and Poisson's ratio to the ground. Tests which operate within the linear, elastic portion of the stress-strain curve for the ground are those usually chosen for analysis.

Static tests

In these a static load is normally applied in one of three ways: over the area of a rigid plate, over the area of a tunnel, and over the area of a bore-hole. This load is increased in increments, the load in each being maintained at a constant value: a new increment commences when deformation under the previous load has ceased.

In *plate bearing* tests the load is applied to a flat surface, usually by an hydraulic jack, and the deformations are recorded. General arrangements are illustrated in Fig. 10.16. In theory the size of loaded area should be related to the rock fabric, but in practice an economic compromise is made and an area 1 m^2 is often used. This is nearly the smallest area desirable for testing most rock masses. The loading pad should be well mated to the test surface.

In tunnel tests part of a circular tunnel is radially loaded by closely spaced sets of jacks. Changes in diameter resulting from this uniformly distributed radial loading are used to define the modulus of elasticity. The volume of rock tested is larger than in a plate bearing test, and more representative values are usually obtained.

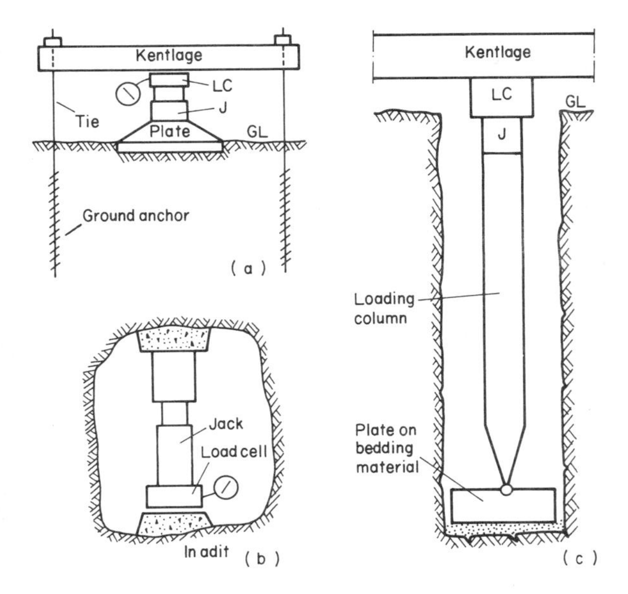

Fig. 10.16 Plate bearing tests performed (**a**) at ground level, (**b**) underground, (**c**) from ground level but on ground at depth, *e.g.* at the level of a future foundation. J = jack. LC = load cell. Drawings are not to scale.

Both plate bearing and pressure-tunnel methods are described by Rocha (1955) and Int. Soc. Rock Mechanics (1978).

Because of the cost of such tests smaller, cheaper methods using hydraulic jacks have been developed for radially loading sections of bore-hole (Fig. 10.17). Many investigators believe that these instruments are best suited for obtaining an index of rock mass deformability, as the technique can induce complicated states of stress and make the analysis of results difficult. All the tests are affected by the disturbance the ground suffers prior to the actual test: see American Society for Testing and Materials (1969).

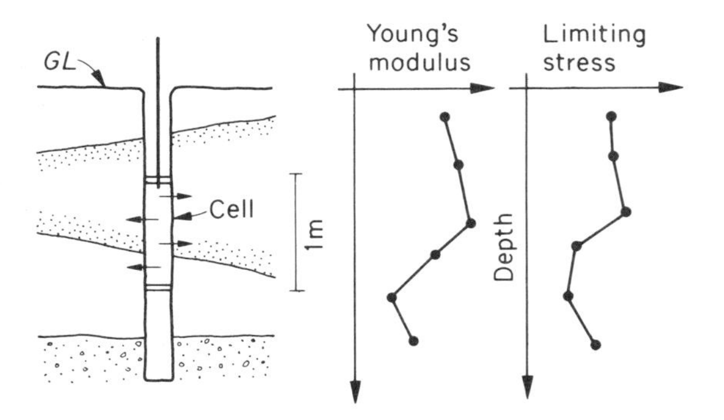

Fig. 10.17 A bore-hole deformation cell for measuring E and strength with examples of the values obtained for these parameters at different depths in a bore-hole (see also Fig. 10.14).

Dynamic tests

These employ the propagation of compressive and shear waves through the ground, their velocity being a function of the elastic moduli of the rock and soil through which they travel. The moduli calculated from them are generally greater than that measured by static tests as the latter often generate non-elastic deformations when pores and fractures close beneath the applied load. Dynamic moduli can be calculated from seismic geophysical surveys conducted at an early stage in a ground investigation and enable differences in the deformability of a large site to be defined, Fig. 10.18 and Meigh (1977).

Dynamic tests can also be used below ground level, in excavations and bore-holes (Bieniawski, 1978). Volumes of ground at depth can be examined by placing the instruments which emit seismic waves in one bore-hole and those which receive the waves in another, the transmission velocity of the waves from bore-hole to bore-hole being a measure of the moduli in the ground between them.

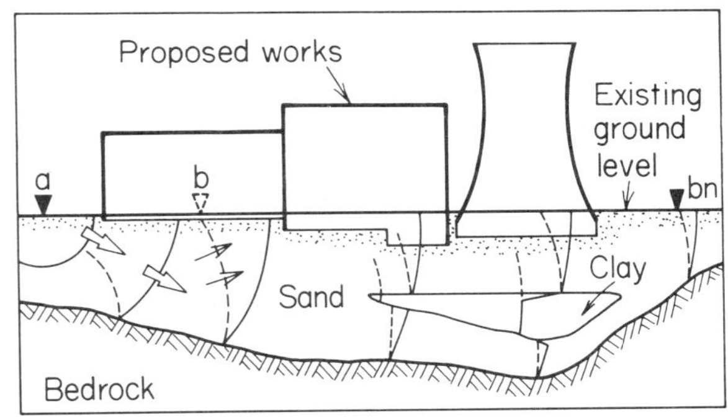

Fig. 10.18 Seismic investigation of ground beneath a proposed power station. A seismic source at *a* emits shock waves whose direct and reflected accelerations (solid lines and broken lines in the figure) are detected by geophones located along a line from *b* to b_n. The dynamic characters of the overburden and bedrock may be deduced from the velocity of the waves, and those returned by bedrock and other surfaces including the boundaries of the buried valley filled with clay, reveal the stratigraphy and structure of the site (p. 176).

Other uses

The geophysical techniques employed to measure dynamic moduli are similar to those needed in certain fields such as blast control and earthquake engineering, where it is necessary to know the speed with which shock waves are propagated through the ground, and the extent to which they will be attenuated. Seismic techniques can be employed to investigate this, and such work is normally conducted by geophysicists. A source of waves is provided at a point on or in the ground by a mechanical device such as a hammer or a vibrator, or by an explosion. Geophones are placed at intervals from this source and record the arrival of waves so produced. For an introduction to the subject see Ambraseys and Hendron (1968); and for a fuller coverage Clarke (1970).

The dynamic behaviour of ground can often be related to physical characters which are of interest to other fields of engineering; e.g. its quality can be assessed. Compressional waves are generated during investigations that use seismic techniques and radiate out from their source, at a velocity (V_f) that is governed by the ground conditions, especially by the amount of fissuring. The effect of discontinuities in the ground can then be estimated by comparing the field velocity (V_f) with the velocity of similar waves through an intact core sample of the same rock (V_c) subjected to stress equal to that *in-situ* and with a similar moisture content. The ratio of the two velocities (V_f/V_c) will approach unity as the rock approaches an unfissured state. Knill (1970) has used this method for studying the grout-take of dam foundations (p. 273).

Seismic velocities have also been correlated to the ease with which ground can be excavated, as described in Chapter 16.

Measurement of shear strength

Three methods are commonly used to measure shear strength *in-situ*.

Shear tests

Shear tests reproduce on a large scale the shearing arrangement used in a laboratory shear box (Fig. 11.9). Figure 10.19 illustrates such a test for use underground,

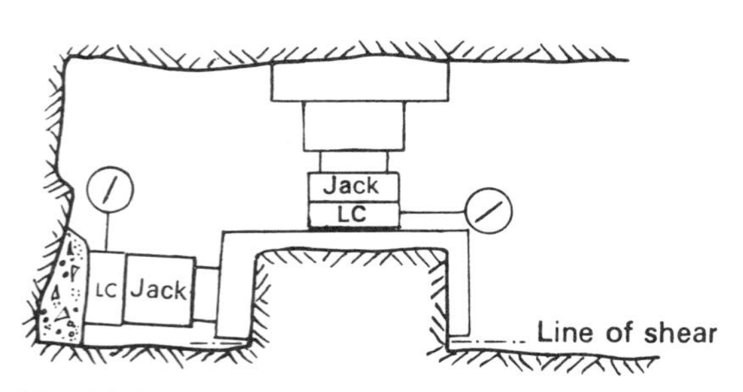

Fig. 10.19 Direct shear test: arrangement for use in adit. Displacement gauges not shown. LC = load cell. Area of shear approximately 1 m².

and Figs 9.24 to 9.26, the form of results such a test would obtain. The test arrangement can be modified for use at ground surface.

Vane tests

These are used in soil. A vane of four thin rectangular blades usually two to four times as long as they are wide, is pressed into the soil and twisted at a uniform rate of about 0.1 degree per second. A cylindrical surface of failure develops at a certain torque the value of which is measured and used to calculate shear strength.

Plate bearing tests

These utilize the test arrangements illustrated in Fig. 10.16*a* and *b* to load the ground until shear failure occurs beneath the plate. This provides a value called the *bearing capacity* of the soil, which is used to assess the maximum load that can be carried by a foundation bearing on the ground.

Other tests

An indication of the relative shear strength of layered soil horizons may be obtained from the resistance they offer to penetration by a rod. A number of simple investigation methods are based on this principle.

Probes are rods used to locate boulders in soft clay and bedrock at depth.

Standard penetration test measures the penetration of a conical probe when it is hammered into the ground with a standard percussive force.

Cone penetrometers are sophisticated and delicate probes which are gently and smoothy forced into the ground; Fig. 10.20 illustrates a *Dutch cone penetrometer*. They

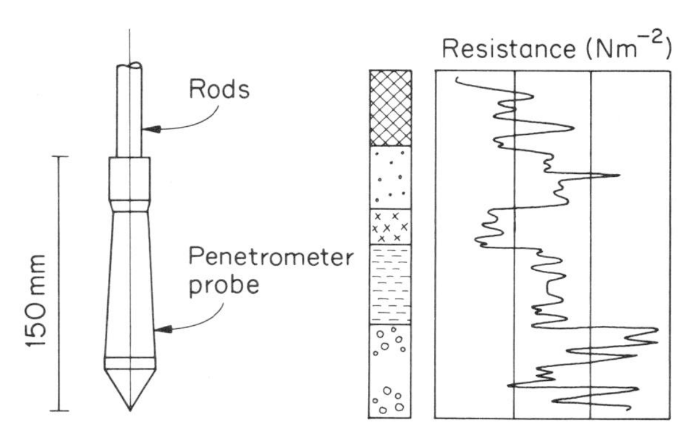

Fig. 10.20 Cone penetrometer and example of continuous log obtained for the soil profile on the left, namely made ground at the top, sand, peat, clay and gravel at the bottom. Pore water pressure may also be measured by some cones (namely piezocones)

may be used in layered sequences containing peat, clay, silt, sand and fine gravel.

These methods have been reviewed by de Mello (1971), Sanglerat (1972), Rodin *et al.* (1974), and de Ruiter (1981). Because of their similarity to pile driving they are often used to investigate soil profiles at sites where driven piles are likely to be used later in construction.

Measurement of hydraulic properties

The two properties most commonly required are the permeability of the ground and its storage; both may be calculated from a *pumping test*.

Pumping test

In this test a well is sunk into the ground and surrounded by observation holes of smaller diameter, which are spaced along lines radiating from the well. Two observation holes are generally held to be a minimum requirement, as in Fig. 10.21. Pumping from the well

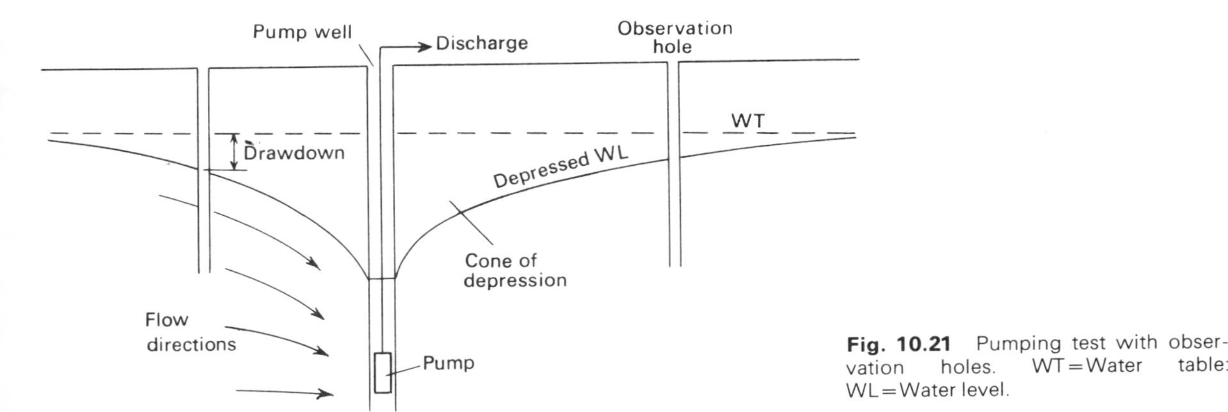

Fig. 10.21 Pumping test with observation holes. WT=Water table: WL=Water level.

lowers the water level in it and in the surrounding ground, so that a *cone of depression* results. By using values for the discharge from the well at given times, and the drawdown measured in the holes at those times, and the known distances of the holes from the centre of the well, the permeability and storage of the ground can be calculated. Values for permeability only can be found from the rise in water levels which occurs when pumping stops: this is called the recovery method. A great variety of tests can be conducted, all of which are based on this fundamental procedure; they are described by Kruseman and de Ridder (1983).

Care should be exercised in choosing the test for a particular site because the value of permeability obtained will depend on the direction of flow from or to the test hole. Different values for permeability will be found in different directions in anisotropic ground, and a test system and orientation should be chosen to simulate as closely as possible the flow regime that will operate when the final engineering structure is complete.

Water engineers use a standard pumping test because the flow around the well when in production will be similar to that in the test. Ground-water lowering and de-watering schemes might use such a test if they consist of series of pumped wells. However, it is unlikely that the permeability derived from the dominantly horizontal and radial flow of these tests would apply to vertical seepage problems beneath large dams or to the flow towards tunnels.

Other tests

Permeability alone may be measured by less expensive tests, using either packers or piezometers, in existing bore-holes that may have been drilled for other ground investigations. Figure 10.22 illustrates such an application: relevant calculations are given by Cedegren (1967).

Packer tests

A packer is an inflatable tube, 1 or 2 m long, that can be lowered into a bore-hole and expanded radially to isolate the length of bore-hole beneath it from that above. Two packers may be used to isolate a section of bore-hole from that beneath and that above. During a packer test water is injected into a section of bore-hole isolated by packers and from this discharge the permeability of the ground in the section can be calculated.

Piezometer tests

When water is injected into piezometers of the type illustrated in Fig. 10.15, it will flow from their tip into the ground. From this flow the permeability of the ground may be assessed.

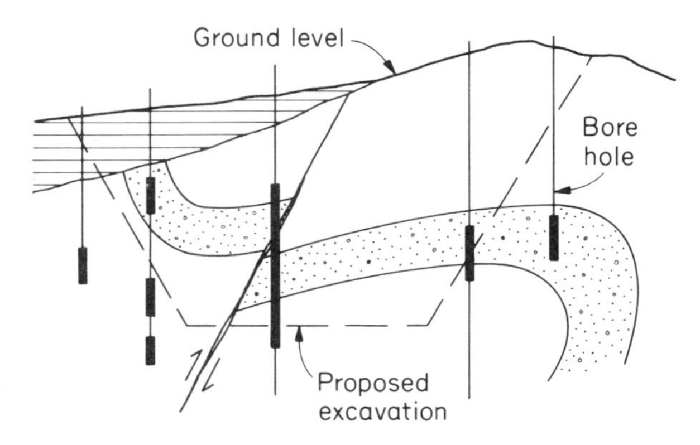

Fig. 10.22 Use of single-hole permeability tests in the investigation bore-holes drilled to study the geology associated with an excavation. Sections tested are indicated.

Selected bibliography

Clayton, C.R.I., Simons, N.E. and Mathews, M.C. (1982). *Site Investigation: a Handbook for Engineers.* Granada Pub. Co. Ltd, London.

British Standard: BS5930 (1981). *Code of Practice for Site Investigation.* British Standards Instn, London.

United States Bureau Reclamation (1974). *Earth Manual.* U.S. Government Printing Office, Washington D.C.

Hvorslev, M.J. (1948). *Sub-surface exploration and sampling of soils for civil engineering purposes.* Amer.

Soc. Civ. Engs. Soil Mechanics and Foundation Division.

American Society Civ. Engs (1972). Report of Task Committee on sub-surface investiagion of foundations of buildings. *Proc. Amer. Soc. Civ. Engs. Jl. Soil Mechanics and Foundation Divison,* **98,** 481–90, 557–78, 749–66, 771–86.

Geological Society of London (1970). The logging of rock cores for engineering purposes. *Q. Jl. Engng Geol,* **3,** 1–24 and **10,** 45–52 (1977).

Geological Society of London (1977). The description of rock masses for engineering purposes. *Q. Jl. Engng Geol.,* **10,** 355–88.

International Society for Rock Mechanics (1980). Basic geotechnical description of rock masses. *Int. Jl. Rock Mech. & Min. Sci. & Geomech. Abstr.,* **18,** 85–110 (1981).

Geological Society of London (1982). Land surface evaluation for engineering purposes. *Q. Jl. Engng Geol.,* **15,** 265–316.

Blaricom, R. Van. (1980). *Practical Geophysics.* Northwest Mining Assocn, Spokane, U.S.A.

Robb, A.D. (1982). *Site Investigation.* ICE Works Construction Guides. Thomas Telford Ltd, London.

Sampling and Testing Residual Soils: A review of international practice (1985). Editors: E. W. Brand and H. B. Phillipson. Scorpion Press, Hong Kong.

Society for Underwater Technology (1985). Offshore site investigation. *Proc. Int. Conference.* London 1985. Graham & Trotman.

11

Laboratory Investigations

Laboratory investigations are normally requested when (*i*) the suitability of a rock or soil for a particular use must be assessed and (*ii*) their composition evaluated.

(*i*) Considerable quantities of rock and soil are used as aggregate and bulk-fill in large engineering schemes, such as in the construction of dams, roads, new towns, etc. The suitability of the natural materials used must be constantly tested during construction and for this a purpose-built laboratory is usually provided as part of the offices on site.

(*ii*) Tests for composition are routine procedure in the commercial extraction of minerals. Miners of metalliferous and non-metalliferous minerals must constantly measure the quality of the ore or stone produced from the mines and quarries. Coal, gas and petroleum engineers are concerned with the calorific value of the fuel won from the ground and water engineers constantly monitor the quality of the ground-water abstracted by water wells.

Soil and rock used for engineering construction is always classified by laboratory tests and this requires the general composition of these materials to be assessed. Their composition is only evaluated in detail when the presence of unstable minerals is suspected. For example, a shale containing the mineral pyrite (p. 84), which oxidizes on exposure, may be unsuitable for use in an embankment, or a basalt that contains the mineral olivine (p. 72), which becomes unstable in the presence of Portland Cement, would be unsuitable as a concrete aggregate. In both these examples the unstable mineral would cause the strength of the rock to decrease with time. Some clay minerals expand, when wetted, especially those of the montmorillonite groupe, and their increase in volume may create unexpected stress and strain in the ground.

Samples and sampling

Care must be taken when sampling soil and rock to ensure that the sample is representative of the particular material to be investigated, and that its disturbance during sampling will not adversely affect the results of the tests to be

Table 11.1 A guide to sample requirements

Material	Bulk or block sample	Core or tube sample (mm) (dia) (length)	Material	Bulk or block sample	Core or tube sample (mm) (dia) (length)
1. *Chemical composition*			4. *Hydraulic characters*		
Clays & silts	0.75–1.0 kg	100 × 300	Clays & silts	$(0.2\ m)^3$	100 × 100
Sands	0.75–1.0 kg	100 × 300	Sands	$(0.2\ m)^3$	100 × 200
Gravels	3.0 kg	100 × 400	Gravels	$(1.0\ m)^3$	100 × 400
Rocks	0.5 kg	40 × 80	Rocks		
Ground-water	2 litres		(coarse grained)	$(0.3\ m)^3$	92 × 90
			(fine grained)	$(0.3\ m)^3$	54 × 50
2. *Comprehensive examination*			5. *Structural characters and classification*		
Clays & silts	100 kg	100 × 600	Clays & silts	2.0 kg	100 × 300
Sands	100 kg	100 × 600	Sands	3.0 kg	100 × 300
Gravels	150 kg	100 × 800	Gravels	30.0 kg	100 × 400
Rocks	100 kg	92 × 500	Rocks	$(0.2\ m)^3$	54 × 120
Ground-water	10 litres				
3. *Strength and deformability including consolidation*					
Clays and silts	$(0.3\ m)^3$	100 × 300			
Sands	$(0.3\ m)^3$	100 × 300			
Gravels	$(0.5\ m)^3$	100 × 400			
Rocks (weathered)	2 off $(0.3\ m)^3$	92 × 240			
Rocks (unweathered)	1 off $(0.3\ m)^3$	54 × 120			

performed upon it (see Broms, 1980 and Amer. Soc. Testing Materials, 1970). No sample can be collected without it being disturbed in some way, for the reasons explained in Chapter 9.

In practice two categories of sample may be obtained:

(*i*) *Routine sample*, i.e. one slightly disturbed but collected with reasonable care by experienced staff who should ensure that it suffers no avoidable change in moisture content, does not lose any of its constituents (e.g. fine sand from mixed gravel) and is not deformed by careless handling if its strength or permeability are to be measured. Routine samples must be correctly stored and adequately labelled. Table 11.1 provides a guide to sample requirements.

(*ii*) *Research sample*, i.e. one collected with the greatest care possible, often using expensive procedures to reduce disturbance: such a sample cannot be economically provided by commercial routine sampling.

The difficulties of obtaining routine samples using conventional methods of commercial sampling are illustrated by Kallstenius (1963) and Rowe (1972).

Guidelines

The following guidelines should be observed when possible and considered the minimum standard required for routine sampling.

Selection of samples

Figure 11.1 illustrates a 20 m deep bore-hole sunk to investigate the ground beneath a foundation 10 m × 10 m square. If samples are taken at 2 m intervals, as in the figure, the volume sampled will be about 1/250 000th of the ground subject to loading. This example demonstrates the care required when selecting samples for testing. Weak strata and zones of weakness in soil and rock should always be tested. Specimens to be tested in a laboratory normally have dimensions which prevent them from containing major surfaces of weakness, such as joints and fissures, and laboratory tests usually over-

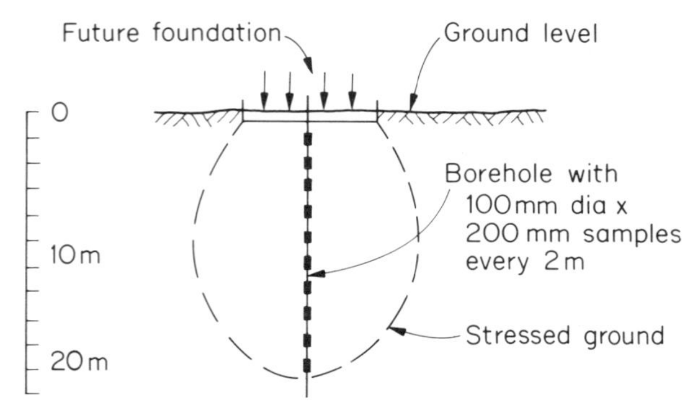

Fig. 11.1 Some limitations with sampling. The total volume of samples is approximately 1/250 000th of the stressed ground. See also Fig. 10.1.

estimate *in-situ* shear strength. Unavoidable deformation of the specimen that occurs when it is being recovered from the ground and prepared for testing, results in most laboratory tests under-estimating *in-situ* moduli: (see Marsland, 1975; and Lessons from Failure, p. 170).

Collection of samples

Changes in moisture content, either by wetting or drying, can be avoided by sealing the sample with an impermeable material such as paraffin wax or polythene (see Stimpson *et al.*, 1970). This should be done as soon as possible after the specimen has been collected; samples should not be collected from areas where the natural moisture content of the ground has been changed by wetting or drying.

Loss of material is a hazard in sands, gravels, and similar deposits which have little or no cohesion and should be avoided. Comparatively undisturbed samples of moist sand may be taken from natural exposures, excavations, and borings, above the water table, by gently forcing a sampling tube into the ground. Samples are less easily obtained below the water table, and if they are to be taken from a bore-hole, the water level in the hole should be kept higher than that in the adjacent ground. The flow of water outward from the bore-hole will then tend to prevent the loss of finer particles which would otherwise be flushed from its sides by inflow of water from the surrounding ground. Having obtained a sample, loss of material from it should be avoided during its transport to the laboratory. These problems are less severe when dealing with rocks than with weak materials and drilling fines, which should be handled with special care.

Local over-stressing often occurs with the trimming and transport of soft materials such as soils. Disturbance from trimming is affected by the sampling and cutting tools that are used, and cutting edges should be sharp so as to cut smoothly and cleanly. They should also be thin so that they cause little displacement when passing through the soil. This is important in the design of tubes that are pushed into the ground to collect a sample; an area ratio of the form

$$\left[\frac{Da^2 - Db^2}{Db^2}\right] \times 100\%$$

should be observed, where Da is the outside diameter of the cutting edge, and Db its inside diameter. The above represents the area of ground displaced by the sampler in proportion to the area of sample. Severe disturbance is likely to occur on the margins of a sample if the ratio is greater than 25%, as illustrated in Fig. 11.2.

Disturbance can also occur during the extrusion of samples either from sample tubes or core barrels, and care should be taken that the friction between the sample and sampler wall is kept as low as possible, and that excessive pressure is avoided during extrusion. Harder materials such as rocks can be sampled by coring, as described in Chapter 9. Hand specimens of rocks should be at least (8 cm)^3 to (10 cm.)^3 and representative of the

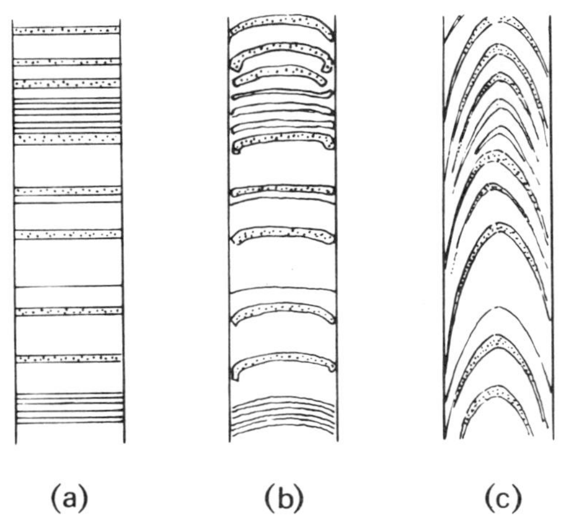

Fig. 11.2 Distortion of bedding by a tube sampler. (**a**) Smooth penetration with correct sampler. (**b**) Jerky penetration with correct sampler. (**c**) Smooth penetration with incorrect sampler.

formation sampled. Ground-water samples can be collected in either glass or polythene screw-cap beakers, the choice depending upon the future analysis that is required.

Table 11.1 indicates the size of samples needed for most of the standard tests made with conventional laboratory apparatus, and Table 11.2 the methods that are normally used for their recovery.

Storage of samples

Certain necessary precautions should be observed when handling and storing samples, particularly those of soils and soft rock. Adequate containers should be provided that will protect them from any further disturbance after collection. At least two large durable labels giving the location and depth from which the samples were taken, their date of collection and serial number should be written for every sample. One label is placed inside the container, the other attached to the outside of it and the sample numbers recorded in a diary. The labels should be written in indelible ink and be able to withstand the wear and tear of site work. It is usually desirable to test soil samples within two weeks of their collection, during which interval they are best stored in a cool room of controlled humidity.

Samples of hard rock are less delicate but still require reasonably careful handling. Individual samples should be labelled securely. This can be done by painting a serial number on the samples. Those which have to be transported should be individually wrapped (newspaper is ideal) and crated in stout boxes. A list of the serial numbers of the specimens should be included in the crate

Table 11.2 A guide to sampling methods

Method	Comments	References
Open drive samplers	Thin walled open tubes from 50 mm to 100 mm diameter (the U-100 sampler in Britain) which are pressed into the ground. Disturbance is common and accepted. Unsuitable for cohesionless materials, hard clays and soft rocks.	Hvorslev (1948) Serota and Jennings (1958)
Piston samplers	Thin walled tubes from 50 mm to 100 mm. diameter, but can be larger. Contains a piston which is withdrawn when sample is collected. Generally less disturbance caused than with an open drive sampler. Normally used for cohesive soils but can cope with granular materials. Special designs are required for the collection of fine sands (Bishop sampler).	Kallstenius (1963) Bishop (1948)
Foil samplers	Special case of piston sampler that protects the sample in a sheath. Gives little disturbance and collects long samples. Can be used in soft sensitive clays and expansive soils. Cannot be used for coarse granular soils but will collect sand samples from above the water table, if moist. 'Delft sampler' is of similar design and capability.	Kjellman, Kallstenius and Wager (1950) Begemann (1966)
Rotary core drilling	For all hard rocks with diameter to suit purpose. Difficult to collect coarse granular soils. Becoming a popular method for sampling stiff clays. Improved sampling obtained using double-tube samplers.	British Standard, 4019 (1974) Earth Manual (1974)
Augers	Will provide disturbed samples of most soils. Extremely cheap.	Earth Manual (1974)
Hand trimming	Least disturbance of all methods in soft and cohesionless materials.	Earth Manual (1974)
Mechanical and manual excavations	For all bulk sampling, either disturbed or undisturbed.	Earth Manual (1974)

and recorded in the site diary. Cores are normally stored and transported in core boxes, which should be clearly labelled with the bore-hole number and site, and the levels from which the cores have come. Further information on the sampling and handling of soils and rocks is given by British Standard 5930 (1981), by Hvorslev (1948), and the *Earth Manual* (U.S. Bur. Reclamation, 1974).

Laboratory tests

The laboratory tests commonly conducted on rock and soil samples are now described: they have been grouped into five categories.

Tests for composition

Analyses for the composition of substances such as coal, petroleum products, ores, fluxes, lime and ground-water, should normally follow prescribed standards. As mentioned earlier, it is occasionally necessary to analyse the composition of other rocks and of soils to solve particular problems when using these materials for construction. In all cases contamination and loss of constituents should be avoided and the standard precautions for chemical analyses observed.

Tests for composition are of three types:

Physical tests: these relate the physical properties of minerals in rock and soil, to their chemistry, as outlined in Table 11.3 (and see Zussman, 1967).

Chemical tests: these include conventional wet methods and dry combustion, and are normally used to determine organic content, sulphates and chlorides. Simple methods that give a qualitative indication of composition are listed in Table 11.4.

Table 11.3 Physical tests

SOLID PHASE	
Hand lens Binocular microscope	Visual assessment of mineralogy and basic composition (F)
Scratch: hardness Colour: streak, lustre Density: specific gravity, bulk, unit weight	Assessment of mineralogy (see Chapter 4, and Tables 4.3, 4.9), weight balance (F)
Activity (involving particle size determination and Atterberg Limits)	Relationship between plasticity and mineralogy: also indicates clay mineralogy (see Fig. 11.13) (F)
Thin section and petrological microscope	Particular mineral assessment (see Chapter 4)
Electron microscope	For assessment of minute characters of particles and crystals
FLUID PHASE	
Water and gas content	Normally by weight loss on drying or vacuation (F)

(F) = can usually be conducted in a site laboratory

Table 11.4 Chemical tests

Odour and taste	Can be extremely characteristic, e.g. for bitumen and rock salt (F)
10% HCl	Reveals carbonates by effervescence (F)
Staining with dyes	Can selectively identify particular organic and inorganic compounds (F)
Combustion	Reveals presence of organic material (F)

(F) = can be undertaken in the field or in the site laboratory

Physico-chemical tests: these employ specialist techniques to measure certain properties of minerals and provide information of diagnostic value. They include electrical conductance, spectroscopy, differential thermal analysis (DTA) and X-ray diffraction (Table 11.5).

Table 11.5 Physico-chemical tests

SOLID PHASE	
Electrical conductance	Non-specific assessment of conductive minerals in a sample (F)
Differential thermal analysis (DTA)	Exothermic and endothermic reactions obtained from a powdered mineral and used for identification
Spectroscopy	Methods that produce emission spectra of measurable radiations that are diagnostic (see text-books of Physical Chemistry) (F*)
X-ray diffraction	Atomic structure diffraction patterns for mineral identification
FLUID PHASE	
In addition to conductance and spectroscopy (above) the following tests can be made:	
pH and Eh	Measurement of ability of fluid to accept or release protons (pH) or electrons (Eh) (F)
Colorimeter	Colour produced on reaction with known reagents (often in tablet form) for indication of the concentration of selected elements and compounds (F)
Chromatography	Analysis of all substances that can be vaporized before separation, for identification of elements

(F) = can usually be conducted in well equipped site laboratory
(F*) special facilities required

Tests for structure

Undisturbed samples that have retained the relative position of grains and hence the shape and size of grain contacts and intervening voids, are required for two of the three tests described here, but are difficult to achieve.

Whole fabric

Whole fabric studies require an undisturbed sample which preserves the shape and distribution of all the voids. Thin sections are commonly used for most rocks (see p. 66). Weaker materials, such as clays and other soft sediments, can be strengthened by impregnation with a wax or an epoxy resin; delicate fabrics can then be observed (Morgenstern and Tchalenko, 1967). A stereo-scanning microscope can be used for small fabrics. Indirect assessments of fabric, usually from its porosity, can be obtained from laboratory measurements of the electrical resistivity of samples. Sonic velocities are also used for this purpose. Both methods are affected by the composition of the sample and by the presence of water in its voids.

Porosity

The structure of voids in a sample is most easily studied by determining the total volume of voids rather than their individual volumes or shape. The latter can be obtained from thin sections as described above (p. 66), but the method is limited to sampling in one plane at a time and is difficult to use for investigating the fine textures found in silts and clays. The volume of voids in rocks is usually assessed either from values of bulk and dry density,

$$\text{Porosity} = 1 - \left[\frac{\gamma}{\gamma_g}\right]$$

where (γ) is bulk density and (γ_g) is the average density of the grains, or from the relative weights of the sample in a saturated and a dry state. The results are expressed as a void ratio (see Chapter 6, p. 112) or as a porosity. These tests require undisturbed samples.

Particle size distribution

Descriptions of the solid part of a fabric are usually restricted to the shape and size of the solids. Particle shape is often considered in classifications for concrete aggregates where it usually applies to the shape of crushed stone rather than to natural aggregates: see for example British Standard, 812 (1975). Grain size is more commonly determined, and for rocks can be obtained from thin sections. In uncemented granular material such as sands, gravels, and clays, it is found either by sieving or by sedimentation methods. In sieving the granular sample is passed through a stack of graded sieves, the largest aperture size being at the top of the stack. The weight of sample retained on any sieve is measured, and expressed as a percentage of the whole sample passing that sieve. The weights retained on successive sieves are then plotted as a cumulative curve against the sieve sizes, as shown in Fig 11.3. This method is employed for grains down to about 0.1 mm diameter, i.e. from cobbles to fine sands. Finer materials such as silt and clay are normally determined by a sedimentation technique: a suspension in water is placed in a cylinder and allowed to settle out, each particle settling at a rate according to its size and specific gravity. The grain size distribution in the suspension can be calculated by taking small samples from the suspension at a given level in the cylinder, over a period of time. Both these techniques, together with a third using an hydrometer, are described in British Standard, 1377 (1975) and by the American Society for Testing and Materials.

The shape of the particle size distribution curves can be expressed approximately by a *uniformity coefficient*, defined as the ration D_{60}/D_{10}, where D_{60} and D_{10} are the particle sizes corresponding to the cumulative percentages 60 and 10 respectively. The uniformity coefficient for the wind blown sand shown in Fig 11.3 is 1.0, that for the sandy alluvium is 9.0. Samples for grain size analysis need not be undisturbed but they should be uncontaminated and complete, i.e. with no fraction of the deposit missing from the sample.

Tests for strength

The strength of specimens tested in a laboratory is affected by the following factors:

(*i*) *Specimen age*. Samples should be tested soon after they have been taken from the ground to avoid unnecessary losses of strength associated with the relief of stress within them, and unknown gains in strength from a reduction in their moisture content (see Chapter 9).

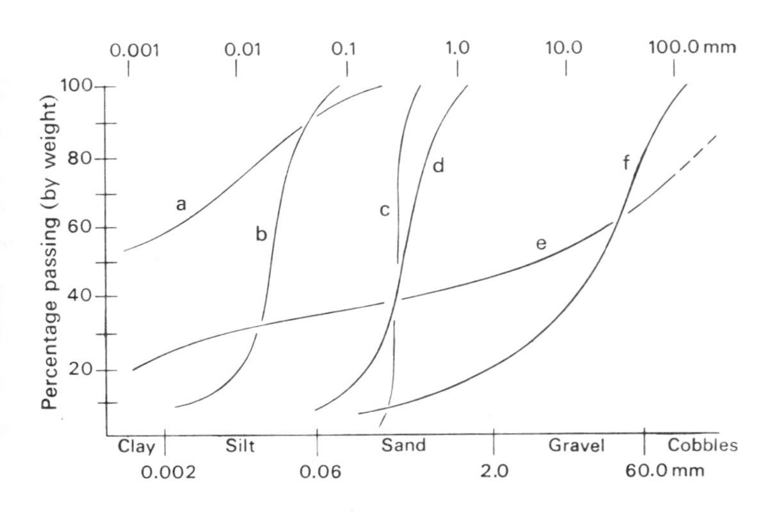

Fig. 11.3 Grading curves for some granular materials: (a) marine clay; (b) loess; (c) wind blown sand; (d) sandy alluvium; (e) boulder clay; (f) gravelly alluvium.

(*ii*) *Specimen size*. Large specimens will contain a greater number of surfaces than smaller samples and exhibit a lower strength (Fig 11.4).

(*iii*) *Speed of testing*. Loads developed during testing create within a specimen a change in pore water pressure, as described in Chapter 9. The strength of a rock or soil will be affected by these changes which may differ in

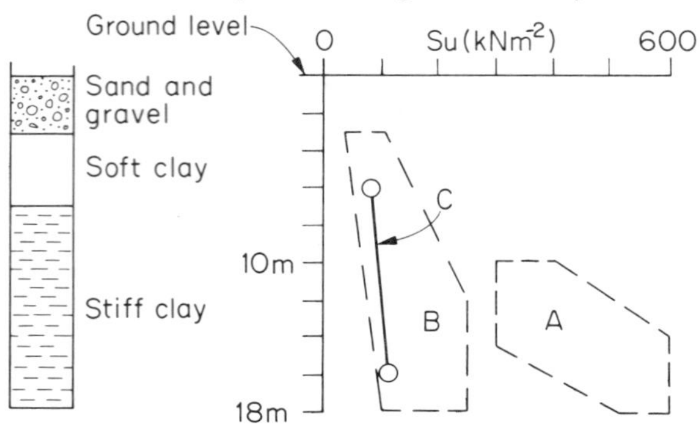

Fig. 11.4 Measurements of undrained shear strength (Su). A = range as measured using bore-hole Penetrometer (*cf.* Fig. 10.20). B = range from 38 mm dia. and 98 mm dia. triaxial samples (*cf.* Fig. 9.11). C = results from 865 mm bore-hole Plate Bearing Test (*cf.* Fig. 10.16) (Marsland, 1971).

magnitude between specimens of the same sample, giving each a different strength. The most useful tests are those conducted at a speed which permits pore pressure changes occurring within the sample to be measured.

(*iv*) *Specimen structure*. Surfaces within a sample, such as bedding and cleavage, will affect its strength, as illustrated in Fig. 11.5.

The loads applied to a specimen should reproduce those expected in the ground, and tests in which the loads are gradually increased reproduce conditions applicable

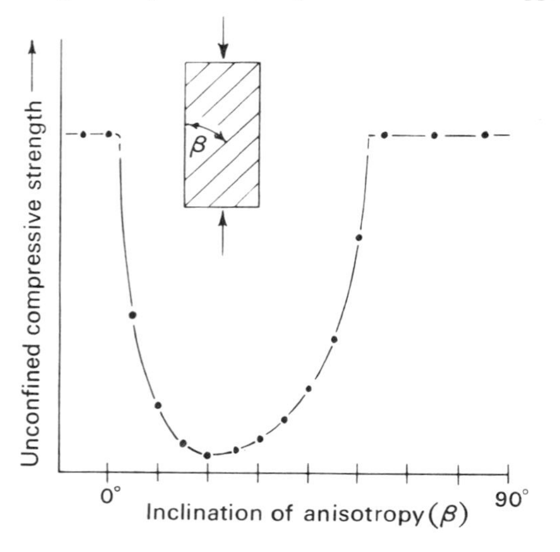

Fig. 11.5 Influence of anisotropy on strength (Donath, 1961).

to most engineering work. Tests to measure soil and rock strength under dynamic conditions experienced during earthquakes or marine loading by waves, must load the specimens in cycles of appropriate periodicity. To measure the creep strength of soil and rock it is necessary to maintain a constant load for long periods.

Elastic moduli

These can be obtained from statically loading cylinders of soil or rock and noting the resulting strains in directions normal and parallel to the direction of applied load. The Modulus of Elasticity and Poisson's Ratio can then be determined. The slope of the initial stress-strain curve in many specimens is less than that of curves obtained in subsequent tests; this is generally attributed to the closing of voids and fissures which have opened during the collection of the sample. It is normal to use the curves from second and subsequent loading cycles for analysis. The elastic constants, under dynamic conditions, can be indirectly determined by measuring the velocity of propagation of compressional and shear waves through the material. (Amer. Soc. Testing Materials, 1975). Undisturbed samples should be used in all these tests.

Consolidation characters

The characters normally required are the *coefficient of compressibility*, which is the change in unit volume that occurs with a change in pressure (used for calculating the magnitude of settlements), and the *coefficient of consolidation*, which is proportional to the ratio of the coefficients of permeability and compressibility (used for calculating the rate of settlement: see Chapter 9). In the two laboratory methods generally used, a sample is compressed with a known load and the resulting changes in its volume with respect to time are measured. The simplest apparatus is an oedometer (Fig. 11.6). In this equipment a sample is axially loaded in a cylindrical container that allows consolidation in one direction only, i.e. vertically. Larger samples can be tested in a Rowe Cell (Fig 11.7). Consolidation in three dimensions can be obtained using triaxial apparatus (Fig 11.8), where a jacketed cylindrical

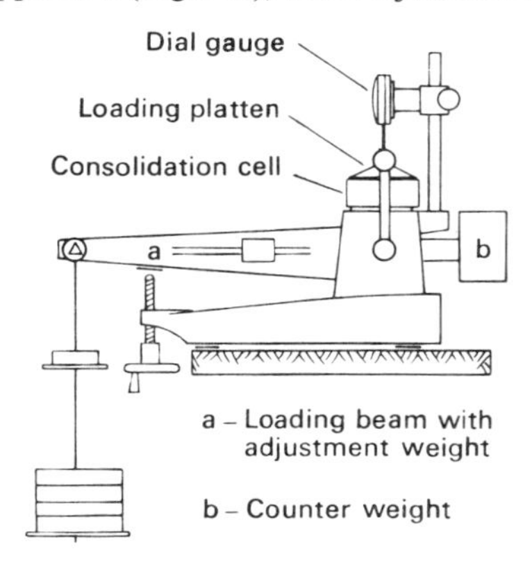

Fig. 11.6 An oedometer. Pore pressure in the sample cannot be measured (see Fig. 11.7).

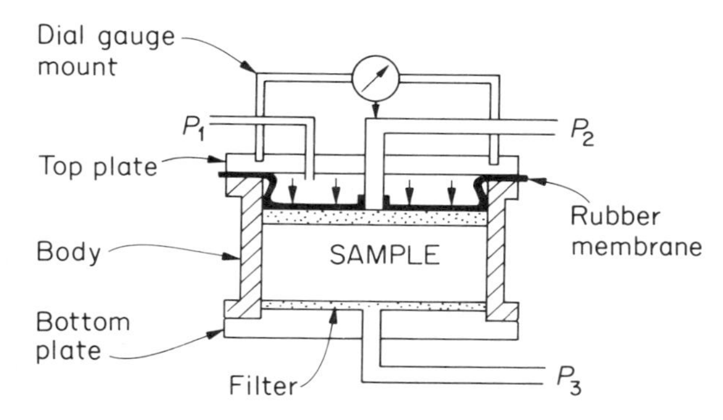

Fig. 11.7 Rudiments of a Rowe Cell (Rowe and Bardon, 1966). The sample is loaded by hydraulic pressure P_1 and consolidated by drainage via P_3. Pore pressures can be measured at both ends of sample via P_2 and P_3. Dial-gauge measures change in sample thickness.

sample can discharge pore water from its upper and lower ends. It is subjected to all-round pressures by pressurizing the fluid in the test cell, so that $\sigma x = \sigma y = \sigma z$ in Fig 11.8. Drainage facilities allow the pore water to escape from the sample, so that its volume decreases and consolidation occurs. Undisturbed samples are normally required.

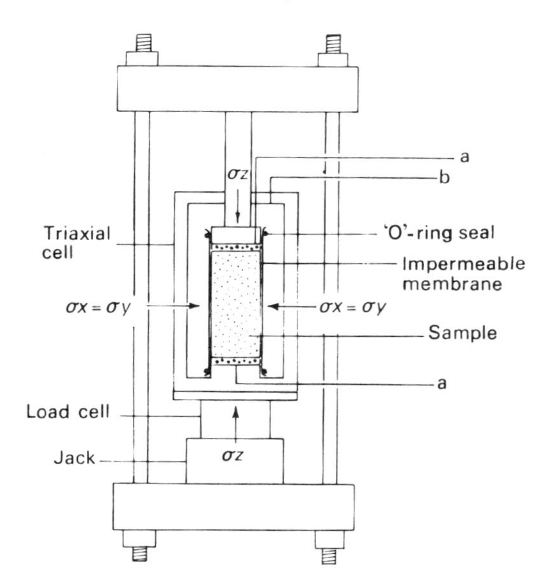

Fig. 11.8 Rudiments of a triaxial cell for soil and weak rock. a = lines for pore fluids. b = line for cell pressure and volume (see also Fig. 9.11).

Swelling characters

The pressure required to prevent the expansion of a specimen may be determined using a oedometer. By this method the swelling pressure generated by the hydration of certain clay minerals (p. 81), or by the hydration of anhydrite to gypsum, or when water in a soil converts to ice (as in permafrost, p. 35) may be measured.

Tensile strength

This can be measured in two ways. The simpler method loads test cylinders in tension until failure occurs, and a standard tensile testing rig is used. The second method loads test discs in compression along a diameter, so inducing tensile failure on a diametral surface. This is sometimes referred to as a Brazilian test. The apparent simplicity of these tests is illusive and results can vary with the specimen preparation, test procedure, and equipment used (Mellor and Hawkes, 1971). Undisturbed samples are required.

Uniaxial (or unconfined) compressive strength

This is normally determined by statically loading a cylinder of rock to failure, the load being applied across the upper and lower faces of the sample. The results obtained are in part a function of the length–breadth ratio of the sample and of the rate of loading. The simplicity of the test is somewhat deceptive (Hawkes and Mellor, 1970). Samples should be undisturbed.

Triaxial compressive strength

A cylindrical sample is placed on a pedestal within the vessel and jacketed with an impermeable membrane. This isolates the sample from the pressurizing fluid which is to surround it (see Fig 9.11). A loading platten is placed on top of the sample. A ram which passes through the roof of the vessel bears against this platten and transmits an axial load from a loading frame to the sample. The load on the sides of the sample is supplied by the pressurized fluid. Figure 11.8 illustrates a simple triaxial apparatus. Hoek and Franklin (1968) describe such a cell for testing rocks. Bishop and Henkel (1962) describe the equipment used for testing soils. In the test a confining load is applied to the sample ($\sigma x = \sigma y = \sigma z$, in Fig 11.8) and then the axial load is increased until failure occurs. Tests can be conducted with or without control of pore pressures, depending upon the sophistication of the equipment. Typical results are shown in Chapter 9 and in Fig 9.19 are illustrated the three tests that have been described above.

Shear strength

Tests of triaxial compressive strength provide a measure of the angle of shearing resistance for soil and rock (Chapter 9). Small versions of the field vane described in Chapter 10 (p. 184) may also be used in the laboratory to obtain a value for soil shear strength. The shear strength of soil and rock surfaces is best evaluated in the laboratory by using a shear box (Fig. 11.9). This houses a normally loaded rectangular sample whose upper and lower halves can be subject to shear displacement about the centrally placed horizontal surface. Typical results are shown in Figs 9.24 to 9.26.

Residual strength. The large displacements requirement to obtain the residual strength of soil (p. 169) can be achieved in a ring-shear apparatus (Fig. 11.10). An annular sample is sheared horizontally into two discs, or rings, and the decrease in frictional resistance that occurs,

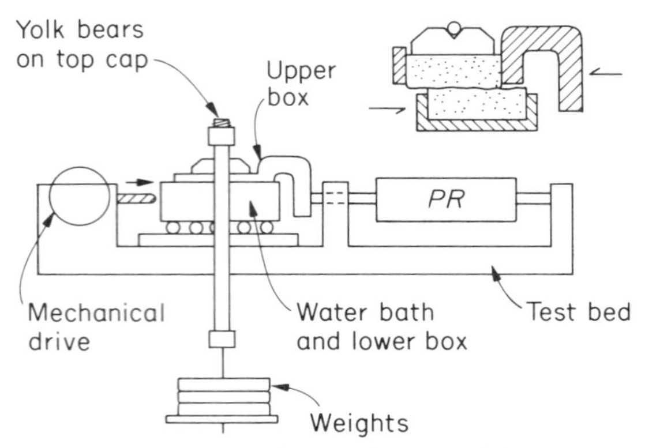

Fig. 11.9 Rudiments of direct shear box. *PR* = Proving Ring: measures the shear load in the sample (Figs 9.24 and 9.25).

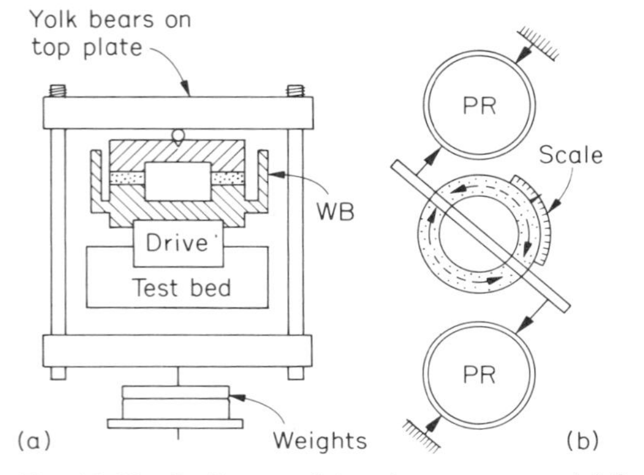

Fig. 11.10 Rudiments of ring shear apparatus. (**a**) Section. Sample turret confines sample between top and bottom plates (WB = Waterbath). (**b**) Plan. Shear load measured by Proving Ring (PR) and displacement by scale. (See Bishop *et al.*, 1971.)

as one disc is rotated relative to the other, can be related to their total displacement.

Tests for hydraulic properties

These measure the conductivity of a soil or rock and the volume of fluid they release by drainage. Field tests give more representative values than laboratory tests (p. 185).

Permeability

Permeability is a measure of the velocity of fluid flow through a porous sample under the hydraulic head operating within the sample; the sample is housed in a permeameter. That shown in Fig. 11.11 is typical of many used for testing sands; silts and clays require slightly different apparatus. Permeability is normally calculated from the following relationship:

$$K = Q/iA$$

where Q = the discharge, i = the hydraulic gradient = $\Delta h/L$, A = cross sectional area of the sample measured at 90° to the general direction of flow, K = coefficient of permeability.

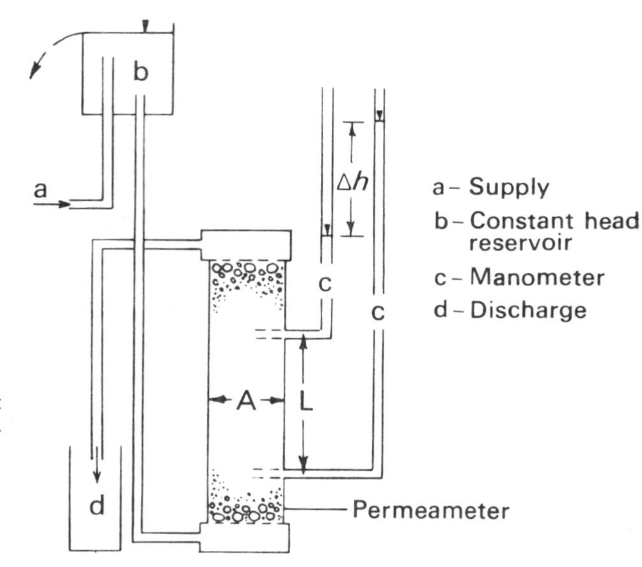

Fig. 11.11 A constant head permeameter for measuring the permeability of granular soils.

This equation is known as Darcy's Law, but being empirical it is not strictly a law. The value of permeability obtained from a test is a function of both the character of the fluid, i.e. its specific weight and viscosity, and the nature of the pore spaces in the solid. Hence the permeability for a sample varies with the character of the fluid used. This limitation can be overcome by considering the fluid characters in a more specific formula and calculating an *intrinsic* permeability, which is a constant for the solid skeleton of the sample. Once calculated it can be used for finding the relationship between Q and i for any fluid passing through the sample, e.g. oil, seawater, fresh water, industrial effluent, etc. Similar tests can be used to find the flow characters of gases through rocks and soils; gas permeameters are used. A theoretical consideration of permeability is provided by Muskat (1937) and test methods are described by Lovelock (1970). The relationship between permeability and stress in rocks can be studied in the laboratory, using the techniques described by Bernaix (1969). That for soils has been discussed by Bishop and Al-Dhahir (1969). In all cases, undisturbed samples are required.

A similar test measures the *thermal conductivity* of rock and soil, i.e. the amount of heat that will flow through them for a given thermal gradient. Such information is needed to calculate the likely temperature in deep mines. Clay-rich materials are poor thermal conductors whereas clay-free materials are good thermal conductors.

Specific yield

This is the volume of fluid released from a porous or fissured material, expressed as a percentage of the total volume of the sample. It need not equal total porosity (p. 112) as not all pores need be interconnected: those which are not can neither release nor receive fluid. A saturated specimen, often that for which permeability has just been tested, is drained either naturally by the force of gravity or by other means. In one method water is forced out of a cylindrical sample by jacketing its walls and blowing air through it from one end. The volume of fluid so released is used to calculate the specific yield of the sample (p. 222).

Effective porosity

This is the volume of fluid accepted by a porous or fissured sample. It need not equal specific yield as capillary tension may retain some of the fluid stored. The difference is called specific retention. Thus specific yield + specific retention = effective porosity. Effective porosity is measured by comparing the weight of a sample in its saturated and dried states, the weight difference being expressed as a volume, 1 g $H_2O = (1\ cm)^3 H_2O$, and this as a percentage of the total volume of the sample.

Index tests

These tests are simple and provide a quick index that may be correlated with design parameters, such as strength and deformation, which must be determined by conventional methods of testing. Index tests do not provide design parameters.

Rock indices

Many indices exist but the following are commonly used.

Point load. This measures the force required to break a piece of rock between converging conical points (Fig. 11.12). The index obtained correlates well with the uniaxial compressive strength of the rock: (Brock and Franklin 1972).

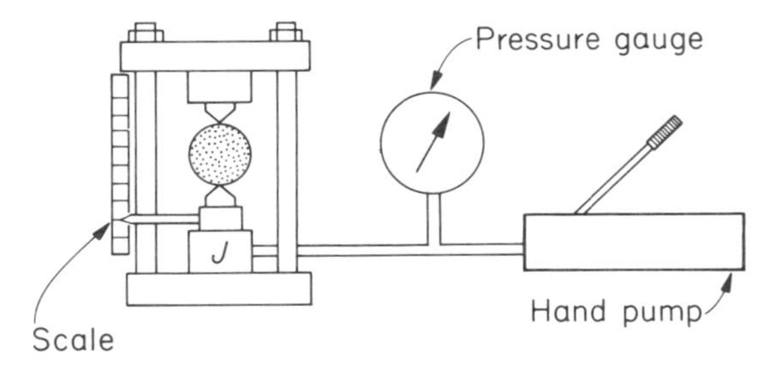

Fig. 11.12 Point load tester. J = jack which closes opposing 60° cones across specimen of rock core. Closure prior to failure measured on scale.

Schmidt Hammer. This instrument measures the rebound of a spring loaded metal piston that is struck against a rock surface. The rebound is an index of the compressive strength of the surface and hence its strength in shear (Int. Soc. Rock Mechanics, 1977*a*). It is also used in tests which assess rock hardness and abrasiveness for drilling purposes.

Slake Durability. In this test a rock is subject to mechanical abrasion, wetting and drying. The index obtained is related to the swelling and shrinkage properties of the rock (Int. Soc. Rock Mechanics, 1977*b*).

Soil indices

The indices commonly used for soil relate some aspect of soil strength to its moisture content.

Atterberg Limits. Atterberg designed two tests which reflect the influence of water content, grain size and mineral composition upon the mechanical behaviour of clays and silts: they are the Liquid and Plastic Limit tests (Fig. 9.22). In the Liquid Limit test, the clayey sample is mixed with water to a creamy paste and placed in a shallow brass dish. A V-groove is then cut through the sample and the dish tapped on its base until the sides of the groove just close. A sample of clay is then taken and its water content measured. The remaining clay is remixed to a new water content and the test repeated. This can be done for a number of values of water content. Water content is plotted against the number of blows; the Liquid Limit is the water content at which the groove is closed on the 25th blow. This, as defined by Atterberg, is the water content above which the remoulded material behaves as a viscous fluid and below which it acts as a plastic solid. An alternative method uses the penetration of a cone, the water content corresponding to a penetration of 20 mm after 5 s being the Liquid Limit. The method is extremely quick.

The Plastic Limit is the water content below which the remoulded sample ceases to behave as a plastic material and becomes friable and crumbly. Atterberg's test consists of finding the water content at which it is no longer possible to roll the clay into an unbroken thread of about 3 mm diameter.

These tests have a definite relationship to mechanical behaviour, they are used for classifying soils for engineering purposes (British Standard, 1377; 1975).

Liquidity Index links the Atterberg Limits of a soil to its natural water content in the ground, i.e.

$$\text{Liquidity Index} = \frac{\textit{in-situ}\text{ water content} - \text{Plastic Limit}}{\text{Plasticity Index}}$$

where the Plasticity Index is the difference between the Plastic and Liquid Limits (see Fig 9.22). The Liquidity Index may be:

(*i*) greater than 1.0; i.e. the *in-situ* water content is greater than the Liquid Limit of the soil, which is thus extremely weak.

(*ii*) between 1.0 and zero; i.e. the *in-situ* water content is at or below the Liquid Limit.

(*iii*) less than zero; i.e. negative when the *in-situ* water content is less than the Plastic Limit and the soil is like a brittle solid.

Activity is the Plasticity Index of a soil, divided by the percentage, by weight, of soil particles within it that are less than 2 μm in size (1 μm = 0.001 mm). Particles of this size are usually clay and Activity provides an index of their plasticity and mineralogy (Fig. 11.13).

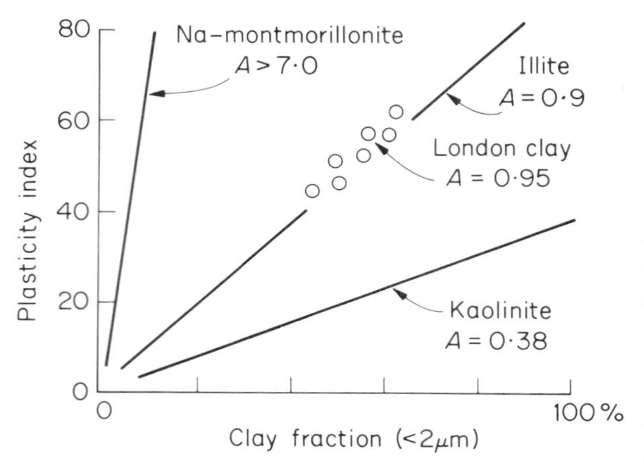

Fig. 11.13 Relationship between plasticity index and proportion of clay fraction in a sample. A=Activity of sample (Skempton, 1950).

Shrinkage measures the reduction in dimensions of a soil specimen with loss of water, and indicates the water content of the Shrinkage Limit (Fig 9.22). Linear and volumetric shrinkage may be measured (Soil Mechanics for Road Engineers, 1952).

Descriptions and classifications

The description and classification of geological materials are related but separate tasks.

If there is only time for one task to be carried out during a site visit then it should be to make a description of the rock or soil present. This will be passed to others to form their own opinion of the site and errors, such as a confusion between silt and clay (see p. 122) can result in serious misconceptions about the performance of a material.

Correct description is so important that strict guidelines for it have been adopted by many countries: see for example British Standard, 5930 (1981). The elements of soil and rock descriptions are given in Tables 10.4 and 10.5.

Few of the descriptive parameters available are used in soil and rock classification but those used are augmented by laboratory measurements of selected mechanical properties. This combination of descriptive and determinative procedures allows a prediction of the likely behaviour of the material to be made; it can also be a link with practical experience in other sites involving soils and rocks that have been similarly classified.

Soil classification

The basic components of a soil classification are shown in Table 11.6. Sub-groups based upon the grading of particle sizes describe the various mixtures that can occur in sediments, e.g. silty clay (precise limits and symbols are shown in Fig. 11.14). Silt and clay are further divided by their plasticity, as in Fig. 11.15. Here Plasticity Index is plotted against Liquid Limit. Silts plot at the bottom and right-hand part of the diagram and clays in the top and left. Also, the products of similar geological processes, e.g. marine sedimentation, or glacial deposition, plot as a diagonal distribution from botton left to top right (Fig.

Table 11.6 Basis for soil classification (from British Standard, 5930)

Category	Soil groups	Sub-groups	Lab. test
INORGANIC	Coarse	Gravel Sand —35% fine	Grain Size (Fig. 11.3)
	Fine	Silt —65% fine Clay	Atterberg Limits (Fig. 11.15)
ORGANIC	Peat, coal, lignite		Botanical and Petrological

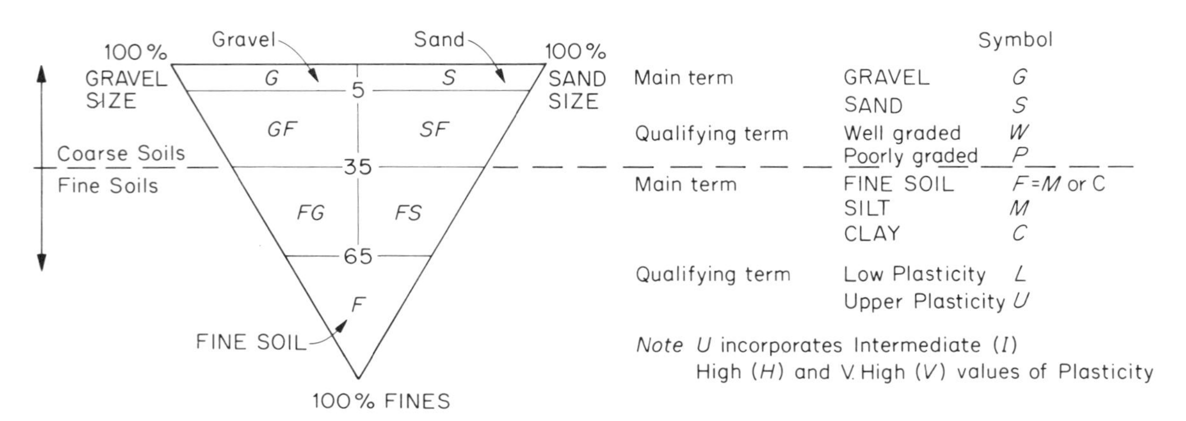

Fig. 11.14 British system for description of soils finer than 60 mm. GF=v. clayey or silty GRAVEL; FG=gravelly FINE SOIL; SF=v. clayey or silty SAND; FS=sandy FINE SOIL.

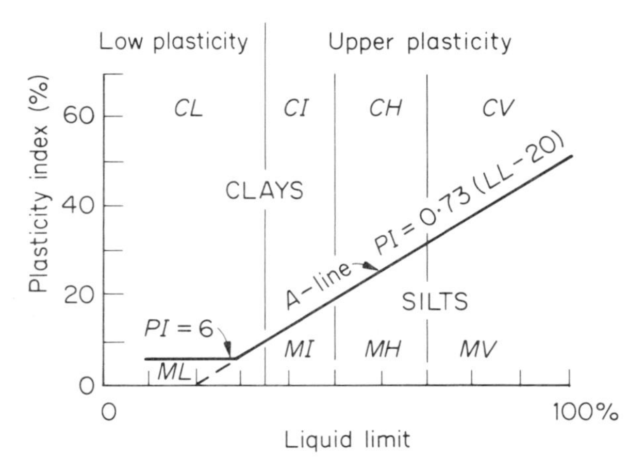

Fig. 11.15 Casagrande's Plasticity Chart. C=clay, M=silt (see Fig. 11.14 for symbols).

11.16). Casagrande (1948) was impressed by this distribution and suggested drawing a line on the graph that separated clays from silts, the A-line: it defines the Plasticity Index as 0.73 (Liquid Limit −20) and has been adopted ever since.

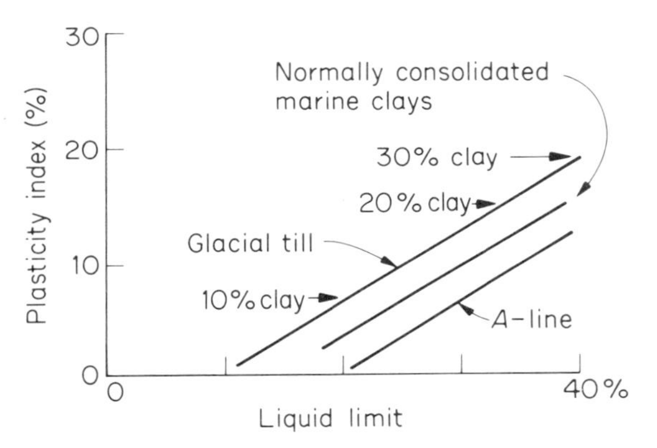

Fig. 11.16 Relationship between plasticity and sediment type (after Boulton and Paul, 1976).

Rock classification

Rock has no classification equivalent to that for soil. Bedding, faulting and jointing in rock create surfaces about which the rock may separate and these have a considerable influence upon its behaviour. For example, open fractures can turn an impermeable shale into an aquifer and a strong granite into a compressible bedrock. Only when such features are closed under pressure at depth will the mechanical characters of a body of rock approach those of the rock itself. Figure 11.17 illustrates a simple classification for rocks based upon the Young's modulus and uniaxial compressive strength of the rock material (Deere, 1968).

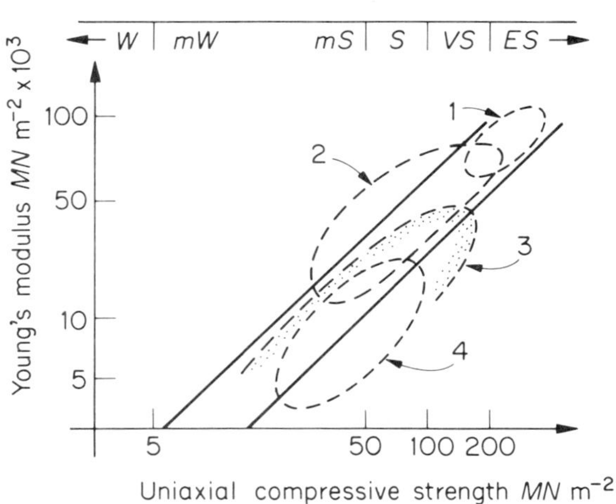

Fig. 11.17 Tendencies of certain rock types. 1. Strong fine grained rocks, e.g. dolerite and limestone. 2. Quartz rich rocks, e.g. granite, gneiss, quartzite. 3. Sandstones. 4. Clay and mica rich rocks, e.g. shale, schist.

Selected bibiography

British Standard: BS1377 (1975). *Code of Practice for Methods of test for soil for civil engineering purposes.* British Standards Instn, London.

American Soc. Testing and Materials: Annual Book of Standards, see current Parts for: Natural Building Stones, Soil and Rock, Peats, Mosses and Humus. (These annuals and Parts are periodically revised) Philadelphia.

Vickers, B. (1982). *Laboratory Work in Soil Mechanics* (2nd Edition). Granada Pub. Co., London, New York.

Liu, C. and Evett, J. (1983). *Soil Properties, Testing, Measurement and Evaluation.* Prentice Hall Int., Hemel Hempstead, England.

Brown, E.T. (1981). *Rock Characterization, Testing and Monitoring.* Int. Soc. Rock Mechanics (ISRM) suggested methods. Pergamon Press, London, New York.

Int. Jl. Rock Mechanics and Mining Sciences and Geomechanics Abstracts (1983). Suggested Methods for Determining the Strength of Rock Materials in Triaxial Compression, **20,** 285–90.

12

Geological Maps

Geological maps record a distribution of features on the ground as if they had been observed from above. Some of the features shown on geological maps may be concealed, e.g. by a cover of vegetation or drift. The maps become less particular in such areas, and this should be borne in mind when reading them; however, it is usually a matter of local detail and the general accuracy of the maps is rarely in question. This is because a distinction is made, on geological maps, between the features which have been seen and those which have been inferred. For example, the *known* position of a boundary is normally shown by a solid line, whereas a broken line is used for its *inferred* position.

It is therefore important to read the symbols that have been used on a map before studying it; they are usually found in the margin, together with a stratigraphical column and other information concerning the geology of the area represented. References may also be given to an explanatory report or memoir in which geological details noted in the field, such as thickness, anisotropy, grading of materials and so on, may be recorded. Maps cannot be interpreted fully without these reports. The colours or symbols shown on standard geological maps, for example, need not necessarily indicate material of uniform character, as such codes are normally used to define stratigraphical boundaries rather than physical characters. Valley deposits are often shown by one colour or one symbol, but reference to a memoir may reveal that they contain a variety of materials, e.g. sands, gravels, clays, and organic matter.

Published geological maps vary in scale and the most detailed map normally produced has a scale of about 1:10 000. In the British Isles the nearest maps to this are 1:10 560 (six inches to one mile and are commonly called the six inch maps), but the 1:10 000 series is now replacing them. These are the base maps that are normally used for detailed field mapping, and the geological maps produced on them should be consulted by engineers whenever possible; many of the smaller scale maps are simply reductions and hence simplifications of these larger scale records. Base maps having a scale that is greater than 1:10 000 are normally used only when it is necessary to map a small area in considerable detail, e.g. a particular foundation or cutting.

The next smaller scale is around 1:50 000. These are useful for assessing the general geology of a region and should be consulted to appreciate the regional setting of an area; this is important in the fields of mining and ground-water. The maps can also be of great assistance to the location of mineral reserves and construction materials.

Maps on scales which are appreciably smaller than 1:50 000 are normally of little use to most engineers. There are three smaller scales which are commonly employed for geological maps, namely 1:200 000, 1:500 000, and 1:1 000 000. These maps illustrate regional tectonic patterns (of interest to engineering seismology) and general geology.

Frequently-used maps

Maps which show the distribution of the various rocks and soils that occur close to the surface of the ground, are most useful for civil engineering work and may be available in 'Solid' and 'Drift' editions. Maps of subsurface geology are needed by mining engineers and others whose engineering work is located at depth.

Solid and Drift editions

The Drift edition is strictly a map of surface geology because it shows the position, and the general character, of all geological materials that occur at ground level. These will include not only the harder rocks, but also such materials as alluvium, glacial drift, mud flows, sand dunes, etc., which conceal the more solid rocks beneath them. It is because these materials have been transported, or drifted, to their present position that the maps which record them are called Drift edition maps.

Not all drift at ground level is thick, indeed much of it can be quite thin, and so it is often easy to construct a map which records the geology beneath the drift, as if the drift had been physically removed. When no drift is shown on a map, only the solid geology, it is called a Solid edition map.

Thus the geology of an area can be studied using two maps, one showing the drift and the solid geology between areas of drift, and another showing only the solid geology as it occurs between, and as it is thought to occur beneath the areas of drift. It is important to check the title of a map, where the edition either Solid or Drift is stated. There is so little drift in some of the areas that the boundaries of both solid and drift geology can be clearly shown on one map which is then called a 'Solid and Drift' edition.

Exposure and outcrop

Rocks are said to be *exposed* when they occur at ground level. An exposure should be distinguished from an outcrop which is the geographical position of a geological unit, regardless of whether or not it is exposed. Figure 12.1 illustrates the two terms and shows that every exposure is part of an outcrop. By studying the geographical distribution of outcrops at ground level, as is shown by a map of surface geology, it is often possible to determine the overall geological structure of the area represented.

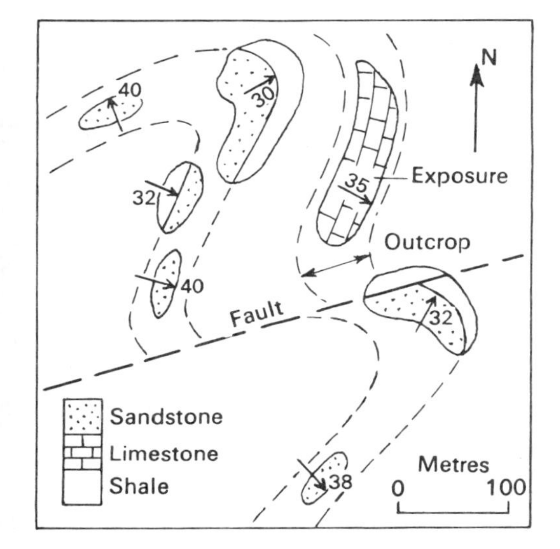

Fig. 12.1 A geological map of exposures and estimated outcrop boundaries (broken lines).

Thickness of strata

Unfortunately the meaning of the word 'thickness' varies with its usage. For example, a mining engineer sinking a vertical shaft through the strata shown in Fig. 12.2*a* would reckon its thickness as t_v, whereas the same engineer driving a horizontal tunnel through the same strata would measure its thickness as being t_h. Geologists usually speak only of *stratigraphic* thickness by which they mean the thickness measured at 90° to the bedding surfaces, t_s in Fig. 12.2*a*. Knowing the stratigraphic thickness and the dip of the strata it is possible to calculate the 'thickness' in any other direction by using the following formula:

$$t_x = t_s . \sec \alpha \qquad \text{(see Fig 12.2}b\text{)}$$

where t_x = the distance required through the stratigraphical unit in a direction x; this need not be the direction of true dip,
t_s = the stratigraphic thickness of the unit,
α = the angle between the direction in which t_s is measured, and the line t_x.

The vertical thickness (t_v), which is encountered in vertical bore-holes and commonly recorded in site investigation reports, is related to the stratigraphical thickness by the formula:

$$t_v = t_s \sec \alpha$$

where α = the angle of dip in the plane containing t_v and t_s, Fig. 12.2*b*).

The thickness of a stratum may also be estimated from its width of outcrop (see Fig. 12.19).

Maps of subsurface geology

The thickness and underground distribution of rock and soil are shown on maps of subsurface geology. These are based upon predictions made from a study of outcrops and of information from bore-holes, well bores, tunnels, mines and geophysical surveys. The maps most frequently used are described below.

Isochore maps

Isochores are lines joining points of equal vertical thickness, so isochore maps record the vertical thickness of geological units. The maps are readily produced using the

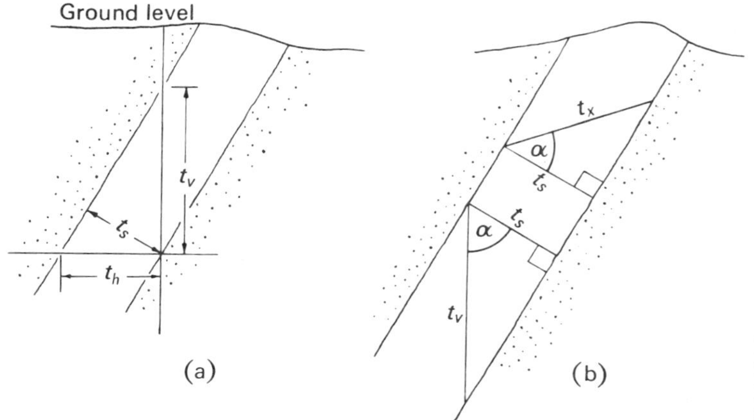

Fig. 12.2 (**a**) Thickness. t_v = vertical; t_h = horizontal; t_s = stratigraphic. (**b**) Relationship between t_s and thickness in other directions.

data obtained from vertical bore-holes which have fully penetrated the units being studied. These maps are often used to illustrate such features as the depth of overburden above some deposit, or the areal variations in the vertical thickness of some concealed unit such as a confined aquifer, a mineral deposit or a zone of weak rock (Fig. 12.3).

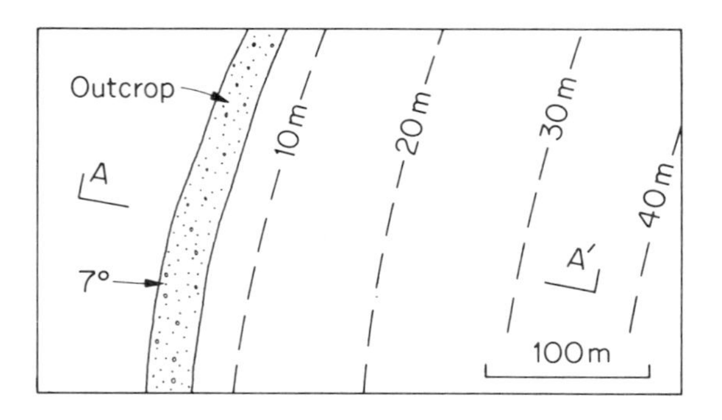

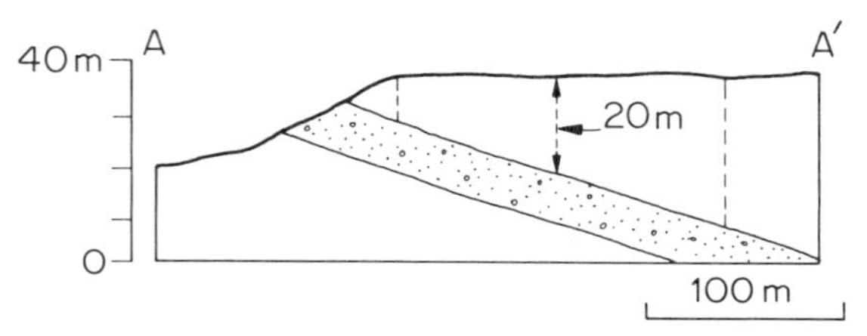

Fig. 12.3 Ischore map for thickness of overburden above a deposit of sand and vertical cross-section.

Isopachyte maps

Isopachyte lines join points of equal stratigraphical thickness and are used to produce maps which are usually of greater interest to the geologist than the engineer. The maps cannot be interpreted as quickly as those of isochores even though the stratigraphic thickness of an horizon can be related to its vertical thickness by the formula

$$t_s = t_v \cos \alpha$$

where α = the true angle of dip (see Fig. 12.2).

Obviously vertical thickness will decrease as dip decreases and will eventually become equal to the stratigraphical thickness when the true dip is zero: Isopachytes then become synonymous with Isochores.

Horizontal-plane maps

These record the geology, as it occurs on a horizontal plane at some level below the surface. The sub-outcrops shown on the maps are influenced by dip and thickness alone, and so dips must be recorded as they cannot be deduced from the sub-outcrop pattern, as with maps of surface geology (see Map interpretation, p. 206). Horizontal plane maps are useful to engineers involved in underground excavations (Fig. 12.4).

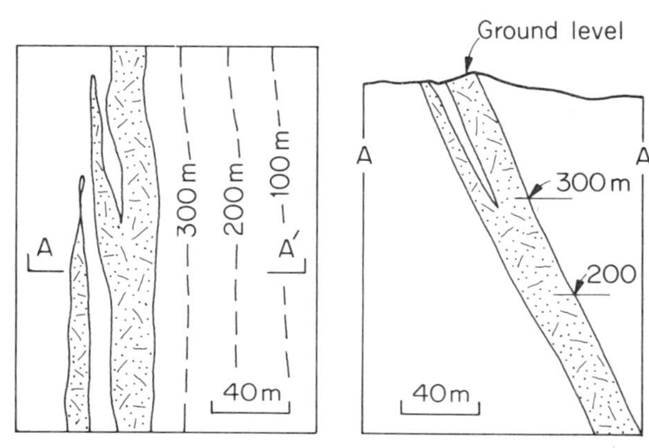

Fig. 12.5 A map showing structural contours (in metres above datum) for the upper surface of a dyke shown in vertical section on the right.

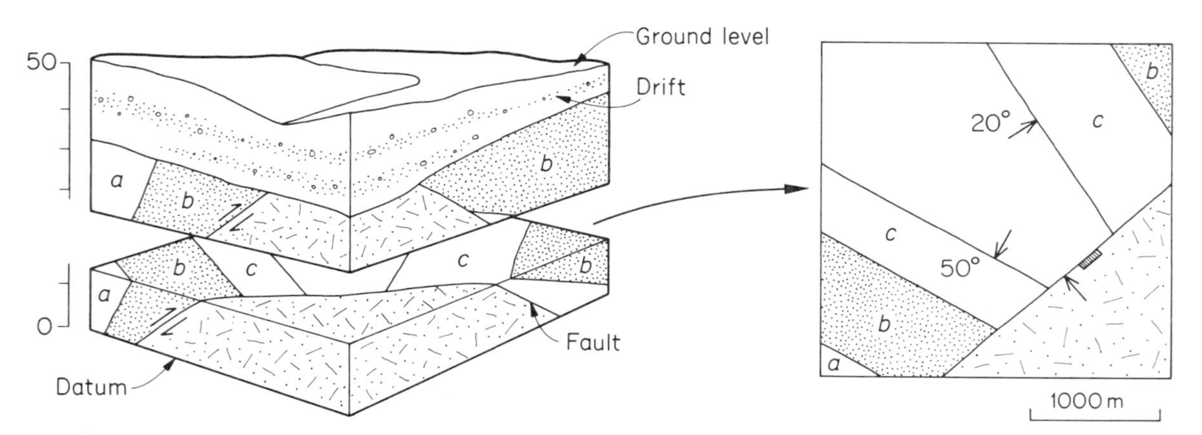

Fig. 12.4 Horizontal-plane map for a surface 15 m above datum (on right) showing an antiform truncated by a fault: note symbol for downthrow side of fault.

Structural contour maps

These record the shape of structural surfaces such as faults, unconformities and the bedding surfaces of folded strata, below ground level. Often they are produced with the aid of bore-hole data. The spacing between the contours is a function of the dip of the surface being illustrated, the contours being closest in those areas where the dip is greatest. The angle of dip can be assessed thus: contour spacing divided by the contour interval = cot d, where d is the dip in the direction in which the contour spacing is measured. These maps are normally used when it is necessary to know the position of a definite structural surface below ground level (Fig. 12.5).

Geophysical maps

The geophysical methods described in Chapter 10 are frequently used to assess subsurface geology and the information obtained from them is often well displayed as a map showing either the physical values obtained, e.g. electrical resistivity, gravitational acceleration, etc., or an interpretation of them. Table 12.1 indicates how seismic velocity may be interpreted. A detailed interpretation of seismic velocity requires prior knowledge of the rock types present: this may be obtained from outcrops and bore-holes. From this data seismic velocity may be correlated with the distribution of rock types and their quality.

Table 12.1 Typical seismic velocities: air and water may fill pores, joints and other voids

Material	Seismic velocity ($m\,s^{-1}$) *Compressional*	*Shear*
Air	330	Zero
Water	1450	Zero
Poor rock	1500	700
Low quality rock	<3000	<1500
Fairly good rock	>3000	>3000
Good rock	>5000	>2500

Figure 12.6 is a map of ground resistivity, high resistivity correlating with poor quality ground and low resistivity (i.e. high conductivity) correlating with better quality ground. By contrast, Fig. 12.7 is an example of a map where the geophysical values are not shown, but an interpretation of them. Each column (1 to 6) represents a particular stratigraphic sequence.

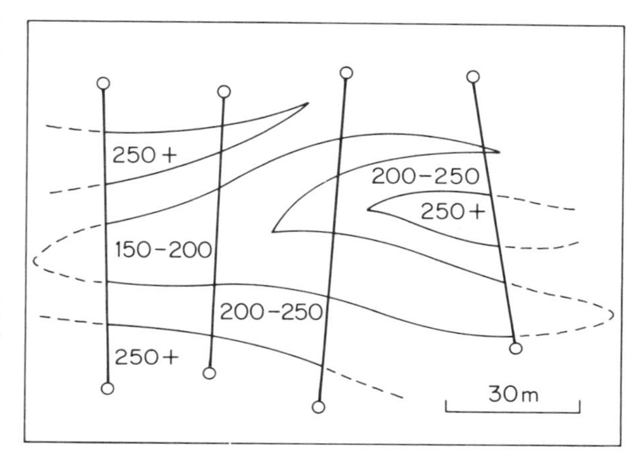

Fig. 12.6 Iso-resistivity map showing four traverse lines along which values for the apparent resistivity of the ground were measured: values in ohm-metres. High resistivity indicates poor rock. (Oliveira *et al.*, 1974.)

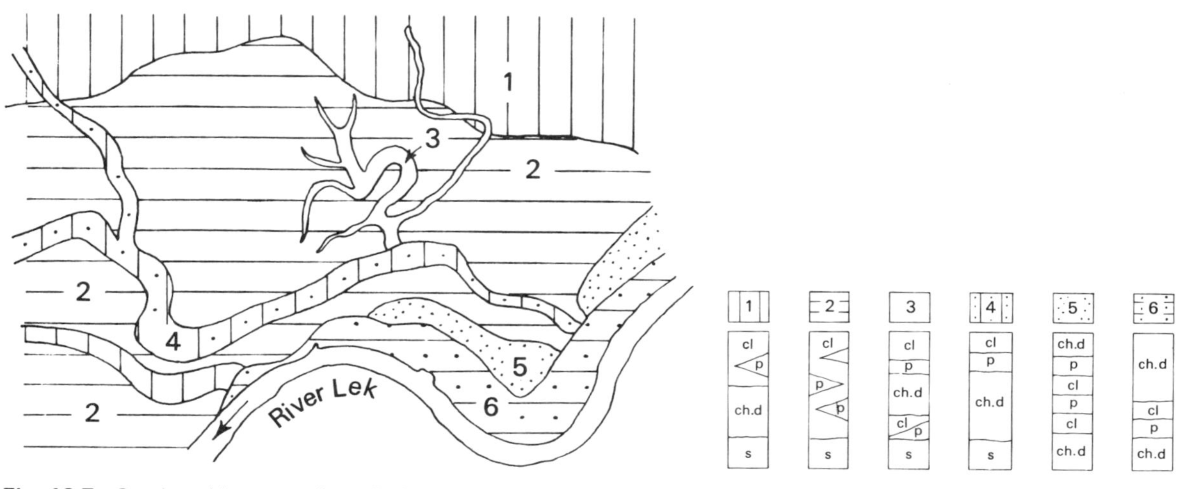

Fig. 12.7 Stratigraphic successions (1–6) exist where shown on the geographical map. cl = clay; p = peat; ch.d = channel deposits; s = sand. (Simplified from a map of the Netherlands, Scale 1:50 000.) (Netherlands Geological Survey.)

Field mapping

Equipment

Normal equipment for mapping includes a topographical map of the area, a map case, soft pencils, compass, clinometer, pocket binoculars and hammer. A hammer head weighing about 0.5 kg is usually sufficient and is made from forged steel so as not to splinter when breaking hard rock (suitable geological hammers can be purchased). The clinometer is used for measuring the dip of surfaces such as bedding and cleavage; it may be a separate instrument, as shown in Fig. 12.8, or incorporated in the compass (as in a Brunton compass). In its simplest form it

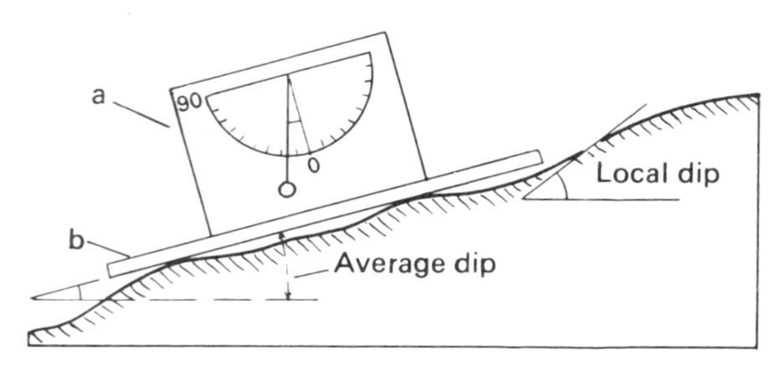

Fig. 12.8 Clinometer (a) with base plate (b) for 'averaging' dips.

consists of a plummet which hangs vertically against a scale, and can be made by mounting a protractor on to a thin plate of material such as plywood or perspex sheeting, with a plum-bob hung from the centre of the protractor circle. The compass is needed for measuring directions of dip or strike; many geologists use a prismatic compass because it also allows the accurate sighting of distant objects, as needed for orienting the map. Other useful equipment includes a surveyor's notebook (although some notes can be written on the map), a straight edge or scale, a pocket lens or magnifier (× 5 is a useful magnification for field work) and a haversack. If air photographs are available, especially when the area to be covered is large, they give a comprehensive view of the ground and are a valuable adjunct. But the geology has to be inspected and mapped on the ground itself.

Mapping

Making the map is usually a painstaking operation, and if it is to be detailed the geologist will walk over the whole area and record every exposure, large or small. When enough information has been obtained it is possible to sketch in the position of geological boundaries, as shown in Fig. 12.1. The map not only records the position of the various rock types, but also the dip of any surfaces exposed. The shape of exposures seen in the field should be represented as closely as possible on the map. All pencil lines and notes are normally inked in each day with a waterproof ink.

The way in which the geologist covers the ground depends on the geology and on the individual. In a complicated area it is often helpful to walk along predetermined traverse lines, noting all that is seen along them, filling in the gaps later whenever appropriate. Such traverse lines are usually taken at right angles to the general strike, when this can be recognized, so that they cross the geological grain of the area. In simpler situations it may be possible to follow a particular geological feature, e.g. a prominent limestone ridge. The exposures seen in the bed and banks of many streams are useful and stream sections are usually studied early in a survey.

Traverse survey

Figure 12.9 is a record of geological features noted by an engineer during a surface traverse along the line of a proposed tunnel. It records the presence of water in the limestone; that rocks of different hardness will have to be excavated; that the dip of the strata changes, indicating that the tunnel will cross either a fault or a fold and that the northern portal may pass through a clay landslide. Such a record is valuable to design engineers when it is available at an early stage in their work.

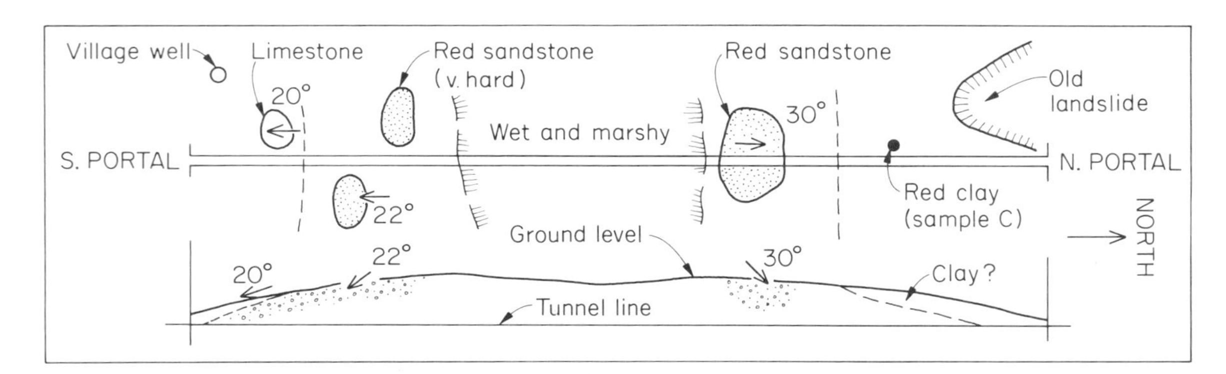

Fig. 12.9 Engineer's record from a surface traverse survey for a tunnel: sketch map, and sketch section below.

Measurement of dip and strike

The true dip of a surface is its maximum angle of inclination from the horizontal. This angle lies in the direction of greatest gradient and dips in any other direction are called *apparent dips* (Fig. 12.10).

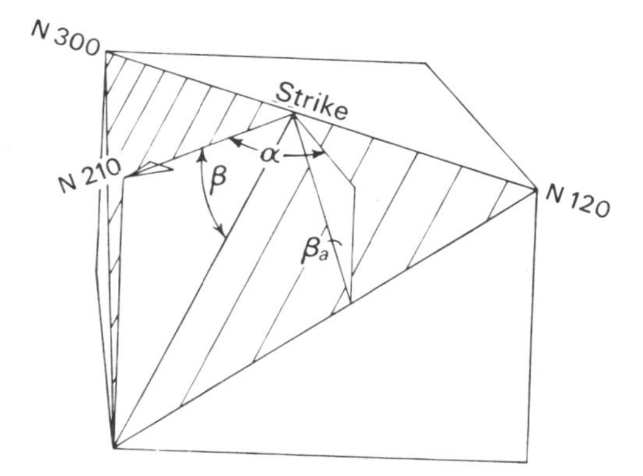

Fig. 12.10 Relationship between true dip and apparent dip. True dip of the shaded surface is β at N 210°. β_a is an apparent dip. Note that $\tan\beta \cos\alpha = \tan\beta_a$.

Dip

Dip is measured with a clinometer (Fig. 12.8), and read to 1° only (West, 1979). The direction of dip is recorded as a compass bearing, e.g. a surface dipping at 60° to the NW, would be recorded in the note book as a dip of 60° at N 315° and shown on the map as an arrow pointing in the direction 315° from North, and having the angle 60° written next to it, as in Fig. 12.1. The true dip of a surface may be used to predict its position at depth (Fig. 12.11).

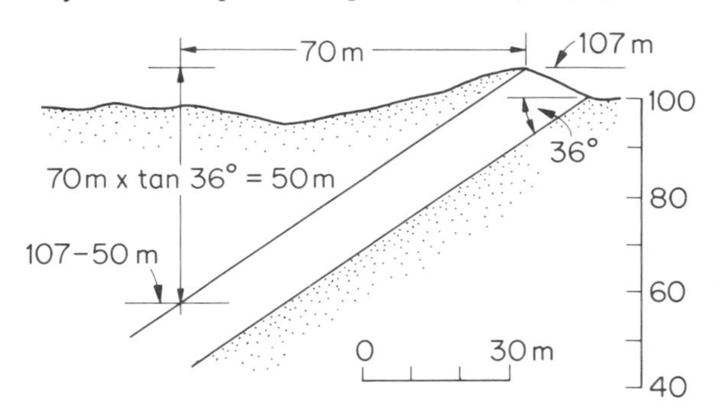

Fig. 12.11 Predicted height above datum for a surface 70 m horizontally from its exposure at 107 m, in a down-dip direction.

Because of the difficulties involved in measuring the true dip of an inaccessible or irregular geological surfaces it is often advisable to use the larger lengths of surface that may be exposed in the sides of pits, quarries, and other excavations; a typical example is illustrated in Fig. 12.12*a*. In the figure an inclined surface intersects two vertical quarry faces, and the inclination of its trace on one face differs from that on the other; the direction of true dip, and hence the angle of true dip, are unknown. At least one of the dips seen on the quarry face must be an apparent dip and it is prudent to assume that both are because the angle and direction of true dip can be calculated from two apparent dips.

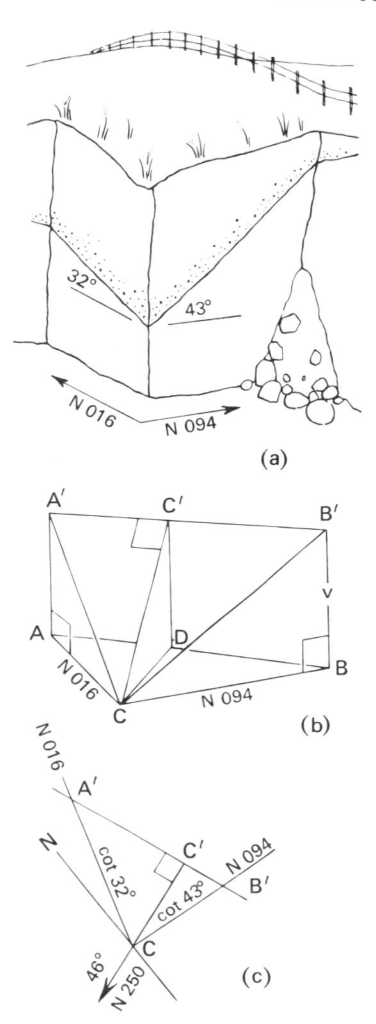

Fig. 12.12 Use of apparent dips for calculation of true dip (see text).

Consider the example shown in Fig. 12.12*a* when redrawn as a geometrical model in Fig. 12.12*b*. To construct this model it is only necessary to measure the two apparent dips, 32° and 43°, and the direction of the surfaces on which they have been measured, namely N 016° and N 094°. Points *A′* and *B′* are both a vertical distance *V* above some horizontal datum, so making the line, *A′B′* a contour for the dipping geological surface. Because the true dip lies in the direction of greatest gradient, lines at 90° to *A′B′* will define that direction; *C′C* is one such line.

The true dip, or inclination of $C'C$ to the horizontal, is conveniently found by making a two-dimensional plan, Fig. 12.12*c*. This diagram can be drawn directly from field data using the following steps:

1 Draw a line to represent the direction of magnetic North.
2 Select a point on this line to represent C.
3 From C draw two lines CA' and CB' so that they are correctly oriented to North. It is then necessary to locate the points A' and B' on these lines.

[One of the commonest mistakes made with this construction is to draw the lines $A'C$ and $B'C$ on the wrong side of the N–S line. The direction of faces A and B in Fig.12.12*a* could equally well have been recorded in the field as N 196° and N 274° respectively. These directions, plotted directly in plan, would have put the triangle $A'B'C$ (to be constructed) on the west side of the N–S line. The angle of dip calculated would be the same as before, but the direction of dip, if measured in the C'–C direction, would be in error by 180°. The construction triangle should point in the same direction as the field exposure.]

4 From Fig. 12.12*b* it can be seen that $AC = AA' \cot 32°$, and $BC = BB' \cot 43°$. Because $AA' = BB' = V$, the actual length of V is immaterial to the construction and can be conveniently taken as unity. The point A' is located a distance equal to cot 32° from C along the line CA', and the point B' a distance equal to cot 43° along the line CB', to some suitable scale.
5 A line drawn at 90° to $A'B'$, i.e. $C'C$, will lie in the direction of true dip, and this direction can be measured directly from the plan, i.e. N 250° in this example.

The angle of true dip is $C'CD$ in Fig. 12.12*b*. The length $DC = DC' \times \cot C'CD$; because $DC' = V = 1$, the equation reduces to $DC = \cot C'CD$. Fig. 12.12*c*. Hence the angle of true dip is that angle whose cotangent is equal to the length CC', i.e. 46° in this example.

This graphical method is usually referred to as the cotangent construction, and it can be used to interpret the inclined surfaces exposed in trial pits, tenches, adits, and similar excavations. Other methods are described by Phillips (1971).

Strike

The strike of a surface is the direction of a horizontal line drawn at 90° to the direction of true dip. The direction N 120° (or N 300°) is the strike of the surface shown in Fig. 12.10 as is the direction of the line $A'B'$ in Fig. 12.12*b*. Because of its horizontality a line drawn in the direction of strike is equivalent to an elevation contour for the surface. Figure 12.13 illustrates how the strike directions of a planar surface can be extended to produce strike lines which are also contours for the surface and are called *structural contours*. However, the majority of geological surfaces are not planar and the unlimited extension of strike lines away from the points at which dip and strike are measured can result in incorrect predictions.

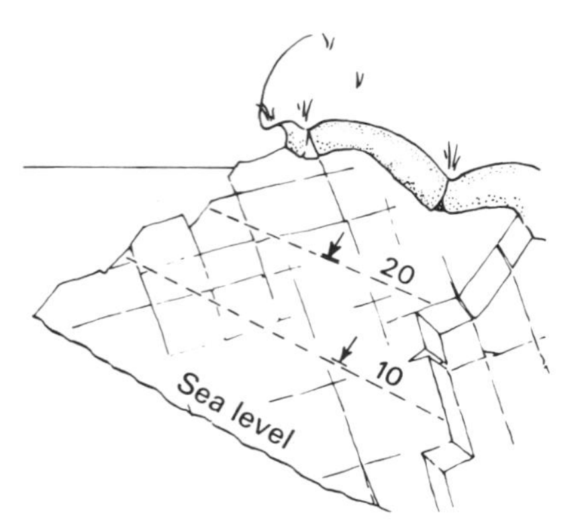

Fig. 12.13 Coastal exposure illustrating that strike lines at chosen elevations can be extended to produce structural contours for a surface.

Constructions for dipping strata

The two constructions described here may be used to predict the likely position of a concealed boundary.

Construction from outcrop

Figure 12.14 illustrates two small valleys between which a hillside exposure reveals the junction of sandstone (dotted) with underlying shales. The dip of this junction at the exposure is 40° at N 160°. In order to predict the approximate position of the junction in the area around the exposure the following construction is used.

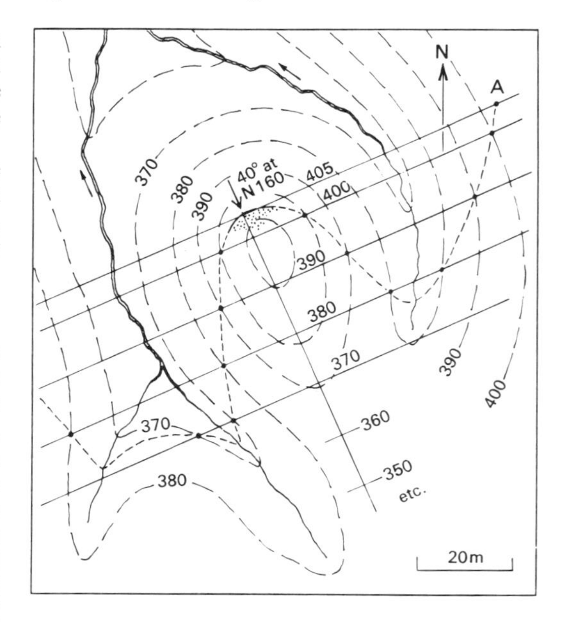

Fig. 12.14 Location of concealed boundary as defined using stratum contours (see text).

A line is drawn on the map in the direction of true dip from the point at which the dip was measured; any lines drawn at 90° to this line will be strike lines for the sandstone-shale junction. The strike line drawn through the point at which the dip was measured will have a value equal to the elevation of the point, in this example 405 m above sea level. By drawing this line it is assumed that the junction between the sandstone and the shale, where present along the line, is 405 m above sea level. It therefore follows that the junction represented by that line will only occur at ground level at points where the strike line intersects the 405 m topographic contour; such a point is shown at location *A* in Fig. 12.14.

It is then necessary to calculate the position of other strike lines and these are usually chosen so that they have the same value as the topographic contours on the base map, i.e. in this example, 400 m, 390 m, 380 m, etc. The separation of the strike lines is calculated from the angle of true dip and the horizontal scale of the map. In this example a dip of 40° is equivalent to a gradient of 1 in 1.19, i.e. 1 m vertical drop in the elevation of the junction occurs in every 1.19 m, measured horizontally in the direction of true dip (see Fig. 12.11); this would be along the bearing N 160°. Hence a drop of 5 m occurs in 5.95 m.

Thus the 400 m strike line can be drawn as a line parallel to the 405 m strike line but separated from it by a horizontal distance equivalent to 5.95 m on the map. Points of outcrop along the 400 m strike line can then be located along the 400 m contour. This construction is continued until the position of the boundary in the area in question is completely predicted. Boundaries which have been located in this manner are usually shown on a map by a broken line to distinguish them from those positions where the boundary was actually seen.

This construction assumes any surface to be *planar* within the area of the map, because the strike lines are shown parallel and the dip is constant. This rarely occurs in practice and the construction should be restricted to small areas. As the predicted position of boundaries should agree with their actual position and hence it is advisable to use all known exposures and dips for the construction.

3-point construction

Figure 12.15*a* shows the position of five 50 m deep bore-holes and Fig. 12.15*b* their representation on a map together with the elevation each penetrated the same stratum, i.e. 15 m above sea level, 37, 42, 26 and 7 m. The dip of the strata may be calculated from this information.

The points are joined by straight lines to produce a pattern of triangles (Fig. 12.15*c*), the elevation of the stratum being known at the corner of each: e.g. triangle ABC intersects the stratum at 15, 37 and 7 m above sea level. It is assumed that the difference in these values is uniform and from this the gradient of the stratum may be found.

For example, the difference in the elevation of the surface in bore-holes *A* and *B* is 22 m, and the horizontal distance separating the bore-holes is 170 m, hence a uniform distribution of this difference along the side *AB* would give a 10 m drop in every 77.3 m, as shown. This process is repeated for each side of every triangle. Contours are then drawn between points of similar elevation on the sides of the triangles, so producing a general contour map of the buried surface.

Having obtained these contours it is possible to calculate the general angle of dip in any direction along the surface, for example along a–b–c Fig. 12.15*c*. Note that the angle of true dip lies in the direction of greatest gradient, i.e. at 90° to the structural contours. The bearing of both true dip, and hence strike, can be measured directly from a correctly orientated map. Obviously the accuracy of any prediction based on this construction is closely related to the number of bore-holes that are used.

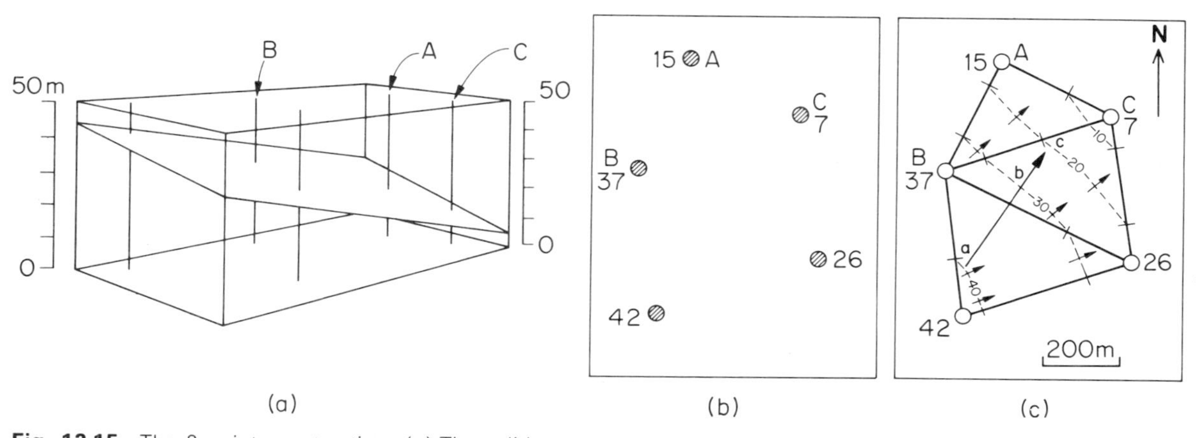

Fig. 12.15 The 3-point construction. (**a**) The solid geometry analysed; (**b**) a map of the bore-hole positions; (**c**) the completed analyses (see text). Broken lines are stratum contours that can be drawn between points of equal elevation. Small arrows = direction of dip. Gradient along a–c = 1 in 20 (2° a–b, 4° b–c).

Map interpretation

The interpretation of a geological map may be made in two stages:

(*i*) identify the relative ages of the rocks (is strata *X* older or younger than strata *Y*?), and
(*ii*) ascertain their structural relationship (is strata *X* above or below *Y*?).

Age relationships

The relative ages of strata shown on a map is given by the list of formations or by the stratigraphic column for the area; this information is normally printed in the margin of the map to a uniform vertical scale. An example of each is illustrated in Fig. 12.16. The divisions in a column are normally based on stratigraphy and lithology, and are arranged so that the oldest strata are always at the base of the column. The thicknesses of lithological formations shown in a column are usually the average stratigraphical thicknesses in the area of the map. Marked local variations are recorded by tapering bands in the column, Fig. 12.16*a*. If there are significant variations in the thickness of formations, over the area of the map, it may be necessary to give more than one stratigraphical column. Stratigraphic columns should not be confused with vertical sections because the latter are obtained from bore-holes, well bores, cliff sections and similar exposures, and the thickness of the strata shown on them is the actual thickness of the strata penetrated at a particular location: the stratigraphic column simply represents the stratigraphical position of all the horizons shown on the map. Vertical sections which reveal gaps in the stratigraphic column

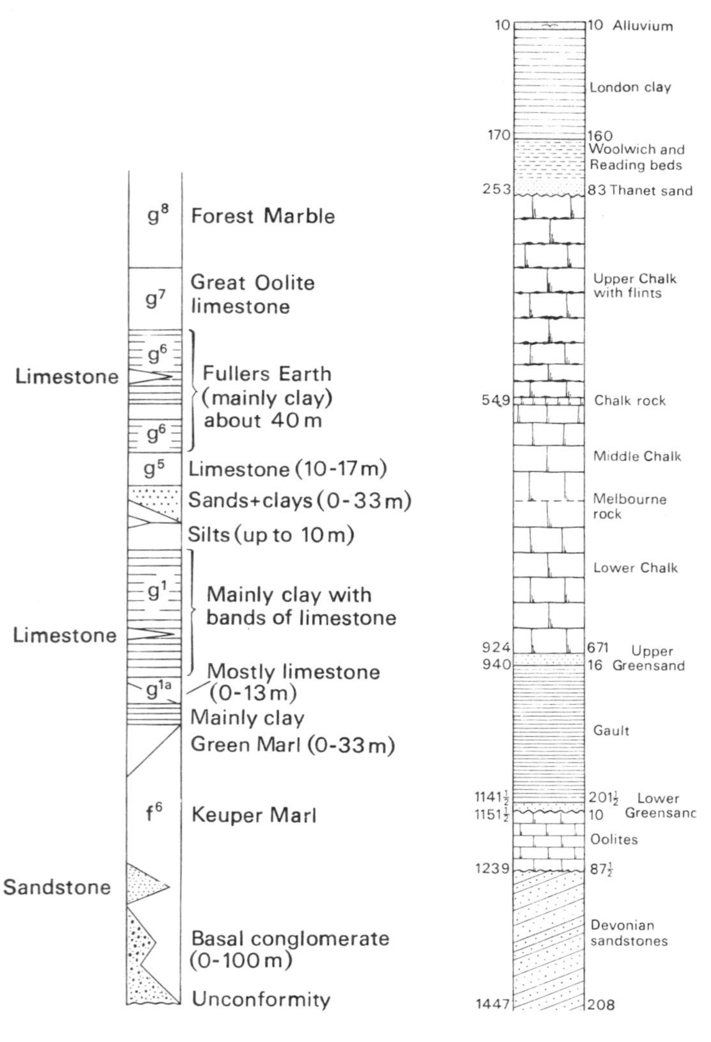

Fig. 12.16 (**a**) Stratigraphic column applicable to an area represented on a map. (**b**) Vertical column recording the strata encountered in a bore-hole: depth below ground level on left, thickness on right. Unconformity represented by wavy line.

indicate the presence of either unconformities or faults. A typical vertical section is illustrated in Fig. 12.16*b*. In these sections it is customary to write the vertical thickness of the strata on one side of the column and the depth below ground level on the other. The positions of recognizable unconformities (e.g. as shown in Fig. 2.2) are usually emphasized by wavy lines.

The relative ages of strata may also be deduced directly from the map by applying the principle of superposition (see Chapter 2). Younger strata will always rest on older strata unless the succession of strata has been inverted by folding. Similarly strata intersected by faults and dykes must be older than these structures: examples of this are shown in Fig. 12.17 where the sequence of events recorded is as follows:

1 Deposition of succession from dolomite (oldest) to tuff.
2 Folding of strata, faulting (Fault 2) and dyke intrusion.
3 Uplift, erosion and deposition of conglomerate.
4 Faulting (Fault 1), the latest geological event that can be deduced from the map.

Unconformities

An uncomfortable surface is a discordance which separates an underlying rock group from an overlying group and represents a period of uplift and erosion (see Chapter 2). When seen on a map an unconformity often appears as a stratigraphical boundary whose course is usually unrelated to the outcrop of the geological structures beneath it. The conglomerate in Fig. 12.17 lies unconformably on the folded rocks beneath and the limit of the conglomerate marks the outcrop of the unconformable surface at ground level. Because unconformities can be considered as surfaces it is possible to determine whether they are either horizontal or inclined. That shown in Fig. 12.17 is horizontal because its outcrop is parallel to the topographic contours; had it been inclined it would V in the direction of dip when crossing valleys and other topographic depressions, such as road cuttings (see Fig. 8.3).

Outliers and inliers

An *outlier* is an outcrop which is completely surrounded (in plan) by rocks of a greater age. The conglomerate within the limits of boundary *B* in Fig. 12.17 is an outlier. As noted in Chapter 8, outliers are usually found close to escarpments (Fig. 8.2). An *inlier* is an outcrop which is completely surrounded by younger rocks. As the dolomite in Fig. 12.17 is older than the surrounding mudstone its outcrop forms a faulted inlier in the core of the antiform. Inliers are sometimes developed in valleys where streams have cut down and locally exposed, in the valley floor, rocks which are older than those forming the sides of the valleys.

Structural relationships

The structure of an area may be deduced from (*i*) the shape of outcrop boundaries, (*ii*) the width of outcrops, (*iii*) the dip of strata, and (*iv*) faults.

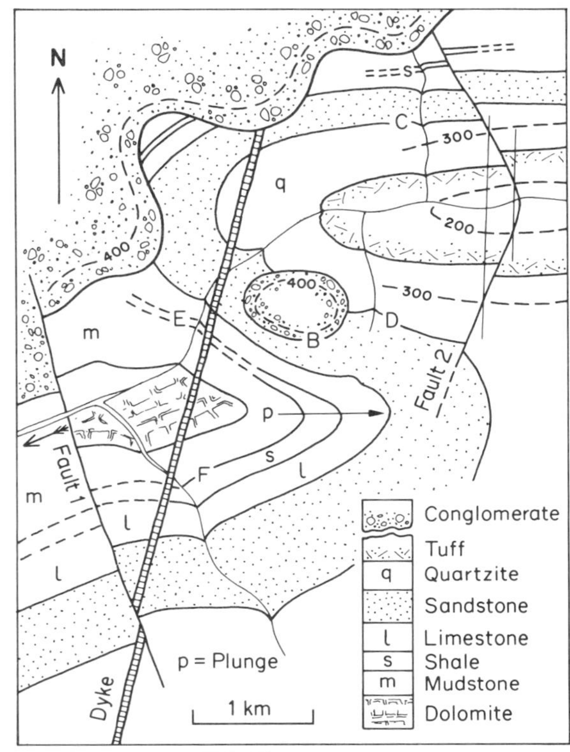

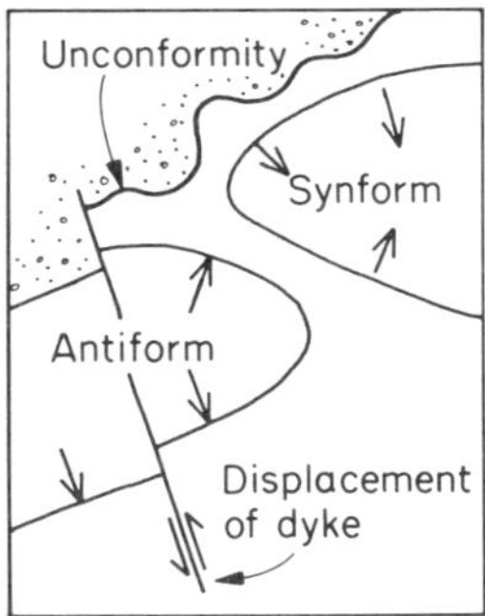

Fig. 12.17 Examples of outcrop patterns commonly seen on maps of surface geology, and sketch of the structural units shown. There are two folds: an antiform and a synform. They have been cut by Fault 2 and intruded by a dyke before being exposed by erosion and covered unconformably by conglomerate. P = plunge of the fold.

Shape of outcrop boundaries

Certain structural characters are easily distinguished; for example, vertical surfaces such as the sides of the dyke in Fig. 12.7 have an outcrop that is unaffected by topography whereas horizontal surfaces, such as boundary *B*, always outcrop parallel to the topographic countours. Vertical and horizontal surfaces can therefore be located by simple inspection.

Many geological surfaces are inclined; the boundary

shown dipping at 40° in Fig. 12.14 is typical. The outcrop pattern of this boundary, as it crosses the valleys, points in the direction of dip. (Check this by looking at the strike line values and the direction of dip as shown by the dip arrow.) This V-ing is characteristic of the outcrop of inclined surfaces intersecting topographic depressions. Fault 2 in Fig. 12.17, is another example. Inclined surfaces and the direction of their inclination can thus be readily seen.

Observations of this kind form the basis for any interpretation of geological structure because once the dip directions are known it is possible to suggest structures that will account for them. For example, at *C* in Fig. 12.17 the dip of the northern outcrop of sandstone is towards the south, whereas at *D*, it is to the north. This sandstone outcrop is part of an east–west trending synform. Similarly the boundary to the outcrop of shale dips to the north-north-east at *E* and to the south-south-east at *F*, and is part of an antiform which is plunging towards the east. Hence if the directions of dip can be deduced from a map the presence and shape of fold structures becomes apparent. Note that isoclinal folds have limbs which dip in the same direction (Fig. 12.18).

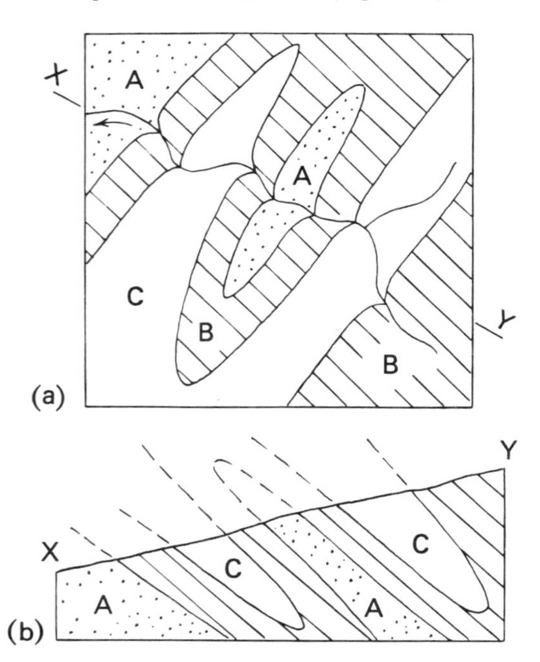

Fig. 12.18 Isoclinal folds in plan (**a**) and section (**b**).

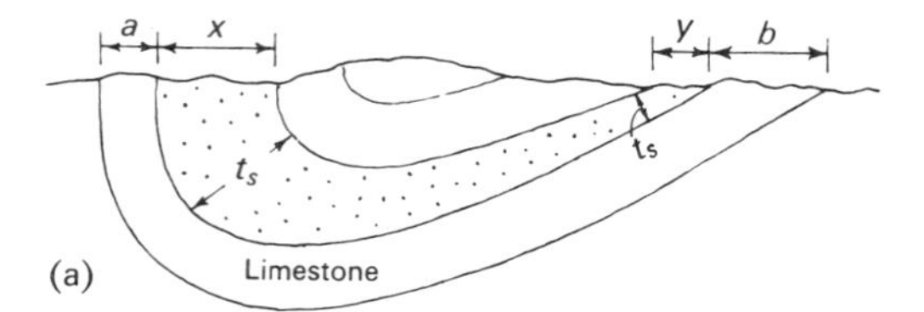

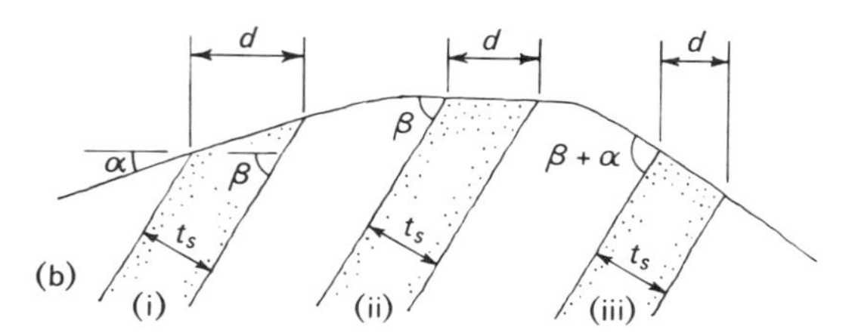

Fig. 12.19 (**a**) Outcrop width as a function of both dip and stratigraphical thickness. (**b**) Outcrop width (d) as a function of geological dip (β) and slope of ground (α).
(*i*) $d=t_s.\text{cosec}(\beta-\alpha).\cos\alpha$. (*ii*) $d=t_s.\text{cosec}\,\beta$.
(*iii*) $d=t_s.\text{cosec}(\beta+\alpha).\cos\alpha$.

Width of outcrop

From Fig. 12.19*a* it is apparent that outcrop width is in some measure a function of the dip of the strata, the wider outcrops occurring in the areas of smaller dip, as is the case with the limestone. If the stratigraphical thickness of the strata is fairly uniform the symmetry of a fold can often be postulated by comparing the widths of the outcrop on the fold limbs. The width of the sandstone outcrop on the limbs of the synform in Fig. 12.17 is fairly constant and it follows that if the sandstone has a uniform stratigraphic thickness the synform must be upright and nearly symmetrical. In contrast, the outcrop of the sandstone on the southern limb of the antiform is considerably wider than on its northern limb, suggesting that the fold is asymmetrical and that its axial surface dips towards the south.

This is a simple and quick way of assessing the fold structures of an area, but there are two possible sources of error, namely excessive variations in topography and stratigraphical thickness.

Considering topography, it is evident from Fig. 12.19*b* that the outcrop of strata of uniform stratigraphical thickness and dip can vary as a function of the shape of the ground surface. Care should therefore be taken when considering the outcrop of geological units whose stratigraphic thickness is appreciably smaller than the variations of topography. Compare this situation with that of the limestone shown in Fig. 12.19*a*. Here the stratigraphical thickness of the limestone is larger than the variation of the topography on its outcrop and outcrop widths (*a*) and (*b*) can be used as an indication of relative dips. The sandstone horizon in Fig. 12.19*a* illustrates the effect of variations in stratigraphic thickness. Here the greater outcrop (*x*) occurs on the steeper limb of the fold. Significant variations of stratigraphical thickness are usually noted in the stratigraphic column for the map.

Faults

Two faults are shown in Fig. 12.17. Faults, as already discussed in Chapter 8, are surfaces or zones about which some displacement has occurred, and the outcrops on either side of a fault are usually shifted relative to one another. This is shown in Fig. 12.17 by the sandstone outcrop in the synform and the faulted unconformity in the region of the antiform; other examples are illustrated in Chapter 8.

Every normal or reverse fault has two characters which should, if possible, be determined, namely its dip and its downthrow side; sometimes both can be assessed from a map. The outcrop of an inclined fault surface will V in the direction of dip when crossing valleys. Strike lines can be sketched, as in Fig. 12.14, and used to confirm both the direction and dip of the fault surface. Two such lines are shown in Fig. 12.17. The outcrop of vertical faults will be unaffected by topography whereas that of horizontal, or low dipping faults (i.e. thrusts) will run parallel or almost parallel to the topographic contours.

When the movement on a fault displaces strata, rocks of one age are brought adjacent to those which are older or younger. Reference to the fault diagrams in Chapter 8 will confirm that the younger rocks of a series are always found on the downthrow side of a fault which cuts uninverted strata and has a vertical component of movement, i.e. a normal or reverse fault.

The fault displacement, e.g. the components of throw and heave and strike slip, can often be calculated by measuring the distance between points which were coincident prior to faulting.

Geological sections

Although a considerable amount can be inferred from maps of surface geology it should be remembered that they are only two dimensional plans and give limited information about the geological variations which occur below the surface. One way of partially overcoming this problem is to construct a number of maps to record the geology which exists at different levels below ground level and to stack them one above the other so as to produce a layered model. Such models are commonly used by mining geologists to assess the size and shape of mineral lodes. A less informative, but much easier method of illustrating subsurface geology consists of drawing one or more geological sections: Figs 8.4 and 10.4 are typical examples.

Drawing a section

The first step in drawing a section is to obtain the ground profile along the line of section; this is done by laying a sheet of section paper along the line of section on the map and marking on its upper edge the separation of the various topographic contours (Fig. 12.20*a*). These points are then plotted on the section paper against a suitable vertical scale and joined by a smooth line which will represent ground level (Fig. 12.20*b*). It may be necessary to extrapolate between these points in order to accommodate local variations in topography which may occur between the contour intervals recorded.

The section paper is then re-positioned on the map as before (Fig. 12.20*c*) and the geological boundaries marked on its edge, with due regard being paid to the direction in which the boundaries are dipping. These positions are then transferred onto the topographic profile and the boundaries extrapolated downwards so producing a sketch of the geological structure as it probably occurs below ground level (Fig. 12.20*d*). The accuracy of such sections can be greatly improved if bore-hole logs and similar data from shafts and well bores can be incorporated.

Special care should be taken with two features when either constructing or interpreting geological sections: (*i*) the orientation of the section with respect to the direction of true dip at any point, and (*ii*) the scale of the section. It is common practice to orientate a line of section so that it shows the geology that would be exposed on a vertical planar surface which lies in the direction of true dip. However, it is often necessary to draw sections in other directions and in these cases they will display apparent dips. True and apparent dips which occur on vertical planar surfaces are related to each other according to the formula given in Fig. 12.10.

The angles of dip shown on sections will also be a function of the horizontal and vertical scales that are

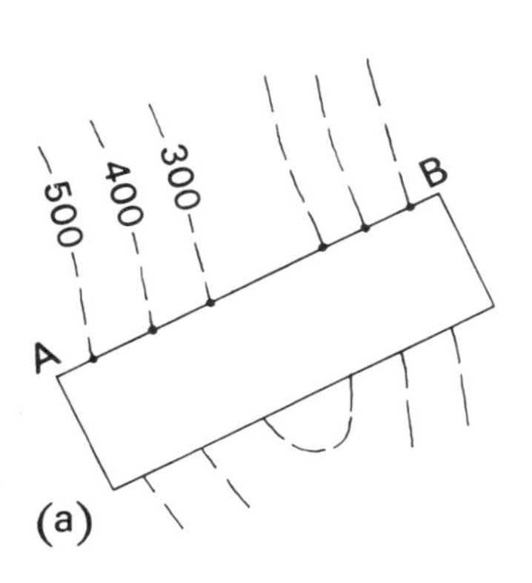

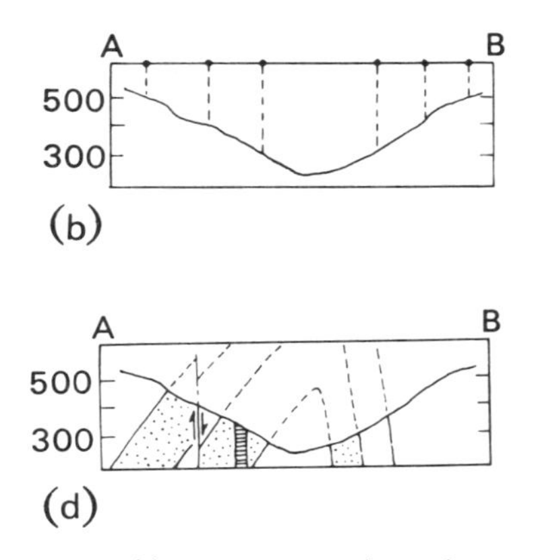

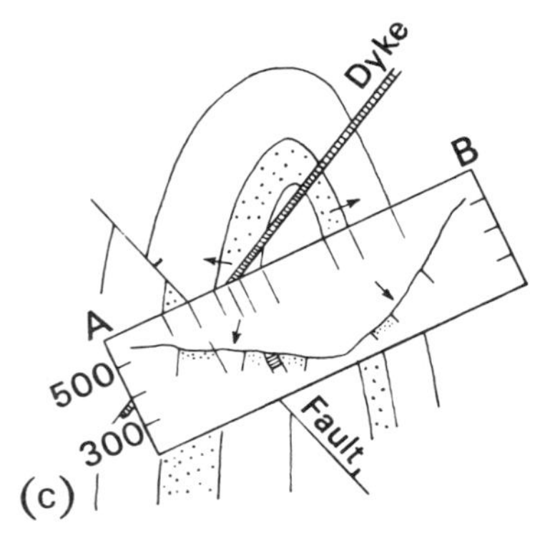

Fig. 12.20 Drawing a geological section: topographic contours not shown in stage (c). See text.

used. It is customary to make the horizontal scale of a section the same as the horizontal scale of the map from which it was constructed; in this way a comparison between the two is facilitated. However, in order to illustrate details such as variations in the thickness of the strata, it is often necessary to make the vertical scale larger than the horizontal. Dips shown on such sections are not those which exist in the ground; they can be calculated by converting the gradient of any surface shown on the section into an equivalent angle. A further consequence of having one scale greater than the other is the convergence or divergence, on the section, of the upper and lower boundaries of geological horizons. This effect varies with the angle the geological horizon makes with the axis of exaggerated scale, as shown in Fig. 12.21.

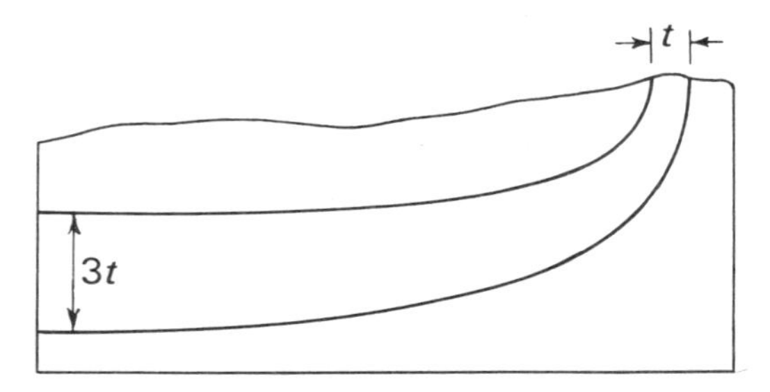

Fig. 12.21 Effect of exaggeration of vertical scale on section. (Here vertical scale = 3 × horizontal scale.) Note apparent change in thickness of a stratum from the position where it is vertical.

Geotechnical sections

Measurements for geotechnical calculations, such as for slope stability, ground treatment, foundation settlement, and volumes of material, are best made on larger sections drawn with the horizontal scale equal to the vertical scale. In this way errors of dip, direction and length arising from scale distortion are avoided.

Thematic maps

A thematic map limits the information presented so that its implications for the particular interests of users are expressed in a form they can appreciate. Thus instead of providing town planners with maps of solid and drift geology, they could be given a map showing where stable ground suitable for conventional foundations, and unstable ground, are situated.

Thematic maps used in geology are of two types: (*i*) those that select some aspect of the geology, this being described as an 'original attribute' (Varnes, 1974), (*ii*) those that show some combination of the original attributes, the characteristic resulting from this summation being called a 'derived attribute'.

The production of thematic maps has been greatly assisted by the use of air surveys, including those from satellites. Sensitive electronic equipment on board these craft detect and measure the various spectra of electromagnetic radiation being naturally emitted from and reflected by the Earth. These can be correlated with such features as rock and soil type and moisture content. The equipment also converts the measurements of each spectra into digits whose numerical combination permits different parts of separate spectra to be added. In this way portions of the great range of spectra measured may be summed to give the most diagnostic value for a particular character of the ground. This has been clearly illustrated by Beaumont (1979) with reference to the search for construction materials. When these values are plotted as a map they produce an *image* of the ground characteristic of interest. Many thematic maps have been prepared in this way. For detailed work aircraft imagery and photography should be at scales of 1:40 000 or larger, preferably at least 1:25 000 (Colwell, 1983).

Maps of resources

These show the distribution of particular minerals, rocks and soils: maps of oil and gas reservoirs and of groundwater supplies are also maps of resources. The earliest example of such a map is that published in 1815 by the engineer and surveyor, William Smith. Smith was commissioned to construct canals in England and to seal these structures he lined them with clay which could be trodden to an impermeable state: this was called puddle-clay. Smith surveyed the neighbouring countryside for clay of this quality and produced a map of its outcrop. From this beginning and from similar mapping elsewhere to supply other commodities, especially limestone, William Smith was eventually able to publish a complete map of the strata of England and Wales.

Modern maps of resources may show the location of rock suitable for crushing and use as aggregate, of supplies of natural aggregate such as sand and gravel, of laterite for road building and of soil for bulk fill. Economic reserves of valuable metalliferous and non-metalliferous rocks will also be mapped. In addition to showing the location of these reserves the maps may also record the quality and quantity of the materials, the thickness of the ground that covers them and has to be penetrated by shafts or removed by stripping to recover them, and their position in relation to ground-water level.

Maps of resources illustrate one of the problems associated with the use of thematic maps, namely that thematic maps are unsuitable for purposes other than those for which they were produced. For example a civil administration may wish to keep areas containing valuable resources of sand and gravel, free from urban development. A thematic map showing the location of these resources is unlikely to reveal that the pits excavated to extract the ballast will require side slopes of 30°, so placing their boundaries beyond the economic limits of the reserve, shown on the map, or that extraction will lower the levels of water in the surrounding ground. The administration requires additional thematic maps.

Derived maps

The development of these maps has been largely inspired by engineering for military purposes. When assessing the strategic and tactical qualities of an area it is necessary to know the response of the ground to vehicle loads, excavation, bombardment and so forth. These, and other qualities, must be derived from a knowledge of the original attributes of the ground, e.g. its strength and permeability. Figure 12.22 illustrates how maps of such derived attributes are compiled.

In this example a map is required that delineates wet weak ground from dry strong ground (Fig. 12.22*a*). A matrix of the characteristics wet, dry, weak and strong shows that six combinations are possible in theory: two are mutually exclusive, two are not required (namely wet strong ground and dry weak ground) and two are those required (Fig. 12.22*b*). A map that illustrates the distribution of these combinations is produced by superimposing one diagram on another (Fig. 12.22*c*): this may be simplified, if desirable, to give the two classes required plus areas not included in the two clases (Fig. 12.22*d*).

Derived maps may be separated by many stages of compilation from the primary geological data on which they have been based and major engineering decisions should not be taken on derivative maps alone: the geological data should always be consulted.

Two types of derived map commonly used in engineering are: (*i*) geomorphological maps and (*ii*) geotechnical maps.

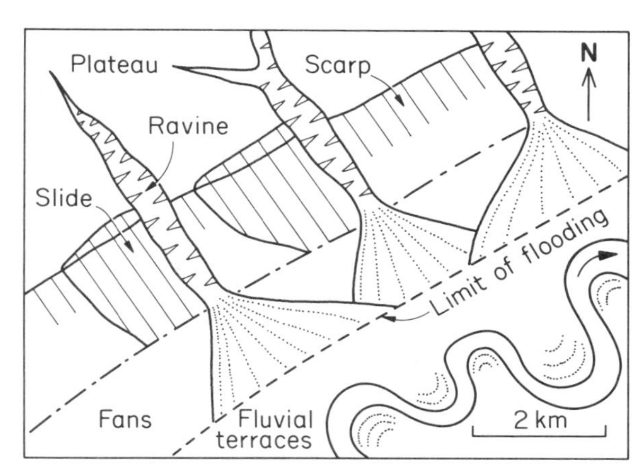

Fig. 12.23 A geomorphological map: contours omitted for clarity.

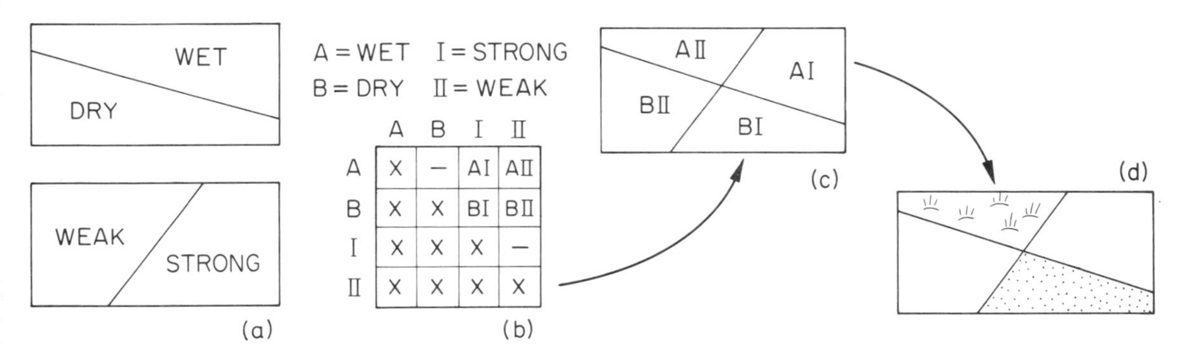

Fig. 12.22 (**a**) Two maps each showing original attributes of an area. (**b**) Contact matrix of characteristics. (**c**) Superposition of the two maps. (**d**) Final transformation. (Varnes, 1974.)

Geomorphological maps

Geomorphology is the study of land surfaces and the process that formed them (see Chapter 3): it is a subject that is relevant to much engineering construction (see Doornkamp *et al.*, 1979).

The International Geographical Union has recommended that aspects of geomorphology should be described on maps under four headings: land appearance, its dimensions and angles, its origin, and its age (see Fig. 10.2*c*). In practice these are usually combined in one of two ways: either as a map of the *products* or geomorphological processes (i.e. topography) or as a map of the *processes* that formed the topography (i.e. weathering, erosion and deposition) (see Fig. 10.2*b*). Figure 12.23 is a map for an area across which a road is proposed, directed from south-west to north-east. It shows the topography, the location of active erosion and slope instability and features of the river catchment especially its liability to flooding. Geomorphological mapping is one of a group of studies collectively described as terrain evaluation; Mitchell (1973), Geological Society of London (1982).

Geotechnical maps

This large group includes maps that predict likely conditions and those that record existing conditions. Recommendations for their preparations are given by the Geological Society of London (1972) and UNESCO (1976). The purpose of such maps is to present the geology of the area in terms that will help in the selection of construction techniques, of ground treatment, and in the prediction of the interaction between an engineering structure and the ground (Dearman and Fookes, 1974).

In the plan shown in Fig. 12.24 five characteristics of

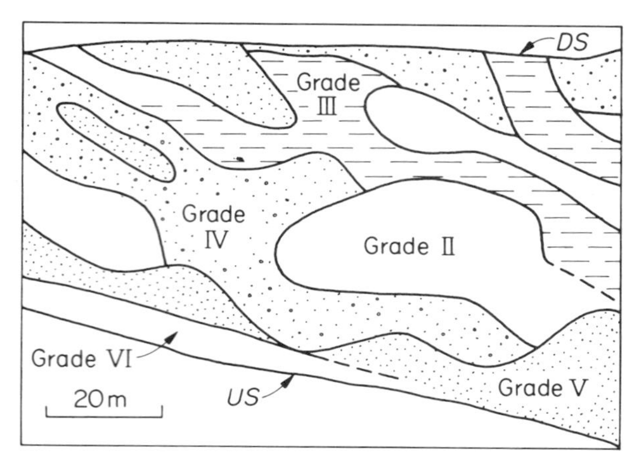

Fig. 12.24 Geotechnical plan of dam foundation (*DS*=downstream boundary; *US*=upstream boundary). Grades: II=bedded solid rock with some shales. III=thinly bedded with some shales. IV=blocky rock with frequent seams of shale and clay-silt, some open joints. V=broken faulted rock. VI=special category poor quality rock. Details from Knill and Jones (Fig. 19) (1965).

the site rocks were considered; namely their state of weathering, compactness, frequency and direction of fractures, cleanness of fractures, and relative proportion of shales. With a procedure of this kind the boundaries drawn on the plan may bear little relationship to those printed on conventional geological maps. For example, a sequence of clays may be subdivided on a geological map according to their age, whereas their fissuring and stiffness may be so similar that they are grouped as one unit on a geotechnical map.

Selected bibliography

Barnes, J.W. (1981). Basic geological mapping. *Geological Society of London Handbook*. The Open University Press, Milton Keynes, and Halstead Press (J. Wiley), New York.

Dackombe, R.V. and Gardiner, V. (1982). *Geomorphological Field Manual*. George Allen & Unwin, London and Boston.

Brink, A.B.A., Partridge, T.C. and Williams, A.A.B. (1982). *Soil Survey for Engineering*. Oxford University Press, London and New York.

Robinson, G.D. and Spieker, A.M. (Editors) (1978). Nature to be commanded. Earth science maps applied to land and water management. *U.S. Geol. Survey Professional Paper*, **950.**

Blyth, F.G.H. (1965). *Geological maps and their Interpretation*. Edward Arnold, London.

Roberts, L.J. (1982). *Introduction to Geological Maps and Structures*. Pergamon Press, Oxford and New York.

Colwell, R. N. (1983). *Manuel of Remote Sensing*. American Society of Photogrammetry. Falls Church, Virginia.

13

Ground-water

Hydrological cycle

Ground-water is the fluid most commonly encountered in engineering construction. It is derived from many sources but most now comes from rainfall and melting snow and is termed *meteoric* ground-water. The passage of water through the surface of the ground is called *infiltration* and its downward movement to the saturated zone at depth is described as *percolation*. Water in the zone of saturated ground moves towards rivers, lakes and the seas, a process known as *ground-water flow*, where it is evaporated and returned to the land as clouds of water vapour which may precipitate as rain or snow. Thus a cycle of events exists, namely precipitation on land, infiltration and percolation, ground-water flow to open water, evaporation and thence precipitation, so starting another cycle. This circulation of water is termed the *hydrological cycle* (Fig. 13.1). The seas and oceans contain approximately 97% of all the water presently involved in the cycle, a little more than 2% is in the form of snow and ice and less than 0.1% is water vapour. The remaining 0.9% is distributed in lakes, rivers and ground-water.

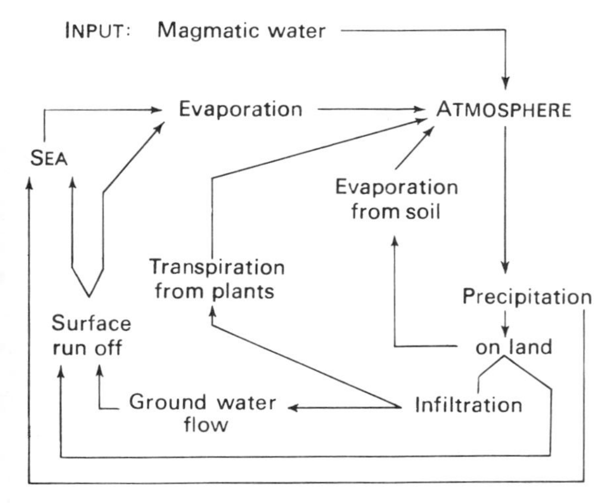

Fig. 13.1 The hydrological cycle.

Infiltration

Infiltration most readily occurs through open fractures such as joints in exposed rock (Fig. 8.15) and the gaping cracks that may develop in zones of tensile strain flanking areas of mining subsidence (Figs 9.28 and 13.13) and around landslides (Fig. 14.7). The superficial deposits that mantle most areas (Fig. 3.1) also permit infiltration through their pores. Deposits of gravel and sand, and screes are able to infiltrate water without difficulty whereas clay-rich mantles retard the ingress of water and characteristically remain wet after periods of rainfall. Vegetation protects the delicate porous structure of many superficial deposits, especially the crumb-structure of top-soil, and ground covered by vegetation has more uniform infiltration than bare ground. In the absence of such protection the impact of raindrops during periods of intense rainfall may disturb the soil structure and render the surface impermeable.

Infiltration ceases once the voids within the ground are full of water, thus if the rate at which water is supplied to the surface exceeds the rate at which it can percolate the volume that can infiltrate will gradually diminish. Horton (1933) defined the maximum rate at which rain can be absorbed by the soil as its *infiltration capacity* and rain falling at a rate that exceeds the infiltration capacity can result in flooding.

Percolation

Once water has penetrated the surface it may commence on a downward journey towards the water table. To do

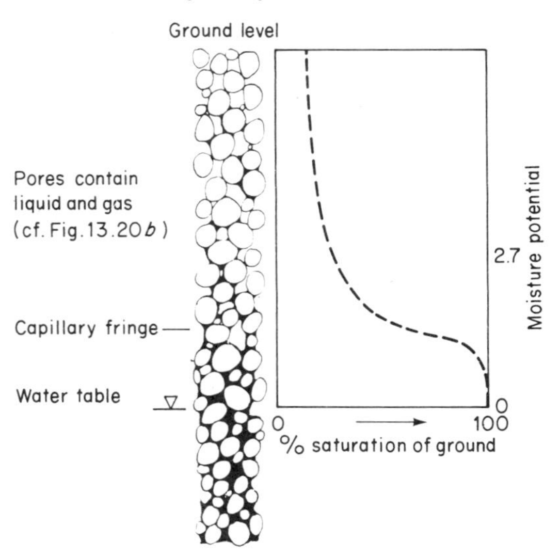

Fig. 13.2 Conditions in pore spaces above the water table (see also Fig. 9.23). (After Meyboom, 1967.)

this the water must first satisfy the capillary requirements of the soil profile and these will depend upon the dryness of the mineral surfaces surrounding the voids. A measure of these requirements is the *moisture potential* which is expressed either as the vacuum necessary to balance soil suction or the logarithm of the head of water (in cm) equivalent to the pressure difference between this vacuum and atmospheric pressure (Fig. 13.2). Fine-grained materials such as clays can develop high potentials when dry which are satisfied only by considerable quantities of water. When no more water can be held by the soil against the pull of gravity, the soil is said to have reached its *field capacity*: the moisture potential is then 2.7 (Fig. 13.2).

Once a profile has reached its field capacity further supplies of water from infiltration drain to deeper levels and this water is described as *gravity water* to distinguish it from the capillary water held on the surface of the soil grains. A wetting front thus moves down through the ground. This water will continue to move down after the cessation of infiltration until the pull of gravity of the pore water is once again balanced by the surface tension on the film of water around the individual soil surfaces, i.e. until the soil is restored to its field capacity.

Capillary fringe and water table

Percolating water may eventually arrive at a depth where all the voids are full of water. A hole drilled to just below this level would not encounter a water-level though its sides would be moist. This is because there still remains in the ground unsatisfied capillary forces. Water would drain from the side of the hole to its base and re-enter the ground. By deepening the hole a level will be reached where water covers its base. This is the *water table* and the pressure of pore water at this level is atmospheric: the moisture potential of the ground is zero (Fig. 13.2). The zone of saturation above the water table is called the *capillary fringe*. The mechanisms involved in percolation are reviewed by Wellings and Bell (1982). A water table that receives water from percolation is said to be *recharged.*

The height to which the capillary fringe rises above the water table depends upon the size of the voids in the ground and the dryness and temperature of the atmosphere. In hot, arid areas the capillary fringe may extend many metres and can reach ground level in low-lying ground. In such areas ground-water is lifted from the water table to the surface and its dissolved constituents precipitated there to form a mineral crust and to fill the pores near ground level with mineral cement.

Ground-water flow

Water moves at various speeds through the ground depending upon its flow path. Near surface flows (described as 'local' in Fig. 13.3) move the fastest and normally supply most of the water that discharges at *springs*. A *spring-line* is defined by the intersection of the water table

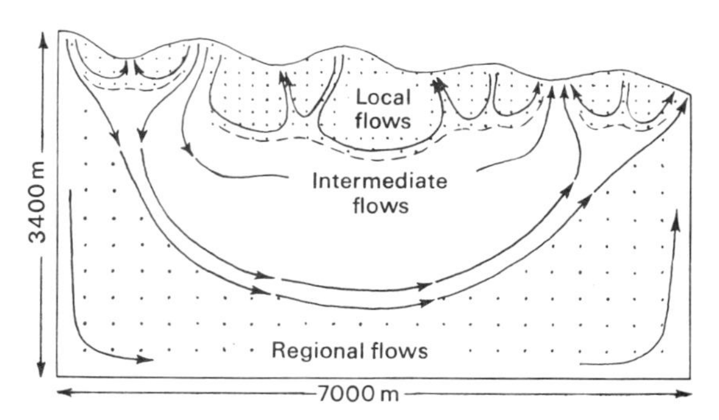

Fig. 13.3 Theoretical distribution of flow (after Toth, 1963): illustrating the relatively rapid circulation of ground-water near surface (local flows), deeper, slower circulation and the very slow circulation of regional flows at great depth.

with ground level (Fig. 13.4): as the water table moves so the spring-line moves, migrating up a topographic slope in response to periods of recharge in wet weather and receding down the slope in periods of dry weather (Fig. 13.20*a*). The headwaters of rivers fed by springs which migrate, will move up and down the valley. Such streams have been called *bournes*: strictly the spring is a bourne and its stream a bourne-flow.

The velocity with which water circulates in the ground gradually decreases with depth and the movement of deep ground-water may be extremely slow (Fig. 13.3). Some of the water resident in deeply buried sediments (i.e. depths exceeding 4 km) of Mesozoic and Cenozoic age (Table 2.1) may be the relic of that in which the sediment accumulated and is called *connate* water, i.e. congenital with the sediment.

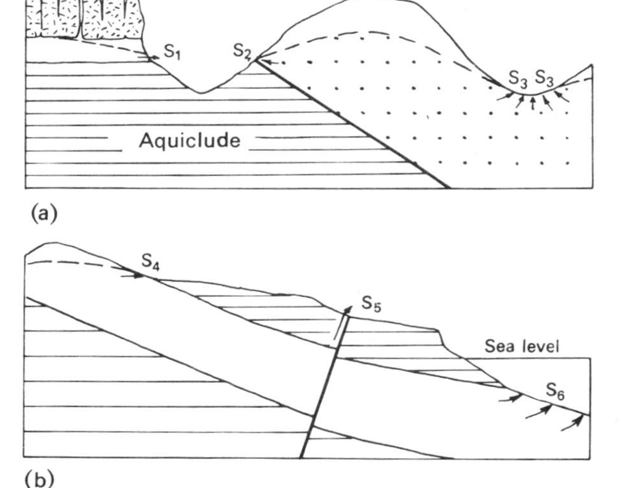

Fig. 13.4 Geological structures commonly associated with springs. S_1 = stratum spring between an aquifer and an aquiclude; S_2 = fault spring; S_3 = valley spring; S_4 = overflow spring; S_5 = artesian spring associated with a fault; S_6 = submarine springs. See also Figs. 13.9, 13.11, 13.16 and 13.25, and the representation of springs on maps, Fig. 13.23.

All water originated from liquids and gases vented through the solidifying crust of the primeval Earth to form the atmosphere. Fluids released this way are described as magmatic (Chapter 5) and small quantities of *magmatic* water continue to enter the hydrological cycle usually in association with volcanic activity (Figs. 13.1 and 5.1).

Character of ground-water

Chemical characters

The water molecule H_2O is strongly polar and thus is a powerful solvent. Ground-water therefore adopts the chemistry of the rock or soil in which it resides and may be described as a solution of H_2O with its dissociated ions, H^+ and OH^-, and ions derived from the dissolution of organic and inorganic solids. Also present are gases, colloidal suspensions, suspended particles, and sometimes viruses and other living organisms which may be dangerous to public health (Fig. 13.5).

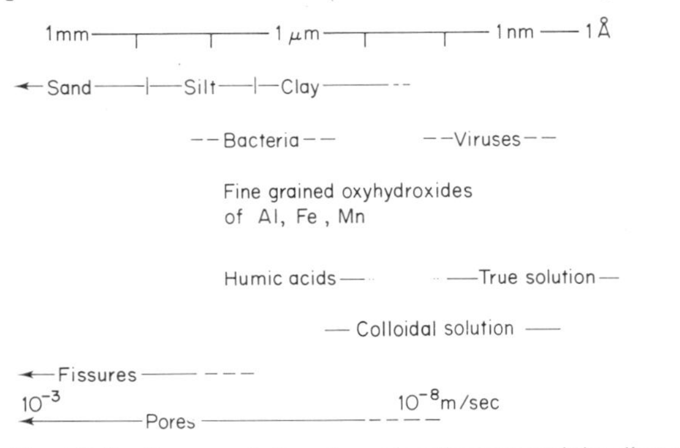

Fig. 13.5 Representative sizes of sediments and the dimensions of solutes and suspended particles in ground-water. Also shown are the dimensions for fissures and pores, and their associated permeability. (From Edmunds, 1981.)

Total dissolved solids (TDS)

The concentration of dissolved constituents may be used to classify ground-water, as in Table 13.1. Ionic concentration may be expressed as TDS and measured in parts per million (ppm) = (grams of solute) ÷ (grams of solu-

Table 13.1 A simple classification of ground-water based upon the concentration of total dissolved solids (TDS): see text. ppm = parts per million

TDS (ppm)	Normal limits (ppm)	Electrical conductivity (µmhos)
0		
Fresh water	500 Drinking water	1000
	700 Irrigation	
1000		2000
Brackish	7000 Farmstock	
10 000		10 000
Saline	30 000 / 40 000 Sea water	50 000
100 000		100 000
Brine > 100 000		> 100 000

Table 13.2 Composition of typical ground-waters (all values in parts per million)

	Likely ionic concentration															pH		
	SiO_2	Al	Fe	Mn	Ca	Mg	Na	K	HCO_3	CO_3	SO_4	Cl	F	NO_3	PO_4	Min	Av	Max
Igneous																		
Acid	37.0	0.11	0.3	0.02	28.0	6.1	11.6	2.3	142.0	0.0	15.7	5.0	0.2	3.8	0.10	6.3	7.2	7.7
Basic	41.0	0.1	0.6	0.06	26.0	14.3	7.7	3.1	202.0	0.0	12.1	22.6	0.2	6.5	0.03	5.6	7.5	8.2
	20.0	0.05	0.1	0.01	27.0	3.1	6.1	1.6	66.0	0.0	31.1	4.4	0.1	2.4	0.03	7.2	7.5	7.7
Sedimentary																		
Arenaceous	23.0	0.1	0.5	0.04	49.0	17.7	19.1	2.8	252.0	2.0	69.0	9.8	0.4	8.6	0.02	6.2	7.3	9.2
Argillaceous	27.0	0.8	1.6	0.06	110.0	51.2	179.0	5.7	330.0	2.8	96.6	121.0	0.6	4.1	0.02	4.0	6.7	8.6
Limestone	13.0	0.1	0.4	0.05	71.0	19.1	12.9	2.2	228.0	0.0	8.8	9.7	0.2	8.9	0.02	7.0	7.5	8.2
Dolomite	14.0	0.2	0.4	0.07	62.0	43.7	13.7	1.1	272.0	0.7	35.0	6.9	0.5	6.3	0.00	7.4	7.7	8.2
Metamorphic																		
Quartzite	10.0	0.07	0.7	0.10	33.0	12.4	6.5	3.3	119.0	0.0	37.5	5.9	0.2	1.7	0.02	6.5	7.1	7.4
Marble	10.0	0.15	0.1	0.03	52.0	10.0	3.0	1.2	192.0	0.0	10.7	4.7	0.1	0.05	0.00	7.6	7.7	7.9
Slate	16.0	0.02	0.5	0.10	39.0	6.0	14.7	3.3	143.0	0.0	40.1	7.8	0.3	4.7	0.03	5.2	7.1	8.0
Schist	16.0	0.02	0.5	0.10	39.0	6.0	14.7	3.3	143.0	0.0	40.1	7.8	0.3	4.7	0.03	5.2	7.1	8.0
Gneiss	33.0	0.1	0.5	0.07	39.0	24.2	29.5	2.5	219.0	0.0	34.3	34.5	0.9	5.8	0.00	5.8	7.1	8.1
Drift																		
Igneous	39.0	0.00	0.2	0.01	18.0	6.2	25.7	3.7	220.0	0.0	74.1	32.6	0.4	5.4	0.00	6.1	7.2	7.9
Sedimentary	19.0	0.00	0.2	0.00	62.0	20.4	53.2	3.9	269.0	0.6	71.1	22.4	0.2	8.6	0.00	7.4	7.8	8.4

tion × 10^6). Because ground-water is an ionic (i.e. electrolyte) solution, its electrical conductivity reflects ionic concentration (see Table 13.1). The composition of typical ground-waters is indicated in Table 13.2.

The quality of ground-water is affected by the period of its contact with the ground. Connate water which may have been in the ground for many millions of years has a high TDS and is often described as a *brine* (Table 13.1). Water circulating closer to ground-level (see Fig. 13.3) moves more quickly, that in fissures travelling much faster than the water in pores (Fig. 13.6) and is generally *fresh*.

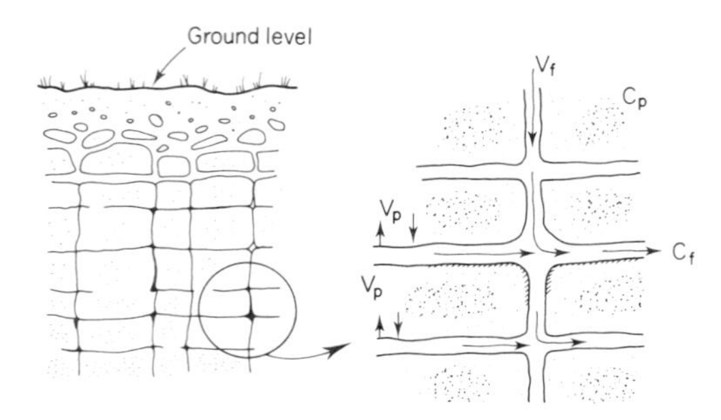

Fig. 13.6 Reaction of ground-water chemistry to host rock environment. Slow moving water in rock pores (velocity Vp, chemical concentration Cp) and fast moving water in fissures (Vf) of composition (Cf). Vf≫Vp and Cp≫Cf.

Eh and pH

The ability of water to react with the ground depends upon its power to supply or receive protons (pH) and supply or receive electrons (Eh). Normal conditions are shown in Fig. 13.7. Infiltrating water rich in O_2 raises the Eh causing dissolved iron to be precipitated as an oxide (often seen as rust-coloured staining on joint surfaces near ground level) and other reactions. Many waters pumped from mines and draining from dumps of industrial waste, have the pH of an acid because of their reaction with the ore.

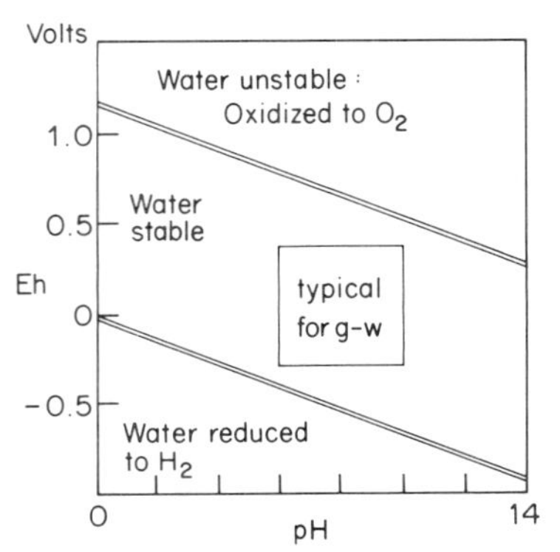

Fig. 13.7 Stability field for water in terms of Eh and pH. (g-w = ground-water) (see Cloke, 1966).

An illustration of the changes that can occur is given in (Fig. 13.8). Recent meteoric water infiltrating through the outcrop has mixed with older saline, perhaps connate water down dip. Note the abruptness of the Eh change. Cores taken from the aquifer at various locations revealed the intimate relationship between rock and water chemistry. Near outcrop the limestone is yellow with occasional grey patches: the yellow represents oxidized limestone, the grey reduced limestone. Cores from locations further east, down dip, revealed limestone that was progressively greyer, the yellow oxidation being restricted to narrow zones that bounded the fractures. The fractures are transmitting the fresher water richer in oxygen.

Cation exchange

Ions in ground-water may be exchanged for others situated either on the surface or within the lattice of clay

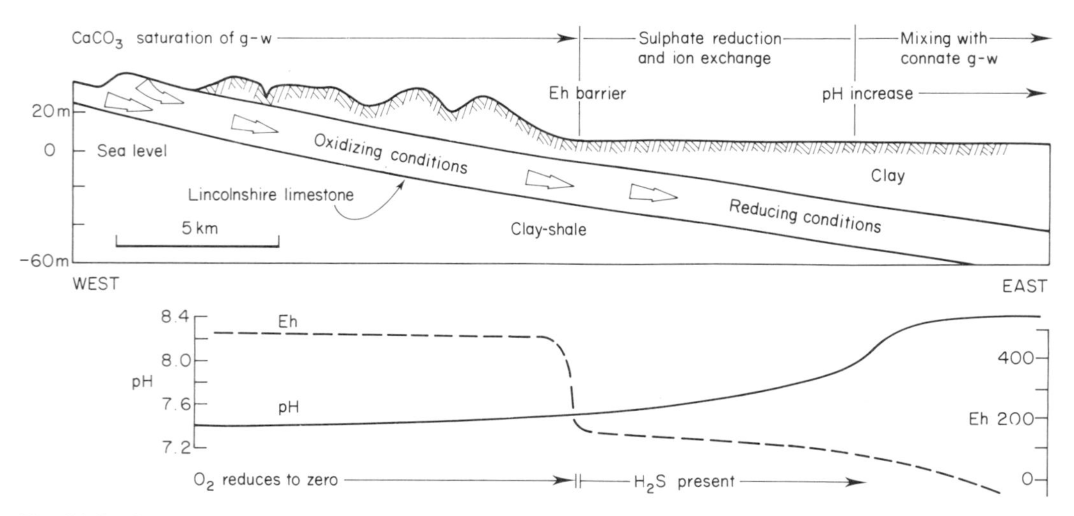

Fig. 13.8 Change in the chemistry of ground-water with distance from its point of infiltration at outcrop: see text for details. Eh measured in millivolts. (From Edmunds, 1973, 1977.)

minerals. For example, ground-water from limestone, rich in calcium (Table 13.2) and described as 'hard', can be 'softened' if it later comes in contact with clays that attract the calcium and release sodium in exchange. Many clay minerals (p. 80), zeolites (p. 80), peat, humus and roots have high exchange capacities.

Dissolved gases

Most ground-water contains gas in solution. Gases commonly derived from the atmosphere are O_2, N_2 and CO_2, the latter being abundant in water that has been in contact with limestone ($CaCO_3$). Methane (CH_4), H_2S and N_2O are produced by the decay of organic matter. Dissolved gases come out of solution when the pressure of ground-water approaches atmospheric pressure and such gas often accumulates in mines, shafts, wells and other poorly ventilated excavations. Their presence may reduce the concentration of O_2 to injurious or even fatal levels. CH_4 is also explosive.

Isotopes

Hydrogen has three naturally-occurring isotopes; ^1H (proteum) and ^2H (deuterium) are stable and have relative abundances of 99.84% and 0.016% respectively. The third isotope is ^3H (tritium) and occurs approximately once in every 10^{-15}H atoms. It is unstable, has a half-life of 12.4 years and permits the age of water to be dated, and its rate of movement through the ground to be calculated. For comparison ^{14}C has a half-life of 5730 years.

Oxygen has three naturally-occurring isotopes, ^{16}O, ^{17}O and ^{18}O; all are stable. Their relative abundance is 99.76%, 0.04% and 0.20% respectively. Stable isotopes enable the history of water to be studied for evaporation preferentially lifts the lighter ^{16}O into the vapour phase thus concentrating ^{18}O in the remaining water. Ratios of ^{18}O:^{16}O (normally about 0.002) and ^{18}O:^2H (normally 6.25:1.0) provide the water molecule with an isotopic signature that can be used to identify, in the ground, water of different histories. Isotope chemistry has permitted the age and stratification of ground-water to be recognized.

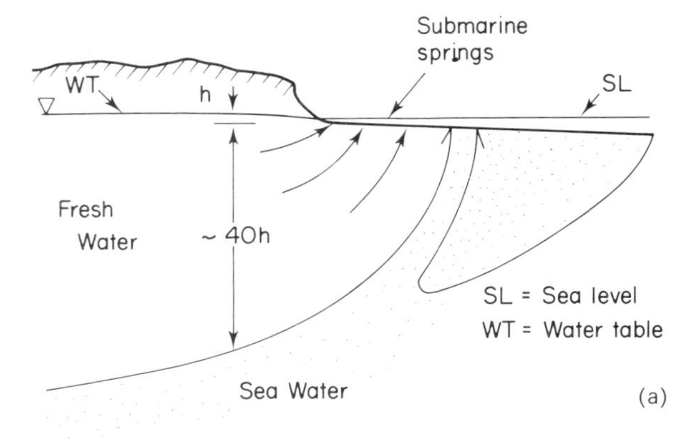

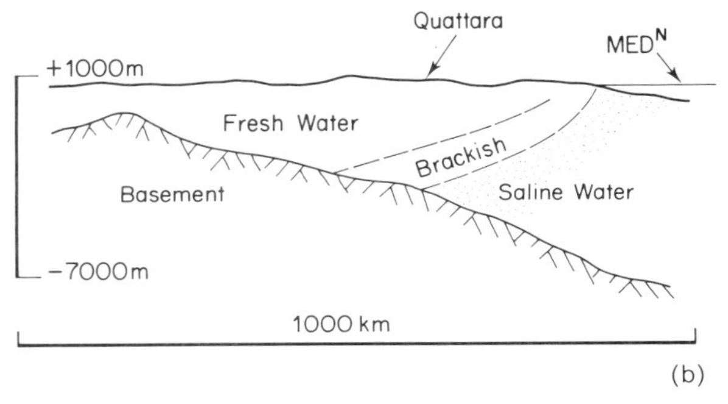

Fig. 13.9 Two examples of the relationship between sea-water and fresh-water at a coast. (**a**) Sea-water supports fresh-water that has a strong coastal discharge; fresh-water–seawater boundary is usually transitional because of diffusion and tidal fluctuations (see Cooper, 1964). (**b**) Section through Egypt: note extensive zone of diffusion (brackish). Med^N = Mediterranean (from Shata, 1982).

Physical characters

Unit weight

The unit weight of fresh ground-water may be taken as $9.81\,kNm^{-3}$ at 20°C and its specific gravity as 1.0. The weight of ground-water increases with TDS, that of sea water being 1.025 times (or 1/40th) greater than fresh water, i.e. $10.05\,kNm^{-3}$. Fresh water floats on saline water, a point of significance in coastal regions (Fig. 13.9). The unit weight decreases with an increase in temperature, being approximately $9.6\,kNm^{-3}$ at 60°C. Unit weight is the force that drives most naturally-occurring flows of ground-water.

Viscosity

The dynamic viscosity of water at 20°C is 1.0×10^{-3} Nsm^{-2} (or 1.0×10^{-3} Pas). It is temperature dependent, rising to $1.3 \times 10^{-3}\,Nsm^{-2}$ at 10°C, to greater than $10^3\,Nsm^{-2}$ as it freezes, being approximately $10^8\,Nsm^{-2}$ when ice. It reduces to $0.3 \times 10^{-3}\,Nsm^{-2}$ at 60°C. Viscosity contributes to the resistance of water to flow; the lower the viscosity of water the less is its resistance to flow through the ground.

Compressibility

The compressibility of water is low and taken as $4.6 \times 10^{-10}\,m^2\,N^{-1}$ at 20°C, varying between $5.0 \times 10^{-10}\,m^2\,N^{-1}$ at 0°C and $4.4 \times 10^{-10}\,m^2\,N^{-1}$ at 60°C, i.e. it is almost constant. Compressibility influences the mass of water stored in the ground for a given pressure and temperature.

Specific heat

Water has the highest value of specific heat of all common liquids and is $4180\,Jkg^{-1}$ degree $Kelvin^{-1}$ (or 1 calorie $gram^{-1}$ per degree °C). Flowing ground-water is thus an important agent for the transfer of heat as witnessed by permafrost phenomena (p. 35) and by its use in geothermal power plants.

Aquifers and aquicludes

Rocks and soils that transmit water with ease through their pores and fractures are called *aquifers* and those that do so with difficulty are *aquicludes*. Typical aquifers are gravel, sand, sandstone, limestone and fractured igneous and metamorphic rocks. Typical aquicludes are clay, mudstone, shale, evaporite and unfractured igneous and metamorphic rock. Table 13.3 lists the void size that may be expected in these materials and their likely permeability. Soluble strata such as limestone and evaporite may have voids created or enlarged by dissolution to form caves and tunnels (Fig. 3.3) and in volcanic rocks lava tunnels may exist (Fig. 5.5). The majority of soils transmit water through their pores (Figs 6.1 to 6.3) whereas transmission through most rocks is by pores and fractures (Figs 5.2, 6.9 to 6.11 and 7.6). Fractures normally transmit more water than pores (Fig. 13.6) as illustrated by the following values for the Chalk of S.E. England, a major aquifer:

Table 13.3 Typical void sizes for granular, fractured and other materials, and associated permeabilities.

Material	Void size (m)	Permeability (ms^{-1})
Clay	$<10^{-6}$ to 10^{-5}	$<10^{-8}$
Silt	10^{-5} to 10^{-4}	10^{-8} to 10^{-6}
Sand	10^{-4} to 10^{-3}	10^{-6} to 10^{-2}
Gravel	$10^{-3}+$	10^{-2} to 10^{0}
Fracture	10^{-3} to 10^{-2}	10^{-3} to 10^{0}
Karst	$10^{-2}+$	$10^{0}+$
Tunnels	$10^{0}+$	$10^{6}+$
Caves	$10^{1}+$	infinite

	Fractures	*Pores*
Void space as	$<2\%$	25–45%
Permeability	$>100\ md^{-1}$	$\sim 1\times 10^{-4}\ md^{-1}$

The size of pathways provided by pores and fractures will reduce if applied loads cause them to close (Fig. 15.13). Shrinkage due to desiccation can open cracks in clays and dissolution can widen voids in soluble materials.

Confinement

Aquifers may be either particular strata (Fig 13.10*a*) or fracture zones (Fig. 13.10*b*). Stratiform aquifers are described as being *confined* if they are buried and *unconfined* if they are exposed (Fig. 13.11). The replenishment of most aquifers by infiltration (a process known as *recharge*, p. 213) occurs over their exposed outcrop and it is within this unconfined portion of an aquifer that the greatest circulation of ground-water occurs.

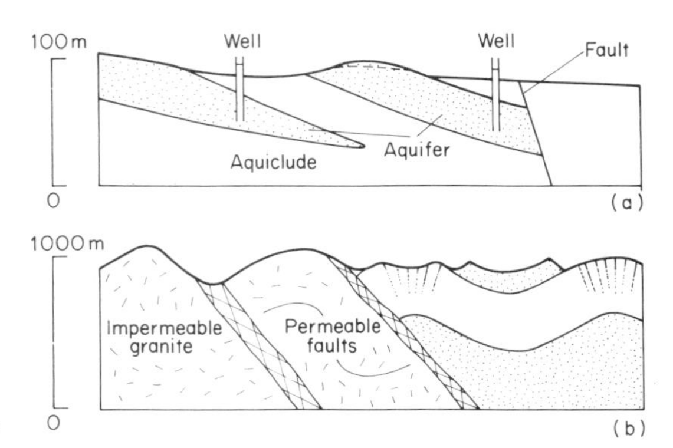

Fig. 13.10 Characteristic aquifers: (**a**) as strata, (**b**) as zones associated with faulting, and tensile fracturing on fold crests.

Isotropy and anisotropy

Igneous, sedimentary and metamorphic rocks, and most soils, are rarely so uniform that their thickness, degree of fracturing, porosity and permeability are the same throughout. Usually these properties vary from place to place, and with direction (see for examples Figs 2.2, 3.1 and 3.36). An important source of variation in sediments is bedding, where the hydraulic properties in the direction parallel to bedding differ from those orthogonal to it. In these respects most aquifers and many aquicludes are anisotropic and calculations to predict ground-water flow must use values of permeability that are appropriate for the direction of flow in the ground.

Hydrogeological boundaries

Geological boundaries define the volume of an aquifer and hydrological boundaries, especially the water table, define the volume of water stored within it. Commonly encountered geological boundaries are stratification of aquifers and of aquifers against aquicludes (Fig. 13.11), the termination of aquifers by faults (Fig. 13.10), unconformities (Fig. 2.2) and igneous intrusions (Figs 5.6, 5.7 and 5.9).

Not every geological boundary need function as a hydrogeological boundary: there must be a noticeable dif-

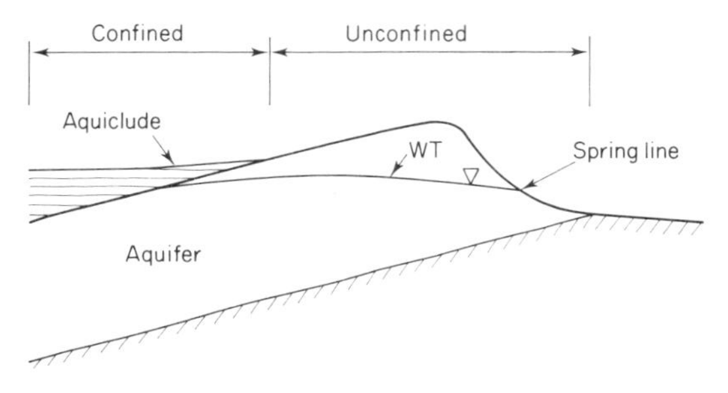

Fig. 13.11 Unconfined outcrop of scarpface with dip-slope confined beneath aquiclude.

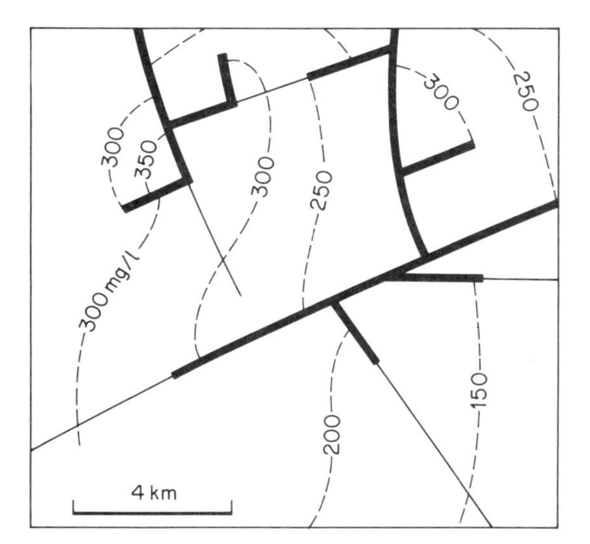

Fig. 13.12 A map of faults that are operating as hydrogeological boundaries and revealed by their interruption of salinity contours. Fault controlled salinity distribution (after Daly *et al.*, 1980) ▬ fault acting as barrier.——— fault not acting as barrier. $\frac{200}{}$ ground-water salinity (mg l^{-1}).

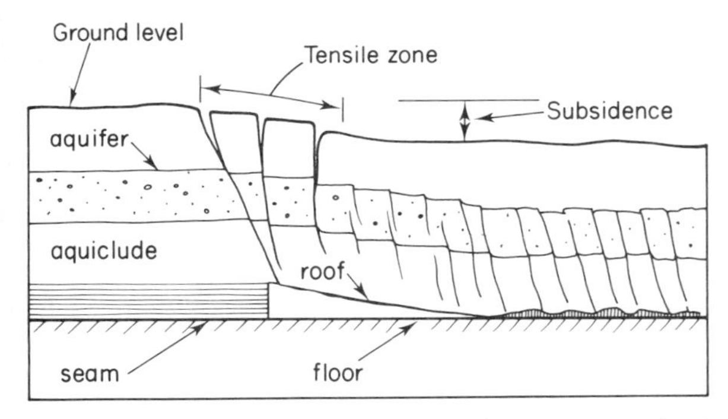

Fig. 13.13 Geological boundary moving above a mined area. Note that the increased fracturing greatly increases permeability.

ference in either the transmissive or storage characters of the ground, or in the chemistry of the ground-water on either side of the boundary. Hydrogeological boundaries are therefore usually revealed by there being across them a change either in water levels, or in the gradient of water levels, or in water quality (Fig. 13.12).

Geological boundaries tend not to change with time, however mining can seriously disrupt the ground (Fig. 13.13). Fractures may open in the zone of tensile strain above a mine (Fig. 9.28), and permit a rapid flow of ground-water in this area. The zone of tensile strain migrates with the mine and is a dynamic geological boundary. Ground broken this way never recovers its original hydraulic characters.

Hydrological boundaries include the water table and spring line (Fig. 13.11), coastal and shore lines (Fig. 13.9), rivers, lakes and reservoirs (Fig. 15.5). Such boundaries usually fluctuate in elevation and are dynamic. The spring line is an important boundary that marks the intersection of the water table with ground level. Ground-water discharges from ground below the spring line (Figs 13.4 and 13.20) thus preventing infiltration from occurring over this area.

Water levels

The various water levels that may be encountered in bore-holes are illustrated in Fig. 13.14. Bore-hole A is unlined and penetrates an impersistent zone of saturation supported on an impermeable zone of limited extent: (*a*) is a *perched* water table. Bore-holes B1 to B4, which are cased only in their upper portion, enter the main zone of saturation and (*b*) is the *main* water-table. Thus a water table is the level of water encountered in a bore-hole that is either unlined, paritally lined, or lined with perforated casing. Bore-hole C penetrates the confined aquifer where ground-water pressure is sufficient to support a column of water in the hole. Unlike the other bores, this hole is lined with casing to its base so that the column of water in it balances the pressure of water in the ground at C′: the level C is called a *piezometric* (or pressure) level for the confined aquifer.

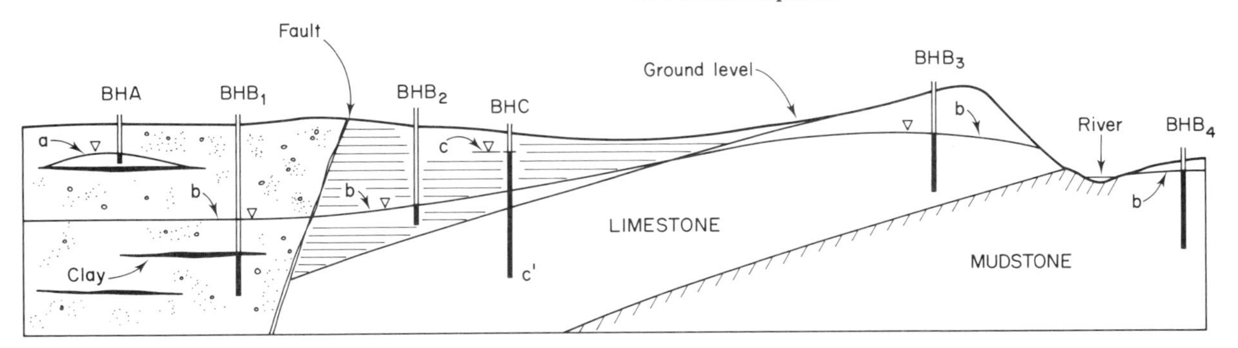

Fig. 13.14 A major sandstone aquifer containing clay lenses, is faulted against a limestone aquifer that is confined by clay and underlain by mudstone. Note: water tables can exist in aquicludes, as shown in the clay and mudstone.

Bores drilled into an aquifer that has its piezometric level above ground-level, will overflow, and are termed *artesian* bores after the French province of Artois where overflowing wells were first recorded. The London Basin (Fig. 2.17*a*) was formerly an *artesian basin* because water levels in the Chalk outcrop north and south of London were higher than ground level in central London where wells drilled through the confining London Clay and into the Chalk, would overflow. Pumping has since reduced the piezometric level to below ground level. The *Great Artesian Basin*, in Queensland, Australia, is a famous example of such a structure, extending 1.56×10^6 km^2 west of its outcrop along the Great Divide, and supplying water, via deep wells, to the arid country of central Queensland. Piezometric levels that are above the top of the aquifer supporting them, but do not reach ground level, are described as *sub-artesian*.

Fluctuation of water levels

Water levels are rarely static for when the rate of recharge to ground-water exceeds its rate of discharge to rivers, lakes and the sea, water levels will rise, and vice versa. Other factors may also cause water levels to change and Fig. 13.15 illustrates the fluctuations that may happen and their common causes.

A uniform distribution of rain rarely occurs thus some areas receive more recharge than others; further, a uniform distribution of rain rarely results in uniform infiltration. Thus a rise in water level will usually be greater in some places than in others. Fluctuations will also vary with proximity of the observation hole in which they are measured to local centres of discharge, such as springs, rivers and engineering works that pump water from the ground. The volume of infiltrated water that can be stored depends upon the volume of voids in the ground: thus ground with little storage may be rapidly filled with and drained of water, and register marked changes in water level in contrast to ground of considerable storage capacity where much water can be stored for little change in water level.

For these reasons, the water level fluctuations measured in one observation hole are rarely those which occur over a large area. Thus such fluctuations should be studied in many holes and the data presented as maps of water level change (Davis & de Weist, 1966).

Ground-water flow

The movement of ground-water and its dissolved constituents may be considered the result of two mechanisms: (1) the physical movement of molecules together with their associated solutes, called *advective flow*, and (2) the chemical movement of solutes, called *diffusive flow*. When rates of advective flow are extremely slow, diffusion be-

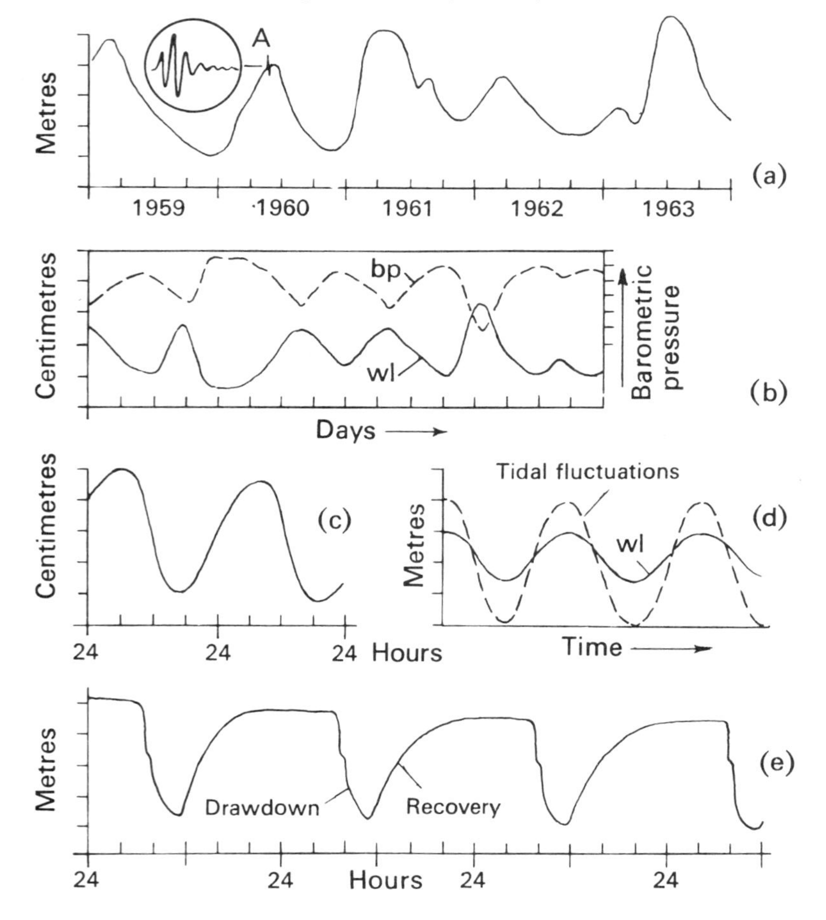

Fig. 13.15 Examples of water table fluctuations. (**a**) Seasonal fluctuations measured in an unconfined aquifer in England: A = short term fluctuations resulting from Chilean earthquake in May 1960. (**b**) Fluctuations in a piezometric level from a confined aquifer reflecting the effects of barometric pressure. (**c**) Daily variations in water table resulting from the transpiration of surrounding vegetation. (**d**) Water levels in a coastal aquifer. (**e**) Periodic fluctuations resulting from the discharge of a nearby well operating between 7 a.m. and noon.

comes the dominant form of mass transfer in response to chemical gradients in the ground. Ground-water that moves sufficiently fast to be of concern to most engineering work is predominantly advective, and may be quantified with reference to the mechanical forms of potential and kinetic energy of the water. These have been defined by Bernoulli and may be termed *elevation* and *pressure* head, both of which are illustrated in Fig. 13.16, and velocity head: the latter is usually so small a component that it can be ignored in most assessments of ground-water flow. Elevation plus pressure head equal *total* head (Fig. 13.16) and flow is always from high to low total head (or water level).

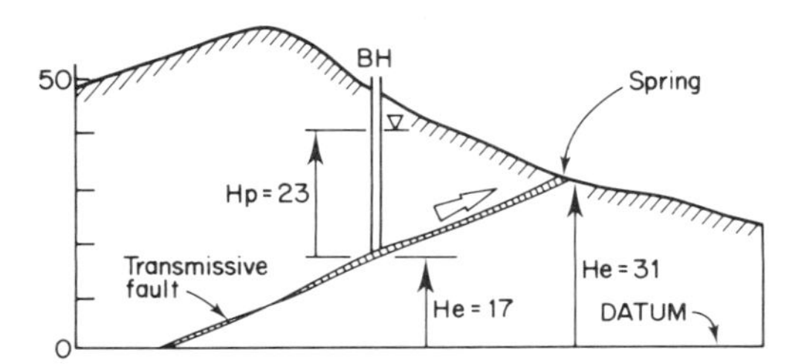

Fig. 13.16 Total head at bore-hole (BH) = 17 + 23 = 40. Total head at spring = 31 + 0 = 31. Flow of ground-water is from high to low total head (or water level). He = elevation head, Hp = pressure head: all measurements are in metres.

As mentioned earlier, water levels rise when the rate of ground-water flow into a volume of ground exceeds its rate of outflow, and vice versa. Under these circumstances, flow is *non-steady* and the mass balance to describe it is:

$$\text{inflow} = \text{outflow} + \text{change in storage}$$

Fig. 13.15 shows that non-steady flow is the normal flow of natural systems. Only when the rate of change of storage equals zero is flow *steady* and under these circumstances water levels *do not* change with time.

Transmission

The discharge of ground-water Q (dimensions L^3T^{-1}) through the aquifer illustrated in Fig. 13.17, of area A (dimensions L^2) at 90° to the direction of flow, is given by the relationship:

$$Q = K\frac{\Delta h}{L}A$$

where $\Delta h/L$ is the loss in total head per unit length of flow (i.e. the hydraulic gradient: dimensionless LL^{-1}) and K, called the *coefficient of permeability*, is the empirically-derived quantity that equates the hydraulic gradient to the discharge per unit area Q/A. Permeability thus has the dimensions of velocity (i.e. LT^{-1}).

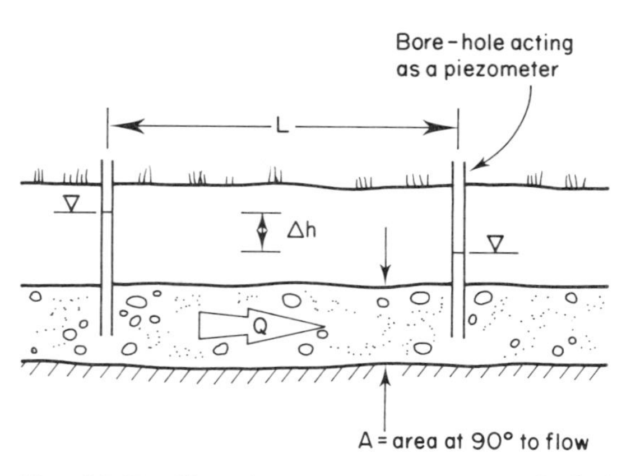

Fig. 13.17. The elementary components of discharge through gravel of permeability (K) (see Fig. 11.11).

Permeability

From the above, it is obvious that permeability is best measured in the field (p. 185) but that small samples may be determined in the laboratory (p. 194). Table 13.3 indicates the range of permeabilities commonly associated with *porous* materials and Fig. 13.18 illustrates the range of values that can operate in *fractured* rock.

Permeability is rarely the same in all directions and the appropriate value to use is that in the direction of ground-water flow. The greatest value of permeability in sediments and sedimentary rocks is usually in the direction of bedding. If ground contains horizons of different permeability an overall value is required and obtained from permeability tests conducted in situ (p. 185).

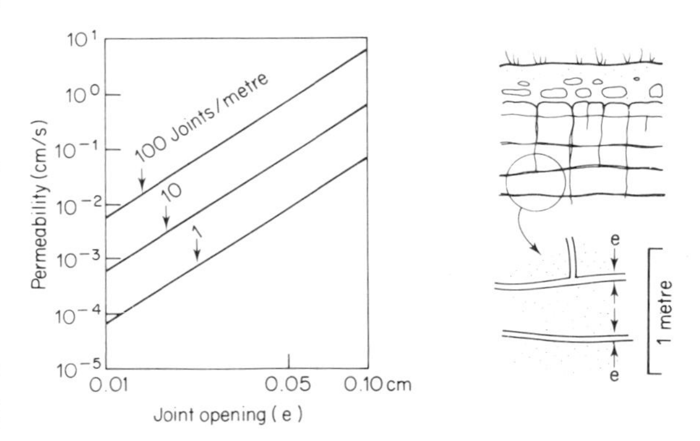

Fig. 13.18 Fracture permeability depends upon both the number of fractures (e.g. joints) and also their openness (e). (From Hoek & Bray, 1977.)

Transmissivity

Water-supply engineers commonly combine the overall value of *in-situ* permeability of strata (K) with its saturated thickness (b: dimensions L) to produce a value called *transmissivity* = Kb (dimensions L^2T^{-1}). As water levels rise and fall so the value for saturated thickness (b) ness (b) will vary (Fig. 13.19). It is also common to find that the value for overall permeability decreases with depth; a most common occurrence in fractured rock. Thus an aquifer may have a range of values for the product Kb depending upon its level of saturation.

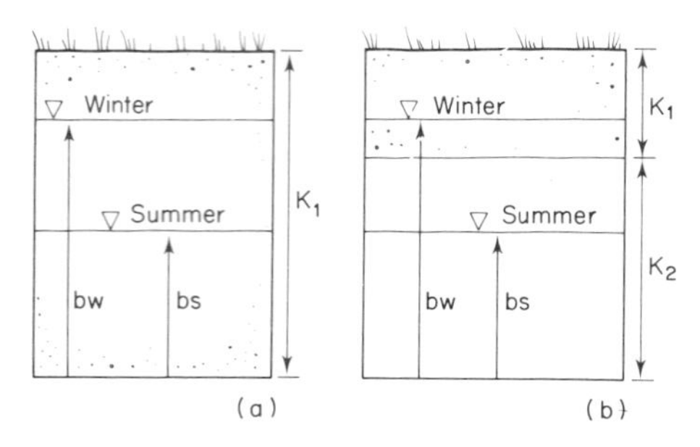

Fig. 13.19 (**a**) Transmissivity = Kb, varies between summer and winter, as b varies. (**b**) Stratification increases variation of transmissivity with variations in b. ∇ = water table. K = permeability. b = saturated thickness.

Velocity of flow

Permeability and hydraulic gradient govern the apparent velocity of flow (v):

$$\frac{Q}{A} = v = \frac{K\Delta h}{L}$$

Thus when the hydraulic gradient is zero, there is no flow, regardless of how permeable the ground may be. The velocity calculated is said to be apparent because the real velocity of flow through the pores and fractures will be larger, as these voids represent only a fraction of the area (A) at 90° to flow (see Fig. 13.17). Absolute measurements of flow velocity can be obtained from the travel time of stable tracers injected into the ground-water.

Seepage force

The loss in total head (Δh) per unit length of flow (L) reflects the transfer of potential *from* the ground-water *to* the soil or rock through which it flows, where it exists as a body force called the *seepage force*. Its magnitude is:

$$\frac{\Delta h}{L} \times \text{unit weight of water}$$

The force operates in the direction of flow and can cause ground failure (Figs. 14.8 and 16.8).

Storage

Pores, fractures and other voids in the ground provide the storage space for ground-water and the amount of water that can be stored may be assessed by observing the volume of water accepted or released from the ground for a given rise or fall in water level.

Specific yield

Figure 13.20 illustrates an unconfined aquifer whose stored ground-water is discharging to a river valley and

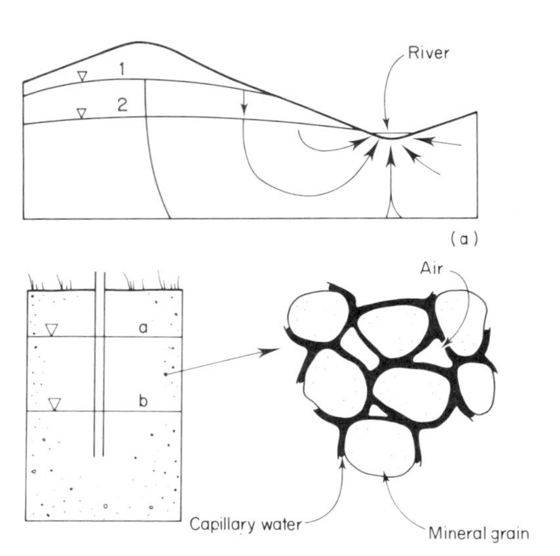

Fig. 13.20 (**a**) River discharge maintained during dry weather by discharge of ground-water from storage as ground-water levels drop from 1 to 2. (**b**) Storage under water table conditions: note that air is present between grains (magnified).

sustaining river flow during periods of dry weather. The water table lowers from *a* to *b* and during the period over which this occurs the river flow is measured so that the volume of ground-water discharged per unit change in water level may be calculated. The specific yield for the aquifer is:

$$\frac{\text{Volume of water released}}{\text{Change in water table} \times \text{area of water table change}}$$

The voids in the aquifer drain and are occupied by air. Typical values for specific yield are in excess of 1% and may be as large as 20%.

Coefficient of storage

Figure 13.21 illustrates a confined aquifer containing water under sufficient pressure to support a sub-artesian

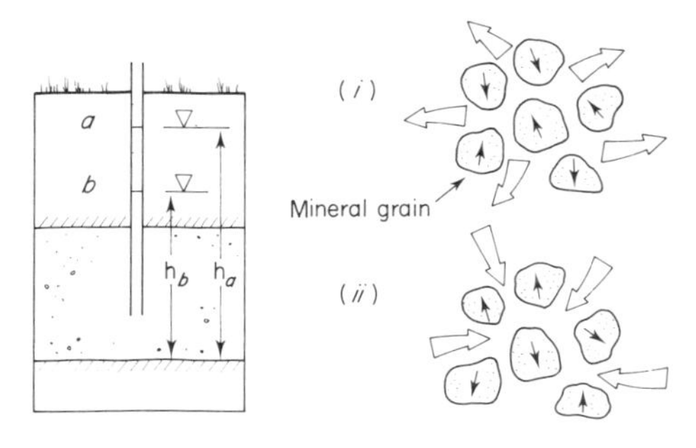

Fig. 13.21 Storage under piezometric conditions with schematic illustration of a granular aquifer; on right, water expansion and compression. (*i*) Level falls from *a* to *b*: water expands and the aquifer also compresses thus expelling water. (*ii*) Level rises from *b* to *a*: water compresses and the aquifer also expands but only very slightly.

piezometric level (see Fig. 13.14). When water pressures in the aquifer are reduced, as when a well pumps from a confined aquifer, the piezometric level falls from *a* to *b*. The aquifer *remains fully saturated* but the reduction in water pressure from $ha\gamma_w$ to $hb\gamma_w$ (where γ_w = unit weight of water) is associated with a slight expansion of the water (compressibility $4.6 \times 10^{-10}\,m^2N^{-1}$) and compression of the aquifer (compressibility usually less than $10^{-7}\,m^2N^{-1}$). Both reactions result in the expulsion of water from the aquifer. The relationship between water level change and change in storage is expressed as a coefficient of storage which equals:

$$\frac{\text{Volume of water released}}{\text{Change in piezometric level} \times \text{area of piezometric change}}$$

Typical values are less than 0.1%; often much less. The compression of geological materials that accompanies a reduction in fluid pressure has been described in Chapter 9. Clays and other highly compressible materials that are interbedded with aquifers suffer irreversible compaction when their water pressure is reduced, and result in consolidation (p. 162) and settlement of ground level (Poland and Davis, 1969; Poland, 1972).

Barometric and tidal efficiency

The interaction of aquifer and water compressibility may also be observed in aquifers that are subjected to a change in applied load. Figure 13.22 illustrates two commonly-occurring conditions. The change in fluid pressure that occurs per unit change in applied stress is the *efficiency* of the system. Wells, shafts and bores should not be entered during periods of falling barometric pressure without first ensuring that the excavation is well ventilated since water tables at depth will be rising and driving air deficient in oxygen towards ground level via the excavation.

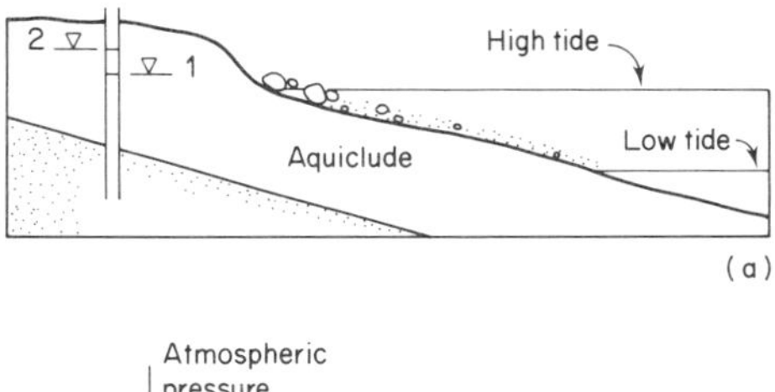

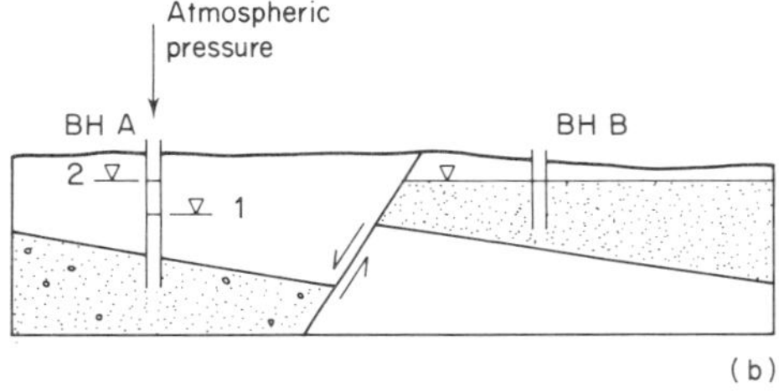

Fig. 13.22 (**a**) Piezometric level 1 at low tide, rises to level 2 at high tide as the weight of sea-water compresses the underlying aquifer. (**b**) Piezometric level 1 rises to level 2 as atmospheric pressure decreases allowing the pressure of water in the aquifer to force water into and up the bore-hole (BH A). (See Fig. 13.15.) Water level in unconfined aquifer (BH B) shows no change.

Calculations of storage

Both the storage coefficient and specific yield are dimensionless quantities thus the volume of water that may be obtained from an aquifer in which there is a water table is (saturated volume × specific yield).

The change in the volume of water stored in an aquifer is either (change in water table × area over which the change applies × specific yield), or (change in sub-artesian or artesian piezometric level × area over which the change applies × coefficient of storage).

These relationships will also provide the change in water level that will accompany any volume of recharge to ground-water or discharge.

Hydrogeological investigations

The hydrogeological investigations that enable predictions to be made either of the influence of ground-water upon engineering works, or its potential as a source for water supply, should be designed to assess the following:

(*i*) the location and thickness of aquifer horizons and zones, their confinement and their hydrogeological boundaries;
(*ii*) the levels of water in the ground, their variation over an area and their fluctuation with time;
(*iii*) the storage and transmissive characters of the ground; and
(*iv*) the quality of the ground-water.

To obtain this information investigations must be conducted at the surface and below ground level.

Surface investigations

Normally there are three objectives for surveys that are conducted at ground level:

(*i*) To make a hydrogeological map of the area so as to show the distribution of aquifers, of geological boundaries, especially those created by stratigraphy and faulting, and of hydrological boundaries such as rivers, lakes and spring lines. Water levels, where known, may also be included (Fig. 13.23). Ground of exceptional permeability may be located at the intersection of faults, or of faults with the axis of an antiform (Fig. 12.17). The mapping of large areas, as needed for large water supply, or hydro-power or irrigation schemes, can be helped greatly by the use of conventional air-photographs and various forms of remote sensing that reveal the presence of water and water-bearing strata (p. 210).
(*ii*) To draw one or more vertical sections across the area to illustrate geological structure, the thickness and confinement of aquifers, and the dip of hydrogeological boundaries (see Figs 13.11, 13.14, 13.20*a* and

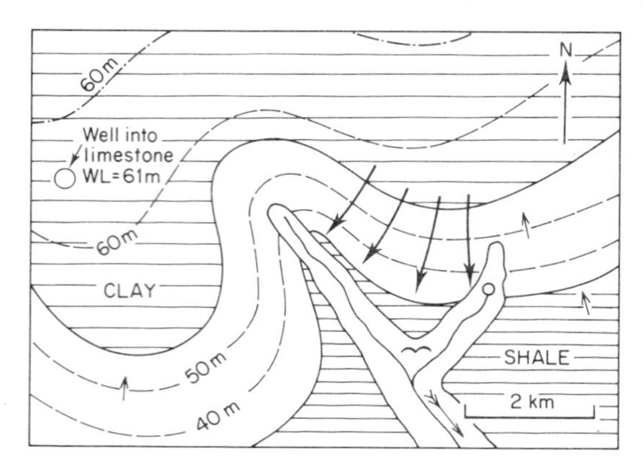

Fig. 13.23 Hydrogeological map. A limestone aquifer lies between two aquicludes: shale below and clay above. The strata dip N. Rivers issue from springs (see UNESCO, 1970). The direction of ground-water flow may be deduced from flow lines drawn orthogonal to the water level contours ‐‐ =water level in limestone. 60 m = height above datum to which water in limestone will rise in a bore. =alluvium. ○ = spring. → = flow line for ground-water.

13.22). Special attention should be given to faults and similar fracture zones that may be important transmitters of water (Figs 13.10 and 13.16). The axes of antiforms are also significant as they tend to be more fractured, and hence more permeable, than the surrounding rocks (Fig. 13.10).

(*iii*) To reconstruct the geological history of the area, particularly for the last 2 my, when sea level has fluctuated in relation to the land (pp. 28 and 39) and ground-water levels were different from those at present. Many rivers have buried valleys (p. 42) in which the alluvium acts as a ribbon-aquifer. Karstic conditions in soluble strata developed at depths that were related to a previous level of the water table, and rising water levels have now flooded the karstic ground (Fig. 13.24). Such areas of palaeo-karst (or 'fossil' karst) provide valuable aquifers for water supply.

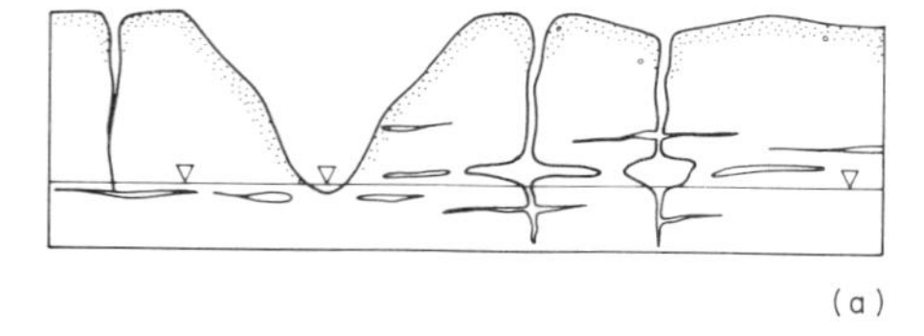

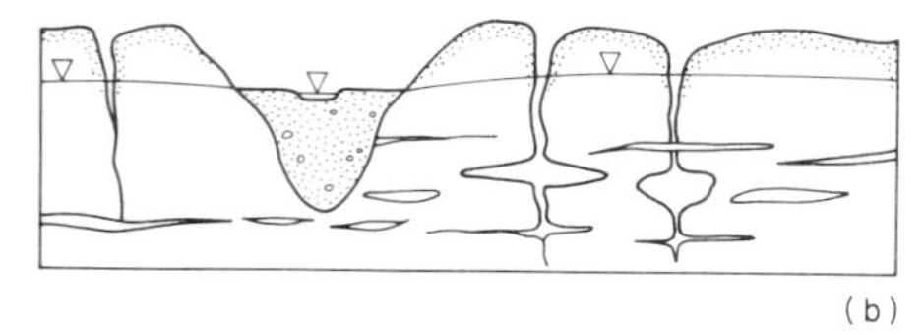

Fig. 13.24 The control of hydrogeological characters by geological history. (**a**) Karst conditions developed during a previous period of low water level. (**b**) Present conditions.

To complete the surface investigations all these data are collated and from them predictions made of the probable location and movement of ground-water (Fig. 13.25). Conceptual models of this nature help identify the type of sub-surface investigation that is needed to provide an adequate hydrogeological study.

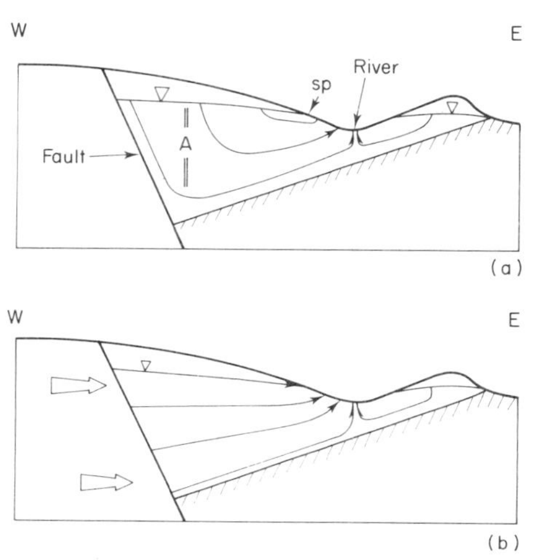

Fig. 13.25 Two models of possible ground-water flow based on available information, viz. a geological cross section, a spring line on both sides of the valley (sp) and a river with dry weather flow (Fig. 13.20). (**a**) No recharge from the west side of the fault. (**b**) Recharge from the west. The measurement of hydraulic head at A would define which system existed.

Sub-surface investigations

These are required to confirm:

(*i*) the level of water in the ground (Fig. 13.14);
(*ii*) the depth, thickness and lateral extent of aquifers and aquicludes;
(*iii*) the permeability of these zones and the storage of aquifers;
(*iv*) the chemistry of the aquifers and their contained water, and their temperature, if required.

The observation and interpretation of water-levels is one of the most important tasks in these investigations. Much may be learnt from the levels of water encountered whilst drilling a bore-hole; Fig. 13.26 illustrates common situations. (1) The level of water in a bore-hole that penetrates an aquifer in which there is no vertical component of ground-water flow, will not change as the bore-hole progresses. (2) When there is a downward component to *in-situ* flow water levels in a bore-hole will lower as the hole progresses: the opposite will happen when there is an upward component to flow (3). (4) A confined aquifer may support a water level that rises up the hole as it

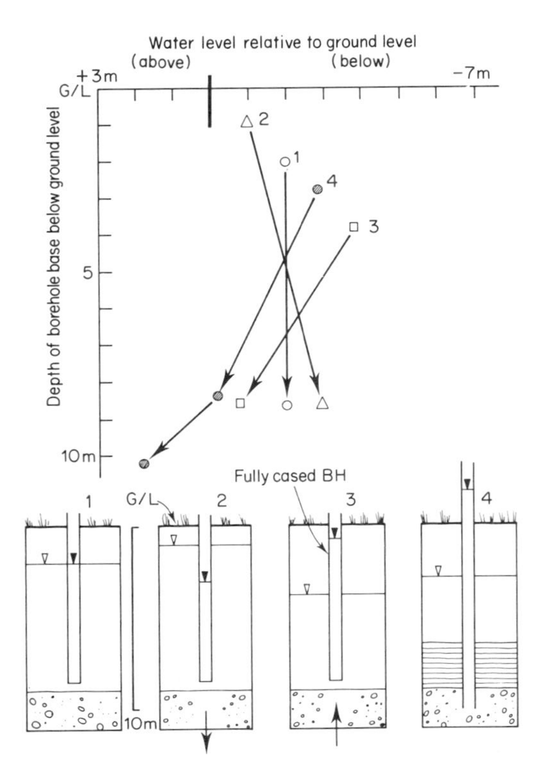

Fig. 13.26 Interpretation of water levels (W/L) during drilling. Compare the depth to water level in a bore-hole with the depth of drilled hole. E.g. (4): W/L encountered at 2.8 m, and rose up the hole as drilling progressed. At 8.5 m the rise in W/L became more rapid as a confined aquifer was penetrated. If the hole is not fully cased the W/L recorded is not an accurate measure of pressure head as water may drain from the hole (as in cases 3 and 4) or towards the hole (as in cases 1 and 2).

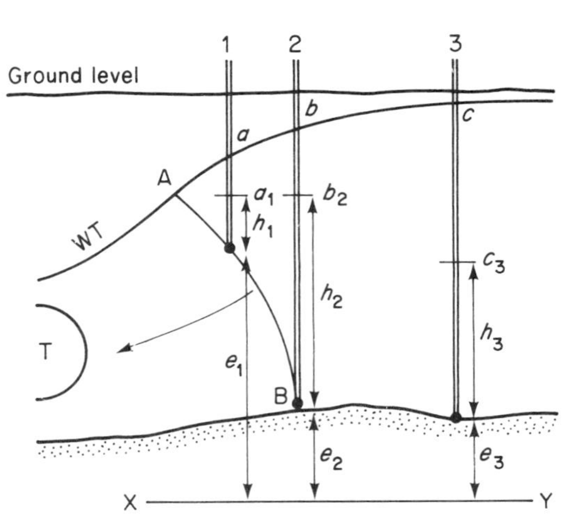

Fig. 13.27 Piezometric levels. h=pressure head; e=elevation head; x–y=datum from which head is measured (Fig. 13.16). A drainage tunnel, T, in clay overlying gravel, depresses the water table, (WT). A–B is line of equal potential, i.e. the total hydraulic head at all points along it is equal, velocity head being ignored. Piezometers 1 and 2 register this equipotential. The water levels encountered during drilling would be a and b respectively falling to a_1 and b_2 once the piezometers were installed (see also Fig. 13.26). Bore-hole 3 enters the gravel. Its water level falls from c to c_3, the piezometric level for the gravel, once its piezometer is installed. The pressure of water at a piezometer tip is=$h\gamma_w$ where γ_w=unit weight of water.

approaches the aquifer: the water is either artesian or sub-artesian (*q.v.*).

A piezometer may be installed into a bore-hole to provide an accurate measure of pressure head at its location in the ground (Fig. 10.15) and enable variations in total head (=water level) to be measured over a period (Fig. 13.15). Figure 13.27 illustrates how such instruments may be used to monitor hydraulic conditions around a tunnel.

The geophysical logging of boreholes provides valuable information that helps to identify permeable zones and to indicate the quality of water in the ground (Table 13.4). Electrical resistivity, caliper and flow logging techniques are popular. Such logging greatly assists engineers investigating the ground for water supplies, shafts, tunnels and large underground storage caverns, and in the task of correlating strata between bore-holes, as each horizon has its own geophysical signature. Beds of similar character may thus be recognized and aquifers and aquicludes delineated.

In-situ assessments of transmissivity (or permeability) and storage are frequently required and may be obtained using the techniques described in Chapter 10 (p. 185).

Samples of water may have to be retrieved for analyses and must be collected with care (see Edmunds, 1981 for guidance). Figure 13.28 illustrates the change in conditions that can occur when sampling at the time of drilling did *not* represent the quality of ground-water under operational conditions.

Table 13.4 Geophysical methods suitable for most bore-holes and commonly used in hydrogeology. (See Keys & McCary, 1971.)

Logs for defining condition of strata	
B.H. diameter	CALIPERS: reveals fractures, cavities, strata boundaries.
Electrical-potential	SELF POTENTIAL: for boundaries and type of strata.
	RESISTIVITY: for strata types and boundaries.
Radiation	NEUTRON: for porosity
	GAMMA: for density
Visual	TELEVISION: for inspection
Logs for defining fluid conditions	
Temperature	reveals permeable horizons intersected by bore-hole.
Flow	as for temperature.
Electrical-potential	CONDUCTIVITY: for fluid composition: salinity.

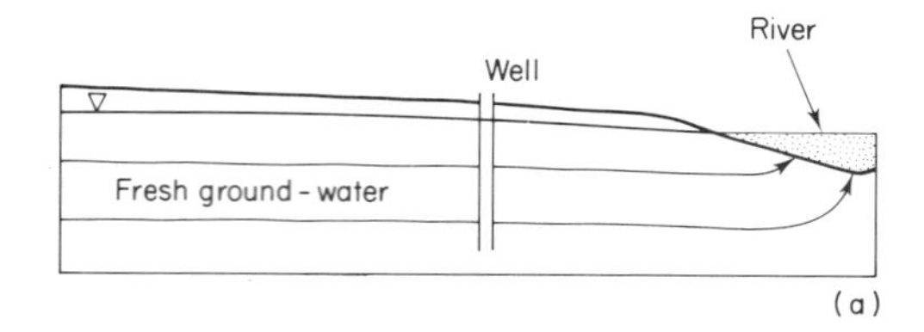

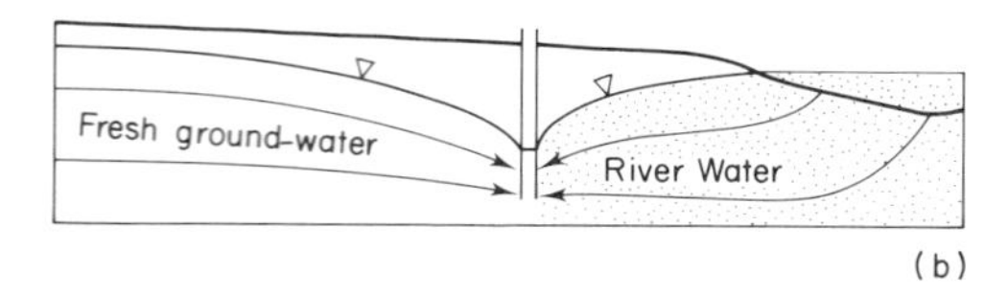

Fig. 13.28 (**a**) Ground-water conditions at time of drilling well. (**b**) Conditions when well is supplying water.

Selected bibliography

General texts

Freeze, R. A. and Cherry, J. A. (1979). *Groundwater.* Prentice-Hall Inc., New Jersey.

Fetter, C. W. (1980). *Applied Hydrogeology.* Charles E. Merrill Pub. Co., Ohio.

Davis, S. N. and de Wiest, R. J. M. (1966). *Hydrogeology.* John Wiley & Sons Inc., New York.

Todd, D. K. (1980). *Ground Water Hydrology.* 2nd Edition. John Wiley & Sons Inc., New York.

Heath, R. C. and Trainer, F. W. (1968). *Introduction to Groundwater Hydrology.* John Wiley & Sons Inc., New York.

Brassington, R. (1983). *Finding Water.* Pelham Books Ltd., London.

Special aspects

Walton, W. C. (1970). *Groundwater Resource Evaluation.* McGraw-Hill Kogakusha Ltd., Tokyo.

Lloyd, J. W. (Ed.) (1981). *Case-Studies in Groundwater Resources Evaluation.* Clarendon Press, Oxford.

Domenico, P. A. (1972). *Concepts and Models in Groundwater Hydrology.* McGraw-Hill, New York.

Cedegren, H. R. (1967). *Seepage, Drainage and Flow Nets.* John Wiley & Sons, New York.

Hem, J. D. (1970). Study and interpretation of chemical characteristics of natural water. *U.S. Geol. Survey. Water Supply Paper*, 1473.

Poland, J. F. (1972). Subsidence and its control: in Underground Waste Management and Environmental implications. *Amer. Assoc. Petrol. Geol. Memoir*, **18,** 50–71.

Drever, J. I. (1982). *The Geochemistry of Natural Waters.* Prentice-Hall, New Jersey.

14

Slope Stability

All slopes have a tendency to move, some more than others. Such movements can vary in origin and magnitude, and range from near-surface disturbances of weathered zones to deep-seated displacements of large rock masses. They occur when the strength of the slope is somewhere exceeded by the stresses within it. Movements that are restricted to surface layers are often controlled by stresses derived from surface or near-surface environments, e.g. precipitation and temperature, whereas movements at depth indicate the presence of adverse stresses at depth. The movements that occur may extend from those small and slow displacements associated with *creep* (p. 160) to the rapid and large displacements of catastrophic slides. Those commonly responsible for engineering problems are illustrated in Fig. 14.1: all are the response to gravity-produced stresses which exist in every slope.

The stability of slopes is thus an important consideration in the design of man-made excavations such as those for an open-cast mine, a quarry, a road cutting, a large foundation or a deep trench. Natural slopes become unstable as a normal phase in slope erosion (Fig. 3.11) and the stability of slopes forming coastal and river cliffs, valley sides, especially where forming reservoir margins, and dam abutments, may have to be assessed if their movement endangers the public or engineering structures.

Slope failure

Slope movement and failure can occur in four ways which may operate either separately or together, as follows:

(*i*) by detachment of rock as rockfalls and topples (Fig. 14.1*a* and *b*);
(*ii*) by shear failure on existing large scale geological surfaces (Fig. 14.1*c*);
(*iii*) by shear failure of rock and soil material, often utilizing weak horizons (Fig. 14.1, *d*, *e*, *f* and *g*);
(*iv*) by gradual adjustments on a microscopic scale as in creep (Fig. 14.1*h*).

An elementary classification of such movements is presented in Table 14.1.

Progressive failure

Failure is usually a gradual event initiated locally at points near the base of the slope. Terzaghi (1950) noted this tendency in clay slopes and termed it *progressive failure* to indicate:

'the spreading of the failure over a potential surface of sliding from a point or line towards the boundaries of the surface. While the stresses in the clay near the periphery of this surface approach the peak value, the shearing resistance of the clay at the area where the failure started is already approaching a much smaller ultimate value' (see Fig. 9.24).

Such failure is suspected to involve the degradation of soil and rock strength by the agents of weathering (Chapter 3) and by creep. A well documented record of progressive failure is that of a slope in N. London (Skempton, 1964): see Fig. 9.27. The slope and its retaining wall, in brown fissured London Clay, were completed in 1912. Movements were at first small and at a fairly constant rate, but gradually increased and eventually culminated in failure; a behaviour typical of slope instability. The slope had taken 29 years to 'fail' and demonstrates that the long-term strength of soils is much lower than their instantaneous strength: a point to remember if designing with values obtained from 'quick' tests. The same is true for rocks (see Radbruck-Hall, 1978).

Because the stability of slopes is time-dependent it is common to consider their stability in the long and the short term, where short term usually means 'for the duration of a contract' and long term 'for the engineering life of the scheme'.

Factor of safety

Analyses for slope stability can determine:

(*i*) the conditions under which movement will commence, and
(*ii*) the amount of deformation that occurs under given conditions.

Analyses for the latter utilize finite element solutions that normally require a much greater knowledge of slope geology and of the non-linear stress-strain behaviour of the materials from which the slope is composed, than is usually available. For these reasons most routine analyses of slope stability concentrate on defining the conditions under which failure will occur.

The commencement of movement can be predicted using *Limit Equilibrium* analyses, where a mechanism, e.g. translation, rotation or toppling (Table 14.1), is postulated to occur about specified surfaces. The resistance required on these surfaces to just prevent failure occurring is then calculated and compared with the available

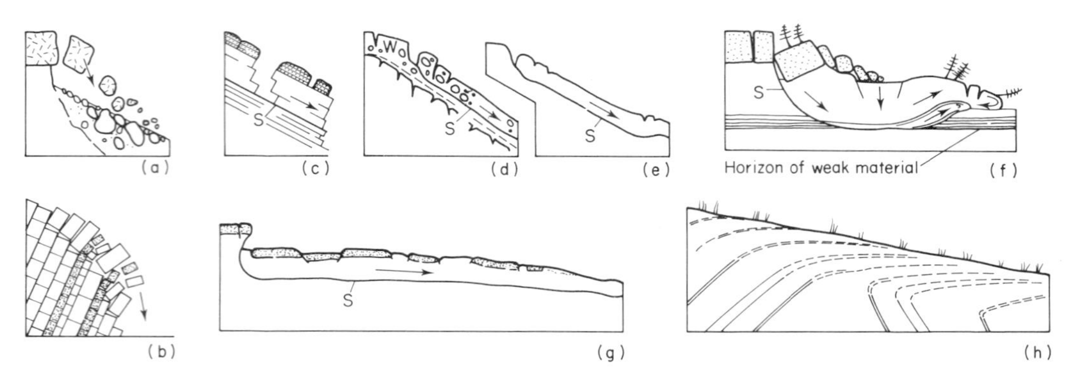

Fig. 14.1 Some common slope movements: S = sliding surface, W = weathered zone (see also Table 14.1 and text).

(*i*) Sevenoaks slip in Weald Clay, an example of failure of a man-made slope (reproduced by permission of the Chief Engineer, British Rail).

Table 14.1 An elementary classification of slope failure and movement (see Varnes, 1978)

A. FEATURES AT TIME OF FAILURE			B. MOVEMENT AFTER FAILURE	
Geometry	*Mechanism*	*Illustration*		
Blocks & columns	Fall & topple	Fig. 14.1*a* Fig. 14.1*b*, 14.4	Rolling & turning of blocks & columns	Fig. 14.1*a*, *b*
Slabs	Slide by translation	Fig. 14.1*c*, *d*, *e*, 14.3, 14.14, 14.16	Sliding of masses: by translation	Fig. 14.1*c*, *d*, *e*
			by rotation	Fig. 14.1*f*, *i*
Sectors	Slide by rotation	Fig. 14.1*f*, *i*, 14.2, 14.19	Flow as a viscous fluid	Fig. 14.1*g*, 14.11,

strength as assessed from laboratory or field tests, or better, the analyses of previous failures in similar materials (p. 171). This comparison defines the *Factor of Safety*, which is the factor by which strength may be reduced to bring the slope to limiting equilibrium (Morgenstern and Sangrey, 1978). When a slope fails the Factor of Safety is 1.0.

Major geological factors

The rocks and soils of which a slope is composed, their geological structure, the influence that ground-water may have upon their strength and upon the forces that operate on them, and the magnitude of stress *in-situ* are factors that influence the stability of a slope.

Types of rock and soil

The strength of rock and soil depends upon its mineralogy and fabric. In Chapter 4 the common minerals are described and in Chapters 5 to 7 their association in rocks and soils are explained and illustrated. It need only be noted here that the clay minerals (p. 80) tend to be the weakest of all.

The smallest fabric commonly of engineering significance to slope stability is that produced by minerals. An example of such a microscopic structure is the delicate flocculated fabric of a Norwegian quick clay (Fig. 6.3*c*): this is most sensitive to strain and capable of rapid collapse which quickly reduces the strength of the clay. Such a clay is *sensitive* (p. 122) and because the remoulding of a quick clay can convert the sediment to a viscous fluid, slopes in such material are liable to flow (Fig. 14.1*g*). Crawford and Eden (1967) describe two such movements in the Leda Clay of the St Lawrence Lowlands of E. Canada; this sediment has an open flocculated structure. One flow covered an area of $283 \times 10^3\,m^2$ to a depth of 5 m and the other involved 25 000 m^3 of clay. Both flows occurred on natural slopes. Similar movements have been described by Aas (1981) and Gregersen (1981) from the quick clays of Norway.

Such movements do not have to be associated with saturated ground. Some very large flow-movements have been recorded for materials that were essentially dry at the time of failure; one such slide, in loess at Kansu is described later, in connection with seismic events.

Mineral fabrics that are anisotropic, such as those of slates, schists and laminated clays, will be weakest in the direction parallel to the fabric, and a careless choice of samples for testing, or of the orientation of specimens tested, can result in an incorrect assessment being made of the strength of the rocks or soils in a slope (Fig. 14.2). Mineral fabrics can occur at different scales: some may exist throughout the slope, as in cleaved rock, whilst others are restricted to shear zones. Thin zones, only a centimetre or so wide, can be crucial to slope stability if they act as surfaces of sliding for large volumes of material (Fig. 14.1*c*).

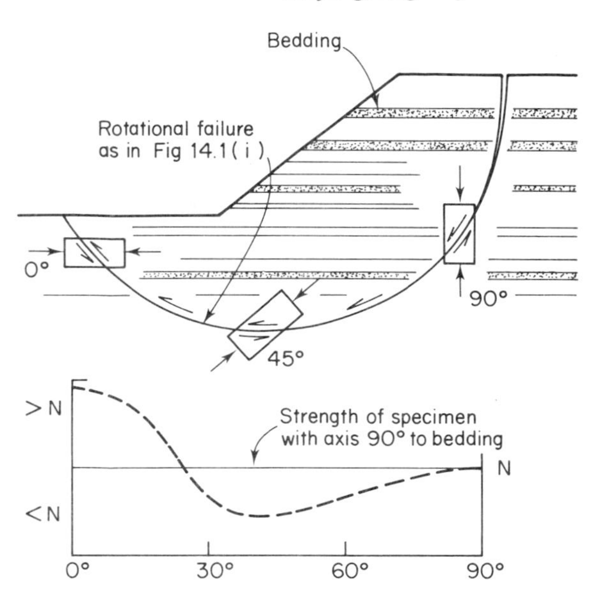

Fig. 14.2 Influences of anisotropy upon strength of a slope (after Skempton & Hutchinson, 1969). See also Fig. 11.5.

Geological structure

Surfaces such as bedding, schistocity, cleavage, faults, joints and fissures in over-consolidated clay, can have a profound influence upon the stability of slopes if their inclination facilitates downhill movement of the slope in which they occur. Examples of this influence are illustrated in Fig. 14.1*b*, *c* and *f*. In addition to this, the strength of joints and other geological surfaces is usually less than that of the intact rock they bound; often they are the weakest component of slope geology.

Figure 14.3 illustrates the influence of a simple geological structure upon the stability of slopes in a cutting. The vertical section (*a*) is drawn in the direction of dip of bedding and in the plane of joint set 2: the N slope of the cutting is controlled by bedding and the S slope by joint set 1. A map of the rock mass is shown in (*b*). The cutting shown in (*a*) is oriented parallel to the strike of bedding (Fig. 8.1) and different orientations, as shown in (*b*) result in slopes of different overall angle of stability, as illustrated by the sections CD and EF.

The manner in which structure can dictate the shape of individual blocks of rock within a slope can be seen in Fig. 14.3, and when these are so proportioned that their centre of gravity is beyond their base, they will topple, as shown in Figs 14.4 and 14.1*b*. Considerable volumes of rock may be moved by toppling (de Freitas and Watters, 1973), and geological structure can be crucial to stability (Goodman *et al.*, 1976; 1982).

Geological structure that permits the agents of weathering (Tables 3.1 and 3.2) to detach rock from the face of a slope will facilitate rockfalls (Fig. 14.1*a*). These sometimes involve large quantities of material and may pose

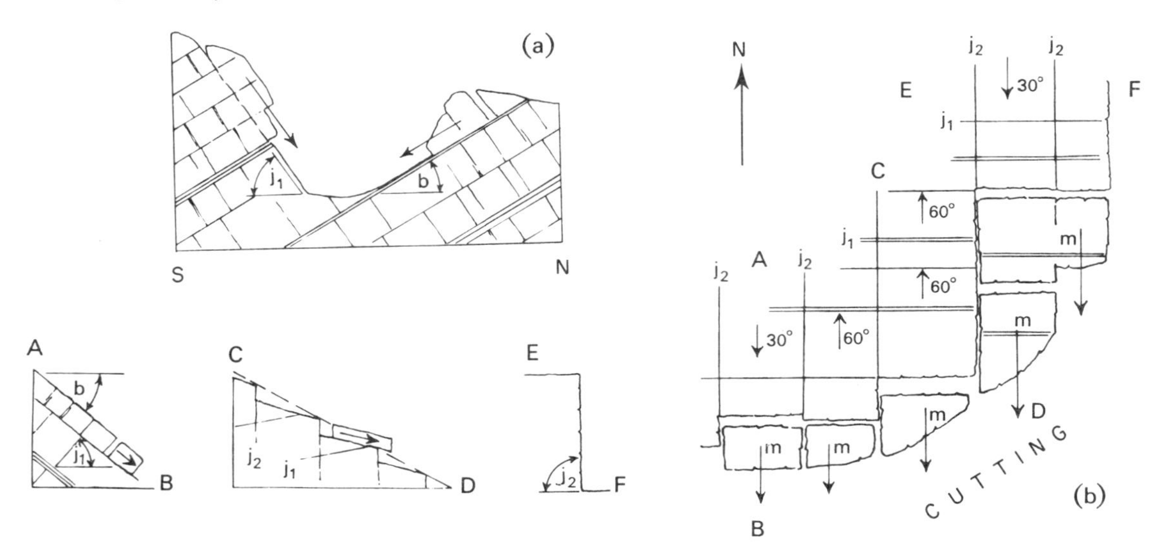

Fig. 14.3 Effect of structure on slope stability. (**a**) In section: j=jointing, b=bedding. (**b**) In plan: dip of bedding=30°S: dip of joint$_1$=60°N; joint$_2$=vertical; m=blocks sliding into the cutting.

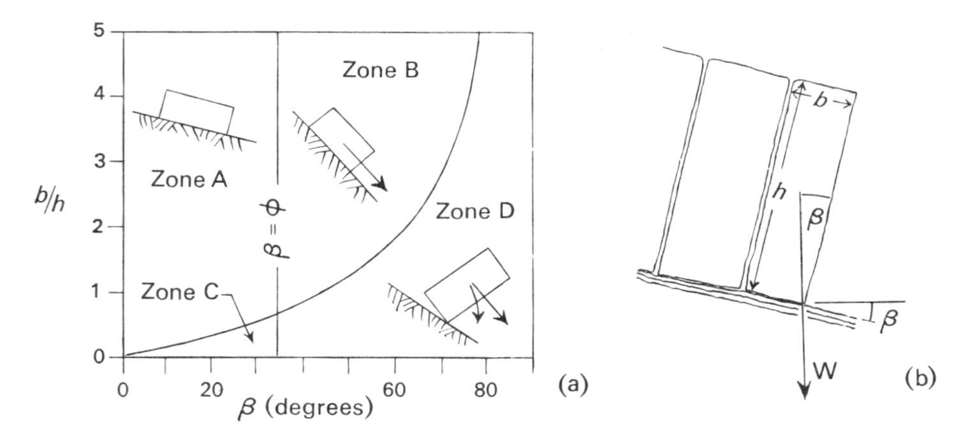

Fig. 14.4 General fields of stability for toppling and translation. ϕ=angle of friction; taken as 35° in this figure, β=dip of basal plane. (**a**) Zone A: b/h>tan β; $\beta<\phi$: a stable zone. Zone B: $b/h>\tan\beta$: $\beta>\phi$: sliding only. Zone C: $b/h<\tan\beta$; toppling only. Zone D: $b/h<\tan\beta$; $\beta>\phi$: sliding and toppling can occur. (**b**) Relevant dimensions for analysis of toppling. (From Ashby, 1971.)

serious engineering problems (Bjerrum and Jorstad, 1968).

Folding, which tilts beds (as in Fig. 14.3), may greatly reduce the shear strength of weaker horizons, for as noted in Chapter 8, certain folds develop by the shearing movement of adjacent layers (Fig. 8.14). This shearing may so re-orient the minerals in the zone of movement that the strength of the surface between adjacent layers is reduced to a residual value (Figs 9.24 to 9.26): such surfaces are often slickensided (Morgenstern and Tchalenko, 1967). A series of surfaces of low shear strength in the folded Siwalik Series of India caused considerable problems in the stability of excavations at the Mangla Dam site, Pakistan (Henkel, 1966). The deformation which occurs in fault zones is similar and often identical in end result. Zones of reduced shear strength are common in the folded strata of coal basins (Fig. 6.19) and represent a widespread and easily overlooked hazard in the stability of slopes in open-pit mines (Stimpson and Walton, 1970).

Ground-water

The self-weight of a dry slope generates stresses that can be modified by the presence and movement of ground-water which can affect slope stability in the following ways:

(*i*) by changing the effective stress (p. 161) and thus the resistance to shear stress,

(*ii*) by generating seepage forces (p. 222 from ground-water flow towards the slope face, which augment those forces tending to destabilize a slope,
(*iii*) by operating as an agent of weathering and erosion to promote dissolution in soluble rocks, swelling in expansive clays and erosion of fine particles from weakly cemented deposits.

The only natural character of a slope that can be changed economically, and on a scale large enough to improve slope stability, is ground-water, because it can be drained by gravity. The alternatives to drainage are expensive regrading of a slope to reduce the gravity induced shear stresses within it, or equally expensive support measures to increase the resistance of the slope to these stresses. Hence the drainage of ground-water is a common component of slope design (Fig. 14.5).

Veder (1981) describes how drainage of ground-water from an unstable soil slope stabilized the ground sufficiently for a road embankment to be founded on it (Fig. 14.6). Lane (1969) describes the stability of walls in a power house built at the base of a 76 m cliff of rock on the side of a valley. The rock contained near vertical joints oriented parallel to the valley side. These permitted the drainage of water from a canal at the top of the gorge to the power station area, where drainage holes had been drilled into the rear wall of the house to provide an exit for the water. The power house was built in 1910 and the drainage system worked successfully for 45 years, but in 1955 it was decided to stop the seepage and the drainage holes were grouted. The cliff collapsed shortly after.

The hydraulic forces created by ground-water that fills a tension crack at the rear of a slope and discharges via a surface of potential sliding are illustrated in Fig. 14.7.

Movement of water through a rock slope creates seepage forces, as illustrated in Fig. 14.8. The unbalanced force produced by the pressure head difference (Δh) acts on each block. Slopes of porous materials such as sands and clays are similarly influenced by seepage forces.

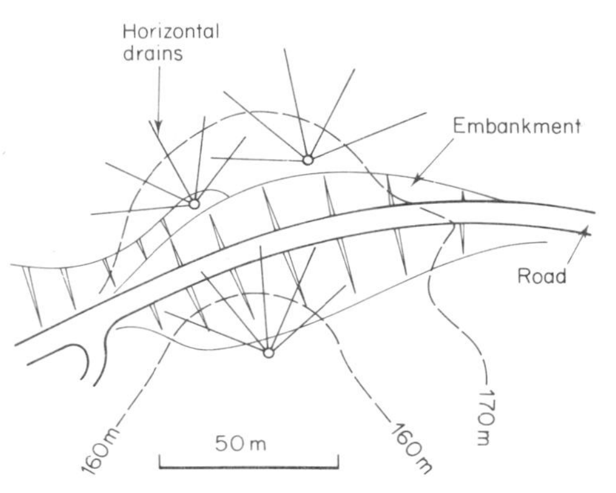

Fig. 14.6 Three fans of horizontal drains to discharge ground-water from the slope to collector wells. Surface drains (not shown) are also present (After Veder, 1981.)

Fig. 14.5 Dewatering of ground around a surface excavation for the construction of an inclined access tunnel to a mine at depth. WP = well points into sand and gravel overlying bedrock. DW = deep wells into bedrock.

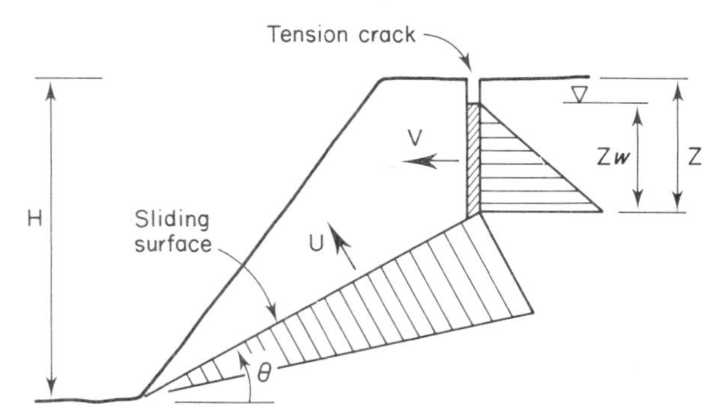

Fig. 14.7 Forces produced by water confined in a tension crack and along a planar sliding surface.
$V = \frac{1}{2}\gamma_w \times Zw^2$ *where* γ_w = unit weight of water.
$U = \frac{1}{2}\gamma_w \times Zw\,(H - Z) \times \mathrm{Cosec}\,\theta$.

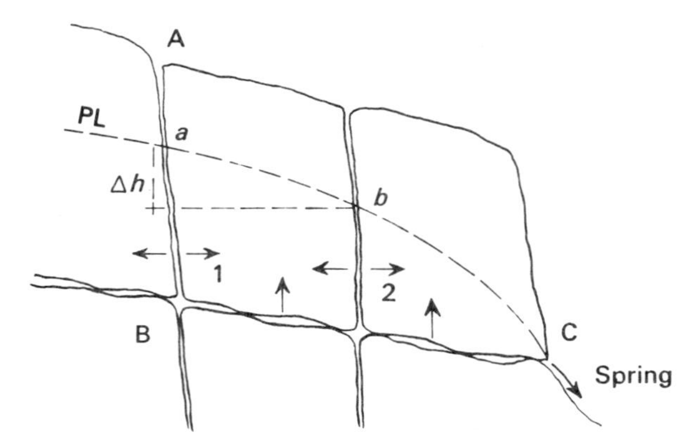

Fig. 14.8 Some water pressures in slopes (after Muller, 1964). PL = piezometric level of water in joint BC which rises as a water level to points *a* and *b* in joints 1 and 2. The thrust on joint 1 is greater than that opposing it on joint 2 by an amount $= \gamma_w \Delta h$, where γ_w = unit weight of water.

Sometimes ground-water flow can be sufficient to dislodge mineral grains from poorly cemented materials such as silts, and wash them out of the ground; a process called *internal erosion*. When this occurs the overlying layers collapse and promote a slope failure. An example is described by Ward (1948) and illustrated in Fig. 14.9. The drainage of surface water from a road into a confined layer of fine sand caused internal erosion and slope failure. Remedial measures included the drainage of road run-off and protection of the seaward exposure of sand with a graded filter of granular material through which the fine sand could not be carried by flowing water.

In-situ stresses

The stresses within a slope are generated by its self-weight, the presence and movement of ground-water, and the geological history of the rock or soil from which the slope is composed: occasionally a slope may also be stressed by accelerations produced from earthquakes.

Stresses produced by the geological history of a rock or soil (p. 159) are embodied in the materials themselves, remaining there after the stimulus which generated them has been removed: for this reason they are called *residual stresses*. The part abnormally high horizontal stress plays in initiating landslides in over-consolidated clay (Fig. 9.6) is an example of their influence (see Bjerrum, 1967). Such residuals decrease in magnitude with time and most of those found in slopes are relics of stress which has been incompletely released by unloading when the slopes were formed. The strain that accompanies the relaxation of such stress is described in Chapter 16 (see for example Fig. 16.12); it augments those strains induced by the self-weight of the slope and with them results in an upward and outward movement of the base of the slope. Hollingworth *et al.* (1944) reported updoming of the floor of clay valleys (Fig. 14.10), stating that:

> '... it appears that the clay has reacted ..., to differential unloading resulting from the erosion of overlying rocks. As the downcutting of streams proceeded, the excess load on either side of the valley could cause a squeezing out of the clay towards the area of minimum load, with consequent forcing up of the rocks in the valley bottom.'

The updoming is called *valley bulging* and has since been recorded at many other localities (Fig. 8.17).

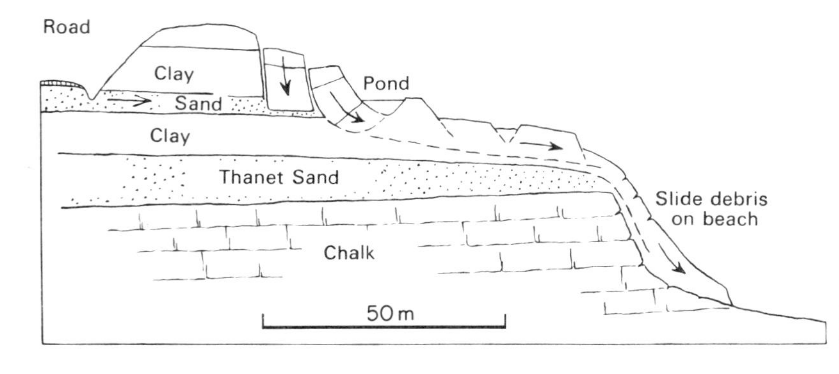

Fig. 14.9 General section through Castle Hill, Newhaven.

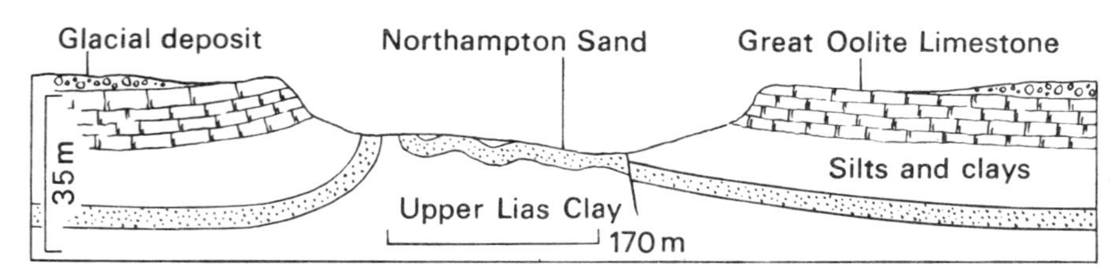

Fig. 14.10 Section across the bulged valley of Slipton in Northamptonshire (after Hollingworth, *et al.*, 1944). See also Horswill & Horton, 1976.

Seismic disturbances

The release of energy from earthquakes results in seismic waves travelling through the ground, which is accelerated by them. Such dynamic loading increases the shear stresses in a slope and decreases the volume of voids within the material of the slope leading to an increase in the pressure of fluids in pores and fractures. Thus shear forces increase and the frictional forces to resist them decrease. The factors which affect the response of a slope are (*i*) the magnitude of the seismic accelerations, (*ii*) their duration, (*iii*) the dynamic strength of the materials affected and (*iv*) the dimensions of the slope.

Seed (1967) describes the coastal landslides that occurred at Turnagain Heights, Anchorage, Alaska. These are composed of outwash sands and gravels overlying 30 m to 50 m of marine Bootlegger Cove clay, a mixture of stiff clay and soft sensitive clay which contains seams of silt and sand. These coastal slopes had withstood earlier earthquakes up to 7.5 intensity (revised Richter Scale). The earthquake of March 27, 1964 was of intensity 8.5 and lasted at least 4 minutes, an unusually long time. After withstanding the movement for 1.5 minutes, the slopes collapsed, in places carrying their seaward edge 600 m into the bay. Landsliding continued for some time after the 'quake. Subsequent tests made on the sediments revealed that the sand lenses within the clay could liquefy within 45 seconds when loaded by vibrations similar to those of the earthquake: the clay itself failed after 1.5 minutes. It is therefore thought that the sand lenses liquefied soon after the 'quake began, their loss of strength promoting eventual failure of the clay. A possible sequence of events is illustrated in Fig. 14.11.

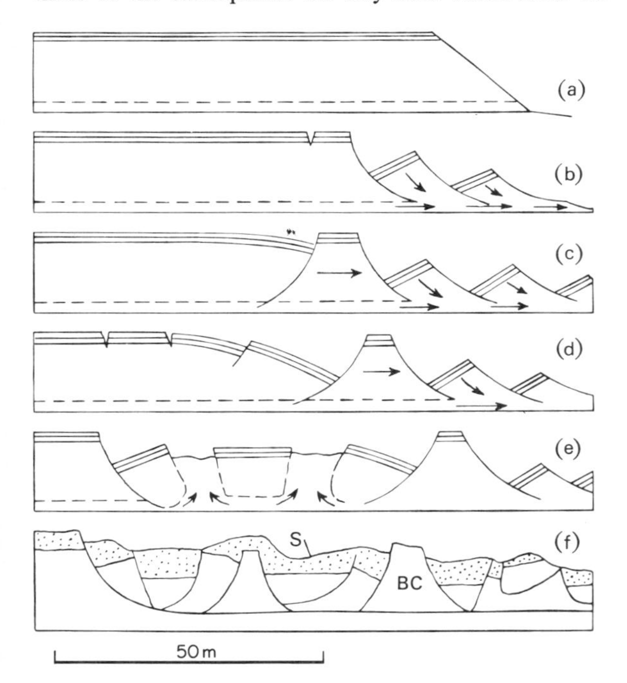

Fig. 14.11 Failure of Turnagain Heights cliffs, Alaska. (**a**)–(**e**), as developed in model tests; (**f**) final situation as observed in the field (after Seed, 1967). S = sands and gravel; BC = Bootlegger Clay.

Failure at Turnagain Heights was associated with liquid pore pressures. Air pore pressures can have the same effect, as illustrated by the landslides in Kansu Province, China, in 1922. Here, thick deposits of loess lay on bedrock: gentle valleys terraced by farming crossed the area. On December 16 an earthquake occurred, producing violent jerking movements, the main period of movement lasting 0.5 minute. The valley sides collapsed and flowed downhill, burying many villages and affecting an area of 272 × 240 km. Close and McCormick (1922) recorded that

> '... the earth ... had shaken loose ... grain from grain, and then cascaded like water, forming vortices, swirls and all the convolutions into which a torrent might shape itself'.

The earthquake had initiated a series of enormous dry flow movements.

Slope history

Natural slopes in which modern landslides occur normally have a geological history of instability, the modern landslide, as in Fig. 14.12, being the latest movement associated with slope erosion. Important aspects of slope history are the previous conditions that affected the slope, and its history of weathering and erosion.

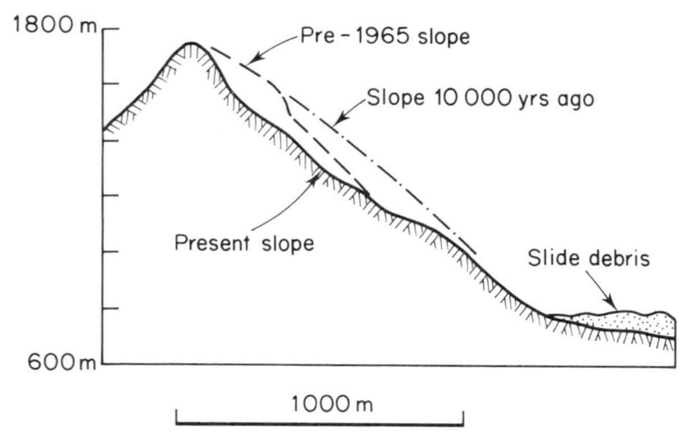

Fig. 14.12 Slope history for Johnson Peak, British Columbia (after Mathews & McTaggart, 1978). Slide debris consists of material from 1965 fall lying stratigraphically above that of a pre-historic slide. (See also Fig. 9.29.)

Previous conditions

Much slope movement that occurred under past periglacial conditions is not immediately apparent at ground level and its presence must be sought with the aid of careful sub-surface investigations (Chapter 10). A number of engineering problems have arisen when excavated slopes cut into the gently and apparently landslide-free sides of valleys, have failed on the buried surfaces of former landslides. Trial pits dug to investigate the ground revealed the presence of ancient mud-flows (Fig. 3.7). Soil testing showed that the angle of residual shearing resistance (ϕ'_r) for the surfaces on which these flows moved, may be *greater* than the present angle of topographic slope. Table 14.2 presents data from over-consolidated, fissured, Cretaceous and Eocene clays in S. England.

Field measurements indicate that in fissured clay a slope of angle greater than approximately $\phi'_r/2$ will ultimately fail, but as seen from Table 14.2, slopes of smaller topographic angle than this may contain surfaces on which sliding has occurred. Obviously the strength of the material has at some time been reduced and as such a condition does not exist at present it must have happened in the past, most probably when the ground was weakened by periglacial conditions during the Pleistocene. Slope movements of this age have been found in many countries (see Fig. 3.8).

Table 14.2 Strength values for ancient slip surfaces (from Weeks, 1969)

Approx. angle of topographic slope (degrees)	ϕ' (degrees)	Depth of surface below existing ground level (metres)
7	15	3.9 (1)
4	15.6	1.8
4	16	1.8 and 3.0
7	12.4	1.6 and 4.5
3	12.7	1.2 to 2.25
5	14	1.3
3½	14	1.6

(1) dated (by ^{14}C) at approximately 10 200 B.C. Trial pits revealed that in some localities a series of mud-flows had moved one over the other, the oldest being at the base. Some flows were associated with soil horizons which were later buried by succeeding flows. The soil horizons contain carbonaceous material from which absolute dates can be obtained.

Weathering

The rate at which chemical changes occur may range from a few days to many years, and can affect both the short term and long term stability of slopes. Soft sediments are most readily involved in these changes as are the soft fillings found in rock fractures, such as the gouge in some fault zones. Hard rocks and those composed of chemically stable minerals are less susceptible to change (see Chemical weathering, p. 31).

Chandler (1969) noted that the weathering of marl increased its liquid limit and natural moisture content and decreased its bulk density, permeability and shear strength (c', ϕ' and $\phi'_{residual}$). Another example is reported by Drouhin *et al.* (1948) from a slope in Algeria where calcareous rocks overlie a glauconitic marl. Calcium, from ground-water percolating through the calcareous strata, was taken by base exchange into the structure of the clay mineral glauconite, which in turn released potassium. The pH of the ground-water increased as its calcium content decreased, causing the marl to deflocculate to a colloidal gel. Slope failure occurs if this process continues until the strength of the marl can no longer support the weight of overlying strata. The mineral glauconite has been found in the lower levels of other large slides, e.g. at Dunedin, New Zealand (Benson, 1946), and at Folkestone Warren, England (p. 238).

Matsuo (1957) describes the influence of exchangeable calcium upon the stability of a railway cutting at Kashio, Japan, which failed after being 'stable' for 10 years. Sliding occurred in a series of clayey sands, clays and gravels, and ground-water issuing from the toe of the slide (at rates between 1.3 and 50 $cm^3 s^{-1}$) was found to have a greater free carbonate content than the rain water infiltrating the slope (2.44 mg l^{-1} and 0.039 mg l^{-1} respectively). It was evident that calcium was being removed from the ground and laboratory tests confirmed that such leaching reduced the strength of the sediments in which the slope was cut.

Chemical changes which can occur rapidly in geological materials are the expansion of anhydrite on hydration and the decay of iron sulphide (pyrite and marcasite) on oxidation. Both changes can occur over short periods (Brune, 1965; Steward *et al.*, 1983). The addition of water may also cause swelling and compressive stress, and its subtraction, shrinkage and tensile stress: mechanical and chemical weathering often coexist (Fig. 3.8).

Erosion

Naturally-eroded slopes are created by the movement of ice, surface water and the sea. Glacial erosion produces U-shaped valleys (Fig. 3.41) whose over-steepened sides became over-stressed when the glaciers melt and the lateral support to them provided by the glaciers is gradually removed. Such slopes eventually fail. Glacial erosion is facilitated by the generation of joint sets a small distance below the base of a glacier: similar joints develop in the rock adjacent to the sides of a glacier. These joints are an addition to those existing prior to glaciation and by stress-relief of the valley sides occurring after glaciation (see Fig. 14.14*b*), and are thought to be produced by the drag of ice against the rock walls and base of a valley. It is common for them to be filled with rock-flour (p. 55) and clay; in this condition they have a low shear strength and are often the surfaces on which slope failure may commence. The lower slopes of many glaciated valleys are covered by glacial deposits which conceal such surfaces (Fig. 3.41) and cut slopes of excavations into the bedrock, these can fail on clay filled joints, causing slope instability that may extend over a large area.

In cold regions, erosion by alternating cycles of freezing and thawing is common and produces *rock-falls*. Bjerrum

(1968) describes the occurrence of rock-falls in Norway (Fig. 14.13). Most occur during the spring thaw in April and the period of maximum precipitation in October. During very wet years, unstable slopes are cleaned off by rock-falls and larger slides, and it may be many years before sufficient material is weakened for further large scale slope instability to develop. The periodicity of slope instability is illustrated in (*c*); the period 1720-60 was exceptionally cold and wet, the incidence of landsliding being 10 times that of the following 50 years. Periodic behaviour of this kind is typical of slope instability.

Erosion of valley slopes by rivers is reviewed in Chapter 3. Unlike glaciers, the level of rivers is related to sea-level and buried valleys are the common legacy of former lower sea-levels (p. 42). Landslides which occurred in the valleys at times of lower sea-level may now be buried by alluvium, but can be re-activated if exposed by engineering excavations.

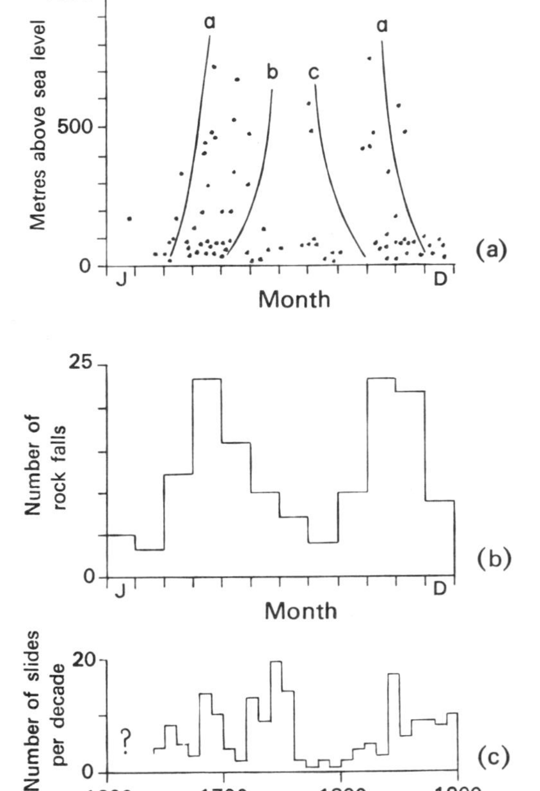

Fig. 14.13 (**a**) 'Rockfalls' (a mixture of falls, topples and small translations, see Fig. 14.1) in E. Norway in relation to altitude, time of fall and temperature. a = average date when the mean daily temperature passes freezing point; b = average last day of frost; c = average first day of frost. (**b**) Number of rockfalls in the years 1951–55 and their montly distribution. (**c**) Number of rockfalls and rock slides in fjords.

Coastal erosion is also considered in Chapter 3. Many coastal areas have raised beaches with the old cliffs forming an inland escarpment whose stability is deteriorating.

Examples of failure

Geological processes rarely work in isolation; most large slope movements result from several processes *collectively* producing an overall condition of critical stability. To illustrate this point three case histories are now described: the Vajont slide at a reservoir; the Turtle Mountain slide, above a mine; and the Folkestone Warren landslide, along a natural coastline.

The Vajont slide

October 9, 1963. (References: Muller, 1964b; Kiersch, 1964; Jaeger, 1969, 1979; Brioli, 1967.)

The river Vajont flows in a steep gorge which cuts through the Alpine folds of north Italy. The broad structure of the valley is a syncline, thought to have been formed in Tertiary times (Fig. 14.14*a*). The rocks involved in the folding are essentially calcareous sediments of Jurassic and Cretaceous age: folding mobilized the shear strength of the weaker clay seams, so that they were at their residual strength when folding ended (see Fig. 8.14).

Pleistocene glaciation removed a considerable volume of rock and scoured out the glacial valley along the axis of the syncline (Fig. 14.14*b*). This unloading of the valley sides promoted the development of stress relief features parallel to the valley itself; in some places new joints were developed, in others conveniently oriented bedding surfaces were opened. Landslides occurred and may have dammed the valley until they were overtopped and eroded away. One slide had actually crossed the valley and its leading edge lay *on* the glacial sands and gravels which locally cap the bedrock. By the time the glaciers had retreated the valley was sufficiently elevated above sea level for its river to cut a gorge some 195 m to 300 m deep in the valley floor.

A hydro-electric scheme for the valley had been planned, and by September 1961 a thin arch concrete dam was constructed. Geological investigations of the valley sides (as distinct from the dam foundations) had been in progress at intervals since 1928 but were intensified in October 1960, when accelerated movement into the reservoir area occurred on the south slope of the valley within 390 m of the dam. The movement was accompanied by a large M-shaped tension gash which extended along the south slope (Fig. 14.14*c* '1960 scar'): approximately 200×10^6 m^3 of rock and was moving on a zone of sliding-surfaces situated about 198 m below ground level (Fig. 14.14*a*). The front of the slide moved at 8–10 cm per day and the remainder at 3–5 cm per day: the movement as a whole suggested that progressive failure and creep were occurring.

The volume of the moving mass precluded all remedial measures other than those which would reduce the pres-

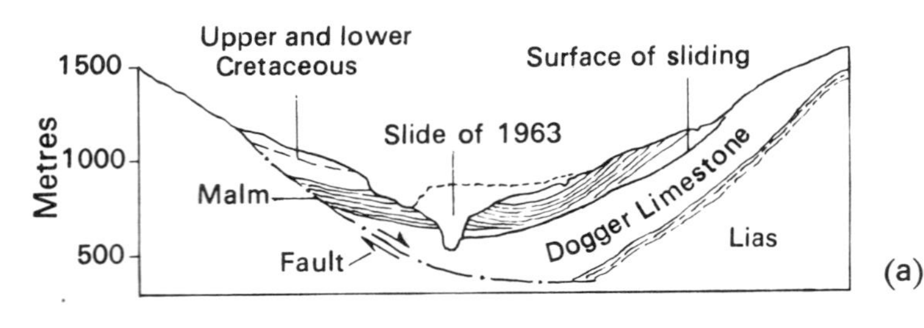

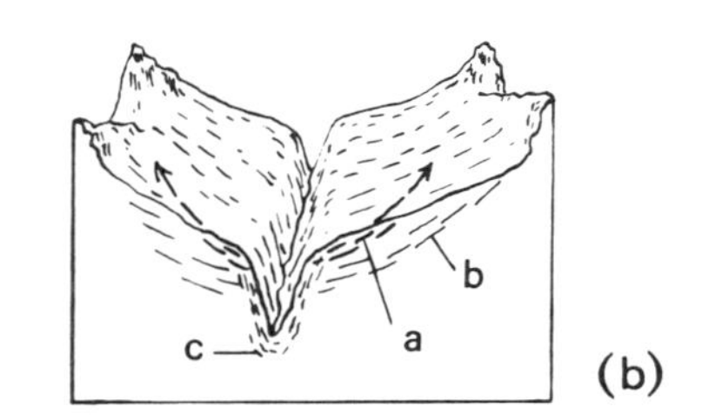

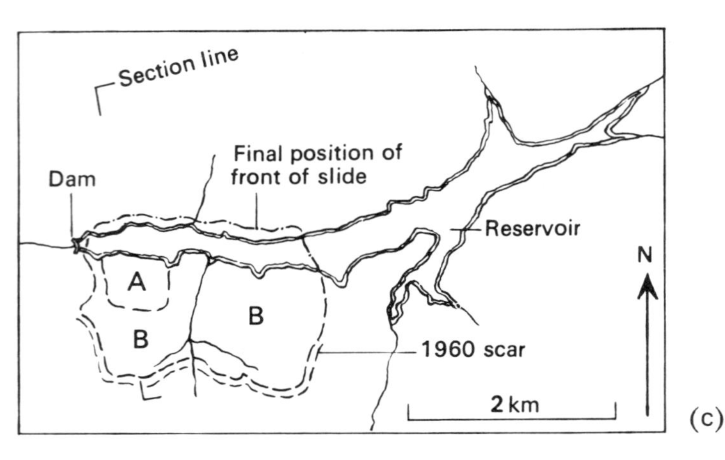

Fig. 14.14 The Vajont Slide. (**a**) General elevation of the geology in metres above sea level. (**b**) Sketch of the gorge. a = joints associated with old glacial valley; b = opening associated with bedding planes; c = joints associated with younger river valley. (**c**) Map of Vajont reservoir. A = 1960 slide; B = 1963 slide. Both slides moved North, into the reservoir.

sure of water within the slide. It was decided to drain the slope by adits and to lower the level of water in the reservoir. The rate of lowering was such that excess water pressures would not develop in the slide, however, the reservoir level rose a further 10 m to 89% of the final height before this drainage programme could begin. On November 4, 7×10^5 m^3 of material slid from the toe of the slide into the reservoir in 10 minutes, and is known as the 1960 slip (Fig. 14.14*c*). The stability of the slope was evidently closely related to the reservoir level. This was

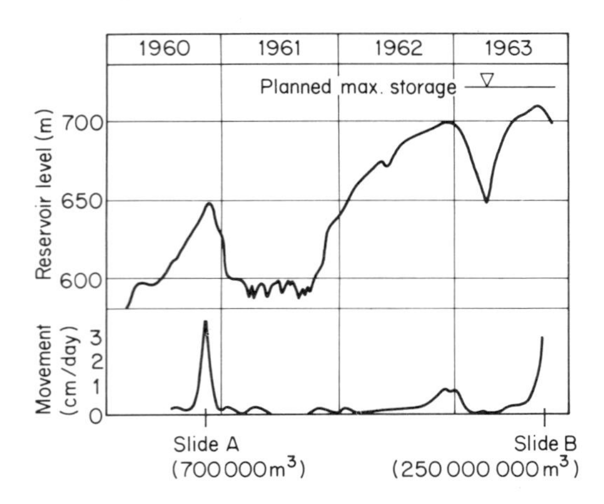

Fig. 14.15 Relationship between movement on the Vajont slope and reservoir level. (Int. Commission on Large Dams, 1979.)

demonstrated throughout 1961, 1962 and 1963 when slope movements accompanying a rise in reservoir level could be halted by a fall in reservoir level (Fig. 14.15). In April 1963 it was considered safe to raise the reservoir level. Slope movement began, slow at first, but increasing: lowering of the reservoir level commenced at the end of September with the intention, based on earlier experience, of bringing the creep to a standstill. Lowering was carried out at a slow rate (15 m per week) and by October 9 the water level had dropped to where it had stood in November 1962 and June 1963 (Fig. 14.15).

That night at 23.38 G.M.T. there was a violent failure that lasted a full minute. The whole of the disturbed mountain-side slid downhill with such momentum that it crossed the river gorge (99 m wide) and rode 135 m up the farther side of the valley (Fig. 14.14*a*). More than 250×10^6 m^3 had moved, at a speed of about 24 ms^{-1}. This sliding lasted for 20 seconds, produced seismic shocks that were recorded throughout Europe, and sent a huge wave over the dam which, however, survived. The wave levelled five villages in the valley below and killed more than 1500 people.

In his analysis of the failure Jaeger simplifies the slide geometry to that given in Fig. 14.16 and shows that the factor of safety for a creeping system of this character decreases with a decrease in the ratio of the angles of friction (ϕ_2/ϕ_1) whatever the values for the ratios W_2/W_1 and α_2/α_1. The value ϕ_2 must have been low to permit the observed deformations in zone *A*, hence ϕ_1, the friction on the 'seat' of the slide became critical to the overall stability. Increase in the uplift forces (U) which occurred

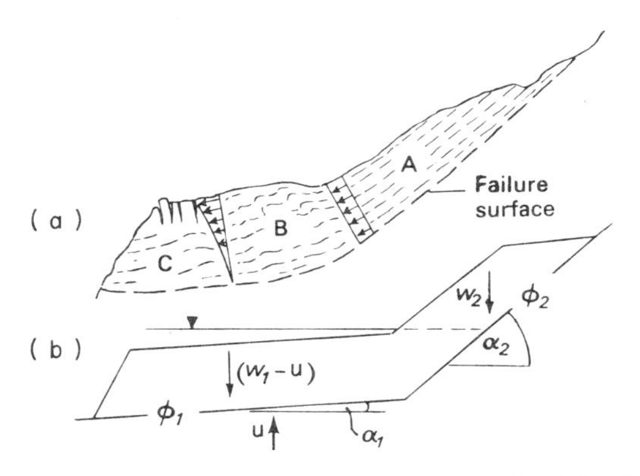

Fig. 14.16 (**a**) General character of the slide (after Muller); arrows show relative velocities of movement. (**b**) Section analysed by Jaeger (1969); W = weight of rock and reservoir water above rock; u = uplift pressures; ϕ = angle of friction; α = dip of sliding surface.

each time the reservoir level was raised, would decrease ϕ_1 and the periodic movements related to rises in reservoir level indicate that the factor of safety must have been hovering between 1.0 and some value above 1.0. The process continued until cohesive forces on the flatter lower section of the slide had been so reduced that rock there failed by rapid fracture. Investigations made after the failure support this theory.

The Turtle Mountain slide

April 29, 1903. (References: McConnell and Brock, 1904; Daly *et al.*, 1912; Cruden and Krahn, 1978.)

Turtle Mountain is a long, narrow, wedge-shaped ridge cut into an anticline that forms part of the front range of the Canadian Rocky Mountains in South Alberta. Its peak rises 945 m above the valley of the Old Man River, and its eastern face overshadows the mining town of Frank. This face has a talus slope at 30 degrees which extends to 245 m above the valley floor where it ends against a precipitous upper cliff. The mountain is composed of Devonian and Carboniferous limestones which have been thrust east over Cretaceous shales, sandstones, and coals (Fig. 14.17). The limestones are an alternating sequence of contrasting beds, some massive and coherent, others flaggy. Bedding plane movement during folding (called flexural slip) has striated the bedding surfaces parallel to the dip direction (Fig. 8.14). Two thrust zones exist, bounding a thickness of contorted strata (T in Fig. 14.17).

Three features of the geological structure are of note:

(*i*) The crest of the anticline, with its propensity for open tensile fractures (see Fig. 8.27) coincides with the mountain crest.
(*ii*) Bedding and jointing dip downslope at angles greater than their angle of friction.
(*iii*) The base of the slope contains sheared and weak rock.

Thus in many ways Turtle Mountain was potentially unstable, and it was surprising that no slide resulted from the severe earthquake in 1901 that was centred in the Aleutian Islands.

Beneath this slope a drift mine was opened in 1901 in the nearly vertical coal seam at the foot of the mountain. From the mouth of the mine, 9 m above river level, a level gangway was driven for 1.6 km along the strike of the coal. A second level was then dug, 9 m below the first, for drainage and ventilation. Large chambers were opened up from the first gangway, each chamber being some 40 m long and 4 m wide and separated from its neighbour by pillars 12 m long; these pillars contained manways 1.5 m square. Most of the coal was stoped down into the chambers, where it was drawn off at chutes.

In October 1902 the miners noticed that the chambers were beginning to squeeze with noticeable severity, particularly between 1 a.m. and 3 a.m. Gangways were being continually re-timbered and manways, driven up to the outcrop, which could not be kept timbered, were abandoned. Early in 1903 coal was being mined with unusual

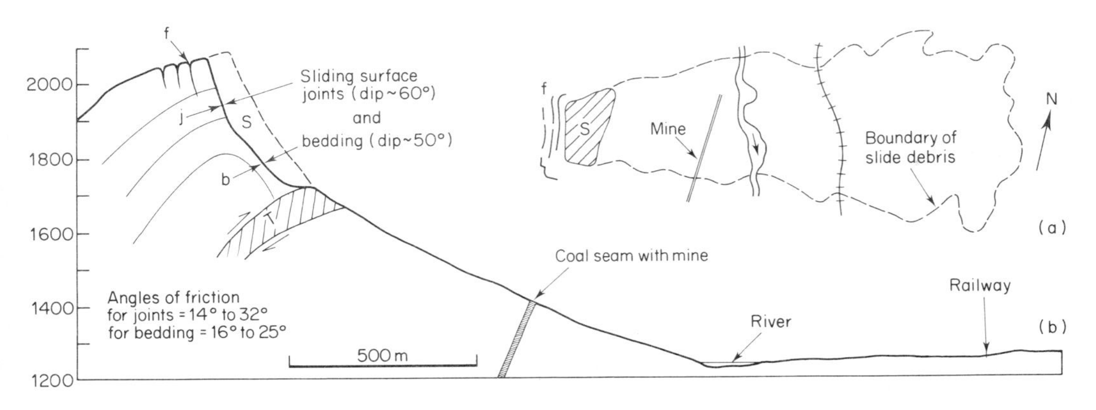

Fig. 14.17 Turtle Mountain slide. (**a**) Sketch map: f = tension fractures, S = slide. (**b**) Section (metres above sea level). T = Thrust zone. (After Cruden & Krahn, 1978.)

ease, and was mining itself over a 457 m length of the working, bringing with it parts of the hanging wall. By April 1903 the mined chambers were up to 122 m high; 187×10^3 m^3 of coal had been removed, with extraction continuing at about 1000 Mg per day.

Above ground level, the countryside was enjoying a warm spell of weather with temperatures up to 23°C, and April 28 in particular was a very warm day. However it was followed by a cold night with heavy frost.

At 4 a.m. on April 29 miners underground noticed that coal was breaking and running down the manways. The men took to the ladders to escape and, continually battered by falling coal, eventually reached the main level. A few minutes later the driver of a shunting engine on the surface heard the cracking of rocks on the mountainside. This was followed suddenly by a sound like an explosion, and 90×10^6 Mg of limestone fell 762 m from the peak of the mountain. The fall of rock hit the ground with a heavy thud that shook the valley, and was deflected into the air by a sandstone ridge just uphill of the coal outcrop, sending it over the outcrop and downwards into the Old Man River. But it did not stop. Instead, it scoured its way through the river and then slid for more than 1.5 km over the rolling hills to the east, making a noise that resembled steam escaping under high pressure. It came to rest abruptly after climbing 120 m up the opposite side of the valley, having covered in its wake an area of over 2.6 km^2. From eye-witness accounts it appears that the whole event took no longer than 100 seconds. The slide killed 70 people as it travelled through Frank.

The behaviour of the slipped mass after its detachment from the mountain is worth noting. Its great speed was assessed at 144–176 km h^{-1}, yet the slide was carrying blocks of limestone 6–12 m long over hummocky ground, and it eventually climbed 120 m. Its character was that of a fluidized sheet of rock debris. These and other points have been studied by Shreve (1968) who concludes that the slide was a highly lubricated moving mass; he puts forward other evidence from similar slides elsewhere, to suggest that the lubricant was nothing more than a layer of compressed air trapped beneath the debris. According to this theory a mass detaches itself, descends a slope, is launched into the air by some topographic feature, and traps on its descent a layer of air on which it rides at high speeds. Voight and Faust (1982) have further demonstrated that frictional heat generated on the sliding surface of an accelerating landslide, is sufficient to cause an expansion of pore fluids. The consequent and rapid loss of the frictional strength of a moving mass may convert sliding at moderate rates to a rapid descent.

The Folkestone Warren slides

(References: Hutchinson, 1969; Hutchinson, *et al.*, 1980.)

Folkestone Warren is a 3.2 km stretch of naturally unstable coastline between the old ports of Folkestone and Dover in south-east England (see Fig. 8.11). It is an area where landsliding is common and has been in progress for centuries (Fig. 14.18).

Some geological details of the Warren are shown in Fig. 14.19. The High Cliff is composed of Cretaceous limestone (the Chalk), with an impervious horizon, the Chalk Marl at its base. The latter lies above 3 m of glauconitic sandy marl known as the 'Chloritic' Marl. Below this is the over-consolidated, fissured and jointed Gault clay, 43–49 m thick. Beneath the Gault lies the Lower Greensand, a coarse yellow sand, highly permeable, containing calcareous and glauconitic horizons.

Three types of movement (Fig. 14.19) are termed by Hutchinson:

(*i*) M-slides: movements in multiple rotational landslips of the undercliff resulting in large seaward displacements; these movements seem to occur on non-circular surfaces.

(*ii*) R-slides: smaller features which are rotational and involve movement only in slipped masses close to the sea cliff.

(*iii*) Chalk falls: large masses of Chalk, which are commonly preceded by downward movements known as 'sets'; these have been associated with subsidence of up to

Fig. 14.18 Folkestone Warren landslip looking east (Crown copyright).

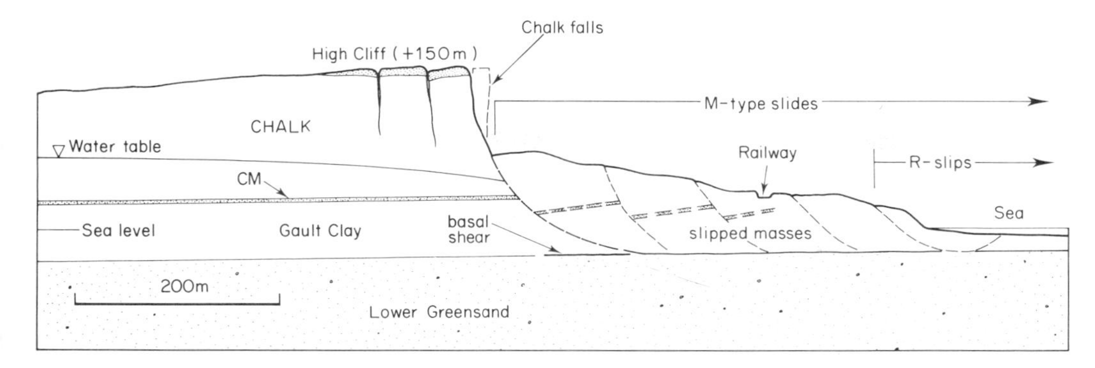

Fig. 14.19 Section through the landslides of Folkestone Warren. CM = Chloritic Marl. (After Hutchinson, 1969; 1980.)

1.5 m at the cliff top and can affect the Chalk for 18 m from the edge of High Cliff.

Hutchinson suggests that the main controls promoting the above movements are the intensity of marine erosion along the toe of the slides and the influence of pore water pressure at potential slip surfaces: there being a recurring relationship between high ground-water levels and slope movement.

Large movements involve failure of the Gault whose effective residual angle of friction (ϕ'_r) ranges between 7° and 15° depending upon fabric and mineralogy, both of which vary through the deposit. The value of (ϕ'_r) for the Chalk ranges from 19° to 35°. The cliffs are therefore composed of contrasting materials, and the following mechanism has been suggested by Hutchinson to explain the retrogression of the slips which has gone on for many years. Considering first the geological setting, the cliffs at Folkestone Warren lie on the northern limb of the Wealden anticline (Fig. 8.11), a structure of Oligocene or early Miocene age. Geological evidence suggests that the upper Gault was consolidated under an effective pressure of some 3.7 to 4.3 MN m^{-2}. Erosion since the formation of the anticline has left the Gault in an over-consolidated state, the present effective pressure on its upper surface being about 2.6MN m^{-2}. Under such conditions the retrogression of the rear scarp of the landslip could have occurred in the following manner.

The over-consolidated Gault, with its content of active clay minerals and their strong diagenetic bonding will, during the unloading by erosion, have released little of the strain energy it accumulated during consolidation. If therefore, as is likely, an expansion potential exists in the Gault, its seaward portion (where marine erosion has removed lateral support) could undergo expansion with the resulting development of a shear surface at or near the base of the formation. This surface would *progress inland* as erosion ate further into the slipped masses. The overlying Chalk could be put into tension as a result of the seaward movement of the Gault, and vertical joints opened up. The local load on the Gault is thus increased, and with reduced lateral support to seaward the clay will fail and the slope subside. After such subsidence, re-engagement of the Chalk could occur and might be sufficient to temporarily stop further movement. Final collapse would coincide with the next M-type slide, which would allow the cycle of regression to be repeated.

These and other examples cited in this Chapter illustrate the crucial role of geology in defining the nature and occurrence of slope movement.

Investigations

It is usually necessary to employ surface and sub-surface investigations to define the stability of a slope. Common techniques for use in these studies are described in Chapters 10 and 11.

Such investigations may commence with an assessment of the shape and geological structure of the slope. Surface mapping and the construction of vertical cross-sections are required (see Chapter 12). The rocks and soils of the slope, together with the dip and strike of surfaces on which movement may occur such as those of bedding, jointing, cleavage and faulting, should be recorded on the map. More than one vertical cross-section may be desirable, each drawn *without* vertical exaggeration so that the dip of surfaces in the plane of section is correctly exhibited.

With these data it should be possible to determine whether the slope has failed before and is thus a slope that contains the remains of a former landslide rather than one in undisturbed ground. Existing failure surfaces may have little shear strength and be easily re-activated by engineering work (Fig. 14.20). It should also be possible to identify either the type of failure that has occurred or is likely to occur, in particular whether movement is translational or rotational (Table 14.1).

Hence the first phase of investigation requires a ground reconnaissance survey and study of air photographs of the slope. Surface geophysical surveys are especially helpful in revealing sub-surface geological structure and ground disturbed by previous movement. Properly supported pits and trenches will expose sliding surfaces close to ground level and may also reveal sufficient recent stratigraphy to permit the history of the slope to be reconstructed (see Fig. 9.29).

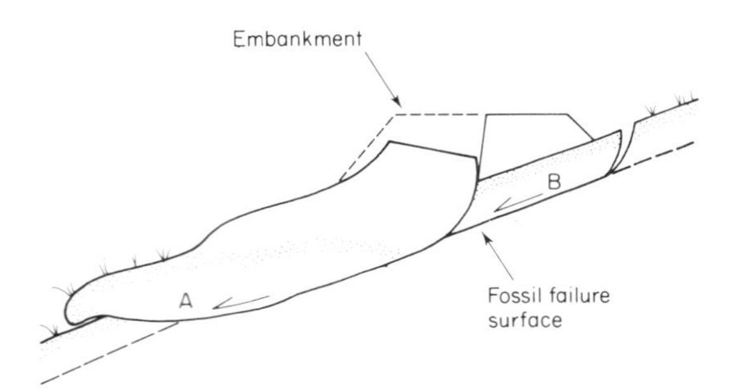

Fig. 14.20 The weight of a new embankment causes movement on a previous failure surface.

The second phase of investigation considers the likely strength of the slope, either by an analysis of a previous slope failure in similar material (see Chapter 9) or from laboratory tests performed on samples collected from the slope. Analyses of previous slides provide values of shearing resistance at failure, because these may be taken to equal the shear forces at failure which can be calculated from the dimensions of the slope and knowledge of the shape of the sliding surface. For such a calculation to be completed it is necessary to know the magnitude of water pressures and seepage forces in the slope. The hydrogeology of the slope must therefore be studied. Aquifers that supply water to a slope and those that drain the slope must be located together with aquicludes which prevent such movement of ground-water. Areas of infiltration and spring seepage should also be identified. Piezometric levels (*q.v.*) should be monitored especially in the vicinity of potential or existing failure surfaces.

The second phase of investigation is therefore conducted to greater depths than the first and normally necessitates the drilling of bore-holes to confirm the internal structure of the slope, and to locate weak zones within it. Into these bore-holes may be installed piezometers to measure ground-water pressure within the slope and devices for monitoring slope failure (e.g. by acoustic emissions) and movement.

In addition to the investigations described above, further studies will be needed in seismic areas to assess the magnitude and frequency of seismic events. This may require a search of historical evidence and extensive field surveys. Many examples of slope failures have been recorded for reference in the Proceedings of the International Symposium on Landslides, Toronto (1984).

Selected bibliography

Hutchinson, J.N. (1968). Mass movement. In *The Encyclopedia of Geomorphology*. Fairbridge, R.W. (Ed.). Reinhold Book Corp., New York.

Schuster, R.l. and Krizek, R.J. (1978). *Landslides: Analysis and Control.* Transportation Research Board Spec. Rep. 176. Nat. Acad. Sci., Washington, D.C.

Hoek, E. and Bray, J.W. (1977). *Rock Slope Engineering* (second edition). Instn. Mining & Metallurgy, London.

Veder, C. (1981). *Landslides and their Stabilization.* Springer-Verlag, New York.

Zaruba, Q. and Mencl, V. (1969). *Landslides and their Control.* Elsevier, Amsterdam, and Academia, Prague.

Voight, B. (Ed.) (1978). *Rock Slides and Avalanches* Vols 1 & 2. Elsevier Sci. Pub. Co., New York.

Embleton, C. and Thornes, J. (Eds.) (1979). *Process in Geomorphology*. Edward Arnold, London.

Selby, M.J. (1982). *Hillslope Materials and Processes.* Oxford University Press, Oxford, London, New York.

15

Reservoirs and Dams

Reservoirs of water may be stored successfully in valleys whose floor and sides will not permit leakage from such artificially created lakes. A dam is required to impound the water, and its design and construction must be capable of preventing the uncontrollable leakage of reservoir water around and beneath its structure. Reservoirs may also be created underground, either by utilizing the natural storage space of pores and fractures in rock, or by excavating caverns of adequate volume for the storage required. The influence of geology upon these schemes is the subject of this chapter.

Surface reservoirs

The volume of a reservoir is reduced by deposition within it of river sediment and landslide debris.

Sedimentation

The work of rivers is described in Chapter 3. Sub-aerial erosion transports debris to river channels, where it is carried as alluvium to lakes and the sea. A dam interrupts this natural sequence of events by preventing alluvium from travelling downstream. Sediment therefore accumulates behind the dam and reduces the volume available for the storage of water. The Sanmenxia reservoir in China was completed in 1960: by 1962 there had accumulated within it 1500×10^6t of sediment, increasing to 4400×10^6t by 1964 (Quian Ning, 1982). The Laoying reservoir, also in China, was completely filled with sediment during a flood, before the dam was completed.

Sediment will also accumulate as deltas at the margins of a reservoir, where rivers discharge into the lake (Fig. 3.20). Around the lake a shoreline develops from the action of waves, generated by wind blowing across the lake. These waves erode the topsoil and underlying profile of superficial deposits, and weathered rock. This erosion contributes sediment to the reservoir.

The rate of denudation within catchments may be calculated from a study of sediment within reservoirs. Numerical models that incorporate the effect of geology, geomorphology, pedology and hydrology of a catchment, have been developed to predict the likely sediment yield of a catchment (Thornes, 1979; Walling, 1983).

In practice, it is important to observe the natural processes of weathering and erosion that are operating within a catchment that is to contain a reservoir (Chee, 1972; U.S. Dept. Agriculture, 1973). A carefully produced map of solid and drift geology, that records the outcrop and exposure of easily eroded strata, is often of considerable aid to the prediction of likely problems arising from sedimentation. It is also instructive to study the recent deposits of alluvium, as they represent the materials being transported by the rivers within a catchment. Such studies complement measurements of the suspended load in rivers and provide valuable information about the nature of transported load when the rivers are in flood (p. 41 and Fig. 3.48).

Landslides

All the types of slope movement described in Chapter 14 may occur on the margins of reservoirs. A record of 500 such movements is shown in Fig. 15.1; 49% accompanied a major rise in the level of the reservoir and 30% followed a major fall in the reservoir level. This case history and others (e.g. Vajont, p. 236) demonstrate that large changes of reservoir level are a significant source of instability in reservoir slopes.

The effect of a rise in reservoir level on the stability of a slope has been likened to the behaviour, on submergence, of a mechanism composed of two blocks linked by a freely pivoting rigid beam; Fig. 15.2 (Kenney, 1967). The upper block, which sits on a surface inclined at its angle of friction (α), represents that part of a slope which tends to drive a landslide forward and the lower block represents that part of the slope which generates most of the resistance to such movement. The mechanism is stable:

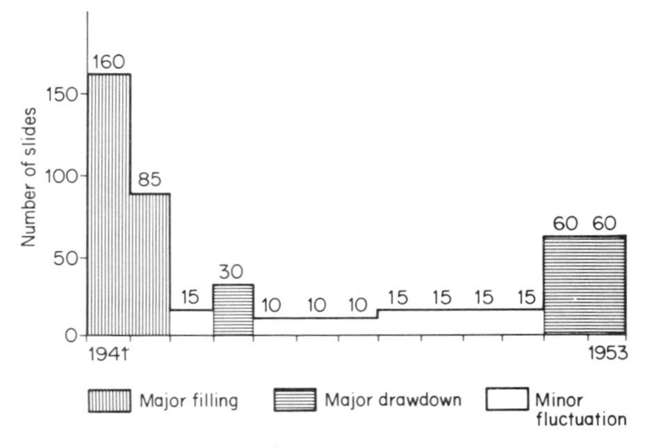

Fig. 15.1 Histogram of 500 slides from one reservoir illustrating their relationship with fluctuations in reservoir level (from Lane, 1967).

i.e. its Factor of Safety is greater than 1.0 (p. 227) when frictional resistance at the base of the lower block is sufficient to hold the upper block in place.

A water level represents the level of the reservoir and is initially lower than the base of the lower block. As this water level rises the lower block is progressively submerged and its weight ($W_{1\ dry}$) thereby reduced to its submerged value. This is accompanied by a similar reduction in the frictional resistance it offers to sliding on its base, the friction on this surface being proportional to the load upon it. The reservoir level at which the Factor of Safety of a slope is either at its lowest value above 1.0 or at 1.0 is called the *critical pool level* (Lane, 1967). Once the level of the reservoir is above this critical elevation further rises in its level help to stabilize a slope, for as the model shows (Fig. 15.2), submergence of the upper block restores the stability of the mechanism to its original value. The moments of a slope when totally submerged are the same as those when it is completely dry.

The model also illustrates the effect of lowering a reservoir level, the sequence of events being in the reverse order to that described above. In addition, the rapid drawdown of a reservoir may leave, within the slopes of its margins, water pressures that are related to its former level (Fig. 15.3). These reduce the effective stress and hence the frictional resistance of surfaces within a slope, and contribute to its instability.

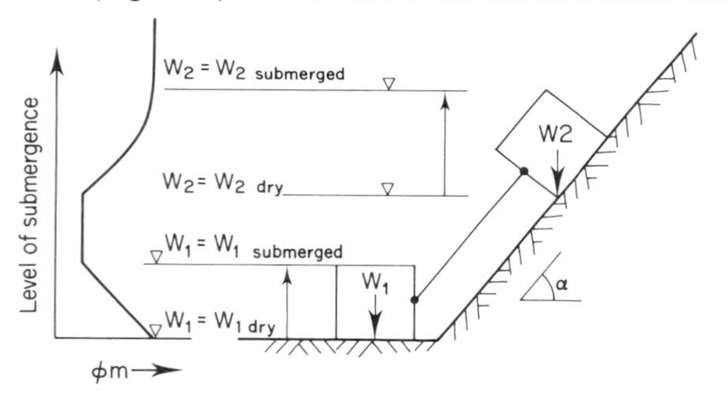

Fig. 15.2 Variation of the mean frictional resistance of this mechanism to sliding (ϕm), with its level of submergence. W=weight of the blocks at different levels of submergence. Compare with Fig. 14.16 where resistance is described in alternative terms of total rock weight and weight of water above rock (when present) – the uplift pressure operating at the base of the slide.

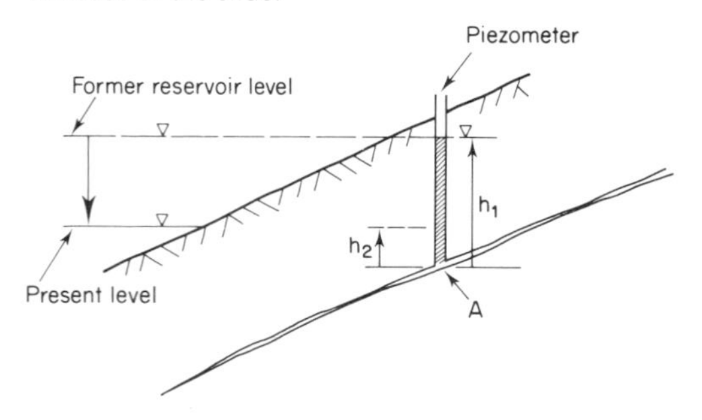

Fig. 15.3 Ground-water pressure at A ($h_1\gamma_w$) is related to the former reservoir level and not that of the present level which would give pressures of ($h_2\gamma_w$). γ_w=unit weight of water.

Geological surveys of reservoir slope stability are commissioned if it is feared that landslides into a reservoir may either significantly reduce its volume available for storage or produce a tidal wave that may over-top the dam and cause flooding downstream. Reservoirs in mountainous terrain are vulnerable to landslides activated by earthquakes. These slides may occur at a considerable distance from the reservoir and travel to it as a flow or avalanche, via a tributary valley (Voight, 1978).

Leakage

There are many recorded examples of reservoirs that have almost uncontrollable leakage of water from their base and sides (Int. Commission on Large Dams, 1973; 1979). To avoid such losses the valleys selected as sites for reservoirs should have at least one of the following geological characteristics; either a floor and margins that contain formations of low permeability, or a natural water level in the valley sides that is higher than the level proposed for the reservoir.

The effect impermeable rocks may have upon restraining leakage is illustrated in Fig. 15.4. To be effective they must create natural barriers that prevent the rapid loss of water away from the sides and base of the reservoir.

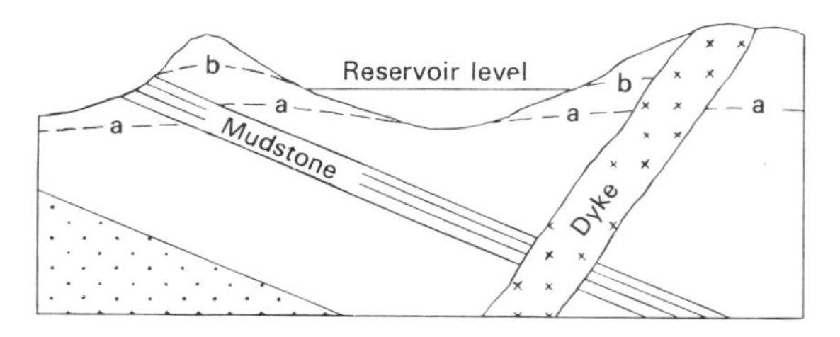

Fig. 15.4 Water-tight reservoir assured by sedimentary and igneous aquicludes. a=original water level; b=water level after impounding.

No such barriers are required to retain water in a reservoir if the level of ground-water (i.e. its total head) is greater than that of the proposed reservoir. This situation is illustrated in Fig. 15.5*a*; no leakage can occur because the natural direction of ground-water flow is from the ground to the reservoir (see also Fig. 13.20*a*). Only when the head of the reservoir exceeds that of the ground-water can leakage commence (Fig. 15.5*b*).

Kennard and Knill (1969) demonstrated how the accurate investigation of head in a valley side permitted a large reservoir to be successfully filled even though the valley contained cavernous limestone and unsealed abandoned mine workings.

Hydrogeological investigations of surrounding water levels are essential to an assessment of likely reservoir leakage. Care must be taken to accurately interpret the significance of water levels encountered in bore-holes and to distinguish perched water tables from the main water

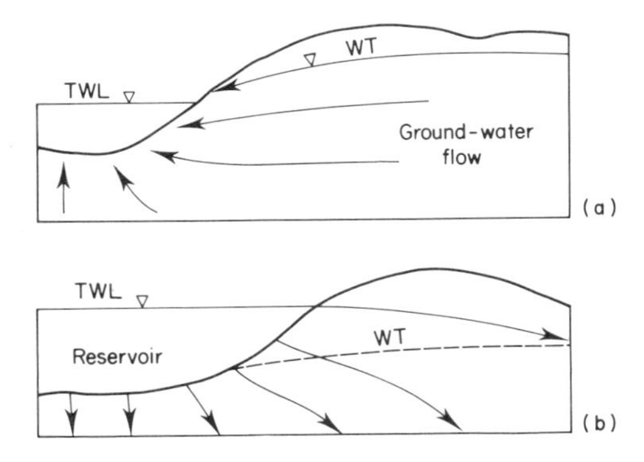

Fig. 15.5 Water-tight and leaky reservoirs. (**a**) The total head of water in the ground exceeds that in the reservoir and there is no leakage of reservoir water. (**b**) The reverse situation, resulting in leakage. WT = water table; TWL = top water level.

table, and other piezometric levels (see Fig. 13.14). Water levels should therefore be monitored diligently during the drilling of bore-holes (Fig. 13.26). Water levels vary (Fig. 13.15); at one time they may be sufficiently high to either prevent or greatly retard reservoir leakage, but may later fall to lower levels, during dry periods, and permit excessive reservoir leakage at a time when the reservoir of water is most needed. For this reason the fluctuation of water levels should be recorded.

In arid regions, where water levels can be low, an artificial impermeable barrier may be created by injecting cement and clay mixtures into the ground so as to reduce unacceptable reservoir leakage (Fig. 15.6). The cost of such work prohibits its extensive use.

Accurate mapping of geological structure and rock types is necessary to reveal the presence of zones and horizons that may either prohibit or permit excessive leakage from a reservoir. The vertical section of Fig. 15.7

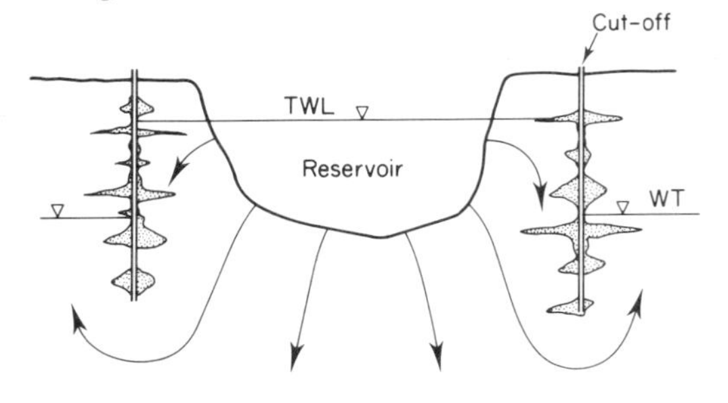

Fig. 15.6 Partially penetrating cut-off to reduce (but not stop) reservoir leakage, shown by flow lines.

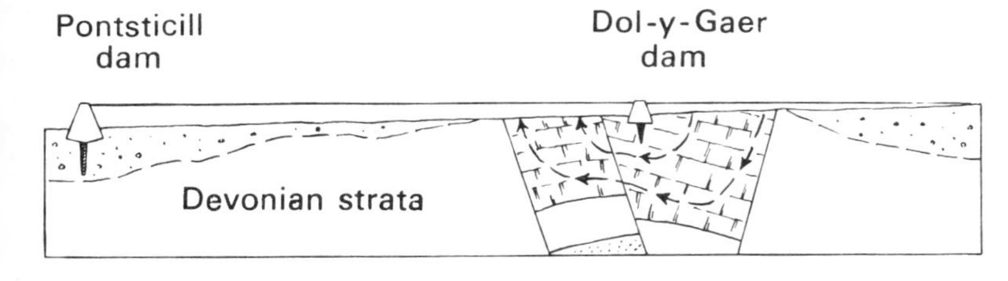

Fig. 15.7 Geological section to illustrate leakage at the Taf Fechan reservoirs, S. Wales.

illustrates the Dol-y-Gaer dam which was built on Carboniferous Limestone, whose extent could easily be mapped and through which serious leakage later occurred. Much of the lost water re-appeared downstream, where it was brought to the surface by the presence of relatively impermeable Devonian strata. Remedial measures failed to control the leakage and the Pontsticill dam was constructed downstream to impound much of the water leaking from the upper reservoir.

The presence of a relatively impermeable layer in the floor of a valley need be no guarantee against leakage when the head of water in the strata beneath the layer is lower than that of the reservoir. Seepage will eventually reach these lower levels and may cause the sealing layer to fail.

Special care should be taken to identify buried valleys (p. 42) as their course need not coincide with that of existing rivers. The map shown in Fig. 15.8 illustrates the

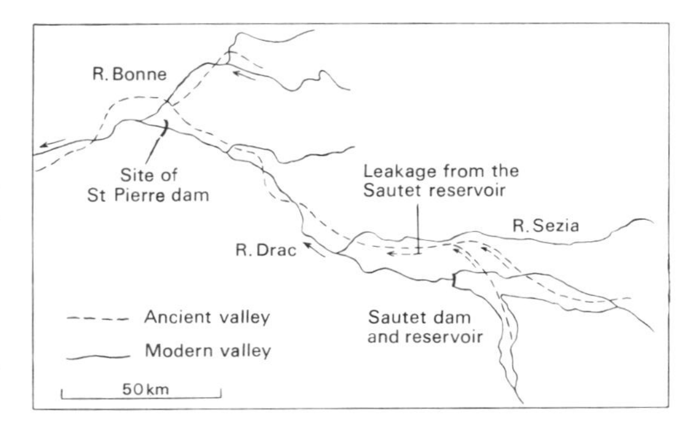

Fig. 15.8 A map of the valleys of the River Drac (from Ailleret, 1951).

valley of the River Drac, in N. France, and its former, now buried, valley along which serious leakage from the Sautet reservoir occurred. The St Pierre dam was constructed only after extensive field tests had confirmed that serious leakage could not occur from its reservoir to the River Bonne.

Water leaking from a reservoir will come into chemical equilibrium with the rocks through which it flows (p. 215) and may dissolve them. Sediments containing the minerals halite (p. 85), gypsum (p. 87), anhydrite (p. 87) and calcite (p. 86) are particularly susceptible to dissolution by flowing ground-water. The failure of the Macmillan reservoir in the U.S.A. was the result of water escaping through a gypsum layer that had been dissolved by leakage. Subramanian and Carter (1969) describe how the dissolution of calcareous sandstones undermined the Clubbiedean Dam in Scotland. Various formulas have been proposed for assessing the solubility of rock in-situ (see Goudie, 1970; Wigley, 1977; James and Kirkpatrick, 1980) although none is precise. Leakage can also erode from joints, other fractures and the cavities of karstic limestones, clayey material which be infilling these voids.

Seismicity

Regions apparently free from earthquakes have experienced seismic activity subsequent to the creation within them of a large reservoir. The term 'induced seismicity' is used to describe these events as they are attributed to changes in the ground-water conditions beneath and around the reservoirs, and from the load of water held within them. The earliest report of note was that of earthquakes associated with the filling of the Boulder Dam reservoir (Carder, 1945) and since then many other similar incidents have been recorded (Housner, 1970; Lane, 1971; National Research Council, 1972; Ackerman *et al.*, 1973).

One of the major difficulties in assessing the cause of induced seismicity is that the natural seismicity of most reservoir sites is not known prior to impounding, hence the level of background seismicity is rarely known for many of the sites where induced seismicity is later reported (UNESCO, 1974). Another difficulty is that much of the small scale seismicity could be associated with the progressive failure of reservoir slopes, which may move as the rising water approaches their critical pool level (p. 242). Unlike normal earthquakes, the more violent shocks associated with impounding are preceded by numerous fore-shocks which progressively increase with time: also, seismic activity characteristically increases at the time of first filling and reduces when the reservoir level is lowered (Rothe, 1973).

Background seismicity and slope failure do not explain the severe shocks which may accompany the filling of deep reservoirs (100 m deep or more). The impounding of the Koyna reservoir in the Deccan of India (p. 25) was followed by a series of earthquakes that culminated in a disastrous event of magnitude M = 6.7 (p. 5). Snow (1982) suggests that large reservoirs may significantly change ground-water pressures to depths of several kilometres. The effect such changes may have upon the crust in the region containing the reservoir can only be appreciated when they are considered in conjunction with former changes that accompanied the uplift and erosion of the region (i.e. regional geological history, especially that of the Cenozoic; Table 2.1).

Dams

The geology of a valley, and the available supplies of suitable construction material, will influence the location of a dam site and the type of dam that is constructed. Beneath every dam is built a *cut-off*. This is a thin barrier that extends into the foundation and either prevents or reduces the leakage of reservoir water under the dam. The cut-off reaches from one abutment to the other and often extends some distance from the abutments into the side slopes of the valley. The depth and lateral extent of a cut-off is governed by the geology of the valley base and sides.

Every dam must be protected against sudden influxes of flood water into its reservoir, by an overflow structure such as a spillway, or other outlet that discharges downstream of the dam. To construct a dam it is necessary to divert the existing river and its flood waters either by retraining it to the side of the valley or diverting it into a tunnel that passes through the abutments and discharges downstream of the dam site. The geology of sites for appurtenance such as overflow structures and river diversion works must be considered.

The magnitude of ground accelerations coming from *natural* seismicity will influence the design of a dam and its auxiliary structures (Instn. Civil Engineers, London, 1981; Oborn, 1979). Geological evidence of past earthquakes may be sought in regions where records are inadequate: see Sherard *et al.* (1974) and the use of stratigraphy in dating events (Fig. 18.15).

Types of dam

There are three types of dam: (*i*) embankments made from sediment and rock (earth-fill and rock-fill are the terms used to describe suitable sedimentary material such as clay and sand, and rock blocks that can be placed to form a safe embankment), (*ii*) concrete dams and their forebears, the masonry dams, and (*iii*) composite dams, which are usually structures composed of more than one type of concrete dam but are occasionally composites that have concrete and embankment sections.

Embankment dams

These consist essentially of a core of impermeable material, such as rolled clay or concrete, supported by permeable shoulders of earth and rock fill. When a clay core is used it is normally flanked by filters of permeable material, such as sand, to protect the core from erosion by the seepage of reservoir water through the dam (Fig. 15.9). Embankment dams, by virtue of the slopes required for their stability, have a broad base and impose lower stresses on the ground than concrete dams of similar height. Their fill is plastic and can accommodate deformations, such as those associated with settlement, more readily than rigid concrete dams. For this reason they can be built in areas where foundation rocks of high strength are not within easy reach of the surface; they are also the safest of all dam types against the risk of damage by earthquake. Their large volume requires copious supplies of suitable materials for earth and rock fill.

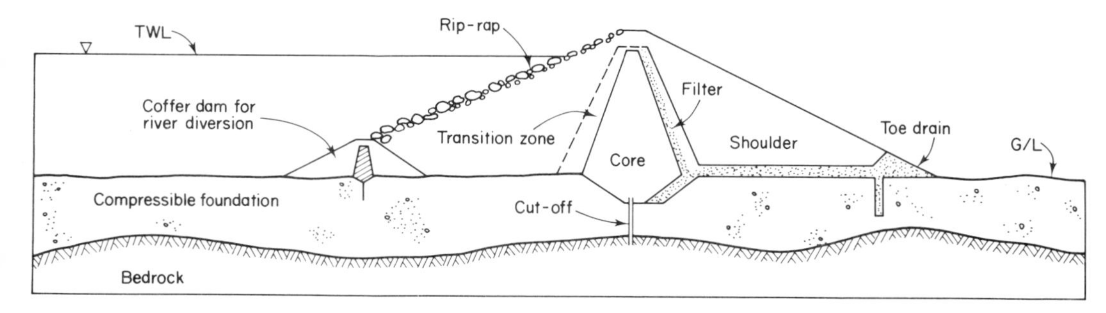

Fig. 15.9 Cross-section through an embankment dam. An impermeable core retains the reservoir and its underground extension (the cut-off) prevents leakage that would otherwise occur beneath the dam. A slow flow of water occurs through these dams and is collected by a filter downstream of their core, which protects the core from erosion (see text). This filter is connected to a basal drainage blanket through which water may readily leave the structure. The blanket also drains water issuing from the foundation strata downstream of the cut-off.

Concrete dams

These assume one of three main designs; gravity, buttress and arch (Fig. 15.10). All require strong foundations that will deform little when loaded by the dam.

A *gravity dam* is a massive impermeable concrete monolith of triangular cross-section, (Fig. 15.10*a*), having dimensions that give it sufficient weight to resist the load from the reservoir. They require ample supplies of concrete aggregate and cement.

A *buttress dam* consists of a series of inclined watertight slabs as its upstream face, supported by a series of buttresses which are triangular in vertical section (Fig. 15.10*b* and *c*). It is less massive than a gravity dam and uses smaller quantities of construction material. The strength of the concrete is used to better effect than in a gravity dam (where concrete weight is the prime factor), as the pressure of water on the upstream face is exploited to load the buttresses and assist the stability of the structure. Foundation loads are high, but may be reduced by widening the base of the buttress (see Fig. 15.14).

An *arch dam* is an impermeable concrete shell shaped as an arch in plan: when curved in vertical section it forms a *cupola* (Fig. 15.10*c*). Many gravity dams are arched in plan but the characteristic of an arch dam is that it takes advantage of the arching effect to reduce its cross-section: arch dams are thin. They require the least volume of construction materials of all designs yet their shape makes them the strongest of all types against overtopping.

Composite dams

These are favoured at sites which are unsuitable for one particular design. They may incorporate features of em-

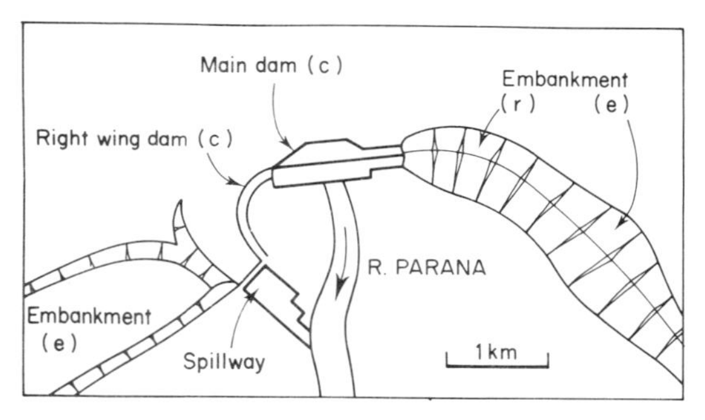

Fig. 15.11 A map of Itaipu Dam spanning the border between Brazil and Paraguay. (c) = concrete; (e) = earth fill; (r) = rock fill.

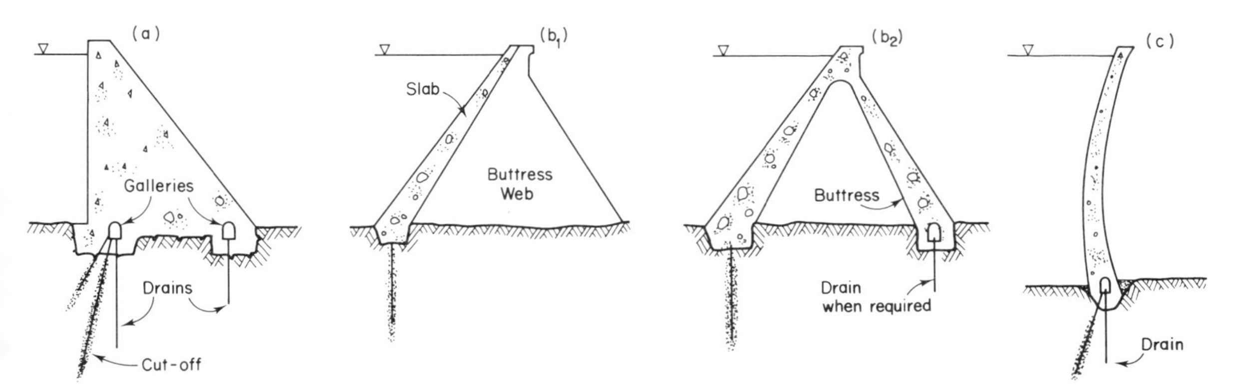

Fig. 15.10 Typical cross-sections for concrete dams. (**a**) Gravity; (**b**) buttress (see Fig. 15.14); (**c**) arch.

bankment, gravity, buttress and arch dams as required by the geology of the foundations, shape of the valley and availability of materials. For example, a broad valley containing reasonably good rock but with a centrally placed zone of deep weathering, may be dammed using gravity structures on its flanks which can support an arch that will span the centre, as at the Stithians Dam in Cornwall (Bainbridge, 1964). Many other examples exist, one of the largest being Itaipu, on the River Parana (Fig. 15.11).

Dam Foundations

Good foundations are highly desirable, but where they are questionable or poor, the location of a dam site becomes an exercise in locating areas where either the rocks and soils can best be improved or the dam most easily designed to compensate for the deficiencies of the ground.

Valleys exist because weathering and erosion have been accentuated at their location; weathered, compressible foundations, and unstable slopes, must be expected near ground level. Stress relief is associated with uplift and erosion (p. 160), and with the excavation of foundations; it can markedly affect the frequency of jointing in a rock valley and the aperture of such fractures (see Grand Coulee Dam, p. 248). Valleys eroded at a time when their rivers were draining to a sea level that was lower than at present (p. 42) are likely to have become a site of deposition following the recovery of sea level to its present elevation. These buried valleys (*q.v.*) may contain thick sequences of weak, compressible sediment such as silt, peat and clay, interbedded with stronger deposits of sand and gravel. Glaciated valleys may contain till beneath their more recent alluvium.

The geology of three sites is illustrated in Fig. 15.12, from which it is apparent that the detailed geology of large dam sites can only be revealed when the river has been diverted and construction has commenced. Weak ground not revealed by pre-construction ground investigations must be expected (Oliveira, 1979). The geology of many other sites is described by Walters (1971), Wahlstrom (1974), Knill (1974), and Legget and Karrow (1983).

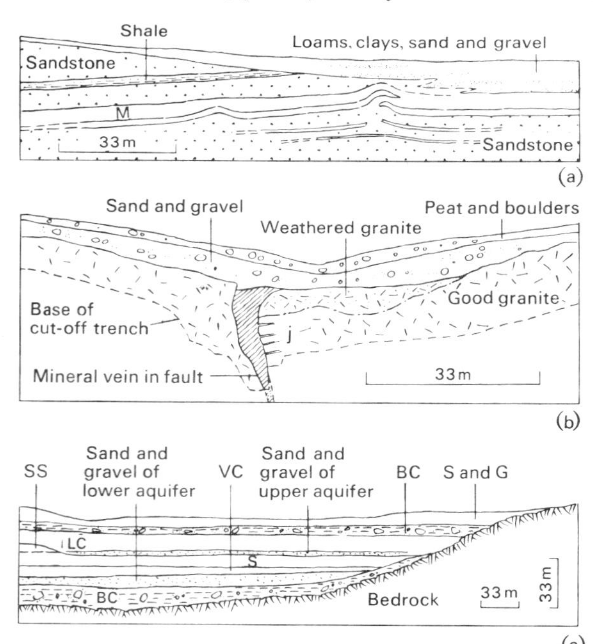

Fig. 15.12 Sections from dam foundations. (**a**) Weir Wood dam, Sussex. M = mudstone. Foundation strata deformed by valley bulge. (**b**) Fernworthy dam, Devon. j = open joints which yield water (after Kennard and Lee, 1947). (**c**) Derwent dam, Northumberland and Durham. SS = sandy silt; LC = laminated clay; VC = varved clay; S = silt; BC = boulder clay; S and G = sand and gravel. Aquifers contained water under artesian pressure (after Ruffle, 1970).

The principal attributes required of a dam foundation are adequate strength, low compressibility and moderate transmissivity.

Strength

The force on a dam foundation must not exceed the strength of the ground (see Fig. 10.1). Those aspects of rock and soil strength reviewed in Chapter 9, are relevant to an assessment of foundation strength. In foundations composed of sediments (Fig. 15.12*c*), compressible and weak strata are either strengthened by drainage and consolidation (p. 274, or removed. Particular care is required with sediments whose pore water cannot drain easily, especially if they rapidly lose their strength when accelerated by earthquake shocks, and liquify (Casagrande, 1971).

Weak and weathered rock is removed from rock foundations and the strength assessed of all rock surfaces on which movement of the foundation rocks could occur (see p. 168). Clayey deposits and residues infilling joints, faults and gaps between separated bedding surfaces, are removed and replaced with cement if the rock is otherwise sound.

Deformation

Every foundation settles beneath the load placed upon it during construction. The ground will be strengthened if this deformation is accompanied by the drainage of ground-water from pores and fractures, and a general increase in effective stress. Deformation should not be excessive, especially for concrete dams which are rigid structures and can tolerate only small differential movement. The centre of many rocky valleys contains zones that are significantly more compressible than their flanks, either by reason of the greater stress relief at the valley floor, with the associated separation of joints and accelerated weathering, or by the presence of faults. Such ground may have to be removed.

Foundations are also loaded by the weight of water the dam retains: this tends to push a dam downstream. The dam is loaded and unloaded as the reservoir level rises and falls, and the deformation of its foundations must be capable of safely transferring these loads to the ground without overstressing the structure of the dam and damaging its cut-off.

Transmissivity

A cut-off is used to control the leakage of reservoir water through permeable zones beneath a dam and may be extended to a depth where permeability is low, to create a full cut-off (Fig. 15.9). This is not always possible, nor desirable, for partial cut-offs may be used to permit some water to leak downstream to maintain water supplies. In these circumstances the cut-off is taken to a depth that increases the flow path beneath the dam (Fig. 15.10) sufficiently to reduce the hydraulic gradient under the dam and the discharge. Most modern cut-offs are created by drilling closely-spaced bore-holes to the required depth and injecting into them grout which permeates the surrounding ground to fill voids, pores and fractures (Figs 17.6 and 17.7).

The pressure head of water (p. 221) beneath a dam is greatly reduced by a cut-off, but *drains* are often provided to ensure that such head is completely dissipated. Unwanted ground-water pressures that may reduce the effective stress in the ground beneath the dam are thus avoided (Figs 15.9 and 15.10). The founding strata for the Derwent Dam are illustrated in Fig. 15.12*c*: the dam is an embankment founded on valley fill. Water under artesian pressure existed in the sand and gravel horizons and an extensive system of vertical sand drains were installed to reduce the pressure head within these horizons, and consolidate (and thus strengthen) the intervening layers of laminated clay. Relief wells were also drilled downstream of the dam to ensure that pressure head in the foundation strata can not increase to a magnitude that endangers the stability of the dam. Similar problems occurred at the Selset Dam, where it was thought that the cut-off would reduce pressure head in the foundations beneath the downstream face of the embankment. The cut-off was unable to do this adequately and relief wells were installed downstream (Bishop *et al.*, 1963).

Joints, and other fractures in rock, can be closed by the load from a foundation, thus trapping within them water which cannot easily drain away, and creating a zone of low transmissivity. If drainage is not possible, undrained conditions develop and the strength of the zone cannot increase with increasing confining pressure. Failure of the Malpasset Dam, in France, has been attributed to this cause. The dam was a cupola, founded on severely broken and faulted rock. It failed in 1967 after its left side had moved two metres downstream. Tests subsequently confirmed that the permeability of the rock on which the dam was founded could reduce to 1/100th of its original value under the loads imposed by the dam. Calculations suggested that this could have created an impermeable barrier (against which would have acted the full head of reservoir water), and in addition could have retarded the dissipation of ground-water pressure sufficiently for the effective stress within the foundation to reduce to a value that permitted local failure to commence. These conditions are thought to have gradually spread through the left abutment (Londe, 1967; 1973). The drainage beneath such dams must be place downstream of their cut-off and upstream of their stressed ground (Fig. 15.13).

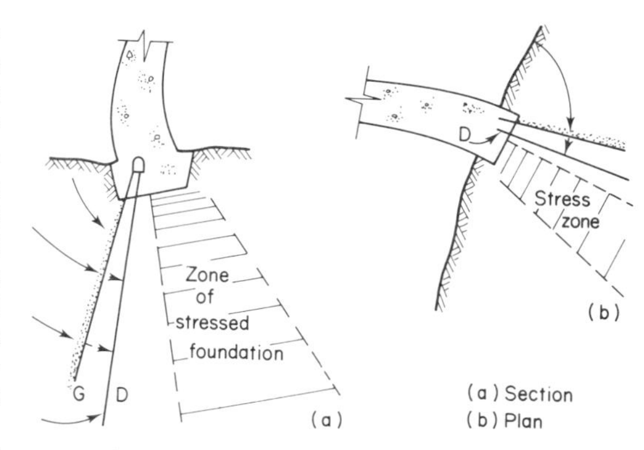

Fig. 15.13 Drainage measures for an arch dam. Drainage curtain: downstream (D) of grout (G) cut-off and upstream of stressed zone. The closure of fractures in the stressed zone *in-situ* reduces permeability and allows development of excessive ground-water pressure (after Londé, 1973).

The contact between a dam and the ground must be protected from the adverse effects of excessive ground-water flow. In many embankments the contact is protected by a filter (Fig. 15.9) which has the dual role of dissipating excessive water pressures and preventing erosion, both of the core and of sediment beneath the dam. Joints and other fractures are usually cleaned and filled with cement before they are covered by the earth fill of an embankment, so that water cannot issue from them at a velocity which can erode fine particles from the fill. Failure, in 1976, of the Teton Dam in Idaho, is suspected to have been associated with erosion of its clay core by seepage from unsealed fractures in the rock on which it was founded; this was an intensely jointed rhyolite (p. 107): U.S. Dept. Interior (1983). Failure of the St Francis gravity dam in California (1928) was attributed to erosion of the foundation sediments: these were conglomerates with some shales and sand which, when wetted, disintegrated to a gritty slurry from which the finer particles could be removed by water flowing beneath the dam. Seepage was observed when the reservoir was first filled and the dam failed in the same month (Amer. Soc. civ. Engrs, 1929).

Materials for dams

The location of suitable materials for construction is an integral part of surveys for a dam, because the design of a dam can be adapted to suit both the foundation conditions and available supplies. Much additional expenditure may be incurred if such supplies run out before the dam and its ancillary works are completed.

Good quality rock aggregate is required in abundance for concrete dams. It should have a tendency to break, when crushed, into the range of particle sizes required to give the concrete good workability and adequate strength on setting. River deposits within the site of the future reservoir are often a source for the finer fractions of a concrete mix. The aggregate must be sufficiently strong to carry the loads imposed on the concrete and free from all contaminates such as clay, shale, mica, pyrite and other weak or reactive material. The aggregate must also be sufficiently stable to withstand the repeated heating and cooling, wetting and drying that the dam will experience during its life.

Embankments also require considerable quantities of material but can be designed in *zones*, where each zone is composed of a different material and allocated a volume that can be satisfied by the quantities of material available on site. The basic configuration is illustrated in Fig. 15.9 but can change to suit available supplies (see Sherard *et al.*, 1963). The zones have different functions and the materials used in them must have certain characters. Clayey, plastic sediment which can be compacted to a density that gives it adequate strength and low permeability, is required for the core. To achieve this the clay is placed at a moisture content suitable for subsequent compaction to generate the undrained shear strength specified for the core. To prevent the core from being eroded by water seeping through the embankment, it is protected on its downstrean face by a filter of sand whose grading and particle size prevents the migration of clay particles through its pores (see filter rules in Vaughan & Soares, 1982). Special care should be taken to avoid the use of dispersive materials in a core. These are clays and clay-sized fragments of other minerals whose composition and structure only provides their surfaces with weak electrostatic forces. The lack of attractive forces prevents these materials from possessing strong cohesion between their particles, making them vulnerable to erosion by flowing water.

Shoulders, whose primary function is to provide an embankment with weight and stability, support the core and its filter. A rip-rap of large armour stones protects the upstream face of the embankment from wave attack and a toe-drain of coarse material saves the downstream face from degradation by wetting from water passing through the dam (Fig. 15.9).

Other aspects of construction materials are considered in Chapter 18.

Six examples

Embankment dam

Mangla Dam N. Pakistan. (Little *et al.*, 1963; Binnie, 1967; Proc. Instn. Civ. Engrs, London, 1967; 1968). Length 3.4 km. Height 116 m. Constructed 1961 to 1967.

Mangla Dam, on the River Jhelum, is one of three large embankments (Jari and Sukian being the other two forming the Indus River Scheme. Foundation rocks consisted of weak interbedded sandstones, siltstones and stiff clays of the Miocene Siwalik Series, overlain by gravelly alluvium. The embankment was made from clay and sandstone obtained from excavations for the spillways, intake and tailrace structures, and gravel won from the alluvium. Excavation for the main works was well advanced when ground investigation for the Jari embankment revealed the presence of low strength shear zones in the Siwalik clays: re-examination of the Mangla and Sukian sites revealed them to be present there also. The foundations were thus weaker than expected. The upstream slope of the dam was therefore supported by a toe-weight of additional fill, and drainage facilities incorporated in the intake works to reduce ground-water pressures and maintain adequate levels of effective stress on the low strength surfaces. These shear zones were attributed to bedding plane slip during folding of the Siwalik Series (p. 149): later movement may also have occurred during uplift and erosion of the site, some 1870 m of strata having been removed by erosion. Difficulties were encountered forming the grout cut-off as moderate injection pressures fractured the rock (p. 272). This failure was attributed to the weakness of the strata, although higher than usual horizontal stresses *in-situ* could be expected on a site that has had 1.8 km of overburden removed, and such stresses would assist injection pressures to cause fracture.

Concrete gravity dam

Grand Coulee Dam, Washington, U.S.A. (Berkey, 1935). Length 1.27 km. Height 168 m. Constructed 1934 to 1942.

Situated in the valley of the Columbia River in the Rocky Mountain range, the dam site consisted of granite beneath a cover of glacial deposits which included river silts and gravels. This cover varied in thickness from 27 to 100 m. Some of the bore-holes used for exploring the site were 0.9 m in diameter, and geologists and engineers could be lowered into the ground to inspect the foundations *in-situ*. The granite was proved to be sound and adequate for bearing a gravity dam. Considerable difficulties were experienced in excavating the granite, because horizontal sheet jointing (Fig. 8.28) was encountered which did not appear to diminish in frequency with depth. It was eventually realized that these joints were being formed by the release of *in-situ* stresses, following the unloading consequent on excavation. At the suggestion of the geologist, Dr C.P. Berkey, excavation was stopped and the dam built, thus replacing the load. The ground was grouted to seal any other joints which may have existed at depth, and the foundations have since proved to be satisfactory. Aggregate for the concrete of the dam was obtained from suitable glacial gravels situated 1.5 km from the site.

Masonry Gravity Dam

Vyrnwy Dam, N. Wales, U.K. (Deacon, 1896). Length 357 m. Height 44 m. Constructed 1887 to 1891.

The upper reaches of the Vyrnwy Valley had been the site of an old glacial lake (p. 54) whose deposits of lacustrine clays, alluvium and peat covered the bedrock to an unknown depth. The engineer put down nearly 200 bore-holes in order to investigate the character of the buried rock surface. These revealed a rock bar at one point across the valley at a depth of 13 m; upstream and downstream of this bar the solid rock surface was deeper. The bar probably represented the barrier of the former glacial lake and the dam was sited at this point. The upper 2-3 m of bedrock were in bad condition and removed from the dam foundations. The final surface was washed with jets of water and scraped clean with wire brushes. No springs issued from the exposed foundation rocks, which were Ordovician slates that dipped upstream, but it was felt desirable to build rubble filled drains into the base of the dam. Building stone was quarried about 1.5 km from the site, and large quantities of stone were rejected because of their poor quarrying properties; more than 700000 t was tipped to waste in order to obtain the necessary rock for the masonry, so highlighting one inherent problem of true masonry structures, namely their enormous demand for good quality dimension stone.

Buttress Dam

Haweswater Dam, Cumbria, U.K. (Davis, 1940; Taylor, 1951). Length 469 m. Height 38 m. Constructed 1934 to 1941.

Located on Ordovician andesites and rhyolites which have *in-situ* seismic velocities in excess of 5484 ms^{-1} and unconfined compressive strengths of 103 $MN\,m^{-2}$ (which is 35 times greater than the maximum stress that the buttresses impose on the ground), the site contained some of the best foundation rocks in Britain. The position of the dam was largely self-selecting, as the valley widens upstream and downstream from the point chosen. Bedrock was covered on one side of the valley by glacial drift to a maximum depth of 9 m and on the other side by a mass of boulders. The bedrock surface was explored with bore-holes and trenches. Natural exposures of rock were penetrated with percussive drills to ensure that they were not the upper surface of large glacially transported boulders. All the superficial material was removed in a trench to expose the bedrock for foundations; rock excavation was confined, as far as possible, to the areas occupied by the buttress units of the dam, the natural rock surface being left undisturbed between them. The cut-off was excavated 1.5 m deeper than the main excavation. A fault zone striking diagonally across the foundations was encountered; it dipped at 60° and contained much broken material but narrowed with depth. The fault was followed down in the cut-off trench to 14 m where it had diminished to a degree that was considered satisfactory. The foundations accepted very little grout.

Arch Dam

Cambambe Dam, N.W. Angola (Sarmento *et al.*, 1964). Length 300 m. Height 68 m. Constructed 1960 to 1963.

This double-curved arch dam is sited in a valley that cuts through Precambrian sandstones and interbedded clay shales. The strata is gently folded in two directions but bedding is everywhere of low dip. Folding has caused bedding plane slip between the sandstones and the shales creating at their interface weak layers of comminuted shale that has a clay-like consistency and variable thickness. In addition, jointing is well developed, two sets being near vertical and a third parallel to bedding. Faulting has also occurred in two directions, one of which produced displacements of 1.5–3.0 m about fault zones filled with mylonite (p. 141) and clay.

Geophysical surveys revealed that the velocity of *P*-waves through the site reduced during the excavation of the foundations and this, together with additional evidence from the exploration galleries, indicated that the rock was sustaining severe relaxation. The opening of vertical joints permitted horizontal relaxation of the valley sides, and movement on the faults allowed the clay layers between bedding to widen thus creating vertical relaxation in the valley floor. Excavation of the foundations proceeded with caution, the last 1.5 m of rock being excavated by hand using crow-bars.

Appreciable ground improvement was required to render the foundation suitable for carrying the thrust of the arch. An extensive programme of ground treatment was conducted from galleries excavated downstream of the dam and within its abutments: these galleries later served as drainage adits for the rock mass downstream of the grout curtain cut-off. Great care was taken to flush the clay from all fractures intersected by the ground treatment holes. Four faults were individually treated, their myolinite and clay being removed by excavation to depths of 15–20 m and replaced with concrete.

Geophysical surveys measured the *P*-wave velocity through the ground between the galleries, and between the galleries and ground level, to monitor the ground improvement attained by the treatment. Instruments within the dam later recorded very small deflections of the dam under the reservoir load (1–2 mm) and confirmed that the foundations had been improved to a satisfactory state.

Composite Dam

Farahnaz Pahlavi Dam, N. Tehran, Iran (Knill and Jones, 1965; Scott *et al.*, 1968). Length 450 m. Height 107 m. Constructed 1962 to 1968.

Situated on the Jaj-e-Rud, the site was considered suitable for either a concrete or an embankment dam. The site geology, which was progressively revealed during the period of ground investigation (1960–61), confirmed the feasibility of a concrete dam of buttress design to assist resistance against earthquakes, the site being close to one of the major seismic belts of the world (p. 4). The complex foundation geology needed extensive exploration for the final design to be completed and an assessment made of the ground improvement required. Tunnels and adits totalling 915 m, in conjunction with several shafts and trenches and 110 bore-holes, were eventually used. Geophysical surveys were made at the surface and in the adits and tunnels; *in-situ* shear strength, deformation and permeability tests, together with grouting trials, were also conducted. The geologists developed a classification based on estimates of the various geological characters which could together affect the engineering behaviour of the rocks, and the foundation area was mapped in these terms (a form of geotechnical zoning: see Fig. 12.24). As the investigations progressed it became apparent that the dam would have to be designed to cope with weak rock and variable conditions in the foundations. Much of the excavation was completed before any concrete was placed, so that a final review of the foundations and alignment could be made. The base of the buttresses was widened to spread the load (Fig. 15.14) and two abutment gravity blocks provided so creating a composite structure.

In addition, an extensive system of subsurface drainage was installed to strengthen the rock by controlling ground-water pressures. Drainage holes were drilled from a concrete-lined drainage tunnel beneath the dam, sited just downstream of a multiple upstream grout-curtain. The left bank buttresses were placed on rock having a seismic velocity in excess of 200 ms^{-1}; the right bank buttresses, which were founded on more deformable material, were supported at their toe by a thrust-block founded on stronger quartzite which outcropped immediately downstream. The straight alignment of the dam was changed so that the right bank buttress thrust into the hillside, and consolidation grouting was used for sealing and strengthening the foundations. Excavated rock was only used for general fill, coarse and fine aggregate being obtained from a limestone quarry opened on the right bank. River gravels and sands were found to be potentially alkali-reactive and were rejected as a source for aggregate.

Underground reservoirs

Underground storage can be obtained either by utilizing the reservoir of natural storage space provided by pores and fractures in rocks underground, or by creating a cavern.

Natural subsurface reservoirs

The geological factors that make a good underground reservoir are similar to those that form a good aquifer.

Fig. 15.14 Concrete buttresses of the Farahnaz Pahlavi Dam. 1 Access to site. 2 Future top water level. 3 Buttresses. 4 Excavation for key into hillside. 5 Spread base for low bearing pressures. 6 River channel. (Photograph by Sir Alexander Gibb and Partners.)

Storage (p. 222) must be sufficient for the volume needed, permeability (p. 221) must permit the easy recharge and discharge of the stored fluid to and from the reservoir, and hydrogeological boundaries (p. 218) must provide satisfactory containment of the stored fluid.

Unconfined reservoirs

These are used for the storage of ground-water in areas where rainfall is infrequent and water stored in surface reservoirs is rapidly returned to the atmosphere by evaporation. The reservoir is created by constructing a cut-off to intercept the natural drainage of ground-water and impound a useful volume of water (Fig. 15.15). The system is best employed in narrow aquifers such as those formed by valley fill. The cut-off requires little or no maintenance and the impounded water is protected from pollution, as all water enters the reservoir by percolation and is thus filtered.

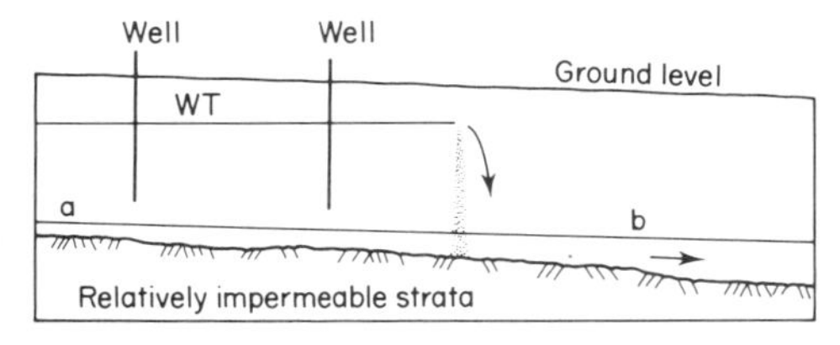

Fig. 15.15 Vertical section illustrating the operation of an underground dam. a = Original water level. b = Unchanged water level downstream of dam. WT = level of water impounded by the dam.

Confined reservoirs

These are normally used for the storage of gas, which is forced through injection wells into aquifer horizons: ethane, propane, methane and butane are commonly stored in this way (Katze *et al.*, 1981). Each gas is injected via separate wells into different parts of the aquifer and saturates the rock in the immediate vicinity, the intervening water saturated volumes of aquifer acting as barriers separating the storage areas. These systems are elastic, because when gas is withdrawn its volume is replaced by ground-water. Between 70% and 80% of the injected product is normally recoverable. The storage of one gas in an anticlinal trap is illustrated in Fig. 15.16. Difficulties can arise when the injected products become contaminated by natural gas in the ground.

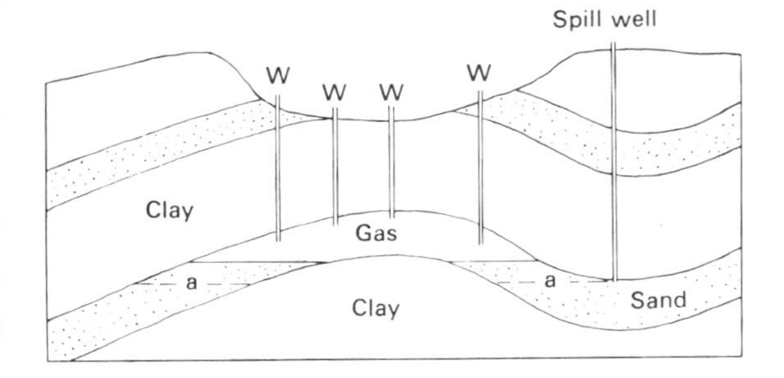

Fig. 15.16 Geological section illustrating underground storage in an anticlinal trap. a = limit of reservoir; w = injection and recovery wells linked to supply lines. Compared this with a naturally formed reservoir shown in Fig. 6.21*a*.

Fluids can only be injected into fully saturated confined reservoirs when the total head at the injection wells exceeds that in the ground. Pressure head (p. 221) must therefore be increased above the natural level and this reduces the effective stress at depth. Such pressures must be carefully controlled if ground failure is to be avoided (see Fig. 18.13).

Chambers

The security offered by a chamber underground has made this form of storage an attractive alternative to storage at the surface, especially when storage is required in proximity to urban areas where land may be required for other uses. The chambers may be excavated in either permeable or impermeable rock, as both can be used to store fluids. The excavation must be stable; a topic considered more fully in Chapter 16.

Impermeable chamber

This isolates the stored fluid from the surrounding ground and may be excavated in permeable rock, and lined, or excavated into impermeable rock.

Naturally impermeable rock has a porosity that retards flow (as might be found in shale, rock salt, or unweathered and non-porous igneous and metamorphic rocks) and a lack of open fractures. Rock free from fractures is rare close to ground level (p. 160) but exists in materials which deform at low *in-situ* stress differences by ductile flow rather than by brittle fracture (Fig. 9.12). Rock salt (p. 85) is both non-porous and ductile, making it a most favourable material in which to excavate impermeable caverns. This sediment occurs as stratified deposits and

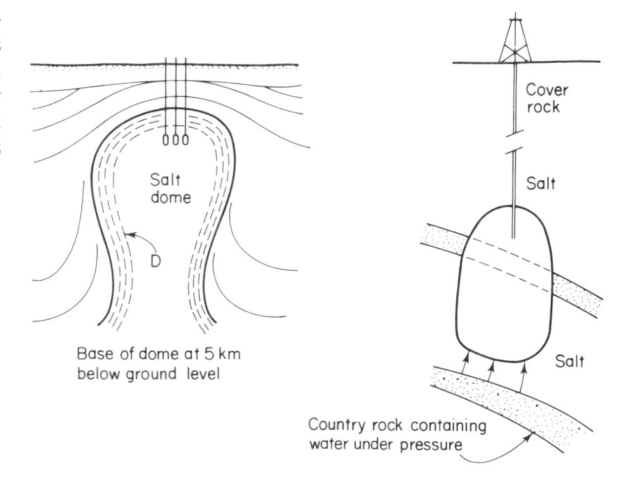

Fig. 15.17 General disposition of storage caverns in a salt dome and (right) detail of problems arising from inclusions of material other than salt (Lechler, 1971 and Hardy *et al.*, 1984). D = most disturbed zone (see also Fig. 6.21*c*).

salt-domes, and the latter, by reason of their upward migration through, and displacement of, overlying strata, can contain within their structure ribbons of country rock, some of which may remain in hydraulic continuity with the surrounding ground (Fig. 15.17). Caverns intersecting such inclusions will not be impermeable, neither will they be entirely ductile, and may become unstable.

Impermeable caverns can also be excavated in brittle rock that contains fractures, if the *in-situ* stresses around the excavation are sufficient to cause the sides, roof and floor of the chamber to move a small distance into the excavation, sufficient to close any fractures that exist. This effect is called 'arching' and will reduce the permeability of the rock in the immediate vicinity of the excavation if the rock is sufficiently strong to transmit, without failing, the stress required to close its fractures (Fig. 15.18).

Fig. 15.18 Underground cavern in gneiss, Scandinavia. (Photograph by J.M. Byggnads och Fastigtighetsab, Stockholm.)

Permeable chamber

Unlined excavations into porous and fractured rock permit liquid stored within them to migrate from the chamber if the total head of the liquid is greater than that of the surrounding ground-water. Conversely, ground-water will migrate into the chamber when its total head is greater than that of the stored liquid. The balance between the head of the two systems is used to confine a liquid in permeable chambers, provided the liquid is immiscible with water and of lower unit weight (Fig. 15.19).

To prevent stored fluids flowing from their chamber to

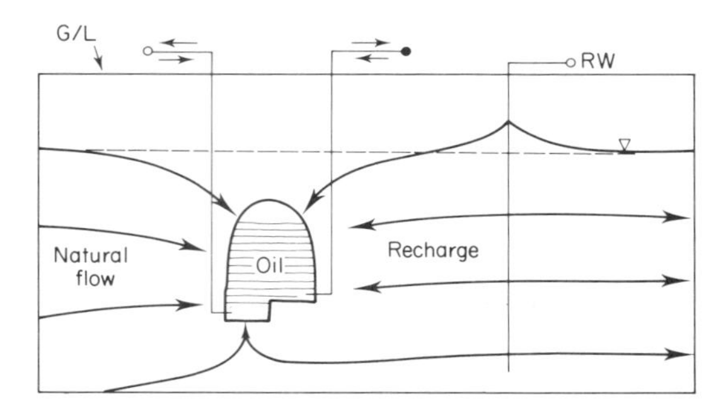

Fig. 15.19 Vertical section showing oil retained by ground-water flow towards a chamber. Below the oil is a water sump. RW = recharge well when required to create a lateral water barrier. Broken line = original water table.

a neighbouring cavern, where the head may temporarily be lower, hydraulic barriers are established between caverns. These consist of water injection points which maintain between the caverns a head that exceeds the head in the caverns (see Fig. 15.19).

Refrigerated chamber

Permeable and impermeable chambers are excellent for storing oil, but difficulties arise when they are used for storing gas. If the temperature of the gas is either at or above 0°C the product must be stored at a depth where its vapour pressure is exceeded by either the strata pressure or the pressure of ground-water. With permeable chambers a downward flow of ground-water into the store is also required to prevent the upward migration of gas bubbles. These problems are greatly reduced by lowering the temperature of the gas so that it can be stored as a liquid at atmospheric pressure. Such low temperatures freeze the surrounding ground and form a refrigerated chamber.

Rock shrinks when it cools. Existing fractures open and new fractures are created as the rock fails in tension under the thermally-induced stresses within it. Unlined refrigerated chambers are therefore rarely impermeable at very low temperatures. Liquid petroleum gas (LPG) can be successfully stored at temperatures as low as −40°C, but liquefied natural gas (LNG), which contains methane, requires temperatures as low as −100°C and this creates severe thermal overstressing in most rocks (Jacobson, 1978).

The presence of ground-water is significant to the performance of refrigerated chambers, as water in the vicinity of the chamber will freeze and expand to seal pores and fractures and make the rock impermeable. But as temperatures continue to lower, ice itself begins to contract and shrinkage cracks develop. Also, the high specific heat of water enables flowing ground-water to retard the growth of a refrigerated zone around such caverns. This may be crucial to the operation of near-surface caverns, where growth of the frozen zone can bring permafrost conditions close to ground-level and affect the foundations of surface structures.

Selected Bibliography

Surface Storage

Thomas, H.H. (1976). *The Engineering of Large Dams.* J. Wiley & Sons, Sydney.

Golze, A.R. (ed.) (1977. *Handbook of Dam Engineering. Van Norstrand Reinhold Co., New York.*

Walters, R.C.S. (1971). *Dam Geology.* Butterworth, London

Wahlstrom, E.E. (1974). *Dams, Dam Foundations and Reservoir Sites.* Elsevier, New York.

International Commission on Large Dams (1973). Lessons from dam incidents. Paris.

Laronne, J.B. and Mosley, M.P. (Eds). (1982). *Erosion and Sediment Yield.* Benchmark Papers in Geology. Hutchinson Ross Pubs. Co., Stroudsburg, Pennsylvania.

Knill, J.L. (1974). The application of engineering geology to the construction of dams in the United Kingdom. *Colloque. Géologie de L'ingénieur, Liege* (1974), p. 113–47.

Underground Storage

Berman, M. (ed.) (1978). *Storage in Excavated Rock Caverns.* Proc. 1st Int. Symp. Stockholm (1977) 3 vols. Pergamon Press, Oxford.

Underground Space. A bi-monthly publication of the American Underground Space. Assoc. University of Minnesota, Minneapolis, U.S.A.

Hardy, H.R. and Langer, M. (eds) (1984). *The Mechanical Behaviour of Salt.* Proc. 1st Conference, Pennsylvania (1981). Trans. Tech. Pub.

16

Excavations

The ease with which rock and soil may be excavated, and the stability of the hole that is thereby created, will be influenced by a number of factors including the following: the strength of the ground and the magnitude of the stresses within it; the geological structure of the ground; the level of ground-water and magnitude of pressure head at depth; the storage of the ground and its permeability. In extremely deep excavations the temperature of the rock and flow of heat through the crust will also be of significance. A knowledge of the geology of ground to be excavated is most desirable and should be considered essential for ensuring the safety of personnel who will work below ground level.

Excavation of rock and soil

The common methods used for creating a hole in the ground are, drilling (for bore-holes and wells), augering (for bores and piles), machine boring (for tunnels), blasting (for excavating strong rock), scraping, ripping and digging (for the surface excavation of soil and weak rock). Each relies for its success on the ability of the method to break and disaggregate rock and soil.

Drilling

The majority of drills are either percussive or rotary. In percussive drilling a chisel-shaped bit is repeatedly struck against the rock so as to form a hole. It pulverizes the rock to a fine debris which can be flushed from the hole by drilling fluid. In rotary drilling the ground is either cut or crushed by tough blades or points which are rotated against the rock under load (Fig. 10.8). The hardness of the mineral constituents, the toughness of the rock, its abrasiveness and geological structure affect the progress of drilling.

Hardness describes the ability of one mineral to scratch another. The Mohs' scale of hardness (p. 62) is based on this criteria and Fig. 16.1 illustrates the relationship between common rock-forming minerals and drilling materials.

Toughness describes the resistance to fracture that comes essentially from the tensile strength of rock. Many drilling bits are designed to induce local tensile failure within rock, as rock is much weaker in tension than it is in compression (Fig. 9.19). Interlocking of mineral grains and strong mineral cement affect toughness, and it is common to find that mineral cleavage is also significant, with coarse-grained rocks, such as gabbros (having large grains and surfaces of cleavage; Fig. 5.19), drilling faster than their finer-grained equivalents, e.g. dolerites (Fig. 5.20). Drilling bits for tough rock have strong shoulders and small, closely-spaced points; those for weaker materials are lighter and carry wider-spaced and pointed teeth.

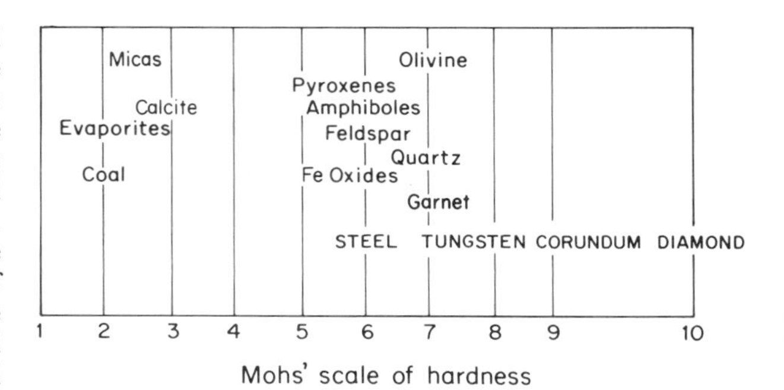

Fig. 16.1 Relative hardness of drill bit materials and minerals.

Abrasiveness describes the ability of rock fragments to wear away the drill and polish its cutting edges (Table 16.1). Rocks composed essentially of quartz are very abrasive.

Table 16.1 Rocks grouped by selected drilling characters (after McGregor, 1967)

IGNEOUS:
Abrasive: Rhyolite, welded tuffs, granite, pegmatite.
Less abrasive: Basalt, dolerite, gabbro
Least abrasive: Weathered intrusive rocks and lavas

METAMORPHIC:
Abrasive (and Hard): Quartzite*, hornfels, gneiss
Less abrasive: Schist
Least abrasive: Phyllite, slate, marble.

SEDIMENTARY:
Abrasive (and Hard): Flint, chert, quartzite, sandstone, quartz-conglomerate
Abrasive (less Hard): Siltstone, siliceous limestone, many sandstones
Abrasive (least Hard): Friable sandstones and grits
Non-abrasive (Hard): Limestone, shale
Non-abrasive (least Hard): Mudstone, marl, coal, oolitic limestone

* usually the most difficult of common rocks to drill.

Structure Broken rock is much more difficult to drill than its intact equivalent, as energy is wasted at every discontinuity. The drilling bit is subjected to uneven and variable loads that damage its cutting edges and cause it to wander from the prescribed course. Some temporary wall support, such as casing, is also required to prevent loose fragments of rock from dropping down the hole where they may jam behind the shoulder of the bit and prevent it from being withdrawn to the surface.

The presence of joints and other discontinuities filled with soft material can also be a source of delay if the softer material binds the bit. Penetration is best in holes which cross discontinuities at a steep angle. To cross them at a low angle leads to deviation of the direction of the hole, which may have to be back-filled and re-drilled on the correct course (Fig. 16.2). Drilling down dip is more difficult than drilling across dip, even though the latter may mean coping with different rock types. Deviation tends to be greater in angled holes than in vertical holes, and tends to increase with distance from the rig. Important holes should be surveyed (see Fig. 16.15).

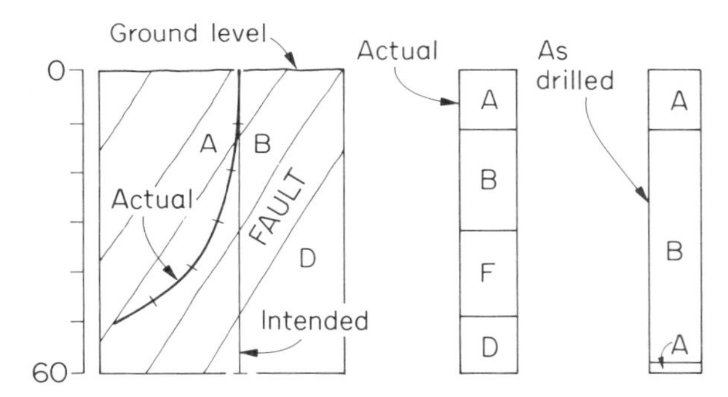

Fig. 16.2 Error in interpretation of subsurface geology created by deviation of drill hole.

Drilling fluids must be selected with care. Air, water, mud, polymers and aerated emulsions (called 'foam') are employed to clean the bit, lift cuttings from the hole, support the sides of the hole, and balance moderate artesian water pressures encountered at depth. Core recovery from weak rocks, such as shale and poorly cemented sandstones, can be severely reduced if the fluid chosen causes the strata to disintegrate (p. 179).

Augering

An auger consists of a drag-bit cutting head connected to a spiral conveyor, or flight. It is most efficient in soft materials such as clay, and can produce holes with diameters up to 1.5 m or more. Augers are frequently used to drill through thick overburden prior to drilling the rock beneath, for drilling weak rocks, and for excavating large holes for cast piles and similar structures. Boulders will severely hinder the penetration of augers if they are greater than one-third the diameter of the tool.

Machine boring

Large diameter machines may be used to bore tunnels. They are an extension of drilling technology and carry, at their head, cutters capable of breaking and excavating the ground (short teeth for sediments, picks for weak rock and discs for strong rock). Although the cutters can be replaced, severe difficulties develop when a machine that is designed for excavating weak rock unexpectedly encounters strong rock. For example, a tunnel bored through ground similar to that illustrated in Figs 5.7 and 5.9 would penetrate moderately strong sedimentary rocks and extremely strong sills and dykes.

To these difficulties are added those of preventing the ground from closing around the machine, or collapsing into it. Machines designed to excavate sediments, such as clay and sand, are enclosed along the length by an outer tube, or *shield*, at the tail of which is erected the permanent tunnel lining needed to support the ground. Stronger materials, particularly sound rock, can often support themselves for a period that is sufficient for the machine to need only a *hood* above it to protect it from occasional falls of rock from the tunnel roof. For these reasons, the homogeneity of the ground affects the performance of boring machines. Examples of ground suitable for tunnel borers are illustrated in Figs 3.36, 6.11, 7.6*b* and 7.10.

Lateral jacks extend from the side of these machines and press against the tunnel wall to generate the reaction the machines require to thrust forward when boring. This reaction can be difficult to obtain in deformable material such as fault gouge.

Blasting

This is extensively used to create surface and underground excavations in rock, and is part of a cycle that includes drilling the holes to house the explosive and clearing away the blasted debris (an activity termed 'mucking').

The structure of rock has a profound influence upon the efficiency of blasting as bedding, jointing, cleavage and other discontinuities reflect the shock waves radiating from a blast and cause the rock to fail in tension. These fractures are also opened by the expansion of gases liberated by the blast, so that the combined effect of an explosion is to break and loosen the rock mass (Fig. 16.3). A ragged profile may thus be formed to the excavation (see Fig. 7.6*b*) unless the blast holes are drilled sufficiently close to each other for there to be created, on detonation, an initial crack that links the holes and forms a boundary that cannot be penetrated by later shock waves. This is called 'smooth-blasting'. A boundary separating rocks of contrasting seismic velocity can refract shock waves and cause severe ground vibrations to be felt some distance from the site of the blast (Fig. 16.4).

The strength of rock also influences the efficiency of blasting. Considerable difficulties are presented by fragmental rocks containing hard and soft material. Agglomerates (p.94), conglomerates (p.118) and breccias (p.118)

Fig. 16.3 A blast (Nobel's Explosives Co. Ltd.).

are particularly difficult when their included fragments are significantly stronger than the matrix which binds them together. Examples of difficult ground are illustrated in Figs 3.4, 6.7 and 6.14.

Stresses within the ground affect the efficiency of a blast. High stresses can increase the difficulty with which fractures may be generated and opened. Tensile stresses resulting from blasting can cause a rock, that is compressed under high stress around a tunnel, to expand and this may be sufficient to lock the rock mass and 'choke' the blast. Hence, the blasting pattern used successfully in an underground excavation, and the associated consumption of explosive per cubic metre of rock excavated, may not be suitable for a similar excavation in comparable rock at a different location (see Hagan, 1983).

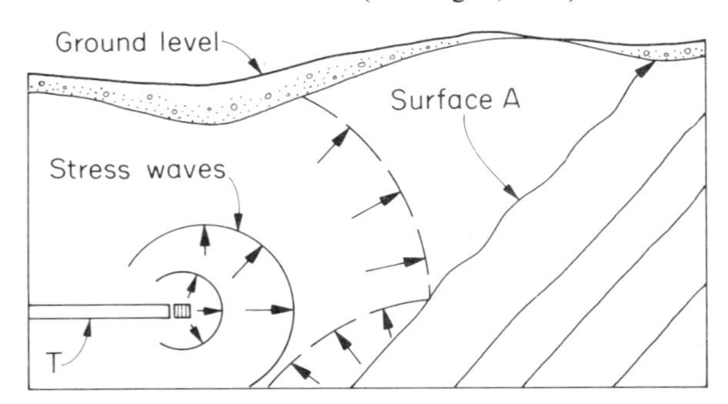

Fig. 16.4 Stress waves from blast in tunnel (T) are refracted to ground level when they reach surface A.

Scraping, ripping and digging

These are the techniques commonly used for removing soft and weak material such as clay, silt, sand, shale, weathered rock and top-soil. The machines employed (scrapers, rippers, dozers, graders and excavators) work best in ground that has a seismic velocity lower than 1000 m s^{-1}. Figure 16.5 is a guide to machine capabilities.

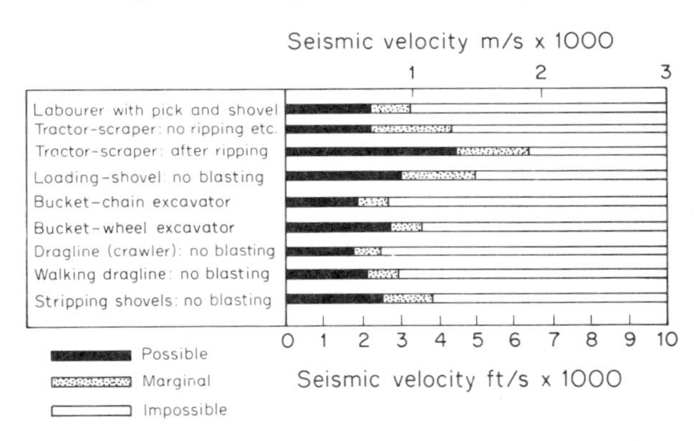

Fig. 16.5 Indication of excavation performance by machine.

A knowledge of the spacing of fractures and other discontinuities, their continuity, and orientation, when combined with a measure of the *in-situ* seismic velocity of the ground (p. 176), can be used to indicate the ease with which a rock or soil may be excavated (Weaver, 1975; Minty and Kearns, 1983). The undrained strength of the ground governs the type and size of machines required, since they must be supported by the load-bearing capacity of the deposit. Care should be taken in selecting equipment for clearing weak material such as peat and recently deposited organic-rich sediments. Many glacial, flood-plain and coastal deposits contain soft horizons interbedded with firmer strata (Fig. 10.20). The Plasticity Index (p. 168) is a useful guide to plant mobility especially as desiccation can convert a soft clay to a material as hard as brick.

Control of ground-water

An excavation below the water table must either be protected against unacceptable inflows of water or be worked by sub-aqueous techniques such as draglines and dredging. The most common reasons for controlling ground-water are to effect a reduction in the flow to an excavation and to lower the pressure of water around it. Commonly occurring patterns of flow are illustrated in Fig. 16.6*a*, and in Fig. 16.6*b* are shown situations where pressure head causes failure. The greatest hazard from ground-water to work underground is an encounter with *unexpected* water-bearing zones and water pressures. Geological investigations are absolutely essential to avoid this hazard and must include sub-surface investigations for surface excavations and probing ahead from excavations in progress underground (see Fig. 17.4).

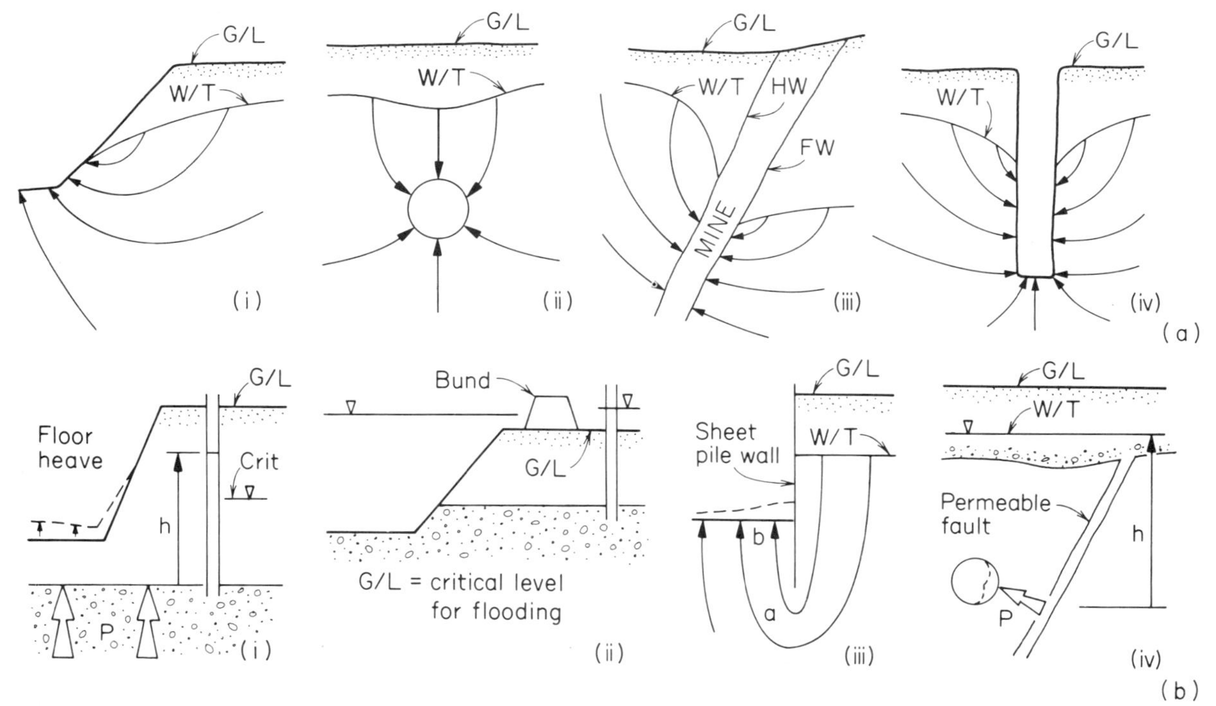

Fig. 16.6 (**a**) Common form of inflow to (*i*) surface excavations, (*ii*) tunnels, (*iii*) mines (HW = Hanging Wall, FW = Foot Wall), (*iv*) shafts. The flow patterns shown are for unsteady conditions where the water table (W/T) lowers with time. (**b**) Failure induced by ground-water. (*i*) Floor heave once h > crit. level; (*ii*) Overspill; (*iii*) Piping; (*iv*) Tunnel failure. P = pressure of water = h × unit weight of water.

Ground-water flow

The mechanics of ground-water flow and the geological factors that influence its rate and duration are described in Chapter 13. To control the ground-water flowing into an excavation it is necessary to know the volume of water that may be drained, the rate at which it will flow and the pressure it will be under.

Volume of water

This is governed by the storage of the ground (p. 222) and by its hydrogeological boundaries (p. 218). Ground-water can flow from a distant source to an excavation and it is therefore necessary to extend beyond the limits of an excavation the ground investigation for relevant boundaries. This point is illustrated in Fig. 16.7 where, in (*a*) both the canal and the sandstone are supplying water to the excavation, and where in (*b*) the tunnel would be dry were it not for a permeable fault in hydraulic continuity with a distant aquifer (the karstic limestone).

Rate of flow

Ground-water discharge is a function of the permeability of the ground and of the hydraulic gradient within it: both parameters are described in Chapter 13. An assessment of permeability is crucial to calculations of inflow and should be obtained from tests conducted *in-situ* (Fig. 10.22). Rocks and soils are generally anisotropic and their permeability varies with direction (p. 185). Care should be taken to measure the permeability of the ground in the direction of ground-water flow; e.g. in the vertical direction beneath the excavation illustrated in Fig. 16.7*a* and in the horizontal direction at its sides. Inspection of the

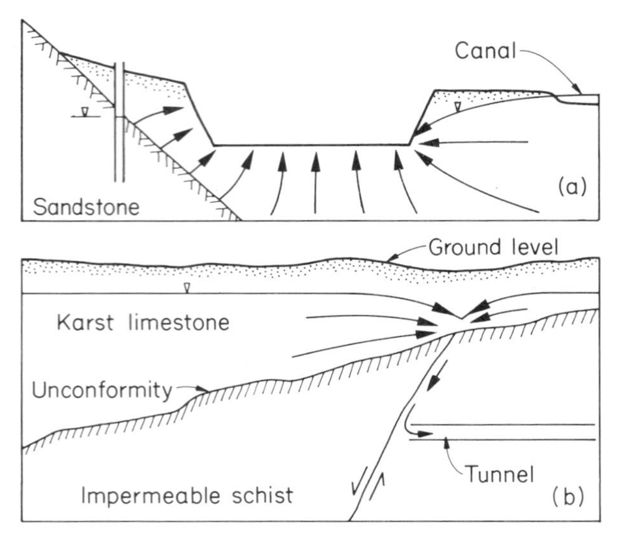

Fig. 16.7 Excavations influenced by remote sources of water.

flow patterns shown in Fig. 16.6 will demonstrate the care required to obtain relevant measurements.

Pressure of water

This may be calculated from an accurate measure of pressure head (p. 221 and will vary from place to place within the ground (Fig. 13.27), and at any one location may be expected to vary with time (Fig. 13.15). As shown in Fig. 16.6*b*, pressure head may be a formidable force to control. Uncontrolled ground-water pressure around a surface excavation can cause the floor of the excavation to fail (see (*i*) in Fig. 16.6*b*) and the excavation to flood (*ii*): it may also cause piping (*iii*) and in underground excavations, can cause failure (*iv*). Piping is a most serious problem in non-cohesive soils such as silt and sand, and occurs when the seepage force (p. 222) associated with an upward flow of ground-water balances the downward force of the sediment grains (i.e. their submerged weight) (Fig. 16.8). Frictional resistance between the grains is reduced to zero and the condition of the deposit converts from that of a stable soil to a fluid. Excavations collapse when this occurs.

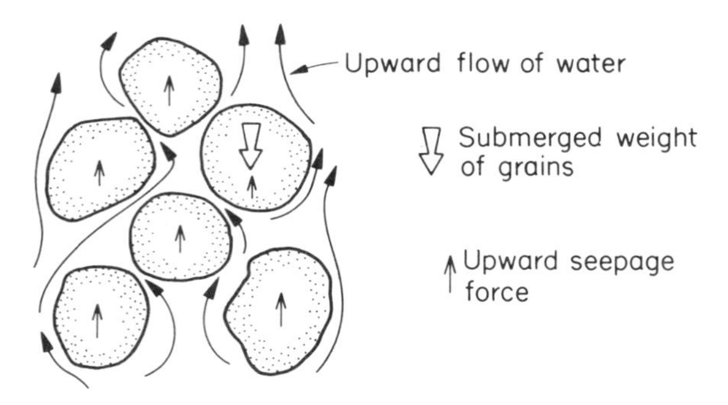

Fig. 16.8 Representation of seepage force in an uncemented granular deposit such as sand or silt.

Control of pressure

Ground-water pressure may be controlled by reducing the pressure head with pumping: a form of de-watering. With reference to Fig. 16.6*b*; pressure head in the underlying deposit of example (*i*) could be reduced to below the critical level at which floor heave commences in the overlying clay. In example (*ii*) the head of the confined aquifer could be reduced to an elevation below that of ground level, and in example (*iii*), head, on the water table side of the sheet pile wall, could be lowered so as to reduce the hydraulic gradient beneath the wall and the magnitude of seepage forces under the excavation. Pressure head may also be isolated by the creation of an artificial barrier such as a cut-off, or grout-curtain (*q.v.*). The permeable fault in example (*iv*) of Fig. 16.6*b* and in Fig. 16.7*b*, could be grouted to isolate the ground near the tunnel from the head of water in the overlying strata. Grout is often injected into the ground immediately behind a tunnel lining so as to create a water-tight barrier that isolates the pressure of water in the ground.

Control of flow

Ground-water flow is sustained by the drainage of water stored in the voids of the ground and from hydrogeological boundaries that recharge water to the ground. Such water may be intercepted before it reaches an excavation and this may be achieved either by pumping water from around the excavation (see Figs 17.2 and 17.4) or by injecting grout to reduce the permeability of the ground (see Fig. 17.8).

Boundaries

The influence of these structures can be gauged from Fig. 16.7 where the excavations shown are being supplied with water from the following recharge boundaries; in (*a*), from the junction between the sandstone and its overlying deposit, and from the base and sides of the canal; and in (*b*), from the unconformity and from the fault. Other examples are illustrated in Figs 13.25 and 13.28*b*.

Flooding of the gold mine at West Driefontein, in South Africa, demonstrates the importance of boundaries. At this mine, impermeable gold-bearing beds are overlain by some 1260 m of cavernous, water-bearing dolomites. Both the gold-bearing rocks and the dolomites are cut by near vertical dykes that are spaced at about 200 m and divide the ground into a series of box-like compartments. To reduce the likelihood of water draining from the dolomite to the mine when fractures behind the hanging wall of a stoped area meet others which connect with the dolomite, the compartment in which the mine is situated was dewatered. The neighbouring compartment was not and a large difference in head existed across the dyke that separated the two. Failure of a hanging wall

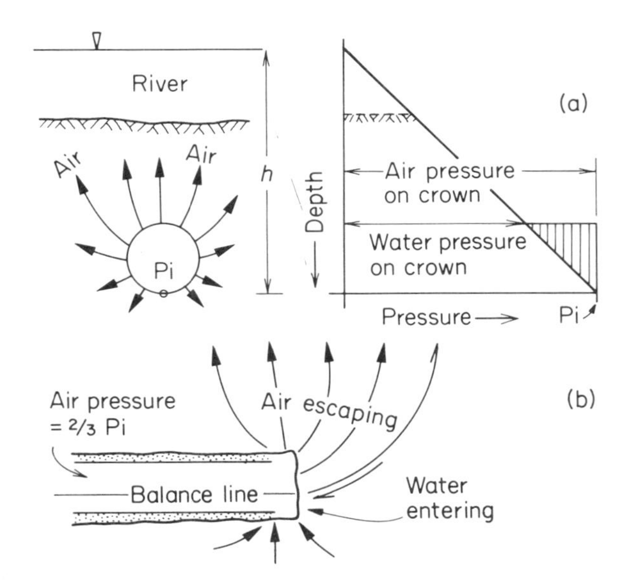

Fig. 16.9 Use of compressed air in a tunnel. (**a**) Hydrostatic pressure at tunnel invert (Pi) balanced by air pressure in tunnel. (**b**) In practice, air pressure is rarely permitted to exceed $\frac{1}{3}$ to $\frac{2}{3}$ Pi. Pi = h × unit weight of water.

near this dyke occurred in October 1968 and water poured into the mine under the head which existed in the dolomite. The incident is described by Cousens and Garrett (1969).

Mines break the ground above them, changing the natural boundaries which exist, and permanently altering the storage and permeability of the ground, and its discharge (Fig. 13.13). When these effects reach the surface (Fig. 9.28), flow to the underground works can be directly influenced by surface hydrology. Broken ground above the hanging wall of the mine illustrated in example (*iii*) of Fig. 16.6*a* will create a zone of enhanced transmissivity and storage immediately above the working area of the mine. Rivers above such areas should be culverted, or lined or diverted before they are able to recharge the ground.

Boundaries also separate waters of different quality and it is common for tunnels and mines in mountainous regions, where geothermal gradients are greater than normal, to encounter hot water in faults and impounded behind aquicludes. During the construction of the Great Appenine Tunnel, ground temperature suddenly increased in a clay shale from 27°C to 45°C and to exceptional temperatures around 63°C after an inrush of methane gas (Szechy, 1966). Temperatures of 65°C were encountered in the Tecolote Tunnel through the Santa Ynez Mountains, California (U.S. Bur. Rec., 1959), and similar temperatures have been measured in tunnels into the Andes.

Compressed air

This may be used in tunnels as an alternative to, or in conjunction with, other forms of ground-water control and ground treatment. The air in the tunnel is pressurized to balance, in part, the pressure head of water in the ground. This reduces the hydraulic gradient within the ground and thus the flow of water to the excavation. The method is illustrated in Fig. 16.9 and is only successful if the geology of the ground above the tunnel and exposed at its face, confines the air to prevent it escaping. The permeability of most soils and rocks with respect to air is approximately 70 times their permeability with respect to water. Considerable caution is required to ensure that a tunnel to be driven beneath a river does not unexpectedly break into coarse and permeable deposits of a buried valley (*q.v.*) as these materials will also allow air to escape, and its pressure to suddenly drop. Water then floods the tunnel and, depending on the air pressure used, some of the workforce will be affected by too rapid a period of decompression.

Surface excavations

Some surface excavations must be stable for long periods, e.g. motorway cuttings and other permanent excavations, whilst others, such as production faces in quarries and many excavations for foundations, need only short term stability. Excavations designed to be stable for short periods may not retain their stability for long periods (see Progressive Failure, p. 227). Of particular importance in this regard is the choice of soil and rock strength values to use in analyses of excavation stability (see Chapter 9), and the change in ground-water pressure that will occur during the lifetime of an excavation, (see p. 220).

Investigations

Ground investigations for a surface excavation must reveal the geology of the area to be excavated and assess the strength, and permeability of the ground (see for example Barton, 1982). Slopes are influenced by the geology of the ground behind them and ground investigations may therefore have to be extended beyond the proposed boundary of the excavation (see Fig. 10.22). Investigations should also be made of the ground beneath the floor of the proposed excavation, and example (*i*) in Fig. 16.6*b* illustrates why this is necessary. Geophysical techniques can be of considerable assistance in these investigations (Table 10.3) especially as the seismic velocity of rocks and soils is also an indication of the ease with which they may be excavated.

Vertical geological sections should be constructed to display the geology (Fig. 10.4) and maps drawn to illustrate the geology when seen in plan (Fig. 10.5). The vertical sections will generally show weathered or weak material overlying relatively stronger material; each may have its own angle of stability in excavation (Fig. 16.10). Typical profiles are shown in Figs 3.1, 3.4 and 3.10; the special problems created by volcanic rocks are illustrated in Fig. 5.5.

Most quarries are able to orient their excavated slopes

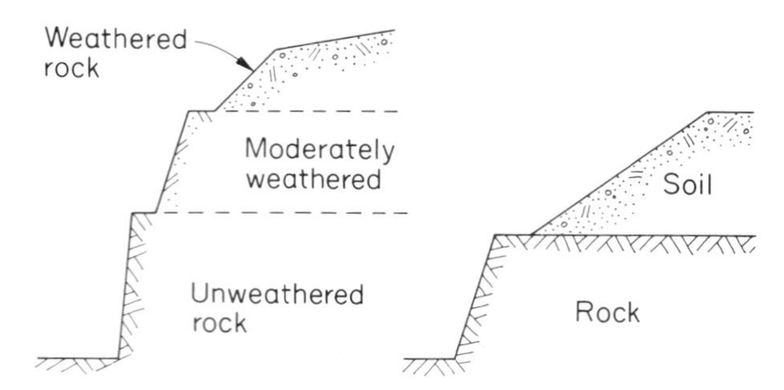

Fig. 16.10 Typical vertical successions in surface excavation.

in relation to bedding and jointing, but for many excavations this is not possible and all the difficulties illustrated in Fig. 14.3 have to be predicted and assessed. Excavation in a direction that is orthogonal to the strike of bedding and sets of major fractures, creates a more stable slope than that for faces advanced in other directions. The common problems of achieving a desired profile in rock are shown in Fig. 16.11. Details of structural geology should therefore be included in ground investigations for an excavation.

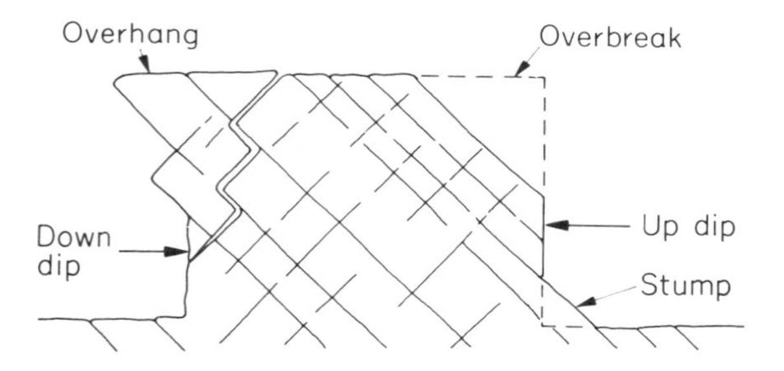

Fig. 16.11 Common problems in achieving a correct profile in a rock excavation.

Deformation and failure

Ground pressures will usually change the shape of an excavation and if deformation is excessive the slope, or floor, of the excavation can fail (Brawner, 1982). An indication of the ground response to be expected is provided by the structure of valleys, for these are the natural excavations produced by the work of rivers and ice. The deformation that can accompany valley erosion is shown in Fig. 14.10, and such a response is seen at a preliminary stage in recently glaciated valleys (Fig. 16.12).

Natural excavations therefore indicate that the slopes of a man-made excavation will tend to move inward and the floor, upward (a rebound termed 'heave'). The amount of movement will reflect the ratio of horizontal to vertical stress *in-situ* and Fig. 16.13 illustrates the displacements which may occur in a homogeneous material. Examples of the influence geological inhomogeneity may have upon this response are described by Lindner (1976) and Nichols (1980).

A typical example is the rebound that occurred in an excavation some 35 × 118 m and 38 m deep in Recent sediments (Bara and Hill, 1967). The deposits consisted of 38 m of alluvium, containing lenticles of sand, silt and clay, above a lacustrine deposit of highly plastic montmorillonitic clay, some 12 m thick. Laboratory tests had

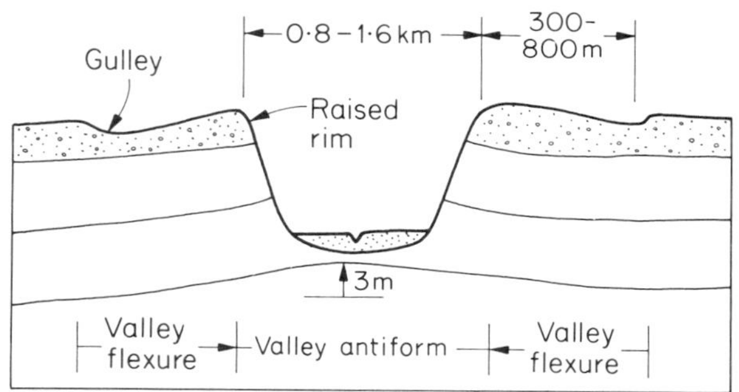

Fig. 16.12 Typical elements of rebound in glaciated valleys. Gulleys lie parallel to the main valley (~60 m deep) and may have depressions 3 m deep (after Matheson and Thomson, 1973): see also Fig. 3.41.

indicated that both the alluvium and the clay would rebound on unloading, and the site was therefore instrumented. Excavation was completed in 5 months and heave developed as predicted, but there were no observed fractures in the excavation surface due to the plastic nature of the deposits.

In contrast to this behaviour, that of the floor of an ore quarry, 300 × 600 m, in Canada, developed a crack 150 m long. Within a few minutes the rock on either side of the crack had risen 2.4 m producing an elongate dome approximately 150 m long and 30 m wide. This upheaval occurred when the quarry had only reached a depth of 15 m: the rock was a thickly-bedded Ordovician limestone which overlay the Precambrian beds containing the ore.

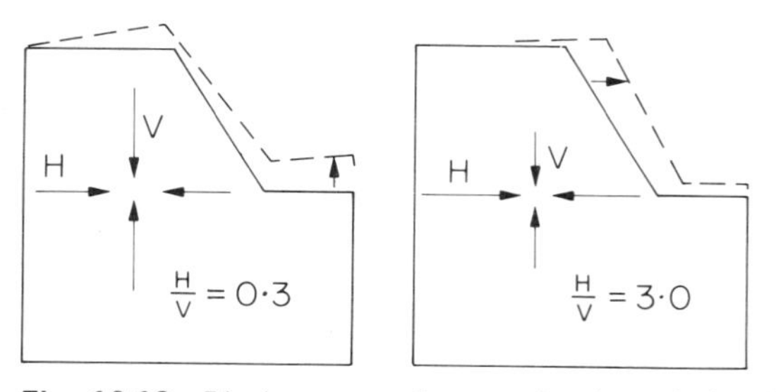

Fig. 16.13 Displacement of excavation boundaries (from Yu and Coates, 1979).

Ground-water

Two problems, peculiar to surface excavations, must be noted; namely the depletion of near surface aquifers and ground-water pollution. Both may occur even though ground-water on site is controlled and of no problem to the site itself.

Excavations act as natural sinks and pumping water out of deep excavations can gradually deplete local ground-water reserves. The water discharged from the excavation may be returned to the general water supply, but the cost of its purification can be greater than that of installing a cut-off to prevent the flow depleting local resources.

Top soil and weathered horizons are important purifiers of water infiltrating into the ground, because they are host to microbiological species that can break down dangerous substances to harmless by-products. The stripping of these horizons precedes surface excavations and once removed, ground-water below the site becomes vulnerable to pollution. Excavations situated above or close to important aquifers should be protected by strict site management, e.g. drains should be installed rather than soak-aways, and special provision made for the storage of petroleum products so that leaks and spills from these products cannot enter the ground. Sand or rolled clay may be placed as a temporary covering over floor areas that are at their formation level.

Underground excavations

Common underground excavations are shafts and drifts (or inclines) for access to underground works, mines for mineral extraction, tunnels and large chambers.

Investigations

The difficulty of adequately predicting geological conditions for such works from surface investigations requires that working practices for underground excavations be quickly adaptable to the actual conditions encountered. Water-bearing zones and fractures are always potentially dangerous. Limestones are particularly difficult materials to investigate if they contain solution cavities. Water-filled fissures and cavities do not have to be intersected by an excavation to endanger it. If the thickness of rock separating them from the excavation is insufficient to withstand the pressure head of water within them, the rock will fail (example (*iv*) in Fig. 16.6*b*). Limestone cavities in the South African dolomites have discharged their content of liquefied mud into shafts, and occasionally this has entrapped workmen. Tunnels present the difficulty of vertical bore-holes intersecting only a small fraction of their total length. Despite this, surface mapping can indicate where bore-holes may best be sunk to provide relevant information for undrilled areas (Fig. 16.14). Shaft locations can be explored with reasonable success by drill-holes sunk from ground level to the full depth of the shaft (Fig. 16.15*a*). Long drill-holes for deep shafts may wander off course and the geology of the shaft may then be misinterpreted (Fig. 16.15*b*). To compensate for these shortcomings it is customary to probe the ground ahead of an excavation (Fig. 16.15*c* and *d*).

In rock

When advancing a tunnel in hard rock, by blasting, it is possible to create a series of benches if difficult ground appears in the tunnel crown. Such drifts can also be advanced, if needed, at the sides, or centre, or invert of the tunnel. In this way difficult ground may be inspected ahead of the main face and the need for support assessed. In moderately hard to firm ground it may be necessary to excavate pilot headings, especially if the ground is badly faulted, or in need of treatment, or beneath a major aquifer or body of water such as a large river. Excavation in moderately hard ground may be influenced by subtle changes in geology such as an increase in the amount of quartz in a mudstone, or the presence of limestone nodules in a shale.

The design of large excavations requires better geological information than can normally be provided by bore-holes drilled from ground level. Bore-holes, however, can enable the feasibility of such an excavation to be assessed, by confirming the presence of rock having adequate mechanical properties, or the absence of unusually diffi-

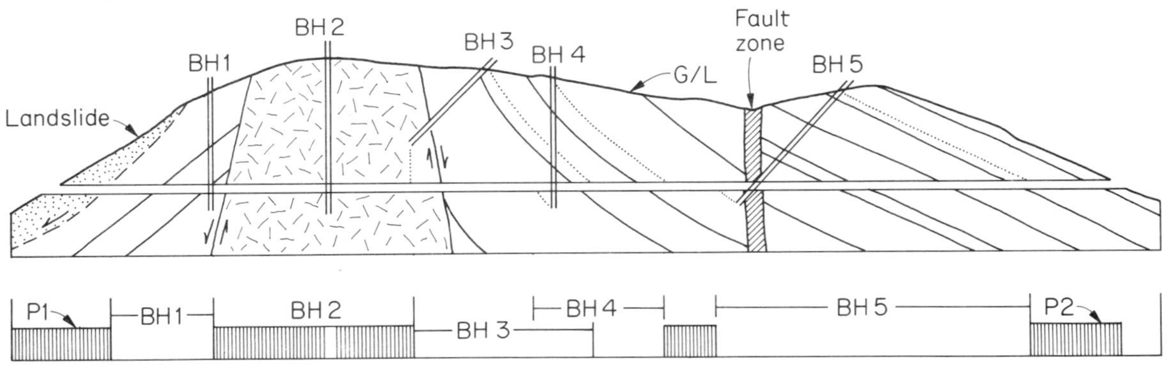

Fig. 16.14 Bore-hole investigation of geology for a tunnel: shaded areas, shown beneath the section, represent tunnel lengths not covered by bore-holes. Area between BH 4 and 5 could have been covered by better positioning of BH 4. Note the value of using inclined holes. Portals (P1 and P2) are the subject of separate investigation (see also Fig. 12.9).

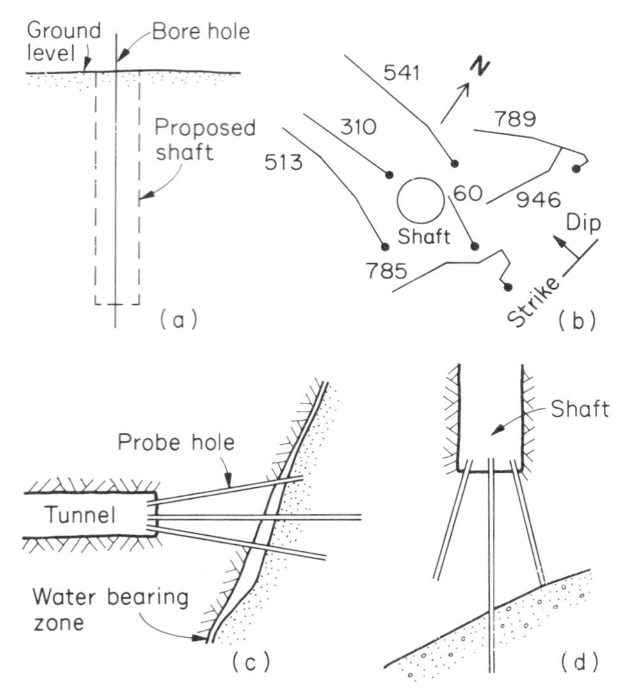

Fig. 16.15 Idealized bore-hole investigation (**a**) for a shaft and a real investigation (**b**). Note how bore-holes have been deflected by dip and strike. Probing ahead (**c**) for a tunnel and (**d**) for a shaft. Most countries have regulations which define the forward distance that must be probed (see also Fig. 17.4).

cult *in-situ* stresses and ground-water pressures. The orientation of major structural characters in the rocks, such as bedding, jointing and cleavage, may similarly be established. Shafts and adits can then be excavated to obtain the detailed geological data required for final design. These smaller, preliminary, excavations can be incorporated into the final excavation and are therefore not wasted.

In sediment

Excavations in gravel, sand, silt and clay, generally need support at all times (p. 279). Investigations are therefore directed towards establishing the stratigraphy of the ground to be excavated, especially the variety of sediments to be encountered. Underground excavations act as drains for surrounding ground-water and can cause consolidation of overlying sediment, and settlement at ground level. All permeable horizons that could drain water to the excavation should be identified. The strength of the sediments must be determined and their *in-situ* moisture content. Special note should be taken of silt occurrences: with too little moisture, silt dries to a powder of little strength, and with too much moisture it flows like a slurry (and see p. 122). The greater Plasticity Index of clays (Fig. 9.22) makes them much easier sediments to excavate especially as they have a low permeability and are less troubled by water flows than other sediments. Shell-bands, and thin seams of limestone and sand, and lenses of gravel, may occur in clay and provide a local ingress of water to the excavation: this will soften the surrounding clay. Organic sediments, such as peat, are weak and often wet: they may also yield methane (see gases).

Gases

Reference has already been made to the presence of gas in the upper levels of the crust (see Figs 6.20, 6.21 and p. 217). Such gas can be lethal when encountered in the enclosed space of an underground excavation. The following gases are of note.

Carbon dioxide (CO_2) 1.53 times heavier than air and accumulates in the bottom of excavations; normally produced by the slow oxidation of coal and sometimes by dissolution of limestone.

Carbon monoxide (CO) 0.97 times the weight of air and rises. Far more toxic than CO_2, is commonly accompanied by methane and is associated with coal-bearing rocks.

Methane (CH_4) 0.55 times the weight of air and rises. An extremely mobile gas that can contaminate strata above its source zone or be carried downwards by water draining to lower excavations. Forms a highly explosive mixture with air and known as 'fire-damp' in coal mines where it originates from the bitumens (Fig. 6.20). May also be found in organic rich sediments where it is described as 'marsh gas'.

Hydrogen sulphide (H_2S) 1.19 times the weight of air, is highly toxic and forms an explosive mixture with air. Arises from the decay of organic substances and is readily absorbed by water.

Other gases Many other gases have been detected underground, including sulphur dioxide, hydrogen, nitrogen and nitrous oxide.

Gas bursts

Most of the gas underground is stored in the pores of sedimentary rocks and is under pressure. It therefore expands when able to do so (1 cm^3 of coal may liberate 0.25 cm^3 or more of gas) and the sudden liberation of a gas rich zone can result in a 'gas-burst' when many hundreds of tonnes of coal can disintegrate into a coal-dust laden cloud of methane. A description of such a burst is provided by Stephenson (1962).

Tunnels above the water table

When water-levels, at depth, rise they drive the air above them towards ground level. This air may be de-oxygenated, the O_2 having been consumed by minerals oxidizing in the ground, and may concentrate in a tunnel that is overlain by impermeable strata.

Stability

When rock or soil is excavated, the load that it formerly carried is transferred to surrounding ground. The stresses *in-situ* therefore change with some increasing and others decreasing in relation to their original value. Failure will occur if the magnitude of the resulting stresses exceeds that which the ground can sustain.

Rockfalls

Blocks of rock may fall or slide into an excavation by moving along surfaces of bedding, jointing and cleavage, when the strength of the surfaces is unable to hold the blocks in place. Such instability readily occurs in an excavation when tension develops in the roof, or when compressive stresses are so reduced that the friction between surfaces is unable to contribute the shear strength they require, above that provided by their cohesion, to maintain stability.

Geological conditions which aggravate this tendency are illustrated in Fig. 16.16.

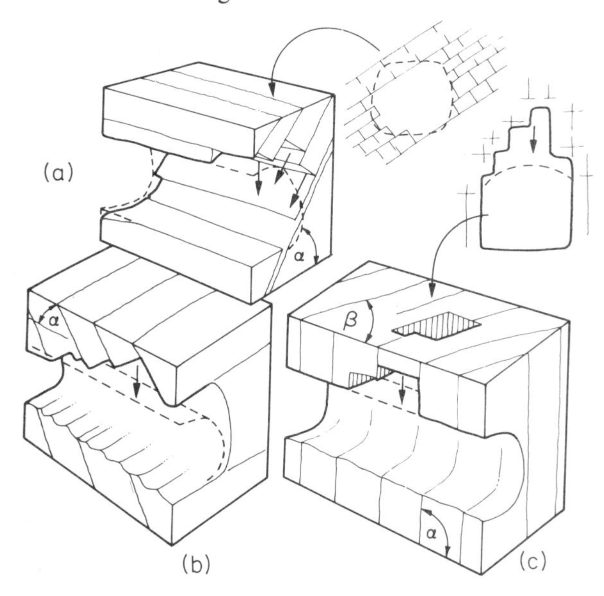

Fig. 16.16 Relations between overbreak and structure: see text for explanation.

(*a*) When the strike of surfaces is parallel to the axis of an excavation, large and often continuous surfaces can form much of the roof: instability may arise when their dip (α) is less than 15° and more than 70°.
(*b*) When the strike of surfaces is orthogonal to the axis of excavation, an uneven profile may develop in the roof when (α) is less than 20°. When (α) is greater than 20° severe falls may occur between weak layers, to produce vaulting in the roof, as shown in (*c*).
(*c*) When the strike of surfaces is oblique to the axis of excavation conditions will be between those described for (*a*) and (*b*) above, depending upon the angles (α) and (β). Such departures from an intended excavation profile are described as *overbreak* and may cause excessive amounts of rock to be removed and the space so formed to be back-filled. A survey of geological structure permits the magnitude of these problems to be predicted (Richards *et al.*, 1972; Shi *et al.*, 1983). Overbreak in the Lochaber Tunnel, cut through schists by drilling and blasting, was small when crossing the strike of the steeply dipping foliation, the cross-section being in good agreement with the outer ring of drill-holes. When the tunnel was aligned to the strike of the foliation considerable overbreak resulted (Peach, 1929; Halcrow, 1930). The tunnel was constructed in the 1920s and since then considerable advances have been made in drilling and blasting practice, which can greatly reduce the incidence of overbreak (p. 225).

Compressive failure

When the compressive stresses around an excavation are high, the surrounding ground may fail in shear. This can occur in shallow shafts and tunnels in soil, at moderate depths in shafts and tunnels in weak rock and at great depths in strong rock. The larger the excavation, the greater will be the change in stress around it and even strong rock can fail at moderate depths if loaded in this way. Excavations in valleys, which are often required in hydroelectric schemes, can be expected to encounter high horizontal stresses originating from the response of the valley to erosion (p. 260). In the excavations for the Snowy Mountains Scheme in S.E. Australia, ratios of horizontal to vertical *in-situ* stress of 1.2, 1.3 and 2.6 were recorded, with the horizontal stresses being considerably greater than that due to the weight of overlying rock (Moye, 1964). The effect of such stresses on the stability of an excavation is illustrated in Fig. 16.17.

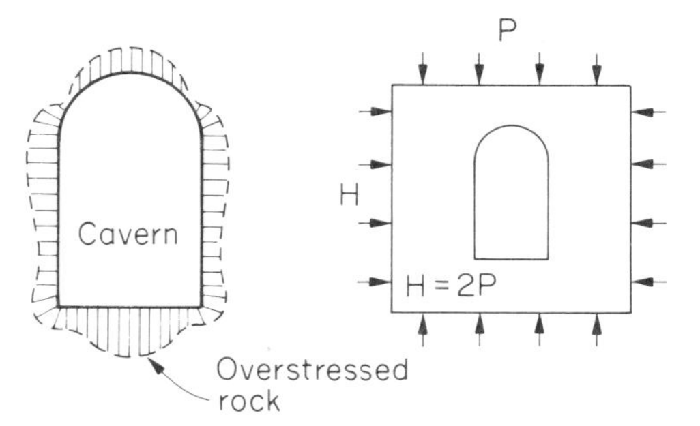

Fig. 16.17 Overstressing capable of causing rock failure. For relevant details see Hoek and Brown (1980).

Three forms of failure in compression are common:
(*i*) *Squeezing ground* describes the effect of overstressing weak ground such as sand, silt, clay, shale and weathered or crushed rock. These soon deform under the surrounding loads and squeeze into the excavation, which must be lined to prevent its total closure. Sometimes the squeezing may grip the shield of a tunnelling machine so securely, that forward progress is prevented. Local swelling can occur when expansive minerals are present, such as anhydrite (p. 87), which raised the floor of a tunnel through the Alps by 25 cm a year (Proctor and White, 1946). Expansive clay minerals, such as montmorillonite (p. 81) can promote serious instability when they occur in thin

seams infilling the joints of strong rock (Brekke *et al.*, 1965).

(*ii*) *Gentle failure* Rock that is not too brittle, with fail gently, relative to that which explodes as a rock-burst (see below). Cracks appear in the surface of an excavation and the ground on either side of them moves into the excavation, sometimes retaining a raised profile, but often spalling to leave a scar (see Fig. 16.18).

Fig. 16.18 Partial closure of an originally horizontal main road (courtesy of the National Coal Board, London).

(*iii*) *Rock bursts* are the sudden and violent detachment of masses of rock from the sides of an excavation. Hundreds of tonnes of rock may be involved, producing ground motions that can be detected at seismic stations many kilometres away (see Fernandez *et al.*, 1982). This failure is a great problem to deep mines in hard, brittle and very strong rock, as found in some of the gold mines of South Africa, and of Canada, and the zinc mines of Idaho in the U.S.A. Bursts are the result of excessive rock stress and to lessen their frequency ground may be de-stressed by local excavation prior to production mining (Jaeger and Cook, 1979). *Bumps* are shocks that originate from failure behind an excavated face, and may dislodge material from the walls and roof of an excavation. The shearing of overlying strata is thought to be a common cause of bumps in coal mines.

Support

A correctly-supported excavation prevents the new conditions of loading in the ground around it from causing either excessive deformation of the excavated profile or failure of the surrounding ground. The movement of ground into an excavated space encourages an arching effect, as illustrated in Fig. 16.19*a*. This strengthens the ground in the vicinity of the excavation and in strong rocks the natural arch may have strength enough to support the surrounding ground: a condition described as 'self-supporting'. This response requires time to occur (Fig. 16.19*a*) which means that there is no unique value for 'ground pressure', and that the response will happen more quickly at some sites than at others, depending upon the strength of the ground, the stresses within it and the size, and shape of the excavation (Brown *et al.*, 1983).

Strong rocks containing weak surfaces such as clay-filled joints, weaker rocks, and all soils, require support

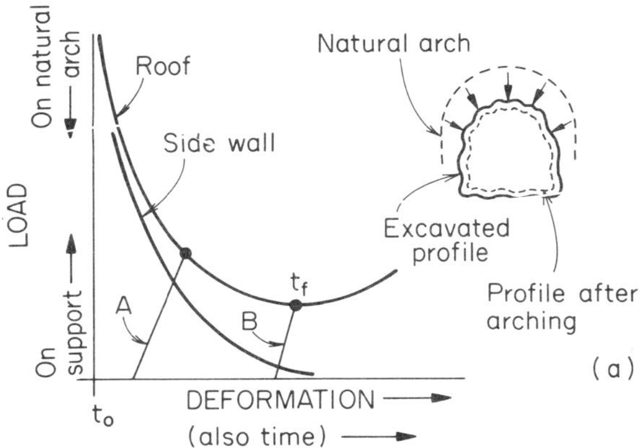

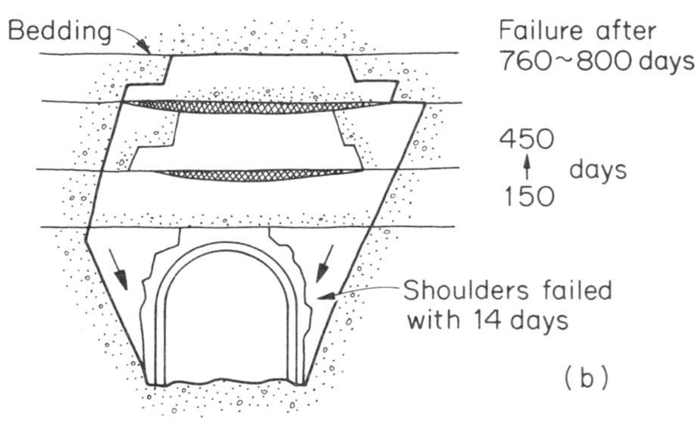

Fig. 16.19 (**a**) Load on excavation boundary. From $t_o \rightarrow t_f$ a ground arch is forming. At t_f the deformation is too great and the ground arch begins to fail. A = support that is applied too early, or is too stiff; B = proper support (Deere *et al.*, 1969). (**b**) Roof sag and final failure recorded during an experimental tunnel collapse (Ward, 1978). This behaviour agrees with the general form of the curves proposed by Deere and shown in (a) above.

because their natural arching is unable to provide long-term stability. The uneveness of failure in these materials is illustrated by the example shown in Fig. 16.19*b*: the walls of this excavation failed in 14 days but the roof requires much longer. In very weak rock and many soils not even short-term stability can be assured and support must be provided quickly. The weakest materials, such as crushed rock in faults, cohesionless sands and silts, may require support *before* they are excavated and this is provided by special excavation techniques, which drive supports into the ground ahead of the excavation, and by ground treatment to increase the strength of the weak material.

The redistribution of load around an excavation can result in a roof being in tension under certain circumstances (usually when the ratio of horizontal to vertical stress in the ground is lower than normal as might occur when the ground is being shaken by an earthquake). Joints, in the roof rocks, open and rock blocks drop from the roof so destroying any natural arch that may exist.

The calculation of support requirements for excavations in soil and rock requires a knowledge of the mechanical properties of these materials, the magnitude of total stress in the ground and of the ground-water pressures that are expected around the excavation. The support of excavations in rock also requires a knowledge of the jointing and other surfaces within the rock mass and these details should be recorded daily in the form of a simple log of the excavation (Fig. 16.20). An indication of the support that will be wanted may be provided by indices which can be calculated from an accurate description of a rock mass. Two indices that are constantly used are those of Barton *et al.* (1974; 1976) and Bieniawski (1974; 1976). These provide an assessment of the period an excavation may remain unsupported (Fig. 16.21*a*), the amount of support it may require (Fig. 16.21*b*: note that the schemes provide detailed guidance on the degree of support that is required), and the likely problems associated with creating a large excavation, as with a cavern, in such ground (Fig. 16.21*c*). The subject is excellently reviewed by Hoek and Brown (1980) and developments with use of these indices has been reviewed by Barton (1983) and Bieniawski (1983).

Bedrock

Shafts and tunnels commonly encounter the sub-surface junction between soft ground and underlying hard ground (the 'bedrock'). This junction is often difficult to penetrate and may require careful support (Fig. 16.22). Its position should be ascertained by exploratory drilling. Unusual conditions may exist in volcanic terrain, with soft horizons of weak ground overlain by harder, more recent, lavas (Fig. 16.23).

Portals

These must usually pass through a surface zone of weathered and weak material that can be of considerable thickness in areas of tropical weathering. They may also have to pass through old landslides on the surface of slopes (Fig. 16.14). For these reasons, portals usually require special support and protection against active forms of slope instability, especially rock falls and avalanches.

Effects at ground level

Underground excavations that are not fully supported invariably result in subsidence at ground level. The strains involved usually have horizontal and vertical components of movement (Fig. 16.24), and may damage buildings, change the permeability of the ground and increase areas liable to flooding adjacent to rivers (Institution of Civil Engineers, 1977). Well stratified deposits of reasonable strength may offer a certain amount of beam support (Fig. 16.19*b*) which may contain the zone of broken rock, but a dangerous situation can develop if strata separation approaches ground level and interferes with the foundations of buildings or the ground where buildings are to be erected. When the ground continues to break under stresses that are too great for it to carry, a front of broken rock will travel up towards ground level. This is described as 'caving' and can result in the sudden appearance of holes at the surface. The drainage of ground-water to

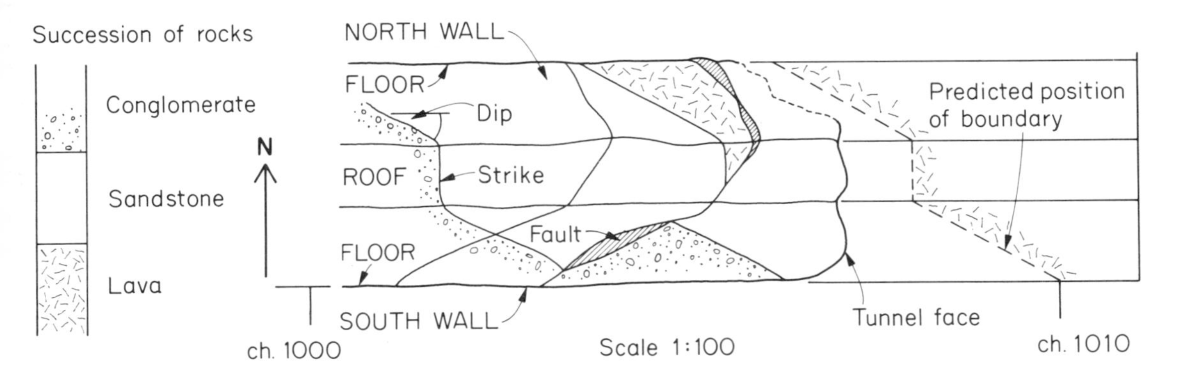

Fig. 16.20 Example of a simple tunnel log. (ch = chainage) and prediction. Such plans can be cut out and the walls folded down to make a simple 3-dimensional model of tunnel geology (see also Face Logs: Fig. 10.7).

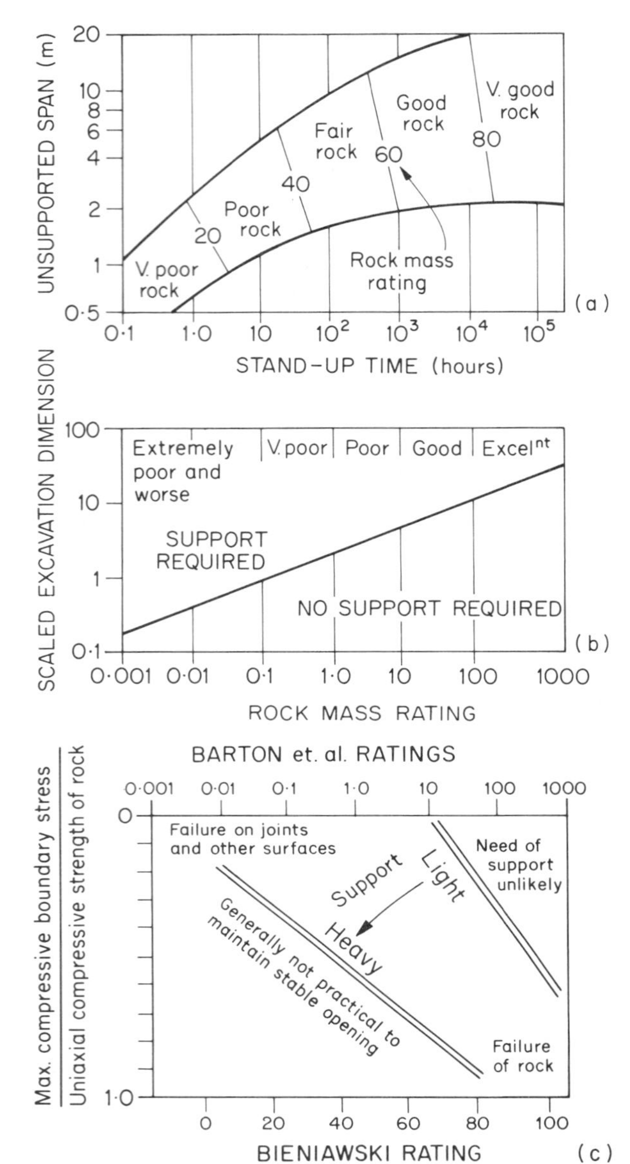

Fig. 16.21 Essential elements of classifications of rock masses by (**a**) Bieniawski, (**b**) Barton *et al.* References cited give details and support recommendations. (**c**) Approximate relationships for large excavations: boundary stress = stress on roof or walls or excavation (from Hoek, 1981).

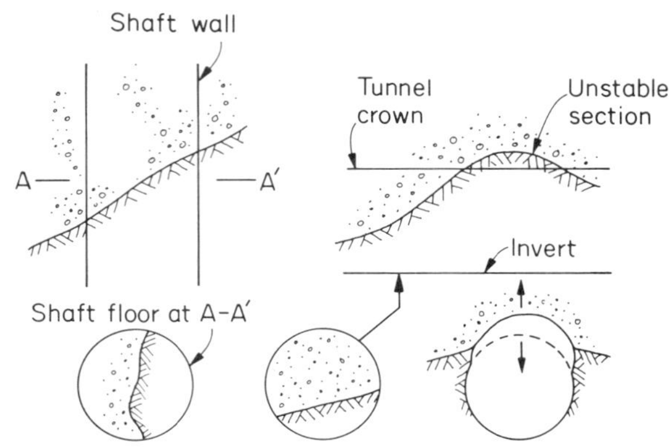

Fig. 16.22 Mixed face in soft and hard (bedrock) materials.

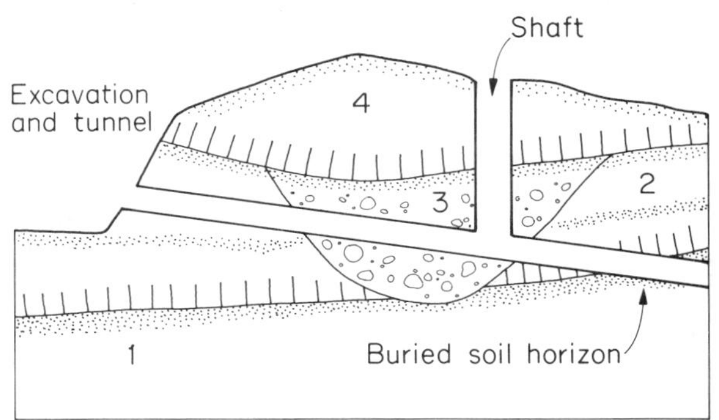

Fig. 16.23 Example of the variations that can be encountered in volcanic rocks. Strong lavas 1, 2 and 4 contain weak zones, profiles of weathered rock, and soil horizons, and moderately strong buried river deposits (3).

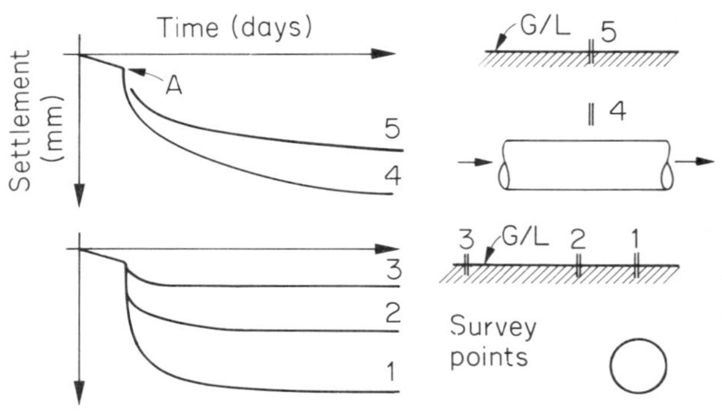

Fig. 16.24 Nature of settlements associated with tunnelling. Prior to 'A' tunnel face is approaching the survey points. At 'A' it is beneath them. After 'A' it is moving away from them.

excavations can result in the consolidation of compressible sediments and settlement, even though the sediments may be a considerable distance above or away from the excavation (de Freitas and Wolmarans, 1978).

When lateral boundaries, such as faults and dykes, do not exist, the areal extent of subsidence becomes largely controlled by the 'angle of draw' (Fig. 16.25) (National Coal Board, 1975). In Coal Measure strata of Britain the angle normally lies between 35° and 38°. In Holland, where the Coal Measures are overlaid by weaker rocks than found in Britain, the angle is around 45°. The 'angle of break' defines a surface of break that joins areas of maximum tensile strain: (Table 16.2).

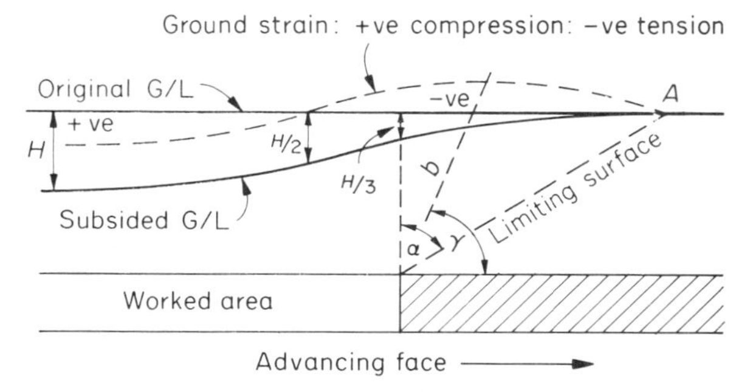

Fig. 16.25 Typical subsidence resulting from the extraction of a level seam. α = angle of draw (measured from the horizontal in some countries); γ = angle of break; H = maximum subsidence. In the typical cycle of ground movement, tension occurs behind the limiting surface and may be associated with the opening of cracks at ground level (Fig. 9.28). Tensile strain reaches a maximum at the surface of break (b). Later the strains become compressive and close cracks. An excavation advancing beneath point A would subject it first to tension and then to compression.

Table 16.2 Typical values for angle of break

	Angle of Break
Weak and generally loose strata	60°–40°
Cohesive materials of clay type	60°
Sands and sandy soils	45°
Shales of medium strength	60°
Hard sandstones and similar rocks	85°

Disposal of excavated material

Bulking

This is the increase in volume which occurs when a material is broken and is additional to volume to be transported and disposed. The increase is usually more than 20% of the original volume, being greatest for the strong

Table 16.3 Swell, as percentage of original volume

Topsoil	20–40
Unconsolidated sediments	10–50
Gravels	15
Sands	12
Silts and clays	10
Sands and gravels, with clay	20–50
Moderately consolidated sediments	20–40
Stiff clays	25
Fissured and hard clays	40
Sedimentary rocks	30–70
Limestone	60–70
Sandstone	40–70
Shale	30–40
Metamorphic and igneous rocks	25–80
Fresh and blocky	50–80
Weathered and blocky	25–40
Slate	30–40

rocks that break into blocks whose edges and corners form resilient protrusions (Table 16.3). The ability of broken ground to flow through chutes and hoppers is an important property that has been investigated by Kvapil (1965). Table 16.4 gives general values of relevance and actual tests should be conducted where possible. The tendency of some materials, particularly the weak rocks, to chip and disintegrate with transport, results in a change of their grading as they travel through the system. This always increases the percentage of fines and generally decreases the mobility of the mass; a point of particular relevance to the efficiency of long conveyor systems (see Kvapil, 1965; Savage *et al.*, 1979).

Table 16.4 Likely angles of friction (from Kvapil, 1965)

Type of material	Wall material	Angle of friction along wall (ϕ_w)
Limestone, dolomite, marble	Steel	30–40°
	concrete	33–43
	wood	37
Granite, gneiss	steel	31–42°
	concrete	35–42
Iron ores	steel	33–42°
	concrete	36–43
	wood	40
Sandstone	steel	32–42°
	concrete	34–42
Shale, mudstone	steel	28–40°
	concrete	29–42

Surface disposal

Waste material that cannot be stored below ground is brought to the surface where its coarse fractions are tipped (Fig. 16.26*a*). Fine material from the mills and processing plant, is transported as a slurry to settling ponds where it is sedimented: the ponds are retained by tailings dams (Fig. 16.26*b*). The design of both tips and tailings dams must obey the principles of soil mechanics because failure of these structures can release waste which may liquefy and engulf all in its path (Klohn, 1980; Barton and Kjaernsli, 1981). Such structures must be founded on ground that is sufficiently strong to carry their weight without failing and should not be placed over zones of natural ground-water discharge without adequate provision being provided for the control and drainage of ground-water (Bishop, 1973).

Two tragic examples of the consequences of foundation failure beneath these structures are recorded by the reports of the Aberfan and Mufulira disasters.

Aberfan (Anon, 1967: 1969)

Aberfan is a small village in the Taff valley of the South Wales coalfield. On its steep slopes, overlooking the village, colliery waste had been tipped since 1914. In 1944 one of the tips slid a small distance downslope: another partially failed in 1947 and 1951. In 1966, tip No. 7 failed, and from it flowed a wave of liquefied tip material which

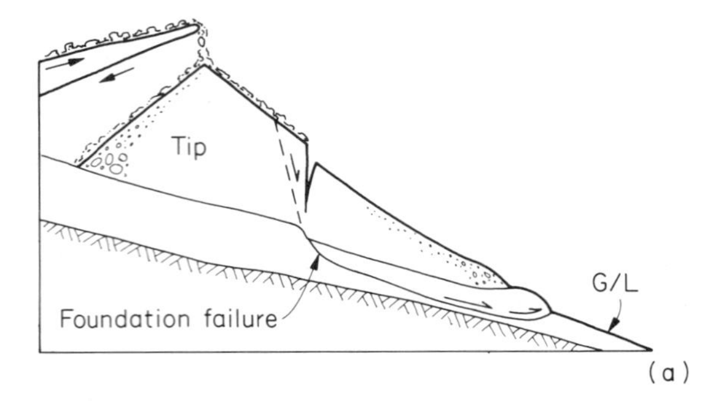

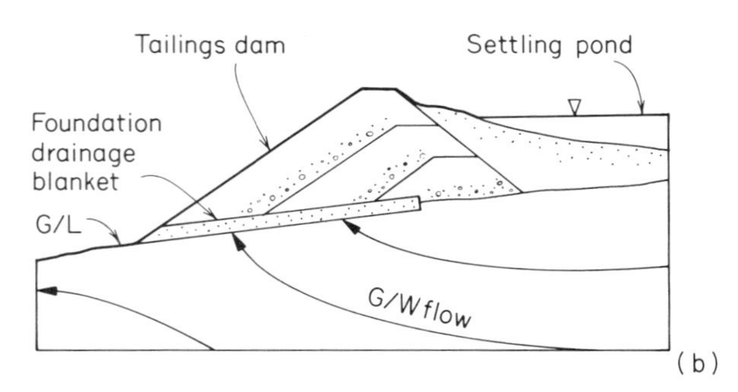

Fig. 16.26 (**a**) Simple tipping and associated danger of foundation failure (see also Fig. 14.20). (**b**) Tailings dam and appropriate drainage for controlling ground-water.

rapidly moved as a flow-slide down the hill into the village, where it engulfed many buildings including a school full of children. 144 people were killed and many others injured.

The enquiry which followed revealed that:

(*i*) the tips had been placed directly on glacial clay that overlies the bedrock;

(*ii*) tip 7 had previously failed in 1963 and that since 1963 the tip had been accumulating on ground underlaid by a sliding surface in clay (see Fig. 16.26*a*);

(*iii*) the toe of the tip had extended over a spring line;

(*iv*) the hydrogeology of the ground beneath the tips had been disturbed by the effects of mining at depth.

The factors combined to cause the 1966 failure to occur following a period of heavy rain. Since this disaster many organizations have revised their codes of practice controlling tipping (see National Coal Board, 1972); unfortunately geology does not follow a code of practice.

Mufulira Mine (Sandy *et al.*, 1976)

At this mine a tailings dump was sited above an area that was to be mined by sub-level caving. A caved zone of broken ground extended to ground level and undermined the tailings dump. Liquid tailings flooded into the void and the mine, filling the western section of the workings from the 1056 m level to the 495 m level; 89 miners were killed.

Selected Bibliography

Hoek, E. and Brown, E.T. (1980). *Underground Excavations in Rock*. The Instn. Mining and Metallurgy, London.

Megaw, T.M. and Bartlett, J.V. (1982). *Tunnels: Planning, Design, Construction.* (in 2 vols). Ellis Horwood Ltd., Chichester, & J. Wiley & Sons, New York.

A.M. Muir-Wood. (1979). Ground behaviour and support for mining and tunnelling. *Trans. Instn. Mining & Metallurgy*, London.

West, G., Carter, P.G., Dumbleton, M.J. and Lake, L.M. (1981). Site investigation for tunnels. *Int. J. Rock. Mech. Min. Sci. & Geomech. Abstr.*, 18, 345–67.

Wahlstrom, E.E. (1973). *Tunnelling in Rock*. Elsevier Pub. Co., New York.

17

Ground Treatment and Support

Considerable economies can be achieved in the design and construction of engineering works if the ground on site can be improved, and supported where necessary.

The extent to which this is necessary should be assessed by ground investigation. Normally it is the weakest ground that is in need of greatest treatment or support and the difficulty of recovering representative samples of such material, especially when it is at depth, frequently results in the nature of such ground being assessed inadequately. It is therefore important to ensure that programmes of bore-hole drilling and sampling report the delays in the progress of work created by weaknesses in the ground. Thus a record should be made of horizons from which samples could not be recovered, or levels at which the bore-hole collapsed, or depths at which the fluid, used in drilling, drained from the hole. Such information implies that the ground in these zones may be in need of treatment and support.

The structure and mineralogy of rocks and soils imposes a limit upon the improvements that can be achieved within them by ground treatment. These limits must be assessed, often by using special tests. For example, an injection test is required to assess the maximum rate at which grout can be injected into the ground; further tests are needed to reveal the maximum pressure that can be used to accomplish this rate of injection, without causing the surrounding ground to fail. Special tests may also be necessary to assess the magnitude of support that is required at a site. One example of such tests is the special survey that may be conducted in bore-holes to determine the dip and strike of surfaces in rock not exposed for mapping. The information from such a survey is used in an assessment of the support required to ensure the stability of an excavation cut into the rock (see Fig. 14.3).

Treatment and support are frequently combined so that a small amount of treatment, which may easily be provided, enables economic support to achieve the overall saving required in either design or construction.

TREATMENT

Dewatering

Ground-water may be controlled either by excluding it from a site or by abstracting it from the ground within and around a site. The techniques commonly employed in these forms of ground treatment are listed in Table 17.1.

Table 17.1 A guide to the application of treatment techniques for the control of ground-water. (1) = mainly chemical grouts.

Method of control	Granular deposits			
	Gravel	Sand	Silt	Clay
ABSTRACTION				
Pumped wells	√	√		
Well points		√	√	
Electro-osmosis				√
Sumps and drains	√	√		
Vertical drains		√	√	
EXCLUSION				
Grouting	√	√	√ (1)	
Compressed air		√	√	√
Freezing	√	√	√	√
METHOD OF CONTROL As listed above	FRACTURED ROCK See Fig. 13.18 for likely values and use the above Table by equating Gravel to $>10^{0}$ cms^{-1}, Sand to 10^{0} to 10^{-4} cms^{-1}, Silt to 10^{-4} to 10^{-6} cms^{-1} and Clay to $<10^{-6}$ cms^{-1}.			

Dewatering describes the abstraction of water, either to reduce the flow of ground-water, or to diminish its pressure. The treatment is influenced by the permeability of the ground, the proximity of hydrogeological boundaries, the storage coefficient of the ground, the pressure head of ground-water to be abstracted and the natural hydraulic gradient within the ground: all these factors are described in Chapter 13.

Dewatering is normally undertaken to improve conditions in surface excavations and to assist the construction of structures founded at, or near ground level. It is not usually employed in underground excavations for fear of consolidating overlying sediments and damaging structures founded upon them (see for example, the drawdown shown in Fig. 13.27). More often, steps are taken to exclude water by grouting, ground freezing or the use of compressed air. The drainage of mines, when required, is achieved using drainage adits or drainage holes. Practical guidelines for the design of dewatering systems are provided by Leonards (1961) and Powers (1981).

Sediments

Thick horizons of porous sediment that have not been sufficiently consolidated to form fractured rock, occur in abundance in strata of Cenozoic age (Table 2.1). Quaternary sediments of differing origin constitute much of the transported material that forms the superficial cover of drift above fractured bedrock. In these strata, clays, silts, sands and gravels may be encountered, occurring as either separate horizons of significant thickness or interbedded. Sometimes these sediments cover a granular and porous zone of weathering at the top of fractured bedrock: such zones may often be treated in the same way as the sediments above them.

Wells may readily abstract water from permeable deposits of sand and gravel. The level of water in the well is reduced, by pumping, below that in the ground thus encouraging a radial flow of ground-water to the well (Fig. 17.1). Sometimes the resulting cone of depression extends further than necessary and dewaters ground that need not be drained; other times it may intersect a hydrogeological boundary that recharges the ground and prevents further drawdown (Fig. 13.28). Such a boundary may be considered a *source*.

Well-points may be used in permeable deposits of silt and sand and in thin deposits of gravel, when small cones of depression are required. Each point sucks water from the ground and has a limited sphere of influence, but when used in banks (Fig. 17.2) it is possible to drain a considerable thickness of strata, and restrict drawdown to the vicinity of the excavation. This is because each well-point acts as a hydrogeological boundary to its neighbour. Such a boundary may be considered a *sink*.

Electro-osmosis is a phenomenon that causes some sediments to expel their pore water when an electric current is conducted through them. It is sometimes used as a special technique for reducing ground-water pressures in fine-grained sediment of low permeability. An electric current is passed through the ground between anodes and cathodes, the latter being a metal well-point. The efficiency of the method decreases with a decrease in the moisture content of the sediment and an increase in the quantity of free-electrolyte in the pore water. Clay mineral activity is also influential, the amount of water being moved per hour per amp, being greatest in inactive clays. For these reasons the method tends to work better in silts than in clays.

Vertical drains achieve much the same effect as electro-osmotic techniques, but for a fraction of the cost. The simplest are bore-holes filled with sand or man-made filter fabrics. They are particularly suitable for draining ground containing horizontal, impersistent permeable horizons, as often found in alluvial, deltaic and estuarine deposits (Fig. 17.3). By providing vertical paths of permeability where only horizontal bands of permeable material existed before, the drains permit layers of silt and sand to drain interbedded clay. In this way the ground may be

Fig. 17.1 Excavation in dewatered gravel. Pipes rising from the excavation carry water being lifted by submersible pumps at depth, which have created a cone of depression (Fig. 10.21) that is greater than the dimensions of the excavation. The site is at Dungeness, close to the sea (Fig. 3.27*b*). Excavation is by digging using drag-lines: a bucket is visible in the bottom left-hand corner. (Photograph by Soil Mechanics Ltd.)

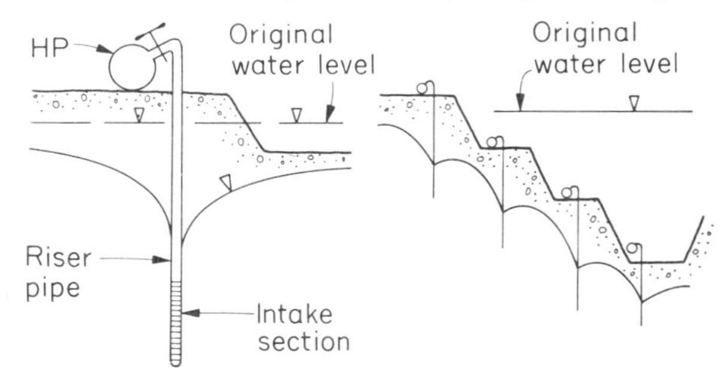

Fig. 17.2 Dewatering point (HP=header pipe) and example of staged excavation.

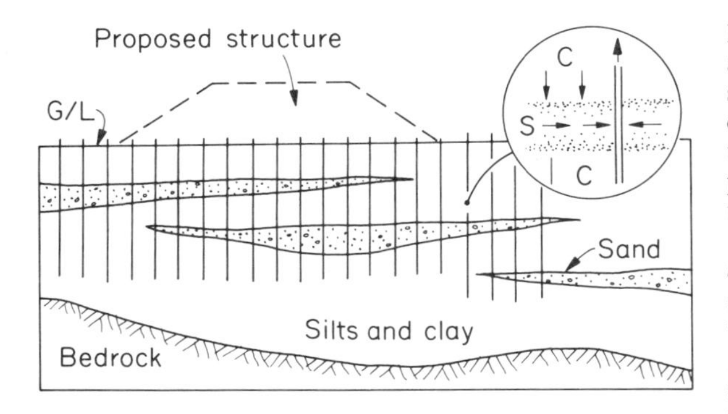

Fig. 17.3 Vertical drains, with detail illustrating how such drains encourage permeable horizons of silt and sand (S) to drain the less permeable material, clay (C).

consolidated and strengthened (see Nicholson and Jardine, 1981).

When assessing the efficiency of a dewatering programme in fine-grained deposits, special care must be taken to correctly determine the *in-situ* permeability of the sediment. Small scale laminae of silt in a clay may have a considerable influence upon the speed with which water may be abstracted from the ground and the water pressures there-in reduced (Rowe, 1968; 1972).

Fractured rock

Open fractures provide rock with a permeability that permits the drainage of water to ditches, wells, drainage holes and adits. The hydraulic conductivity of fractures is extremely sensitive to their aperture and open joints can carry considerably more water than their neighbours, which might be open by a slightly lesser amount. Therefore it is the large fractures that should be sought because they will drain the myriad of smaller fractures in hydraulic continuity with them and effect an overall drainage of the rock mass. This action can create an apparently haphazard response to the performance of drainage holes drilled out from underground excavations. Those for the adit shown in Fig. 17.4 illustrate this behaviour. Drainage

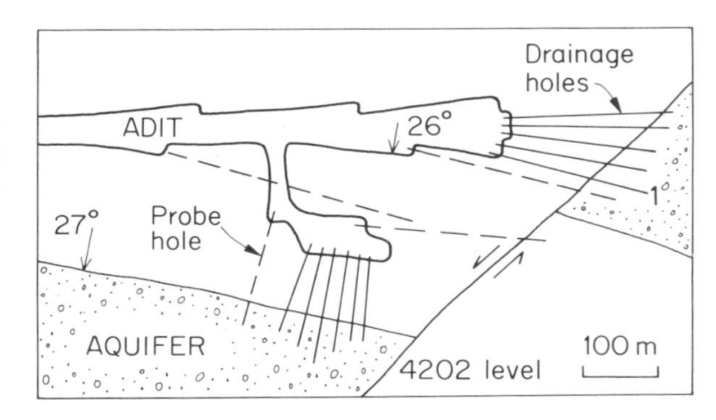

Fig. 17.4 Example of drainage measures used underground. Probe holes shown by broken lines.

hole No. 1 will intersect the fault before the other four and will discharge water whilst the remaining four may still have much smaller yields. To understand the significance of flow from drainage holes, it is necessary to study the geological structure of the ground, and in this example probe holes had been drilled ahead of the drainage holes to confirm the presence of a fault.

Joints are the most common type of fracture in rock and commonly occur in sets (p. 155). Drainage holes that intersect many sets are likely to yield more water than those whose orientation parallels a major set (see Fig. 10.3). Some fractures may be filled with clay and similar material which prevents them from conducting water until such time as the filling is eroded by flowing water. This can occur in ground surrounding a recently formed excavation if the change in stress within it causes the clay-filled fractures to open. Drainage holes that were formerly dry, and fractured rock that previously appeared to be impermeable, may then discharge water.

Adits may be driven to drain water from underground. They are commonly used in mines to relieve the pressure of water in adjacent aquifers. Sometimes they are used to drain hillsides so as to stabilize an overlying slope. *Drainage holes* may be driven from adits to extend the influence of these drainage measures.

Wells operate in the manner described on p. 270 and the structures most commonly used from ground-level to dewater a considerable thickness of fractured rock.

Ditches may successfully drain surface excavations when the permeability of the fractured rock permits draw-down to occur at a rate that does not hinder the advance of the excavation (Fig. 17.5). The excavation should be remote from a local source of recharge, such as a river or lake. Aquifers beneath the excavation, containing water under pressure (see Fig. 16.6*b*, example (*i*)), may have to be dewatered by wells.

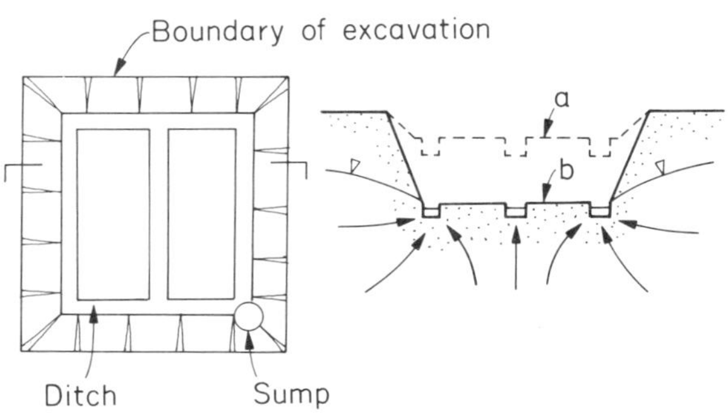

Fig. 17.5 Lowering of water levels within and around a surface excavation, by natural drainage to ditches. a = intermediate position; b = final level.

Grouting

This process injects material of liquid consistency into fractured rocks and porous sediments where it sets after a period as a permanent inclusion in pores and fissures.

The materials used may be suspensions of cement or clay or other fine particles, mixed with water; or combinations of these materials; or emulsions such as bitumen in water; or true solutions which form insoluble precipitates after injection (the 'chemical' grouts). Grouting is normally used either to reduce the permeability of the ground or to increase its strength, and often consists of repeated injections of grout via holes specially drilled for the purpose: examples are shown in Fig. 17.6. Initial injections are

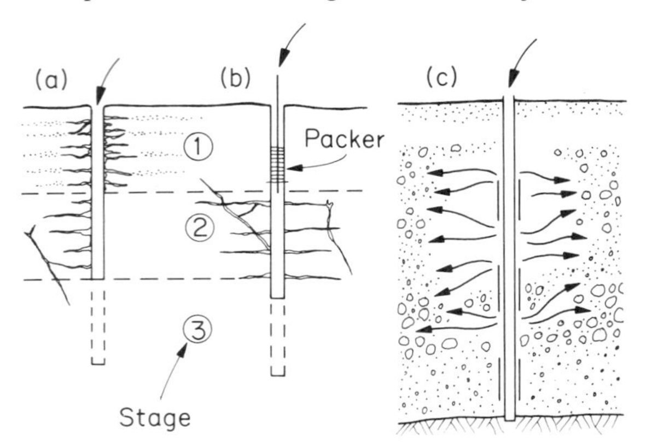

Fig. 17.6 Basic method of grouting for rock (**a**), (**b**), and unconsolidated sediments (**c**). (**a**) An open hole is drilled to the base of stage 1, grouted and then redrilled to the next stage, and so on to the desired depth of treatment. (**b**) Grout injection via packers where sections of hole can also be used in stages and grouted. (**c**) Grout injection using the tube-a-manchette technique.

made through primary holes spaced at regular intervals, with this treatment being supplemented where necessary by subsequent injections from secondary and tertiary holes (Fig. 17.7). Reviews of grouting procedures are presented by Glossop (1960), Little (1975), Scott (1975), Skipp (1975) and Camberfort (1977).

Geology influences the choice of grouts to be used because the size of pores and fractures to be filled limits the size of particles that may be used in a suspension.

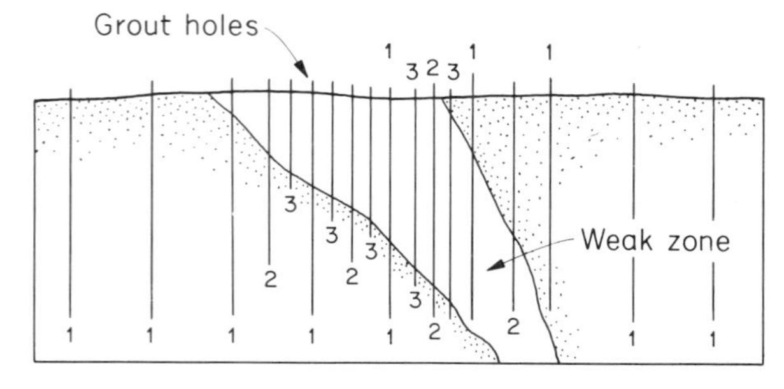

Fig. 17.7 Primary (1), secondary (2) and tertiary (3) grout holes used to treat ground of variable quality. The weak zone, in need of more treatment than the better ground on either side, is grouted using a grid of more closely spaced injection holes.

Also, the hydraulic resistance of these voids defines the viscosity a grout must possess to be easily injected, and the stresses *in-situ* regulate the maximum pressure that can be used to effect this injection. Migration of grout is controlled by the permeability of the ground and restrained by boundaries separating ground of different hydrogeological character. A knowledge of the geology of the ground is thus an essential requirement for attaining a successful result from grouting.

Sediments

Treatment of unconsolidated porous sediments such as sand and gravel, requires the permeation of grout through their delicate porous network. There is a limit to the size of pore that can be invaded by suspensions of cement and clay, and silts are usually beyond this limit: in these sediments, chemical grouts may have to be used (see example described later, Fig. 17.21). Stratified deposits of sand and gravel, interbedded with silt and clay, can be extremely difficult sequences to treat. If excessive injection pressures are used in areas where the pore size of the strata is small, the ground around the hole will fracture and each fracture will be penetrated by a sheet-like intrusion of grout. When these intrusions set they will retard the migration of grout from subsequent injections thereby isolating volumes of ground from further treatment.

In weak sediments it is not possible to drill unlined grout holes that will remain sufficiently stable to accept repeated injections of grout. The holes must be supported by a lining through which grout can pass. This lining contains a series of perforated sections each surrounded by a flexible sheath. Successive injections of grout are achieved by lowering packers into the hole, to isolate a length of section that bridges the stratum to be treated, and injecting grout into the section. When the pressure of grout in the section is sufficient to distend the flexible sheath, grout will flow through the annulus between it and the lining tube, and into the ground. The method is known as 'tube-a-manchette' and is commonly used for grouting alluvium (Fig. 17.6*c*).

Weak rock

Rock may be weak because it is a sediment that has not been well cemented, or consolidated, (many limestones and sandstones of Cretaceous and Tertiary age have this character: Table 2.1), or because it is weathered (as in almost completely weathered granite), or intensely fractured (as in major fault zones). In most cases the dominant voids will be small pores and fractures whose dimensions resist the permeation of grout suspensions. They are therefore difficult to treat. Excessive injection pressures will fracture the material, as it is weak. This process (termed 'claquage') can cause the ground to be veined by a network of grout filled fractures that collectively reduce its overall permeability. Unfortunately the weak rock may possess sufficient brittleness for claquage to create both large and small fractures. The large fractures can be filled with grout but the smallest fractures may be too small to accept a grout suspension. Incorrect injection pressures can sometimes increase the permeability of weak rock by creating too many ungrouted micro-fractures.

Fractured strong rock

Ground containing non-porous, fractured strong rock is normally the easiest of all types to grout. Rocks containing sets of large fractures and many smaller fractures, are more difficult to treat. Significant reductions in permeability can be achieved in these rocks by grouting only the sets of large fractures, if by doing so the network of smaller fractures is divided into isolated blocks, each of which is bounded by a grouted fracture (Fig. 17.8). Grout would have to penetrate the network of small fissures if the deformability of the ground had to be improved.

Clay-filled fractures will not accept grout and must be flushed with water until all erodable infilling is removed (see Lane, 1963; and Cambambe Dam, p. 249).

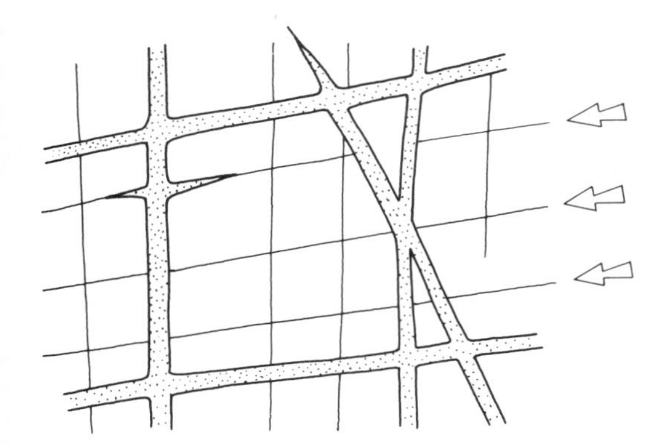

Fig. 17.8 The interruption of water flow through small fractures by grout injected into large fractures that are more easily treated.

Investigations

Attention should be given to the character of the voids to be filled and the geological structure of the ground to be treated, prior to the design and commencement of a programme of grouting. Rarely is it possible to predict with any certainty the volume of grout that rocks and soils will accept, but indications of likely quantities can be gained by observing the following geological characters.

(*i*) The amount of water that can be discharged into the ground from a bore-hole. In soils an *in-situ* test of permeability will often provide the necessary indication, but in rocks a special test may be used which measures the volume of water absorbed per metre length of hole and describes the result in units of Lugeon: 1 Lugeon is approximately equivalent to a permeability of 1.0×10^{-5} cm s^{-1}. The variety of responses that can be obtained from such a test are illustrated in Fig. 17.9 and discussed by Houlsby (1976). Injection creates a hydraulic gradient that radiates from the hole and increases with increasing pressure of injection. Fractured rock may respond to this by permitting flow which behaves as if laminar conditions

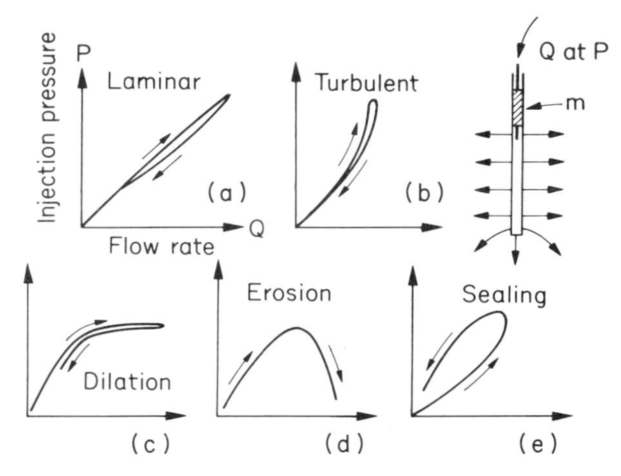

Fig. 17.9 Form of ground response to injection, as recorded from a water absorption test: in each case the ordinate represents injection pressure and the abscissa, flow rate into the ground: based on Houlsby (1976). The test is conducted over a section of bore-hole isolated by a packer (m): see text.

operate (Fig. 17.9*a*); in others, it appears that turbulent conditions develop with increasing pressure (*b*). When pressures are too great, the rock fractures are forced apart and the ground dilates (*c*). Weak material, infilling fractures, may be eroded when the velocity of flow reaches a certain value, so that the ground becomes more permeable than it was originally (*d*). Movement of erodable material may block pores and fractures, sealing the ground and making it difficult to permeate (*e*).

(*ii*) The amount of water drained from a hole: this is used in underground excavations to assess the presence of voids in need of treatment.

(*iii*) The ratio between the seismic velocity of the rock *in-situ* and of the intact material as measured on a sample of the rock tested in the laboratory. This ratio reflects the openness of the rock mass and has been used effectively by Knill (1970) to indicate grout take. (Fig. 17.10.)

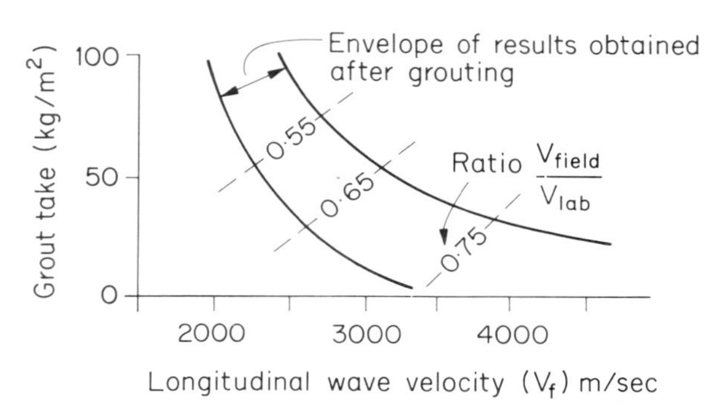

Fig. 17.10 Relationship between grout take of grout curtains in U.K. dams, and seismic velocity of the dam foundations: V_f = velocity through rock mass in field, V_{lab} = velocity through rock core from the rock mass (from Knill, 1970).

(*iv*) The length of core recovered from drill holes, expressed as a percentage of the length of hole drilled, the RQD of the core and a log of its fractures (see Chapter 10).

Other features to note are the presence of large voids such as solution cavities, lava tunnels (and underground workings), that may require more than normal treatment (Khan *et al.*, 1970), and ground strain that may have opened joint sets; e.g. at the crest of a slope or above mined areas. Many of the uncertainties associated with predicting the response of rocks and soils to grouting, especially with foreseeing the pressures that may be used, can be removed by conducting a grout test (Little *et al.*, 1963).

Ground-water The natural rate of ground-water flow should be studied as this may be sufficient to transport unset grout away from areas of treatment. Zones of high velocity flow may need special attention. The chemistry of ground-water is also relevant as it affects the type of grout that may be used, its viscosity and its setting time (Glossop, 1960). The properties of many chemical grouts are particularly sensitive to the quality of the ground-water with which they mix.

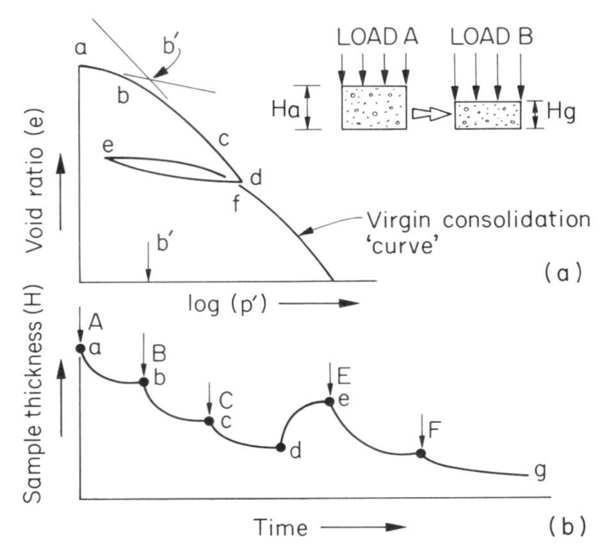

Fig. 17.12 The consolidation of a sediment and its response to unloading, (**a**) with respect to effective stress and (**b**), with respect to time (see also Fig. 9.8).

Consolidation

This form of treatment is appropriate for strengthening soft to firm lightly consolidated sediments and for reducing the settlement that will occur within them, when later loaded by an engineering structure. The treatment of these sediments seeks to continue the consolidation that occurred during their deposition. At that time, the particles of sediment were progressively loaded by the weight of grains settling above them and moved closer together, provided water could escape from the intervening pore space (Fig. 17.11).

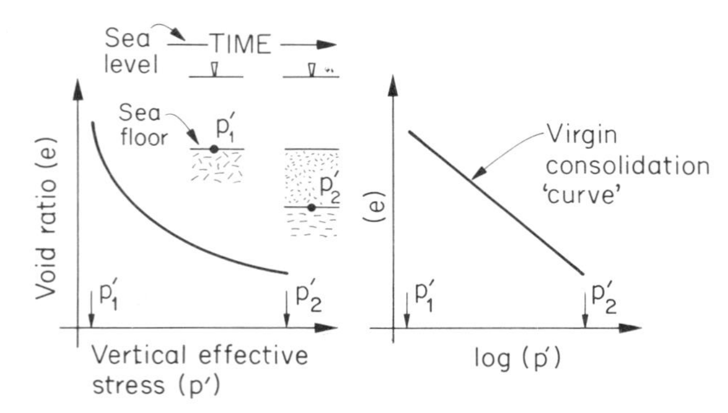

Fig. 17.11 Relationship between the void ratio of a sediment at depth (p′) and the vertical effective stress within it during gravitational consolidation accompanying burial from depth p'_1 to p'_2 (see also Fig. 6.4).

Stratigraphic history

The behaviour of the sediments under load may be studied in the laboratory and is illustrated in Fig. 17.12. An undisturbed sample of sediment is placed into an oedometer (p. 192), loaded (see A in Fig. 17.12) and permitted to consolidate (a–b), and then loaded again, in increments (B, C, E, F) until the required magnitude of effective stress within it has been attained. Initial consolidation is usually non-linear (a–b–c in 17.12*a*). The reason for this can be explained by observing the behaviour of the sample when it is unloaded (d–e): it expands, but does not recover all the deformation it has sustained. Reloading (e–f) returns the behaviour of the sample to the original (virgin) curve. Hence the similar behaviour seen at (a–b–c) may be taken to represent the return of the sample to its virgin curve; i.e. it is a record of previous unloading. The load represented by position b′ is called the pre-consolidation load to signify that a continuation of the gravitational consolidation started by nature will only occur once this magnitude of effective stress has been exceeded.

To extend this behaviour to field practice, it is necessary to increase the effective stress within a sedimentary deposit to a value that is greater than the pre-consolidation load of the succession to promote a continuation of its consolidation. This can be achieved by draining pore water from the strata; a process that may take many months to produce a widespread reduction in pore water pressures, particularly in successions containing strata of low permeability, such as clay. Consolidation can be accelerated by loading the sediment with a surcharge placed at ground level (for example, by building an embankment, or stockpile above the area to be consolidated). Vertical drains, as shown in Fig. 17.3, greatly assist the drainage of water from horizontally-stratified successions of sediment.

Shallow water sediments

Many of the deposits which are suitable for ground treatment by consolidation, accumulated under shallow water alluvial, estuarine and deltaic environments (Chapter 3), and contain lateral and vertical variations of sediment type (facies). Lenses of gravel can be interbedded with clay and peat. The time required for a suitable amount of consolidation to occur in these variable deposits can only be assessed when the stratigraphy of the succession is accurately known. Sediments most susceptible to improvement will have a moisture content that is between the Liquid and Plastic Limit for the material (Fig. 9.22): strata with a moisture content that is less than that for its Plastic Limit, are unlikely to be much improved by this form of treatment. Undisturbed samples of the compressible layers should therefore be obtained and if the boreholes, from which these samples are taken, are later logged using geophysical methods, it is possible to gain an accurate record of the stratigraphy on site.

Peat Special care should be taken to identify deposits of peat and similar organic material because they are normally the most compressible of all the strata. Organic deposits, by reason of their fibrous structure, have non-linear consolidation characters, the general character of which is illustrated in Fig. 17.13. The first phase of consolidation is controlled by the network of plant fibres, but as time progresses, water within the fibres begins to be expelled, giving the material a marked phase of secondary consolidation (Berry, 1983).

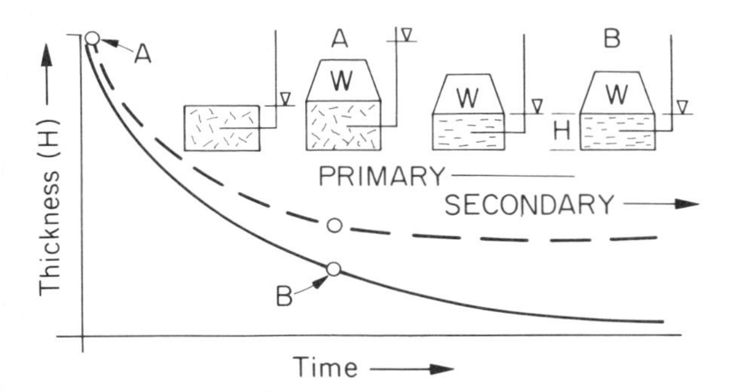

Fig. 17.13 Consolidation of peat. Primary consolidation accompanies the gradual increase in effective stress as pore pressures within the peat dissipate (represented in the figure by the pressure head in manometers). Secondary consolidation is that which can be seen to occur under an apparently constant level of effective stress. The mechanisms responsible for primary and secondary consolidation, overlap, with primary consolidation dominating the initial response to load. Broken line = normal performance of a clay (see Figs 9.7, 9.8): solid line = that for organic deposits.

Sub-aerial sediments

Not all compressible weak sediments accumulated under water: for example, colluvial deposits and loess formed in sub-aerial conditions, often in a dry climate. These sediments may possess a high void ratio and have a very open

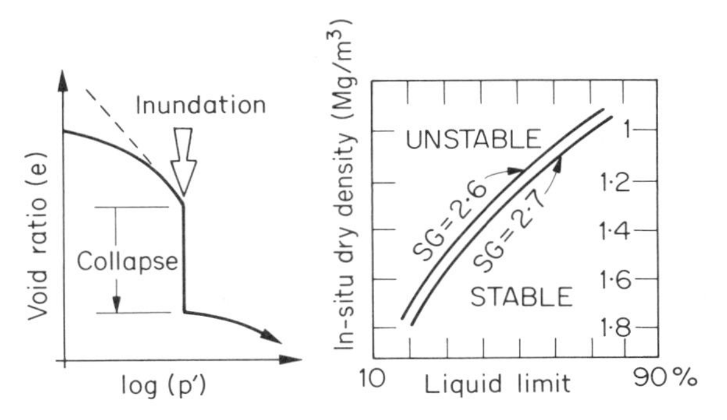

Fig. 17.14 Collapsing soils and their characters, SG = specific gravity of soil grains (after Hall *et al.*, 1965).

micro-structure which can collapse when saturated. Such deposits are described as 'collapsing soils' and are treated by inundating them with water (Fig. 17.14).

Sandy sediments

The configuration of particles of sediment within a compressible deposit, such as a lightly-consolidated clay, is unstable (see Fig. 6.3 *a–d*); hence the deposit compresses when the magnitude of effective stress within it is changed. The packing of particles in sands and sandy sediments is much more stable (Fig. 6.8). Such deposits are not easily consolidated by inducing a drainage of their pore water, especially if they exist above the water table. Methods for compacting these materials use dynamic forces. They impart energy to the ground by dropping a weight onto its surface (tamping), or by penetrating the thickness of the deposit to be treated, with a large diameter vibrating poker, often aided in its penetration by simultaneously jetting water into the ground (vibroflotation) (see Charles *et al.*, 1982; Institution of Civil Engineers, 1976).

Thermal treatment

Freezing

In northern latitudes, bordering the permafrost (p. 35) of the Arctic circle, ground containing water freezes natur-

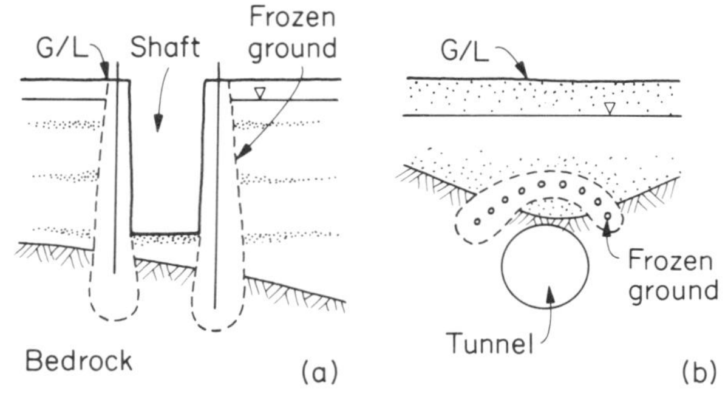

Fig. 17.15 Examples of ground that has been frozen to assist excavation (**a**) through, and (**b**) beneath weak and saturated deposits of drift.

ally during the winter and converts even the weakest sediments to a condition that is hard and strong. Excavations that must penetrate weak ground are completed during the winter, when the sediments are frozen solid, so saving the need to provide the excavations with drainage facilities and temporary support.

The condition of freezing may be artificially produced by pumping a refrigement, usually calcium chloride ('brine') but sometimes liquid nitrogen, through tubes that have been inserted into the ground. When the tubes are placed sufficiently close to each other for the columns of frozen ground, which grow around them, to coalesce, a continuous frozen zone may be formed (Fig. 17.15). Relevant reviews and case histories are provided by Frivik *et al.* (1981) and Schmid (1981). Freezing is generally a more expensive method of controlling ground-water and of improving strength than either grouting or de-watering. It is usually considered only if these methods are thought to be unsuitable or have been used, and have failed to provide the desired improvement. Unlike hydraulic conductivity (i.e. permeability), which may vary over many orders of magnitude, the thermal conductivity of most geological materials lies within the narrow range of 0.1 to 10.0 $Wm^{-1}K$. Thus, ground-freezing techniques, when used in mixed strata, produce much more predictable results than those of either dewatering or grouting.

The gain in strength obtained by freezing is remarkable, and is illustrated in Fig. 17.16. Voids, such as pores and fractures, do not have to be saturated, but must contain at least a film of water on their surfaces (see, for example, Fig. 9.23).

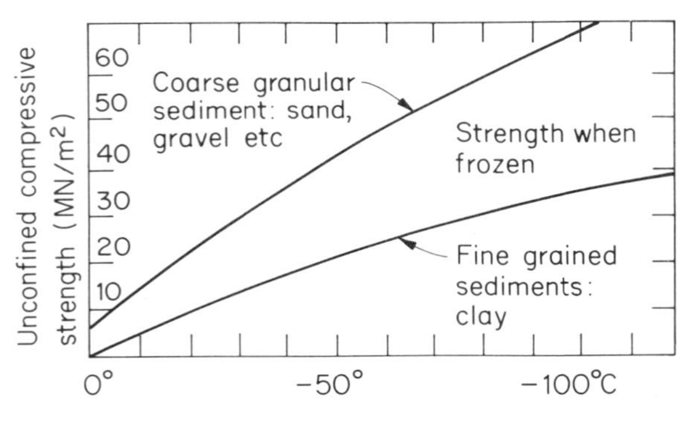

Fig. 17.16 Likely strength for frozen ground. The temperature for brines ranges from −30° to −50°C and that for liquid nitrogen is −196°C. (Data from British Oxygen Co., U.K.)

Frozen sediments

Gravel, sand and silt These sediments can be successfully frozen: indeed, freezing is an excellent method for improving the strength of silt, whose fine grain size renders it difficult to dewater and almost impossible to permeate without recourse to costly chemical solutions.

Clayey sediments Fine-grained argillaceous deposits (Table 6.2) can be frozen but are the most difficult materials in which to obtain completely frozen conditions. This is thought to result from the concentration of ions in the pore water adjacent to the surface of the clay minerals. Too great a concentration of ions prevents the pore water from completely converting to a crystalline form. Consequently, clays are often only partially frozen and tend to creep under load.

Organic deposits Peat and other organic deposits, will freeze, but the process damages the cells of the fibres and on thawing the mechanical properties of the deposits may be significantly different from their original values.

Frozen rocks

It is not usually necessary to freeze rocks, although water in the joints and pores of rock may be frozen. Rocks are much more brittle than soils and are badly affected by the volumetric changes that accompany severe freezing. Weak rocks can be seriously damaged by the expansion of water in their pores and on thawing, may disaggregate to a slurry.

The surfaces of fractures in strong rock may be weakened by freezing, particularly if they are weathered, thus reducing their shear strength once thawed. Strong rocks containing fractures filled with clay and the fine-grained products of weathering, may be even more substantially altered by freezing. The material infilling the fractures increases in porosity as its pore water expands, enabling it to accept more water (which will be readily available from the reservoir of surrounding fractures) when a thaw commences, thus increasing its moisture content and decreasing its strength.

Volumetric changes

Sediments exhibit approximately 9% expansion when artificially frozen and sometimes heave can be observed in treated ground. (Much greater expansion can occur when *ice lenses* form and separate bedding. This is a common feature of naturally-frozen ground: see Fig. 3.6). As temperatures continue to drop, volumetric contraction develops. Sediments shrink and form vertical columns, often polygonal in cross section, separated from each other by fine cracks. On thawing, these vertical cracks do not properly heal and the vertical permeability of sediments is often found to have increased from its original value (Chamberlain and Gow, 1978). Rocks may fracture under the tensile stresses induced by contraction and it is often discovered that their *in-situ* permeability has been increased by ground freezing.

Investigations

The presence and movement of ground-water is of greatest importance to the efficiency of artificial ground freezing and should be investigated. Ground-water provides a positive heat flux and flow rates exceeding 2.0 m day^{-1} can swamp the negative heat flux from the freezing tubes. Under these conditions ground-water may not freeze.

The chemistry of ground-water is also relevant (p. 215) as high salinities act as an antifreeze, but this rarely creates a serious problem. Sea-water freezes at 3°C below the temperature of fresh water and a saturated solution of NaCl, as might be found near a deposit of salt, will freeze at −21°C. Cryogenic fluids used for freezing can lower temperatures considerably beyond these levels (see Fig. 17.16).

The position and character of hydrogeological boundaries needs to be known, especially if a frozen zone is to key into such a boundary (see Fig. 17.15). Fluctuations of water level, to be expected outside the frozen zone, should also be noted as they will influence the load that must be carried by the frozen ground.

Heating

An improvement in the strength of weak sediments has been attained by raising their temperature to 600°C and higher. The treatment relies on the irreversible changes that occur in clays when heated beyond 400°C: at 100°C adsorbed water is driven off from clay surfaces and by 900°C certain clays begin to fuse to a brick-like material. The technique requires liquid or gas fuels to be burnt under pressure in a sealed bore-hole and is claimed, in some circumstances, to compete economically with piled foundations in thick sedimentary successions. It has been extensively used for improving ground prior to construction and for stabilizing landslides. The method was developed, and is largely exploited, in E. Europe (Litvinov *et al.*, 1961).

SUPPORT

Aspects of the support of underground excavations have been described in Chapter 16, where the purpose of its installation was explained: namely to prevent ground around an excavation from becoming unstable. Support may be provided by rods, in the form of dowels, bolts and anchors; by arches and rings; by a complete lining of metal or concrete; and by retaining walls (Fig. 17.17). Ground treatment is frequently used in conjunction with support; thus a rock slope may be supported by bolts and drained by holes at its base. Sometimes more than one support method is employed: a slender retaining wall may be anchored into good ground some distance behind the wall, when this combination is more economic than a substantial retaining wall.

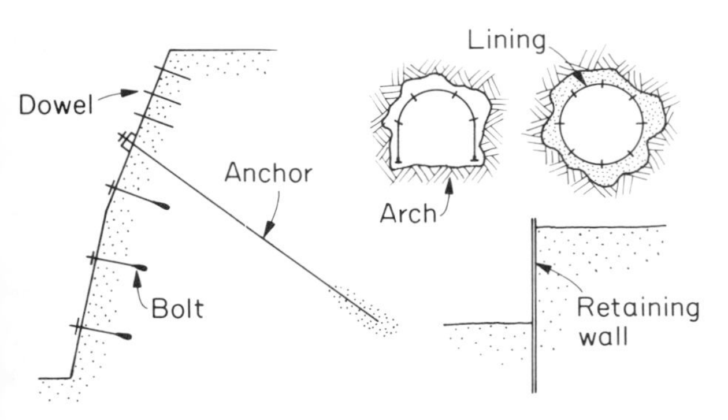

Fig. 17.17 Examples of commonly-used systems of ground support: not to scale.

Most systems of support require the strength of the ground to contribute to their effect. Bolts must react with good ground at depth and arches must be supported by the floor on which they stand. The interaction between ground and its support is not a phenomenon that is amenable to prediction by theory, especially in weak rock and in ground where the variation of mechanical properties, from one location to another, cannot be predicted. It is therefore prudent to install instruments that will measure the response of the ground. Ward (1976) measured the reaction of fissile mudstone to various methods of tunnelling and types of support, and illustrative results are shown in Fig. 17.18. The magnitude of the response of the ground was influenced by both the method of excavation used and the type of support provided: it also changed with time (see also Fig. 16.19). Drill and blast methods of excavation, which can greatly disturb the surrounding ground, resulted in a noticeably different ground reaction from that associated with less violent excavation by machine bore.

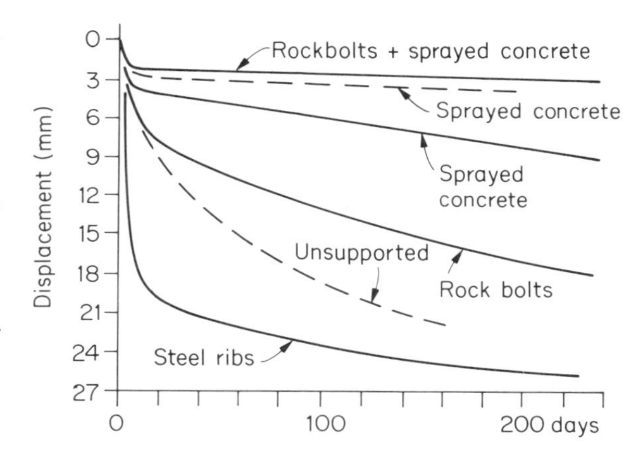

Fig. 17.18 Displacements measured in a rock tunnel (from Ward *et al.*, 1976). Solid liine = drill and blast excavations; Broken line = machine excavation. Sprayed concrete = shotcrete lining (see text).

Rods, bolts and anchors

The simplest form of rod support is a dowel, placed with or without grout, in a hole drilled to accept it. Bolts have their far end secured in rock and on their exposed length is placed a face plate and nut. Anchors are long bolts or cables which, once grouted into their hole, can carry considerable load. Reviews of these systems, which are illustrated in Fig. 17.17, are provided by Littlejohn *et al.* (1977; 1980); Hobst *et al.* (1983).

Rock bolts and anchors

The structural geology of the ground around an excavation should be known to determine the inclination (θ) of

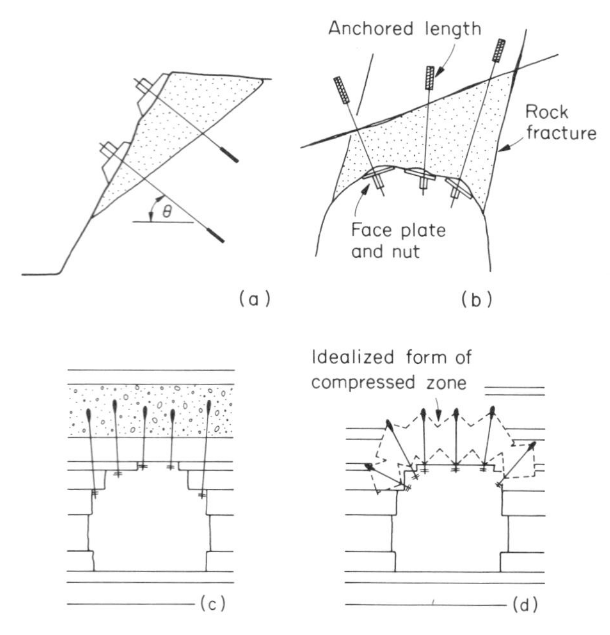

Fig. 17.19 Rock supported by bolts (**a**) on a slope, (**b**) within a tunnel (unstable ground is stippled). Rock roof (**c**) suspended from stable, strong strata, and (**d**) compressed to form a stable arch of rock.

bolts and anchors that provides the most desirable resultant force on the rock surfaces whose friction resistance must be increased (Fig. 17.19*a*). The disposition of fractures will also define the size of the blocks in need of support (Fig. 17.19*b*).

Rock strength limits the load that can be carried by the ground around the anchored length of these supports and beneath their face plate. Slow failure of rock will gradually lessen the support with which it was provided.

Rock strength and structure must be considered together to assess whether the desired support can be obtained by suspending rock from firmer ground beyond the excavation (Fig. 17.19*b* and *c*). If this is not possible, the strength of the rock may have to be increased by using bolts to create a zone of compression in which joints and bedding are pressed together, so that the frictional resistance of their surfaces is increased sufficiently to provide stable conditions (Fig. 17.19*d*).

Clay-filled joints, and fractures containing soft weathered material, will deform under the load imposed by a bolt or anchor. Such deformation slackens the tension within these supports. Ground-water flow, capable of eroding the infilling from joints and fractures, aggravates this condition. Because water softens weak rock, especially lightly cement (friable) sandstones, and mudstones, and erodes clay-filled fractures, its use as a drilling fluid should be avoided when boring the holes to house bolts and anchors in these materials.

Soil anchors

Anchors may be used in sediments when the deposits can provide sufficient reaction to the loads carried by the cable. The performance and installation of soil anchors is much influenced by the stratigraphy of the sediment lying above and around the anchor. When the sediment has an undrained *in-situ* strength greater than 100 kNm^{-2}, it is usually possible to under-ream the hole so as to improve the reaction of the anchorage (Fig. 17.20). Holes for anchors should be drilled as quickly as possible and the cables installed, and grouted immediately, because the condition of the soil around the hole will deteriorate rapidly and this can reduce the strength of the bond between the anchor and the ground. Water softens sediments and should not be used as a drilling fluid.

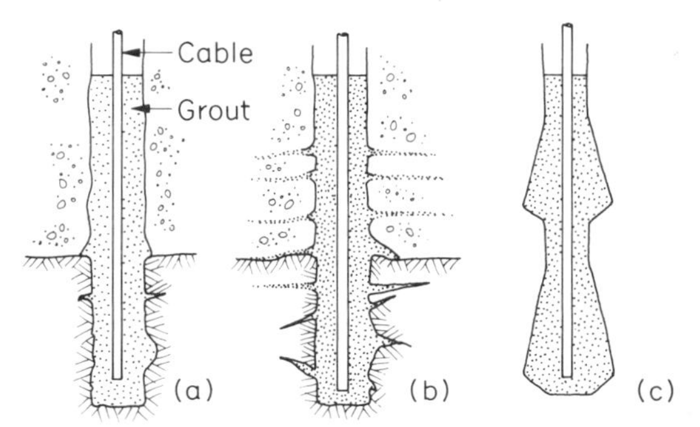

Fig. 17.20 Anchorages for cables: (**a**) grouted bore-hole, (**b**) bore-hole grouted under pressure to fill fractures in rock and coarse permeable horizons in the overlying drift, (**c**) under-reamed. (**a**) and (**b**) are suitable for use in soil and rock, (**c**) is suitable for stiff to very stiff clays and weak rock.

Arches, rings and linings

Arches may be erected at regular intervals to produce 'sets' that are normally linked by horizontal bars to form a supported corridor. The space between the arches and the excavation must be packed, to prevent excessive deformation of the ground, and the roof and walls between the arches may be lined with 'lagging' to prevent the ingress of rock. The floor on which the arches are erected must be able to carry the load transmitted to it by the arch legs. When the floor is weak, as when crossing a fault zone, it may be necessary to change the cross-section of the excavation to a circle, and use a complete ring.

Squeezing ground

Arches and rings are particularly useful in ground that is susceptible to severe deformation around an excavation: a condition that is described as 'squeezing ground' (p. 263). Such deformation can be encountered in crushed rock under stress, as in major and recently active fault zones, and in weak and moderately strong rocks that are severely overstressed by the excavation (as illustrated in Fig. 16.18 where squeezing accompanies gentle failure).

The compressibility of arches and rings can be designed to absorb strain energy at the rate it is being released by the ground and can be deliberately designed to permit the occurrence of large deformations in a moderately controlled manner. In this way, stresses in excess of those that could reasonably be restrained by linings, may be safely relaxed.

Weak and variable ground

Tunnelling in consolidated sediments such as clay, sand and gravel, and in soft sedimentary rocks, i.e. soft-ground tunnelling, uses linings to provide support at all times. The lining consists of pre-formed segments that are locked together to form a complete circular section: voids between the lining and the ground are filled with grout (Fig. 17.21).

Fig. 17.21 Tunnelling in chemically grouted gravel. The tunnel is circular in section, excavated with pneumatic spades and supported by a complete lining of segments bolted together to form rings. Ahead of the lining is excavated gravel: note its stability after treatment. A heading leads off into the distance and was the pilot tunnel from which was undertaken treatment of the surrounding ground. Treatment was accomplished by drilling grout holes radially from the walls of the heading into the gravel. (Photograph by Soil Mechanics Ltd.)

Shotcrete is a form of concrete that can be sprayed onto the face of an excavation where it builds up to form a lining, usually a few centimetres thick. It may be mixed with fibres of metal or glass and thereby reinforced. The method was developed in continental Europe and is the basis of an excavation system described as the New Austrian Tunnelling Method (or NATM). The speed and ease with which shotcrete can be applied enables the surface of an excavation to be protected and supported almost as soon as it is exposed, if this is required. Such action greatly reduces the deterioration that can rapidly occur in weak and variable ground (see Fig. 17.18). Consequently, large excavations have been opened in moderately weak rock, using shotcrete linings (John, 1981).

Shotcrete is particularly useful for protecting argillaceous rocks from absorbing moisture in the tunnel atmosphere, and softening: much of their *in-situ* strength can thus be retained. It also protects weak material, infilling fractures, from erosion by ground-water flowing towards the excavation: the sprayed lining provides a barrier to such flow. By controlling the flow of ground-water, the risk of settlement and collapse of ground level above the excavation is also avoided.

Shotcrete permits the amount of support provided to the roof and sides of an excavation, to vary with local changes in conditions. A thick lining may be formed in areas of weak ground and a thinner lining in stronger areas. This if of great value in ground where conditions cannot be predicted with accuracy, as in the weathering zone of igneous rocks (especially of granites: p. 33), in volcanic accumulations (Fig. 16.23) and in intensely-folded strata.

Other forms of support may be used in conjunction with shotcrete, such as metal meshes, rock bolts and arches, when further support is required. In extremely poor ground a conventional full concrete lining may be necessary (Fig. 17.17).

Retaining walls

It is not possible to excavate stable vertical slopes in sediments and weak rock. When such a profile is required, it is customary to form a retaining wall *in-situ*, prior to excavation, so that excavation can take place against this boundary without causing the ground behind to become unstable. The wall is normally built by driving a line of

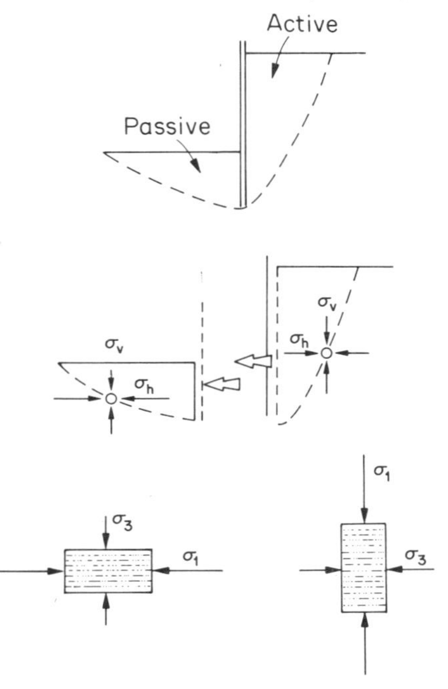

Fig. 17.22 Elements of earth pressure around a retaining wall. Note that in the passive zone bedding is close to the orientation of σ_1 (and see Fig. 14.2).

sheet-piles, or by boring a line of secant piles, or by excavating a slurry supported trench later to be filled with concrete, to a predetermined depth that is greater than the depth of the proposed excavation. The resulting structure forms a cantilever that retains the ground adjacent to the excavation: it may be either free standing (Fig. 17.22) or supported also by anchors (Fig. 17.23).

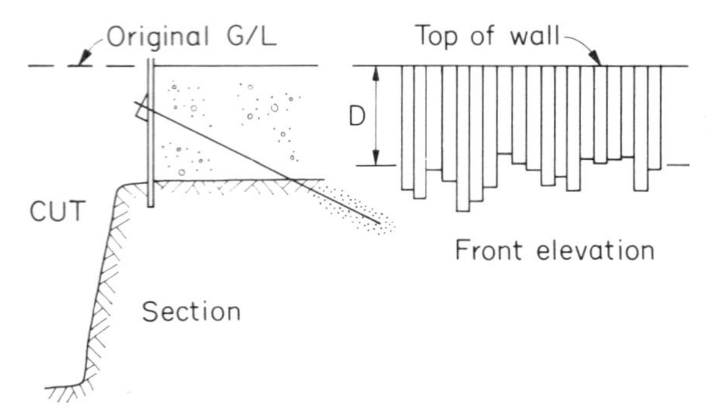

Fig. 17.23 Retaining wall to bed-rock. D=anticipated depth to bed-rock.

Earth pressures

Equilibrium relies upon the passive pressure in front of the wall being sufficient to balance the active pressure behind the wall (see Fig. 17.22). If such a wall were to move forward, the horizontal stress in the ground behind the wall would decrease. Ground failure would occur when the difference in principal stresses reaches a magnitude that cannot be sustained. Similarly, for the wall to move forward the horizontal stress in the ground in front of the wall must increase: too great an increase can result in ground failure. The ratio of horizontal to vertical stress when failure commences behind the wall, is called the *coefficient of active earth pressure* (K_A). The same ratio when failure commences in front of the wall, is called the *coefficient of passive earth pressure* (K_P).

These coefficients are required for design, and can be obtained from suitable tests of sediment strength. They are not constants; their magnitude will change with changing ground conditions. Hence ground investigations should carefully establish the correct values of effective stress that will operate on site. To do this, attention must be given to ascertaining the value of the *coefficient of earth pressure at rest* (K_o), which is the ratio of horizontal to vertical stress that naturally exists in the ground prior to excavation (p. 161). Over-consolidated deposits will have a higher ratio than normally consolidated sediments, and the ratio generally increases with increasing over-consolidation. The top of the London Clay has a horizontal effective stress that is 2.5 to 3.0 times the vertical effective stress.

Investigations

The stratigraphy of the site should be established in some detail and note taken of the stratification, and presence of weak horizons, and laminae. Shallow-water sediments and colluvium are liable to contain lateral variations in their succession, so that the strata encountered in one bore need not exist over the entire site. Ground in front of the wall should therefore be investigated as well as the ground behind the wall. Reference to Fig. 17.22 illustrates that failure at the base of the wall will be much influenced by horizontal displacements which are likely to be parallel to bedding in sediments. Where the wall is to be founded on rock at depth, the weathering at bedrock surface should be investigated, especially if pockets of clay-rich weathered debris are likely to exist there, buried beneath more granular drift. Permeable, free-draining deposits, such as sand and gravel, should be identified, together with layers of low permeability in which changes of pore pressure could not be expected to dissipate readily during periods of deformation. The shear strength of the materials should be measured in terms of effective stress.

Attention must also be given to existence of over-consolidated deposits and presence of fissures in cohesive soils. Heavily over-consolidated clays (which include those that have been frozen by permafrost, perhaps during the last ice-age: see Fig. 3.8) contain fissures which will tend to open in the zone of active pressure and change the distribution of load against a wall. If rainfall fills these fissures with water a wall can be subjected to hydrostatic pressures. The content of expandable (or soluble) minerals should be determined in ground that has had a considerable history of dryness, if excavation work is to expose it to rainfall, or wetting from other sources.

Ground-water conditions must also be investigated, for as shown in Fig. 16.6*b*, uncontrolled seepage can undermine the stability of retaining walls. It is often prudent to locate the depth of an impermeable horizon into which the wall may be keyed so as to control such problems of ground-water.

In all these investigations it is helpful to note the resistance the ground may offer to pile driving and pile formation. The density of the ground, and the difficulty with which borehole casing may be advanced during drilling, are good indicators of likely problems to be encountered during wall construction. Many glacial deposits of sand and gravel are extremely difficult to penetrate. In tropical and arid regions layers of iron-pan and other forms of duricrust (p. 38) may exist, buring beneath weaker overlying sediments. Boulders should always be noted, when present.

Ground investigation for retaining walls that are to be founded on bedrock at depth, should reveal the topography of the bedrock surface. This is often irregular and differences must be expected between the predicted and actual base of such walls (Fig. 17.23).

Selected bibliography

American Society of Civil Engineers. (1982). *Grouting in Geotechnical Engineering*. (Ed. W.H. Baker). Special Publication American Society Civil Engineers.
Institution of Civil Engineers. (1964). *Grouts and Drilling*

Muds in Engineering Practice. Proceedings of a Symposium. Butterworths, London.

Institution of Civil Engineers. (1976). *Ground Treatment by Deep Compaction*. Proceedings of a Symposium in Print.

Powers, J.P. (1981). *Construction Dewatering: A Guide to Theory and Practice*. J. Wiley and Sons, New York.

Institution of Civil Engineers. (1982). *Vertical Drains*. Third Geotechnical Symposium in print. Thomas Telford, London.

Bell, F.C. (Ed.) (1975). *Methods of Treatment of Unstable Ground*. Newnes-Butterworths, London.

Hoek, E. and Brown, E.T. (1980). *Underground Excavations in Rock*. Institution Mining & Metallurgy, London.

Hobst, L. & Zajic, J. (1983). Anchoring in Rock and Soil. *Developments in Geotechnical Engineering*. No. **33.** Elsevier Scientific Publishing Co., Amsterdam.

British Standards Institution. (1982). *Ground Anchorages* (Draft): full report in preparation. British Standards Institution, London.

18

Development and Redevelopment

The development of a community may be influenced by many geological phenomena, for example, earthquakes (Chapter 1), weathering and erosion (Chapter 3), volcanoes (Chapter 5), landslides and avalanches of rock and soil (Chapter 14). To these obstacles may also be added the problems of engineering with the materials and landscape created by geological processes, for example, the difficulty of creating reservoirs (Chapter 15), and of excavating routes for communications (Chapter 14 and 16). These, and other examples, are mentioned in previous Chapters, as indicated.

In this Chapter we wish to describe those geological factors that affect (*i*) the provision of an adequate supply of water from underground sources; (*ii*) the suitability of minerals, sediments and rocks as materials for the construction of engineering structures; (*iii*) the design of foundations that will safely transfer to the ground the load of the structure they must carry, and (*iv*) the safe and controlled disposal of waste, on land.

Water supplies

Most ground-water supplies tap local flows of circulating meteoric water (p. 213 and Fig. 13.3) that exist within the catchment over which rain has fallen.

Catchments

Two types of catchment must be recognized, namely surface and underground.

Surface catchments are bounded by the highest ground separating neighbouring drainage systems. This high ground is called the *surface-divide* or *watershed.* Rain falling on an impermeable catchment will drain over its surface to the river within the catchment.

Underground catchments are defined by water levels, not ground levels, and are bounded by a line that joins the highest water levels beneath a surface catchment. This line is called the *ground-water divide*. Often, the boundary of the underground catchment lies almost vertically beneath that of the surface catchment, and rain percolating to the water table will be carried by ground-water flow towards the river draining the surface catchment (see Figs 13.20*a* and 13.23). When the position of the divides does not coincide, rainwater which percolates to the water table will be carried, by ground-water flow, *beneath* the divide of the surface catchment in which it fell: this is called *underflow* (Fig. 18.1).

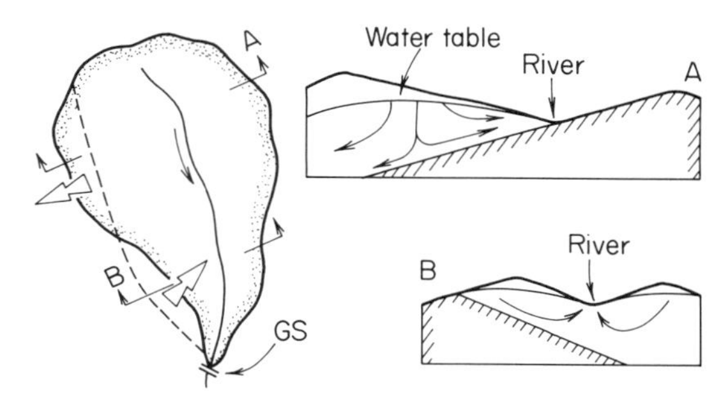

Fig. 18.1 On the left, a map of a catchment: solid line = topographic divide; broken line = ground-water divide; arrows indicate *underflow.* Two vertical cross-sections illustrate the geological structures that result in this underflow; arrows indicate movement of ground-water. GS = gauging station for measuring river flow.

Water budgets

If more water is taken from the ground than enters it from recharge (p. 214), water levels will fall and, in theory, the ground will eventually be drained. The water budget

Table 18.1 Examples of water budgets

Mass balance for a budget		
INFLOW	= OUTFLOW	+ CHANGE IN STORAGE
For a catchment (simple budget)		
Precipitation	= Evaporation + Transpiration + River run-off	+ Infiltration
For a catchment (more complex budget)		
Precipitation + +ve underflow + imports of water	= Evaporation + transpiration + River run-off + Pumped groundwater + −ve under-flow + exports of water	+ Infiltration + net change in lake levels + artificial recharge
For an aquifer		
Recharge	= Ground-water + discharge	Changes in water level × coef. of storage

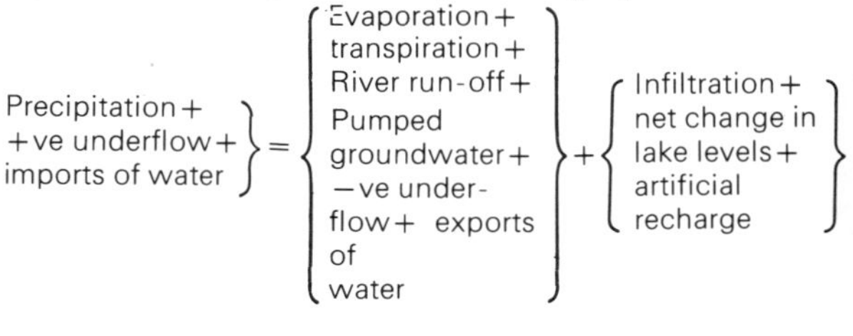
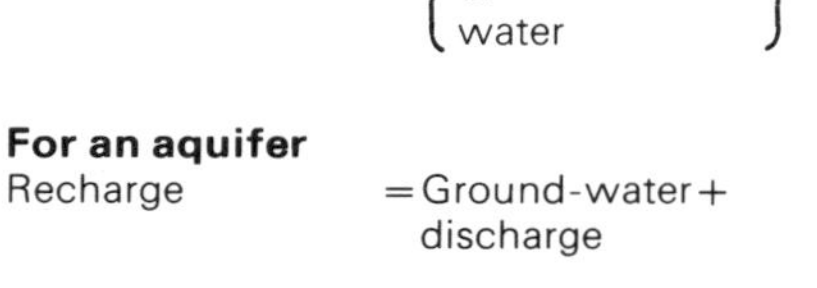

of an underground reservoir must therefore be calculated, and Table 18.1 shows how such a budget may be written. Infiltration, inflow and outflow (i.e. ground-water flow) and changes in storage are described in Chapter 13.

Precipitation is all forms of rainfall and snow, and is measured by rain and snow gauges as a depth of water falling over a unit area.

Evaporation is the return of water to the atmosphere, as a vapour from the surface of bodies of water. The maximum rate of evaporation (i.e. *potential evaporation*) is measured as a depth of water evaporated from an exposed evaporating pan of unit area.

Transpiration is the return of water to the atmosphere by plants, which draw water from the soil. In dry weather this can produce a deficit in the soil moisture required for percolation to occur, and the depth of rainfall that is required to restore the ground to its field capacity (p. 214) is called the *soil moisture deficit*.

Run-off is the river discharge measured at a gauging station (Fig. 18.1). Rivers which flow during periods of dry weather are sustained by the drainage of ground-water from underground storage (Fig. 13.20). Those which flow only in wet weather are usually carrying water which has not infiltrated, but remained on the surface. The typical behaviour of a river flowing through a catchment that contains permeable rocks and soils near the surface, is shown in Fig. 18.2(*a*). Rain which falls at a rate that exceeds the infiltration capacity of the ground (p. 213), will leave the catchment rapidly and produce a 'flash' of discharge (Fig. 18.2*b*), commonly called a *flood*.

To complete a budget it is also necessary to calculate the change in volume of reservoirs, lakes and other bodies of water, the gains or losses from underflow, the amount of water abstracted by wells and from rivers, and the volume of this abstracted water that is returned to the catchment. All the components are then summed to obtain the *net* change in storage, which if negative indicates that more water is leaving the ground than is entering.

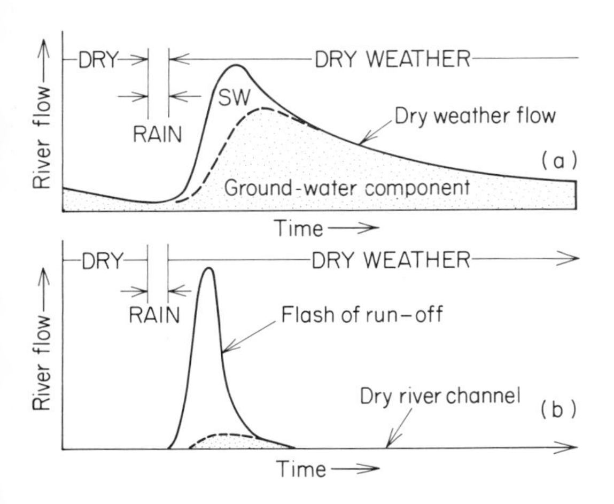

Fig. 18.2 River run-off hydrographs. (**a**) For a river sustained by ground-water flow from aquifers within the catchment. Total run-off contains a surface water component (SW) and a ground-water component. (**b**) Flash run-off from a catchment that cannot absorb rainfall: in this case the response of a river in a hot, arid climate is illustrated.

Location of sources

Physical geology provides the most immediate information of relevance to the location of sources. River patterns indicate the relative permeability of the ground and the controls exerted by structural geology and rock type upon infiltration and the movement of ground-water (Figs 3.3 and 3.15). The shape of valleys records their history and the likely presence of permeable, and saturated, deposits overlying the concealed base of a valley, or stranded as terraces above the valley floor (Figs 3.17 and 3.41). Levees and flood plains (Fig. 3.17) are witness to the existence of flashy run-off.

To these observations should also be added a study of river hydrographs and in particular the river discharge during dry weather, as this is supplied by ground-water. Because dry-weather flow is sustained by water draining from storage, its recession is characteristically exponential. Thus, when the hydrograph is plotted as a semi-log graph, the linear portion of its tail may be taken to represent the contribution from ground-water; called the *ground-water component of discharge*. Typical trends are illustrated in Fig. 18.3; low values for the slope of the recession indicate the existence of substantial volumes of ground-water somewhere upstream.

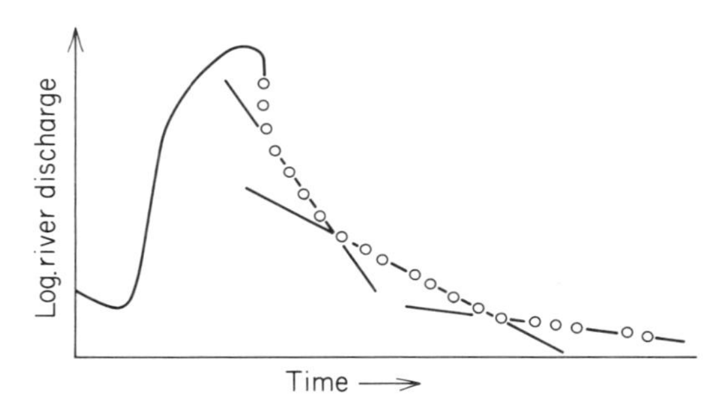

Fig. 18.3 Semi-log plot of the recession of a river. Three components of recession can be recognized by their different gradient. The number of components present is governed by the hydrology of the catchment and the duration of the recession.

Having reviewed the general hydrology of the area, it is necessary to focus attention on specific locations, to search for a supply. In doing this, the hydrogeological investigations described in Chapter 13 (p. 223–6) must be completed, bearing in mind that the source should yield a supply that is economic for the user. Thus, an aquifer which satisfies the needs of a farmer may be quite inadequate for the needs of a township.

The volume of supply required will indicate the size of the aquifer sought (see calculation of storage, p. 223) and the magnitude of the surplus that must be indicated by the water budget. Sources close to ground level are usually capable of yielding only small supplies, drained from impersistent permeable horizons and perched (*q.v.*) resources within the drift overlying bedrock (see for example Figs 3.1, 3.4 and 3.10). More substantial supplies usually came from larger geological structures such as the extensive network of voids in porous, or fractured bedrock beneath the drift (Figs 3.1 and 3.10), and the widespread system of fractures provided by many faults. Large supplies may be associated with favourable hydrogeological boundaries (p. 218). Impermeable barriers, such as dykes (Fig. 5.7) can act as underground dams storing, on their upstream side, large volumes of water. Permeable boundaries, such as open faults, can cross the countryside and act as drains for the aquifers they intersect (see for example Fig. 12.17).

In arid regions (e.g. North Africa) suitable large sources may only be found at great depth and consist of water trapped in the ground since the wetter climates of the ice age. These aquifers are not now being recharged with fresh water, and the withdrawal of their supplies is being balanced by the invasion of saline waters (brines *q.v.*) from depth. Eventually the waters from these aquifers will become badly contaminated.

Spring supplies

Supplies of ground-water may easily be obtained when the level of the water table is sufficiently close to ground level to have created a spring line (Fig 13.11). Shallow pits excavated below the spring line will fill with water and act as a cistern from which supplies can be tapped. The concentrated and local discharge of ground-water below the spring line, produces a *spring*. Springs may occur at the heads of valleys where they result from the convergence of ground-water flow paths (Fig 13.23). Many dykes and vertical faults are hydrogeological boundaries (*q.v.*) and deflect ground-water along their length towards their valley-side exposure, where it emerges as a spring. Volcanic rocks (especially large flows of basalt, much jointed by cooling) and limestones (especially when karstic: Fig 3.2 and 3.3) are capable of providing substantial supplies of spring water. Spring flow may increase in wet weather and decrease in dry weather, and will cease when the water table falls below ground level.

Well supplies

Most of the supplies that have to be raised from underground are tapped by wells. The wells may range in diameter from 100 cm to more than a metre and are bored by well drilling machines. Hand-dug wells may be excavated when machines are not available and when the water table is within reasonable distance of ground level.

Wells The cone of depression created by a pumping well is illustrated in Figs 10.21 and 17.1. The size of a cone of depression created by pumping depends upon the size of the well, the transmissivity and storage of the ground, and the proximity of the hydrogeological boundaries. Wells pumping from ground of low transmissivity and storage will produce a steep cone of depression that does not extend an appreciable distance into the surrounding ground. A well will not yield water when its cone of depression reaches the level of the pump intake, and must be rested to recover its water level. Better aquifiers produce a shallow cone of depression of large diameter that cannot be lowered to the level of the pump intake by maximum pump discharge. The volume of water discharged from a well divided by the drawdown of water level in the well, describes the *specific capacity* of the well.

In porous and permeable sediments, such as gravelly alluvium, the yield of a well can be proportional to its depth. This is not the case with fractured-rock aquifers where a large percentage of the yield may be supplied by a limited number of water-bearing fractures. A well that does not intersect these fractures will be dry; similarly, the yield of a well will not be improved by extending the well beyond the zone of fractures into unfractured ground at depth. Fractured rocks are the most difficult of all geological materials in which to locate a well supply, particularly as a well is a vertical structure and not able

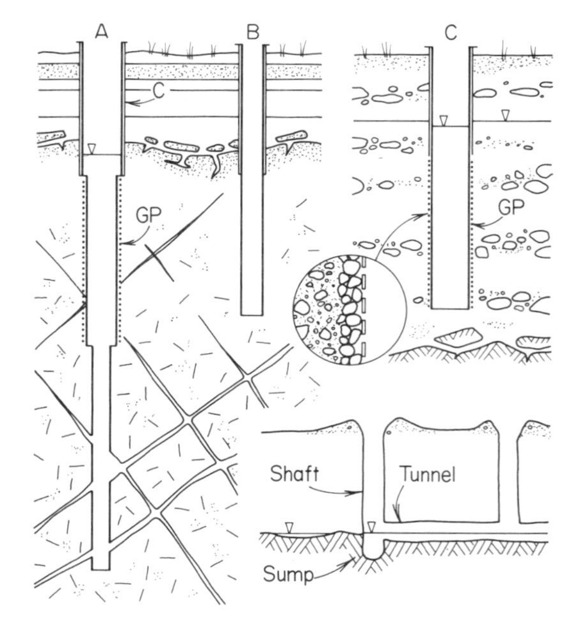

Fig. 18.4 (A) Well supply from fractured rock concealed by superficial deposits: C=casing, GP=gravel pack. The well is productive because it intersects water-bearing fractures: its lower portion is unscreened. (B) A dry well in the same rock. (C) Well supply from unconsolidated sediments above bedrock. GP=gravel pack placed around the well screen to prevent grains of sediment, below a certain size, from entering the wall. Inset shows the pack separating the ground from the well screen. Bottom right: a type of quanat developed from shafts. These drawings are not at the same scale.

to intersect vertical fractures that do not lie on its line (see Fig 10.3). Commonly encountered situations are illustrated in Fig 18.4.

Well screens are used to support the sides of a well in ground that would collapse under the force of converging radial flow towards the well, during pumping. The maximum aperture within a screen is defined by the size of particle that must be retained in the ground. To design a screen for use in unconsolidated sediment, it is necessary to measure the range of particle size within the deposit (p. 191) since too large an aperture will permit fine particles to be carried into the well, by the flow of water. This may clog the well, damage the pump, put suspended sediment into the supply, and cause settlement at ground level around the well.

Well tests determine the drawdown that accompanies a rate of discharge and period of pumping. Important wells are often tested for many weeks. In areas where a large fluctuation in water level is expected it may be necessary to test either for months, or at intervals at different seasons. The transmissivity and storage of many aquifers decreases with depth, so that a shallow cone of depression may accompany a test when water levels are high. When water levels are low, and the well must be supplied from zones of low transmissivity and storage, the cone of depression becomes steep and narrow, and the well may be pumped dry. Well tests also permit the quality of the water to be monitored for any deterioration in standard (Fig 18.5).

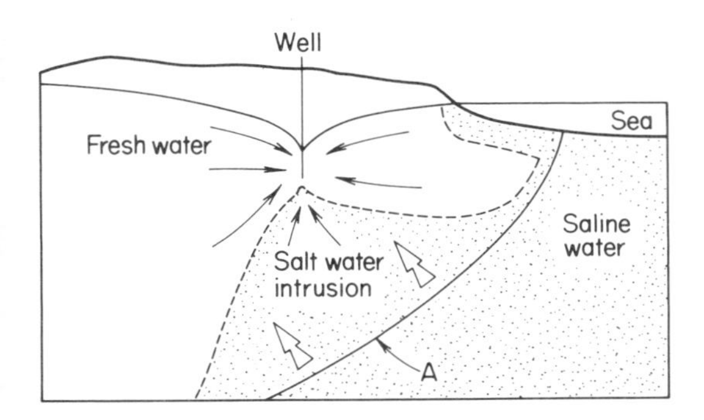

Fig. 18.5 A discharging well situated adjacent to a coast will cause salt water intrusion that can eventually pollute the well. A = the original position of the transition zone separating fresh from salt water (see also Figs. 13.9 and 13.28).

Well development The yield from most wells can be improved by removing any fine-grained debris that may have entered the pores and fractures of the aquifer during drilling operations. Wells in limestone are often injected with acid to purge their fractures. Screens in unconsolidated aquifers should be designed to permit fine particles of the deposit to pass through their aperture and into the well where they may be removed. This leaves the surrounding ground more porous and more permeable than it was originally. Pumping water from a well, surging water up and down a well, and playing jets of water against the walls of a well, are methods used to develop wells.

Adit supplies

Adits (also called *infiltration galleries*) may be excavated from the base of large diameter wells to tap additional supplies of water. They are either horizontal or gently inclined, and in fractured rock are oriented to intersect the maximum number of water-bearing fractures per metre of adit length (Fig 18.6). Many hand-dug wells have been extended in this way to collect meagre supplies that become perched above impermeable bedrock at depth (Fig 18.4). Adits may also be excavated into hillsides to skim water from the top of the water table and lead it towards the valley: this is the basis of the ancient system of quanats used throughout the Middle East for irrigation and supply.

Fig. 18.6 Adit in chalk for water collection (photograph by George Stow & Co. Ltd.)

Construction materials

Sand, gravel, clay, aggregate and dimension stone, are examples of a wide range of materials that may be used in their raw state for construction. When available close to site they are normally much cheaper to use than a manufactured alternative such as concrete, and as noted when discussing the materials used in dams (p. 247), it is advantageous to change the design of the structure so that available materials may be utilized. This is common practice in the construction of large structures such as embankments, roads and airstrips. To correctly assess the potential of economically available raw materials it is necessary to appreciate both their geological character and their mode of formation.

Types of material

Unconsolidated sediments provide the most immediate source of easily excavated gravel and sand. The form of these deposits is described in Chapter 3 and their composition is reviewed in Chapter 6 (Tables 6.2, 6.3 and 6.5). Many of the deposits closest to the surface will be recent drift and have a variable thickness, infilling valleys and other depressions in the land surface. They will usually be shallow water and sub-aerial deposits and are likely to be extremely variable, requiring careful investigation prior to exploitation. Surveys using trial pits, bore-holes and geophysical methods are invaluable for this purpose. Unwanted deposits such as peat, may be interbedded with the sediments. Hard layers, where pores have been filled with mineral cement, may also be encountered; *iron-pan* may be found in areas that have been water-logged and duricrusts (p. 38) in deposits exposed to weathering in arid and semi-arid climates.

Consolidated sediments and sedimentary rocks that exist as uplifted bodies of strata of appreciable thickness can form a reliable and plentiful source of material. Their characters are described in Chapter 6. They will have a preferential weakness along bedding and tend to excavate as tabular and slabby blocks if well stratified. Blocks of argillaceous material (Table 6.2) may break down on exposure and many sediments described as mudstone and shale, will absorb water, expand and flake. Porous sediments may not be durable, particularly if composed of weak minerals such as calcite, as repeated wetting and drying, or freezing and thawing within their pores can readily destroy their strength. Sandstones of a given porosity therefore tend to be stronger and more durable than limestones of similar porosity. Weak sediments, of any composition, will disintegrate to a powder, or slurry when disturbed. Special care must be taken with deposits containing chemical sediments such as anhydrite, gypsum, the evaporites, the coals and pyrite. These are unstable materials that may dissolve, burn, or ozidize to release acid, and should be discarded as materials for use in construction.

Igneous rocks, when fresh and composed of stable minerals, provide excellent sources of strong and durable material. Ultrabasic rocks (Fig. 5.17) are usually avoided because many of their minerals weather easily and are much altered to clays. Granite, gabbro, basalt and dolerite are normally most suitable. All but the latter occur as large bodies sufficient for the needs of most sites and their characters are described in Chapter 5. Dolerite forms as a minor intrusion (Fig. 5.7) and can be impersistent, but frequently occurs on a large enough scale for most requirements. All igneous rocks preferentially weather along joints (Fig. 3.4) and granitic intrusions may contain zones much altered by mineralization and hydrothermal processes.

Metamorphic rocks can be extremely strong but many are adversely affected by foliation (Table 7.3). This causes them to split into elongate fragments and to have markedly anisotropic characters. Seams of mica commonly contaminate what would otherwise be excellent rock, and large pods of micaceous material, some metres long, may also occur. Metamorphic rocks tend to be severely deformed and good rock may terminate at a fault. Their basic characters are described in Chapter 7 and many forms of deformation which affect them are illustrated in Chapter 8.

Volumes of material

When borrow pits and quarries are opened they often reveal that the volume of suitable material available is much smaller than was originally estimated to exist. To ensure that the requirements for raw materials can be supplied without interruption, it is prudent to locate a volume that is one-third greater than that needed. Commonly occurring boundaries which limit the volumes available are the base of the zone of weathering (or the thickness of overburden), the water table and faulting. The nature of weathering is described in Chapter 3, the location and characters of the water table are illustrated in Chapter 13, and the effects of faulting are explained in Chapter 8. Surface excavations are discussed in Chapter 16. The example shown in Fig 18.7 illustrates a quarry that contains less acceptable material than was anticipated because of the irregular base to the zone of weathering. To obtain the required volume the quarry must either extend towards the fault and create a high back face, in which slope failure may occur and fill the quarry, or deepen and excavate below the water table.

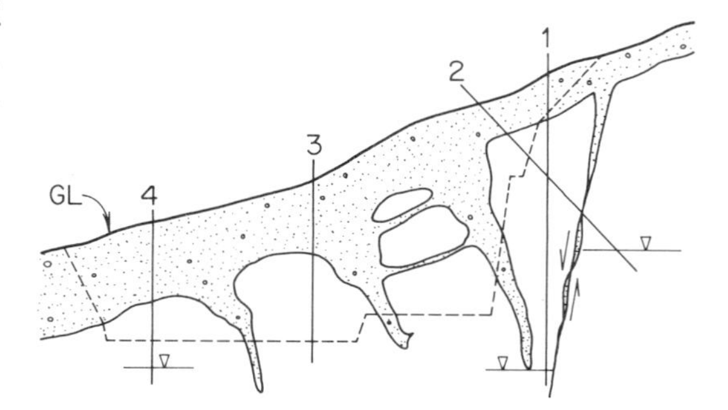

Fig. 18.7 Vertical section through a hillside and outline of proposed quarry (broken line). The ground was investigated using four bore-holes and bedrock was assumed to have a surface that followed the general shape of ground level. See text.

Aggregates

These are rock fragments which may be combined (or *aggregated*) to produce a mixture that can be used in construction. The mixture is usually based upon fragment size, similar sizes being used when an open, porous aggregate is required and a range of sizes when the voids between fragments must be filled with smaller particles. The aggregate may be bonded with cement or bitumen, as in concrete and macadam, or used unbonded as in embankment dams (p. 248). Sand and gravel are naturally-deposited aggregates of sediment: they are extracted from pits and passed over screens which separate their fragments into the range of particle sizes required. Rocks are quarried, fragmented by crushers and then screened. The very fine fractions screened from sands and gravels, and from crushed rock, often contain an unacceptable concentration of clay and mineral flour, and are discarded: a process called *scalping*.

Trade names are descriptions adopted by users of aggregate for groups of rocks of related origin. The British trade names are quoted here as an example: selected examples of rocks included within the groups are named in brackets and described in Chapters 5, 6 and 7.

Artificial i.e. not natural, e.g. slag.
Basalt Group (inc. andesite and dolerite).
Flint Group (inc. chert).
Gabbro Group (inc. picrite, peridotite and diorite).
Granite Group (inc. syenite, pegmatite, granulite and gneiss).
Gritstone Group (inc. sandstone and pyroclastics).
Hornfels Group (inc. all thermally metamorphosed rocks with the exception of marble).
Limestone Group (inc. marble).
Porphyry Group (inc. rhyolite and aplite).
Quartzite Group (inc. ganister).
Schist Group (inc. phyllite and slate).

Suitability A material may be unsuitable for use as an aggregate if the *shape* of its fragments are flaky and elongate: slates and schists are often rejected for this reason (Fig. 7.1). Deposits of sand and gravel will not be suitable sources of aggregate when they contain *contaminating substances* such as clay (either on the grains or as pellets), mica, shells, and fragments of coal, lignite or peat, in quantities that will adversely affect its performance. Rocks that do not provide *durable* fragments which retain their mass when exposed to a variety of conditions, will not form a suitable source of aggregate and many weak limestones and sandstones, and numerous argillaceous rocks, fail to provide good aggregates for this reason. Sand, gravel and rock which contain minerals that *react* with cement paste are unsuitable for use as concrete aggregate. The suitability of a material for use as an aggregate is assessed under the following headings:

Grading for sands and gravels, and for rock when crushed.
Shape and texture of particles; many smooth particles of gravel, if lightly crushed, can be given a shape and surface texture that provide an aggregate that has a large angle of friction and good bonding with cement and bitumen.
Specific gravity and bulk density of fragments and of aggregates respectively.
Petrographic examination by viewing a thin section of the rock with a petrological microscope. The following materials are undesirable and rocks containing them in concentrations greater than 0.5 to 1.0%, should not be used as aggregates without further careful testing: organic fragments, the evaporites (p. 127), hydrated Fe-oxides (p. 85), sulphides and sulphates (p. 84–7), opal and chalcedony (p. 78), many zeolites (p. 80) and olivine (p. 72).
Mechanical tests which determine the strength of the rock (see Table 18.2) and of its aggregate, and include such tests as 'aggregate impact value' and 'aggregate crushing value', and measure the tendency of the rock to shrink and swell with changes in its moisture content.

Table 18.2 Examples of the unconfined compressive strength that may be expected from samples of fresh rock taken from potential sources of aggregate; i.e. excluding obviously weak and weathered material.

Unconfined strength (MN m^{-2})			
Material	Lowest	Likely	Highest
Granite	100	230	350
Gabbro	150	280	350
Dolerite	150	310	550
Basalt	130	300	500
Gneiss	100	230	350
Quartzite	150	200	310
Sandstone	75	150	350
Limestone	75	145	250
Hornfels	120	300	400

Durability tests which measure resistance to *mechanical abrasion* (e.g. Los Angeles Abrasion Value, and the Polished Stone Value); break-down under repeated cycles of wetting and drying using water and other solvents (e.g. Slake Durability Index, Sulphate Soundness test and other *physico-chemical tests*); and the content of organics, chlorides and sulphates by *chemical tests*. The reaction between rock fragments and cement is measured by *alkali-reactivity tests*.

Standards which specify the physical and mineralogical requirements of aggregates used for specific purposes, have been defined by organizations such as the American Society for Testing and Materials, the British Standards Institution, and others. The standards vary in detail from country to country and reflect the national sources of aggregate and the weathering the aggregates must resist. Aggregates used in high latitudes must withstand freezing and thawing, and those in hot and low latitudes must resist salt attack. Other standards specify the terminology to be used when aggregates are described and the tests required to assess their suitability.

Bound aggregates

Mineral–cement reactions

Cement is provided with free lime (CaO) which (*i*) reacts with the CO_2 of the atmosphere to precipitate $CaCO_3$ around the cement crystals and protect them from atmospheric weathering, and (*ii*) creates a level of alkalinity that protects the reinforcing steel from corrosion. If the alkalinity produced by the free lime induces the minerals within the aggregate to liberate their Si (and associated Al), new minerals will form within an alkali-silica gel that develops as a reaction rim around the unstable aggregate (Fig 18.8). The gel absorbes water from its immediate

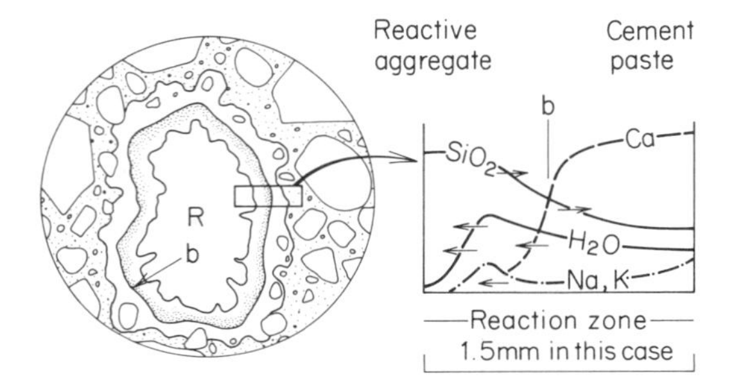

Fig. 18.8 Concrete viewed in thin section. A piece of reactive aggregate (R), bound in cement, is surrounded by healthy aggregate. Its original boundary (b) has been invaded and destroyed (stippled zone) and around the boundary a clear gel has developed. The graph (right) illustrates the relative concentration of constituents (concentration increases upwards) and their movement during reaction. (Reaction boundary details from French, 1980.)

surroundings and swells, creating internal pressures that can exceed the strength of the concrete. Micro-cracks are formed which permit water to migrate into the concrete, enabling the gel to expand even further. This water also dilutes the alkalinity of the concrete and permits the reinforcing bars to rust (another process involving expansion). Repeated wetting and drying causes the gel to expand and contract and subjects the concrete to cyclic loading which eventually destroys its structure. The rate of destruction may vary: some structures are severely damaged within 3 years and others show their first sign of damage after 30 years.

Most alkali-reactions involve silicate minerals that have a poorly-developed crystal structure, which releases Si readily in an alkaline environment (and with it, associated Al, Na and Ca). These minerals characteristically contain *amorphous* silica. Opal and chalcedony are examples: they may be found as a volcanic glass filling vesicles in lavas such as rhyolite and andesite, and can exist as fragments in sands derived from these sources. Chert is another example, and is often found associated with limestones. Once silica is locked into a crystal structure, even if it is a microcrystalline structure as in flint (SiO_2), it is stable and almost unreactive. Olivine is an example of a mineral which, when formed in rapidly-cooling lava, may contain an unbound silicate structure (see Fig 4.20*a*) that is capable of responding to an alkaline environment. A reaction zone of serpentine can form around it, and cause a 50% expansion in volume.

Mineral–bitumen reactions

Bitumen has a slight negative charge and is attracted to the positive charge of rock surfaces rich in ferromagnesian minerals. Basaltic and gabbroic aggregates therefore bind well with bitumen; dolomite and bauxite behave in a similar manner. Acid igneous rocks tend to be negatively charged and aggregates containing considerable amounts of feldspar and quartz in the form of large crystals, do not bind well with bitumen. Flint and quartzite are particularly troublesome, even with tar, which has slightly better binding qualities.

Unbound aggregates

These materials are conventionally used to construct the pavements of highways and airfields and to provide filter material, and ballast. The aggregates should be composed of fragments of strong rock that will withstand crushing and abrasion. This may exclude many of the weaker sediments from consideration. Potential materials should be tested under both wet and dry conditions, and should also be exposed to an environment that is representative of the conditions they will encounter in service. An apparently strong aggregate (a coarse-grained gneiss) from the Shai Hills, Nigeria, was used as armour stone to protect marine works on the coast. Heavy blocks of the rock began to break up after a short time due to a small content of the mineral prehnite. Prehnite is a secondary mineral (p. 79), sometimes classed with the zeolites: its presence was detected in thin sections cut from the rock. Although only present in small amounts, its rapid weathering weakened the rock, resulting in lower resistance to the impact forces which the armour stone had to withstand from the waves.

Weak rocks may decompose within a few years of being exposed within a structure and should only be considered when their performance over many years has been established. The Balderhead Dam, in Yorkshire, is an embankment composed of a clay core supported by shoulders of compacted shale of Carboniferous age. Extensive tests were conducted upon the shale to assess its suitability as a construction material. Much valuable information concerning the rate of weathering of the shale was also obtained from field observations. A similar shale had been used 25 years before, to build the Burnhope Dam, and its condition was investigated by excavating a trial pit into the downstream side of the dam: the shale was found to be unaltered. Spoil tips from a tunnel driven through the shale some 50 years earlier, were also examined and confirmed the very slow rate of alteration of the shale. Many weak rocks, including many shales, are not as suitable as this particular shale: see Kennard *et al.* (1967).

Earthfill

This is often derived from unconsolidated sediment; examples of its use are described on p. 247 (materials for embankment dams). The permeability of placed fill may be reduced by compaction, and its strength thereby increased, as particles of sediment are forced into closer contact with each other. Construction machinery passing over placed fill may provide sufficient weight for the required compaction to be achieved, and when extra loads are required, rollers and vibrators may be used. The density to which fill may be compacted varies with moisture content, and by using a Proctor Test (Proctor, 1933 and subsequent Standards) the *optimum moisture content* of a sedimentary material may be defined, this being the moisture content that permits a given compactive effort to achieve a maximum dry unit weight, i.e. dry density (Fig. 18.9). The mineralogy of a sediment, the shape of its

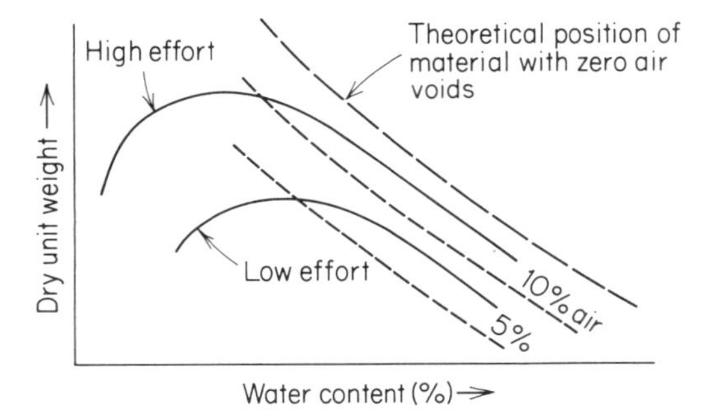

Fig. 18.9 Variation of dry unit weight with compactive effort and quantity of moulding water. Best compaction, and peak dry weight, occur at the *optimum moisture content*.

grains and the size of its particles (i.e. their grading: Fig. 11.3) influence the compaction of a granular material and govern its optimum moisture content.

Residual soils (p. 1) will not have been subject to the mechanical sorting that occurs when a sediment is transported by wind and by water. Consequently, residual soils may contain a great variety of weathering products some of which can have unusual properties.

At the Sasumua Dam site, in Kenya, lavas and tuffs had been reduced to a residual soil, known as the Sasumua Clay. The plasticity index of the clay (p. 168) varied with the chemistry of the dispersing agent used in preparing specimens and further investigation revealed the residual soil to be composed of the clay minerals halloysite (60%) and kaolinite (4%), in conjunction with geothite (16%), quartz (6%) and other constituents.

Most clays have a considerable range of moisture contact separating their liquid and plastic limits (Fig. 9.22), but the Sasumua clay had a much lower range than normal. It also had a higher angle of friction and a higher permeability than most clays of similar liquid limit. This behaviour was explained by the nature of the halloysite, which forms as small tubes (0.5×10^{-6} m in length) that hold water, and combine as clusters to form porous grains with rough surfaces. This structure has a low plasticity index whilst having a high liquid limit, and accounts for the higher than usual values of strength and permeability. The material compacted well and the dam performs satisfactorily, seepage losses being low. Investigations undertaken at the same time, revealed that other dams had been constructed from similar material and had worked well: Tjipanoendjang Dam, 1927 (Java) and Silvan Dam, 1931 (Australia). The materials are described by Terzaghi (1958) and Dixon *et al.* (1970).

Lateritic soils may develop when rocks are weathered and leached by a tropical climate (p. 38). Bases are removed first and followed by silica released from decomposing silicate minerals, with the exception of quartz. The combination of silica with alumina (also freed by weathering) produces kaolinite. Iron oxides accumulate under the conditions of rapid oxidation and the colour of lateritic soils varies from dark red to yellow, depending upon their content of iron and the state of its hydration. With prolonged weathering even the kaolinite begins to decompose, releasing silica into solution and increasing the concentration of alumina within the weathering profile, to produce a *bauxitic soil.* When the water table falls, iron becomes oxidized and fixed as haematite or geothite. Fluctuations in the water table cause this deposit to gradually accumulate forming either a honeycombed and cemented mass of *laterite* or layers of cemented nodules known as *laterite gravel.*

Lateritic soils form a valuable source of construction material that is extensively used in tropical countries. The properties of the soils which are not self-hardening can be improved, if inadequate, by mixing into them additives such as lime (a process called *soil stabilization* and suitable for use with many types of soil: Ingles and Metcalf, 1972). Some materials (for which the word *laterite* was originally used: p. 38) can be cut into bricks and other shapes, which irreversibly harden on drying in the air. Laterites possessing this property are now called *plinthites*, to distinguish them from other lateritic materials. Cemented laterite (e.g. laterite gravel) is frequently used as an unbound aggregate for road construction.

Dimension stone

Strong, durable stone, cut to size and used as slabs, blocks and columns, has been a building material for a greater period than has concrete, and where the rock has been correctly selected it has stood the test of time. The use of dimension stone has now been superseded by concrete, although it continues to be employed for facing and other ornamental work, and for providing resistance to abrasion: Leary (1983) provides an excellent account of its use in practice. Hard stone may often be used to protect the base of bridge piers, the sills of hydraulic structures and other parts of engineering works that will be exposed to severe wear. Igneous and metamorphic rocks are favoured, the finer-grained varieties being better than those with coarse-grained structures. Sediments are weakened by their porosity, which enables water and its dissolved constituents to invade the stone and cause decay. Such damage is evident in the stonework of most cities where sulphurous acid reacts with the $CaCO_3$ of the rock to form $CaSO_4$, which on hydration becomes crystalline gypsum, occupying a greater volume. A sulphate skin forms on the surface of the stone and gradually splits off and falls away, a process called *exfoliation*. Sediments also suffer from the presence of bedding, which if placed vertically (called 'face bedding') will spall off. Sedimentary rocks should always be laid with their bedding horizontal.

Foundations

The purpose of a foundation is to transfer the load of a structure to the ground without causing the ground to respond with uneven and excessive movement. Most buildings are supported on one of four types of foundation, viz. pads, strips, rafts and piles: these may be modified and combined to form a suitable foundation for the ground conditions that exist. Examples of dam foundations have been described in Chapter 15.

Bearing capacity

The intensity of loading that causes shear failure to occur beneath a foundation is termed the *bearing capacity* of the ground. This capacity is governed by the fabric of the rocks and soils beneath a foundation and by the reaction of this fabric to changes in effective stress. Such changes will accompany periods of construction that change the total stress on the ground and are described in Chapter 9. Thus the behaviour of a soft to firm compressible clay will not only be that of an inherently weak sediment but also that of an undrained material: the clay will have a low bearing capacity. A hard boulder clay will be stronger and have a higher bearing capacity, but will also behave as an undrained material. Dense deposits of sand and gravel have their particles packed closely together yet retain a permeability that is sufficient to readily dissipate any increase in pore pressure that may accompany an increase in total stress. These sediments can have a high bearing capacity and behave as a drained material. Loose sands and gravels have an open texture and a lower bearing capacity than their denser varieties. Strong rocks have a saturated strength that exceeds the safe working stress of concrete (around $4000\,kN\,m^{-2}$) and are therefore not loaded to their bearing capacity. But even strong rocks contain weak surfaces, such as joints and soft horizons interbedded with stronger strata, and the load from a large foundation can cause movement to occur along these planes. The bearing capacity of weak rocks, especially porous sedimentary rocks and weathered igneous and metamorphic rocks, can be exceeded by foundation loads from structures of moderate size.

Bearing capacity failure is avoided by ensuring that a foundation never transmits an intensity of load that causes the ground to exceed its range of elastic behaviour.

Later movements

Most foundations settle because rocks and soils respond to loads placed upon them. The total settlement that results normally consists of three components:

(*i*) a reversible settlement which occurs immediately the net pressure on the ground is increased and is attributable to the elastic deformation of the ground;
(*ii*) an irreversible settlement attributable to consolidation and occurring when the fluid pressures increased by the increase in net foundation load, begin to dissipate from pores and fractures;
(*iii*) an irreversible settlement that occurs during the life of the foundation and attributable to imperceptible creep under conditions of constant effective stress.

Sometimes settlement results from other causes, such as the collapse of unstable soils (p. 275) and of ground above a concealed cavity (such as a solution hole, or a mine). Many foundations rise because they are constructed on soils and rocks which expand when wetted: some are lifted by the growth of ice or the crystallization of gypsum, or the hydration of anhydrite in the rocks on which they are founded. Foundations located on landslides may move laterally and vertically, usually at different speeds.

Investigations

Two types of investigation may be required: a general survey which identifies the most suitable areas in which to found buildings (see for example Price *et al.*, 1969; and Taylor, 1969) and a particular study of the geology beneath the site of proposed buildings: see Chapter 10. Every investigation must establish the vertical sequence of soil and rock, and its lateral variations. Measurements of deformation, strength and *in situ* permeability are of special relevance, as they permit the depth and type of most suitable foundation to be assessed, and provide values that may be used in an analysis of foundation stability and settlement. Foundations for light buildings are often designed using values obtained from index tests conducted in the field (such as the standard penetration tests and cone penetrometers, p. 184), and from laboratory tests performed on samples retrieved from boreholes. Laboratory measurements of strength, consolidation and permeability are described in Chapter 11: special note should be taken of the selection of samples for testing and the description and classification of untested samples.

Many foundations are constructed in excavations which have to be designed so as to remain dry and stable during the period of foundation construction. Relevant aspects of the investigations required to assess the stability and drainage of these excavations are described in Chapters 14 and 16.

Six examples of vertical profiles illustrated in Fig. 18.10 and described in turn.

Six examples

(*a*) The sequence at Pisa is typical of recently-deposited shallow water sediments and of thick successions of weak

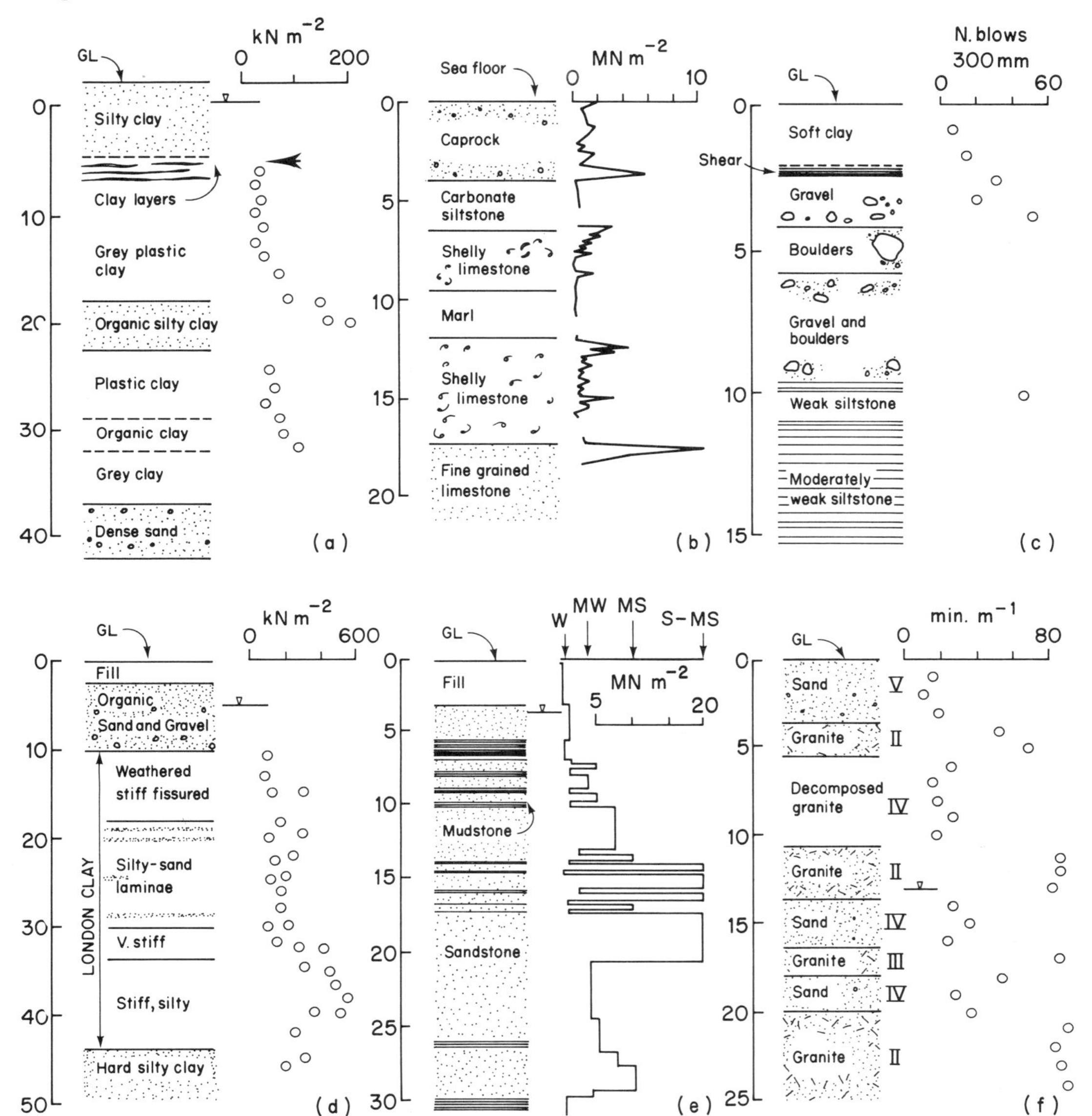

Fig. 18.10 Six vertical profiles, recording the geology beneath foundations and illustrating the variety of techniques that may be used to describe the strength of the ground (see text).
(**a**) Results from *in-situ* vane tests (Mitchell *et al.*, 1977): arrow indicates level of suspected shear failure. (**b**) Results of point load tests on cores (Dennis, 1978). (**c**) Results from standard penetration tests. (**d**) Results from laboratory tests of strength on samples taken from cores (Burland *et al.*, 1977). (**e**) Description of cores: W=weak, MW=moderately weak, MS=moderately strong, S=strong (Geol. Soc. London, 1970) and scale of strength implied by the use of these terms (Cole *et al.*, 1977). (**f**) Description of cores in terms of grades of weathering in Hong Kong (VI=soil, V=completely decomposed, IV=highly decomposed, III=moderately decomposed, II=slightly decomposed, I=freshrock; see Ruxton *et al.*, 1957) and rate of drilling progress (mins per metre).

and compressible detritus which can overlie strata of adequate strength.
(*b*) Coastal sediments often reveal variations in strength that can be attributed to former changes in sea level. Such a change was recorded at Umm Shaif, in the Arabian Gulf. Here weak carbonate sands, silts and clays (=marls) are interbedded with stronger horizons which have gained their strength from the precipitation of mineral cement in their pores during periods of brief exposure after deposition. The mineral cement is believed to have been carried upwards from underlying sediments, by the evaporation of water from the exposed sediment. Note that a hard caprock overlies weaker material: similar situations can be encountered on continents where duricrusts (p. 38), formed by cementation during weathering, overlie weaker horizons from which their mineral cement has been leached.
(*c*) Glacial drift can contain horizons of greatly differing character if they have been formed by different processes of deposition (p. 56). At the S. Wales site, palaeo-shear surfaces existed in the soft clay as a relic of former slope movements that had translated the overlying strata down hill.
(*d*) Older and apparently uniform deposits, such as the London Clay, may be more variable than they appear, particularly if affected by the agents of weathering and erosion. The behaviour of such sequences beneath a foundation will be sensitive to the presence of pre-existing fractures (e.g. fissures in the clay) and permeable laminae of silt and sand.
(*e*) Sedimentary rocks characteristically exhibit stratification which in this case consists of alternating layers of sandstone, mudstone and siltstone. Considerable differences exist in the strength of the strata, and even within the sandstones. Such a sequence will also contain pre-existing fractures (joints). These surfaces, and weak layers of sediment, permit heavy foundations to settle, and sometimes allow bearing capacity to be exceeded.
(*f*) The *in-situ* weathering of igneous rock can produce a complicated profile of hard and soft zones (see Fig. 3.4). Acid igneous rocks, such as granite, often produce a quartz rich residue that becomes progressively richer in clay, with depth. Basic igneous rocks, such as basalt and dolerite (=diabase), convert readily to clay which can expand on wetting and shrink on drying: see also Sasumua clay (p. 289).

These examples demonstrate the prudence of investigating to depths greater than proposed foundation level. Shallower investigations may suffice for light structures and Sowers (1979) suggests that 2–3 m per storey (or equivalent load) usually provides a satisfactory limit to the depth of drilling, but this requires a knowledge of local conditions (see below: special problems).

Mechanisms of failure

Vertical profiles permit the likely mode of failure to be established of a foundation placed at a particular level below the surface. A variety of such mechanisms is illustrated in Fig. 18.11.

The near-circular form of failure is normal for soils loaded beyond their bearing capacity. A famous example of this mechanism is the failure that tilted the tower of Pisa: the failure surface is thought to extend into a weak horizon at the top of the grey plastic clay (Fig. 18.10*a*). The following characters may be taken as a guide to soil behaviour:

Sands and gravels, when dense, have a high bearing capacity and low compressibility, but when loose they have a low bearing capacity.

Glacial clays, if heavily over-consolidated, have a high bearing capacity, but can contain weak and compressible horizons (Fig. 18.10*c*) and varved clays (*q.v.*) which are often troublesome. Water-bearing sands and gravels, and large boulders, must also be expected (Fig. 3.42).

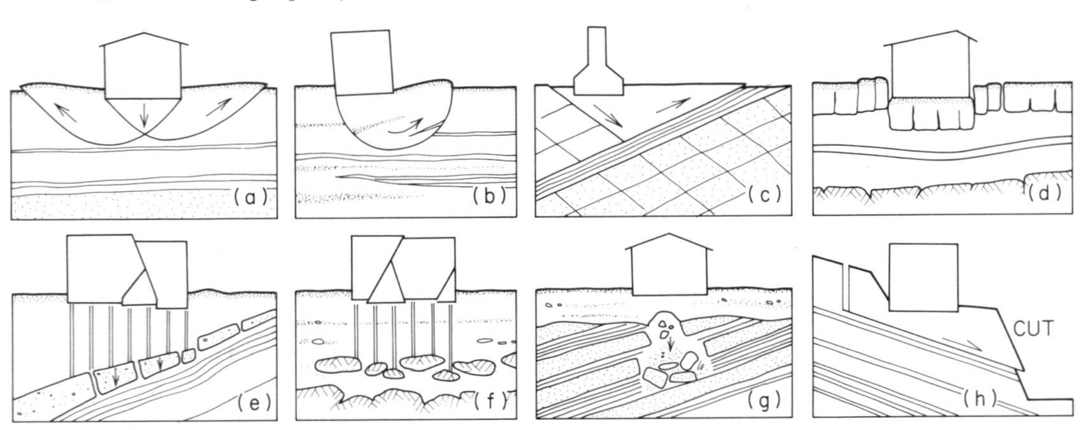

Fig. 18.11 Examples of failure mechanisms beneath foundations. (**a**) and (**b**) are typical of sedimentary formations. (**c**) Illustrates movement of rock. (**d**) and (**e**) demonstrate the effect of overloading strong strata underlain by weak strata. (**f**) Represents the problems that arise when boulders are mistaken for bedrock. (**g**) Shows the upward growth of cavities that are not filled by bulking of the fallen debris, and (**f**) records one of the difficulties of foundations on slopes.

Stiff clays, such as London Clay (Fig. 18.10*d*), have a reasonable bearing capacity and moderate compressibility, but are liable to soften when wetted.

Soft clays, as in Fig. 18.10*a* and carbonate muds (Fig. 18.10*b*) are weak. The clays can be extremely sensitive (*q.v.*). Their bearing capacity is low and their compressibility is high.

Silts are usually soft and weak, especially when saturated, and have a low bearing capacity.

Loess and other cemented, but porous sediments (as in Fig. 18.10*b*) may have a high bearing capacity and low compressibility, but be liable to collapse. Loess is susceptible to collapse when saturated (Fig. 17.14).

Organic sediments have a high compressibility and low strength often over-estimated by vane tests when the vane becomes entwined in the organic fibres (see Fig. 18.10*a*).

Tropical soils Red sandy soils normally have a high bearing capacity. Black soils (e.g. adobe) are extremely hard materials when dry and appear as attractive founding material, but when saturated they lose their strength and often expand.

The failure mechanisms that can occur in rocks usually depend upon movement along pre-existing surfaces, these being the weakest part of many rocks masses. Problems created by strong rock underlain by weak material are illustrated in Fig. 18.11 (and see profiles *b*, *c*, *e* and *f*, Fig. 18.10).

Special problems

The problems arising from local conditions can often be grouped under one of the following headings.

Swelling ground This can occur when ice lenses grow between bedding, so jacking up the overlying strata (a cold regions phenomenon: see Johnston, 1981). Heave can also result when the moisture content of argillaceous sediments is increased. Sometimes it is attributed to the adsorption of water, but other times it may be due to the conversion of sulphides to sulphates and the hydration of anhydrite (p. 87) to gypsum (p. 87). Many argillaceous soils expand when trees that were rooted in them, are felled prior to construction. If the roots had dried the soil then removal of the trees will permit soil moisture to increase, and the ground to swell (Samuels *et el.*, 1974). Swelling clays are the source of much damage to foundations throughout the world and regions badly affected by their presence include W. Canada, Colorado, Texas and Wyoming, India (where the black cotton soils exist in abundance) and Nigeria, Israel, South Africa and to some extent, S. Australia. California, Utah, Nebraska and S. Dakota are also troubled by these materials (Chen, 1975).

Shrinking ground Expansive soils shrink when dried, and the removal of soil moisture by vegetation and by drought, can seriously damage shallow foundations (Tomlinson *et al.*, 1978). Some foundations collapse because they are on unstable soils (Fig. 17.14) or have been built over cavernous ground. This is a severe problem in South Africa where lives have been lost when complete structures collapsed into solution cavities. All soluble strata is suspect, and areas where mining has occurred should be treated with the same caution.

Ground-water The movement of water levels and the chemistry of the ground-water are characteristics that should be studied at every site. When water levels are lowered consolidation may occur. The spectacular problems of settlement in Venice and Mexico City, caused by the extraction of water from aquifers at depth, are well-known examples of consolidation. Many other examples are described by Holzer (1984). A shaft of an end-bearing pile can be loaded to failure by the downward drag against it of consolidating sediment. Uneven consolidation beneath a building will result in differential settlement and structural failure.

Some ground-waters have a concentration of sulphate that is aggressive to concrete. The sulphate comes from soluble sulphate minerals in the ground, and for this reason the sulphate content of both soil and water should be measured; it is useful to measure their pH at the same time. Sulphates are common in deposits of clay, evaporite and peat. Oxidation of pyrite can produce free sulphuric acid and this occurrence is indicated by a low pH for the soil (less than 4.3) and a high content of sulphate. Metal piles can corrode in saline ground and the electrical conductivity of the soil will indicate the severity of this effect. Deposits which accumulate in salt-flats (p. 48) and sabkhas (p. 49) can be very saline.

Waste disposal on land

March harmful waste degrades to harmless substances that eventually enter the atmosphere, as a gas, or the hydrosphere, as a solute in ground-water. Safe waste disposal ensures that dangerous products cannot travel far from their repository until they have degraded to a safe condition. To predict the rate of migration of waste products that have been buried in the ground, it is necessary to study the geology of each disposal site. Special attention must be directed to the movement of ground-water. Some waste contains materials that do not degrade and do not decay: cadmium and mercury are examples. The disposal of these wastes on land may be achieved by burying them beyond reach of the agents of weathering and erosion. Underground disposal sites are thus an attractive location for these materials. Chapters 10 and 11 outline the standard forms of investigation that will be used to study the geology of disposal sites. Maps and vertical sections, similar to those described in Chapter 12, are valuable aids to understanding the likely movement of a contaminant through the ground. Chapter 13 explains the principles which will govern most of the ground-water flow that affects waste disposal sites.

Landfill

Much domestic and building waste, and certain industrial wastes, degrade quite rapidly to safe by-products and can be disposed of in surface excavations. Abandoned quarries and pits offer favourable sites for this waste, which is used to fill them in and restore ground level to its former elevation. Waste stored in this manner is called *landfill*. Rain that percolates through the waste whilst it is still degrading, may emerge from the base of the fill as an unpleasant leachate, rich in organic and inorganic constituents. Ground-water supplies in aquifers that extend beneath a landfill site, would be contaminated if such a leachate were to reach them. Some sites are lined to contain the leachate and later covered to prevent the ingress of rain. Other sites are not lined and in these it is customary to ensure that a thickness of unsaturated, porous granular ground separates the base of the fill from the underlying water table. The quality of a leachate is often improved by percolation through this unsaturated zone, but not always by an adequate amount. Any improvement that does occur retards the migration of constituents from the fill; a process called *attenuation*. Carbon dioxide and methane, which are both usually generated in landfill, are extremely difficult to retard. Methane can be lethal if it accumulates in buildings founded above landfill or is carried by ground-water to areas where it may accumulate in wells and cisterns. Some

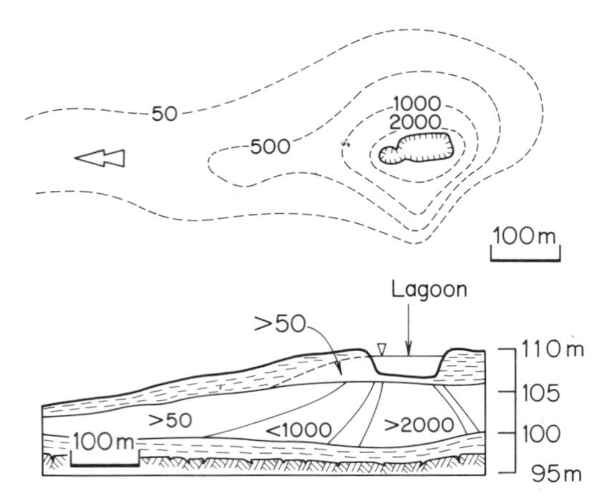

Fig. 18.12 Map of a landfill site in which liquid waste is also disposed, forming a lagoon. A leaky, disused clay-pit overlies unsaturated sands which are separated from bedrock by another clay horizon. The main water table is in bedrock. The concentration and distribution of total organic carbon (in mgl^{-1}) is illustrated in plan and section, and outlines a *plume* of pollution. Ground-water that may occasionally flow through the sand will tend to *dilute* and disperse the plume: (based on Williams, 1984).

sediments, residual soils and weathering profiles, have better attenuation characters than others and this is thought to reflect their mineralogy and microscopic texture. Two of the processes which are considered to make an important contribution to the improvement of leachate quality during its migration through the ground are, (*i*) buffering of the pH of the leachate, by calcareous minerals and (*ii*) removal of solutes from the leachate by their adsorption onto the surface of clay minerals, and by their exchange for ions within the lattice of the clay minerals. Slow rates of percolation favour better cleansing of the leachate and the speed with which water may flow through fractures usually makes jointed rock a poor retarder of pollutants (see Figs 13.5 and 13.6).

It is common to find that sites do not behave in the same manner. Leachate that becomes incorporated into ground-water flow, is *diluted* by the ground-water and *dispersed* by constant sub-division through the anastomosing network of pores and fractures in the ground. A *plume* of dilted leachate thus develops. The *plume* of leachate emerging from beneath a disposal site, and the attenuation of some of its constituents, are illustrated in Fig. 18.12.

Injection

Liquid waste may easily enter aquifers and be transported a considerable distance from the disposal site, possibly to emerge in a public supply. To avoid this occurrence, producers of large quantities of liquid waste have attempted to inject the waste into ground at great depth. At such depths, flows, if any, would be of a regional nature and extremely slow (Fig. 13.3) and the waste would be isolated from local flows that sustain water supplies. Deep injection wells are used for this purpose.

One such well was sunk at Denver, Colorado: it was 3658 m deep and injected waste into rocks of Precambrian age at a depth of 3642 m. The waste was mainly a concentrated solution of NaCl, having an average pH of 8.5. Injection commenced in 1962 and was accompanied by a spate of small earth tremors (Fig. 18.13): none had been reported from the region before. Injection continued, each period of operation being followed by local earthquakes. The focus of these 'quakes was calculated to be at a depth of between 3220 m and 4830 m below ground level: their epicentre, which was close to the well, was gradually migrating away from the well. Injection ceased in 1966 but the tremors continued for three more years with events of magnitude 4.8, 5.1 and 5.8 occurring in 1967, by which time the epicentre had migrated 6 to 8 km from the well. Various explanations for these events have been forwarded (van Poolen and Hoover, 1970):

(*i*) Injection pressures had reduced the effective stress at depth and caused slip to occur on a pre-existing fault.
(*ii*) Thermal stresses produced by the difference in temperature between the waste and the rock had caused rock failure (but unlikely to affect rock 6 to 8 km away).

(*iii*) The chemistry of the waste weakened the ground. It was calculated that 213 Mg of rock had been dissolved and that Na–K exchange had involved 325 Mg of K-feldspar in the Precambrian rock.

These effects are illustrated in Fig. 18.13.

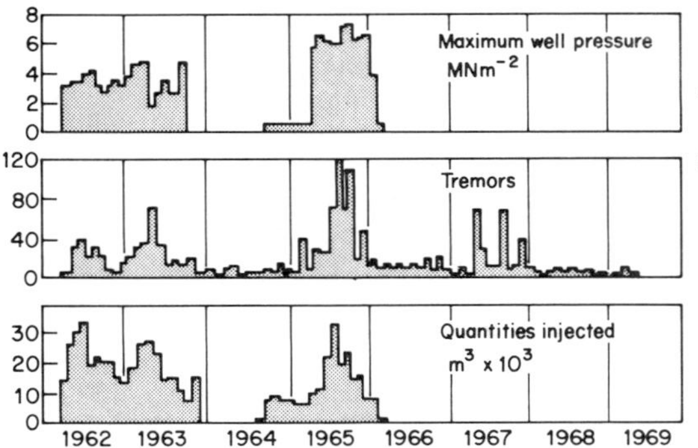

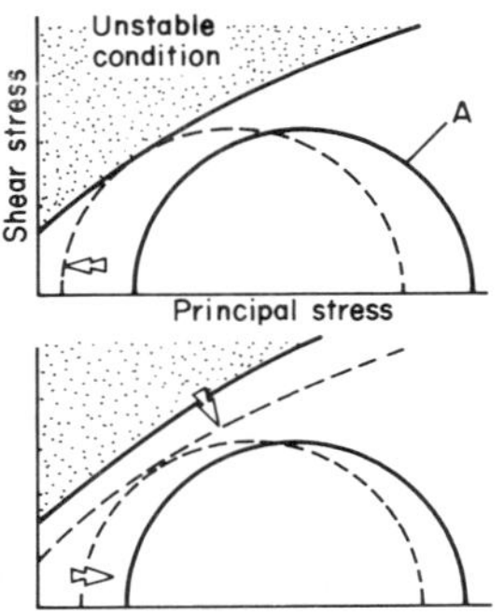

Fig. 18.13 Injection and earthquake history for the Rocky Mountain Arsenal well, Colorado (from Poolen and Hoover, 1970). Top right: Mohr circles for decreasing effective stress with increasing injection pressure (circle A = original condition). Botom right: effective stress increases as injection pressure decreases but shear strength reduces with dissolution of the rock. See text, and Figs 9.17 and 9.19.

High pressure recharge wells often generate ground movement. The failure of Baldwin Hills Reservoir, approximately 13 km from Los Angeles, was attributed to injection causing movement on a pre-existing fault. The reservoir ruptured and water destroyed 277 homes in Baldwin Hills (Hamilton and Mechan, 1971; and Castle *et al.*, 1973). Much smaller movements, of similar origin, have been associated with fluid injection at depths of 2 km, for the extraction of geothermal energy from the hot dry rock of the Carnmenellis Granite in Cornwall (Fig. 2.22).

Nuclear waste

Radioactive waste may be divided into low, intermediate and high level categories on the basis of its level of activity. Low and intermediate wastes are mainly composed of ordinary materials, such as paper and metal, which have been contaminated by small amounts of radioactivity. Low level waste may be diluted and dispersed into the sea and air: it may also be buried in shallow trenches. Intermediate waste is buried. The short-lived intermediate wastes, which decay to very low levels of activity over 200 to 300 years, are sealed in canisters and buried in deep, concrete lined repositories that are excavated into sediments of low permeability, such as overconsolidated clays. The longer-lived intermediate wastes are also sealed into containers but stored in specially constructed underground cavities deep below ground level: an example is illustrated in Fig. 18.14; for a repository in bedded salt see Weart (1979).

High level waste contains spent fuel and generates heat. It requires at least 600 years before it decays to safe levels and about 10^6 years before its activity becomes comparable to that of the original uranium ore from which the nuclear fuel was obtained. This waste must be contained for thousands of years. It may be fused with other substances to form a boro-silicate glass, or a ceramic called supercalcine, composed of crystalline phases, or a synthetic igneous rock called synrock, which is also composed of crystalline phases. Once in this form, the waste is sealed into long-lasting containers and buried at great

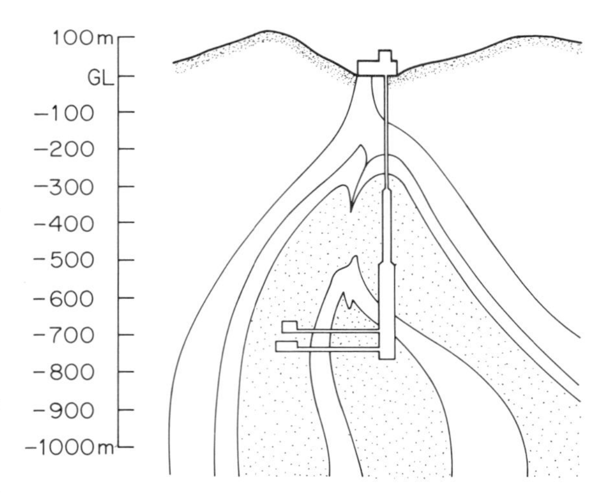

Fig. 18.14 Sketch section of the salt mine at Asse in W. Germany. Nuclear waste could be lowered from the 700 level to cavern storage at 750 level. Since 1967 experimental storage has been monitored at the 511 level. Note shape of the salt dome: see also Figs 6.21c and 8.18.

depths. This process is designed to separate the high level waste from man by three barriers: (*i*) that created by immobilizing the waste in glass, or in the lattice or crystalline phases of ceramic or synthetic rock, (*ii*) that created by the container, and (*iii*) that created by the geology of the ground around the container.

Contributions from geology

Radioactive elements, glass and metal alloys are formed by natural geological processes and those produced millions of years ago have been studied in the laboratory and in the field to guide the design of disposal systems for high level waste. For example, volcanic glass such as obsidian and pitchstone (p. 107) of Tertiary age and older (i.e. between 2 and 65 my: see Table 2.1) may be found in many parts of the world and provides evidence that glass-like substances have the durability required to retain waste. Ringwood (1978) has presented much geological and geochemical evidence to demonstrate that radioactive materials can be safely locked into the lattice of minerals where they will reside for millions of years.

Lattices

All igneous rocks contain small amounts of radioactive elements (e.g. U, Th, K, Sr, Cs, Rb) which have become firmly incorporated into the lattice of the rock-forming minerals. The biotite mica illustrated in Fig. 4.17 contains two small crystals of zircon which were enveloped by the growth of the mica crystal. The zircon contains uranium and thorium which has radiated the mica lattice in its immediate vicinity, creating a visible 'halo'. This phenomenon can be observed in the micas of most granites. Sphene (Fig. 4.34) contains strontium. Such minerals may be dated by their radioactivity and from this evidence it is concluded their lattice can resist leaching over extensive periods of geological time, and all other changes that may cause them to release their radioactive elements. Zircon crystals recovered from Proterozoic rocks which are over 2000 my old (Table 2.1) have retained the uranium in their lattice despite the damage their lattice has sustained by radiation. Caesium has been found in micas that are millions of years old. Strontium and rubidium are known to be immobilized in the lattice of feldspars for very long periods. These and other natural examples demonstrate that radioactive waste can be contained.

Radiation damage

Alpha particles emitted from actinide elements are the principle cause of damage to the lattice of host minerals. The crystal lattice becomes disordered and up to 10% volumetric expansion may occur: crystals in this condition are described as being in a metamict state. Zircon is most susceptible to radiation damage (see examples in Fig. 4.34) and its volumetric expansion may exceed 10% (this contributes to the halo seen in micas and mentioned earlier). Zircon crystals that were 500 my old and in a metamict state, have been subjected to prolonged periods of heating and leaching in the laboratory but tenaciously retained their uranium. Such treatment did remove, from the crystals, lead that had been produced by the decay of uranium, the lead not being incorporated into the lattice of the zircon by reason of its incompatible ionic radius and charge. Ringwood quotes other examples: crystals of uraninite retrieved intact from conglomerates of the Witwatersrand in S. Africa (2000 my old); thorianite derived from weathered pegmatites, occurs in 500 my old deposits of alluvial gravel, in Sri Lanka.

Geological barrier

The geology of the ground around a container of waste provides the thickest and final barrier between the waste and man. Ground-water is the most likely agent to transport waste across this barrier and hence high level waste will be stored in rocks of very low permeability and remote from concentrated flows of ground-water, as may be found in some faults. Thick deposits of rock salt and argillaceous sediment are considered to be suitable in many respects, and large intrusions of igneous rock, such as granite, are also possible repositories for such waste. The waste would be placed at a depth where ground stresses have permitted only a few fractures to open (p. 160). Many rock-forming minerals are able to absorb radioactive species from ground-water: clays and argillaceous sediments in general exhibit a remarkable capacity for such absorption (Fried, 1979). Similar geological evidence from a variety of rocks and sediments supports the conclusion that unfractured geological barriers are able to prevent the rapid release of radioactive substances from deep repositories.

By far the most spectacular example of such evidence is the uranium deposit at Oklo in Gabon, W. Africa. This deposit is of Precambrian age and is approximately 1800 my old (Table 2.1). It has been exposed by an open pit mine that was developed to extract the uranium ore, and the excavation has uncovered a fossil nuclear power station. Several large bodies of uranium ore accumulated and became a natural reactor which is calculated to have released 100×10^9 kW h^{-1} of energy. Like all reactors, it produced plutonium which, like the uranium, has retained its position and character so perfectly over this immense passage of time, that reaction rates across the surrounding formations can be interpreted in terms of neutron physics (Int. Atomic Energy Agency, 1975).

Geological history

Waste disposal sites will be located in areas where the geological record proves a long history of stability. A region containing relatively undeformed rocks of Mesozoic age lying unconformably on folded Palaeozoic strata records that for the past 65 to 248 my the area has been dominated to gentle vertical movements (see Table 2.1 and Fig. 2.17*a*). Older examples are plentiful (Fig. 2.2). In this way, stable areas can be identified. Most stable

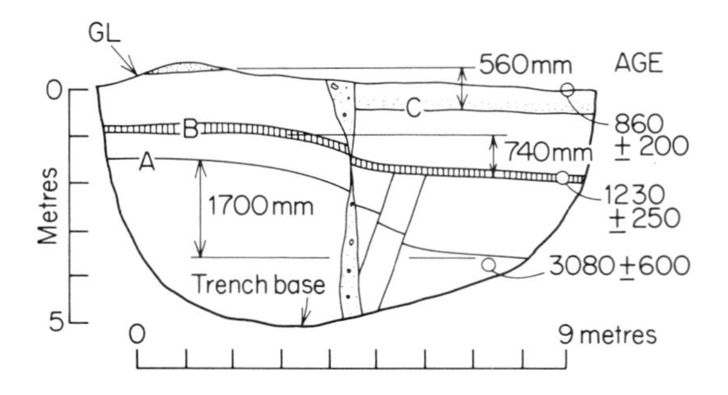

Fig. 18.15 Trial pit in Recent lacustrine sediments where fault displacements record former earthquakes. Dates provided by C^{14} and given in years since present. The event that broke layer A displaced it by 1700–740 mm, i.e. 960 mm; that for B by 740–560 mm (from Clark *et al.*, 1972).

areas contain faults whose occasional movement can be expected to produce earthquakes. A detailed reconstruction of past events is sometimes possible if stratigraphic evidence is preserved. An example of the remarkable record that may be obtained is illustrated in Fig. 18.15. From these records the frequency and magnitude of events may be determined for periods in excess of 2 my; a far greater period than that covered by historic documents.

The safe disposal of nuclear waste will be one of the most important technical achievements of the 21st century. All branches of geology are being used to support the studies required for this task and no clearer example could be provided of the importance of geology to society and the engineer.

Selected bibliography

Watson, J. (1983) *Geology and Man*. George Allen and Unwin, London and Boston.

Legget, R.F. and Karrow, P.F. (1983) *Handbook of Geology in Civil Engineering*. McGraw-Hill Book Co., New York, London.

Cooke, R.U., Brunsden, D., Doornkamp, J.C. and Jones, D.K.C. (1982). *Urban Geomorphology in Dry Lands*. Oxford University Press, London and New York, on behalf of The United Nations University.

Holzer, T.L. (Ed.) (1984). *Man-induced Land Subsidence*. Reviews in Engineering Geology. Vol. VI. Geological Society of America, Boulder, Colorada.

Water supplies

Lloyd, J.W. (Ed.) (1981). *Case-studies in Groundwater Resources Evaluation*. Clarendon Press, Oxford.

Walton, W.C. (1970). *Groundwater Resource Evaluation*. McGraw-Hill Pub. Co., New York.

American Society for Testing and Materials (1981). *Water for sub-surface injection*. Special Technical Publication (STP) 735. ASTM, Philadelphia.

United States Bureau Reclamation (1977). *Groundwater Manual: a guide for the investigation, development and management of groundwater resources*. U.S. Bur. Rec. Denver, Coldorado.

Construction materials

American Society for Testing and Materials (Published Annually). Section 4, (for 1984) Construction: with special reference to Vol. 04.01 *Cement; Lime; Gypsum*. Vol. 04.02 *Concrete and Mineral aggregates*. Vol. 04.03 *Road and Pavement Materials*. Vol. 04.08 *Natural Building Stones, Soil and Rock. ASTM, Philadelphia.*

British Standards Institution (1975). *Methods for Sampling and testing of mineral aggregates, sands and fillers*. British Standard BS 812:1975. British Standards Institution, London.

British Standards Institution (1973) *Aggregates from natural sources for concrete*. British Standard BS 822:1973. British Standards Institution, London.

Schaffer, R.J. (1972) (A facsimile reprint). *The Weathering of Natural Building Stones*. Dept. Sci. Industrial Research. Special Report No 18. H.M. Stationary Office. London.

McFarlane, M.J. (1976). *Laterite and Landscape*. Academic Press, London and New York.

Foundations

Tomlinson, M.J. (1980). *Foundation Design and Construction*, 4th Edition. Pitman Publishing Ltd, London.

British Standards Institution (1972). *Code of Practice for Foundations*. CP2004:1972. British Standards Institution, London.

Bell, F.G. (Ed.) (1978). *Foundation Engineering in Difficult Ground*. Newnes-Butterworths, London and Boston.

Pells, P.J.N. (Ed.) (1980). *Structural Foundations on Rock*. Proc. Int. Conf. Sydney, 1980. A A Balkema, Rotterdam. Vols I and II.

Waste disposal

American Society for Testing and Materials (1981). *Permeability and Groundwater Contaminant Transport*. Special Technical Publication (STP) 746. ASTM, Philadelphia.

United Nations Educational, Scientific and Cultural Organization (1980). *Aquifer contamination and protection*. Project 8.3 of the Int. Hydrological Programme. UNESCO, Paris.

International Atomic Energy Agency (1981). *Underground disposal of Radioactive Waste: Basic Guidance*. I.A.E.A., Vienna.

International Atomic Energy Agency (1981). *Site Investigations for Repositories for Solid Radioactive Waste in Deep Continental Geological Formations*. I.A.E.A., Vienna.

International Atomic Energy Agency (1981). *Site Investigations for Repositories for Solid Radioactive Waste in Shallow Ground*. I.A.E.A., Vienna.

Milnes, A.G. (1985). *Geology and Radwaste*. Academic Press, London.

SI Units

The International System of Units has seven standards of measurement of which the following are commonly used in geology and geotechnical engineering.

Fundamental mechanical units

Length

The standard of length is a *metre* (m): dimensions [L]. The following submultiple or multiple may be used: millimetre (mm) = 0.001 m, and kilometre (km) = 1000 m

Conversions

Length [L]

1 ft	= 0.3048 m	1 m	=	3.2808 ft
1 mile	= 1.6093 km	1 km	=	0.6214 mile

Area [L^2]

1 ft^2	= 0.0929 m^2	1 m^2	=	10.7639 ft^2
1 $mile^2$	= 2.5899 km^2	1 km^2	=	0.3861 $mile^2$

Volume [L^3]

1 ft^3	= 0.0283 m^3	1 m^3	=	35.3147 ft^3
1 yd^3	= 0.7646 m^3	1 m^3	=	1.3079 yd^3

Fluid vol [L^3]

1 UK gal	= 0.0045 m^3	1 m^3	= 219.9692 UK gal
1 US gal	= 0.0038 m^3	1 m^3	= 264.1721 US gal

Note that 1 m^3 = 1000 litres = 1 000 000 cm^3

Mass

The standard of mass is the *kilogram* (kg), i.e. 1000 grams: dimensions [M]. The following submultiple or multiple may be used:

gram (g) = 0.001 kg, and megagram (Mg) = 1000 kg

Conversions

Mass [M]

1 lb mass	= 0.4536 kg	1kg	= 2.2046 lb mass
1 UK ton	= 1.0160 Mg	1 Mg	= 0.9842 UK ton
1 US ton	= 0.9072 Mg	1 Mg	= 1.1023 US ton

Note that 1 Mg = 1 tonne

Mass density In SI units, mass density is quoted in (kg m^{-3}): dimensions [ML^{-3}]. In geology and geotechnical engineering it is common to measure in quantities of 1000 kg m^{-3} = 1 Mg m^{-3}.

Note that 1 Mg m^{-3} = 1 tonne m^{-3} = mass density of fresh water.

Time

The SI standard of time is the *second* (s): dimensions [T].

Note that 1 hour = 3600 s, 1 day = 86400 s.

Fundamental Heat Units

The SI base unit of thermodynamic temperature is the *Kelvin* (K) which has Absolute Zero as zero and the triple point of water (i.e. the temperature at which ice, water and water vapour are in equilibrium) as 273.15 K. The Celsius scale uses this point as its zero and because its scale divisions are the same as those of the Kelvin scale (i.e. a change of 1 K is the same as 1°C), degrees Celsius = temperature in Kelvin minus 273.15.

Note that °C = 5(°F − 32) ÷ 9
°F = (°C × 9 ÷ 5) + 32

Derived units

Commonly used multiples of the fundamental units, such as are needed for force, pressure and stress, have been given names. Two such derived units, which are commonly used in geology and geotechnical engineering, are the *Newton* and the *Pascal*.

Force

Force is the product of mass × acceleration, as defined by Newton in his first law of motion, and is given the derived unit of the *Newton* (N): dimensions [MLT^{-2}]. A force of 1 Newton will accelerate a mass of 1 kg by 1 m s^{-2}.

Weight is a force produced by the action of Earth's gravitational acceleration (g) upon mass. The natural variations that occur in the value of gravitational acceleration over the surface of the globe, can be ignored for the purposes of geotechnical engineering, and the standard acceleration used, namely 9.806 m s^{-2}. Thus as mass of 1000 kg has a weight on Earth of 9806 N; usually stated as 9.806 kN.

Note that 9.806 m s^{-2} = 32.174 ft s^{-2}.

Conversions

Weight [MLT^{-2}]

1 lbf	= 4.4482 N	1 N	= 0.2248 lbf
1 tonf (UK)	= 9.9639 kN	1 kN	= 0.1004 tonf (UK)
1 tonf (US)	= 8.8959 kN	1 kN	= 0.1124 tonf (US)

Unit weight In SI units, unit weight is quoted in N m^{-3} or in the following multiples, kN m^{-3}, MN m^{-3}.

Note that unit weight of fresh water = 62.43 lbf ft^{-3} = 9.806 kNm^{-3}

Pressure and stress

These are both measures of force per unit area and are given the derived unit of the *Pascal* (Pa): dimensions [$ML^{-1}T^{-2}$].

1 Pascal = 1 Newton m^{-2}

The following multiples may be used:

kilopascal (KPa) = 1000 Pa
Megapascal (MPa) = 1 000 000 Pa

Conversions

Pressure and stress [$ML^{-1}T^{-2}$]:

1 lbf in^{-2} = 6.8948 kPa
1 kPa = 0.1450 lbf in^{-2}

1 lbf ft^{-2} = 0.0479 kPa
1 kPa = 20.8768 lbf ft^{-2}

1 UK tonf ft^{-2} = 0.1073 MN m^{-2}
1 MN m^{-2} = 9.3197 UK tonf ft^{-2}

1 US tonf ft^{-2} = 0.0958 MN m^{-2}
1 MN m^{-2} = 10.4384 US tonf ft^{-2}

Other measures of pressure:

1 standard atmosphere = 101.325 kPa = 1.01325 bar
1 ft head of water = 2.989 kPa

Other quantities

From the standards and derived units described above, measurements of the following quantities may be obtained.

Elastic moduli [$ML^{-1}T^{-2}$], quoted in Pascals.
Coef. volume compressibility [L^2/MLT^{-2}], quoted in m^2 $Newton^{-1}$.
Coef. consolidation [L^2T^{-1}], quoted in m^2 s^{-1}.
Coef. permeability [LT^{-1}], quoted in m s^{-1}.

References

British Standards Institution (1964). *The International System (SI) of Units*. British Standard: BS 3763. London.

American Society for Testing and Materials (1982). *Metric Practice Guide*. Compilation of ASTM Standards. E380-82. ASTM, Philadelphia.

Addenda

Typical unit weights for various dry unweathered rocks and wet and dry soils (kNm^{-3}): min or max = unlikely to be less or more than; mean = most frequently; N = no value can be recommended; — = unknown.

ROCK TYPE	min	common range	mean	max
Granite	23.7	25.1 to 27.6	26.8	29.2
Syenite	24.6	25.1 to 26.7	25.8	28.3
Diorite	26.7	26.7 to 28.3	26.8	29.8
Gabbro	26.7	28.9 to 29.8	29.5	30.1
Porphyry	23.1	23.4 to N	24.8	28.1
Dolerite	22.5	25.4 to 27.6	27.1	28.1
Rhyolite	19.5	N	N	26.8
Andesite	19.5	21.7 to 22.5	N	26.8
Basalt	22.0	25.1 to 27.2	27.1	27.9
* " (vesicular)	N	13.5 to 18.9	N	N
*Agglomerate	N	17.9 to 25.3	N	N
*Tuff	11.9	13.3 to 14.9	N	27.5
*Sandstone	14.9	18.7 to 26.8	20.9 to 23.7	29.8
*Limestone	11.9	19.5 to 26.8	N	28.4
*Oolitic Lst.	17.9	20.1 to 23.6	N	26.5
*Chalk	14.6	18.2 to 22.5	19.5	26.1
*Shale	17.6	18.0 to 22.6	19.2	26.5
*Laterite	6.4	9.0 to 16.6	N	20.9
Gneiss	23.8	25.4 to 26.8	N	27.6
Schist	18.2	26.7 to 27.6	N	27.9
Slate	26.5	27.3 to 28.3	N	N
Marble	23.9	26.7 to 27.6	26.8	28.4

SOIL TYPE	min†	common range	mean	max†
Quartz sand (dry)	11.9	13.0 to 19.7	N	21.7
" " (sat.)	17.3	17.9 to 21.6	N	22.1
Coral sand (dry)	8.2	10.8 to 12.8	N	14.3
" " (sat.)	15.0	16.7 to 17.9	N	18.9
Silt (dry)	11.3	12.6 to 18.5	16.9	20.9
" (sat.)	16.5	17.4 to 19.8	18.2	21.5
Sandy gravel (dry)	15.4	17.7 to 19.3	N	21.3
" " (sat.)	19.2	20.4 to 21.6	N	22.4
Soft clay(sat.)	12.6	13.9 to 18.4	16.5	19.6
Stiff clay (sat.)	13.2	17.4 to 21.6	20.1	23.5
Peat	7.8	8.3 to 10.7	9.3 to 10.2	12.7
Loess (dry)	13.7	14.5 to 16.4	15.5	17.7
" (sat.)	15.7	16.7 to 19.2	17,2 to 18.6	21.6

* These rocks can readily accept water and their weight will increase in accordance with their water content.
† Broadly correlates with loosest and densest state of compaction most likely to be expected.

References

(References added for reprints are given on p. 315.)

Aas, G. (1981) Stability of natural slopes in quick-clays. *Norwegian Geotechnical Institute Bulletin*, **135.**

Ackerman, W.C., White, G.F. and Worthington, E.B. (1973). *Man-made Lakes; their Problems and Environmental Effects*. Geophysical Monograph 17. American Geophysical Union. Washington D.C.

Adams, J.W. (1978). Lessons learned at Eisenhower Memorial Tunnel. *Tunnels and Tunnelling*, **10:** 4, 20-3.

Ager, D.V. (1975). *Introducing Geology*. Faber and Faber, London.

Ager, D.V. (1975). The geological evolution of Europe. *Proceedings of the Geologists' Association*, **86,** 127-54.

Ailleret, M.P. (1951). Estimation of leakage from an impounding reservoir before construction. *Journal of the Institution of Water Engineers*, **5,** 98-108. (See also *Journal of the Institution of Water Engineers* (1958), **12,** 144. Note after impounding of reservoir.)

Ambraseys, N. and Hendron, A. (1968). Dynamic behaviour of rock masses. In: Stagg, K.G. and Zienkiewicz, O.C. (Eds). *Rock Mechanics and Engineering Practice*. J. Wiley and Sons, London.

American Society for Testing and Materials. (1969) *Determination of the* in-situ *Modulus of Deformation of Rock*. Special Technical Publication (STP) 477. A.S.T.M. Philadelphia.

American Society for Testing and Materials. (1970). *The Sampling of Soil and Rock*. Special Technical Publication (STP) 483. A.S.T.M. Philadelphia.

American Society for Testing Materials. (1981). *Acoustic Emissions in Geotechnical Engineering Practice*. Special Technical Publication (STP) 750. A.S.T.M. Philadelphia.

American Society for Testing and Materials. (1981). *Water for Sub-surface Injection*. Special Technical Publication (STP) 735. A.S.T.M. Philadelphia.

American Society for Testing and Materials. (1981). *Permeability and Groundwater Contaminant Transport*. Special Technical Publication (STP) 746. A.S.T.M. Philadelphia.

American Society for Testing and Materials. (1982). *Metric Practice Guide*. Compilation of A.S.T.M. Standards. E380-82. A.S.T.M. Philadelphia.

American Society for Testing and Materials (Annual). (1984). Published annually with Section devoted to Construction (Section 4 for 1984). A.S.T.M. Philadelphia.

American Society of Civil Engineers. (1929). Essential facts concerning the failure of the St. Francis Dam. Report of Committee of Board of Direction. *American Society of Civil Engineers Proceedings*, **55,** 2147-63; and subsequent discussions in **55,** and **56.**

American Society of Civil Engineers. (1972). Report of Task Committee on sub-surface investigation of foundations of buildings. *Proceedings American Society of Civil Engineers. Journal of the Soil Mechanics and Foundations Division*, **98,** 481-90, 557-78, 749-66, 771-86.

American Society of Civil Engineers (1982). *Grouting in Geotechnical Engineering*. (Ed.) W.H. Baker. Special Publication of the American Society of Civil Engineers.

Anderson, E.M. (1951). *Dynamics of Faulting* (2nd Edition). Oliver and Boyd, Edinburgh.

Anderson, J.G.C., Arthur, J. and Powell, D.B. (1977). The engineering geology of the Dinorwic underground complex and its approach tunnels. *Proceedings of British Geotechnical Society Conference on Rock Engineering, Newcastle-upon-Tyne*, 491-510.

Andric, M., Roberts, G.T. and Tarvydas, R.K. (1976). Engineering geology of the Gordon Dam, S.W. Tasmania. *Quarterly Journal of Engineering Geology*, **9,** 1-24.

Anon. (1967). *Report of Tribunal appointed to inquire into the Aberfan disaster of October 21, 1966*. H.M. Stationery Office, London.

Anon. (1969). *Technical Reports submitted to the Tribunal appointed to inquire into the Aberfan disaster of October 21, 1966*. (Items 1-8). H.M. Stationery Office, London.

Anon. (1980). *Evaporite Deposits. Illustration and Interpretation of some Environmental Sequences*. Chambre syndicale de la recherche et de la production du petrole et du gaz naturel, Paris.

Anttikoski, U.V. and Saraste, A.E. (1977). The effect of horizontal stress in the rock on the construction of the Salmisaari oil caverns. In: Bergman, M. (Ed.). *Storage in Excavated Rock Caverns* (Pergamon Press, Oxford), **3,** 593-8.

Antevs, E. (1929). Maps of the Pleistocene glaciations. *Bulletin of the Geological Society of America*, **40,** 631-720.

Ashby, J.P. (1971). *Sliding and Toppling Modes of Failure in Models and Jointed Rock Slopes*. M.Sc. Thesis, Imperial College, University of London. (See also Goodman, R.E. and Bray, J.W., 1976.)

Bagnold, R.A. (1940). Beach formation by waves: some model experiments in a wave tank. *Journal of the Institution of Civil Engineers*, **15,** 27–52.

Bagnold, R.A. (1941). *The Physics of Blown Sand and Desert Dunes*. Methuen, London.

Bainbridge, C.G. (1964). The Stithians Reservoir scheme in S.W. Cornwall. *Water and Water Engineering*, **68,** 309–14.

Baker, B.H., Mohr, P.A. and Williams, L.A. (1972). Geology of the eastern rift system of Africa. *Geological Society of America, Special Paper* 136.

Balk, R. (1937). Structural Behaviour of igneous rocks. *Geological Society of America Memoir*, **5.**

Bara, J.P. and Hill, R.R. (1967). Foundation rebound at the Dos Amingos pumping plant. Proceedings of the American Society of Civil Engineers. *Journal of the Soil Mechanics and Foundations Division*, **93,** SM5, 153–68.

Barden, K. and Sides, G. (1970). The influence of weathering on the microstructure of Keuper Marl. *Quarterly Journal of Engineering Geology*, **3,** 259–60.

Barnes, J.W. (1981). *Basic Geological Mapping*. Geological Society of London Handbook, The Open University Press, Milton Keynes, UK, and Halsted Press, J. Wiley, New York.

Barton, N. (1976). Recent experiences with the Q-system of tunnel support design. *Proceedings of the Symposium on Exploration for Rock Engineering, Johannesburg*, **1,** 107–17.

Barton, N. (1982). Shear strength investigations for surface mining. In: *Third International Conference on Stability on Surface Mining (June 1981)*. Vancouver, Canada. Society of Mining Engineers and American Institute of Mining, Metallurgy and Petroleum Engineers Inc., New York.

Barton, N. (1983). Application of Q-system and index tests to estimate shear strength and deformability of rock masses. Proceedings of the International Symposium on Engineering Geology and Underground Construction. *International Association of Engineering Geology*, Lisbon, **2,** 11.51–11.70.

Barton, N. and Choubey, V. (1977). The shear strength of rock joints in theory and practice. *Rock Mechanics*, **10,** 1–54.

Barton, N. and Kjaernsli, B. (1981). Shear strength of rock fill. *Proceedings of the American Society of Civil Engineers' Journal of the Geotechnical Engineering Division*, **GT7,** 873–91.

Barton, N., Lien, R. and Lunde J. (1974). Engineering classification of rock masses for the design of tunnel support. *Rock Mechanics*, **6,** 189–236.

Bateman, A.M. (1950). *Economic Mineral Deposits* (2nd Edition). J. Wiley and Son, New York.

Bates, R.L. (1969). *Geology of the Industrial Rocks and Minerals*. Dover, New York.

Battey, M.H. (1972). *Mineralogy for Students*. Oliver and Boyd, Edinburgh.

Beaumont, T.E. (1979). Remote sensing for location and mapping of engineering construction materials in developing countries. *Quarterly Journal of Engineering Geology*, **12,** 147–58.

Begemann, H.K.S. (1966). *A New Approach for Taking a Continuous Soil Sample*. Laboratorium voor Grondwechanicia, Paper 4.

Bell, F.G. (Ed.). (1975) *Methods of Treatment of Unstable Ground*. Newnes-Butterworths, London.

Bell, F.G. (Ed.). (1978). *Foundation Engineering in Difficult Ground*. Newnes-Butterworths, London and Boston.

Benson, W.N. (1946). Landslides and their relation to engineering in the Dunedin district New Zealand. *Economic Geology*, **41,** 328–47.

Bergman, M. (Ed.) (1978). Rockstore '77. *Storage in Excavated Rock Caverns*. Proceedings 1st International Symposium, Stockholm, 1977. 3 vols. Pergamon Press, Oxford.

Berkey, C.P. (1935). Foundation conditions for Grand Coulee and Bonneville Projects. *Civil Engineering (New York)*, **5,** No. 2, 67–71.

Bernaix, J. (1969). New laboratory methods for studying the mechanical properties of rocks. *International Journal of Rock Mechanics and Mining Sciences*, **6,** 43–90.

Berry, P.L. (1983). Application of consolidation theory for peat to the design of a reclamation scheme by preloading. *Quarterly Journal of Engineering Geology*, **16,** 103–12.

Best, M.G. (1982). *Igneous and Metamorphic Petrology*. W.H. Freeman and Co., San Francisco.

Best, R. and Fookes, P.G. (1970). Some geotechnical and sedimentary aspects of ball-clays from Devon. *Quarterly Journal of Engineering Geology*, **3,** 207–39.

Bieniawski, Z.T. (1969). *In-situ* large scale testing of coal. *Proceedings of the Conference on* in-situ *Investigations of Soils and Rocks*, 1969. British Geotechnical Society, London.

Bieniawski, Z.T. (1974). Geomechanics classification of rock masses and its application to tunnelling. *Proceedings of the Third International Congress of Rock Mechanics*. International Society of Rock Mechanics, Denver, **11A,** 27–32.

Bieniawski, Z.T. (1976). Rock mass classification in rock engineering. *Proceedings of the Symposium on Exploration for Rock Engineering, Johannesburg*, **1,** 97–106.

Bieniawski, Z.T. (1978). Determining rock mass deformability. *International Journal of Rock Mechanics and Mining Sciences, and Geomechanics Abstracts*, **15,** 237–47.

Bieniawski, Z.T. (1983). The Geomechanics Classification (RMR System) in design applications to underground excavations. Proceedings of the International Symposium on Engineering Geology and Underground Construction. *International Association of Engineering Geology, Lisbon*, **2,** 11.33–11.47.

Billings, M.P. (1925). On the mechanics of dyke emplacement. *Journal of Geology*, **33,** 140–50.

Binnie, G.M. (1967). The Mangla Dam project. *Proc*

eedings of the Institution of Civil Engineers, **36,** 213–17.

Bird, J.M. and Dewey, J.F. (1970). Lithosphere plate continental margin tectonics and the evolution of the Appalachian orogen. *Geological Society of America, Bulletin*, **81,** 1031–59.

Bishop, A.W. (1948). A new sampling tool for use in cohesionless sands below ground-water level. *Geotechnique*, **1,** 125–31.

Bishop A.W. (1973). The stability of tips and spoil heaps. *Quarterley Journal of Engineering Geology*, **6,** 335–77.

Bishop, A.W. and Al-Dhahir, Z. (1969). Some comparisons between laboratory tests, *in-situ* tests, and full-scale performance, with special reference to permeability and coefficient of consolidation. In: *Conference on* in-situ *Investigations in Soils and Rocks*. British Geotechnical Society, London, 1969.

Bishop, A.W., Green, G.E., Garga, V.K., Andresen, A. and Brown, J.D. (1971). A new ring-shear apparatus and its application to the measurement of residual strength. *Geotechnique*, **21,** 273–328.

Bishop, A.W. and Henkel D.J. (1962). *The Measurement of Soil Properties in the Triaxial Test*. (2nd Edition). Edward Arnold, London.

Bishop, A.W., Kennard, M.F. and Vaughan, P.R. (1963). The development of uplift pressures downstream of a grouted cut-off during the impounding of Selset reservoir. *Grouts and Drilling Muds in Engineering Practice*. Butterworth, London.

Bjerrum, L. and Jorstad, F.A. (1968). Stability of rock slopes in Norway. *Norwegian Geotechnical Institute Publication*, **79.**

Bjerrum, L. (1967). Mechanism of progressive failure in slopes of overconsolidated plastic clay and clay shales. *Proceedings of the American Society of Civil Engineers, Journal of the Soil Mechanics and Foundations Division*, **93,** 1–49.

Blyth, F.G.H. (1954). The southern margin of the Cairnsmore of Fleet granite at the Clints of Dromore, Galloway. *Proceedings Geologists' Association*, **65,** 224–50.

Blyth, F.G.H. (1976). *Geological Maps and their Interpretation* (2nd Edition). Edward Arnold, London.

Bott, M.H.P. (1982). *The Interior of the Earth* (2nd Edition). Edward Arnold, London and Elsevier, New York.

Boulton, G.S., Jones, A.S., Clayton, K.M. and Kenning, M.J. (1977). A British ice-sheet model and patterns of glacial erosion and deposition in Britain. In: Shotton, F.W. (Ed.) *British Quarternary Studies: Recent Advances*. Oxford University Press.

Boulton, G.S. (1980). Classification of till. *Quarternary Newsletter*, No. **31,** Quarternary Research Association, Birkbeck College, London.

Bott, M.H.P. (1982). *The Interior of the Earth*. Edward Arnold, London.

Boulton, G.S. and Paul, M.A. (1976). The influence of genetic processes on some geotechnical properties of glacial tills. *Quarterly Journal of Engineering Geology*, **9,** 159–94.

Brawner, C.O. (Ed.). (1982). *Third International Conference on Stability in Surface Mining (1–3 June, 1981)*. Vancouver, Canada. Society of Mining Engineers and American Institute of Mining, Metallurgical and Petroleum Engineers Inc., New York.

Brekke, T. and Selmer-Olsen, R. (1965). Stability problems in underground construction caused by montmorillonite carrying joints and faults. *Engineering Geology*, **1,** 3–19.

Bridges, E.M. (1970). *World Soils*. Cambridge University Press, London.

British Geotechnical Society. (1973). *Field Instrumentation*. Proceedings of Symposium, London, 1973. Newnes-Butterworths, London.

British Standards Institution (1964). *The International (SI) System of Units*. British Standard: BS 3763. B.S.I. London.

British Standards Institution (1972). *Foundations*. Code of Practice: CP 2004: 1972. B.S.I. London.

British Standards Institution (1973). *Aggregates from Natural Sources for Concrete*. British Standard: BS 822: 1973. B.S.I. London.

British Standards Institution. (1974). *Core Drilling Equipment*. Code of Practice: B.S. 4019. B.S.I. London.

British Standards Institution (1975). *Methods for Sampling and Testing of Mineral Aggregates, Sands and Fillers*. British Standard: BS 812: 1975. B.S.I. London.

British Standards Institution (1975). *Methods of Testing for Soil for Civil Engineering Purposes*. Code of Practice: BS 1377. B.S.I. London.

British Standards Institution (1981). *Site Investigations*. Code of Practice: BS 5930. B.S.I. London.

British Standards Institution. (1981). *Earthworks*. Code of Practice: BS 6031: B.S.I. London.

British Standards Institution. (1982). *Recommendations for Ground Anchorages*. (Draft DD81): full report in preparation. B.S.I. London.

Brock, E. and Franklin, J.A. (1972). The point load strength test. *International Journal of Rock Mechanics and Mining Sciences*, **9,** 669–97.

Broms, B.B. (1980). Soil sampling in Europe: state-of-the-art. *American Society of Civil Engineers, Journal of the Geotechnical Engineering Division*, **106,** GT1; 65–98.

Brioli, L. (1967). New knowledges of the geomorphology of the Vajont slide slip surface. *Rock Mechanics and Engineering Geology*, **5,** 38–88.

Brink, A.B.A., Partridge, T.C. and Williams, A.A.B. (1982). *Soil Survey for Engineering*. Oxford University Press, London and New York.

Brown, E.T. (1981) *Rock Characterization, Testing and Monitoring*. International Society of Rock Mechanics (ISRM) suggested methods. Pergamon Press, London.

Brown, E.T., Bray, J.W., Ladanyi, B. and Hoek, E. (1983). Ground response curve for rock tunnels. *Proceedings of American Society of Civil Engineers. Journal of Geotechnical Engineering Division*, **109,** 15–39.

Brown, R.J.E. (1970). *Permafrost in Canada, its Influence on Northern Development*. University of Toronto Press, Toronto.

Brune, G. (1965). Anhydrite and gypsum problems in engineering geology. *Bulletin Association of Engineering Geologists*, **2,** 26–33.

Bullard, E.C., Everett, J.E. and Smith, A.G. (1965). The fit of the continents around the Atlantic. *Philosophical Transactions of the Royal Society*, **A258,** 41–51.

Burland, J.B. and Hancock, R.J.R. (1977). Underground car park at the House of Commons, London. Geotechnical aspects. *The Structural Engineer*, **55,** 87–100.

Bussell, M.A., Pitcher, W.S. and Wilson, P.A. (1976). Ring-complexes of the Peruvian coastal batholith. *Canadian Journal of Earth Sciences*, **13,** 1020–30.

Cambefort, H. (1977). The principles and applications of grouting. *Quarterly Journal of Engineering Geology*, **10,** 57–95.

Carder, D.S. (1945). Seismic investigations in the Boulder Dam area, 1940–44, and the influence of reservoir loading on local earthquake activity. *Bulletin Seismological Society of America*, **35,** 175–92.

Carey, S.W. (1976). *The Expanding Earth*. Elsevier Scientific Publishing Company, Amsterdam.

Casagrande, A. (1948). Classification and identification of soils. *Transactions American Society of Civil Engineers*, **113,** 901–30.

Casagrande, A. (1971). The liquefaction phenomena. *Geotechnique*, **21,** 197–202.

Castle, R., Yerkes, T. and Yould, T. (1973). Ground rupture in the Baldwin Hills – an alternative explanation. *Bulletin of the Association of Engineering Geologists*, **10,** 21–49.

Cedegren, H.R. (1967). *Seepage, Drainage and Flow-Nets*. J. Wiley and Sons, New York.

Chamberlain, E.J. and Gow, A.J. (1978). Effect of freezing and thawing on the permeability and structure of soils. Proceedings of the International Symposium on Ground-Freezing. Ruhr University, Bochum, 31–44. (Published in *Engineering Geology*, **13,** 73–92. (1979).)

Chandler, R.J. (1969). The effect of weathering on the shear strength properties of Keuper Marl. *Geotechnique*, **19,** 321–34.

Charles, J.A. and Watts, K.S. (1982). A field study of the use of the dynamic consolidation ground treatment technique on soft alluvial soil. *Ground Engineering*, **15,** (5), 17–25.

Chee, S.P. and Sweetman, A.P. (1972). Sedimentation characteristics of gorge-type reservoirs. *Water Resources Bulletin*, **8,** 881–6.

Chen, F.H. (1975). *Foundations on Expansive Soils*. Elsevier Scientific Publishing Company, Amsterdam and New York.

Clark, M.M., Grantz, A. and Rubin, M. (1972). Holocene activity of the Coyote Creek Fault as recorded in sediments of Lake Cahuilla. *United States Geological Survey Professional Paper 787*. U.S. Department of the Interior.

Clarke, G. (Ed.). (1970). *Dynamic Rock Mechanics*. Proceedings of the 12th Symposium on Rock Mechanics, University of Missouri. American Institute of Mining Engineers, New York.

Clayton, C.R.I., Simons, N.E. and Mathews, M.C. (1982). *Site Investigation. A Handbook for Engineers*. Granada Publishing, London.

Clifford, T.N. (1971). Location of mineral deposits. In: Gass, I.G., Smith, P.J. and Wilson, R.C.L. (Eds.). *Understanding the Earth*. Published for the Open University Press by the Artemis Press, Sussex.

Cloke, P.L. (1966). The geochemical application of Eh-pH diagrams. *Journal of Geological Education*, **4,** 140–8.

Close, U. and McCormick, E. (1922). Where the mountains walked. *National Geographic Magazine*, **41,** 445–64.

Coates, D.F. (1964). Some cases of residual stress effects in engineering work. In: Judd, W. (Ed.). *State of Stress in the Earth's Crust*: 679–88. Proceedings of the International Conference in Santa Monica, California 1963. Elsevier Publishing Company, New York.

Cocks, L.R.M. and Fortey, R.A. (1982) Faunal evidence for oceanic separations in the Palaeozoic of Britain. *Journal of the Geological Society*, **139,** 467–80.

Cole, K.W. and Stroud, M.A. (1971). Rock socket piles at Coventry Point, Market Way, Coventry. In: *Piles in Weak Rock*. Institution of Civil Engineers, London.

Cook, R.U., Brunsden, D., Doornkamp, J.C. and Jones, D.K.C. (1982). *Urban Geomorphology in Dry Lands*. Oxford University Press, London and New York, on behalf of the United Nations University.

Cook, R.U. and Warren, A. (1973). *Geomorphology in Deserts*. B.T. Batsford, London.

Cooper, H.H., Kohout, F.A., Henry, H.R. and Glover, R.E. (1964). Sea water in coastal acquifers. *United States Geological Survey, Water Supply Paper*, 1613-C.

Costa, J.E. and Baker, V.R. (1981). *Surficial Geology*. John Wiley and Sons, New York.

Cousens, R.R.M. and Garrett, W.S. (1969). The flooding at the West Driefontein Mine. *9th Commonwealth Mining and Metallurgical Congress, London*, **1,** 931–72.

Cox, K. (1971). Minerals and Rocks. In: Gass, I.G., Smith, P.J., and Wilson, R.C.L. (Eds.). *Understanding the Earth*. Published for the Open University Press by the Artemis Press, Sussex.

Crawford, C. and Eden J. (1967). Stability of natural slopes in sensitive clay. *Proceedings of the American Society of Civil Engineers. Journal of Soil Mechanics and Foundations Division*, **93,** 419–37.

Cruden, D.M. and Krahn, J. (1978). Frank rockslide, Alberta, Canada. In: Voight, B. (Ed.). *Rockslides and Avalanches*. Vol. 1. Elsevier Scientific Publishing Company, New York.

Dakombe, R.V. and Gardiner, V. (1982). *Geomorphol-*

ogical Field Manual. George Allen and Unwin, London and Boston.

Daly, D., Lloyd, J.W., Misstear, B.D.R. and Daly, E.D. (1980). Fault control of ground-water flow and hydrochemistry in the aquifer system of the Castlecomer Plateau, Ireland. *Quarterly Journal of Engineering Geology*, **13,** 167–76.

Daly, R.A., Miller, W.G. and Rice, G.S. (1912). Report of the Commission appointed to investigate Turtle Mountain, Frank, Alberta. *Canadian Geological Survey, Memoir*, **27.**

Davies, D.G. (1940). The Haweswater Dam. *Transactions of the Liverpool Engineering Society*, **61,** 79–104.

Davies, T.A. and Gorsline, D.S. (1976). Oceanic sediments and sedimentary processes. In: Riley, J.R. (Ed.). *Chemical Oceanography*. R. Chester Academic Press, London.

Davis, S.N. and de Wiest, R.J.M. (1966). *Hydrogeology*. J. Wiley and Sons, New York.

Deacon, G.F. (1896). The Vyrnwy Works for the water supply of Liverpool. *Journal of the Institution of Civil Engineers, London*, **126,** 24–125.

Dearman, W.R. and Fookes, P.G.F. (1974). Engineering geological mapping for civil engineering practice in the United Kingdom. *Quarterly Journal of Engineering Geology*, **7,** 223–56.

Deere, D.U. (1968). Geological considerations. In: Stagg, K.G. and Zienkiewicz, O.C. (Eds). *Rock Mechanics and Engineering Practice*. J. Wiley and Sons, London.

Deere, D.U., Peck, R.B., Monsees, J.E. and Schmidt, B. (1969). Design of tunnel liners and support systems. Final Report. Office of High Speed Ground Transportation. *United States Department of Transportation Contract, No.* **3-0152.**

Dennis, J.A.N. (1978). Offshore structures. *Quarterly Journal of Engineering Geology*, **11,** 79–80.

Dewey, J.F. (1982). Plate tectonics and the evolution of the British Isles. *Journal of the Geological Society*, **139,** 371–414.

Dewey, J.F. and Pankhurst, R.J. (1970). The evolution of the Scottish Caledonides in relation to their isotopic age pattern. *Transactions of the Royal Society of Edinburgh*, **68,** 361–87.

Dixon, C.J. (1979). *Atlas of Economic Mineral Deposits*. Chapman and Hall, London.

Dixon, H.H. and Robertson, R.H.S. (1970). Some engineering experiences in tropical soils. *Quarterly Journal of Engineering Geology*, **3,** 137–50.

Domenico, P.A. (1972). *Concepts and Models in Groundwater Hydrology*. McGraw Hill, New York.

Donath, F.A. (1961). Experimental study of shear failure in anisotropic rocks. *Bulletin of the Geological Society of America*, **72,** 985–90.

Doornkamp, J.C., Brunsden, D., Jones, D.K.C., Cooke, R.U. and Bush, P.R. (1979). Rapid geomorphological assessments for engineering. *Quarterly Journal of Engineering Geology*, **12,** 189–204.

Drever, J.I. (1982). *The Geochemistry of Natural Water*. Prentice Hall, New Jersey.

Drouhin, G., Gautier, M. and Dervieux, F. (1948). Slide and subsidence of the hills of St. Raphael: Telemy. *Proceedings 2nd International Conference on Soil Mechanics and Foundation Engineering, Rotterdam*, **5,** 104–6.

Dumbleton, M.J. and West, G. (1976). Preliminary sources of information for site investigation in Britain. *Road Research Laboratory, Report LR 403*. Department of the Environment. H.M. Stationery Office, London.

Dunning, F. (1970). *Geophysical Exploration*. Institute of Geological Sciences, H.M. Stationery Office, London.

Dury, G.H. (1951). A 400-foot beach in South-eastern Warwickshire. *Proceedings of the Geologists' Association*, **62,** 167–73.

Edmunds, W.M. (1973). Trace element variations across an oxidation-reduction barrier in a limestone aquifer. *Proceedings of Symposium on Hydrogeochemistry and Biogeochemistry, Tokyo, (1970)*, 500–26. Clarke Co., Washington, D.C.

Edmunds, W.M. (1977). Ground-water geochemistry: controls and processes. *Proceedings of Symposium on Ground-water Quality. Water Research Centre, Reading*, (1976), 115–47.

Edmunds, W.M. (1981). Hydrochemical investigations. In: Lloyd, J.W. (Ed.). *Case Studies in Groundwater Resources Evaluation*, 87–112. Clarendon Press, Oxford.

Embleton, C. and Thornes, J. (Eds.). (1979). *Process in Geomorphology*. Edward Arnold, London.

Evans, A.M. (1980). *An Introduction to Ore Geology*. Blackwell Scientific, Oxford.

Farmer, I.W. (1983). *Engineering Behaviour of Rocks*. Chapman and Hall, London, New York.

Faul, H. (1977). A history of geologic time. *American Science*, **66,** 159–65.

Fernandez, L.M. and van der Heever, P.K. (1982). Ground movement and damage accompanying a large seismic event in the Klerksdorp district. *Proceedings of 1st International Symposium on Seismicity in Mines, Johannesburg, 1982.*

Fetter, C.W. (1980). *Applied Hydrogeology*. Charles E. Merrill Publishing Company, Ohio.

Fleming, G. (1969). Design curves for suspended load estimation. *Proceedings of Institution of Civil Engineers*, **43,** 1–9.

Flint, R.R. (1957). *Glacial and Pleistocene Geology*. J. Wiley and Sons, New York.

Flint, R.F. (1971). *Glacial and Quarternary Geology*. J. Wiley and Sons, New York.

Fiske, R.S., Hopson, C.A. and Waters, A.C. (1963). Geology of Mount Rainier National Park, Washington. *United States Geological Survey, Professional Paper*, **444.**

Francis, P. (1976). *Volcanoes*. Penguin Books, London and New York.

Freeze, R.A. and Cherry, J.A. (1979). *Groundwater*. Prentice-Hall Inc., New Jersey.

Freire, F.C.V. and Souza, R.J.B. (1979). Lining, support and instrumentation of the cavern for the Paulo Alfonso IV Power Station, Brazil. In: *Tunnelling '79*. Institution of Mining and Metallurgy, London, 182–92.

de Freitas, M.H. and Watters, R.J. (1973). Some field examples of toppling failure. *Geotechnique*, **23,** 495–514.

de Freitas, M.H. and Wolmarans, J.F. (1978). Dewatering and settlement in the Bank Compartment of the Far West Rand, South Africa. *Water and Mining in Underground Works, (SIAMOS) Granada* **1,** 619–35.

de Freitas, M.H. (Ed.). (1981). Mudrocks of the United Kingdom. *Quarterly Journal of Engineering Geology*, **14,** 241–372.

French, W.J. (1980). Reactions between aggregates and cement paste: an interpretation of the pessimum. *Quarterly Journal of Engineering Geology*, **13,** 231–47.

Fried, S.M. (Ed.). (1979). Radioactive waste in geologic storage. *American Chemical Society, Symposium Series* **100.** Washington D.C.

Frivik, P.E., Janbu, N., Saetersdal, R. and Finborud, L.I. (Eds). (1981). Proceedings of 2nd International Symposium on Ground Freezing. (1980), Trondheim, Norway. Published in *Engineering Geology*, **18,** Special Issue, parts 1–4.

Gass, I.G., Smith, P.J. and Wilson, R.C.L. (1971). *Understanding the Earth*. Artemis Press, Sussex and M.I.T. Press, Cambridge, Massachusetts.

Geological Society of London. (1970). The logging of rock cores for engineering purposes. *Quarterly Journal of Engineering Geology*, **3,** 1–24 and **10,** 45–52 (1977).

Geological Society of London. (1972). The preparation of maps and plans in terms of engineering geology. *Quarterly Journal of Engineering Geology*, **5,** 293–382.

Geological Society of London. (1977). The description of rock masses for engineering purposes. *Quarterly Journal of Engineering Geology*, **10,** 355–88.

Geological Society of London, (1982). Land surface evaluation for engineering purposes. *Quarterly Journal of Engineering Geology*, **15,** 265–316.

Geological Survey of Great Britain. (1961). British Regional Geology: Scotland. *The Tertiary Volcanic Districts* (3rd Edition). H.M. Stationery Office, London.

Gillen, C. (1982). *Metamorphic Geology*. George Allen and Unwin, London.

Gillot, J.E. (1968). *Clay in Engineering Geology*. Elsevier Publishing Company, London, New York.

Glossop, R. (1960). The invention and development of injection processes. *Geotechnique*, **11,** 91–100 and 225–279.

Golze, A.R. (Ed.). (1977). *Handbook of Dam Engineering*. Van-Norstrand Reinhold Company, New York.

Goodman, R.E. and Bray, J.W. (1976). Toppling of rock slopes. In: *Rock Engineering for Foundations and Slopes*. University of Colorado, Boulder, Colorado. **2,** 201–34. American Society of Civil Engineers.

Goodman, R.E. and Shi, G.H. (1982). Geology and rock slope stability: application of a 'Keyblock' concept for rock slopes. In: *3rd International Conference on Stability in Surface Mining (June 1981)*. Vancouver, Canada. Society of Mining Engineers and American Institute of Mining, Metallurgical and Petroleum Engineers, Inc., New York.

Goudge, M.F., Haw, V.A. and Hewitt, D.F. (Eds.). (1957). *The Geology of Canadian Industrial Mineral Deposits*. Proceedings of the 6th Commonwealth Mining and Metallurgy Congress, Montreal.

Goudie, A.S. (1970). Input and output considerations in estimating rates of chemical denudation. *Earth Science Journal*, **4,** 60–6.

Green, J. and Short, N.M. (1971). *Volcanic Landforms and Surface Features*. Springer-Verlag, New York.

Green, R.P. and Gallagher, J.M. (Eds.). (1980). *Future Coal Prospects: Country and Regional Assessments*. The second and final WOCOL report. World Coal Study. Ballinger Publishing Company (Harper and Row), Cambridge, Massachusetts.

Greensmith, J.T. (1978). *Petrology of the Sedimentary Rocks* (6th Edition), George Allen and Unwin, London.

Griffiths, D. and King, R. (1969). *Applied Geophysics for Engineers and Geologists*. Pergamon Press, Oxford.

Gregersen, P. (1981). The quick clay landslide in Rissa, Norway. *Norwegian Geotechnical Institute Bulletin*, **135.**

Gregory, K.J. and Walling, D.E. (1973). *Drainage Basin Form and Process*. Edward Arnold, London.

Grim, R.E. (1962). *Applied Clay Mineralogy*. McGraw Hill Book Co. New York and London.

Gustafson, L.B. and Hunt, J.P. (1975). The porphyry copper deposit at El Salvador, Chile. *Economic Geology*, **70,** 857–912.

Hagan, T.N. (1983). The influence of rock properties in the design, and results of, blasts in underground construction. Proceedings of the International Symposium on Engineering Geology and Underground Construction. *International Association of Engineering Geologists, Lisbon*, **1,** III. 57–66.

Haimson, B.C. (1978). The hydrofracturing stress measuring method and recent field results. *International Journal of Rock Mechanics and Mining Sciences, and Geomechanics Abstracts*, **15,** 167–78. (See also Haimson, B.C. (1981). Measuring rock stress for hydroprojects. *Water Power and Dam Construction (Oct)*. 37–41.)

Halbouty, M.T. (1967). *Salt Domes: Gulf Region, United States and Mexico*. Gulf Publishing Company, Houston, Texas.

Halcrow, W.T. (1930). The Lochaber Water Power Scheme. *Proceedings of the Institution of Civil Engineers, London*, **231,** 54–63.

Hall, A.L. (1932). The Bushveld igneous complex of the central Transvaal. *South African Geological Survey Memoir*, **28.**

Hall, C.E. and Carlson, J.W. (1965). Stabilization of soils subject to hydrocompaction. *Engineering Geology, Bulletin of the Association of Engineering Geologists*, **2,** 47-58.

Hamilton, D. and Meehan, R. (1971). Ground rupture in Baldwin Hills. *Science*, **172,** 333-44.

Hamilton, W. and Meyers, B. (1967). The nature of batholiths. *United States Geological Survey. Professional Paper* **554-C.**

Harrison, J.V. and Falcon, N.L. (1936). Gravity collapse structures and mountain ranges in south-western Iran. *Quarterly Journal of the Geological Society, London*, **92,** 91-102.

Harland, W.B., Cox, A.V., Llewellyn, P.G., Pickton, C.A.G., Smith, A.G. and Walters, R. (1982). *A Geological Time Scale*. Cambridge University Press, Cambridge.

Hawkes, I. and Mellor, M. (1970). Uniaxial testing in rock mechanics laboratories. *Engineering Geology*, **4,** 117-285.

Heath, R.C. and Trainer, F.W. (1968). *Introduction to Groundwater Hydrology*. J. Wiley and Sons, New York.

Heezen, B.C. and Menard, H.W. (1963). Topography of the deep sea floor. In: Hill, M.N. (Ed.). *The Sea*. J. Wiley and Sons, New York.

Heezen, B.C., Tharp, M. and Ewing, M. (1959). The floors of the oceans. *Geological Society of America Special Publications*, **65.**

Hem, J. (1970). Study and interpretation of the chemical characters of natural waters. *United States Geological Survey Water Supply Paper*, **1473.**

Henkel, D.J. (1966). The stability of slopes in the Siwalik rocks in India. *Proceedings of 1st International Congress of the International Society of Rock Mechanics, Lisbon*, **2,** 161-5.

Hobson, G.D. and Tiratsoon, E.N. (1981). *Introduction to Petroleum Geology* (2nd Edition). Scientific Press, Beaconsfield, England.

Hobst, L. and Zajic, J. (1983). Anchoring in rock and soil. *Developments in Geotechnical Engineering*, No. **33.** Elsevier Scientific Publishing Company, Amsterdam.

Hoek, E. (1981). Geotechnical design of large openings at depth. In: *Proceedings of the Rapid Excavation and Tunnelling Conference, California*, **2,** 1167-80. Society of Mining Engineers and American Institution of Mining, Metallurgical and Petroleum Engineers, New York.

Hoek, E. and Bray, J.W. (1977). *Rock Slope Engineering* (2nd Edition). Institution of Mining and Metallurgy, London.

Hoek, E. and Brown, E.T. (1980). *Underground Excavations in Rock*. Institution of Mining and Metallurgy, London.

Hollingworth, S.E., Taylor, J.H. and Kellaway, G.A. (1944). Large scale superficial structures in the Northamptonshire Ironstone field. *Quarterly Journal of the Geological Society of London*, **100,** 1-44.

Holmes, A. (1978). *Principles of Physical Geology* (3rd Edition). Nelson, London.

Horswill, P. and Horton, A. (1976). Cambering and valley bulging in the Gwash Valley at Empingham, Rutland (including Discussion). *Philosophical Transactions of the Royal Society* **A283,** 427-62.

Horton, R.E. (1933). Role of infiltration in the hydrological cycle. *Transactions of the American Geophysical Union*, **14,** 446-60.

Hosking, K.F.G. and Shrimpton, G.J. (Eds). (1964). *Present views of Some Aspects of the Geology of Cornwall and Devon*. 150th Anniversary Edition of the Royal Geological Society of Cornwall, Penzance.

Hoskins, E. (1966). An investigation of the flat jack method of measuring stress. *International Journal of Rock Mechanics and Mining Sciences and Geomechanics Abstracts*, **3,** 249-64.

Houlsby, A.C. (1976). Routine interpretation of the Lugeon water-test. *Quarterly Journal of Engineering Geology*, **9,** 303-13.

Housner, G.W. (1970). Seismic events at Koyna Dam. In: *Proceedings of 11th Symposium on Rock Mechanics*. American Institute of Mining, Metallurgy, and Petroleum Engineers, New York.

Hutchinson, J.N. (1968). Mass Movement. In: Fairbridge, R.W. (Ed.). *The Encyclopaedia of Geomorphology*. Reinhold Book Corporation, New York.

Hutchinson, J.N. (1969). A reconsideration of the coastal landslides at Folkestone Warren, Kent. *Geotechnique*, **19,** 6-38.

Hutchinson, J.N., Bromhead, E.N. and Lupini, J.F. (1980). Additional observations on the Folkestone Warren landslides. *Quarterly Journal of Engineering Geology, London*, **13,** 1-31.

Hutchinson, J.N. and Bhandari, R.K. (1971). Undrained loading, a fundamental mechanism of mudflows and other mass movements. *Geotechnique*, **21,** 353-58.

Hvorslev, M.J. (1948). *Sub-surface Exploration and Sampling of Soils for Civil-Engineering Purposes*. Waterways Experiment Station, Vicksberg, Mississippi. Reprinted by Engineering Foundation, United Engineering Centre, New York. 1962 and 1965. Sometimes classified under American Society of Civil Engineers. Soil Mechanics and Foundations Division Committee on Sampling and Testing, 1949.

Irving, E. (1979). Pole positions and continental drift since the Devonian. In: McEthinny, M.W. (Ed.). *The Earth, its Origin, Structure and Evolution*. Academic Press, New York.

Ingles, O.G. and Metcalf, J.B. (1972). *Soil Stabilization*. Butterworths, Sydney.

International Atomic Energy Agency. (1975). *The Oklo Phenomenon*. Proceedings of the Symposium of International Atomic Energy Agency, French Atomic Energy Commission, Government of the Republic of Gabon, Libreville, Gabon, I.A.E.A. Vienna.

International Atomic Energy Agency. (1981). *Underground Disposal of Radioactive Waste: Basic Guidance*. I.A.E.A., Vienna.

International Atomic Energy Agency. (1982). *Site Investigations for Repositories for Solid Radioactive Waste in Deep, Continental Geological Formations*. I.A.E.A., Vienna.

International Atomic Energy Agency. (1982). *Site Investigations for Repositories for Solid Radioactive Waste in Shallow Ground*. I.A.E.A., Vienna.

International Commission on Large Dams. (1973). *Lessons from Dam incidents*. I.C.O.L.D., Paris.

International Commission on Large Dams. (1979). *Deterioration Cases Collected and Their Preliminary Assessment*. Committee on the Deterioration of Dams and Reservoirs. Vol. 1. I.C.O.L.D., New Delhi, India.

International Society for Rock Mechanics. (1977). Suggested methods for monitoring rock movements using borehole extensenometers. *International Journal of Rock Mechanics and Mining Sciences, and Geomechanics Abstracts*, **15**, 305–17.

International Society for Rock Mecanics. (1977). Suggested methods for determining hardness and abrasiveness of rocks. *International Journal of Rock Mechanics and Mining Sciences, and Geomechanics Abstracts*, **15**, 89–97

International Society for Rock Mechanics. (1977). Suggested methods for determining water content, porosity, density, absorption and related properties and swelling and slake-durability index properties. *International Journal of Rock Mechanics and Mining Sciences, and Geomechanics Abstracts*, **16**, 141–56.

International Society for Rock Mechanics. (1978). Suggested methods for determining *in-situ* deformability of rock. *International Journal of Rock Mechanics and Mining Sciences, and Geomechanics Abstracts*, **16**, 195–214.

International Society for Rock Mechanics. (1980). Basic geotechnical description of rock masses. *International Journal of Rock Mechanics and Mining Sciences and Geomechanics Abstracts*, **18**, 85–110.

International Society for Rock Mechanics. (1981). Suggested methods for geophysical logging of bore holes. *International Journal of Rock Mechanics and Mining Sciences, and Geomechanics Abstracts*, **18**, 67–83.

Institution of Civil Engineers. (1964). *Grouts and Drilling Muds in Engineering Practice*. Butterworths, London.

Institution of Civil Engineers. (1976). *Ground Treatment by Deep Compaction*. Thomas Telford Ltd, London.

Institution of Civil Engineers. (1977). *Ground Subsidence*. Thomas Telford Ltd, London.

Institution of Civil Engineers. (1981). *Dams and Earthquakes*. Proceedings of Conference (1980). Thomas Telford Ltd., London.

Institution of Civil Engineers. (1982). *Vertical Drains*. Thomas Telford, London.

Institute of Geological Sciences. (1977). *The Story of the Earth*. H.M. Stationery Office, London.

Jacobson, R.R.E., MacLeod, W.N. and Black, R. (1958). Ring-complexes in the younger granite province of northern Nigeria. *Geological Society of London Memoir*, **1.**

Jacobsson, U. (1978). Storage for liquified gases in unlined refrigerated rock caverns. In: Rockstore 77. *Proceedings of the 1st International Symposium, Stockholm*, **2,** 449–58.

Jaeger, J.C. (1969). The stability of partly immersed fissured rock masses and the Vajont rock slide. *Civil Engineering and Public Works Review*, 1204–7.

Jaeger, J.C. (1979). *Rock Mechanics and Engineering*. Cambridge University Press, Cambridge.

Jaeger, J.C. and Cook, N.G.W. (1979). *Fundamentals of Rock Mechanics* (3rd Edition). Chapman and Hall, London.

James, A.N. and Kirkpatrick, I.M. (1980). Design of foundations of dams containing soluble rocks and soils. *Quarterly Journal of Engineering Geology*, **13,** 189–98.

Jenkyns, H.C. (1980). Tethys; past and present. *Proceedings of the Geologists' Association*, **91,** 107–18.

John M. (1981). Application of the new Austrian tunnelling method under various rock conditions. In: *Proceedings of the Rapid Excavation and Tunnelling Conference, San Francisco*, **1,** 409–26. American Institute of Mining, Metallurgical and Petroleum Engineers Inc., New York.

Johnston, G.H. (Ed.). (1981). Permafrost-Engineering Design and Construction. National Research Council, Canada. J. Wiley and Sons Ltd., Toronto.

Kallstenius, T. (1963). Studies on clay samples taken with the standard piston sampler. *Royal Swedish Institute Proceedings*, No. **21.**

Katz, D.L. and Tek, M.R. (1981). Overview of underground storage of natural gas. *Journal of Petroleum Technology*, **33,** 943–51.

Kellog, C.E. (1950). *Soil*. Scientific American, **183,** 1, 30–39.

Kennard, J. and Lee, J.J. (1947). Some features of the construction of Fernworthy Dam. *Journal of the Institution of Water Engineers*, **1,** 11–38.

Kennard, M.F. and Knill, J.L. (1969). Reservoirs on limestone, with particular reference to the Cow Green Scheme. *Journal of the Institution of Water Engineers*, **23,** 87–113.

Kennard, M.F. and Wakeling, R.T.M. (1979). Site investigations for reservoirs. *Journal of the Institution of Water Engineers and Scientists*, **33,** 363–76.

Kennard, M.F., Knill, J.L. and Vaughan, P.R. (1967). The geotechnical properties and behaviour of Carboniferous shale at the Balderhead Dam. *Quarterly Journal of Engineering Geology*, **1,** 3–23.

Kenney, T.C. (1964). Sea level movements and the geologic histories of post glacial marine soils at Boston, Ottawa, and Oslo. *Geotechnique*, **14,** 203–30.

Kenney, T.C. (1967). Stability of the Vajont valley slope. *Felsmechanik and Ingenieur-geologic*, **5**, 10–16.

Keys, W.S. and McCary, L.M. (1971). *Techniques of Water Resources Investigations of the United States Geological Survey*. Chapter E1. Application of borehole geophysics to water resources investigation. U.S. Government Printing Office, Washington.

Khan, S. and Alinaqui, S. (1970). Foundation treatment for underseepage control at Tarbela Dam project. *Transactions of the International Congress on Large Dams*, Montreal, **2**, Paper 60.

Kiersch, G.A. (1964). Vajont reservoir disaster. *Civil Engineering* (*March*), 32–9.

King, Cuchlaine, A.M. (1972). *Beaches and Coasts* (2nd Edition). Edward Arnold, London.

Kjellman, W., Kallstenius, T., and Wager, O. (1950). Soil sampler with metal foils. Device for taking undisturbed samples of very great length. *Proceedings of the Royal Swedish Geotechnical Institute*, No. **1.**

Klohn, E.J. (1980). Current tailings dam design and construction methods. *Mining Engineer*, **33**, 798–808.

Knill, J.L. (1970). The application of seismic methods in the prediction of grout take in rock. In: Proceedings of the Conference. In-situ *Investigations in Soils and Rocks*. British Geotechnical Society of London, 93–100.

Knill, J.L. (1974). The application of engineering geology to the construction of dams in the United Kingdom. In: *Colloque Geologie de L'ingenieur, Liege, 1974*, 113–147. (Centenaire de la Societe Geologique de Belgique.)

Knill, J.L. and Jones, K.S. (1965). The recording and interpretation of geological conditions in the foundations of the Roseires, Kariba, and Latiyan dams. *Geotechnique*, **15**, 94–124.

Kruseman, G. and de Ridder, N. (1970). *Analysis and Evaluation of Pumping Test Data, Bulletin* **11.** International Institute of Land Reclamation. The Netherlands.

Kuesel, T.R. and King, E.H. (1979). Marta's Peachtree Centre Station. Proceedings of the 1979 Rapid Excavation and Tunnelling Conference. In: Maevis, A.C. and Hustrulid, W.A. (Eds.). *American Institution of Mining Engineers, New York*, **2**, 1521–44.

Kvapil, R. (1965). Gravity flow of granular materials in hoppers and bins in mines. *International Journal of Rock Mechanics and Mining Sciences*, **2**, 35–41 and 277–304.

Lane, K.S. (1969). Engineering problems due to fluid pressure in rock. *Proceedings of the 11th Symposium on Rock Mechanics, Berkeley, California*, 501–40.

Lane, K.S. (1967). Stability of reservoir slopes. In: Fairhurst, C. (Ed.). *Failure and Breakage of Rock*. Proceedings of the 8th Symposium on Rock Mechanics, Minnesota.

Lane, R.G.T. (1963). The jetting and grouting of fissured quartzite at Kariba. In: *Grouts and Drilling Muds in Engineering Practice*. Butterworths, London.

Lane, R.G.T. (1971). Seismic activity at man made reservoirs. *Proceedings of the Institution of Engineers, London*, **50**, 15–24 (and see discussion, **51**).

Leary, E. (1983). *The Building Limestones of the British Isles*. Building Research Establishment. H.M. Stationery Office, London.

Lechler, S. (1971). Storage caverns in German salt domes. *Petroleum and Petrochemical International*, **11**, (12), 64–73 and 80c.

Legget, R.F. (1976). Glacial Till: an interdisciplinary study. *Royal Society of Canada, Special Publication*, **12.**

Legget, R.F. and Karrow, P.F. (1983). *Handbook of Geology in Civil Engineering*. McGraw-Hill Book Company, New York, London.

Leonards, G.A. (1962). *Foundation Engineering*. McGraw-Hill Book Company, New York.

Lindner, E. (1976). Swelling rock: a review. In: *Rock Engineering for Foundations and Slopes*. Proceedings of a Conference, University of Colorado, **1**, 141–81. American Society of Civil Engineers, New York.

Linton, D.L. (1955). The problem of tors. *Geographical Journal*, **121**, 470–86.

Little, A.L. (1975). Ground-water control by exclusion. In: Bell, F.G. (Ed.). *Methods of Treatment of Unstable Ground.* Newnes-Butterworths, London.

Little, A.L., Stewart, J.C. and Fookes, P.G. (1963). Bedrock grouting tests at Mangla Dam, West Pakistan. In: *Grouts and Drilling Muds in Engineering Practice*. Butterworths, London.

Littlejohn, G.S. (1980). Design estimation of the ultimate load-holding capacity of ground anchors. *Ground Engineering* (*Nov.*), 25–39.

Littlejohn, G.S. and Bruce, D.A. (1977). *Rock Anchors: State of the Art*. Foundation Publications Ltd., Brentwood, Essex.

Litvinov, I.M., Rzhanitzn, B.A. and Bezruk, V.M. (1961). Stabilisation of soil for constructional purposes. *Proceedings of the 5th International Congress on Soil Mechanics and Foundation Engineering*, **2**, 775–80.

Lloyd, J.W. (Ed.). (1981). *Case-Studies in Groundwater Resources Evaluation*. Clarendon Press, Oxford.

Londe, P. (1967). Discussion of Theme 6. *Proceedings of the 1st International Congress of the International Society of Rock Mechanics, Lisbon*, **3**, 449–53.

Londe, P. (1973). Water seepage in rock slopes. *Quarterly Journal of Engineering Geology*, **6**, 75–92.

Lovelock, P. (1970). The laboratory measurement of soil and rock permeability. *Water Supply Papers Technical Communication*, No. **2.** Institute of Geological Sciences. H.M. Stationery Office, London.

Marsland, A. (1971). Large *in-situ* tests to measure the properties of stiff, fissured clays. *Proceedings of the 1st Australia–New Zealand Conference on Geomechanics, Melbourne*, **1**, 180–9.

Marsland, A. (1975). *In-situ* and laboratory tests on glacial clays at Redcar. *Proceedings of a Symposium on the Behaviour of Glacial Materials.* Midlands Soil Mechanics and Foundation Society.

Matheson, D.S. and Thomson, S. (1973). Geological implications of valley rebound. *Canadian Journal of Earth Science*, **10,** 961–78.

Mathews, W.H. and McTaggart, K.C. (1978). Hope rockslides, British Columbia, Canada. In: Voight, B. (Ed.). *Rockslides and Avalanches*, Vol. 1. Elsevier Scientific Publishing Company, New York.

Matsuo, S. (1957). A study of the effect of cation exchange on the stability of slopes. *Proceedings of the 4th International Conference on Soil Mechanics and Foundation Engineering, London*, **2,** 330–3.

Mayer, C., Shea, E.P., Goddard, C.C. *et al.* (1868). Ore deposits at Butte, Montana. In Ridge, J.D. (Ed.). *Ore Deposits in the United States, 1933–67.* American Institute of Mining and Petroleum Engineering Inc., New York.

MacKenzie, W.S. and Guilford, C. (1980). *Atlas of Rock-forming Minerals in Thin Section.* Longman, London.

McConnell, R.G. and Brock, R.W. (1904). Report on the great landslide at Frank, Alberta, Canada. *Canadian Parliament Session Papers*, **38,** Paper 25, Part 8.

McFarlane, M.J. (1976). *Laterite and Landscape.* Academic Press. London and New York.

McGregor, K. (1967). *The Drilling of Rock.* C.R. Books Ltd. London.

Megaw, T.M. and Bartlett, J.V. (1982). *Tunnels: Planning, Design, Construction*, (in 2 vols). Ellis Harwood Ltd., Chichester, and J. Wiley and Sons, New York.

Meigh, A.C. (1977). General Report, Session 3, Site Investigation Proceedings of the International Symposium. The Geotechnics of Structurally Complex Formations. *Associato Geotechnico Itali, Capri*, **1.**

Meigs, P. (1953). World distribution of arid and semi-arid homoclimates. *Review of Research in Arid Zone Hydrology.* 203–209. UNESCO, Paris.

de Mello, V.F.B. (1971). The standard penetration test. *Proceedings of the 4th Pan-american Conference on Soil Mechanics and Foundation Engineering, San Juan, Puerto-Rico*, **1,** 1–86.

Mellor, M. and Hawkes, I. (1971). Measurement of tensile strength by diametral compression of discs and annuli. *Engineering Geology*, **5,** 173–225.

Meyboom, P. (1967). Hydrogeology. In: *Groundwater in Canada.* Geological Survey of Canada. *Economic Geology Report*, **24.**

Miller, J.P. (1961). Solutes in small streams draining single rock types, Sangre de Cristo Range, New Mexico. *United States Geological Survey Water Supply Paper 1535 F.*

Millot, G. (1970). *Geology of Clays.* Springer-Verlag, New York, London.

Minty, E.J. and Kearns, G.K. (1983). Rock Mass Workability. In: *Collected Case Studies in Engineering Geology, Hydrogeology Environmental Geology.* Geological Society of Australia. Sydney, 59–81.

Mitchell, J.K. (1970). In-place treatment of foundation soils. *Proceedings of the American Society of Civil Engineers. Journal of the Soil Mechanics and Foundations Division*, **96,** 73–110.

Mitchell, C.W. (1973). *Terrain Evaluation.* Longman Group, London.

Mitchell, J.K. (1976). *Fundamentals of Soil Behaviour.* J. Wiley and Sons. New York and London.

Mitchell, J.K., Vivatrat, V., and Lambe, T.W. (1977). Foundation performance of the Tower of Piza. *Proceedings of the American Society of Civil Engineers. Journal of the Geotechnical Engineering Division*, **103,** GT3, 227–49, and see discussion, **104,** GT1, 95–106.

Miyashiro, A., Shido, F. and Ewing, M. (1970). Petrologic models for the mid-Atlantic ridge. *Deep-Sea Research*, **17,** 109–123.

Morgenstern, N.R. and Sangrey, D.A. (1978). Methods of stability analysis. In: Schuster, R.L. and Krizek, R.J. (Eds.). *Landslides: Analysis and Control.* Transport Research Board Special Report 176. National Academy of Science, Washington D.C.

Morgenstern, N.R. and Tchalenko, J.S. (1967). Microstructural observations on shear zones from slips in natural clays. *Proceedings of the Geotechnical Conference, Oslo*, **1,** 147–52.

Moye, D. (1964). Rock mechanics in the investigation and construction of TUMUT 1 Underground Power Station, Snowy Mountains, Australia. In: *Engineering Geology Case Histories*, No. **3,** 123–54. Geological Society of America.

Muir-Wood, A.M. (1979). Ground behaviour and support for mining and tunnelling. *Transactions of the Institution of Mining and Metallurgy, London*, **88,** A23–A34.

Müller, L. (1964). The stability of rock bank slopes and the effect of rock water on the same. *International Journal of Rock Mechanics and Mining Science*, **1,** 475–504.

Müller, L. (1964). The rock slide in the Vajont valley. *Rock Mechanics and Engineering Geology*, **2,** 148–212.

Murchison, D. and Westoll, T.S. (Eds.). (1968). *Coal and Coal-bearing Strata.* Oliver and Boyd, London.

Muskat, M. (1937). *The Flow of Homogeneous Fluids through Porous Media.* McGraw-Hill Publishing Company, London.

National Coal Board. (1972). *Technical Handbook for Spoil Heaps and Lagoons.* National Coal Board, London.

National Coal Board. (1975). *Subsidence Engineers' Handbook. Mining Department.* National Coal Board, London.

National Research Council. (1972). *Earthquakes Related to Reservoir Filling.* Report of Joint Panel on Problems Concerning Seismology and Rock Mechanics. National Research Council, Washington, D.C.

Nichols, T.C. (1980). Rebound, its nature and effect on engineering works. *Quarterly Journal of Engineering Geology*, **13,** 133–52.

Nicholson, D.P. and Jardine, R.J. (1981). Performance of vertical drains at Queenborough by-pass. *Geotechnique*, **31**, 67–90.

Oborn, L.E. (1979). Seismotectonics and dam construction. General Report for Theme 3. Symposium on Engineering Geology, Problems in Hydrotechnical Construction. Tbilisi, USSR, September 1979. *Bulletin of the International Association of Engineering Geology*, **20**, 94–105.

Oliveira, R. (1979). Engineering geological problems related to the study, design and construction of dam foundations. Panel Report on Theme 1 of Symposium on Engineering Geological Problems in Hydrotechnical Construction. Tbilisi (USSR) 1979. *Bulletin of the International Association of Engineering Geology*, **20**, 1–6.

Oliveira, R., Esteves, J.M., Rodrigues, L.F. and Vieira, A.M. (1974). Geotechnical studies of the foundation rock mass of Valhelkas dam, Portugal. *Proceedings of the 2nd International Congress of Engineering Geology, Sao Paulo, Brazil.*

Ollier, C.D. (1969). *Weathering*. Longman, London.

Oxburgh, E.R. (1974). The plain-man's guide to plate tectonics. *Proceedings of the Geologists' Association*, **85**, 299–358.

Parasnis, D.S. (1973). *Mining Geophysics*. Elsevier Publishing Company, New York.

Park, R.G. (1983). *Foundations of Structural Geology*. Blackie, London.

Peach, B.N. (1929). The Lochaber Water-Power Scheme and its geological aspect. *Transactions of the Institution of Mining Engineers*, **78**, 212–25.

Pells, P.J.N. (Ed.). (1980). Structural Foundations on Rock. *Proceedings of the International Conference Sydney 1980*. A.A. Balkema, Rotterdam, Vols. I and II.

Peters, W.C. (1978). *Exploration and Mining Geology*. J. Wiley and Sons, New York.

Phillips, F.C. (1971). *The Use of Stereographic Projection in Structural Geology* (3rd Edition). Edward Arnold London.

Poland, J.F. (1972). Subsidence and its control. In: *Underground Waste Management and Environmental Implications. American Association of Petroleum Geologists' Memoir*, **18**, 50–71.

Poland, J.F. and Davis, G.H. (1969). Land subsidence due to withdrawal of fluids. Geological Society of America. *Reviews in Engineering Geology*, **2**, 187–269.

Pomerol, C. (1982). *The Cenozoic Era: Tertiary and Quarternary*. Ellis Horwood, Pub. and J. Wiley and Sons, Chichester.

Poolen, H. van and Hoover, O. (1970). Waste disposal and earthquakes at Rocky Mountain Arsenal, Denver, Colorado. *Journal of Petroleum Technology*, **22**, 983–93.

Powers, J.P. (1981). *Construction Dewatering: A Guide to Theory and Practice*. J. Wiley and Sons, New York.

Press, F. and Siever, R. (1982). *Earth* (3rd Edition). W.H. Freeman and Company Ltd., Oxford.

Price, D.G., Malkin, A.B. and Knill, J.L. (1969). Foundations of multi-storey blocks and the Coal Measures, with special reference to old mine workings. *Quarterly Journal of Engineering Geology*, **1**, 271–322.

Price, N.J. (1975). Rates of deformation. *Journal of the Geological Society of London*, **131**, 553–76.

Price, N.J. (1981). *Fault and Joint Development in Brittle and Semi-brittle Rock*. Pergamon Press, Oxford.

Priest, S.D. and Hudson, J.A. (1976). Discontinuity spacings in rock. *International Journal of Rock Mechanics and Mining Sciences*, **13**, 134–53.

Proceedings of the Institution of Civil Engineers (1967, 1968), Mangla Dam. *Proceedings of the Institution of Civil Engineers*, **38**, 337–576; and **41**, 119–203.

Proctor, R.R. (1933). Fundamental principles of soil compaction. *Engineering News* – Record. August, September.

Proctor, R.V. and White, T. (1946). *Rock Tunnelling with Steel Supports* (2nd Edition, 1968). Commercial Shearing and Stamping Company, Youngstown.

Quian-Ning, (1982). General Report for Question 54 (presented by C. Shen). *Proceedings of the XIV International Congress on Large Dams. Rio de Janeiro*, **V**, 399–402. (See also **3**, 255–690 devoted to reservoir sedimentation.)

Radbruch-Hall, D.H. (1978). Gravitational creep of rock masses on slopes. In: Voight, B. (Ed.). *Rockslides and Avalanches*, Elsevier Scientific Publishing Company, New York.

Ragan, D.M. (1968). *Structural Geology, an Introduction to Geometrical Techniques* (2nd Edition). J. Wiley and Sons, New York.

Read, H.H. (1955). Granite Series in Mobile Belts. *Geological Society of America Special Paper*, **62.**

Read, H.H. and Watson J. (1971). *Introduction to Geology*, **1**, (2nd Edition). Macmillan, London.

Reedman, J.H. (1979). *Techniques in Mineral Exploration*. Applied Science Publishers, London.

Richards, L.R., Sharp, J.C. and Pine, R.J. (1978). Design considerations for large unlined caverns at shallow depths in jointed rock. In: Storage in Excavated Rock Caverns. *Proceedings of the 1st International Symposium, Stockholm, 1977*, **2**, 239–46.

Richey, J.E., MacGregor, A.G. and Anderson, F.W. (1961). *Scotland: The Tertiary Volcanic Districts*. British Regional Geology.

Ringwood, A.E. (1978). *Safe Disposal of High Level Nuclear Reactor Wastes: a New Strategy*. Australian National University Press, Canberra.

Roberts, J.L. (1982). *Introduction to Geological Maps and Structures*. Pergamon Press, Oxford.

Robinson, G.D. and Spieker, A.M. (Eds.). (1978). Nature to be commanded. Earth-science maps applied to land and water management. *United States Geological Survey. Professional Paper* **950.**

Robinson, L.H. (1959). Effect of pore and confining pressure on the failure process in sedimentary rocks. *Colorado School of Mines, Quarterly*, **54,** 177–99.

Rocha, M. (1971). A method of integral sampling of rock masses. *Rock Mechanics*, **3,** 1–12.

Rocha, M., Serafim, J. and da Silveira, A. (1955). Deformability of foundation rocks. *Proceedings of 5th International Congress on Large Dams, Paris*, **3,** 531–7.

Rodin, S., Corbeth, B.O., Sherwood, D.E. and Thorburn, S. (1974). Penetration testing in the UK: State of the Art Report. *Proceedings of the European Symposium on Penetration Testing 1974*, **1.** Swedish Geotechnical Society, Stockholm.

Rothe, J.P. (1973). Summary: Geophysics Report. In: Ackerman, W.C., White, G.F. and Worthington, E.B. (Eds.). *Man-made Lakes; their Problems and Environmental Effects. Geophysical Monograph* **17.** American Geophysical Union, Washington, D.C.

Rowe, P.W. (1968). The influence of geological features of clay deposits on the design and performance of sand drains. *Proceedings of the Institution of Civil Engineers, London*, **39,** 465–6.

Rowe, P.W. (1972). The relevance of soil fabric to site investigation practice. *Geotechnique*, **27,** 195–300.

Ruffle, N. (1970). The Derwent Dam: design considerations. *Proceedings of the Institution of Civil Engineers, London*, **45,** 479–521. (See also, in same volume, papers on the construction and stability of the dam.)

de Ruiter, J. (1981). Current penetrometer practice. *American Society of Civil Engineers Convention, St. Louis. State of the Art Report, Session* **35.**

Ruxton, B.P. and Berry, L. (1957). Weathering of granite and associated erosional features in Hong Kong. *Bulletin of the Geological Society of America*, **68,** 1263–92.

Samuels, S.G. and Cheney, J.E. (1974). Long-term heave of a building on clay due to tree removal. *Proceedings of the British Geotechnical Society conference, Settlement of Structures, Cambridge 1974*, Paper III/8, 212–20.

Sandy, J.D., Piesold, D.D.A., Fleischer, V.D. and Forbes, P.J. (1976). Failure and subsequent stabilisation of No. 3 dump at Mufulira Mine, Zambia. *Transactions of the Institution of Mining and Metallurgical Engineers*, A144–62.

Sanglerat, G. (1972). The penetrometer and soil exploration. Development in Geotechnical Engineering, Elsevier Scientific Publishing Company, New York.

Sarmento, G. and Vaz, L. (1964). Cambambe Dam. Problems posed by the foundation ground and their solution. *Transactions of the 8th International Congress on Large Dams, Edinburgh*, **1,** 443–64.

Savage, S.B. and Sayed, M. (1979). In: Conwin, S.C. (Ed.). *Mechanics Applied to the Transport of Bulk Materials*. American Society of Mechanical Engineers, New York.

Schaffer, R.J. (1972). (A facsimile reprint) The weathering of Natural Building Stones. *Department of Scientific and Industrial Research Special Report* **18.** H.M. Stationery Office, London.

Schmid, L. (1981). Milchbuck Tunnel. Application of the freezing method to drive a 3-lane highway tunnel close to the surface. In: *Proceedings of the Rapid Excavation Tunnelling Conference, San Francisco, 1981*, **1,** 427–45. American Institution of Mining, Metallurgical and Petroleum Engineers, Inc., New York.

Schuster, R.L. and Krizek, R.J. (1978). Landslides, Analysis and Control. *Transportation Research Board Special Report* **176.** National Academy of Science, Washington, D.C.

Scott, K.F., Reeve, W.T.N. and Germond, J.P., (1968). Farahnaz Pahlavi Dam at Latiyan. *Proceedings of the Institution of Civil Engineers, London*, **39,** 353–95.

Scott, R.A. (1975). Fundamental conditions governing the penetration of grouts. In: Bell, F.G. (Ed). *Methods of Treatment of Unstable Ground.* Newnes-Butterworths, London.

Seed, H.B. (1976). The Turnagain Heights landslide. *Journal of the Soil Mechanics and Foundations Division American Society of Civil Engineers*, **93,** 325–53.

Selley, R.C. (1981). *An Introduction to Sedimentology* (2nd Edition). Academic Press, London.

Shata, A.A. (1982). Hydrogeology of the Great Nubian Sandstone Basin, Egypt. In: Middle East Hydrogeology. *Quarterly Journal of Engineering Geology*, **15,** 127–34.

Sherard, J.L., Cluff, L.S. and Allen, C.R. (1974). Potentially active faults in dam foundations. *Geotechnique*, **24,** 367–428.

Sherard, J.L., Woodward, R.J., Gizienski, S.F. and Clevenger, W.A. (1963). *Earth and Earth-Rock Dams.* J. Wiley and Sons, New York.

Shi, G.H. and Goodman, R.E. (1983). Underground support design using block theory to determine keyblock bolting requirements. *Proceedings of the Symposium on Rock Mechanics in the Design of Tunnels. International Society for Rock Mechanics (South African Group)*. August 1983.

Shotton, F.W. (1953). Pleistocene deposits of the area between Coventry and Rugby. *Philosophical Transactions of the Royal Society* **B.237,** 209–60.

Shreve, R.L. (1968). The Blackhawk landslide. *Geological Society of America Special Paper* **108.**

Simons, N.E. and Menzies, B.K. (1977). *A Short Course in Foundation Engineering*. Newnes-Butterworths, London.

Siva Subramanian, A. and Carter, A. (1969). Investigation and treatment of leakage through Carboniferous rocks at Clubbieden Dam, Midlothian. *Scottish Journal of Geology*, **5,** 207–23.

Skempton, A.W. (1950). Soil mechanics in relation to geology. *Proceedings of the Yorkshire Geological Society*, **29,** 33–62.

Skempton, A.W. (1954). The pore pressure coefficients *A* and *B*. *Geotechnique*, **4,** 143–7.

Skempton, A.W. (1961). Effective stress in soils, concrete and rocks. In: *Pore Pressure and Suction in Soils.* Butterworths, London.

Skempton, A.W. (1964). Long term stability of clay slopes. *Geotechnique*, **14,** 77–101.

Skempton, A.W. (1966). Some observations on tectonic shear zones. *Proceedings of the 1st International Congress of the International Society of Rock Mechanics, Lisbon*, **1,** 329–35.

Skempton, AW., (1970). The consolidation of clays by gravitational compaction. *Quarterly Journal of the Geological Society of London*, **125,** 373–411.

Skempton, A.W. and Hutchinson, J.N. (1969). Stability of natural slopes and embankment foundations. *Proceedings of the 7th International Conference of Soil Mechanics and Foundation Engineering, Mexico City, State of the Art*, Vol. 291–340.

Skempton, A.W. and Northey, R.D. (1952). The sensitivity of clays. *Geotechnique*, **3,** 30–53.

Skipp, B.O. (1975). Clay grouting and alluvial grouting. In: Bell, F.G. (Ed.). *Methods of Treatment of Unstable Ground.* Newnes-Butterworths, London.

Smith, A.G. and Hallam, A. (1970). The fit of the southern continents. *Nature*, **225,** 139–44.

Smith, A.G., Hurley, A.M. and Briden, J.C. (1981). *Phanerozoic Palaeocontinental World Maps.* Cambridge University Press, London.

Smith, G.N. (1982). *Elements of Soil Mechanics for Civil and Mining Engineers* (5th Edition). Granada Publishing Ltd., London.

Smith, P.J. (1973). *Topics in Geophysics.* The Open University Press, Milton Keynes.

Snow, D.T. (1982). Case Histories of Kariba and Koyna. In: Hydrogeology of Induced Seismicity and Tectonism. *Geological Society of America, Special Paper*, **189,** 317–60.

Soil Mechanics for Road Engineers. (1952). *Department of Scientific and Industrial Research Road Research Laboratory.* H.M. Stationery Office, London.

Sowers, G.F. (1979). *Introductory Soil Mechanics and Foundations* (4th Edition). Collier-Macmillan International Editions. New York and London.

Stearns, H.T. (1966). *Geology of the State of Hawaii.* Pacific Books, Palo Alto, California.

Stearns, S.R. (1966). Permafrost (perennially frozen ground). *U.S. Army Cold Regions Research and Engineering Laboratory, Monograph*, **1,** -A2.

Stephenson, H.S. (1962). Report on the explosion at Hampton Valley Colliery, Lancashire. *Hampton Valley Report* No. **4.** Examination of affected area with special reference to the nature and spread of the explosion. Safety in Mines Research Establishment, Ministry of Power.

Stevenson, T. (1844–1849). Account of experiment upon the force of waves of the Atlantic and German Oceans. *Transactions of the Royal Society of Edinburgh*, **16,** 23–32.

Steward, H.E. and Cripps, J.C. (1983). Some engineering implications of chemical weathering of pyritic shale. *Quarterly Journal of Engineering Geology, London*, **16,** 281–90.

Stimpson, B., Metcalfe, R.G. and Walton, G. (1970). A new field technique for sealing and packing rock and soil samples. *Quarterly Journal of Engineering Geology*, **3,** 127–33.

Stimpson, B. and Walton, G. (1970). Clay mylonites in English Coal Measures: their significance in opencast slope stability. *Proceedings of the 1st International Congress of the International Association of Engineering Geology, Paris*, **2,** 1388–93.

Strakhov, N.M. (1967). *Principles of Lithogenesis.* Consultants Bureau, New York, **1,** Oliver and Boyd, London.

Sugden, D.E. and John, B.S. (1976). *Glaciers and Landscape. A geomorphological approach.* Edward Arnold, London.

Szechy, K. (1966). *The Art of Tunnelling.* Akademai Kiado, Budapest.

Taylor, G.E. (1951). The Haweswater Reservoir. *Journal of the Institution of Water Engineers*, **5,** 355–92.

Taylor, R.K. (1969). Site investigations in coalfields; the problem of shallow mine workings. *Quarterly Journal of Engineering Geology*, **1,** 115–34.

Terzaghi, K. (1950). Mechanism of landslides. In: *Application of Geology to Engineering Practice; Berkey Volume.* Geological Society of America.

Terzaghi, K. (1958). Design and Performance of the Sasumua Dam. *Proceedings of the Institution of Civil Engineers*, London, **9,** 369–94.

Terzaghi, R. (1965). Sources of error in joint surveys. *Geotechnique*, **15,** 287–304.

Thackray, J. (1980). *The Age of the Earth.* H.M. Stationery Office, London.

The Dynamic Earth (1983). A Scientific American Book. W.H. Freeman & Company Ltd., Oxford.

Thomas, H.H. (1976). *The Engineering of Large Dams.* Parts 1 and 2. J. Wiley and Sons, Sydney.

Thornbury, W.D. (1969). *Principles of Geomorphology.* J. Wiley and Sons, New York.

Thornes, J. (1979). Fluvial processes. In: Embleton, C. and Thornes, J. (Eds.). *Process in Geomorphology.* Edward Arnold, London.

Times Newspapers Ltd. (1980). *The Times Concise Atlas of the World.* Times Books Ltd., London.

Tissot, B.P. and Welte, D.H. (1978). *Petroleum Formation and Occurrence.* Springer-Verlag, Berlin.

Todd, D.K. (1980). *Groundwater Hydrology* (2nd Edition). J. Wiley and Sons, New York.

Toit, A.L. du (1937). *Our Wandering Continents.* Oliver and Boyd, Edinburgh.

Tomlinson, M.J. (1980). *Foundation Design and Construction* (4th Edition). Pitman Publishing Ltd., London.

Tomlinson, M.J., Driscoll, R. and Burland, J.B. (1978). Foundations for low-rise buildings. *The Structural Engineer*, Part A. **56A,** 161–73.

Toth, J. (1963). A theoretical analysis of groundwater

flow in small drainage basins. *Journal of Geophysical Research*, **68,** 4795–812.

Trudgill, S.T. (1983). *Weathering and Erosion*. Butterworths, London.

UNESCO (1974). Seismic effects of reservoir impounding. *Proceedings of the International Colloquium, London, 1973*. Judd, W. (Ed.). Elsevier Publishing Company, Amsterdam.

UNESCO in conjunction with International Association of Engineers Geology (1976). *Engineering Geological Maps*. A guide to their preparation. The UNESCO Press, Paris.

UNESCO (1980). *Aquifer Contamination and Protection*. Project 8.3 of the International Hydrological Programme. The UNESCO Press, Paris.

UNESCO in conjunction with the Institute of Geological Sciences (London) and the International Association of Engineering Hydrologists (1970). *International Legend for Hydrogeological Maps*. Cook, Hammond and Knell, London.

United States Bureau of Reclamation (1959). *Technical Record of the Design and Construction of the Tecolote Tunnel*. U.S. Bureau of Reclamation, Denver, Colorado.

United States Bureau of Reclamation (1974). *Earth Manual*. U.S. Government Printing Office, Washington, D.C.

United States Department of Agriculture (1973). *National Engineering Handbook*. U.S. Department of Agriculture. Soil Conservation Service, Washington, D.C.

United States Department of the Interior (1983). Failure of the Teton Dam. Final *Report of the Failure Review Group*. U.S. Government Printing Office, Washington, D.C.

Varnes, D.J. (1974). The logic of geological maps with reference to their interpretation and use for engineering purposes. *U.S. Geological Survey Professional Paper* **873.**

Varnes, D.J. (1978). Slope movement types and processes. In: Schuster, R.L. and Kruzek, R.J. (Eds). *Landslides, Analysis and Control*. Special Report 176, Transport and Road Reserch Board. National Academy of Science, Washington, D.C.

Vaughan, P.R. and Soares, H.F. (1982). Design of filters for clay cores of dams. Journal of Geotechnical Division, *Proceedings American Society of Civil Engineers*, **108,** GT1, 17–31.

Veder, C. (1981). *Landslides and their Stabilization*. Springer Verlag, New York.

Vickers, B. (1982). *Laboratory Work in Soil Mechanics* (2nd Edition). Granada Publishing, London.

Voight, B. (Ed.). (1978). *Rock Slides and Avalanches*. Volumes 1 and 2. Elsevier Scientific Publishing Company, New York.

Voight B. and Faust, C. (1982). Frictional heat and strength loss in some rapid landslides. *Geotechnique*, **32,** 43–54.

Wager, L.R. and Brown, G.M. (1968). *Layered Igneous Rocks*. Oliver and Boyd, London.

Wahlstrom, E.E. (1973). *Tunnelling in Rock*. Elsevier Publishing Company, New York.

Wahlstrom, E.E. (1974). *Dams, Dam Foundations and Reservoir Sites*. Elsevier, New York.

Walling, D.E. (1983). The sediment delivery problem. *Journal of Hydrology*, **65,** 209–37.

Walters, R.C.S. (1971). *Dam Geology*. Butterworth, London.

Walton, G. and Coates, H.(1980). Some footwall failure modes in South Wales open cast workings. *Proceedings of the 2nd Conference on Ground Movements and Structures*.

Walton, K. (1969). *The Arid Zone*. Hutchinson, London.

Walton, W.C. (1970). *Groundwater Resource Evaluation*. McGraw-Hill Publishing Company, New York.

Ward, W.H. (1948). A coastal landslip. *Proceedings of the 2nd International Conference on Soil Mechanics and Foundation Engineering, Rotterdam*, **2,** 33–7.

Ward, W.H. (1978). Ground supports for tunnels in weak rocks. *Geotechnique*, **28,** 133–71.

Ward, W.H., Coates, D.J. and Tedd, P. (1976). Performance of tunnel support systems in the Four Fathom Mudstone. *Proceedings of 'Tunnelling '76'*, 329–40. Institution of Mining and Metallurgy, London, 1976.

Washburn, A.L. (1979). *Periglacial Processes and Environments* (2nd Edition). Edward Arnold, London.

Washburn, A.L. (1979). *Geocryology. A survey of periglacial processes and environments*. Edward Arnold, London.

Watson, J. (1983). *Geology and Man*. George Allen and Unwin, London and Boston.

Weart, W.D. (1979). WIPP: A bedded salt respository for defence radioactive waste in southeastern New Mexico. In: Fried, S. (Ed.). *Radioactive Waste in Geologic Storage*. American Chemical Society Symposium Series 100. Washington, D.C.

Weaver, J.M. (1975). Geological factors significant in the assessment of rippability. *Transactions of the South African Institute of Civil Engineers*, **17,** 313–6.

Weeks, A.G. (1969). The stability of natural slopes in south-east England as affected by periglacial activity. *Quarterly Journal of Engineering Geology*, **2,** 49–62.

Wellings, S.R. and Bell, J.P. (1982). Physical controls of water movement in the unsaturated zone. *Quarterly Journal of Engineering Geology*, **15,** 235–41.

West, G. (1979). *A Preliminary Study of the Reproducibility of Joint Measurements in Rock. Report SR* **488.** Transport and Road Research Laboratory, Crowthorn, UK.

West, G., Carter, P.G., Dumbleton, M.J. and Lake, L.M. (1981). Site investigation for tunnels. *International Journal of Rock Mechanics and Mining Sciences and Geomechanics Abstracts*, **18,** 345–67.

Wigley, T.M.L. (1977). WATSPEC: A computer program for determining the equilibrium specification of aqueous solutions. *British Geomorphological Research Group, Technical Bulletin, No.* **20.**

Williams, D., Stanton, R.L. and Rambaud, F. (1975). The Planes – San Antonio pyritic deposit of Rio

Tinto, Spain: its nature, environment and genesis. *Transactions of the Institution of Mining and Metallurgy London*, **84,** B73–82.

Williams, G.M., Ross, C.A.M., Stuart, A., Hitchman, S.P. and Alexander, L.S. (1984). Controls on contaminant migration at the Villa Farm Lagoons. *Quarterly Journal of Engineering Geology*, **17,** 39–56.

Williams, P.J. (1982). *The Surface of the Earth: an Introduction to Geotechnical Science*. Longman, London and New York.

Wills, L.J. (1973). A palaeogeographical map of the Palaeozoic floor below the Permian and Mesozoic formations in England and Wales. *Geological Society of London, Memoir*, **7.**

Wilson, I.G. (1972). Aeoloian bedforms, their development and origins. *Sedimentology*, **19,** 173–210.

Woldstedt, P. (1954). *Das Eiszeitalter, Volume 1* (2nd Edition). Enke, Stuttgart.

Wright, W.B. (1937). *The Quarternary Ice Age*. Macmillan, London.

Wroth, C.P. and Hughes, J.M.O. (1973). An instrument for the *in-situ* measurement of the properties of soft clays. *Proceedings of the 8th International Conference of Soil Mechanics and Foundations. Moscow, 1973*, **1.2,** 487–94.

Wyllie, P.J. (1971). *The Dynamic Earth*. J. Wiley and Sons, New York.

Yu, Y.S. and Coates, D.F. (1979). Canadian experience in simulating pit slopes by the finite element method. In: *Rockslides and Avalanches*, Vol. 2. Voight, B. (Ed.). Elsevier Scientific Publishing Company, New York.

Zussman, J. (1967). *Physical Methods in Determinative Mineralogy*. Academic Press, London.

Additional references for reprints

Adams, A.E., MacKenzie, W.S. and Guilford, C. (1984). *Atlas of Sedimentary Rocks under the Microscope*. Longman, Harlow, England.

Colwell, R.N. (1983). *Manual of Remote Sensing*, 2nd edition. Vols I and II. American Society of Photogrammetry, Falls Church, Virginia.

Hardy, H.R. and Langer, M. (Eds) (1984). *The Mechanical Behaviour of Salt*. Proc. 1st Conference. Pennsylvania (1981). Trans. Tech. Pub.

Holzer, T.L. (Ed.) (1984). *Man-induced Land Subsidence*. Reviews in Engineering Geology. Vol. VI. Geological Society of America, Boulder, Colorada.

Liu, C. and Evett, J. (1983). *Soil Properties, Testing, Measurement and Evaluation*. Prentice Hall Int., Hemel Hempstead, England.

Milnes, A.G. (1985). *Geology and Radwaste*. Academic Press, London.

Mitchell, J.K. (1986). Practical Problems from Surprising Soil Behaviour. Proceedings of the American Society of Civil Engineers. *Journal of Geotechnical Engineering Division*, **112,** 225–89.

Newmark, N.M. and Rosenblueth, E. (1971). *Fundamentals of Earthquake Engineering*. Prentice-Hall, New Jersey.

Proceedings IV International Symposium on Landslides (1984). Toronto. Vols I, II and III.

Robb, A.D. (1982). *Site Investigation*. ICE Work Construction Guides. Thomas Telford Ltd, London.

Rosenblueth, E. (ed.) (1980) *Design of Earthquake Resistant Structures*. Pentech Press, London.

Sampling and Testing Residual Soils: A review of international practice (1985). Editors: E.W. Brand and H.B. Phillipson. Scorpion Press, Hong Kong.

Society for Underwater Technology (1985). *Offshore Site Investigation*. Proc. Int. Conference, London, 1985. Graham & Trotman.

Index